Basic Mathematics

Eighth Edition

Marvin L. Bittinger

Indiana University—Purdue University
at Indianapolis

ADDISON-WESLEY

An imprint of Addison Wesley Longman, Inc.

Reading, Massachusetts • Menlo Park, California • New York • Harlow, England
Don Mills, Ontario • Sydney • Mexico City • Madrid • Amsterdam

Publisher	Jason A. Jordan
Project Manager	Christine Poolos
Assistant Editor	Michelle Fowler
Managing Editor	Ron Hampton
Production Supervisor	Kathleen A. Manley
Production Assistant	Jane Estrella
Design Direction	Susan Carsten
Text Designer	Rebecca Lloyd Lemna
Editorial and Production Services	Martha Morong/Quadrata, Inc.
Art Editor	Janet Theurer
Marketing Managers	Craig Bleyer and Mark Harrington
Illustrators	Scientific Illustrators, Gayle Hayes, and Rolin Graphics
Compositor	The Beacon Group
Cover Designer	Jeannet Leendertse
Cover Photographs	© Photodisc, 1998; © TSM/T. Kevin Smyth, 1998
Manufacturing Supervisor	Ralph Mattivello

PHOTO CREDITS

1, Pascal Rondeau/Tony Stone Images **4,** Copyright 1997, USA TODAY. Reprinted with permission **69,** Pascal Rondeau/Tony Stone Images **83,** Greg Probst/Tony Stone Images **108,** John Coletti/Tony Stone Images **126,** AP/Wide World Photos **130,** Greg Probst/Tony Stone Images **135,** Kirsten Willis **169,** Jeff Greenberg/Photo Researchers **173,** Kirsten Willis **193,** Mark Richards/PhotoEdit **236,** Gary Gay/The Image Bank **241,** Mark Richards/PhotoEdit **252,** Focus on Sports **261, 283,** Nicholas Devore III/Photographers Aspen **286,** Jim Zuckerman/Westlight **288,** AP/Wide World Photos **301,** Brian Spurlock **303,** Kenneth Chen/Envision **331,** Reuters/Jeff Christensen/Archive Photos **336,** Brian Spurlock **363, 373, 392,** Steven Needham/Envision **397,** Scott Goodman/Envision **401,** Tom Stratman **417,** Eric Neurath/Stock Boston **430,** Owen Franken/Stock Boston **435,** Eric Neurath/Stock Boston **471, 478,** © 1998 Al Satterwhite **494,** Henryk T. Kaiser/Envison **513, 516,** AP/World Wide Photos **551, 588,** © 1998 Jeff Schultz/Alaska Stock **603,** Bruce Hands/Tony Stone Images

LIBRARY OF CONGRESS CATALOGING-IN-PUBLICATION DATA

Bittinger, Marvin L.
Basic mathematics—8th ed./by Marvin L. Bittinger.
p. cm.
Includes index.
ISBN 0-201-95958-5
1. Arithmetic. I. Title.
QA107.K43 1998
513'.1—dc21 98-4836
CIP

3 4 5 6 7 8 9 10—VH—010099

Contents

Preface

This text is the first in a series of texts that includes the following:

Bittinger: *Basic Mathematics,* Eighth Edition
Bittinger: *Fundamental Mathematics,* Second Edition
Bittinger: *Introductory Algebra,* Eighth Edition
Bittinger: *Intermediate Algebra,* Eighth Edition
Bittinger/Beecher: *Introductory and Intermediate Algebra: A Combined Approach*

Basic Mathematics, Eighth Edition, is a significant revision of the Seventh Edition, particularly with respect to design, art program, pedagogy, features, and supplements package. Its unique approach, which has been developed and refined over eight editions, continues to blend the following elements in order to bring students success:

- ***Writing style.*** The author writes in a clear easy-to-read style that helps students progress from concepts through examples and margin exercises to section exercises.
- ***Problem-solving approach.*** The basis for solving problems and real-data applications is a five-step process (*Familiarize, Translate, Solve, Check,* and *State*) introduced early in the text and used consistently throughout. This problem-solving approach provides students with a consistent framework for solving applications. (See pages 61, 173, and 583.)
- ***Real data.*** Real-data applications aid in motivating students by connecting the mathematics to their everyday lives. Extensive research was conducted to find new applications that relate mathematics to the real world.

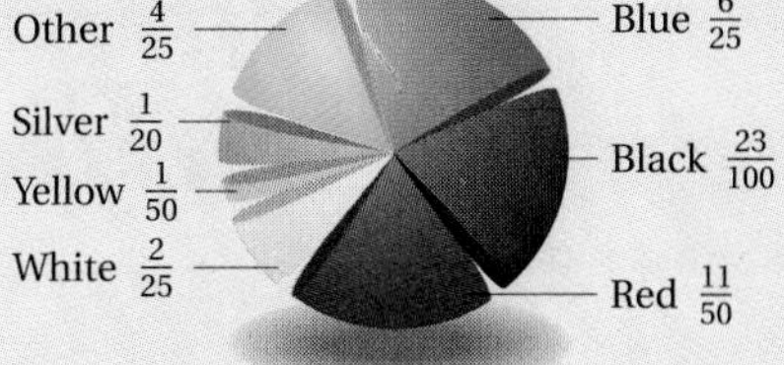

Source: Bicycle Market Research Institute

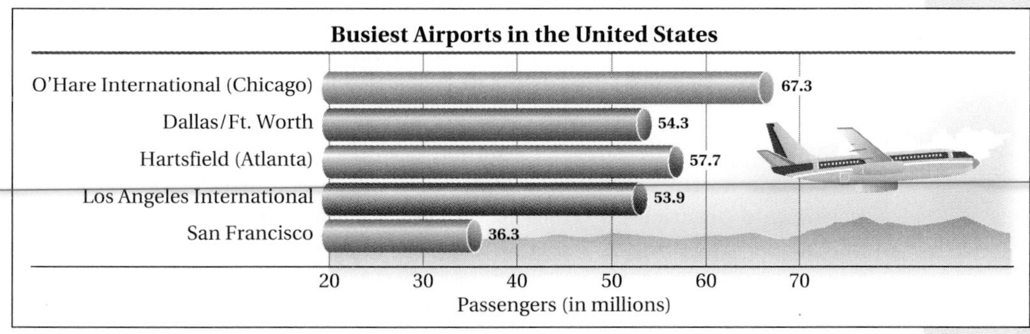

Source: Air Transport Association of America

- ***Art program.*** The art program has been expanded to improve the visualization of mathematical concepts and to enhance the real-data applications.

- ***Reviewer feedback.*** The author solicits feedback from reviewers and students to help fulfill student and instructor needs.
- ***Accuracy.*** The manuscript is subjected to an extensive accuracy-checking process to eliminate errors.
- ***Supplements package.*** All ancillary materials are closely tied with the text and created by members of the author team to provide a complete and consistent package for both students and instructors.

What's New in the Eighth Edition?

The style, format, and approach of the Seventh Edition have been strengthened in this new edition in a number of ways.

Updated Applications Extensive research has been done to make the applications in the Eighth Edition even more up to date and realistic. A large number of the applications are new to this edition, and many are drawn from the fields of business and economics, life and physical sciences, social sciences, and areas of general interest such as sports and daily life. To encourage students to understand the relevance of mathematics, many applications are enhanced by graphs and drawings similar to those found in today's newspapers and magazines. Many applications are also titled for quick and easy reference, and most real-data applications are credited with a source line. (See pages 247, 337, 374, 375, and 386.)

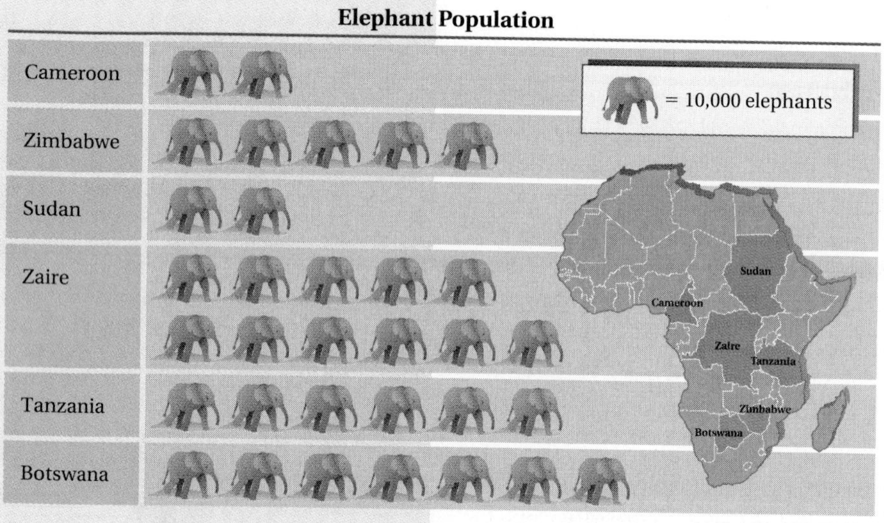

Elephant Population

Cameroon
Zimbabwe
Sudan
Zaire
Tanzania
Botswana

= 10,000 elephants

Source: National Geographic

Improving Your Math Study Skills Occurring at least once in every chapter, and referenced in the table of contents, these mini-lessons provide students with concrete techniques to improve studying and test-taking. These features can be covered in their entirety at the beginning of the course, encouraging good study habits early on, or they can be used as they occur in the text, allowing students to learn them gradually. These features can also be used in conjunction with Marvin L. Bittinger's "Math Study Skills" Videotape, which is free to adopters. Please see your Addison Wesley Longman sales consultant for details on how to obtain this videotape. (See pages 6, 216, and 344.)

Calculator Spotlights Designed specifically for the basic mathematics student, these optional features include scientific-calculator instruction and practice exercises (see pages 75, 309, 456, and 481). Answers to all Calculator Spotlight exercises appear at the back of the text.

New Art and Design To enhance the greater emphasis on real data and applications, we have extensively increased the number of pieces of technical and situational art (see pages 3, 242, 388, and 583). The use of color has been carried out in a methodical and precise manner so that its use carries a consistent meaning, which enhances the readability of the text. For example, when perimeter is considered, figures have a red border to emphasize the perimeter. When area is considered, figures are outlined in black and screened with amber to emphasize the area. Similarly, when volume is considered, figures are three-dimensional and air-brushed blue. When fractional parts are illustrated, those parts are shown in purple. In addition, the margin color has been improved so that student annotations can be read more clearly. Answer lines have also been deleted from all section exercise sets to allow room for more exercises and additional art to better illustrate the exercises.

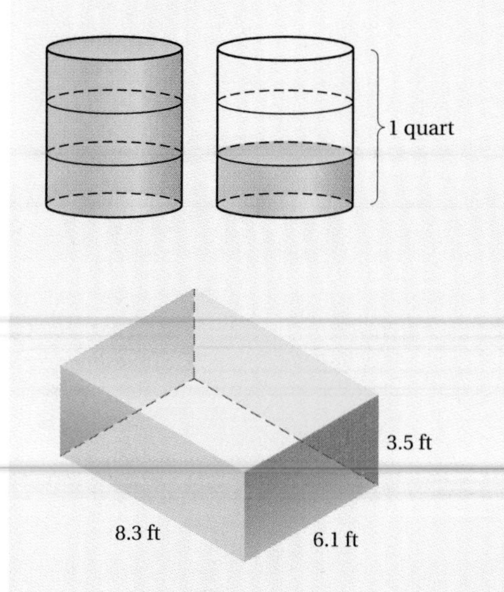

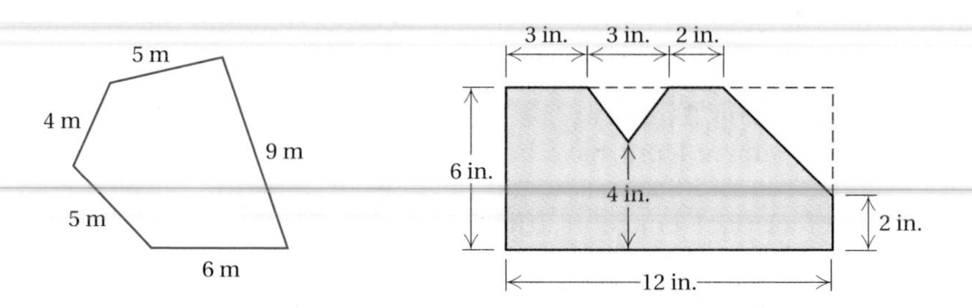

World Wide Web Integration The World Wide Web is a powerful resource available to more and more people every day. In an effort to get students more involved in using this source of information, we have added a World Wide Web address (www.mathmax.com) to every chapter opener (see pages 83, 261, and 301). Students can go to this page on the World Wide Web to further explore the subject matter of the chapter-opening application. Selected exercise sets, marked on the first page of the exercise set with an icon (see pages 155, 353, and 531), have additional practice-problem worksheets that can be downloaded from this site. Additional, more extensive, Summary and Review pages for each chapter, as well as other supplementary material, can also be downloaded for instructor and student use.

Collaborative Learning Features An icon located at the end of an exercise set signals the existence of a Collaborative Learning Activity correlating to that section in Irene Doo's *Collaborative Learning Activities Manual* (see pages 72, 156, and 348). Please contact your Addison Wesley Longman sales consultant for details on ordering this supplement.

Exercises The deletion of answer lines in the exercise sets has allowed us to include more exercises in the Eighth Edition. Exercises are paired, meaning that each even-numbered exercise is very much like the odd-numbered one that precedes it. This gives the instructor several options: If an instructor wants the student to have answers available, the odd-numbered exercises are assigned; if an instructor wants the student to practice (perhaps for a test), with no answers available, then the even-numbered exercises are assigned. In this way, each exercise set actually serves as two exercise sets. Answers to all odd-numbered exercises, with the exception of the Thinking and Writing exercises, and *all* Skill Maintenance exercises are provided at the back of the text. If an instructor wants the student to have access to all the answers, a complete answer book is available.

Skill Maintenance Exercises The Skill Maintenance exercises have been enhanced by the inclusion of 60% more exercises in this edition. These exercises focus on the four Objectives for Retesting listed at the beginning of each chapter, but they also review concepts from other sections of the text in order to prepare students for the Final Examination. Section and objective codes appear next to each Skill Maintenance exercise for easy reference. Answers to all Skill Maintenance exercises appear at the back of the book (see pages 110, 228, and 316).

Synthesis Exercises These exercises now appear in every exercise set, Summary and Review, Chapter Test, and Cumulative Review. Synthesis exercises help build critical thinking skills by requiring students to synthesize or combine learning objectives from the section being studied as well as preceding sections in the book. In addition, the Critical Thinking feature from the Seventh Edition has been incorporated into the Synthesis exercises for the Eighth Edition (see pages 122, 288, and 532).

Thinking and Writing Exercises ◈ Two Thinking and Writing exercises (denoted by the maze icon) have been added to the Synthesis section of every exercise set and Summary and Review. Designed to develop comprehension of critical concepts, these exercises encourage students to both think and write about key mathematical ideas in the chapter (see pages 170, 218, 272, and 592).

Content We have made the following improvements to the content of *Basic Mathematics.*

- The topic of estimation is spiraled throughout the text. It is used to approximate the results of operations, to check possible solutions to applied problems, and as a basic skill. Estimation has been expanded in this edition to include fractions and mixed numerals. (See pages 27, 181, and 235.)
- Chapters 4 (*Addition and Subtraction: Decimal Notation*) and 5 (*Multiplication and Division: Decimal Notation*) from the Seventh Edition have been combined into one chapter (Chapter 4, *Decimal Notation*) to allow more time for expanded coverage of other topics.
- In Chapter 7 (*Data Analysis, Graphs, and Statistics*), a new section on data analysis and predictions has been added. Students will learn to compare two sets of data using their means and to make predictions from a set of data using interpolation and extrapolation. This chapter is also filled with new real-data applications.

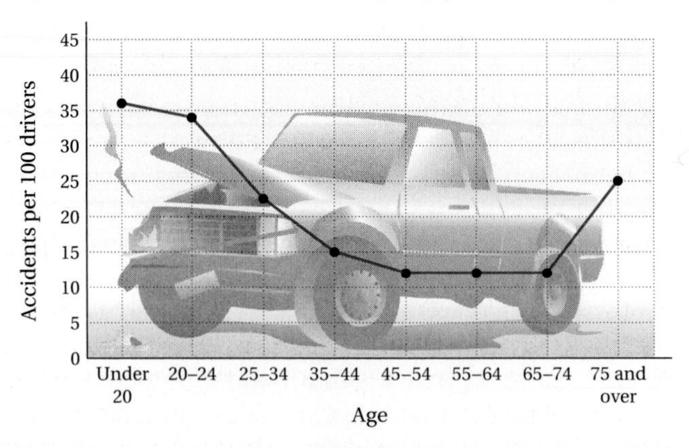

- In Chapter 9 (*More Geometry and Measures*), a new section on angles and triangles has been added. Students will learn to name a given angle in four different ways and to measure it with a protractor; classify a given angle as right, straight, acute, or obtuse; classify a triangle as equilateral, isosceles, or scalene, and as right, acute, or obtuse; and given two angle measurements of a triangle, find the third angle measurement.

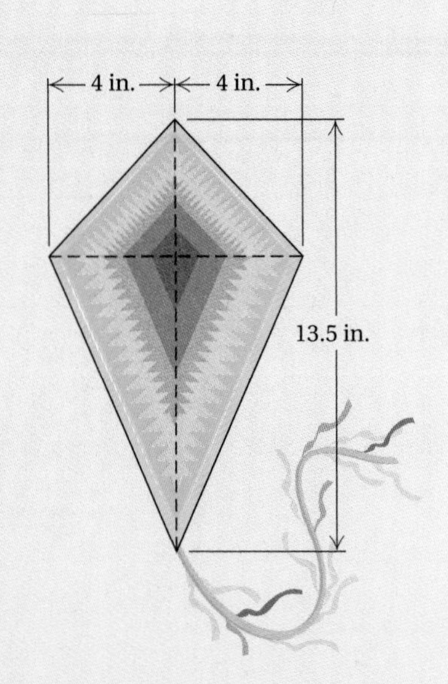

Learning Aids

Interactive Worktext Approach The pedagogy of this text is designed to provide an interactive learning experience between the student and the exposition, annotated examples, art, margin exercises, and exercise sets. This approach provides students with a clear set of learning objectives, involves them with the development of the material, and provides immediate and continual reinforcement and assessment.

Section objectives are keyed by letter not only to section subheadings, but also to exercises in the exercise sets and Summary and Review, as well as answers to the Pretest, Chapter Test, and Cumulative Review questions. This enables students to easily find appropriate review material if they are unable to work a particular exercise.

Throughout the text, students are directed to numerous *margin exercises,* which provide immediate reinforcement of the concepts covered in each section.

Review Material The Eighth Edition of *Basic Mathematics* continues to provide many opportunities for students to prepare for final assessment.

Now in a two-column format, the *Summary and Review* appears at the end of each chapter and provides an extensive set of review exercises. Reference codes beside each exercise or direction line preceding it allow the student to easily return to the objective being reviewed (see pages 129, 357, and 463).

Also included at the end of every chapter but Chapter 1 is a *Cumulative Review,* which reviews material from all preceding chapters. At the back of the text are answers to all Cumulative Review exercises, together with section and objective references, so that students know exactly what material to study if they miss a review exercise (see pages 191, 511, and 597).

Objectives for Retesting are covered in each Summary and Review and Chapter Test, and are also included in the Skill Maintenance exercises and in the Printed Test Bank (see pages 302 and 472).

For Extra Help Many valuable study aids accompany this text. Below the list of objectives found at the beginning of each section are references to appropriate videotape, audiotape, tutorial software, and CD-ROM programs to make it easy for the student to find the correct support materials.

Objectives

a Multiply and simplify using fractional notation.

b Solve applied problems involving multiplication.

For Extra Help

TAPE 4   TAPE 4B   MAC WIN   CD-ROM

Testing The following assessment opportunities exist in the text.

The *Diagnostic Pretest,* provided at the beginning of the text, can place students in the appropriate chapter for their skill level by identifying familiar material and specific trouble areas (see page xxi).

Chapter Pretests can then be used to place students in a specific section of the chapter, allowing them to concentrate on topics with which they have particular difficulty (see pages 262, 418, and 472).

Chapter Tests allow students to review and test comprehension of chapter skills, as well as the four Objectives for Retesting from earlier chapters (see pages 81, 411, and 467).

Answers to all Diagnostic Pretest, Chapter Pretest, and Chapter Test questions are found at the back of the book, along with appropriate section and objective references.

Supplements for the Instructor

Annotated Instructor's Edition

0-201-33874-2

The *Annotated Instructor's Edition* is a specially bound version of the student text with answers to all margin exercises, exercise sets, and chapter tests printed in a special color near the corresponding exercises.

Instructor's Solutions Manual

0-201-43407-5

The *Instructor's Solutions Manual* by Judith A. Penna contains brief worked-out solutions to all even-numbered exercises in the exercise sets and answers to all Thinking and Writing exercises.

Printed Test Bank/Instructor's Resource Guide

by Donna DeSpain

0-201-43408-3

The test-bank section of this supplement contains the following:

- Three alternate test forms for each chapter, with questions in the same topic order as the objectives presented in the chapter
- Five alternate test forms for each chapter, modeled after the Chapter Tests in the text
- Three alternate test forms for each chapter, designed for a 50-minute class period
- Two multiple-choice versions of each Chapter Test
- Two cumulative review tests for each chapter, with the exception of Chapter 1
- Eight final examinations: three with questions organized by chapter, three with questions scrambled as in the Cumulative Reviews, and two with multiple-choice questions
- Answers for the Chapter Tests and Final Examination

The resource-guide section contains the following:

- A conversion guide from the Seventh Edition to the Eighth Edition
- Extra practice exercises (with answers) for 40 of the most difficult topics in the text
- Critical Thinking exercises and answers
- Black-line masters of grids and number lines for transparency masters or test preparation
- Indexes to the videotapes and audiotapes that accompany the text
- Three-column chapter Summary and Review listing objectives, brief procedures, worked-out examples, multiple-choice problems similar to the example, and the answers to those problems
- Instructor support material for the CD-ROM

Collaborative Learning Activities Manual

0-201-34574-9

The *Collaborative Learning Activities Manual,* written by Irene Doo of Austin Community College, features group activities that are tied to sections of the text via an icon Collaborative Learning Manual. Instructions for classroom setup are also included in the manual.

Answer Book

0-201-43409-1

The *Answer Book* contains answers to all exercises in the exercise sets in the text. Instructors can make quick reference to all answers or have quantities of these booklets made available for sale if they want students to have access to all the answers.

TestGen-EQ

0-201-38133-8 (Windows), 0-201-38140-0 (Macintosh)

This test generation software is available in Windows and Macintosh versions. TestGen-EQ's friendly graphical interface enables instructors to easily view, edit, and add questions, transfer questions to tests, and print tests in a variety of fonts and forms. Search and sort features help the instructor quickly locate questions and arrange them in a preferred order. Six question formats are available, including short-answer, true–false, multiple-choice, essay, matching, and bimodal formats. A built-in question editor gives the instructor the ability to create graphs, import graphics, insert mathematical symbols and templates, and insert variable numbers or text. Computerized testbanks include algorithmically defined problems organized according to each textbook. An "Export to HTML" feature lets instructors create practice tests for the World Wide Web. TestGen-EQ is free to qualifying adopters.

QuizMaster-EQ

0-201-38133-8 (Windows), 0-201-38140-0 (Macintosh)

QuizMaster-EQ enables instructors to create and save tests and quizzes using TestGen-EQ so students can take them on a computer network. Instructors can set preferences for how and when tests are administered. QuizMaster-EQ automatically grades the exams and allows the instructor to view or print a variety of reports for individual students, classes, or courses. This software is available for both Windows and Macintosh and is fully networkable. QuizMaster-EQ is free to qualifying adopters.

Supplements for the Student

Student's Solutions Manual

0-201-34022-4

The *Student's Solutions Manual* by Judith A. Penna contains fully worked-out solutions with step-by-step annotations for all the odd-numbered exercises in the exercise sets in the text, with the exception of the Thinking and Writing exercises. It may be purchased by your students from Addison Wesley Longman.

"Steps to Success" Videotapes

0-201-30359-0

Steps to Success is a complete revision of the existing series of videotapes, based on extensive input from both students and instructors. These videotapes feature an engaging team of mathematics teachers who present comprehensive coverage of each section of the text in a student-interactive format. The lecturers' presentations include examples and problems from the text and support an approach that emphasizes visualization and problem solving. A video icon at the beginning of each section references the appropriate videotape number. The videotapes are free to qualifying adopters.

"Math Study Skills for Students" Videotape

0-201-84521-0

Designed to help students make better use of their math study time, this videotape help students improve retention of concepts and procedures taught in classes from basic mathematics through intermediate algebra. Through carefully-crafted graphics and comprehensive on-camera explanation, Marvin L. Bittinger helps viewers focus on study skills that are commonly overlooked.

Audiotapes

0-201-43404-0

The audiotapes are designed to lead students through the material in each text section. Bill Saler explains solution steps to examples, cautions students about common errors, and instructs them at certain points to stop the tape and do exercises in the margin. He then reviews the margin-exercise solutions, pointing out potential errors. An audiotape icon at the beginning of each section references the appropriate audiotape number. The audiotapes are free to qualifying adopters.

InterAct Math Tutorial Software

0-201-38077-3 (Windows), 0-201-38084-6 (Macintosh)

InterAct Math Tutorial Software has been developed and designed by professional software engineers working closely with a team of experienced developmental-math teachers. This software includes exercises that are linked one-to-one with the odd-numbered exercises in the text and require the same computational and problem-solving skills as their companion exercises in the text. Each exercise has an example and an interactive guided solution that are designed to involve students in the solution process and to help them identify precisely where they are having trouble. In addition, the software recognizes common student errors and provides students with appropriate customized feedback. With its sophisticated answer recognition capabilities, *InterAct Math Tutorial Software* recognizes equivalent forms of the same answer for any kind of input. It also tracks for each section student activity and scores that can then be printed out. A disk icon at the beginning of each section identifies section coverage. Available for Windows and Macintosh computers, this software is free to qualifying adopters or can be bundled with books for sale to students.

World Wide Web Supplement (www.mathmax.com)

This on-line supplement provides additional practice and learning resources for the student of basic mathematics. For each book chapter, students can find additional practice exercises, Web links for further exploration, and expanded Summary and Review pages that review and reinforce the concepts and skills learned throughout the chapter. In addition, students can download a plug-in for Addison Wesley Longman's *InterAct Math Tutorial Software* that allows them to access additional tutorial problems directly through their Web browser. Students and instructors can also learn about the other supplements available for the

MathMax series via sample audio clips and complete descriptions of other services provided by Addison Wesley Longman.

MathMax Multimedia CD-ROM for Basic Mathematics

0-201-39734-x (Windows), 0-201-44456-9 (Macintosh)

The Basic Math CD provides an active environment using graphics, animations, and audio narration to build on some of the unique and proven features of the MathMax series. Highlighting key concepts from the book, the content of the CD is tightly and consistently integrated with the *Basic Mathematics* text and retains references to the *Basic Mathematics* numbering scheme so that students can move smoothly between the CD and other *Basic Mathematics* supplements. The CD includes Addison Wesley Longman's *InterAct Math Tutorial Software* so that students can practice additional tutorial problems. An interactive Summary and Review section allows students to review and practice what they have learned in each chapter; and multimedia presentations reiterate important study skills described throughout the book. A CD-ROM icon at the beginning of each section indicates section coverage. The Basic Math CD is available for both Windows and Macintosh computers. Contact your Addison Wesley Longman sales consultant for a demonstration.

Your author and his team have committed themselves to publishing an accessible, clear, accomplishable, error-free book and supplements package that will provide the student with a successful learning experience and will foster appreciation and enjoyment of mathematics. As part of our continual effort to accomplish this goal, we welcome your comments and suggestions at the following email address:

Marv Bittinger
exponent@aol.com

Acknowledgments

Many of you have helped to shape the Eighth Edition by reviewing, participating in telephone surveys and focus groups, filling out questionnaires, and spending time with us on your campuses. Our deepest appreciation to all of you and in particular to the following:

Courtney Adams, *Louisiana State University—Eunice*
Andrea Adlman, *Ventura College*
Michelle Bach, *Kansas City Kansas Community College*
Linda Balfour, *Schoolcraft College*
Ted Bright, *College of Marin*
Kim Brown, *Tarrant County Junior College, Northeast*
Tyrone Clinton, *St. Petersburg Junior College*
Barbara Conway, *Berkshire Community College*
Dan Cortney, *Glendale Community College*
Sherry Crabtree, *Northwest Shoals Community College*
Nancy Desilet, *Carroll Community College*
Irene Doo, *Austin Community College*
Peter Embalabala, *Lincolnland Community College*
Rebecca Farrow, *John Tyler Community College*
Kathy Fenimore, *Frederick Community College*
Jacqueline Fesq, *Raritan Valley Community College*
Margaret Finster, *Erie Community College—South Campus*
David Frankenreiter, *Jefferson College*
Jimy Fulford, *Gulf Coast Community College*
Dewey Furnass, *Ricks College*
Bob Grant, *Mesa Community College*
Janet Guynn, *Blue Ridge Community College*
Linda Hurst, *Central Texas College*
Charlotte Hutt, *Rogue Community College*
Kathy Janoviak, *Mid-Michigan Community College*
Yvonne Jessee, *Mountain Empire Community College*
Kathy Johnson, *Volunteer State Community College*
Robert Kaiden, *Lorain County Community College*
Joanne Kendall, *College of the Mainland*
Roxann King, *Prince Georges Community College*
Steve Kinholt, *Green River Community College*
Paulette Kirkpatrick, *Wharton County Junior College*
Jeanne Kubier, *Oklahoma State University—Oklahoma City*
Lynn Maracek, *Rancho Santiago College*
Lawrence Marler, *Olympic College*
Bob Martin, *Tarrant County Junior College, Northeast*
Molly Misko, *Gadsden State Community College*
Frank Mulvaney, *Delaware County Community College*

Ethel Muter, *Raritan Valley Community College*
Ellen O'Connell, *Triton College*
Julie Pendleton, *Brookhaven College*
Charles Perry, *Southern Union State Junior College*
George Podorski, *Jefferson College—Hillsboro*
Elizabeth Polen, *County College of Morris*
Mary Pusch, *Rogue Community College*
Patricio Rojas, *New Mexico State University—Grants*
James Ronner, *Southwestern Michigan College*
Pat Roux, *Delgado Community College*
Martin Sade, *Pima Community College*
Radha Shrinivas, *St. Louis Community College—Forest Park*
Maxine Smith, *Greenville Technical College*
Larry Smyrski, *Henry Ford Community College*
Tom Swiersz, *St. Petersburg Junior College*
Sharon Testone, *Onondaga Community College*
Jane Thieling, *Dyersburg State Community College*
Victor Thomas, *Holyoke Community College*
Joyce Wellington, *Southeastern Community College*
Lily Yang, *Brevard Community College*
Ben Zandy, *Fullerton College*

We also wish to recognize the following people who wrote scripts, presented lessons on camera, and checked the accuracy of the videotapes:

Beth Burkenstock
Donna DeSpain
Margaret Donlan, *University of Delaware*
David J. Ellenbogen, *Community College of Vermont*
Barbara Johnson, *Indiana University—Purdue University of Indianapolis*
Judith A. Penna, *Indiana University—Purdue University of Indianapolis*
Patricia Schwarzkopf, *University of Delaware*
Clen Vance, *Houston Community College*

I wish to thank Jason Jordan, my publisher and friend at Addison Wesley Longman, for his encouragement, for his marketing insight, and for providing me with the environment of creative freedom. The unwavering support of the Developmental Math group and the endless hours of hard work by Martha Morong and Janet Theurer have led to products of which I am immensely proud.

I also want to thank Judy Beecher, my co-author on many books and my developmental editor on this text. Her steadfast loyalty, vision, and encouragement have been invaluable. In addition to writing the Student's Solutions Manual, Judy Penna has continued to provide strong leadership in the preparation of the printed supplements, videotapes, and Interactive CD-ROM. Other strong support has come from Donna DeSpain for the Printed Test Bank; Bill Saler for the audiotapes; Irene Doo for the Collaborative Learning Activities Manual; and Irene Doo, Pat Ewert, Linda Long, and Patty Slipher for their accuracy checking.

M.L.B.

Diagnostic Pretest

Chapter 1

1. Add: $1425 + 382$.

2. Solve: $32 + x = 61$.

3. Multiply: $\begin{array}{r} 321 \\ \times\ 47 \\ \hline \end{array}$

4. A thermos contains 128 oz of juice. How many 6-oz cups can be filled from the thermos? How many ounces will be left over?

Chapter 2

5. Solve: $\frac{5}{8} \cdot x = \frac{3}{16}$.

6. Multiply and simplify: $4 \cdot \frac{3}{8}$.

7. Find the prime factorization of 144.

8. A recipe calls for $\frac{3}{4}$ cup of flour. How much is needed to make $\frac{2}{3}$ of a recipe?

Chapter 3

9. Find the LCM of 12 and 16.

10. Add: $\frac{3}{8} + \frac{1}{6}$.

11. Multiply and simplify: $4\frac{1}{5} \cdot 3\frac{2}{3}$.

12. A 3-m fence post was set $1\frac{2}{5}$ m in the ground. How much was above the ground?

13. A car travels 249 mi on $8\frac{3}{10}$ gal of gas. How many miles per gallon did it get?

Chapter 4

14. Which number is larger, 0.00009 or 0.0001?

15. Round to the nearest tenth: 25.562.

16. Add: $12.035 + 0.08 + 27.7$.

17. A driver bought gasoline when the odometer read 68,123.2. At the next gasoline purchase, the odometer read 68,310.1. How many miles had been driven?

18. Multiply: 0.012×2.5.

19. Find decimal notation: $\frac{7}{3}$.

20. Solve: $1.5 \times t = 3.6$.

21. What is the cost of 5 tee shirts at $23.99 each?

22. Estimate the product 4.68×32.431 by rounding to the nearest one.

Chapter 5

23. Solve: $\frac{1.2}{x} = \frac{0.4}{1.5}$.

24. If 3 cans of green beans costs $1.49, how many cans of green beans can you buy for $5.96?

25. *Unit Price.* It costs $3.79 for a 22-oz bag of tortilla chips. Find the unit price in cents per ounce. Round to the nearest tenth of a cent.

Chapter 6

26. Find percent notation: $\frac{1}{8}$.

27. Find decimal notation: 1.35%.

28. *Percent of Decrease.* The price of a pair of running shoes was reduced from \$45 to \$27. Find the percent of decrease in price.

29. *Simple Interest.* What is the simple interest on \$230 principal at the interest rate of 8.5% for one year?

Chapter 7

30. Find the average, the median, and the mode of the following set of numbers:

22, 25, 27, 25, 22, 25.

31. A car traveled 296 mi on 16 gal of gasoline. What was the average number of miles per gallon?

32. *Test Score.* In order to get a B in math, a student must average 80 on four tests. Scores on the first three tests were 85, 72, and 78. What is the lowest score the student can get on the last test and still get a B?

Chapter 8

Complete.

33. 2 yd = ______ in.

34. 4 cm = ______ km

35. Find the area and the circumference of a circle with a diameter of 12 cm. Use 3.14 for π.

36. Find the area and the perimeter of a rectangle with length 3 ft and width 2.5 ft.

37. Simplify: $\sqrt{121}$.

Chapter 9

Complete.

38. 2 min = ______ sec

39. 5 mg = ______ g

40. 2 yd^2 = ______ ft^2

41. 10 qt = ______ gal

42. The diameter of a ball is 18 cm. Find the volume. Use 3.14 for π.

Chapter 10

43. Find the absolute value: $|-4.2|$.

44. Find decimal notation: $-\frac{4}{9}$.

Compute and simplify.

45. $-2 - (-1.9)$

46. $\frac{5}{6}\left(-\frac{1}{10}\right)$

Chapter 11

Solve.

47. $2x - 1 = 4x + 5$

48. $\frac{1}{3}x + \frac{1}{5} = \frac{2}{3} - \frac{1}{4}x$

49. A student bought a sweater and a pair of jeans. The sweater cost \$59.95. This was \$8.39 more than the cost of the jeans. How much did the jeans cost?

50. A 22-oz box of cereal costs \$3.49. How many boxes of cereal can you buy for \$27.92?

Basic Mathematics

Eighth Edition

1

Operations on the Whole Numbers

An Application

Total sales, in millions of dollars, of bicycles and related sporting supplies were $2973 in 1992, $3534 in 1993, $3470 in 1994, and $3435 in 1995 (***Source***: National Sporting Goods Association). Find the total sales for the entire four-year period.

This problem appears as Exercise 4 in Exercise Set 1.8.

The Mathematics

We let T = the total sales. Since we are combining sales, addition can be used. We translate the problem to the equation

$$\underbrace{2973 + 3534 + 3470 + 3435 = T.}$$

This is how addition can occur in applications and problem solving.

Introduction

In this chapter, we consider addition, subtraction, multiplication, and division of whole numbers. Then we study the solving of simple equations and apply all these skills to the solving of problems.

World Wide Web For more information, visit us at www.mathmax.com

Pretest: Chapter 1

1. Write a word name: 3,078,059.

2. Write expanded notation: 6987.

3. Write standard notation: Two billion, forty-seven million, three hundred ninety-eight thousand, five hundred eighty-nine.

4. What does the digit 6 mean in 2,967,342?

5. Round 956,449 to the nearest thousand.

6. Estimate the product $594 \cdot 126$ by first rounding the numbers to the nearest hundred.

7. Add.

$$\begin{array}{r} 7\,3\,1\,2 \\ +\,2\,9\,0\,4 \\ \hline \end{array}$$

8. Subtract.

$$\begin{array}{r} 7\,0\,1\,2 \\ -\,2\,9\,0\,4 \\ \hline \end{array}$$

9. Multiply: $359 \cdot 64$.

10. Divide: $23{,}149 \div 46$.

Use either $<$ or $>$ for ▒ to write a true sentence.

11. 346 ▒ 364

12. 54 ▒ 45

Solve.

13. $326 \cdot 17 = m$

14. $y = 924 \div 42$

15. $19 + x = 53$

16. $34 \cdot n = 850$

Solve.

17. Anna weighs 121 lb and Kari weighs 109 lb. How much more does Anna weigh?

18. How many 12-jar cases can be filled with 1512 jars of spaghetti sauce?

19. *Population.* The population of Illinois is 11,830,000. The population of Ohio is 11,151,000. (***Source:*** U.S. Bureau of the Census) What is the total population of Illinois and Ohio?

20. A lot measures 48 ft by 54 ft. A pool that is 15 ft by 20 ft is constructed on the lot. How much area is left over?

Evaluate.

21. 5^2

22. 4^3

Simplify.

23. $8^2 \div 8 \cdot 2 - (2 + 2 \cdot 7)$

24. $108 \div 9 - \{4 \cdot [18 - (5 \cdot 3)]\}$

1.1 Standard Notation

We study mathematics in order to be able to solve problems. In this chapter, we learn how to use operations on the whole numbers. We begin by studying how numbers are named.

Objectives

a Convert from standard notation to expanded notation.

b Convert from expanded notation to standard notation.

c Write a word name for a number given standard notation.

d Write standard notation for a number given a word name.

e Given a standard notation like 278,342, tell what 8 means, what 3 means, and so on; identify the hundreds digit, the thousands digit, and so on.

For Extra Help

TAPE 1

TAPE 1A

MAC WIN

CD-ROM

a From Standard Notation to Expanded Notation

To answer questions such as "How many?", "How much?", and "How tall?", we use whole numbers. The set, or collection, of **whole numbers** is

0, 1, 2, 3, 4, 5, 6, 7, 8, 9, 10, 11, 12, ...

The set goes on indefinitely. There is no largest whole number, and the smallest whole number is 0. Each whole number can be named using various notations. The set 1, 2, 3, 4, 5, . . . , without 0, is called the set of **natural numbers**.

As examples, we use data from the bar graph shown here.

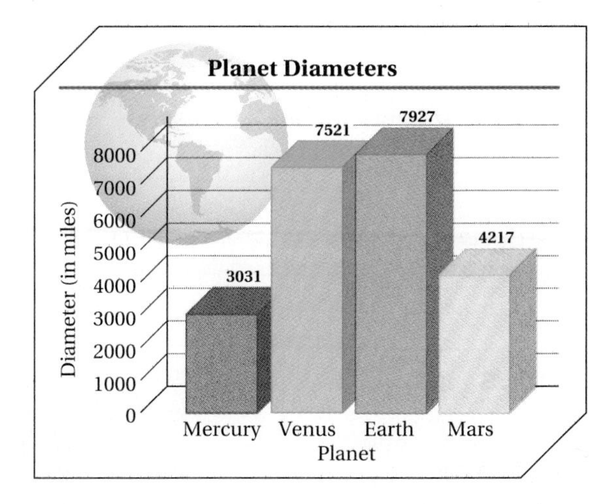

Note that the diameter of Mars is 4217 miles (mi). **Standard notation** for this number is 4217. We find **expanded notation** for 4217 as follows:

4217 = 4 thousands + 2 hundreds + 1 ten + 7 ones.

Example 1 Write expanded notation for 3031 mi, the diameter of Mercury.

3031 = 3 thousands + 0 hundreds + 3 tens + 1 one

Example 2 Write expanded notation for 54,567.

54,567 = 5 ten thousands + 4 thousands + 5 hundreds + 6 tens + 7 ones

Do Exercises 1 and 2 (in the margin at the right).

Write expanded notation.

1. 1805

2. 36,223

Answers on page A-1

Example 3 Write expanded notation for 3400.

3400 = 3 thousands + 4 hundreds + 0 tens + 0 ones, or
3 thousands + 4 hundreds

Do Exercises 3–5.

b From Expanded Notation to Standard Notation

Example 4 Write standard notation for 2 thousands + 5 hundreds + 7 tens + 5 ones.

Standard notation is 2575.

Example 5 Write standard notation for 9 ten thousands + 6 thousands + 7 hundreds + 1 ten + 8 ones.

Standard notation is 96,718.

Example 6 Write standard notation for 2 thousands + 3 tens.

Standard notation is 2030.

Do Exercises 6–8.

c Word Names

"Three," "two hundred one," and "forty-two" are **word names** for numbers. When we write word names for two-digit numbers like 42, 76, and 91, we use hyphens. For example, U.S. Olympic team pitcher Michelle Granger can pitch a softball at a speed of 72 mph. A word name for 72 is "seventy-two."

Examples Write a word name.

7. 43 Forty-three

8. 91 Ninety-one

Do Exercises 9–11.

For large numbers, digits are separated into groups of three, called **periods**. Each period has a name: *ones, thousands, millions, billions,* and so on. When we write or read a large number, we start at the left with the largest period. The number named in the period is followed by the name of the period; then a comma is written and the next period is named. Recently, the U.S. national debt was $5,103,040,000,000. We can use a **place-value** chart to illustrate how to use periods to read the number 5,103,040,000,000.

Write expanded notation.

3. 3210

4. 2009

5. 5700

Write standard notation.

6. 5 thousands + 6 hundreds + 8 tens + 9 ones

7. 8 ten thousands + 7 thousands + 1 hundred + 2 tens + 8 ones

8. 9 thousands + 3 ones

Write a word name.

9. 57

10. 29

11. 88

Answers on page A-1

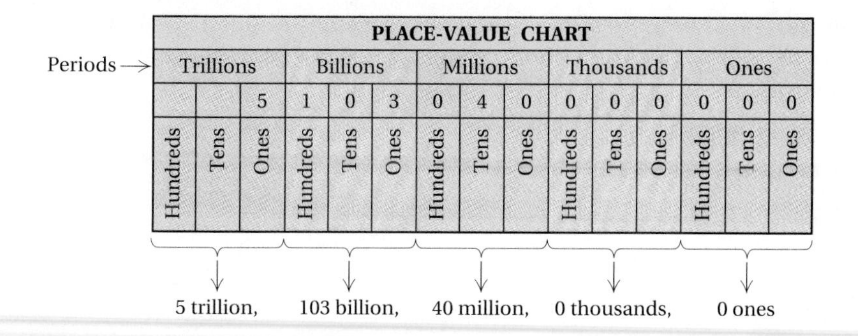

PLACE-VALUE CHART														
Trillions			Billions			Millions			Thousands			Ones		
		5	1	0	3	0	4	0	0	0	0	0	0	0
Hundreds	Tens	Ones	Hundreds	Tens	Ones	Hundreds	Tens	Ones	Hundreds	Tens	Ones	Hundreds	Tens	Ones

Example 9 Write a word name for 46,605,314,732.

Forty-six billion,

six hundred five million,

three hundred fourteen thousand,

seven hundred thirty-two

The word "and" *should not* appear in word names for whole numbers. Although we commonly hear such expressions as "two hundred *and* one," the use of "and" is not, strictly speaking, correct in word names for whole numbers. For decimal notation, it is appropriate to use "and" for the decimal point. For example, 317.4 is read as "three hundred seventeen *and* four tenths."

Do Exercises 12–15.

d From Word Names to Standard Notation

Example 10 Write standard notation.

Five hundred six million,

three hundred forty-five thousand,

two hundred twelve

Standard notation is 506,345,212.

Do Exercise 16.

e Digits

A **digit** is a number 0, 1, 2, 3, 4, 5, 6, 7, 8, or 9 that names a place-value location.

Examples What does the digit 8 mean in each case?

11. 278,342 — 8 thousands

12. 872,342 — 8 hundred thousands

13. 28,343,399,223 — 8 billions

Do Exercises 17–20.

Write a word name.

12. 204

13. 79,204

14. 1,879,204

15. 22,301,879,204

16. Write standard notation.

Two hundred thirteen million,
one hundred five thousand,
three hundred twenty-nine

What does the digit 2 mean in each case?

17. 526,555

18. 265,789

19. 42,789,654

20. 24,789,654

Answers on page A-1

Golf Balls. On an average day, Americans buy 486,575 golf balls. In 486,575, what digit tells the number of:

21. Thousands?

22. Ten thousands?

23. Ones?

24. Hundreds?

Example 14 *Dunkin Donuts.* On an average day about 2,739,726 Dunkin Donuts are served in the United States. In 2,739,726, what digits tells the number of:

a) Hundred thousands 7

b) Thousands 9

Do Exercises 21–24.

Improving Your Math Study Skills

Tips for Using This Textbook

Throughout this textbook, you will find a feature called "Improving Your Math Study Skills." At least one such topic is included in each chapter. Each topic title is listed in the table of contents beginning on p. iii.

One of the most important ways to improve your math study skills is to learn the proper use of the textbook. Here we highlight a few points that we consider most helpful.

- **Be sure to note the special symbols [a], [b], [c], and so on, that correspond to the objectives you are to be able to perform.** They appear in many places throughout the text. The first time you see them is in the margin at the beginning of each section. The second time is in the subheadings of each section, and the third time is in the exercise set. You will also find them next to the skill maintenance exercises in each exercise set and in the review exercises at the end of the chapter, as well as in the answers to the chapter tests and the cumulative reviews. These objective symbols allow you to refer back whenever you need to review a topic.
- **Note the symbols in the margin under the list of objectives at the beginning of each section.** These refer to the many distinctive study aids that accompany the book.
- **Read and study each step of each example.** The examples include important side comments that explain each step. These carefully chosen examples and notes prepare you for success in the exercise set.
- **Stop and do the margin exercises as you study a section.** When our students come to us troubledabout how they are doing in the course, the first question we ask is "Are you doing the margin exercises when directed to do so?" This is one of the most effective ways to enhance your ability to learn mathematics from this text. Don't deprive yourself of its benefits!
- **When you study the book, don't mark the points that you think are important, but mark the points you do not understand!** This book includes many design features that highlight important points. Use your efforts to mark where you are having trouble. Then when you go to class, a math lab, or a tutoring session, you will be prepared to ask questions that home in on your difficulties rather than spending time going over what you already understand.
- **If you are having trouble, consider using the *Student's Solutions Manual,* which contains worked-out solutions to the odd-numbered exercises in the exercise sets.**
- **Try to keep one section ahead of your syllabus.** If you study ahead of your lectures, you can concentrate on what is being explained in them, rather than trying to write everything down. You can then take notes only of special points or of questions related to what is happening in class.

Answers on page A-1

Exercise Set 1.1

Always review the objectives before doing an exercise set. See page 3. Note how the objectives are keyed to the exercises.

a Write expanded notation.

1. 5742

2. 3897

3. 27,342

4. 93,986

5. 5609

6. 9990

7. 2300

8. 7020

b Write standard notation.

9. 2 thousands + 4 hundreds + 7 tens + 5 ones

10. 7 thousands + 9 hundreds + 8 tens + 3 ones

11. 6 ten thousands + 8 thousands + 9 hundreds + 3 tens + 9 ones

12. 1 ten thousand + 8 thousands + 4 hundreds + 6 tens + 1 one

13. 7 thousands + 3 hundreds + 0 tens + 4 ones

14. 8 thousands + 0 hundreds + 2 tens + 0 ones

15. 1 thousand + 9 ones

16. 2 thousands + 4 hundreds + 5 tens

c Write a word name.

17. 85

18. 48

19. 88,000

20. 45,987

21. 123,765

22. 111,013

23. 7,754,211,577

24. 43,550,651,808

Write a word name for the number in the sentence.

25. *NBA Salaries.* In a recent year, the average salary of a player in the NBA was $1,867,000.

26. The area of the Pacific Ocean is about 64,186,000 square miles.

27. *Population.* The population of South Asia is about 1,583,141,000.

28. *Monopoly.* In a recent Monopoly game sponsored by McDonald's restaurants, the odds of winning the grand prize was estimated to be 467,322,388 to 1.

d Write standard notation.

29. Two million, two hundred thirty-three thousand, eight hundred twelve

30. Three hundred fifty-four thousand, seven hundred two

31. Eight billion

32. Seven hundred million

Write standard notation for the number in the sentence.

33. *Light Distance.* Light travels nine trillion, four hundred sixty billion kilometers in one year.

34. *Pluto.* The distance from the sun to Pluto is three billion, six hundred sixty-four million miles.

35. *Area of Greenland.* The area of Greenland is two million, nine hundred seventy-four thousand, six hundred square kilometers.

36. *Memory Space.* On computer hard drives, one gigabyte is actually one billion, seventy-three million, seven hundred forty-one thousand, eight hundred twenty-four bytes of memory.

e What does the digit 5 mean in each case?

37. 235,888

38. 253,888

39. 488,526

40. 500,346

In 89,302, what digit tells the number of:

41. Hundreds?

42. Thousands?

43. Tens?

44. Ones?

Synthesis

Exercises designated as *Synthesis exercises* differ from those found in the main body of the exercise set. The icon ◆ denotes synthesis exercises that are writing exercises. Writing exercises are meant to be answered in one or more complete sentences. Because answers to writing exercises often vary, they are not listed at the back of the book.

Exercises marked with a 🖩 are meant to be solved using a calculator. These and the other synthesis exercises will often challenge you to put together two or more objectives at once.

45. ◆ Write an English sentence in which the number 260,000,000 is used.

46. ◆ Explain why we use commas when writing large numbers.

47. 🖩 What is the largest number that you can name on your calculator? How many digits does that number have? How many periods?

48. How many whole numbers between 100 and 400 contain the digit 2 in their standard notation?

1.2 Addition

Objectives

a Write an addition sentence that corresponds to a situation.

b Add whole numbers.

For Extra Help

TAPE 1 | TAPE 1A | MAC WIN | CD-ROM

a Addition and the Real World

Addition of whole numbers corresponds to combining or putting things together. Let's look at various situations in which addition applies.

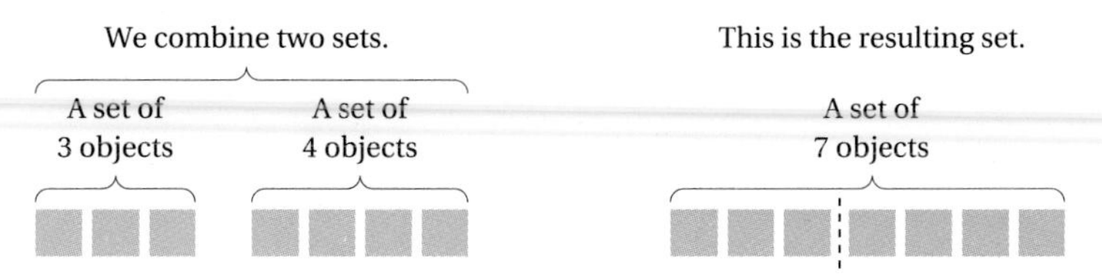

The addition that corresponds to the figure above is

$3 + 4 = 7.$

The number of objects in a set can be found by counting. We count and find that the two sets have 3 members and 4 members, respectively. After combining, we count and find that there are 7 objects. We say that the **sum** of 3 and 4 is 7. The numbers added are called **addends**.

$$3 + 4 = 7$$

Addend, Addend, Sum

Example 1 Write an addition sentence that corresponds to this situation.

A student has \$3 and earns \$10 more. How much money does the student have?

An addition that corresponds is \$3 + \$10 = \$13.

Do Exercises 1 and 2.

Write an addition sentence that corresponds to the situation.

1. John has 8 music CD-ROMs in his backpack. Then he buys 2 educational CD-ROMs at the bookstore. How many CD-ROMs does John have in all?

2. Sue earns \$45 in overtime pay on Thursday and \$33 on Friday. How much overtime pay does she earn altogether on the two days?

Addition also corresponds to combining distances or lengths.

Example 2 Write an addition sentence that corresponds to this situation.

A car is driven 44 mi from San Francisco to San Jose. It is then driven 42 mi from San Jose to Oakland. How far is it from San Francisco to Oakland along the same route?

44 mi + 42 mi = 86 mi

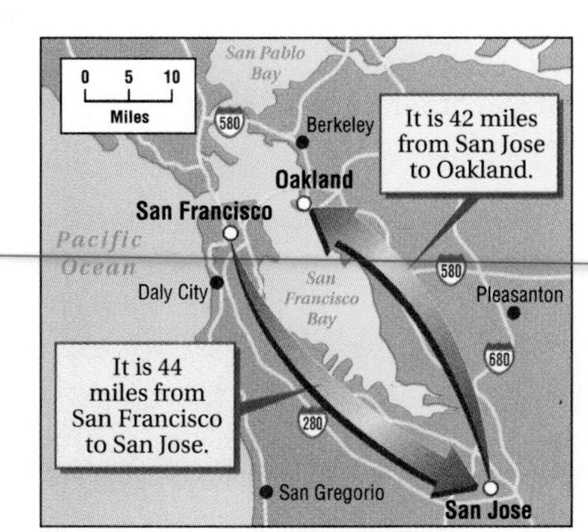

Do Exercises 3 and 4.

Write an addition sentence that corresponds to the situation.

3. A car is driven 100 mi from Austin to Waco. It is then driven 93 mi from Waco to Dallas. How far is it from Austin to Dallas along the same route?

4. A coaxial cable 5 ft (feet) long is connected to a cable 7 ft long. How long is the resulting cable?

Answers on page A-1

Write an addition sentence that corresponds to the situation.

5. Find the perimeter of (distance around) the figure.

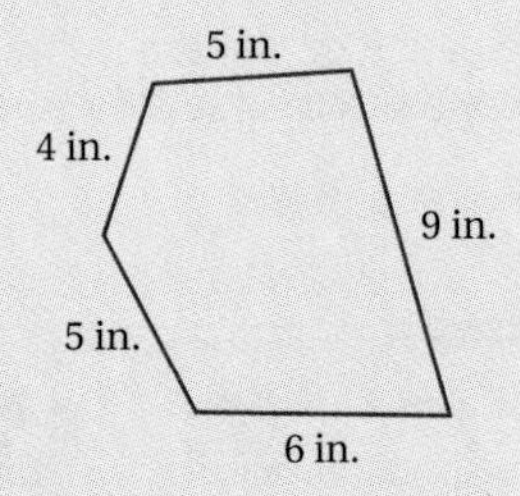

6. Find the perimeter of (distance around) the figure.

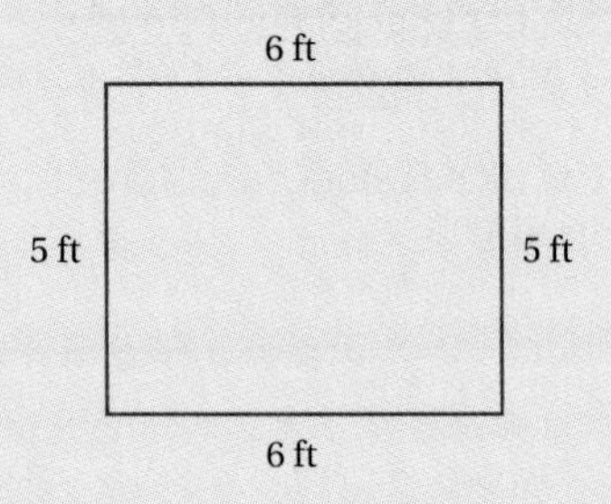

Write an addition sentence that corresponds to the situation.

7. The front parking lot of Sparks Electronics contains 30,000 square feet (sq ft) of parking space. The back lot contains 40,000 sq ft. What is the total area of the two parking lots?

8. You own a small rug that contains 8 square yards (sq yd) of fabric. You buy another rug that contains 9 sq yd. What is the area of the floor covered by both rugs?

Answers on page A-1

When we find the sum of the distances around an object, we are finding its **perimeter.**

Example 3 Write an addition sentence that corresponds to this situation.

A computer sales rep travels the following route to visit various electronics stores. How long is the route?

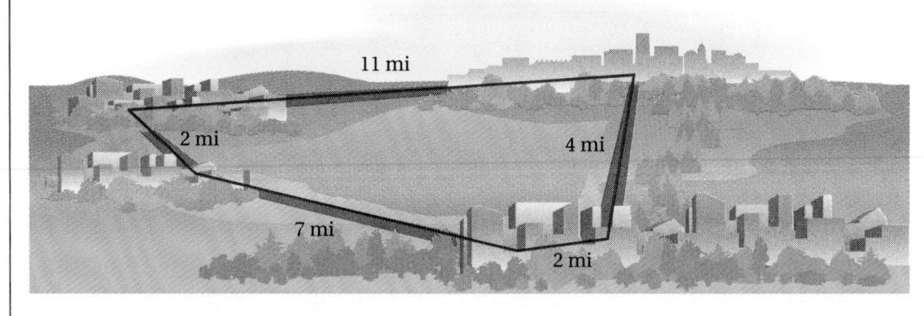

$$2 \text{ mi} + 7 \text{ mi} + 2 \text{ mi} + 4 \text{ mi} + 11 \text{ mi} = 26 \text{ mi}$$

Do Exercises 5 and 6.

Addition also corresponds to combining areas.

Example 4 Write an addition sentence that corresponds to this situation.

The area of a standard large index card is 40 square inches (sq in.). The area of a standard small index card is 15 sq in. Altogether, what is the total area of a large and a small card?

The area of the large index card is 40 sq in.		The area of the small index card is 15 sq in.		The total area of the two cards is 55 sq in.
40 sq in.	+	15 sq in.	=	55 sq in.

Do Exercises 7 and 8.

Addition corresponds to combining volumes as well.

Example 5 Write an addition sentence that corresponds to this situation.

Two trucks haul dirt to a construction site. One hauls 5 cubic yards (cu yd) and the other hauls 7 cu yd. Altogether, how many cubic yards of dirt have they hauled to the site?

$$5 \text{ cu yd} + 7 \text{ cu yd} = 12 \text{ cu yd}$$

Do Exercises 9 and 10.

b Addition of Whole Numbers

To add numbers, we add the ones digits first, then the tens, then the hundreds, and so on.

Example 6 Add: 7312 + 2504.

Place values are lined up in columns.

```
  7 3 1 2     Add ones.
+ 2 5 0 4
---------
        6

  7 3 1 2     Add tens.
+ 2 5 0 4
---------
      1 6

  7 3 1 2     Add hundreds.
+ 2 5 0 4
---------
    8 1 6

  7 3 1 2     Add thousands.
+ 2 5 0 4
---------
  9 8 1 6
```

We show you this for explanation.

You need write only this.

```
  7 3 1 2  <- Addends
+ 2 5 0 4  <-
---------
  9 8 1 6  <- Sum
```

Do Exercise 11.

Example 7 Add: 6878 + 4995.

```
      1
  6 8 7 8     Add ones. We get 13 ones, or 1 ten + 3 ones.
+ 4 9 9 5     Write 3 in the ones column and 1 above the tens.
---------     This is called carrying, or regrouping.
        3

    1 1
  6 8 7 8     Add tens. We get 17 tens, or 1 hundred + 7 tens.
+ 4 9 9 5     Write 7 in the tens column and 1 above the hundreds.
---------
      7 3

  1 1 1
  6 8 7 8     Add hundreds. We get 18 hundreds, or 1 thousand +
+ 4 9 9 5     8 hundreds.
---------     Write 8 in the hundreds column and 1 above the thousands.
    8 7 3

  1 1 1
  6 8 7 8     Add thousands. We get 11 thousands.
+ 4 9 9 5
---------
1 1 8 7 3
```

Do Exercises 12 and 13.

Write an addition sentence that corrresponds to the situation.

9. Two trucks haul sand to a construction site to use in a driveway. One hauls 6 cu yd and the other hauls 8 cu yd. Altogether, how many cubic yards of sand have they hauled to the site?

10. A football fan drives to all college football games using a motor home. On one trip the fan buys 80 gallons (gal) of gasoline and on another, 56 gal. How many gallons were bought in all?

11. Add.

```
  6 2 0 3
+ 3 5 4 2
```

Add.

12.

```
  7 9 6 8
+ 5 4 9 7
```

13.

```
  9 8 0 4
+ 6 3 7 8
```

Answers on page A-1

Add from the top.

14.

$$\begin{array}{r} 9 \\ 9 \\ 4 \\ +\ 5 \\ \hline \end{array}$$

15.

$$\begin{array}{r} 8 \\ 6 \\ 9 \\ 7 \\ +\ 4 \\ \hline \end{array}$$

16. Add from the bottom.

$$\begin{array}{r} 9 \\ 9 \\ 4 \\ +\ 5 \\ \hline \end{array}$$

Answers on page A-1

How do we do an addition of three numbers, like $2 + 3 + 6$? We do so by adding 3 and 6, and then 2. We can show this with parentheses:

$$2 + (3 + 6) = 2 + 9 = 11.$$ **Parentheses tell what to do first.**

We could also add 2 and 3, and then 6:

$$(2 + 3) + 6 = 5 + 6 = 11.$$

Either way we get 11. It does not matter how we group the numbers. This illustrates the **associative law of addition,** $a + (b + c) = (a + b) + c$. We can also add whole numbers in any order. That is, $2 + 3 = 3 + 2$. This illustrates the **commutative law of addition,** $a + b = b + a$. Together the commutative and associative laws tell us that to add more than two numbers, we can use any order and grouping we wish.

Example 8 Add from the top.

$$\begin{array}{r} 8 \\ 9 \\ 7 \\ +\ 6 \\ \hline \end{array}$$

We first add 8 and 9, getting 17; then 17 and 7, getting 24; then 24 and 6, getting 30.

$$\begin{array}{r} 8 \\ 9 \\ 7 \\ +\ 6 \\ \hline 30 \end{array} \quad 8 + 9 \rightarrow 17, \quad 17 + 7 \rightarrow 24, \quad 24 + 6 \rightarrow 30$$

You write only this.

Do Exercises 14 and 15.

Example 9 Add from the bottom.

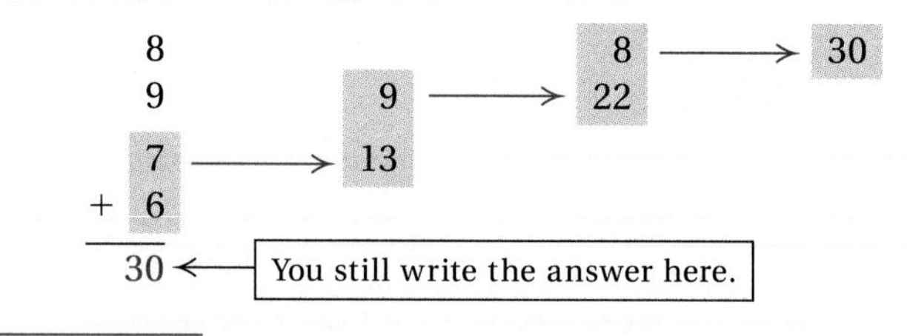

Do Exercise 16.

Sometimes it is easier to look for pairs of numbers whose sums are 10 or 20 or 30, and so on.

Examples Add.

10.

$$\begin{array}{r} 14 \\ 8 \\ 6 \\ 2 \\ +\ 9 \\ \hline 39 \end{array} \qquad \begin{array}{r} 20 \\ \\ \\ 10 \\ 9 \\ \hline 39 \end{array}$$

You should write only the answer in the position shown.

11. $23 + 19 + 7 + 21 + 4 = 74$

$30 + 40 + 4$

74

Do Exercises 17–19.

Example 12 Add: $2391 + 3276 + 8789 + 1498$.

$$\begin{array}{r} \scriptstyle 2 \\ 2\,3\,9\,1 \\ 3\,2\,7\,6 \\ 8\,7\,8\,9 \\ +\ 1\,4\,9\,8 \\ \hline 4 \end{array}$$

Add ones: We get 24, so we have 2 tens + 4 ones. Write 4 in the ones column and 2 above the tens.

$$\begin{array}{r} \scriptstyle 3\ 2 \\ 2\,3\,9\,1 \\ 3\,2\,7\,6 \\ 8\,7\,8\,9 \\ +\ 1\,4\,9\,8 \\ \hline 5\,4 \end{array}$$

Add tens: We get 35 tens, so we have 30 tens + 5 tens. This is also 3 hundreds + 5 tens. Write 5 in the tens column and 3 above the hundreds.

$$\begin{array}{r} \scriptstyle 1\ 3\ 2 \\ 2\,3\,9\,1 \\ 3\,2\,7\,6 \\ 8\,7\,8\,9 \\ +\ 1\,4\,9\,8 \\ \hline 9\,5\,4 \end{array}$$

Add hundreds: We get 19 hundreds, or 1 thousand + 9 hundreds. Write 9 in the hundreds column and 1 above the thousands.

$$\begin{array}{r} \scriptstyle 1\ 3\ 2 \\ 2\,3\,9\,1 \\ 3\,2\,7\,6 \\ 8\,7\,8\,9 \\ +\ 1\,4\,9\,8 \\ \hline 1\,5\,9\,5\,4 \end{array}$$

Add thousands: We get 15 thousands.

Do Exercise 20.

Add. Look for pairs of numbers whose sums are 10, 20, 30, and so on.

17.
$$\begin{array}{r} 15 \\ 7 \\ 5 \\ 3 \\ +\ 8 \\ \hline \end{array}$$

18. $6 + 12 + 14 + 8 + 7$

19. $27 + 8 + 13 + 2 + 11$

20. Add.
$$\begin{array}{r} 1\,9\,3\,2 \\ 6\,7\,2\,3 \\ 9\,8\,7\,8 \\ +\ 8\,9\,4\,1 \\ \hline \end{array}$$

To the instructor and the student: This section presented a review of addition of whole numbers. Students who are successful should go on to Section 1.3. Those who have trouble should study developmental unit A and then repeat Section 1.2.

Answers on page A-1

Improving Your Math Study Skills

Getting Started in a Math Class: The First-Day Handout or Syllabus

There are many ways in which to improve your math study skills. We have already considered some tips on using this book (see Section 1.1). We now consider some more general tips.

- **Textbook.** On the first day of class, most instructors distribute a handout that lists the textbook and other materials needed in the course. If possible, call the instructor or the department office before the term begins to find out which textbook you will be using and visit the bookstore to pick it up. This way, you can purchase the book before class starts and be ready to begin studying. Delay in obtaining a copy of the textbook may cause you to fall behind in your homework.
- **Attendance.** The handout may also describe the attendance policy for your class. Some instructors take attendance at every class, while others use different methods to track students' attendance. Regardless of the policy, you should plan to attend class every time. Missing even one class can cause you to fall behind. If attendance counts toward your course grade, find out if there is a way to make up for missed days. In general, missing a class is not as catastrophic if you put in the effort to catch up by studying the material on your own.

  If you do miss a class, call the instructor as soon as possible to find out what material was covered and what was assigned for the next class. If you have a study partner, call this person; ask if you can make a copy of his or her notes and find out what the homework assignment was. It is a good idea to meet with your instructor in person to clarify any concepts that you do not understand. This way, when you do return to class, you will be able to follow along with the rest of the group.
- **Homework.** The first-day handout may also detail how homework is handled. Find out when, and how often, homework will be assigned, whether homework is collected or graded, and whether there will be quizzes over the homework material.

  If the homework will be graded, find out what part of the final grade it will determine. Ask what the policy is for late homework: Some instructors are willing to accept homework after the deadline, while others are more strict. If you do miss a homework deadline, be sure to do the assigned homework anyway, as this is the best way to learn the material.
- **Grading.** The handout may also provide information on how your grade will be calculated at the end of the term. Typically, there will be tests during the term and a final exam at the end of the term. Frequently, homework is counted as part of the grade calculation, as are the quizzes. Find out how many tests will be given, if there is an option for make-up tests, or if any test grades will be dropped at the end of the term.

  Some instructors keep the class grades on a computer. If this is the case, find out if you can receive current grade reports throughout the term. This will help you focus on what is needed to obtain the desired grade in the course. Although a good grade should not be your only goal in this class, most students find it motivational to know what their grade is at any time during the term.
- **Get to know your classmates.** It can be a big help in a math class to get to know your fellow students. You might consider forming a study group. If you do so, find out their phone numbers and schedules so that you can coordinate study time for homework or tests.
- **Get to know your instructor.** It can, of course, help immensely to get to know your instructor. Trivial though it may seem, get basic information like his or her name, how he or she can be contacted outside of class, and where the office is.

  Learn about your instructor's teaching style and try to adapt your learning to it. Does he or she use an overhead projector or the board? Will there be frequent in-class questions?

Exercise Set 1.2

a Write an addition sentence that corresponds to the situation.

1. Isabel receives 7 e-mail messages on Tuesday and 8 on Wednesday. How many e-mail messages did she receive altogether on the two days?

2. At a construction site, there are two gasoline containers to be used by earth-moving vehicles. One contains 400 gal and the other 200 gal. How many gallons do both contain altogether?

3. A builder buys two parcels of land to build a housing development. One contains 500 acres and the other 300 acres. What is the total number of acres purchased?

4. During March and April, Deron earns extra money doing income taxes part time. In March he earned $220, and in April he earned $340. How much extra did he earn altogether in March and April?

Find the perimeter of (distance around) the figure.

5.

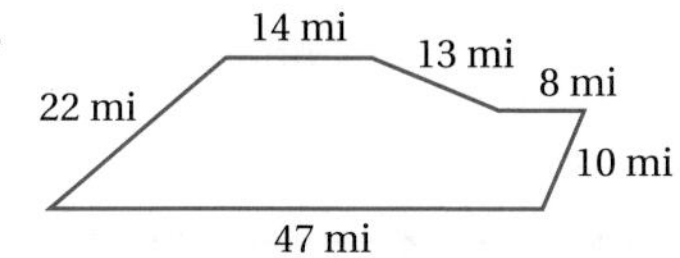

6.

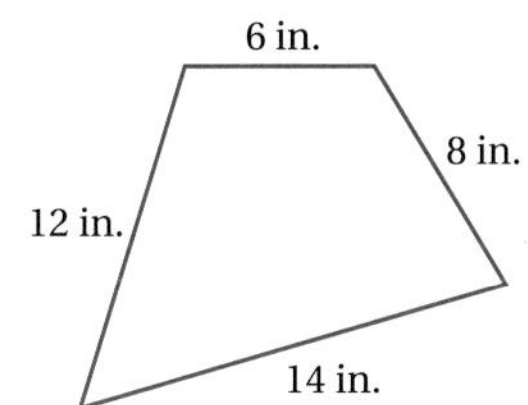

7.

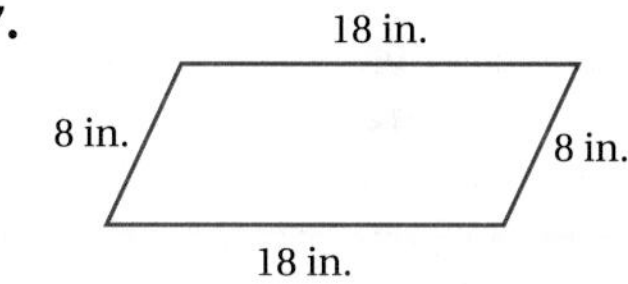

8.

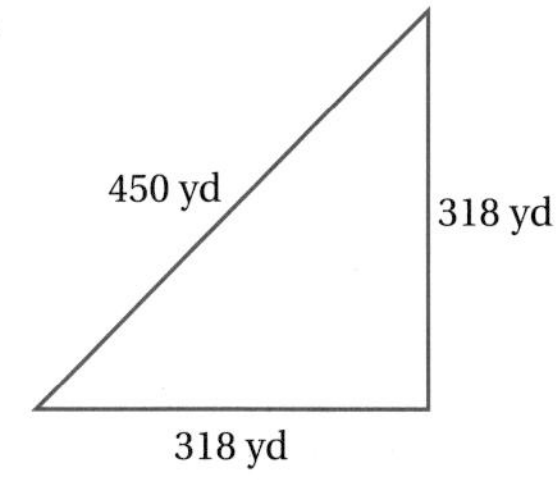

9.

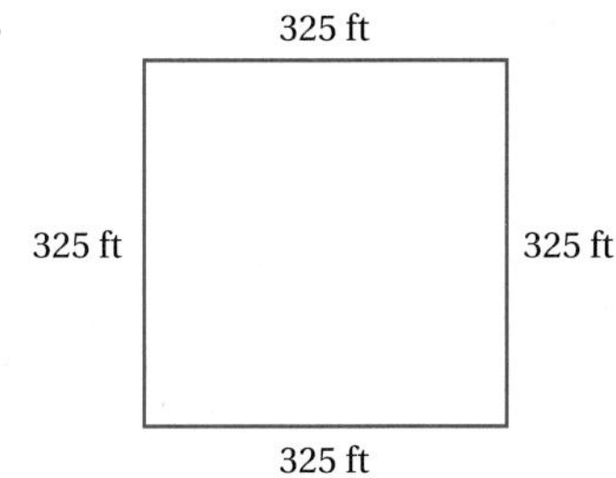

10.

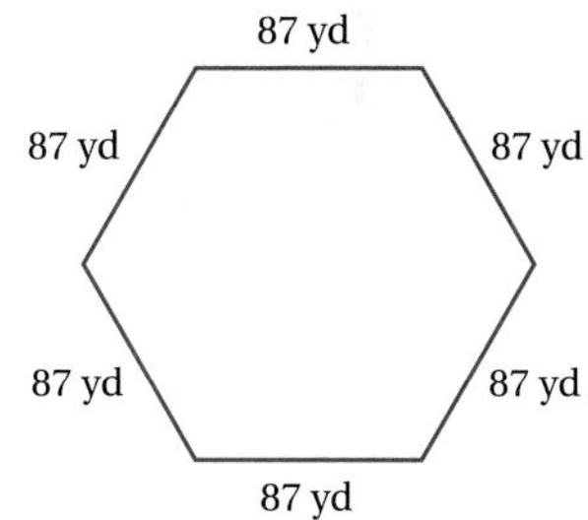

b Add.

11. $\begin{array}{r} 364 \\ +\ \ 23 \\ \hline \end{array}$

12. $\begin{array}{r} 1521 \\ +\ \ 348 \\ \hline \end{array}$

13. $\begin{array}{r} 1716 \\ +3482 \\ \hline \end{array}$

14. $\begin{array}{r} 7503 \\ +2683 \\ \hline \end{array}$

15. $\begin{array}{r} 86 \\ +\ 78 \\ \hline \end{array}$

16. $\begin{array}{r} 73 \\ +\ 69 \\ \hline \end{array}$

17. $\begin{array}{r} 99 \\ +\ \ 1 \\ \hline \end{array}$

18. $\begin{array}{r} 999 \\ +\ \ 11 \\ \hline \end{array}$

19. 789 + 111

20. 839 + 386

21. 909 + 101

22. 707 + 909

23. 8113 + 390

24. 271 + 3338

25. 356 + 4910

26. 280 + 34,702

27. 3870 + 92 + 7 + 497

28. 10,120 + 12,989 + 5738

29. $\begin{array}{r} 5093 \\ +\ 3217 \\ \hline \end{array}$

30. $\begin{array}{r} 3654 \\ +\ 2700 \\ \hline \end{array}$

31. $\begin{array}{r} 4825 \\ +\ 1783 \\ \hline \end{array}$

32. $\begin{array}{r} 6775 \\ +\ 1432 \\ \hline \end{array}$

33. $\begin{array}{r} 9999 \\ +\ 6785 \\ \hline \end{array}$

34. $\begin{array}{r} 45{,}879 \\ +\ 21{,}786 \\ \hline \end{array}$

35. $\begin{array}{r} 23{,}443 \\ +\ 10{,}989 \\ \hline \end{array}$

36. $\begin{array}{r} 67{,}654 \\ +\ 98{,}786 \\ \hline \end{array}$

37. $\begin{array}{r} 77{,}543 \\ +\ 23{,}767 \\ \hline \end{array}$

38. $\begin{array}{r} 44{,}654 \\ +\ \ 4{,}765 \\ \hline \end{array}$

39. $\begin{array}{r} 99{,}999 \\ +\ \ \ \ 112 \\ \hline \end{array}$

40. $\begin{array}{r} 127{,}556 \\ +\ \ 68{,}766 \\ \hline \end{array}$

Add from the top. Then check by adding from the bottom.

41. $\begin{array}{r} 7 \\ 9 \\ 4 \\ +\ 8 \\ \hline \end{array}$

42. $\begin{array}{r} 4 \\ 3 \\ 9 \\ 1 \\ +\ 8 \\ \hline \end{array}$

43. $\begin{array}{r} 8 \\ 6 \\ 2 \\ 3 \\ +\ 7 \\ \hline \end{array}$

44. $\begin{array}{r} 9 \\ 4 \\ 7 \\ 8 \\ +\ 7 \\ \hline \end{array}$

Add. Look for pairs of numbers whose sums are 10, 20, 30, and so on.

45. $\begin{array}{r} 7 \\ 18 \\ 3 \\ 37 \\ +\ 2 \\ \hline \end{array}$

46. $\begin{array}{r} 23 \\ 16 \\ 11 \\ 18 \\ +\ 19 \\ \hline \end{array}$

47. $\begin{array}{r} 45 \\ 25 \\ 36 \\ 44 \\ +\ 80 \\ \hline \end{array}$

48. $\begin{array}{r} 38 \\ 27 \\ 32 \\ 14 \\ +\ 76 \\ \hline \end{array}$

Add.

49. $\begin{array}{r} 23 \\ 62 \\ +\ 45 \\ \hline \end{array}$

50. $\begin{array}{r} 43 \\ 11 \\ +\ 37 \\ \hline \end{array}$

51. $\begin{array}{r} 451 \\ 36 \\ +\ 862 \\ \hline \end{array}$

52. $\begin{array}{r} 31 \\ 753 \\ +\ 924 \\ \hline \end{array}$

53. $\begin{array}{r} 2,603 \\ 28,214 \\ +\ 6,109 \\ \hline \end{array}$

54. $\begin{array}{r} 93,249 \\ 1,268 \\ +\ 74,823 \\ \hline \end{array}$

55. $\begin{array}{r} 12,070 \\ 2,954 \\ +\ 3,400 \\ \hline \end{array}$

56. $\begin{array}{r} 42,487 \\ 83,141 \\ +\ 36,712 \\ \hline \end{array}$

57. $\begin{array}{r} 327 \\ 428 \\ 569 \\ 787 \\ +\ 209 \\ \hline \end{array}$

58. $\begin{array}{r} 989 \\ 566 \\ 834 \\ 920 \\ +\ 703 \\ \hline \end{array}$

59. $\begin{array}{r} 4835 \\ 729 \\ 9204 \\ 8986 \\ +\ 7931 \\ \hline \end{array}$

60. $\begin{array}{r} 5,946 \\ 834 \\ 12,956 \\ 928,342 \\ 34,901 \\ +\ 56,000 \\ \hline \end{array}$

61.
$$\begin{array}{r} 2037 \\ 4923 \\ 3471 \\ +1248 \\ \hline \end{array}$$

62.
$$\begin{array}{r} 4567 \\ 1023 \\ 4821 \\ +3683 \\ \hline \end{array}$$

63.
$$\begin{array}{r} 3420 \\ 8719 \\ 4312 \\ +6203 \\ \hline \end{array}$$

64.
$$\begin{array}{r} 2003 \\ 149 \\ 58 \\ +3426 \\ \hline \end{array}$$

65.
$$\begin{array}{r} 5{,}678{,}987 \\ 1{,}409{,}312 \\ 898{,}888 \\ +4{,}777{,}910 \\ \hline \end{array}$$

66.
$$\begin{array}{r} 78{,}899{,}311 \\ 6{,}784{,}170 \\ 11{,}541{,}913 \\ +\quad 100{,}817 \\ \hline \end{array}$$

Skill Maintenance

The exercises that follow begin an important feature called *skill maintenance exercises.* These exercises provide an ongoing review of any preceding objective in the book. You will see them in virtually every exercise set. It has been found that this kind of extensive review can significantly improve your performance on a final examination.

67. Write standard notation for 7 thousands + 9 hundreds + 9 tens + 2 ones. [1.1b]

68. Write a word name for the number in the following sentence: [1.1c]

In a recent year, the gross revenue of the NBA was $924,600,000 (***Source:*** *Wall Street Journal*).

69. What does the digit 8 mean in 486,205? [1.1e]

70. Write standard notation: [1.1d]

Twenty-three million.

Synthesis

71. ◆ Describe a situation that corresponds to the addition 80 sq ft + 140 sq ft. (See Examples 2–5.)

72. ◆ Explain in your own words what the associative law of addition means.

Add.

73. ▦ 5,987,943 + 328,959 + 49,738,765

74. ▦ 39,487,981 + 8,709,486 + 989,765

75. A fast way to add all the numbers from 1 to 10 inclusive is to pair 1 with 9, 2 with 8, and so on. Use a similar approach to add all the numbers from 1 to 100 inclusive.

1.3 Subtraction

Objectives

a. Write a subtraction sentence that corresponds to a situation involving "take away."

b. Given a subtraction sentence, write a related addition sentence; and given an addition sentence, write two related subtraction sentences.

c. Write a subtraction sentence that corresponds to a situation involving "how much more."

d. Subtract whole numbers.

For Extra Help

TAPE 1 TAPE 1B MAC WIN CD-ROM

a Subtraction and the Real World: Take Away

Subtraction of whole numbers corresponds to two kinds of situations. The first one is called "take away."

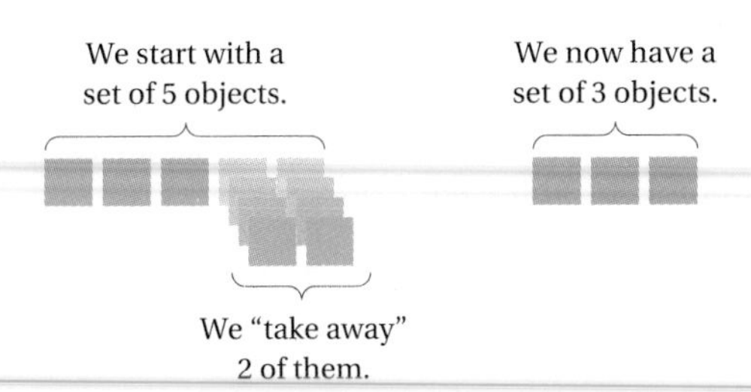

The subtraction that corresponds to the figure above is as follows.

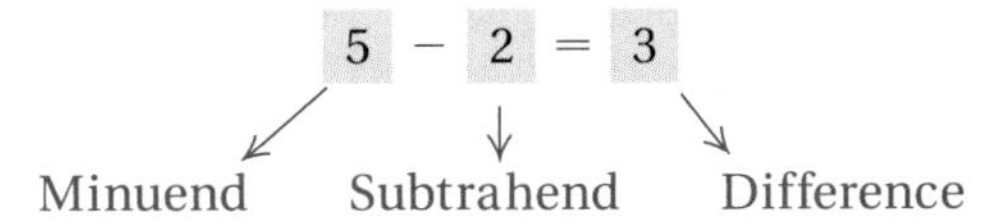

A **subtrahend** is the number being subtracted. A **difference** is the result of subtracting one number from another. That is, it is the result of subtracting the subtrahend from the **minuend**.

Examples Write a subtraction sentence that corresponds to the situation.

1. Juan goes to a music store and chooses 10 CDs to take to the listening station. He rejects 7 of them, but buys the rest. How many CDs did Juan buy?

2. A student has $300 and spends $85 for office supplies. How much money is left?

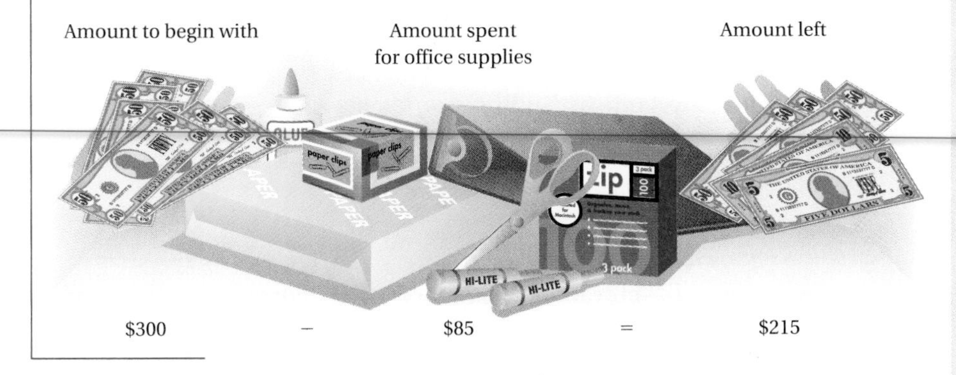

Do Exercises 1 and 2.

Write a subtraction sentence that corresponds to the situation.

1. A contractor removes 5 cu yd of sand from a pile containing 67 cu yd. How many cubic yards of sand are left in the pile?

2. Sparks Electronics owns a field next door that has an area of 20,000 sq ft. Deciding they need more room for parking, the owners have 12,000 sq ft paved. How many square feet of field are left unpaved?

Answers on page A-2

b Related Sentences

Subtraction is defined in terms of addition. For example, $5 - 2$ is that number which when added to 2 gives 5. Thus for the subtraction sentence

$5 - 2 = 3,$ **Taking away 2 from 5 gives 3.**

there is a *related addition* sentence

$5 = 3 + 2.$ **Putting back the 2 gives 5 again.**

In fact, we know answers to subtractions are correct only because of the related addition, which provides a handy way to check a subtraction.

Example 3 Write a related addition sentence: $8 - 5 = 3$.

$8 - 5 = 3$

↑

This number gets added (after 3).

↓

$8 = 3 + 5$

By the commutative law of addition, there is also another addition sentence:

$8 = 5 + 3.$

The related addition sentence is $8 = 3 + 5$.

Do Exercises 3 and 4.

Example 4 Write two related subtraction sentences: $4 + 3 = 7$.

$4 + 3 = 7$

This addend gets subtracted from the sum.

$4 = 7 - 3$

(7 take away 3 is 4.)

$4 + 3 = 7$

This addend gets subtracted from the sum.

$3 = 7 - 4$

(7 take away 4 is 3.)

The related subtraction sentences are $4 = 7 - 3$ and $3 = 7 - 4$.

Do Exercises 5 and 6.

c How Much More?

The second kind of situation to which subtraction corresponds is called "how much more"? We need the concept of a missing addend for "how-much-more" problems. From the related sentences, we see that finding a *missing addend* is the same as finding a *difference*.

Missing addend: $12 = 3 + \square$

Difference: $12 - 3 = \square$

Examples Write a subtraction sentence that corresponds to the situation.

5. A student has \$47 and wants to buy a graphing calculator that costs \$89. How much more is needed to buy the calculator?

Write a related addition sentence.

3. $7 - 5 = 2$

4. $17 - 8 = 9$

Write two related subtraction sentences.

5. $5 + 8 = 13$

6. $11 + 3 = 14$

Answers on page A-2

To find the subtraction sentence, we first consider addition.

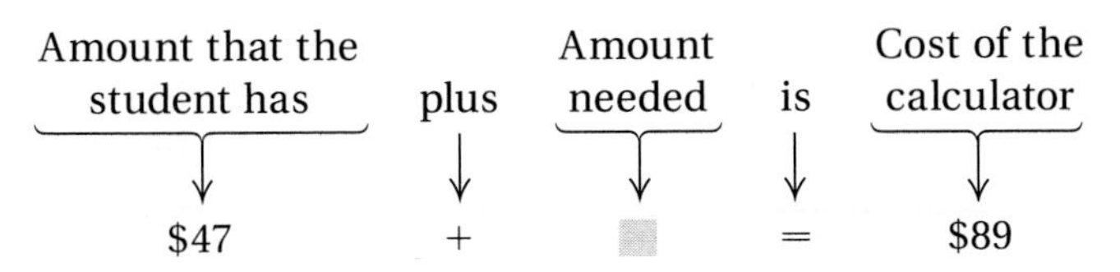

Now we write a related subtraction sentence:

$47 + \square = 89$

$\square = 89 - 47.$ **The addend 47 gets subtracted.**

6. Cathy is reading *True Success: A New Philosophy of Excellence,* by Tom Morris, as part of her philosophy class. It contains 288 pages. She has read 126 pages. How many more pages must she read?

Now we write a related subtraction sentence:

$126 + \square = 288$

$\square = 288 - 126.$ **126 gets subtracted.**

Do Exercises 7 and 8.

d Subtraction of Whole Numbers

To subtract numbers, we subtract the ones digits first, then the tens, then the hundreds, and so on.

Example 7 Subtract: $9768 - 4320$.

$$\begin{array}{r} 9\,7\,6\,8 \\ -\;4\,3\,2\,0 \\ \hline 8 \end{array}$$ **Subtract ones.**

$$\begin{array}{r} 9\,7\,6\,8 \\ -\;4\,3\,2\,0 \\ \hline 4\,8 \end{array}$$ **Subtract tens.**

$$\begin{array}{r} 9\,7\,6\,8 \\ -\;4\,3\,2\,0 \\ \hline 4\,4\,8 \end{array}$$ **Subtract hundreds.**

$$\begin{array}{r} 9\,7\,6\,8 \\ -\;4\,3\,2\,0 \\ \hline 5\,4\,4\,8 \end{array}$$ **Subtract thousands.**

This is for explanation.

$$\begin{array}{r} 9\,7\,6\,8 \\ -\;4\,3\,2\,0 \\ \hline 5\,4\,4\,8 \end{array}$$

You should write only this.

Do Exercise 9.

Write an addition sentence and a related subtraction sentence corresponding to the situation. You need not carry out the subtraction.

7. It is 348 mi from Miami to Jacksonville. Alice has driven 67 mi from Miami to West Palm Beach on the way to Jacksonville. How much farther does she have to drive to get to Jacksonville?

8. A bricklayer estimates that it will take 1200 bricks to complete the side of a building but he has only 800 bricks on the job site. How many more bricks will be needed?

9. Subtract.

$$\begin{array}{r} 7\,8\,9\,3 \\ -\;4\,0\,9\,2 \\ \hline \end{array}$$

Answers on page A-2

Subtract. Check by adding.

10. $$\begin{array}{r} 8\,6\,8\,6 \\ -\,2\,3\,5\,8 \\ \hline \end{array}$$

11. $$\begin{array}{r} 7\,1\,4\,5 \\ -\,2\,3\,9\,8 \\ \hline \end{array}$$

Subtract.

12. $$\begin{array}{r} 7\,0 \\ -\,1\,4 \\ \hline \end{array}$$

13. $$\begin{array}{r} 5\,0\,3 \\ -\,2\,9\,8 \\ \hline \end{array}$$

Subtract.

14. $$\begin{array}{r} 7\,0\,0\,7 \\ -\,6\,3\,4\,9 \\ \hline \end{array}$$

15. $$\begin{array}{r} 6\,0\,0\,0 \\ -\,3\,1\,4\,9 \\ \hline \end{array}$$

16. $$\begin{array}{r} 9\,0\,3\,5 \\ -\,7\,4\,8\,9 \\ \hline \end{array}$$

To the instructor and the student: This section presented a review of subtraction of whole numbers. Students who are successful should go on to Section 1.4. Those who have trouble should study developmental unit S and then repeat Section 1.3.

Answers on page A-2

Sometimes we need to borrow.

Example 8 Subtract: 6246 − 1879.

$$\begin{array}{r} 3\;16 \\ 6\,2\,\not{4}\,\not{6} \\ -\,1\,8\,7\,9 \\ \hline 7 \end{array}$$

We cannot subtract 9 ones from 6 ones, but we can subtract 9 ones from 16 ones. We borrow 1 ten to get 16 ones.

$$\begin{array}{r} 13 \\ 1\;\not{3}\;16 \\ 6\,\not{2}\,\not{4}\,\not{6} \\ -\,1\,8\,7\,9 \\ \hline 6\,7 \end{array}$$

We cannot subtract 7 tens from 3 tens, but we can subtract 7 tens from 13 tens. We borrow 1 hundred to get 13 tens.

$$\begin{array}{r} 11\;13 \\ 5\;\not{1}\;\not{3}\;16 \\ \not{6}\,\not{2}\,\not{4}\,\not{6} \\ -\,1\,8\,7\,9 \\ \hline 4\,3\,6\,7 \end{array}$$

We cannot subtract 8 hundreds from 1 hundred, but we can subtract 8 hundreds from 11 hundreds. We borrow 1 thousand to get 11 hundreds.

We can always check the answer by adding it to the number being subtracted.

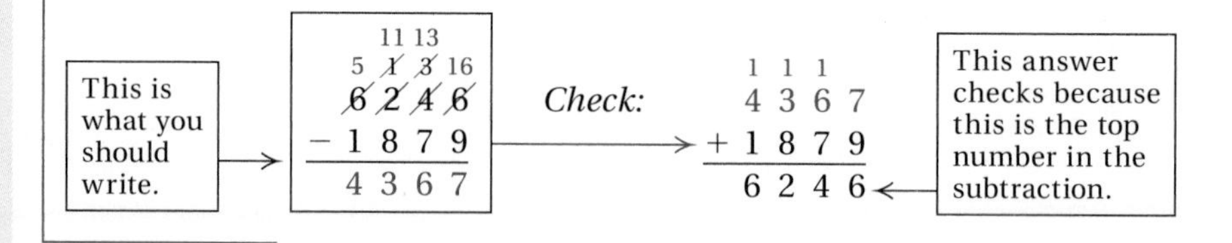

Do Exercises 10 and 11.

Example 9 Subtract: 902 − 477.

$$\begin{array}{r} 8\;\;9\;12 \\ \not{9}\,\not{0}\,\not{2} \\ -\,4\,7\,7 \\ \hline 4\,2\,5 \end{array}$$

We cannot subtract 7 ones from 2 ones. We have 9 hundreds, or 90 tens. We borrow 1 ten to get 12 ones. We then have 89 tens.

Do Exercises 12 and 13.

Example 10 Subtract: 8003 − 3667.

$$\begin{array}{r} 7\;\;9\;\;9\;13 \\ \not{8}\,\not{0}\,\not{0}\,\not{3} \\ -\,3\,6\,6\,7 \\ \hline 4\,3\,3\,6 \end{array}$$

We have 8 thousands, or 800 tens.
We borrow 1 ten to get 13 ones. We then have 799 tens.

Examples

11. Subtract: 6000 − 3762.

$$\begin{array}{r} 5\;\;9\;\;9\;10 \\ \not{6}\,\not{0}\,\not{0}\,\not{0} \\ -\,3\,7\,6\,2 \\ \hline 2\,2\,3\,8 \end{array}$$

12. Subtract: 6024 − 2968.

$$\begin{array}{r} 11 \\ 5\;\;9\;\;\not{1}\;14 \\ \not{6}\,\not{0}\,\not{2}\,\not{4} \\ -\,2\,9\,6\,8 \\ \hline 3\,0\,5\,6 \end{array}$$

Do Exercises 14–16.

Exercise Set 1.3

a Write a subtraction sentence that corresponds to the situation. You need not carry out the subtraction.

1. Jeanne has $1260 in her college checking account. She spends $450 for her food bill at the dining hall. How much is left in her account?

2. *Frozen Yogurt.* A dispenser at a frozen yogurt store contains 126 ounces (oz) of strawberry yogurt. A 13-oz cup is sold to a customer. How much is left in the dispenser?

3. A chef pours 5 oz of salsa from a jar containing 16 oz. How many ounces are left?

4. *Chocolate Cake.* One slice of chocolate cake with fudge frosting contains 564 calories. One cup of hot cocoa made with skim milk contains 188 calories. How many more calories are in the cake than in the cocoa?

b Write a related addition sentence.

5. $7 - 4 = 3$

6. $12 - 5 = 7$

7. $13 - 8 = 5$

8. $9 - 9 = 0$

9. $23 - 9 = 14$

10. $20 - 8 = 12$

11. $43 - 16 = 27$

12. $51 - 18 = 33$

Write two related subtraction sentences.

13. $6 + 9 = 15$

14. $7 + 9 = 16$

15. $8 + 7 = 15$

16. $8 + 0 = 8$

17. $17 + 6 = 23$

18. $11 + 8 = 19$

19. $23 + 9 = 32$

20. $42 + 10 = 52$

c Write an addition sentence and a related subtraction sentence corresponding to the situation. You need not carry out the subtraction.

21. *Kangaroos.* There are 32 million kangaroos in Australia and 17 million people. How many more kangaroos are there than people?

22. *Interstate Speeds.* Recently, speed limits on interstate highways in many Western states were raised from 65 mph to 75 mph. By how many miles per hour were they raised?

23. A set of drapes requires 23 yards (yd) of material. The decorator has 10 yd of material in stock. How much more must be ordered?

24. Marv needs to bowl a score of 223 in order to beat his opponent. His score with one frame to go is 195. How many pins does Marv need in the last frame to beat his opponent?

d Subtract.

25. $\begin{array}{r} 16 \\ -\ \ 4 \\ \hline \end{array}$

26. $\begin{array}{r} 86 \\ -13 \\ \hline \end{array}$

27. $\begin{array}{r} 65 \\ -21 \\ \hline \end{array}$

28. $\begin{array}{r} 87 \\ -34 \\ \hline \end{array}$

29. $\begin{array}{r} 866 \\ -333 \\ \hline \end{array}$

30. $\begin{array}{r} 526 \\ -323 \\ \hline \end{array}$

31. $\begin{array}{r} 4547 \\ -3421 \\ \hline \end{array}$

32. $\begin{array}{r} 6875 \\ -2111 \\ \hline \end{array}$

33. 86 − 47

34. 73 − 28

35. 625 − 327

36. 726 − 509

37. 835 − 609

38. 953 − 246

39. 981 − 747

40. 887 − 698

41. $\begin{array}{r} 7769 \\ -2387 \\ \hline \end{array}$

42. $\begin{array}{r} 6431 \\ -2896 \\ \hline \end{array}$

43. $\begin{array}{r} 3982 \\ -2489 \\ \hline \end{array}$

44. $\begin{array}{r} 7650 \\ -1765 \\ \hline \end{array}$

45. $\begin{array}{r} 5046 \\ -2859 \\ \hline \end{array}$

46. $\begin{array}{r} 6308 \\ -2679 \\ \hline \end{array}$

47. $\begin{array}{r} 7640 \\ -3809 \\ \hline \end{array}$

48. $\begin{array}{r} 8003 \\ -\ \ 599 \\ \hline \end{array}$

49. $\begin{array}{r} 12,647 \\ -\ 4,899 \\ \hline \end{array}$

50. $\begin{array}{r} 16,222 \\ -\ 5,888 \\ \hline \end{array}$

51. $\begin{array}{r} 46,771 \\ -\ 12,977 \\ \hline \end{array}$

52. $\begin{array}{r} 95,654 \\ -\ 48,985 \\ \hline \end{array}$

53. 10,002 − 7834

54. 23,048 − 17,592

55. 90,237 − 47,209

56. 84,703 − 298

57. $\begin{array}{r} 80 \\ -\ 24 \\ \hline \end{array}$

58. $\begin{array}{r} 40 \\ -\ 37 \\ \hline \end{array}$

59. $\begin{array}{r} 90 \\ -\ 54 \\ \hline \end{array}$

60. $\begin{array}{r} 90 \\ -\ 78 \\ \hline \end{array}$

61. $\begin{array}{r} 140 \\ -\ 56 \\ \hline \end{array}$

62. $\begin{array}{r} 470 \\ -\ 188 \\ \hline \end{array}$

63. $\begin{array}{r} 690 \\ -\ 236 \\ \hline \end{array}$

64. $\begin{array}{r} 803 \\ -\ 418 \\ \hline \end{array}$

65. $\begin{array}{r} 903 \\ -\ 132 \\ \hline \end{array}$

66. $\begin{array}{r} 6408 \\ -\ 258 \\ \hline \end{array}$

67. $\begin{array}{r} 2300 \\ -\ 109 \\ \hline \end{array}$

68. $\begin{array}{r} 3506 \\ -\ 1293 \\ \hline \end{array}$

69. $\begin{array}{r} 6808 \\ -\ 3059 \\ \hline \end{array}$

70. $\begin{array}{r} 7840 \\ -\ 3027 \\ \hline \end{array}$

71. $\begin{array}{r} 8092 \\ -\ 1073 \\ \hline \end{array}$

72. $\begin{array}{r} 6007 \\ -\ 1589 \\ \hline \end{array}$

73. 5843 − 98

74. 10,002 − 398

75. 101,734 − 5760

76. 15,017 − 7809

77. 10,008 − 19

78. 21,043 − 8909

79. 83,907 − 89

80. 311,568 − 19,394

81. $\begin{array}{r} 7000 \\ -\ 2794 \\ \hline \end{array}$

82. $\begin{array}{r} 8001 \\ -\ 6543 \\ \hline \end{array}$

83. $\begin{array}{r} 48,000 \\ -\ 37,695 \\ \hline \end{array}$

84. $\begin{array}{r} 17,043 \\ -\ 11,598 \\ \hline \end{array}$

Skill Maintenance

85. What does the digit 7 mean in 6,375,602? [1.1e]

86. Write a word name for 6,375,602. [1.1c]

87. Write standard notation for 2 ten thousands + 9 thousands + 7 hundreds + 8 ones. [1.1b]

88. Add: 9807 + 12,885. [1.2b]

Synthesis

89. ◆ Is subtraction commutative (is there a commutative law of subtraction)? Why or why not?

90. ◆ Describe a situation that corresponds to the subtraction $20 − $17. (See Examples 2 and 5.)

Subtract.

91. 🖩 3,928,124 − 1,098,947

92. 🖩 21,431,206 − 9,724,837

93. Fill in the missing digits to make the equation true:
9, ▒48,621 − 2,097▒81 = 7,251,140.

1.4 Rounding and Estimating; Order

a Rounding

We round numbers in various situations if we do not need an exact answer. For example, we might round to check if an answer to a problem is reasonable or to check a calculation done by hand or on a calculator. We might also round to see if we are being charged the correct amount in a store.

To understand how to round, we first look at some examples using number lines, even though this is not the way we normally do rounding.

Example 1 Round 47 to the nearest ten.

Here is a part of a number line; 47 is between 40 and 50.

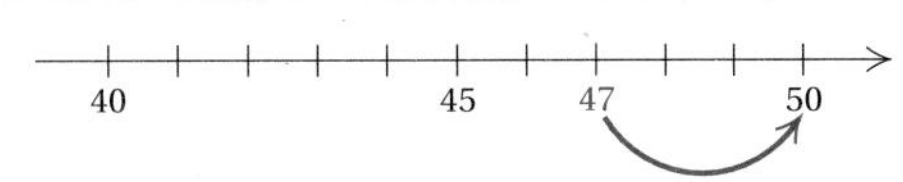

Since 47 is closer to 50, we round up to 50.

Example 2 Round 42 to the nearest ten.

42 is between 40 and 50.

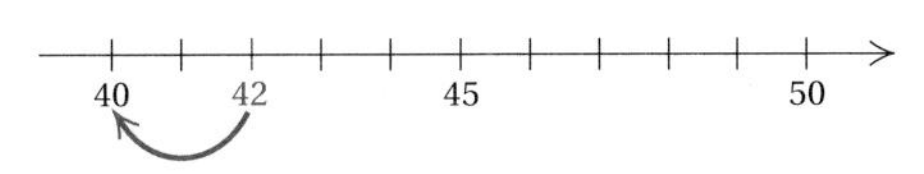

Since 42 is closer to 40, we round down to 40.

Do Exercises 1–4.

Example 3 Round 45 to the nearest ten.

45 is halfway between 40 and 50.

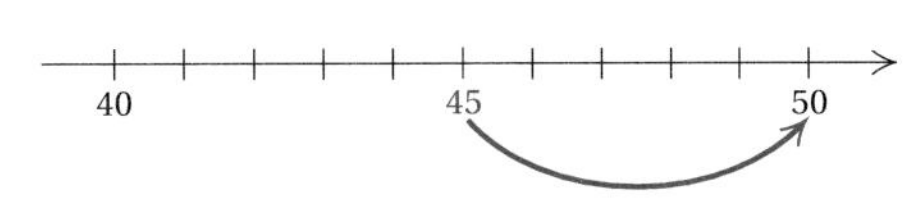

We could round 45 down to 40 or up to 50. We agree to round up to 50.

> When a number is halfway between rounding numbers, round up.

Do Exercises 5–7.

Here is a rule for rounding.

> To round to a certain place:
> **a)** Locate the digit in that place.
> **b)** Consider the next digit to the right.
> **c)** If the digit to the right is 5 or higher, round up; if the digit to the right is 4 or lower, round down.
> **d)** Change all digits to the right of the rounding location to zeros.

Objectives

a Round to the nearest ten, hundred, or thousand.

b Estimate sums and differences by rounding.

c Use < or > for ▒ to write a true sentence in a situation like 6 ▒ 10.

For Extra Help

TAPE 1 | TAPE 1B | MAC WIN | CD-ROM

Round to the nearest ten.

1. 37

2. 52

3. 73

4. 98

Round to the nearest ten.

5. 35

6. 75

7. 85

Answers on page A-2

Round to the nearest ten.

8. 137

9. 473

10. 235

11. 285

Round to the nearest hundred.

12. 641

13. 759

14. 750

15. 9325

Round to the nearest thousand.

16. 7896

17. 8459

18. 19,343

19. 68,500

Answers on page A-2

Example 4 Round 6485 to the nearest ten.

a) Locate the digit in the tens place.

6 4 8 5
↑ (under 8)

b) Consider the next digit to the right.

6 4 8 5
↑ (under 5)

c) Since that digit is 5 or higher, round 8 tens up to 9 tens.

d) Change all digits to the right of the tens digit to zeros.

6 4 9 0 ← **This is the answer.**

Example 5 Round 6485 to the nearest hundred.

a) Locate the digit in the hundreds place.

6 4 8 5
↑ (under 4)

b) Consider the next digit to the right.

6 4 8 5
↑ (under 8)

c) Since that digit is 5 or higher, round 4 hundreds up to 5 hundreds.

d) Change all digits to the right of hundreds to zeros.

6 5 0 0 ← **This is the answer.**

Example 6 Round 6485 to the nearest thousand.

a) Locate the digit in the thousands place.

6 4 8 5
↑ (under 6)

b) Consider the next digit to the right.

6 4 8 5
↑ (under 4)

c) Since that digit is 4 or lower, round down, meaning that 6 thousands stays as 6 thousands.

d) Change all digits to the right of thousands to zeros.

6 0 0 0 ← **This is the answer.**

CAUTION! 7000 is not a correct answer to Example 6. It is incorrect to round from the ones digit over, as follows:

6485, 6490, 6500, 7000.

Do Exercises 8–19.

There are many methods of rounding. For example, in computer applications, the rounding of 8563 to the nearest hundred might be done using a different rule called **truncating**, meaning that we simply change all digits to the right of the rounding location to zeros. Thus, 8563 would round to 8500, which is not the same answer that we would get using the rule discussed in this section.

b Estimating

Estimating is used to simplify a problem so that it can then be solved easily or mentally. Rounding is used when estimating. There are many ways to estimate.

Example 7 Michelle earned $21,791 as a consultant and $17,239 as an instructor in a recent year. Estimate Michelle's yearly earnings.

There are many ways to get an answer, but there is no one perfect answer based on how the problem is worded. Let's consider a couple of methods.

METHOD 1. Round each number to the nearest thousand and then add.

$$\begin{array}{r} 21{,}791 \\ +\ 17{,}239 \\ \hline \end{array} \qquad \begin{array}{r} 22{,}000 \\ +\ 17{,}000 \\ \hline 39{,}000 \end{array} \leftarrow \text{Estimated answer}$$

METHOD 2. We might use a less formal approach, depending on how specific we want the answer to be. We note that both numbers are close to 20,000, and so the total is close to 40,000. In some contexts, such as retirement planning, this might be sufficient.

The point to be made is that estimating can be done in many ways and can have many answers, even though in the problems that follow we ask you to round in a specific way.

Example 8 Estimate this sum by first rounding to the nearest ten:

78 + 49 + 31 + 85.

We round each number to the nearest ten. Then we add.

$$\begin{array}{r} 78 \\ 49 \\ 31 \\ +\ 85 \\ \hline \end{array} \qquad \begin{array}{r} 80 \\ 50 \\ 30 \\ +\ 90 \\ \hline 250 \end{array} \leftarrow \text{Estimated answer}$$

Do Exercise 20.

Example 9 Estimate this sum by first rounding to the nearest hundred:

850 + 674 + 986 + 839.

We have

$$\begin{array}{r} 850 \\ 674 \\ 986 \\ +\ 839 \\ \hline \end{array} \qquad \begin{array}{r} 900 \\ 700 \\ 1000 \\ +\ \ 800 \\ \hline 3400 \end{array}$$

Do Exercise 21.

20. Estimate the sum by first rounding to the nearest ten. Show your work.

$$\begin{array}{r} 74 \\ 23 \\ 35 \\ +\ 66 \\ \hline \end{array}$$

21. Estimate the sum by first rounding to the nearest hundred. Show your work.

$$\begin{array}{r} 650 \\ 685 \\ 238 \\ +\ 168 \\ \hline \end{array}$$

Answers on page A-2

22. Estimate the difference by first rounding to the nearest hundred. Show your work.

$$\begin{array}{r} 9285 \\ -\,6739 \\ \hline \end{array}$$

23. Estimate the difference by first rounding to the nearest thousand. Show your work.

$$\begin{array}{r} 23{,}278 \\ -\,11{,}698 \\ \hline \end{array}$$

Use $<$ or $>$ for $\square$ to write a true sentence. Draw a number line if necessary.

24. 8 $\square$ 12

25. 12 $\square$ 8

26. 76 $\square$ 64

27. 64 $\square$ 76

28. 217 $\square$ 345

29. 345 $\square$ 217

Answers on page A-2

Example 10 Estimate the difference by first rounding to the nearest thousand: 9324 − 2849.

We have

$$\begin{array}{r} 9324 \\ -\,2849 \\ \hline \end{array} \qquad \begin{array}{r} 9000 \\ -\,3000 \\ \hline 6000 \end{array}$$

Do Exercises 22 and 23.

The sentence $7 - 5 = 2$ says that $7 - 5$ is the same as 2. Later we will use the symbol $\approx$ when rounding. This symbol means **"is approximately equal to."** Thus, when 687 is rounded to the nearest ten, we may write

$$687 \approx 690.$$

c Order

We know that 2 is not the same as 5. We express this by the sentence $2 \neq 5$. We also know that 2 is less than 5. We can see this order on a number line: 2 is to the left of 5. The number 0 is the smallest whole number.

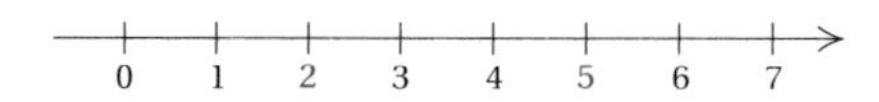

For any whole numbers a and b:

1. $a < b$ (read "a is less than b") is true when a is to the left of b on a number line.
2. $a > b$ (read "a is greater than b") is true when a is to the right of b on a number line.

We call $<$ and $>$ **inequality symbols.**

Example 11 Use $<$ or $>$ for $\square$ to write a true sentence: 7 $\square$ 11.

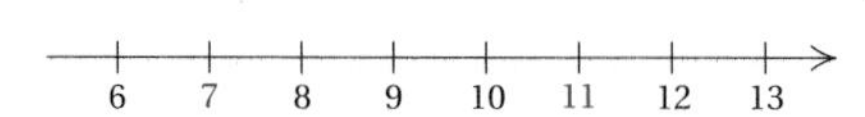

Since 7 is to the left of 11, $7 < 11$.

Example 12 Use $<$ or $>$ for $\square$ to write a true sentence: 92 $\square$ 87.

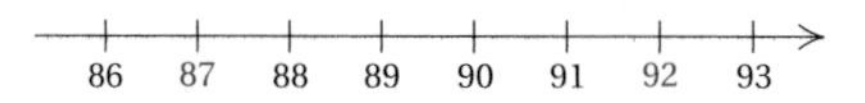

Since 92 is to the right of 87, $92 > 87$.

A sentence like $8 + 5 = 13$ is called an **equation**. A sentence like $7 < 11$ is called an **inequality**. The sentence $7 < 11$ is a true inequality. The sentence $23 > 69$ is a false inequality.

Do Exercises 24–29.

Exercise Set 1.4

a Round to the nearest ten.

1. 48 **2.** 17 **3.** 67 **4.** 99

5. 731 **6.** 532 **7.** 895 **8.** 798

Round to the nearest hundred.

9. 146 **10.** 874 **11.** 957 **12.** 650

13. 9079 **14.** 4645 **15.** 32,850 **16.** 198,402

Round to the nearest thousand.

17. 5876 **18.** 4500 **19.** 7500 **20.** 2001

21. 45,340 **22.** 735,562 **23.** 373,405 **24.** 6,713,855

b Estimate the sum or difference by first rounding to the nearest ten. Show your work.

25.
$$\begin{array}{r} 78 \\ +\ 97 \\ \hline \end{array}$$

26.
$$\begin{array}{r} 62 \\ 97 \\ 46 \\ +\ 88 \\ \hline \end{array}$$

27.
$$\begin{array}{r} 8074 \\ -\ 2347 \\ \hline \end{array}$$

28.
$$\begin{array}{r} 673 \\ -\ \ \ 28 \\ \hline \end{array}$$

Estimate the sum by first rounding to the nearest ten. Do any of the given sums seem to be incorrect when compared to the estimate? Which ones?

29.
$$\begin{array}{r} 45 \\ 77 \\ 25 \\ +\ 56 \\ \hline 343 \end{array}$$

30.
$$\begin{array}{r} 41 \\ 21 \\ 55 \\ +\ 60 \\ \hline 177 \end{array}$$

31.
$$\begin{array}{r} 622 \\ 78 \\ 81 \\ +\ 111 \\ \hline 932 \end{array}$$

32.
$$\begin{array}{r} 836 \\ 374 \\ 794 \\ +\ 938 \\ \hline 3947 \end{array}$$

Estimate the sum or difference by first rounding to the nearest hundred. Show your work.

33.
$$\begin{array}{r} 7348 \\ +\ 9247 \\ \hline \end{array}$$

34.
$$\begin{array}{r} 568 \\ 472 \\ 938 \\ +\ 402 \\ \hline \end{array}$$

35.
$$\begin{array}{r} 6852 \\ -\ 1748 \\ \hline \end{array}$$

36.
$$\begin{array}{r} 9438 \\ -\ 2787 \\ \hline \end{array}$$

Estimate the sum by first rounding to the nearest hundred. Do any of the given sums seem to be incorrect when compared to the estimate? Which ones?

37. $\begin{array}{r} 216 \\ 84 \\ 745 \\ +\,595 \\ \hline 1640 \end{array}$

38. $\begin{array}{r} 481 \\ 702 \\ 623 \\ +\,1043 \\ \hline 1849 \end{array}$

39. $\begin{array}{r} 750 \\ 428 \\ 63 \\ +\,205 \\ \hline 1446 \end{array}$

40. $\begin{array}{r} 326 \\ 275 \\ 758 \\ +\,943 \\ \hline 2302 \end{array}$

Estimate the sum or difference by first rounding to the nearest thousand. Show your work.

41. $\begin{array}{r} 9643 \\ 4821 \\ 8943 \\ +\,7004 \\ \hline \end{array}$

42. $\begin{array}{r} 7648 \\ 9348 \\ 7842 \\ +\,2222 \\ \hline \end{array}$

43. $\begin{array}{r} 92{,}149 \\ -\,22{,}555 \\ \hline \end{array}$

44. $\begin{array}{r} 84{,}890 \\ -\,11{,}110 \\ \hline \end{array}$

c Use $<$ or $>$ for ▢ to write a true sentence. Draw a number line if necessary.

45. 0 ▢ 17

46. 32 ▢ 0

47. 34 ▢ 12

48. 28 ▢ 18

49. 1000 ▢ 1001

50. 77 ▢ 117

51. 133 ▢ 132

52. 999 ▢ 997

53. 460 ▢ 17

54. 345 ▢ 456

55. 37 ▢ 11

56. 12 ▢ 32

Skill Maintenance

Add. [1.2b]

57. $\begin{array}{r} 67{,}789 \\ +\,18{,}965 \\ \hline \end{array}$

58. $\begin{array}{r} 9002 \\ +\,4587 \\ \hline \end{array}$

Subtract. [1.3d]

59. $\begin{array}{r} 67{,}789 \\ -\,18{,}965 \\ \hline \end{array}$

60. $\begin{array}{r} 9002 \\ -\,4587 \\ \hline \end{array}$

Synthesis

61. ◆ When rounding 748 to the nearest hundred, a student rounds to 750 and then to 800. What mistake is the student making?

62. ◆ Explain how estimating and rounding can be useful when shopping for groceries.

63.–66. 🖩 Use a calculator to find the sums or differences in Exercises 41–44. Since you can still make errors on a calculator—say, by pressing the wrong buttons—you can check your answers by estimating.

1.5 Multiplication

Objectives

a Write a multiplication sentence that corresponds to a situation.

b Multiply whole numbers.

c Estimate products by rounding.

For Extra Help

TAPE 1 TAPE 2A MAC WIN CD-ROM

a Multiplication and the Real World

Multiplication of whole numbers corresponds to two kinds of situations.

Repeated Addition

The multiplication 3×5 corresponds to this repeated addition:

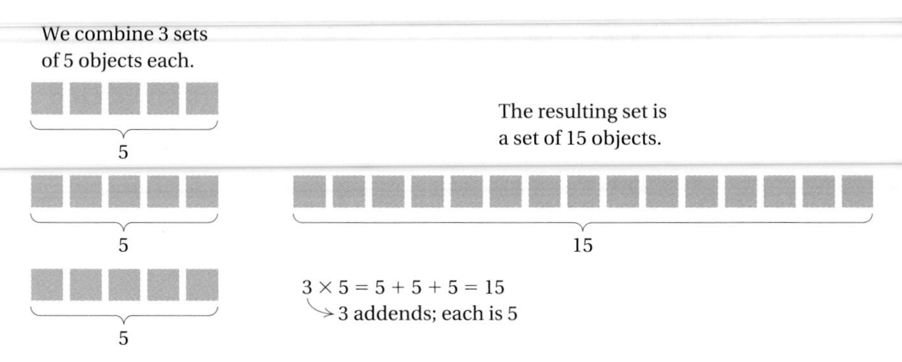

We say that the *product* of 3 and 5 is 15. The numbers 3 and 5 are called *factors.*

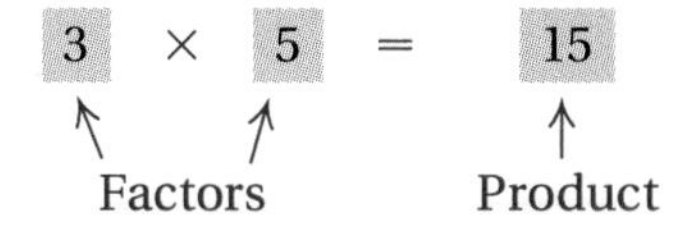

The numbers that we multiply can be called **factors**. The result of the multiplication is a number called a **product**.

Rectangular Arrays

The multiplication 3×5 corresponds to this rectangular array:

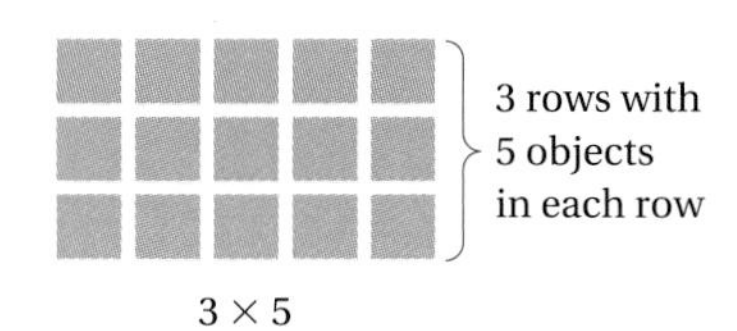

When you write a multiplication sentence corresponding to a real-world situation, you should think of either a rectangular array or repeated addition. In some cases, it may help to think both ways.

We have used an "×" to denote multiplication. A dot "·" is also commonly used. (Use of the dot is attributed to the German mathematician Gottfried Wilhelm von Leibniz in 1698.) Parentheses are also used to denote multiplication—for example, $(3)(5) = 15$, or $3(5) = 15$.

Write a multiplication sentence that corresponds to the situation.

1. Marv practices for the U.S. Open bowling tournament. He bowls 8 games a day for 7 days. How many games does he play altogether for practice?

2. A lab technician pours 75 milliliters (mL) of acid into each of 10 beakers. How much acid is poured in all?

3. *Checkerboard.* A checkerboard consists of 8 rows with 8 squares in each row. How many squares in all are there on a checkerboard?

Answers on page A-2

Examples Write a multiplication sentence that corresponds to the situation.

1. It is known that Americans drink 24 million gal of soft drinks per day (*per day* means *each day*). What quantity of soft drinks is consumed every 5 days?

   We draw a picture or at least visualize the situation. Repeated addition fits best in this case.

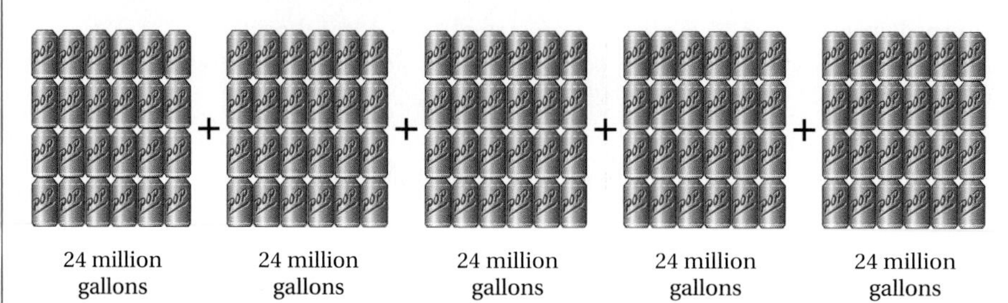

2. One side of a building has 6 floors with 7 windows on each floor. How many windows are there on that side of the building?

   We have a rectangular array and can easily draw a sketch.

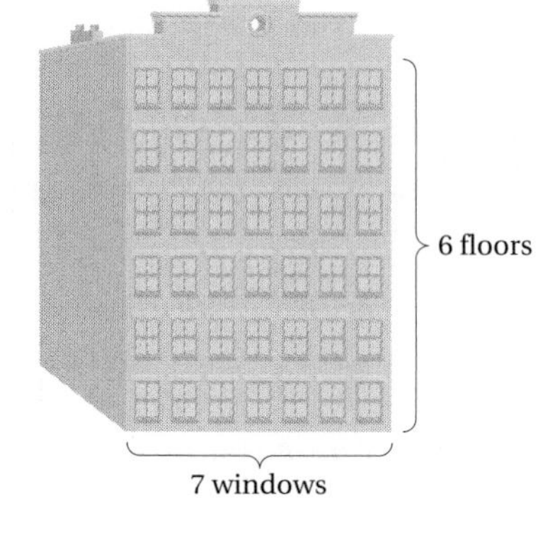

$6 \cdot 7 = 42$

Do Exercises 1–3.

Area

The area of a rectangular region is often considered to be the number of square units needed to fill it. Here is a rectangle 4 cm (centimeters) long and 3 cm wide. It takes 12 square centimeters (sq cm) to fill it.

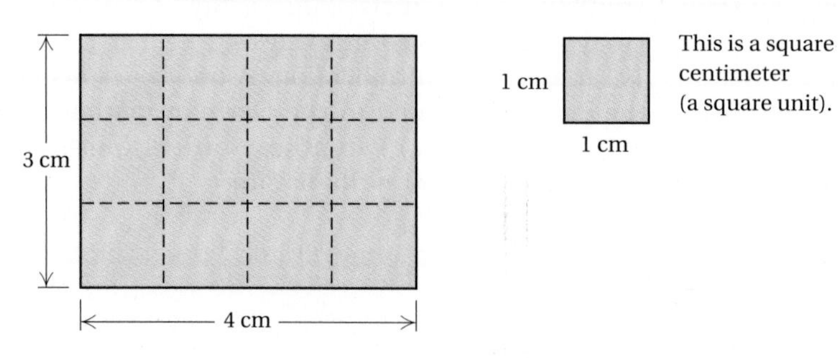

In this case, we have a rectangular array. The number of square units is $3 \cdot 4$, or 12.

Example 3 Write a multiplication sentence that corresponds to this situation.

A rectangular floor is 10 ft long and 8 ft wide. Find its area.

We draw a picture.

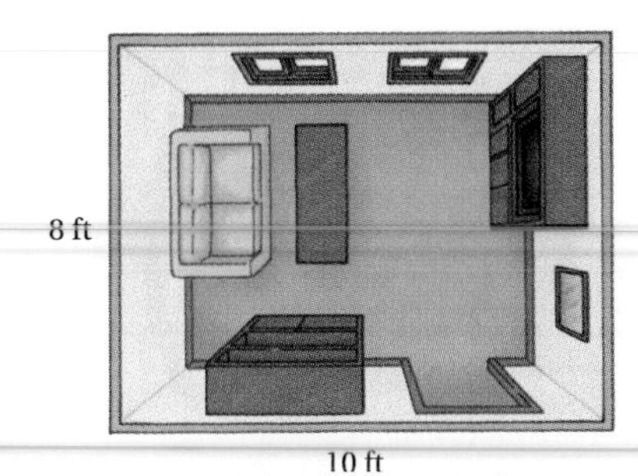

If we think of filling the rectangle with square feet, we have a rectangular array. The length $l = 10$ ft, and the width $w = 8$ ft. The area A is given by the formula

$$A = l \cdot w = 10 \times 8 = 80 \text{ sq ft}.$$

Do Exercise 4.

4. What is the area of this pool table?

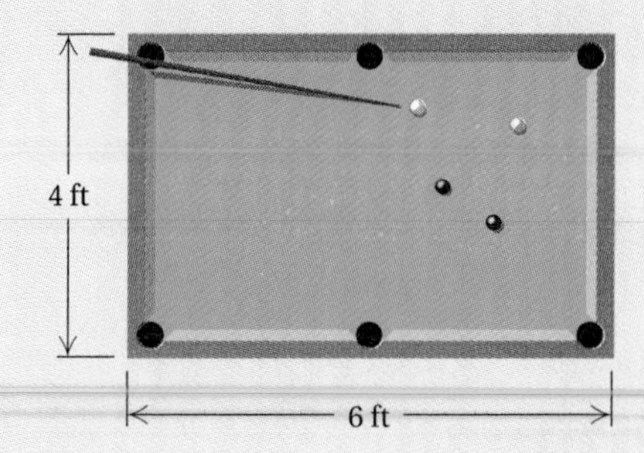

b Multiplication of Whole Numbers

Let's find the product

$$\begin{array}{r} 54 \\ \times\ 32 \\ \hline \end{array}$$

To do this, we multiply 54 by 2, then 54 by 30, and then add.

$$\begin{array}{r} 54 \\ \times\ \ 2 \\ \hline 108 \end{array} \qquad \begin{array}{r} {}^{1}\ \ \\ 54 \\ \times\ \ 30 \\ \hline 1620 \end{array}$$

Since we are going to add the results, let's write the work this way.

$$\begin{array}{rl} 54 & \\ \times\ 32 & \\ \hline 108 & \text{Multiplying by 2} \\ 1620 & \text{Multiplying by 30} \\ \hline 1728 & \text{Adding to obtain the product} \end{array}$$

The fact that we can do this is based on a property called the **distributive law.** It says that to multiply a number by a sum, $a \cdot (b + c)$, we can multiply each part by a and then add like this: $(a \cdot b) + (a \cdot c)$. Thus, $a \cdot (b + c) = (a \cdot b) + (a \cdot c)$. Applied to the example above, the distributive law gives us

$$54 \cdot 32 = 54 \cdot (30 + 2) = (54 \cdot 30) + (54 \cdot 2).$$

Answer on page A-2

Multiply.

5.
```
  4 5
× 2 3
```

6. 48 × 63

Multiply.

7.
```
  7 4 6
×   6 2
```

8. 245 × 837

Answers on page A-2

Example 4 Multiply: 43 × 57.

```
   2
   5 7
×  4 3
 1 7 1      Multiplying by 3

   2
   2
   5 7
×  4 3
 1 7 1      Multiplying by 40. (We write a 0
2 2 8 0     and then multiply 57 by 4.)
```

You may have learned that such a 0 does not have to be written. You may omit it if you wish. If you do omit it, remember, when multiplying by tens, to put the answer in the tens place.

```
   2
   2
   5 7
×  4 3
 1 7 1
2 2 8 0
2 4 5 1     Adding to obtain the product
```

Do Exercises 5 and 6.

Example 5 Multiply: 457 × 683.

```
   5 2
   6 8 3
×  4 5 7
 4 7 8 1       Multiplying 683 by 7

   4 1
   5 2
   6 8 3
×  4 5 7
 4 7 8 1
3 4 1 5 0      Multiplying 683 by 50

   3 1
   4 1
   5 2
   6 8 3
×  4 5 7
    4 7 8 1
  3 4 1 5 0
2 7 3 2 0 0    Multiplying 683 by 400
3 1 2,1 3 1    Adding
```

Do Exercises 7 and 8.

Zeros in Multiplication

Example 6 Multiply: 306×274.

Note that $306 = 3$ hundreds $+ 6$ ones.

```
    2 7 4
  × 3 0 6
  -------
  1 6 4 4     Multiplying by 6
8 2 2 0 0     Multiplying by 3 hundreds. (We write 00
                and then multiply 274 by 3.)
---------
8 3,8 4 4     Adding
```

Do Exercises 9–11.

Example 7 Multiply: 360×274.

Note that $360 = 3$ hundreds $+ 6$ tens.

```
    2 7 4     Multiplying by 6 tens. (We write 0
×   3 6 0       and then multiply 274 by 6.)
---------
1 6 4 4 0     Multiplying by 3 hundreds. (We write 00
8 2 2 0 0       and then multiply 274 by 3.)
---------
9 8,6 4 0     Adding
```

Do Exercises 12–15.

Note the following.

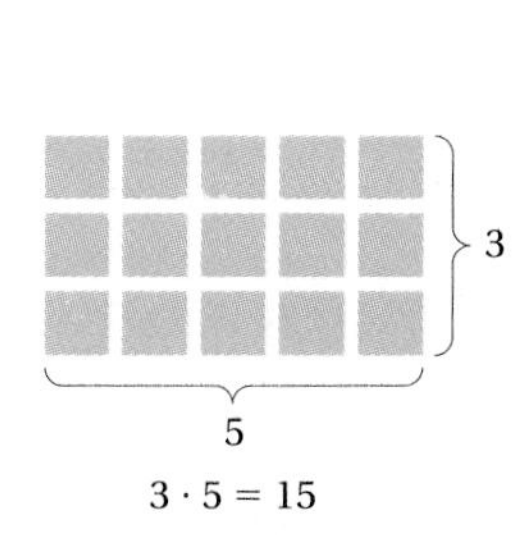

$3 \cdot 5 = 15$

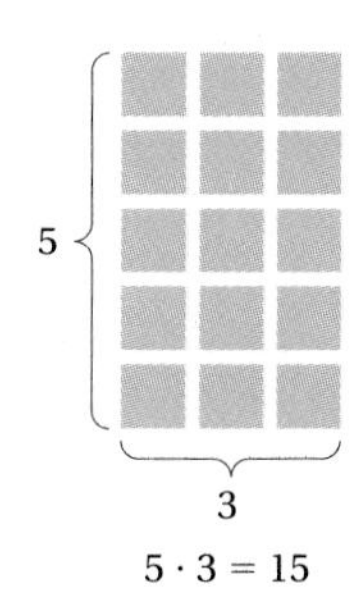

$5 \cdot 3 = 15$

If we rotate the array on the left, we get the array on the right. The answers are the same. This illustrates the **commutative law of multiplication.** It says that we can multiply two numbers in any order, $a \cdot b = b \cdot a$, and still get the same answer.

Do Exercise 16.

Multiply.

9.
```
  4 7 2
× 3 0 6
```

10. 408×704

11.
```
  2 3 4 4
× 6 0 0 5
```

Multiply.

12.
```
  4 7 2
× 8 3 0
```

13.
```
  2 3 4 4
× 7 4 0 0
```

14. 100×562

15. 1000×562

16. a) Find $23 \cdot 47$.

b) Find $47 \cdot 23$.

c) Compare your answers to parts (a) and (b).

Answers on page A-2

Multiply.

17. $5 \cdot 2 \cdot 4$

18. $5 \cdot 1 \cdot 3$

19. Estimate the product by first rounding to the nearest ten and the nearest hundred. Show your work.

$$\begin{array}{r} 837 \\ \times\, 245 \\ \hline \end{array}$$

To the instructor and the student: This section presented a review of multiplication of whole numbers. Students who are successful should go on to Section 1.6. Those who have trouble should study developmental unit M and then repeat Section 1.5.

Answers on page A-2

To multiply three or more numbers, we usually group them so that we multiply two at a time. Consider $2 \cdot (3 \cdot 4)$ and $(2 \cdot 3) \cdot 4$. The parentheses tell what to do first:

$2 \cdot (3 \cdot 4) = 2 \cdot (12) = 24.$ **We multiply 3 and 4, then 2.**

We can also multiply 2 and 3, then 4:

$(2 \cdot 3) \cdot 4 = (6) \cdot 4 = 24.$

Either way we get 24. It does not matter how we group the numbers. This illustrates that **multiplication is associative**: $a \cdot (b \cdot c) = (a \cdot b) \cdot c$. Together the commutative and associative laws tell us that to multiply more than two numbers, we can use any order and grouping we wish.

Do Exercises 17 and 18.

c Rounding and Estimating

Example 8 Estimate the following product by first rounding to the nearest ten and to the nearest hundred: 683×457.

Nearest ten	*Nearest hundred*	*Exact*
680	700	683
× 460	× 500	× 457
40800	350000	4781
272000		34150
312800		273200
		312131

Note in Example 8 that the estimate, having been rounded to the nearest ten, is

312,800.

The estimate, having been rounded to the nearest hundred, is

350,000.

Note how the estimates compare to the exact answer,

312,131.

Why does rounding give a larger answer than the exact one?

Do Exercise 19.

Exercise Set 1.5

a Write a multiplication sentence that corresponds to the situation.

1. The *Los Angeles Sunday Times* crossword puzzle is arranged rectangularly with squares in 21 rows and 21 columns. How many squares does the puzzle have altogether?

2. *Pixels.* A computer screen consists of small rectangular dots called *pixels.* How many pixels are there on a screen that has 600 rows with 800 pixels in each row?

3. A typical beverage carton contains 8 cans, each of which holds 12 oz. How many ounces are there in the carton?

4. There are 7 days in a week. How many days are there in 18 weeks?

What is the area of the region?

5.

3 ft

6 ft

6.

7 mi

7 mi

7.

11 yd

11 yd

8.

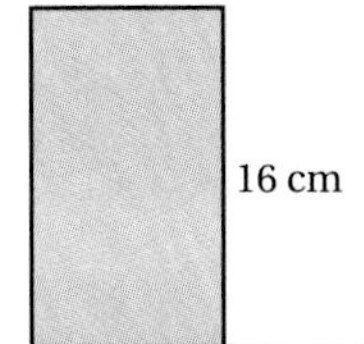

16 cm

9 cm

9.

3 mm

48 mm

10.

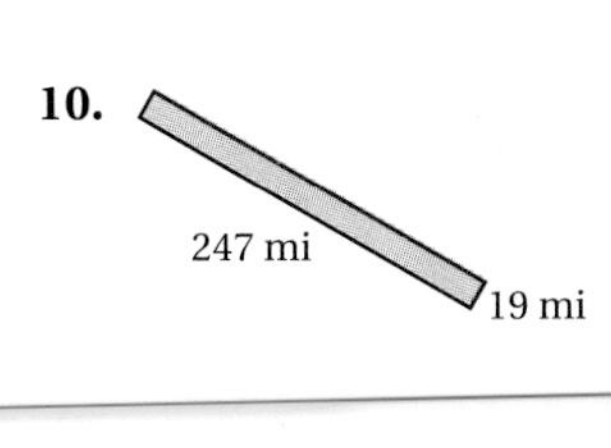

247 mi

19 mi

b Multiply.

11. $\begin{array}{r} 87 \\ \times 10 \\ \hline \end{array}$

12. $\begin{array}{r} 100 \\ \times 96 \\ \hline \end{array}$

13. $\begin{array}{r} 2340 \\ \times 1000 \\ \hline \end{array}$

14. $\begin{array}{r} 800 \\ \times 70 \\ \hline \end{array}$

15. $\begin{array}{r} 65 \\ \times 8 \\ \hline \end{array}$

16. $\begin{array}{r} 87 \\ \times 4 \\ \hline \end{array}$

17. $\begin{array}{r} 94 \\ \times 6 \\ \hline \end{array}$

18. $\begin{array}{r} 76 \\ \times 9 \\ \hline \end{array}$

19. $\begin{array}{r} 652 \\ \times 100 \\ \hline \end{array}$

20. $\begin{array}{r} 652 \\ \times 10 \\ \hline \end{array}$

21. $\begin{array}{r} 4371 \\ \times 1000 \\ \hline \end{array}$

22. $\begin{array}{r} 4371 \\ \times 100 \\ \hline \end{array}$

23. $3 \cdot 509$

24. $7 \cdot 806$

25. 7(9229)

26. 4(7867)

27. 90(53)

28. 60(78)

29. (47)(85)

30. (34)(87)

31. $\begin{array}{r} 640 \\ \times 72 \\ \hline \end{array}$

32. $\begin{array}{r} 666 \\ \times 66 \\ \hline \end{array}$

33. $\begin{array}{r} 444 \\ \times 33 \\ \hline \end{array}$

34. $\begin{array}{r} 509 \\ \times 88 \\ \hline \end{array}$

35. $\begin{array}{r} 509 \\ \times 408 \\ \hline \end{array}$

36. $\begin{array}{r} 432 \\ \times 375 \\ \hline \end{array}$

37. $\begin{array}{r} 853 \\ \times 936 \\ \hline \end{array}$

38. $\begin{array}{r} 346 \\ \times 650 \\ \hline \end{array}$

39. $\begin{array}{r} 489 \\ \times 340 \\ \hline \end{array}$

40. $\begin{array}{r} 7080 \\ \times \quad 160 \\ \hline \end{array}$

41. $\begin{array}{r} 4378 \\ \times 2694 \\ \hline \end{array}$

42. $\begin{array}{r} 8007 \\ \times \quad 480 \\ \hline \end{array}$

43. $\begin{array}{r} 6428 \\ \times 3224 \\ \hline \end{array}$

44. $\begin{array}{r} 8928 \\ \times 3172 \\ \hline \end{array}$

45. $\begin{array}{r} 3482 \\ \times \quad 104 \\ \hline \end{array}$

46. $\begin{array}{r} 6408 \\ \times 6064 \\ \hline \end{array}$

47. $\begin{array}{r} 5006 \\ \times 4008 \\ \hline \end{array}$

48. $\begin{array}{r} 6789 \\ \times 2330 \\ \hline \end{array}$

49. $\begin{array}{r} 5608 \\ \times 4500 \\ \hline \end{array}$

50. $\begin{array}{r} 4560 \\ \times 7890 \\ \hline \end{array}$

51. $\begin{array}{r} 876 \\ \times 345 \\ \hline \end{array}$

52. $\begin{array}{r} 355 \\ \times 299 \\ \hline \end{array}$

53. $\begin{array}{r} 7889 \\ \times 6224 \\ \hline \end{array}$

54. $\begin{array}{r} 6501 \\ \times 3449 \\ \hline \end{array}$

55. $\begin{array}{r} 555 \\ \times \quad 55 \\ \hline \end{array}$

56. $\begin{array}{r} 888 \\ \times \quad 88 \\ \hline \end{array}$

57. $\begin{array}{r} 734 \\ \times 407 \\ \hline \end{array}$

58. $\begin{array}{r} 5080 \\ \times \quad 302 \\ \hline \end{array}$

c Estimate the product by first rounding to the nearest ten. Show your work.

59. $\begin{array}{r} 45 \\ \times 67 \\ \hline \end{array}$

60. $\begin{array}{r} 51 \\ \times 78 \\ \hline \end{array}$

61. $\begin{array}{r} 34 \\ \times 29 \\ \hline \end{array}$

62. $\begin{array}{r} 63 \\ \times 54 \\ \hline \end{array}$

Estimate the product by first rounding to the nearest hundred. Show your work.

63. $\begin{array}{r} 876 \\ \times 345 \\ \hline \end{array}$

64. $\begin{array}{r} 355 \\ \times 299 \\ \hline \end{array}$

65. $\begin{array}{r} 432 \\ \times 199 \\ \hline \end{array}$

66. $\begin{array}{r} 789 \\ \times 434 \\ \hline \end{array}$

Estimate the product by first rounding to the nearest thousand. Show your work.

67. $\begin{array}{r} 5608 \\ \times 4576 \\ \hline \end{array}$

68. $\begin{array}{r} 2344 \\ \times 6123 \\ \hline \end{array}$

69. $\begin{array}{r} 7888 \\ \times 6224 \\ \hline \end{array}$

70. $\begin{array}{r} 6501 \\ \times 3449 \\ \hline \end{array}$

Skill Maintenance

71. Add. [1.2b]

$\begin{array}{r} 20 \\ 850 \\ +3500 \\ \hline \end{array}$

72. Subtract. [1.3d]

$\begin{array}{r} 6003 \\ -2894 \\ \hline \end{array}$

73. Round 2345 to the nearest ten, then to the nearest hundred, and then to the nearest thousand. [1.4a]

Synthesis

74. ◆ Explain in your own words what it means to say that multiplication is commutative.

75. ◆ Describe a situation that corresponds to the multiplication 4 · $150. (See Examples 1 and 2.)

76. 🖩 An 18-story office building is box-shaped. Each floor measures 172 ft by 84 ft with a 20-ft by 35-ft rectangular area lost to an elevator and a stairwell. How much area is available as office space?

1.6 Division

a Division and the Real World

Division of whole numbers corresponds to two kinds of situations. In the first, consider the division $20 \div 5$, read "20 divided by 5." We can think of 20 objects arranged in a rectangular array. We ask "How many rows, each with 5 objects, are there?"

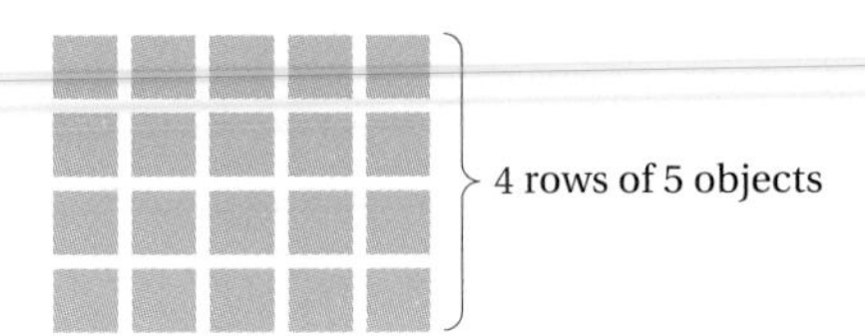

Since there are 4 rows of 5 objects each, we have

$$20 \div 5 = 4.$$

In the second situation, we can ask, "If we make 5 rows, how many objects will there be in each row?"

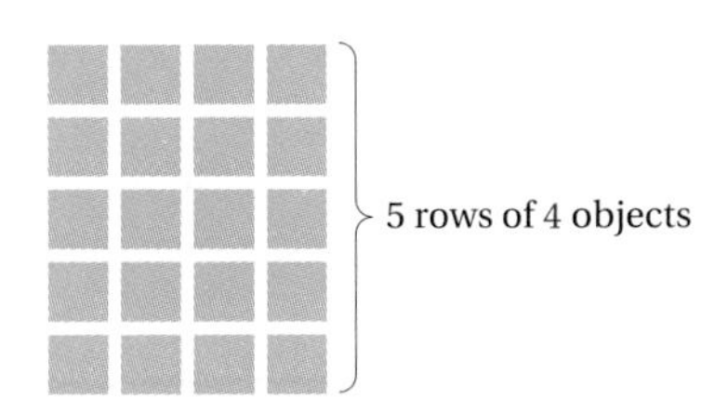

Since there are 4 objects in each of the 5 rows, we have

$$20 \div 5 = 4.$$

We say that the **dividend** is 20, the **divisor** is 5, and the **quotient** is 4.

$$\underset{\text{Dividend}}{\underset{\downarrow}{20}} \div \underset{\text{Divisor}}{\underset{\downarrow}{5}} = \underset{\text{Quotient}}{\underset{\downarrow}{4}}$$

The *dividend* is what we are dividing into. The result of the division is the *quotient.*

We also write a division such as $20 \div 5$ as

$$20/5 \quad \text{or} \quad \frac{20}{5} \quad \text{or} \quad 5\overline{)20}.$$

Example 1 Write a division sentence that corresponds to this situation.

A parent gives $24 to 3 children, with each child getting the same amount. How much does each child get?

We think of an array with 3 rows. Each row will go to a child. How many dollars will be in each row?

$$24 \div 3 = 8$$

Objectives

a Write a division sentence that corresponds to a situation.

b Given a division sentence, write a related multiplication sentence; and given a multiplication sentence, write two related division sentences.

c Divide whole numbers.

For Extra Help

TAPE 2 | TAPE 2A | MAC WIN | CD-ROM

Write a division sentence that corresponds to the situation. You need not carry out the division.

1. There are 112 students in a college band, and they are marching with 14 in each row. How many rows are there?

2. A college band is in a rectangular array. There are 112 students in the band, and they are marching in 8 rows. How many students are there in each row?

Answers on page A-2

Example 2 Write a division sentence that corresponds to this situation. You need not carry out the division.

How many mailboxes that cost $45 each can be purchased for $495?

We think of an array with 45 one-dollar bills in each row. The money in each row will buy a mailbox. How many rows will there be?

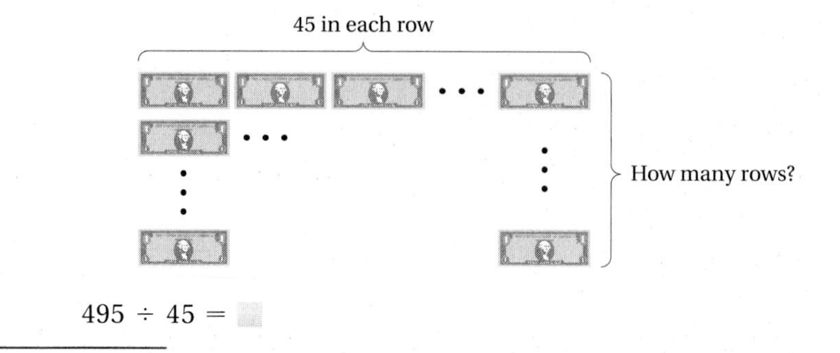

$495 \div 45 = \square$

> Whenever we have a rectangular array, we know the following:
>
> (The total number) ÷ (The number of rows) = (The number in each row).
>
> Also:
>
> (The total number) ÷ (The number in each row) = (The number of rows).

Do Exercises 1 and 2.

b Related Sentences

By looking at rectangular arrays, we can see how multiplication and division are related. The following array shows that $4 \cdot 5 = 20$.

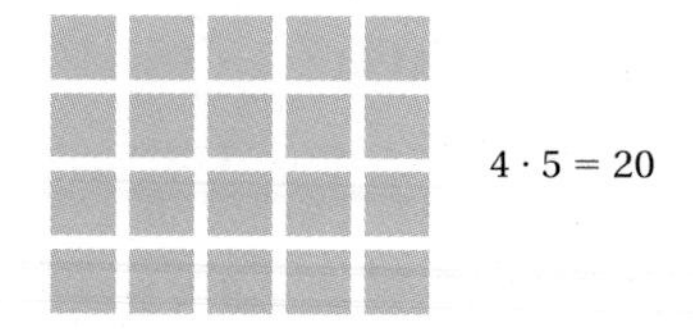

The array also shows the following:

$20 \div 5 = 4$ and $20 \div 4 = 5$.

Division is actually defined in terms of multiplication. For example, $20 \div 5$ is defined to be the number that when multiplied by 5 gives 20. Thus, for every division sentence, there is a related multiplication sentence.

$20 \div 5 = 4$ **Division sentence**

$20 = 4 \cdot 5$ **Related multiplication sentence**

To get the related multiplication sentence, we use
Dividend = Quotient · Divisor.

Example 3 Write a related multiplication sentence: $12 \div 6 = 2$.

We have

$12 \div 6 = 2$ **Division sentence**

$12 = 2 \cdot 6$. **Related multiplication sentence**

The related multiplication sentence is $12 = 2 \cdot 6$.

By the commutative law of multiplication, there is also another multiplication sentence: $12 = 6 \cdot 2$.

Do Exercises 3 and 4.

Write a related multiplication sentence.

3. $15 \div 3 = 5$

4. $72 \div 8 = 9$

For every multiplication sentence, we can write related divisions, as we can see from the preceding array.

Example 4 Write two related division sentences: $7 \cdot 8 = 56$.

We have

$7 \cdot 8 = 56$ — This factor becomes a divisor. — $7 = 56 \div 8$.

$7 \cdot 8 = 56$ — This factor becomes a divisor. — $8 = 56 \div 7$.

The related division sentences are $7 = 56 \div 8$ and $8 = 56 \div 7$.

Do Exercises 5 and 6.

Write two related division sentences.

5. $6 \cdot 2 = 12$

6. $7 \cdot 6 = 42$

c Division of Whole Numbers

Multiplication can be thought of as repeated addition. Division can be thought of as repeated subtraction. Compare.

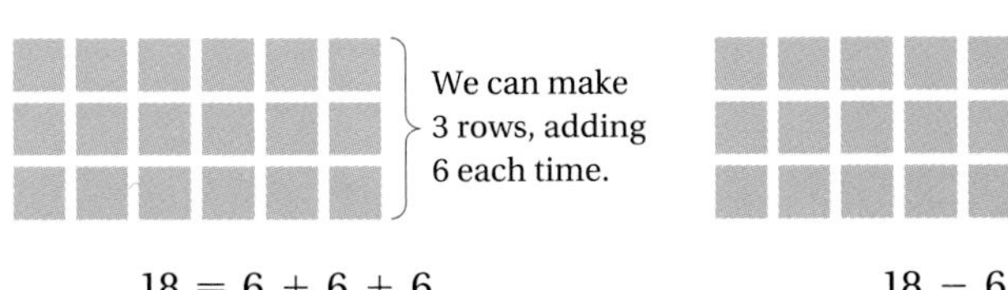

If we take away 6 objects at a time, we can do so 3 times.

$18 = 6 + 6 + 6$
$= 3 \cdot 6$

$18 - 6 - 6 - 6 = 0$
3 times

$18 \div 6 = 3$

Answers on page A-2

Divide by repeated subtraction. Then check.

7. $54 \div 9$

8. $61 \div 9$

9. $53 \div 12$

10. $157 \div 24$

Answers on page A-2

To divide by repeated subtraction, we keep track of the number of times we subtract.

Example 5 Divide by repeated subtraction: $20 \div 4$.

$$\begin{array}{r} 20 \\ -\ 4 \\ \hline 16 \\ -\ 4 \\ \hline 12 \\ -\ 4 \\ \hline 8 \\ -\ 4 \\ \hline 4 \\ -\ 4 \\ \hline 0 \end{array}$$

We subtracted 5 times, so $20 \div 4 = 5$.

Example 6 Divide by repeated subtraction: $23 \div 5$.

$$\begin{array}{r} 23 \\ -\ 5 \\ \hline 18 \\ -\ 5 \\ \hline 13 \\ -\ 5 \\ \hline 8 \\ -\ 5 \\ \hline 3 \end{array}$$

We subtracted 4 times.

We have 3 left. This number is called the *remainder.*

We write

$$23 \div 5 = 4 \text{ R } 3$$

Dividend (23), Divisor (5), Quotient (4), Remainder (3)

CHECKING DIVISIONS. To check a division, we multiply. Suppose we divide 98 by 2 and get 49:

$$98 \div 2 = 49.$$

To check, we think of the related multiplication sentence $49 \cdot 2 = \square$. We multiply 49 by 2 and see if we get 98.

If there is a remainder, we add it after multiplying.

Example 7 Check the division in Example 6.

We found that $23 \div 5 = 4$ R 3. To check, we multiply 5 by 4. This gives us 20. Then we add 3 to get 23. The dividend is 23, so the answer checks.

Do Exercises 7–10.

When we use the process of long division, we are doing repeated subtraction, even though we are going about it in a different way.

To divide, we start from the digit of highest place value in the dividend and work down to the lowest through the remainders. At each step we ask if there are multiples of the divisor in the quotient.

Example 8 Divide and check: $3642 \div 5$.

```
      ?
5 ) 3 6 4 2
```

1. **We start with the thousands digit in the dividend. Are there any thousands in the thousands place of the quotient? No; $5 \cdot 1000 = 5000$, and 5000 is larger than 3000.**

```
        7
5 ) 3 6 4 2
    3 5 0 0
    -------
      1 4 2
```

2. **Now we go to the hundreds place in the dividend. Are there any hundreds in the hundreds place of the quotient? Think of the dividend as 36 hundreds. Estimate 7 hundreds. Write 7 in the hundreds place, multiply 700 by 5, write the answer below 3642, and subtract.**

```
        7 3   <- No!
5 ) 3 6 4 2
    3 5 0 0
    -------
      1 4 2
      1 5 0  <- Can't subtract
```

3. **a) We go to the tens place of the first remainder. Are there any tens in the tens place of the quotient? To answer the question, think of the first remainder as 14 tens. Estimate 3 tens. When we multiply, we get 150, which is too large.**

b) We lower our estimate to 2 tens. Write 2 in the tens place, multiply 20 by 5, and subtract.

```
        7 2
5 ) 3 6 4 2
    3 5 0 0
    -------
      1 4 2
      1 0 0
      -----
        4 2
```

4. **We go to the ones place of the second remainder. Are there any ones in the ones place of the quotient? To answer the question, think of the second remainder as 42 ones. Estimate 8 ones. Write 8 in the ones place, multiply 8 by 5, and subtract.**

```
        7 2 8
5 ) 3 6 4 2
    3 5 0 0
    -------
      1 4 2
      1 0 0
      -----
        4 2
        4 0
        ---
          2
```

You may have learned to divide like this, not writing the extra zeros. You may omit them if desired.

```
      7 2 8
5 ) 3 6 4 2
    3 5
    ---
      1 4
      1 0
      ---
        4 2
        4 0
        ---
          2
```

The answer is 728 R 2. To check, we multiply the quotient 728 by the divisor 5. This gives us 3640. Then we add 2 to get 3642. The dividend is 3642, so the answer checks.

Do Exercises 11–13.

We can summarize our division procedure as follows.

> To do division of whole numbers:
>
> a) Estimate.
>
> b) Multiply.
>
> c) Subtract.

Divide and check.

11. $4 \overline{)\,239}$

12. $6 \overline{)\,8855}$

13. $5 \overline{)\,5075}$

Answers on page A-2

Divide.

14. $45\overline{)6030}$

15. $52\overline{)3288}$

Sometimes rounding the divisor helps us find estimates.

Example 9 Divide: 8904 ÷ 42.

We mentally round 42 to 40.

```
           2
      ________
42 ) 8 9 0 4   <— Think: 89 hundreds ÷ 40.
     8 4 0 0       Estimate 2 hundreds, but write
     -------       2 × 42 = 84.
       5 0 4
```

```
           2 1
      ________
42 ) 8 9 0 4
     8 4 0 0
     -------
       5 0 4   <— Think: 50 tens ÷ 40.
       4 2 0       Estimate 1 ten, but write
       -----       1 × 42 = 42.
         8 4
```

```
           2 1 2
      ________
42 ) 8 9 0 4
     8 4 0 0
     -------
       5 0 4
       4 2 0
       -----
         8 4   <— Think: 84 ones ÷ 40.
         8 4       Estimate 2 ones, but write
         ---       2 × 42 = 84.
           0
```

CAUTION! Be careful to keep the digits lined up correctly.

The answer is 212. *Remember*: If after estimating and multiplying you get a number that is larger than the divisor, you cannot subtract, so lower your estimate.

Do Exercises 14 and 15.

Calculator Spotlight

Calculators usually provide division answers in decimal notation. (Decimal notation will be considered in Chapter 4.) There are calculators that give quotients and remainders directly. One such calculator, the *TI Math Explorer,* uses a special division key [INT÷]. An I indicator is displayed when [INT÷] is pressed. Then a [Q] and an [R] in brackets indicate the quotient and the remainder.

Example Divide: 3642 ÷ 5.

Press	**Display**
3642 [INT÷]	I 3642
5 =	728 2
	[Q] [R]

The quotient is 728 and the remainder is 2.

Exercises

Use a calculator that finds quotients and remainders to do the following divisions.

1. 8855 ÷ 6

2. 9724 ÷ 27

3. 44,847 ÷ 56

4. 6030 ÷ 45

Answers on page A-2

Zeros in Quotients

Example 10 Divide: 6341 ÷ 7.

```
        9
    ______
7 ) 6 3 4 1  <— Think: 63 hundreds ÷ 7.
    6 3 0 0       Estimate 9 hundreds.
    -------
        4 1
```

```
        9 0
    ________
7 ) 6 3 4 1
    6 3 0 0
    -------
        4 1  <— Think: 4 tens ÷ 7. There are no tens
                 in the quotient (other than the tens in 900).
                 We write a 0 to show this.
```

```
        9 0 5
    _________
7 ) 6 3 4 1
    6 3 0 0
    -------
        4 1  <— Think: 41 ones ÷ 7.
        3 5      Estimate 5 ones.
        ---
          6
```

The answer is 905 R 6.

Do Exercises 16 and 17.

Example 11 Divide: 8889 ÷ 37.

We round 37 to 40.

```
            2
      _______
3 7 ) 8 8 8 9  <— Think: 37 ≈ 40; 88 hundreds ÷ 40.
      7 4 0 0      Estimate 2 hundreds, but write
      -------      2 × 37 = 74.
      1 4 8 9
```

```
          2 4
      _______
3 7 ) 8 8 8 9
      7 4 0 0
      -------
      1 4 8 9  <— Think: 148 tens ÷ 40.
      1 4 8 0      Estimate 4 tens, but write
      -------      4 × 37 = 148.
            9
```

```
          2 4 0
      _________
3 7 ) 8 8 8 9
      7 4 0 0
      -------
      1 4 8 9
      1 4 8 0
      -------
            9  <— Think: 9 ones ÷ 40.
                   There are no ones in the quotient.
```

The answer is 240 R 9.

Do Exercises 18 and 19.

Divide.

16. $6\overline{)4846}$

17. $7\overline{)7616}$

Divide.

18. $27\overline{)9724}$

19. $56\overline{)44{,}847}$

To the instructor and the student: This section presented a review of division of whole numbers. Students who are successful should go on to Section 1.7. Those who have trouble should study developmental unit D and then repeat Section 1.6.

Answers on page A-2

Improving Your Math Study Skills

Homework

Before Doing Your Homework

- **Setting.** Consider doing your homework as soon as possible after class, before you forget what you learned in the lecture. Research has shown that after 24 hours, most people forget about half of what is in their short-term memory. To avoid this "automatic" forgetting, you need to transfer the knowledge into long-term memory. The best way to do this with math concepts is to perform practice exercises repeatedly. This is the "drill-and-practice" part of learning math that comes when you do your homework. It cannot be overlooked if you want to succeed in your study of math.

  Try to set a specific time for your homework. Then choose a location that is quiet and uninterrupted. Some students find it helpful to listen to music when doing homework. Research has shown that classical music creates the best atmosphere for studying: Give it a try!
- **Reading.** Before you begin doing the homework exercises, you should reread the assigned material in the textbook. You may also want to look over your class notes again and rework some of the examples given in class.

  You should not read a math textbook as you would a novel or history textbook. Math texts are not meant to be read passively. Be sure to stop and do the margin exercises when directed. Also be sure to reread any paragraphs as you see the need.

While Doing Your Homework

- **Study groups.** For some students, forming a study group can be helpful. Many times, two heads are better than one. Also, it is true that "to teach is to learn again." Thus, when you explain a concept to your classmate, you often gain a deeper understanding of the concept yourself. If you do study in a group, resist the temptation to waste time by socializing.

  If you work regularly with someone, be careful not to become dependent on that person. Work on your own some of the time so that you do not rely heavily on others and are able to learn even when they are not available.
- **Notebook.** When doing your homework, consider using notebook paper in a spiral or three-ring binder. You want to be able to go over your homework when studying for a test. Therefore, you need to be able to easily access any problem in your homework notebook. Write legibly in your notebook so you can check over your work. Label each section and each exercise clearly, and show all steps. Your clear writing will also be appreciated by your instructor should your homework be collected. Also, tutors and instructors can be more helpful if they can see and understand all the steps in your work.

  When you are finished with your homework, check the answers to the odd-numbered exercises at the back of the book or in the *Student's Solutions Manual* and make corrections. If you do not understand why an answer is wrong, put a star by it so you can ask questions in class or during the instructor's office hours.

After Doing Your Homework

- **Review.** If you complete your homework several days before the next class, review your work every day. This will keep the material fresh in your mind. You should also review the work immediately before the next class so that you can ask questions as needed.

Exercise Set 1.6

a Write a division sentence that corresponds to the situation. You need not carry out the division.

1. *Canyonlands.* The trail boss for a trip into Canyonlands National Park divides 760 pounds (lb) of equipment among 4 mules. How many pounds does each mule carry?

2. *Surf Expo.* In a swimwear showing at Surf Expo, a trade show for retailers of beach supplies, each swimsuit test takes 8 minutes (min). If the show runs for 240 min, how many tests can be scheduled?

3. A lab technician pours 455 mL of sulfuric acid into 5 beakers, putting the same amount in each. How much acid is in each beaker?

4. A computer screen is made up of a rectangular array of pixels. There are 480,000 pixels in all, with 800 pixels in each row. How many rows are there on the screen?

b Write a related multiplication sentence.

5. $18 \div 3 = 6$

6. $72 \div 9 = 8$

7. $22 \div 22 = 1$

8. $32 \div 1 = 32$

9. $54 \div 6 = 9$

10. $72 \div 8 = 9$

11. $37 \div 1 = 37$

12. $28 \div 28 = 1$

Write two related division sentences.

13. $9 \times 5 = 45$

14. $2 \cdot 7 = 14$

15. $37 \cdot 1 = 37$

16. $4 \cdot 12 = 48$

17. $8 \times 8 = 64$

18. $9 \cdot 7 = 63$

19. $11 \cdot 6 = 66$

20. $1 \cdot 43 = 43$

c Divide.

21. $277 \div 5$

22. $699 \div 3$

23. $864 \div 8$

24. $869 \div 8$

25. $4\overline{)1\,2\,2\,8}$

26. $3\overline{)2\,1\,2\,4}$

27. $6\overline{)4\,5\,2\,1}$

28. $9\overline{)9\,1\,1\,0}$

29. $297 \div 4$

30. $389 \div 2$

31. $738 \div 8$

32. $881 \div 6$

33. $5\overline{)8\,5\,1\,5}$

34. $3\overline{)6\,0\,2\,7}$

35. $9\overline{)8\,8\,8\,8}$

36. $8\overline{)4\,1\,3\,9}$

37. $127{,}000 \div 10$

38. $127{,}000 \div 100$

39. $127{,}000 \div 1000$

40. $4260 \div 10$

41. $7\,0\overline{)3\,6\,9\,2}$

42. $20\overline{)5798}$

43. $30\overline{)875}$

44. $40\overline{)987}$

45. $852 \div 21$

46. $942 \div 23$

47. $85\overline{)7672}$

48. $54\overline{)2729}$

49. $111\overline{)3219}$

50. $102\overline{)5612}$

51. $8\overline{)843}$

52. $7\overline{)749}$

53. $5\overline{)8047}$

54. $9\overline{)7273}$

55. $5\overline{)5036}$

56. $7\overline{)7074}$

57. $1058 \div 46$

58. $7242 \div 24$

59. $3425 \div 32$

60. $48\overline{)4899}$

61. $24\overline{)8880}$

62. $36\overline{)7563}$

63. $28\overline{)17{,}067}$

64. $36\overline{)28{,}929}$

65. $80\overline{)24{,}320}$

66. $90\overline{)88{,}560}$

67. $285\overline{)999{,}999}$

68. $306\overline{)888{,}888}$

69. $456\overline{)3{,}679{,}920}$

70. $803\overline{)5{,}622{,}606}$

Skill Maintenance

71. Write expanded notation for 7882. [1.1a]

72. Use < or > for ▒ to write a true sentence: [1.4c]

888 ▒ 788.

Write a related addition sentence. [1.3b]

73. $21 - 16 = 5$

74. $56 - 14 = 42$

Write two related subtraction sentences. [1.3b]

75. $47 + 9 = 56$

76. $350 + 64 = 414$

Synthesis

77. ◆ Describe a situation that corresponds to the division $1180 \div 295$. (See Examples 1 and 2.)

78. ◆ Is division associative? Why or why not?

79. A group of 1231 college students is going to take buses for a field trip. Each bus can hold only 42 students. How many buses are needed?

80. ▦ Fill in the missing digits to make the equation true:

$34{,}584{,}132 \div 76\square = 4\square{,}386.$

1.7 Solving Equations

a Solutions by Trial

Let's find a number that we can put in the blank to make this sentence true:

$$9 = 3 + \square.$$

We are asking "9 is 3 plus what number?" The answer is 6.

$$9 = 3 + \boxed{6}$$

Do Exercises 1 and 2.

A sentence with = is called an **equation**. A **solution** of an equation is a number that makes the sentence true. Thus, 6 is a solution of

$9 = 3 + \square$ because $9 = 3 + \boxed{6}$ is true.

However, 7 is not a solution of

$9 = 3 + \square$ because $9 = 3 + \boxed{7}$ is false.

Do Exercises 3 and 4.

We can use a letter instead of a blank. For example,

$$9 = 3 + x.$$

We call x a **variable** because it can represent any number.

> A **solution** is a replacement for the variable that makes the equation true. When we find all the solutions, we say that we have **solved** the equation.

Example 1 Solve $x + 12 = 27$ by trial.

We replace x with several numbers.

If we replace x with 13, we get a false equation: $13 + 12 = 27$.
If we replace x with 14, we get a false equation: $14 + 12 = 27$.
If we replace x with 15, we get a true equation: $15 + 12 = 27$.

No other replacement makes the equation true, so the solution is 15.

Examples Solve.

2. $7 + n = 22$
(7 plus what number is 22?)
The solution is 15.

3. $8 \cdot 23 = y$
(8 times 23 is what?)
The solution is 184.

Note, as in Example 3, that when the variable is alone on one side of the equation, the other side shows us what calculations to do in order to find the solution.

Do Exercises 5–8.

Objectives

a Solve simple equations by trial.

b Solve equations like $t + 28 = 54$, $28 \cdot x = 168$, and $98 \div 2 = y$.

For Extra Help

TAPE 2 TAPE 2B MAC WIN CD-ROM

Find a number that makes the sentence true.

1. $8 = 1 + \square$

2. $\square + 2 = 7$

3. Determine whether 7 is a solution of $\square + 5 = 9$.

4. Determine whether 4 is a solution of $\square + 5 = 9$.

Solve by trial.

5. $n + 3 = 8$

6. $x - 2 = 8$

7. $45 \div 9 = y$

8. $10 + t = 32$

Answers on page A-3

Solve.

9. $346 \times 65 = y$

10. $x = 2347 + 6675$

11. $4560 \div 8 = t$

12. $x = 6007 - 2346$

Solve.

13. $x + 9 = 17$

14. $77 = m + 32$

Answers on page A-3

b Solving Equations

We now begin to develop more efficient ways to solve certain equations. When an equation has a variable alone on one side, it is easy to see the solution or to compute it. For example, the solution of

$$x = 12$$

is 12. When a calculation is on one side and the variable is alone on the other, we can find the solution by carrying out the calculation.

Example 4 Solve: $x = 245 \times 34$.

To solve the equation, we carry out the calculation.

$$\begin{array}{r} 245 \\ \times \quad 34 \\ \hline 980 \\ 7350 \\ \hline 8330 \end{array}$$

The solution is 8330.

Do Exercises 9–12.

Look at the equation

$$x + 12 = 27.$$

We can get x alone on one side of the equation by writing a related subtraction sentence:

$x = 27 - 12$ **12 gets subtracted to find the related subtraction sentence.**

$x = 15.$ **Doing the subtraction**

It is useful in our later study of algebra to think of this as "subtracting 12 *on both sides.*" Thus,

$x + 12 - 12 = 27 - 12$ **Subtracting 12 on both sides**

$x + 0 = 15$ **Carrying out the subtraction**

$x = 15.$

> To solve $x + a = b$, subtract a on both sides.

If we can get an equation in a form with the variable alone on one side, we can "see" the solution.

Example 5 Solve: $t + 28 = 54$.

We have

$t + 28 = 54$

$t + 28 - 28 = 54 - 28$ **Subtracting 28 on both sides**

$t + 0 = 26$

$t = 26.$

The solution is 26.

Do Exercises 13 and 14.

Example 6 Solve: $182 = 65 + n$.

We have

$$182 = 65 + n$$
$$182 - 65 = 65 + n - 65 \quad \textbf{Subtracting 65 on both sides}$$
$$117 = 0 + n \quad \textbf{65 plus } n \textbf{ minus 65 is } 0 + n.$$
$$117 = n.$$

The solution is 117.

Do Exercise 15.

15. Solve: $155 = t + 78$.

Example 7 Solve: $7381 + x = 8067$.

We have

$$7381 + x = 8067$$
$$7381 + x - 7381 = 8067 - 7381 \quad \textbf{Subtracting 7381 on both sides}$$
$$x = 686.$$

The solution is 686.

Do Exercises 16 and 17.

Solve.

16. $4566 + x = 7877$

17. $8172 = h + 2058$

We now learn to solve equations like $8 \cdot n = 96$. Look at

$$8 \cdot n = 96.$$

We can get n alone by writing a related division sentence:

$$n = 96 \div 8 = \frac{96}{8} \quad \textbf{96 is divided by 8.}$$
$$= 12. \quad \textbf{Doing the division}$$

Note that $n = 12$ is easier to solve than $8 \cdot n = 96$. This is because we see easily that if we replace n on the left side with 12, we get a true sentence: $12 = 12$. The solution of $n = 12$ is 12, which is also the solution of $8 \cdot n = 96$.

It is useful in our later study of algebra to think of the preceding as "dividing by 8 *on both sides*." Thus,

$$\frac{8 \cdot n}{8} = \frac{96}{8} \quad \textbf{Dividing by 8 on both sides}$$
$$n = 12. \quad \textbf{8 times } n \textbf{ divided by 8 is } n.$$

> To solve $a \cdot x = b$, divide by a on both sides.

Answers on page A-3

Solve.

18. $8 \cdot x = 64$

19. $144 = 9 \cdot n$

20. Solve: $5152 = 8 \cdot t$.

21. Solve: $18 \cdot y = 1728$.

22. Solve: $n \cdot 48 = 4512$.

Answers on page A-3

Example 8 Solve: $10 \cdot x = 240$.

We have

$$10 \cdot x = 240$$

$$\frac{10 \cdot x}{10} = \frac{240}{10} \quad \text{Dividing by 10 on both sides}$$

$$x = 24.$$

The solution is 24.

Do Exercises 18 and 19.

Example 9 Solve: $5202 = 9 \cdot t$.

We have

$$5202 = 9 \cdot t$$

$$\frac{5202}{9} = \frac{9 \cdot t}{9} \quad \text{Dividing by 9 on both sides}$$

$$578 = t.$$

The solution is 578.

Do Exercise 20.

Example 10 Solve: $14 \cdot y = 1092$.

We have

$$14 \cdot y = 1092$$

$$\frac{14 \cdot y}{14} = \frac{1092}{14} \quad \text{Dividing by 14 on both sides}$$

$$y = 78.$$

The solution is 78.

Do Exercise 21.

Example 11 Solve: $n \cdot 56 = 4648$.

We have

$$n \cdot 56 = 4648$$

$$\frac{n \cdot 56}{56} = \frac{4648}{56} \quad \text{Dividing by 56 on both sides}$$

$$n = 83.$$

The solution is 83.

Do Exercise 22.

Exercise Set 1.7

a Solve by trial.

1. $x + 0 = 14$

2. $x - 7 = 18$

3. $y \cdot 17 = 0$

4. $56 \div m = 7$

b Solve.

5. $13 + x = 42$

6. $15 + t = 22$

7. $12 = 12 + m$

8. $16 = t + 16$

9. $3 \cdot x = 24$

10. $6 \cdot x = 42$

11. $112 = n \cdot 8$

12. $162 = 9 \cdot m$

13. $45 \times 23 = x$

14. $23 \times 78 = y$

15. $t = 125 \div 5$

16. $w = 256 \div 16$

17. $p = 908 - 458$

18. $9007 - 5667 = m$

19. $x = 12{,}345 + 78{,}555$

20. $5678 + 9034 = t$

21. $3 \cdot m = 96$

22. $4 \cdot y = 96$

23. $715 = 5 \cdot z$

24. $741 = 3 \cdot t$

25. $10 + x = 89$

26. $20 + x = 57$

27. $61 = 16 + y$

28. $53 = 17 + w$

29. $6 \cdot p = 1944$

30. $4 \cdot w = 3404$

31. $5 \cdot x = 3715$

32. $9 \cdot x = 1269$

33. $47 + n = 84$

34. $56 + p = 92$

35. $x + 78 = 144$

36. $z + 67 = 133$

37. $165 = 11 \cdot n$

38. $660 = 12 \cdot n$

39. $624 = t \cdot 13$

40. $784 = y \cdot 16$

41. $x + 214 = 389$

42. $x + 221 = 333$

43. $567 + x = 902$

44. $438 + x = 807$

45. $18 \cdot x = 1872$

46. $19 \cdot x = 6080$

47. $40 \cdot x = 1800$

48. $20 \cdot x = 1500$

49. $2344 + y = 6400$

50. $9281 = 8322 + t$

51. $8322 + 9281 = x$

52. $9281 - 8322 = y$

53. $234 \times 78 = y$

54. $10{,}534 \div 458 = q$

55. $58 \cdot m = 11{,}890$

56. $233 \cdot x = 22{,}135$

Skill Maintenance

57. Write two related subtraction sentences: $7 + 8 = 15$. [1.3b]

58. Write two related division sentences: $6 \cdot 8 = 48$. [1.6b]

Use $>$ or $<$ for ▒ to write a true sentence. [1.4c]

59. 123 ▒ 789

60. 342 ▒ 339

Divide. [1.6c]

61. $1283 \div 9$

62. $17\overline{)5689}$

Synthesis

63. ◆ Describe a procedure that can be used to convert any equation of the form $a + b = c$ to a related subtraction equation.

64. ◆ Describe a procedure that can be used to convert any equation of the form $a \cdot b = c$ to a related division equation.

Solve.

65. ▦ $23{,}465 \cdot x = 8{,}142{,}355$

66. ▦ $48{,}916 \cdot x = 14{,}332{,}388$

1.8 Applications and Problem Solving

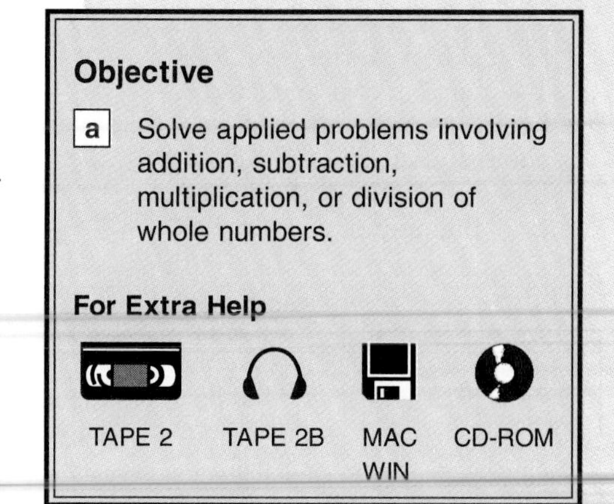
Objective

a Solve applied problems involving addition, subtraction, multiplication, or division of whole numbers.

For Extra Help

TAPE 2 TAPE 2B MAC WIN CD-ROM

a Applications and problem solving are the main uses of mathematics. To solve a problem using the operations on the whole numbers, we first look at the situation. We try to translate the problem to an equation. Then we solve the equation. We check to see if the solution of the equation is a solution of the original problem. Thus we are using the following five-step strategy.

FIVE STEPS FOR PROBLEM SOLVING

1. *Familiarize* yourself with the situation. If it is described in words, as in a textbook, *read carefully.* In any case, think about the situation. Draw a picture whenever it makes sense to do so. Choose a letter, or *variable,* to represent the unknown quantity to be solved for.
2. *Translate* the problem to an equation.
3. *Solve* the equation.
4. *Check* the answer in the original wording of the problem.
5. *State* the answer to the problem clearly with appropriate units.

Example 1 *Minivan Sales.* Recently, sales of minivans have soared. The bar graph at right shows the number of Chrysler Town & Country LXi minivans sold in recent years. Find the total number of minivans sold during those years.

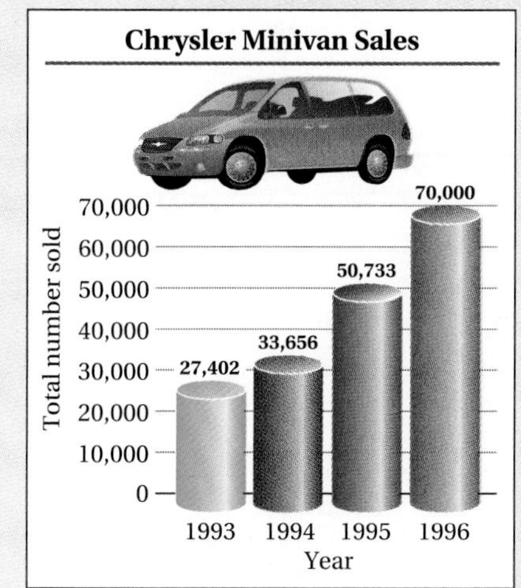

Source: Chrysler Corporation

1. **Familiarize.** We can make a drawing or at least visualize the situation.

27,402	+	33,656	+	50,733	+	70,000	=	n
in 1993		in 1994		in 1995		in 1996		Total sold

Since we are combining objects, addition can be used. First we define the unknown. We let n = the total number of minivans sold.

2. **Translate.** We translate to an equation:

$$27{,}402 + 33{,}656 + 50{,}733 + 70{,}000 = n.$$

3. **Solve.** We solve the equation by carrying out the addition.

$$\begin{array}{r} \scriptstyle 1\ 1\quad 1\ \ \ \ \\ 27{,}402 \\ 33{,}656 \\ 50{,}733 \\ +\ \ 70{,}000 \\ \hline 181{,}791 \end{array}$$

Thus, $181{,}791 = n$, or $n = 181{,}791$.

4. **Check.** We check 181,791 in the original problem. There are many ways in which this can be done. For example, we can repeat the calculation. (We leave this to the student.) Another way is to check the reasonableness of the answer. In this case, we would expect the answer to be larger than the sales in any of the individual years, which it is. We can also estimate by rounding. Here, we round to the nearest thousand:

$$\begin{aligned} &27{,}402 + 33{,}656 + 50{,}733 + 70{,}000 \\ &\qquad \approx 27{,}000 + 34{,}000 + 51{,}000 + 70{,}000 \\ &\qquad = 182{,}000. \end{aligned}$$

1. *Teacher needs in 2005.* The data in the table show the estimated number of new jobs for teachers in the year 2005. The reason is an expected boom in the number of youngsters under the age of 18. Find the total number of jobs available for teachers in 2005.

Type of Teacher	Number of New Jobs
Secondary	386,000
Aides	364,000
Childcare workers	248,000
Elementary	220,000
Special education	206,000

Source: Bureau of Labor Statistics

2. *Checking Account.* You have \$756 in your checking account. You write a check for \$387 to pay for a VCR for your campus apartment. How much is left in your checking account?

Answers on page A-3

Since $181{,}791 \approx 182{,}000$, we have a partial check. If we had an estimate like 236,000 or 580,000, we might be suspicious that our calculated answer is incorrect. Since our estimated answer is close to our calculation, we are further convinced that our answer checks.

5. **State.** The total number of minivans sold during these years is 181,791.

Do Exercise 1.

Example 2 *Hard-Drive Space.* The hard drive on your computer has 572 megabytes (MB) of storage space available. You install a software package called Microsoft® Office, which uses 84 MB of space. How much storage space do you have left after the installation?

1. **Familiarize.** We first make a drawing or at least visualize the situation. We let $M =$ the amount of space left.

2. **Translate.** We see that this is a "take-away" situation. We translate to an equation.

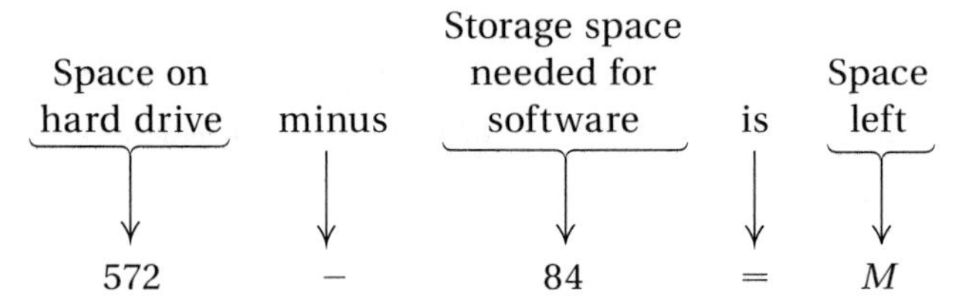

3. **Solve.** This sentence tells us what to do. We subtract.

$$\begin{array}{r} {\scriptstyle 16} \\ {\scriptstyle 4\ \not{6}\ 12} \\ \not{5}\,\not{7}\,\not{2} \\ -\quad 8\ 4 \\ \hline 4\ 8\ 8 \end{array}$$

Thus, $488 = M$, or $M = 488$.

4. **Check.** We check our answer of 488 MB by repeating the calculation. We note that the answer should be less than the original amount of memory, 572 MB, which it is. We can also add the difference, 488, to the subtrahend, 84: $84 + 488 = 572$. We can also estimate:

$$572 - 84 \approx 600 - 100 = 500 \approx 488.$$

5. **State.** There is 488 MB of memory left.

Do Exercise 2.

In the real world, problems may not be stated in written words. You must still become familiar with the situation before you can solve the problem.

Example 3 *Travel Distance.* Vicki is driving from Indianapolis to Salt Lake City to work during summer vacation. The distance from Indianapolis to Salt Lake City is 1634 mi. She travels 1154 mi to Denver. How much farther must she travel?

1. **Familiarize.** We first make a drawing or at least visualize the situation. We let $x =$ the remaining distance to Salt Lake City.

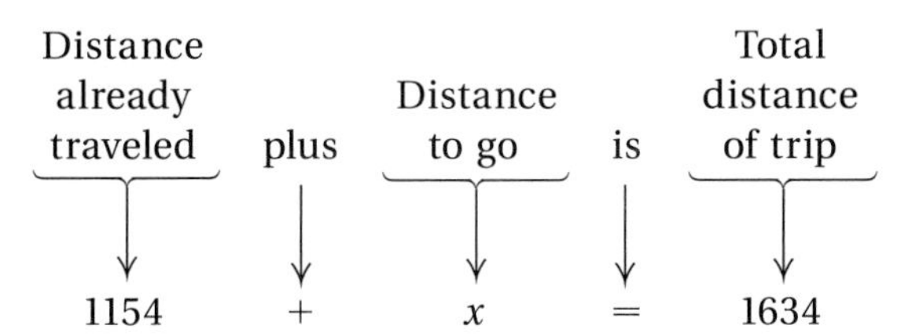

2. **Translate.** We see that this is a "how-much-more" situation. We translate to an equation.

Distance already traveled	plus	Distance to go	is	Total distance of trip
↓	↓	↓	↓	↓
1154	+	x	=	1634

3. **Solve.** We solve the equation.

$$1154 + x = 1634$$

$$1154 + x - 1154 = 1634 - 1154 \quad \text{Subtracting 1154 on both sides}$$

$$x = 480$$

$$\begin{array}{r} \overset{5\ 13}{1\,\cancel{6}\,\cancel{3}\,4} \\ -\ 1\,1\,5\,4 \\ \hline 4\,8\,0 \end{array}$$

4. **Check.** We check our answer of 480 mi in the original problem. This number should be less than the total distance, 1634 mi, which it is. We can add the difference, 480, to the subtrahend, 1154: $1154 + 480 = 1634$. We can also estimate:

$$1634 - 1154 \approx 1600 - 1200$$

$$= 400 \approx 480.$$

The answer, 480 mi, checks.

5. **State.** Vicki must travel 480 mi farther to Salt Lake City.

Do Exercise 3.

3. *Calculator Purchase.* Bernardo has $76. He wants to purchase a graphing calculator for $94. How much more does he need?

Answer on page A-3

4. *Total Cost of Laptop Computers.* What is the total cost of 12 laptop computers with CD-ROM drives and matrix color if each one costs $3249?

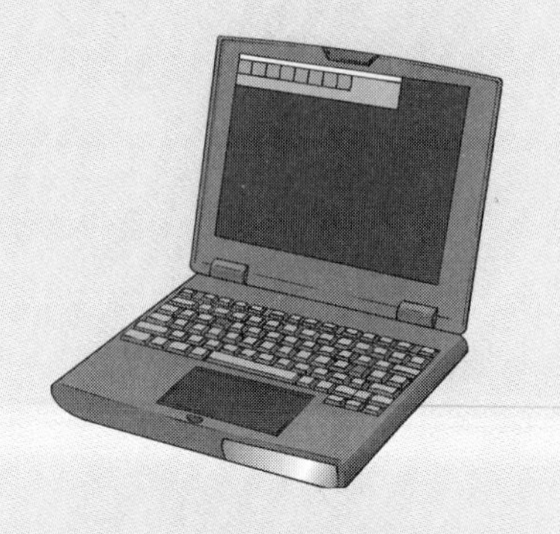

Example 4 *Total Cost of VCRs.* What is the total cost of 5 four-head VCRs if each one costs $289?

1. **Familiarize.** We first make a drawing or at least visualize the situation. We let n = the cost of 5 VCRs. Repeated addition works well here.

2. **Translate.** We translate to an equation.

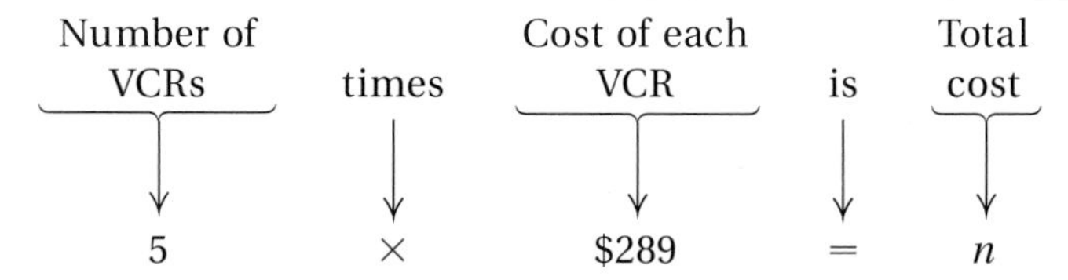

3. **Solve.** This sentence tells us what to do. We multiply.

$$\begin{array}{r} {\scriptstyle 4\ 4\ } \\ 2\ 8\ 9 \\ \times \quad\ \ 5 \\ \hline 1\ 4\ 4\ 5 \end{array}$$

Thus, $n = 1445$.

4. **Check.** We have an answer that is much larger than the cost of any individual VCR, which is reasonable. We can repeat our calculation. We can also check by estimating:

$$5 \times 289 \approx 5 \times 300 = 1500 \approx 1445.$$

The answer checks.

5. **State.** The total cost of 5 VCRs is $1445.

Do Exercise 4.

Answer on page A-3

Example 5 *Bed Sheets.* The dimensions of a sheet for a king-size bed are 108 in. by 102 in. What is the area of the sheet? (The dimension labels on sheets list width × length.)

1. Familiarize. We first make a drawing. We let A = the area.

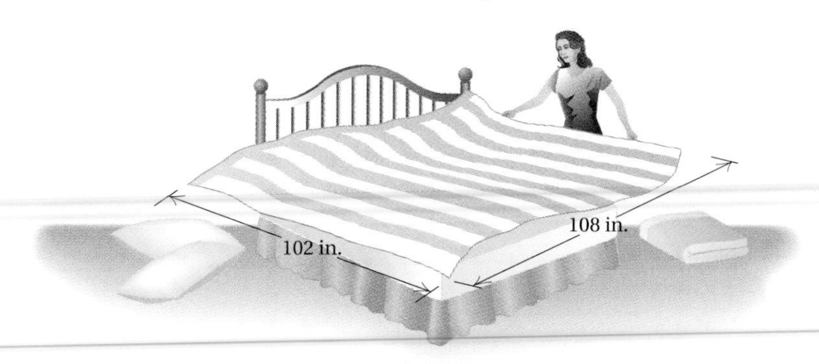

2. Translate. Using a formula for area, we have

$$A = \text{length} \cdot \text{width} = l \cdot w = 102 \cdot 108.$$

3. Solve. We carry out the multiplication.

$$\begin{array}{r} 1\ 0\ 8 \\ \times\quad 1\ 0\ 2 \\ \hline 2\ 1\ 6 \\ 1\ 0\ 8\ 0\ 0 \\ \hline 1\ 1\ 0\ 1\ 6 \end{array}$$

Thus, $A = 11{,}016$.

4. Check. We repeat our calculation. We also note that the answer is larger than either the length or the width, which it should be. (This might not be the case if we were using decimals.) The answer checks.

5. State. The area of a king-size bed sheet is 11,016 sq in.

Do Exercise 5.

Example 6 *Diet Cola Packaging.* Diet Cola has become very popular in the quest to control our weight. A bottling company produces 2203 cans of cola. How many 8-can packages can be filled? How many cans will be left over?

1. Familiarize. We first draw a picture. We let n = the number of 8-can packages to be filled.

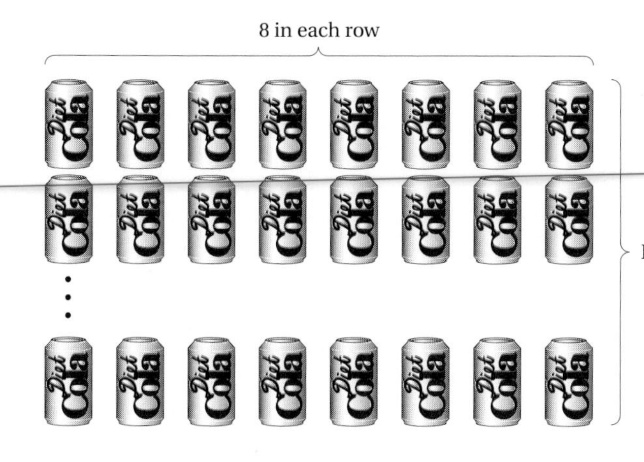

5. *Bed Sheets.* The dimensions of a sheet for a queen-size bed are 90 in. by 102 in. What is the area of the sheet?

Answer on page A-3

6. *Diet Cola Packaging.* The bottling company also uses 6-can packages. How many 6-can packages can be filled with 2269 cans of cola? How many cans will be left over?

Answer on page A-3

2. **Translate.** We can translate to an equation as follows.

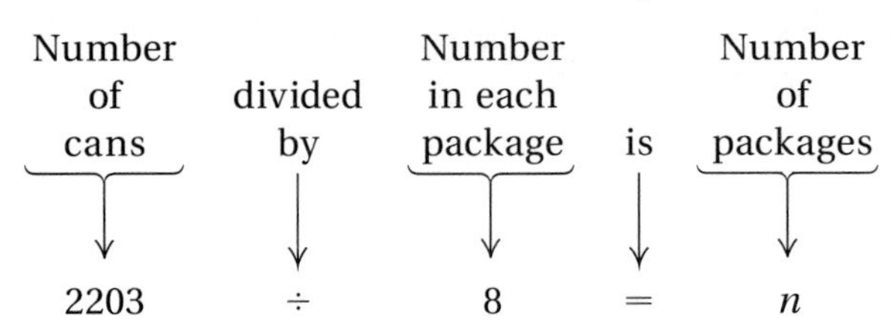

3. **Solve.** We solve the equation by carrying out the division.

$$\begin{array}{r} 275 \\ 8\,\overline{)\,2203} \\ \underline{1600} \\ 603 \\ \underline{560} \\ 43 \\ \underline{40} \\ 3 \end{array}$$

4. **Check.** We can check by multiplying the number of packages by 8 and adding the remainder, 3:

$8 \cdot 275 = 2200, \qquad 2200 + 3 = 2203.$

5. **State.** Thus, 275 8-can packages can be filled. There will be 3 cans left over.

Do Exercise 6.

Example 7 *Automobile Mileage.* The Chrysler Town & Country LXi minivan featured in Example 1 gets 18 miles to the gallon (mpg) in city driving. How many gallons will it use in 4932 mi of city driving?

1. **Familiarize.** We first make a drawing. It is often helpful to be descriptive about how you define a variable. In this example, we let g = the number of gallons (g comes from "gallons").

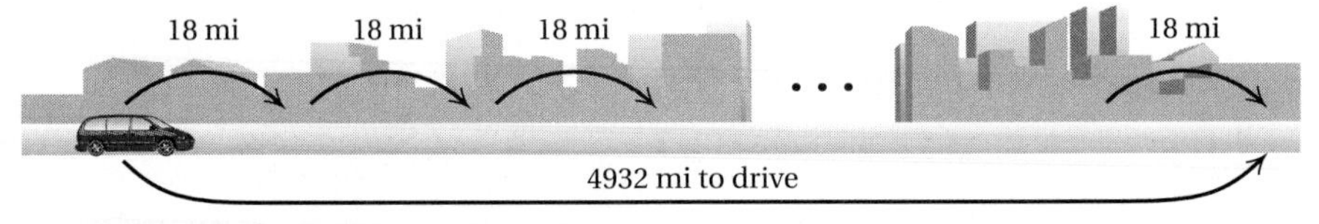

2. **Translate.** Repeated addition applies here. Thus the following multiplication corresponds to the situation.

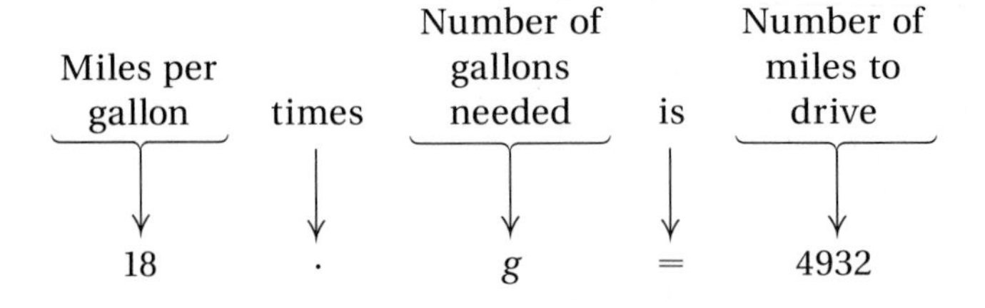

3. **Solve.** To solve the equation, we divide by 18 on both sides.

$$18 \cdot g = 4932$$
$$\frac{18 \cdot g}{18} = \frac{4932}{18}$$
$$g = 274$$

$$\begin{array}{r} 274 \\ 18\,\overline{)\,4932} \\ 3600 \\ \hline 1332 \\ 1260 \\ \hline 72 \\ 72 \\ \hline 0 \end{array}$$

4. **Check.** To check, we multiply 274 by 18: $18 \cdot 274 = 4932$.
5. **State.** The minivan will use 274 gal.

Do Exercise 7.

Multistep Problems

Sometimes we must use more than one operation to solve a problem, as in the following example.

Example 8 *Weight Loss.* Many Americans exercise for weight control. It is known that one must burn off about 3500 calories in order to lose one pound. The chart shown here details how many calories are burned by certain activities. How long would an individual have to run at a brisk pace in order to lose one pound?

1. **Familiarize.** We first make a chart.

ONE POUND			
3500 calories			
100 cal 8 min	100 cal 8 min	... ...	100 cal 8 min

2. **Translate.** Repeated addition applies here. Thus the following multiplication corresponds to the situation. We must find out how many 100's there are in 3500. We let x = the number of 100's in 3500.

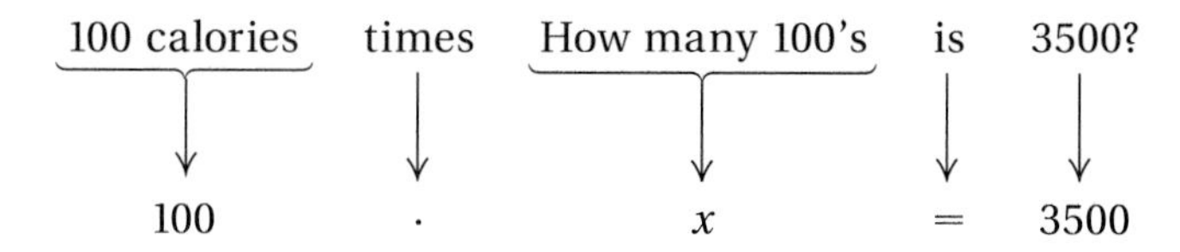

$$100 \cdot x = 3500$$

7. *Automobile Mileage.* The Chrysler Town & Country LXi minivan gets 24 miles to the gallon (mpg) in country driving. How many gallons will it use in 888 mi of country driving?

Answer on page A-3

8. *Weight Loss.* Using the chart for Example 8, determine how long an individual must swim in order to lose one pound.

9. *Bones in the Hands and Feet.* There are 27 bones in each human hand and 26 bones in each human foot. How many bones are there in all in the hands and feet?

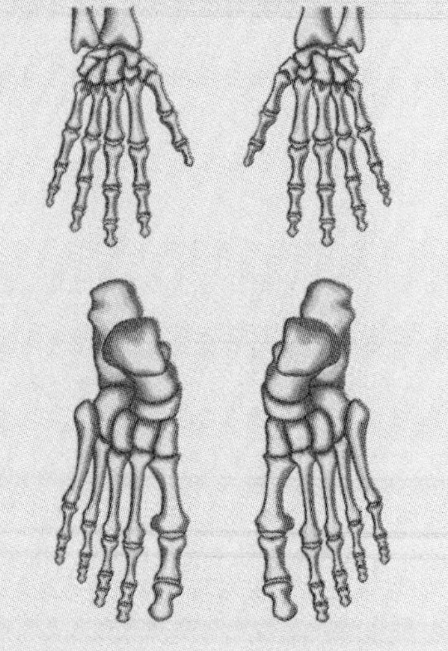

Answers on page A-3

3. Solve. To solve the equation, we divide by 100 on both sides.

$$100 \cdot x = 3500$$

$$\frac{100 \cdot x}{100} = \frac{3500}{100}$$

$$x = 35$$

$$\begin{array}{r} 35 \\ 100\overline{)3500} \\ \underline{3000} \\ 500 \\ \underline{500} \\ 0 \end{array}$$

We know that running for 8 min will burn off 100 calories. To do this 35 times will burn off one pound, so you must run for 35 times 8 minutes in order to burn off one pound. We let t = the time it takes to run off one pound.

$$35 \times 8 = t$$

$$280 = t$$

$$\begin{array}{r} 35 \\ \underline{\times \quad 8} \\ 280 \end{array}$$

4. Check. Suppose you run for 280 min. If we divide 280 by 8, we get 35, and 35 times 100 is 3500, the number of calories it takes to lose one pound.

5. State. It will take 280 min, or 4 hr, 40 min, of running to lose one pound.

Do Exercises 8 and 9.

As you consider the following exercises, here are some words and phrases that may be helpful to look for when you are translating problems to equations.

Addition:	sum, total, increase, altogether, plus
Subtraction:	difference, minus, how much more?, how many more?, decrease, deducted, how many left?
Multiplication:	given rows and columns, how many in all?, product, total from a repeated addition, area, of
Division:	how many in each row?, how many rows?, how many pieces?, how many parts in a whole?, quotient, divisible

Exercise Set 1.8

a Solve.

1. During the first four months of a recent year, Campus Depot Business Machine Company reported the following sales:

January	$3572
February	2718
March	2809
April	3177

What were the total sales over this time period?

2. A family travels the following miles during a five-day trip:

Monday	568
Tuesday	376
Wednesday	424
Thursday	150
Friday	224

How many miles did they travel altogether?

Bicycle Sales. The bar graph below shows the total sales, in millions of dollars, for bicycles and related supplies in recent years. Use this graph for Exercises 3–6.

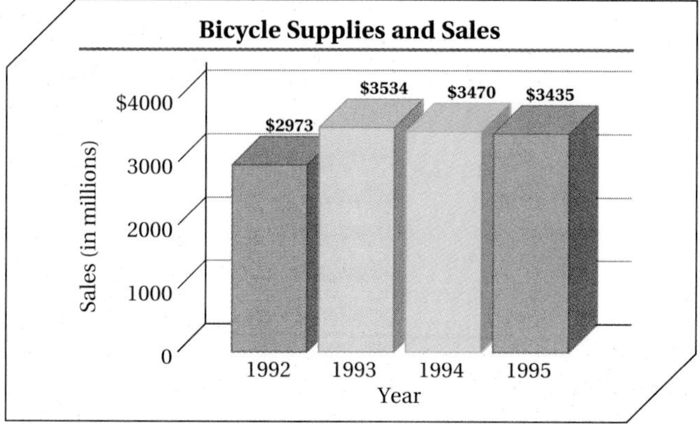

Source: National Sporting Goods Association

3. What were the total sales for 1993 and 1994?

4. What were the total sales for 1992 through 1995?

5. How much more were the sales in 1993 than in 1994?

6. How much more were the sales in 1994 than in 1992?

7. *Longest Rivers.* The longest river in the world is the Nile, which has a length of 4145 mi. It is 138 mi longer than the next longest river, which is the Amazon in South America. How long is the Amazon?

8. *Largest Lakes.* The largest lake in the world is the Caspian Sea, which has an area of 317,000 square kilometers (sq km). The Caspian is 288,900 sq km larger than the second largest lake, which is Lake Superior. What is the area of Lake Superior?

Amazon
?
Coffee beans
South America

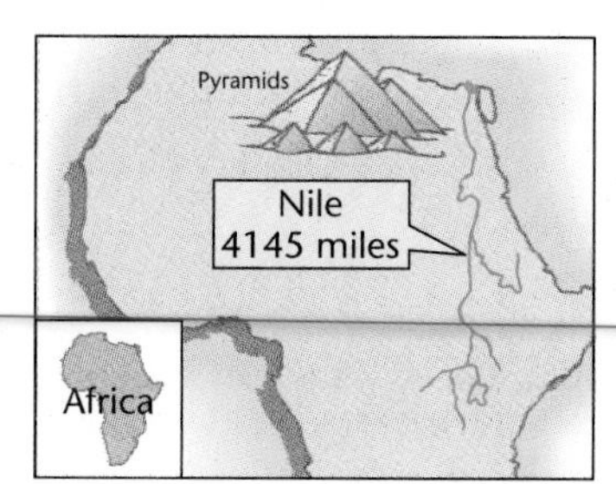

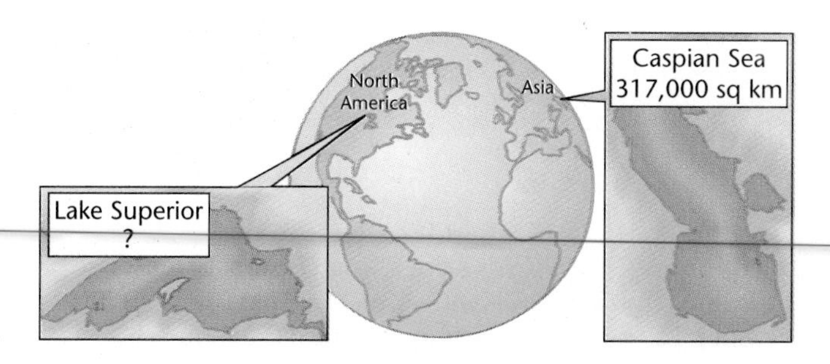

9. *Sheet Perimeter.* The dimensions of a sheet for a queen-size bed are 90 in. by 102 in. What is the perimeter of the sheet?

10. *Sheet Perimeter.* The dimensions of a sheet for a king-size bed are 108 in. by 102 in. What is the perimeter of the sheet?

11. *Paper Quantity.* A ream of paper contains 500 sheets. How many sheets are in 9 reams?

9 reams

500 sheets in each

12. *Reading Rate.* Cindy's reading rate is 205 words per minute. How many words can she read in 30 min?

13. *Elvis Impersonators.* When Elvis Presley died in 1977, there were already 48 professional Elvis impersonators (***Source:*** *Chance Magazine* 9, no. 1, Winter 1996). In 1995, there were 7328. How many more were there in 1995?

14. *LAV Vehicle.* A combat-loaded U.S. Light Armed Vehicle 25 (LAV-25) weighs 3930 lb more than its empty curb weight. The loaded LAV-25 weighs 28,400 lb. (***Source:*** *Car & Driver* 42, no. 1, July 1996: 153–155) What is its curb weight?

15. Dana borrows $5928 for a used car. The loan is to be paid off in 24 equal monthly payments. How much is each payment (excluding interest)?

16. A family borrows $4824 to build a sunroom on the back of their house. The loan is to be paid off in equal monthly payments of $134 (excluding interest). How many months will it take to pay off the loan?

17. *Cheers Episodes.* *Cheers* is the longest-running comedy in the history of television, with 271 episodes created. A local station picks up the syndicated reruns. If the station runs 5 episodes per week, how many full weeks will pass before it must start over with past episodes? How many episodes will be left for the last week?

18. A lab technician separates a vial containing 70 cubic centimeters (cc) of blood into test tubes, each of which contain 3 cc of blood. How many test tubes can be filled? How much blood is left over?

19. There are 24 hours (hr) in a day and 7 days in a week. How many hours are there in a week?

20. There are 60 min in an hour and 24 hr in a day. How many minutes are there in a day?

21. You have $568 in your checking account. You write checks for $46, $87, and $129. Then you deposit $94 back in the account upon the return of some books. How much is left in your account?

22. The balance in your checking account is $749. You write checks for $34 and $65. Then you make a deposit of $123 from your paycheck. What is your new balance?

23. *NBA Court.* The standard basketball court used by college and NBA players has dimensions of 50 ft by 94 ft (***Source*:** National Basketball Association).

a) What is its area?
b) What is its perimeter?

24. *High School Court.* The standard basketball court used by high school players has dimensions of 50 ft by 84 ft.

a) What is its area? What is its perimeter?
b) How much larger is the area of an NBA court than a high school court? (See Exercise 23.)

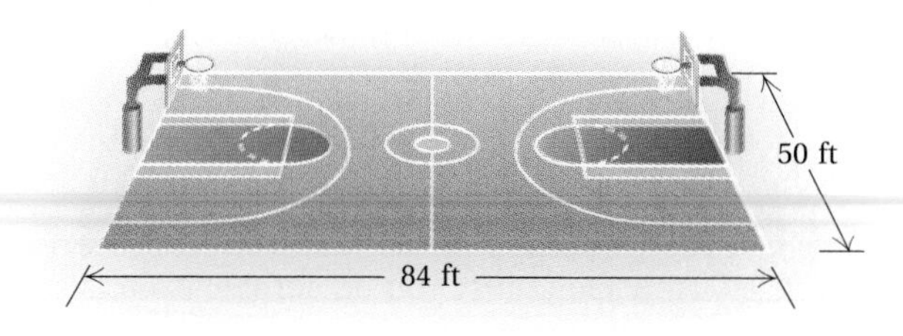

25. Copies of this book are generally shipped from the warehouse in cartons containing 24 books each. How many cartons are needed to ship 840 books?

26. Sixteen-ounce bottles of catsup are generally shipped in cartons containing 12 bottles each. How many cartons are needed to ship 528 bottles of catsup?

27. Copies of this book are generally shipped from the warehouse in cartons containing 24 books each. How many cartons are needed to ship 1355 books? How many books are left over?

28. Sixteen-ounce bottles of catsup are generally shipped in cartons containing 12 bottles each. How many cartons are needed to ship 1033 bottles of catsup? How many bottles are left over?

29. *Map Drawing.* A map has a scale of 64 mi to the inch. How far apart *in reality* are two cities that are 25 in. apart on the map? How far apart *on the map* are two cities that, in reality, are 1728 mi apart?

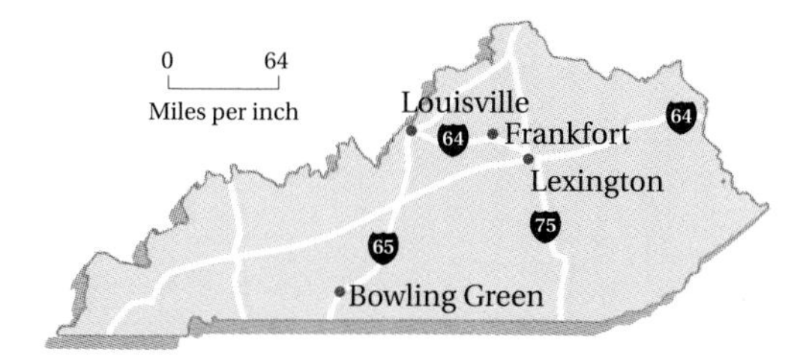

30. *Map Drawing.* A map has a scale of 25 mi to the inch. How far apart *on the map* are two cities that, in reality, are 2200 mi apart? How far apart *in reality* are two cities that are 13 in. apart on the map?

31. A carpenter drills 216 holes in a rectangular array in a pegboard. There are 12 holes in each row. How many rows are there?

32. Lou works as a CPA. He arranges 504 entries on a spreadsheet in a rectangular array that has 36 rows. How many entries are in each row?

33. Elaine buys 5 video games at $44 each and pays for them with $10 bills. How many $10 bills did it take?

34. Lowell buys 5 video games at $44 each and pays for them with $20 bills. How many $20 bills did it take?

35. Before going back to college, David buys 4 shirts at \$59 each and 6 pairs of pants at \$78 each. What was the total cost of this clothing?

36. Ann buys office supplies at Office Depot. One day she buys 8 reams of paper at \$24 each and 16 pens at \$3 each. How much did she spend?

37. *Weight Loss.* Use the information from the chart on page 67. How long must you do aerobic exercises in order to lose one pound?

38. *Weight Loss.* Use the information from the chart on page 67. How long must you bicycle at 9 mph in order to lose one pound?

39. *Index Cards.* Index cards of dimension 3 in. by 5 in. are normally shipped in packages containing 100 cards each. How much writing area is available if one uses the front and back sides of a package of these cards?

40. An office for adjunct instructors at a community college has 6 bookshelves, each of which is 3 ft long. The office is moved to a new location that has dimensions of 16 ft by 21 ft. Is it possible for the bookshelves to be put side by side on the 16-ft wall?

Skill Maintenance

Round 234,562 to the nearest: [1.4a]

41. Hundred.

42. Thousand.

Estimate the computation by rounding to the nearest thousand. [1.4b]

43. $2783 + 4602 + 5797 + 8111$

44. $28{,}430 - 11{,}977$

Estimate the product by rounding to the nearest hundred. [1.5c]

45. $787 \cdot 363$

46. $887 \cdot 799$

Synthesis

47. ◆ Of the five problem-solving steps listed at the beginning of this section, which is the most difficult for you? Why?

48. ◆ Write a problem for a classmate to solve. Design the problem so that the solution is "The driver still has 329 mi to travel."

49. 🖩 *Speed of Light.* Light travels about 186,000 miles per second (mi/sec) in a vacuum as in outer space. In ice it travels about 142,000 mi/sec, and in glass it travels about 109,000 mi/sec. In 18 sec, how many more miles will light travel in a vacuum than in ice? in glass?

50. Carney Community College has 1200 students. Each professor teaches 4 classes and each student takes 5 classes. There are 30 students and 1 teacher in each classroom. How many professors are there at Carney Community College?

Make a budget for a road trip to your favorite destination.

1.9 Exponential Notation and Order of Operations

Objectives

a Write exponential notation for products such as $4 \cdot 4 \cdot 4$.

b Evaluate exponential notation.

c Simplify expressions using the rules for order of operations.

d Remove parentheses within parentheses.

For Extra Help

TAPE 2 | TAPE 3A | MAC WIN | CD-ROM

a Exponential Notation

Consider the product $3 \cdot 3 \cdot 3 \cdot 3$. Such products occur often enough that mathematicians have found it convenient to create a shorter notation, called **exponential notation,** explained as follows.

$$\underbrace{3 \cdot 3 \cdot 3 \cdot 3}_{\text{4 factors}} \text{ is shortened to } 3^4 \leftarrow \text{exponent (3 is the base)}$$

We read 3^4 as "three to the fourth power," 5^3 as "five to the third power," or "five cubed," and 5^2 as "five squared." The latter comes from the fact that a square of side s has area A given by $A = s^2$.

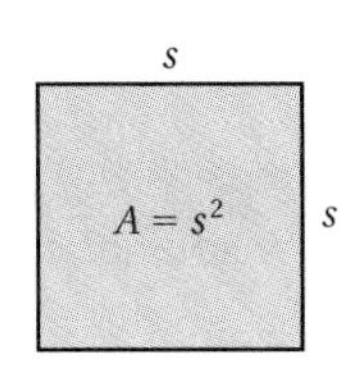

Example 1 Write exponential notation for $10 \cdot 10 \cdot 10 \cdot 10 \cdot 10$.

Exponential notation is 10^5. **5 is the *exponent*. 10 is the *base*.**

Example 2 Write exponential notation for $2 \cdot 2 \cdot 2$.

Exponential notation is 2^3.

Do Exercises 1–4.

Write exponential notation.

1. $5 \cdot 5 \cdot 5 \cdot 5$

2. $5 \cdot 5 \cdot 5 \cdot 5 \cdot 5$

3. $10 \cdot 10$

4. $10 \cdot 10 \cdot 10 \cdot 10$

b Evaluating Exponential Notation

We evaluate exponential notation by rewriting it as a product and computing the product.

Example 3 Evaluate: 10^3.

$10^3 = 10 \cdot 10 \cdot 10 = 1000$

Example 4 Evaluate: 5^4.

$5^4 = 5 \cdot 5 \cdot 5 \cdot 5 = 625$

Caution! 5^4 does not mean $5 \cdot 4$.

Do Exercises 5–8.

Evaluate.

5. 10^4

6. 10^2

7. 8^3

8. 2^5

Answers on page A-3

Simplify.

9. $93 - 14 \cdot 3$

10. $104 \div 4 + 4$

11. $25 \cdot 26 - (56 + 10)$

12. $75 \div 5 + (83 - 14)$

Simplify and compare.

13. $64 \div (32 \div 2)$ and $(64 \div 32) \div 2$

14. $(28 + 13) + 11$ and $28 + (13 + 11)$

Answers on page A-3

c Simplifying Expressions

Suppose we have a calculation like the following:

$$3 + 4 \cdot 8.$$

How do we find the answer? Do we add 3 to 4 and then multiply by 8, or do we multiply 4 by 8 and then add 3? In the first case, the answer is 56. In the second, the answer is 35. We agree to compute as in the second case.

Consider the calculation

$$7 \cdot 14 - (12 + 18).$$

What do the parentheses mean? To deal with these questions, we must make some agreement regarding the order in which we perform operations. The rules are as follows.

RULES FOR ORDER OF OPERATIONS

1. Do all calculations within parentheses ( ), brackets [ ], or braces { } before operations outside.
2. Evaluate all exponential expressions.
3. Do all multiplications and divisions in order from left to right.
4. Do all additions and subtractions in order from left to right.

It is worth noting that these are the rules that a computer uses to do computations. In order to program a computer, you must know these rules.

Example 5 Simplify: $16 \div 8 \times 2$.

There are no parentheses or exponents, so we start with the third step.

$16 \div 8 \times 2 = 2 \times 2$ **Doing all multiplications and divisions in order from left to right**

$= 4$

Example 6 Simplify: $7 \cdot 14 - (12 + 18)$.

$7 \cdot 14 - (12 + 18) = 7 \cdot 14 - 30$ **Carrying out operations inside parentheses**

$= 98 - 30$ **Doing all multiplications and divisions**

$= 68$ **Doing all additions and subtractions**

Do Exercises 9–12.

Example 7 Simplify and compare: $23 - (10 - 9)$ and $(23 - 10) - 9$.

We have

$$23 - (10 - 9) = 23 - 1 = 22;$$
$$(23 - 10) - 9 = 13 - 9 = 4.$$

We can see that $23 - (10 - 9)$ and $(23 - 10) - 9$ represent different numbers. Thus subtraction is not associative.

Do Exercises 13 and 14.

Example 8 Simplify: $7 \cdot 2 - (12 + 0) \div 3 - (5 - 2)$.

$7 \cdot 2 - (12 + 0) \div 3 - (5 - 2) = 7 \cdot 2 - 12 \div 3 - 3$ Carrying out operations inside parentheses

$= 14 - 4 - 3$ Doing all multiplications and divisions in order from left to right

$= 7$ Doing all additions and subtractions in order from left to right

Do Exercise 15.

Example 9 Simplify: $15 \div 3 \cdot 2 \div (10 - 8)$.

$15 \div 3 \cdot 2 \div (10 - 8) = 15 \div 3 \cdot 2 \div 2$ Carrying out operations inside parentheses

$= 5 \cdot 2 \div 2$ Doing all multiplications and divisions in order from left to right

$= 10 \div 2$

$= 5$

Do Exercises 16–18.

Example 10 Simplify: $4^2 \div (10 - 9 + 1)^3 \cdot 3 - 5$.

$4^2 \div (10 - 9 + 1)^3 \cdot 3 - 5$

$= 4^2 \div (1 + 1)^3 \cdot 3 - 5$ Carrying out operations inside parentheses

$= 4^2 \div 2^3 \cdot 3 - 5$ Adding inside parentheses

$= 16 \div 8 \cdot 3 - 5$ Evaluating exponential expressions

$= 2 \cdot 3 - 5$

$= 6 - 5$ } Doing all multiplications and divisions in order from left to right

$= 1$

Do Exercises 19–21.

Calculator Spotlight

Calculators often have an [x^y], [a^x], or [^] key for raising a base to a power. To find 3^5 with such a key, we press [3] [x^y] [5] [=]. The result is 243.

1. Find 4^5. **2.** Find 7^9. **3.** Find 2^{20}.

To determine whether a calculator is programmed to follow the rules for order of operations, press [3] [+] [4] [×] [2] [=]. If the result is 11, that particular calculator follows the rules. If the result is 14, the calculator performs operations as they are entered. To compensate for the latter case, we would press [4] [×] [2] [=] [+] [3] [=].

4. Find $84 - 5 \cdot 7$. **5.** Find $80 + 50 \div 10$.

When a calculator has parentheses, [(] and [)], expressions like $5(4 + 3)$ can be found without first entering the addition. We simply press [5] [×] [(] [4] [+] [3] [)] [=]. The result is 35.

6. Find $9(7 + 8)$. **7.** Find $8[4 + 3(7 - 1)]$.

15. Simplify:

$9 \times 4 - (20 + 4) \div 8 - (6 - 2)$.

Simplify.

16. $5 \cdot 5 \cdot 5 + 26 \cdot 71 - (16 + 25 \cdot 3)$

17. $30 \div 5 \cdot 2 + 10 \cdot 20 + 8 \cdot 8 - 23$

18. $95 - 2 \cdot 2 \cdot 2 \cdot 5 \div (24 - 4)$

Simplify.

19. $5^3 + 26 \cdot 71 - (16 + 25 \cdot 3)$

20. $(1 + 3)^3 + 10 \cdot 20 + 8^2 - 23$

21. $95 - 2^3 \cdot 5 \div (24 - 4)$

Answers on page A-3

22. *NBA Tall Men.* The heights, in inches, of several of the tallest players in the NBA are given in the bar graph below. Find the average height of these players.

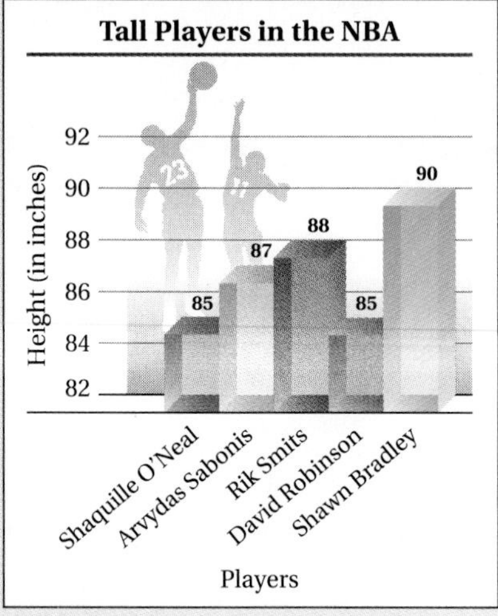

Source: NBA

Simplify.

23. $9 \times 5 + \{6 \div [14 - (5 + 3)]\}$

24. $[18 - (2 + 7) \div 3] - (31 - 10 \times 2)$

Answers on page A-3

Example 11 *Average Height of Waterfalls.* The heights of the four highest waterfalls in the world are given in the bar graph at right. Find the average height of all four. To find the **average** of a set of numbers, we first add the numbers and then divide by the number of addends.

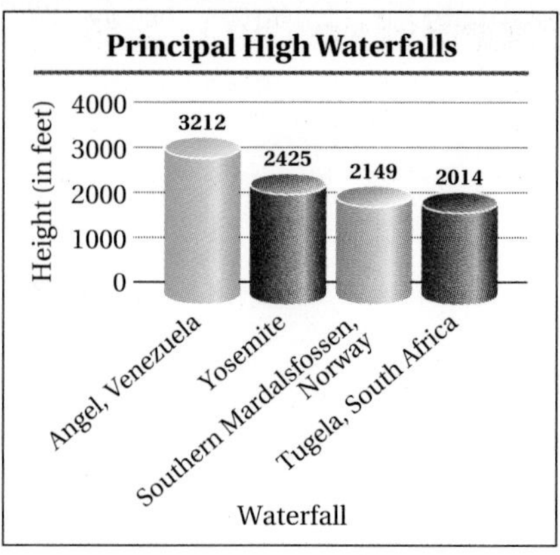

Source: World Almanac

The average is given by

$$(3212 + 2425 + 2149 + 2014) \div 4.$$

To find the average, we carry out the computation using the rules for order of operations:

$$\begin{aligned}(3212 + 2425 + 2149 + 2014) \div 4 &= 9800 \div 4 \\ &= 2450.\end{aligned}$$

Thus the average height of the four highest waterfalls is 2450 ft.

Do Exercise 22.

d Parentheses Within Parentheses

When parentheses occur within parentheses, we can make them different shapes, such as [ ] (also called "brackets") and { } (also called "braces"). All of these have the same meaning. When parentheses occur within parentheses, computations in the innermost ones are to be done first.

Example 12 Simplify: $16 \div 2 + \{40 - [13 - (4 + 2)]\}$.

$16 \div 2 + \{40 - [13 - (4 + 2)]\}$

$= 16 \div 2 + \{40 - [13 - 6]\}$ **Doing the calculations in the innermost parentheses first**

$= 16 \div 2 + \{40 - 7\}$ **Again, doing the calculations in the innermost parentheses**

$= 16 \div 2 + 33$

$= 8 + 33$ **Doing all multiplications and divisions in order from left to right**

$= 41$ **Doing all additions and subtractions in order from left to right**

Example 13 Simplify: $[25 - (4 + 3) \times 3] \div (11 - 7)$.

$$\begin{aligned}[25 - (4 + 3) \times 3] \div (11 - 7) &= [25 - 7 \times 3] \div (11 - 7) \\ &= [25 - 21] \div (11 - 7) \\ &= 4 \div 4 \\ &= 1\end{aligned}$$

Do Exercises 23 and 24.

Exercise Set 1.9

a Write exponential notation.

1. $3 \cdot 3 \cdot 3 \cdot 3$

2. $2 \cdot 2 \cdot 2 \cdot 2 \cdot 2$

3. $5 \cdot 5$

4. $13 \cdot 13 \cdot 13$

5. $7 \cdot 7 \cdot 7 \cdot 7 \cdot 7$

6. $10 \cdot 10$

7. $10 \cdot 10 \cdot 10$

8. $1 \cdot 1 \cdot 1 \cdot 1$

b Evaluate.

9. 7^2

10. 5^3

11. 9^3

12. 10^2

13. 12^4

14. 10^5

15. 11^2

16. 6^3

c Simplify.

17. $12 + (6 + 4)$

18. $(12 + 6) + 18$

19. $52 - (40 - 8)$

20. $(52 - 40) - 8$

21. $1000 \div (100 \div 10)$

22. $(1000 \div 100) \div 10$

23. $(256 \div 64) \div 4$

24. $256 \div (64 \div 4)$

25. $(2 + 5)^2$

26. $2^2 + 5^2$

27. $(11 - 8)^2 - (18 - 16)^2$

28. $(32 - 27)^3 + (19 + 1)^3$

29. $16 \cdot 24 + 50$

30. $23 + 18 \cdot 20$

31. $83 - 7 \cdot 6$

32. $10 \cdot 7 - 4$

33. $10 \cdot 10 - 3 \cdot 4$

34. $90 - 5 \cdot 5 \cdot 2$

35. $4^3 \div 8 - 4$

36. $8^2 - 8 \cdot 2$

37. $17 \cdot 20 - (17 + 20)$

38. $1000 \div 25 - (15 + 5)$

39. $6 \cdot 10 - 4 \cdot 10$

40. $3 \cdot 8 + 5 \cdot 8$

41. $300 \div 5 + 10$

42. $144 \div 4 - 2$

43. $3 \cdot (2 + 8)^2 - 5 \cdot (4 - 3)^2$

44. $7 \cdot (10 - 3)^2 - 2 \cdot (3 + 1)^2$

45. $4^2 + 8^2 \div 2^2$

46. $6^2 - 3^4 \div 3^3$

47. $10^3 - 10 \cdot 6 - (4 + 5 \cdot 6)$

48. $7^2 + 20 \cdot 4 - (28 + 9 \cdot 2)$

49. $6 \cdot 11 - (7 + 3) \div 5 - (6 - 4)$

50. $8 \times 9 - (12 - 8) \div 4 - (10 - 7)$

51. $120 - 3^3 \cdot 4 \div (5 \cdot 6 - 6 \cdot 4)$

52. $80 - 2^4 \cdot 15 \div (7 \cdot 5 - 45 \div 3)$

53. Find the average of \$64, \$97, and \$121.

54. Find the average of four test grades of 86, 92, 80, and 78.

d Simplify.

55. $8 \times 13 + \{42 \div [18 - (6 + 5)]\}$

56. $72 \div 6 - \{2 \times [9 - (4 \times 2)]\}$

57. $[14 - (3 + 5) \div 2] - [18 \div (8 - 2)]$

58. $[92 \times (6 - 4) \div 8] + [7 \times (8 - 3)]$

59. $(82 - 14) \times [(10 + 45 \div 5) - (6 \cdot 6 - 5 \cdot 5)]$

60. $(18 \div 2) \cdot \{[(9 \cdot 9 - 1) \div 2] - [5 \cdot 20 - (7 \cdot 9 - 2)]\}$

61. $4 \times \{(200 - 50 \div 5) - [(35 \div 7) \cdot (35 \div 7) - 4 \times 3]\}$

62. $\{[18 - 2 \cdot 6] - [40 \div (17 - 9)]\} + \{48 - 13 \times 3 + [(50 - 7 \cdot 5) + 2]\}$

Skill Maintenance

Solve. [1.7b]

63. $x + 341 = 793$

64. $7 \cdot x = 91$

Solve. [1.8a]

65. *Colorado.* The state of Colorado is roughly the shape of a rectangle that is 270 mi by 380 mi. What is its area?

66. On a long four-day trip, a family bought the following amounts of gasoline for their motor home:

23 gallons, 24 gallons,
26 gallons, 25 gallons.

How much gasoline did they buy in all?

Synthesis

67. ◆ The expression $9 - (4 \times 2)$ contains parentheses. Are they necessary? Why or why not?

68. ◆ The expression $(3 \cdot 4)^2$ contains parentheses. Are they necessary? Why or why not?

Simplify.

69. ▦ $15(23 - 4 \cdot 2)^3 \div (3 \cdot 25)$

70. ▦ $(19 - 2^4)^5 - (141 \div 47)^2$

Each of the expressions in Exercises 71–73 is incorrect. First find the correct answer. Then place as many parentheses as needed in the expression in order to make the incorrect answer correct.

71. $1 + 5 \cdot 4 + 3 = 36$

72. $12 \div 4 + 2 \cdot 3 - 2 = 2$

73. $12 \div 4 + 2 \cdot 3 - 2 = 4$

74. Write out the symbols +, −, ×, ÷, and ( ) and one occurrence each of 1, 2, 3, 4, 5, 6, 7, 8, and 9 to represent 100.

Collaborative Learning Manual — Use the order of operations to simplify expressions.

Summary and Review Exercises: Chapter 1

The review exercises that follow are for practice. Answers are given at the back of the book. If you miss an exercise, restudy the objective indicated in blue next to the exercise or direction line that precedes it.

Write expanded notation. [1.1a]

1. 2793

2. 56,078

Write standard notation. [1.1b]

3. 8 thousands + 6 hundreds + 6 tens + 9 ones

4. 9 ten thousands + 8 hundreds + 4 tens + 4 ones

Write a word name. [1.1c]

5. 67,819

6. 2,781,427

Write standard notation. [1.1d]

7. Four hundred seventy-six thousand, five hundred eighty-eight

8. *San Francisco International.* The total number of passengers passing through San Francisco International Airport in a recent year was thirty-six million, two hundred sixty thousand, sixty-four.

9. What does the digit 8 mean in 4,678,952? [1.1e]

10. In 13,768,940, what digit tells the number of millions? [1.1e]

11. Write an addition sentence that corresponds to the situation. [1.2a]

Tony has $406 in her checking account. She is paid $78 for a part-time job and deposits that in her checking account. How much is then in the account?

12. Find the perimeter. [1.2a]

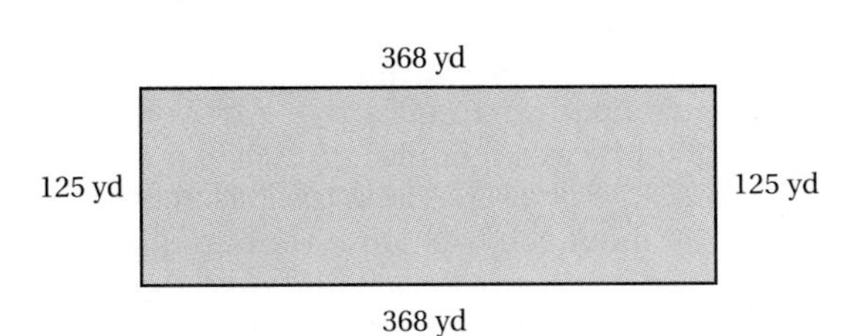

Add. [1.2b]

13. 7304 + 6968

14. 27,609 + 38,415

15. 2743 + 4125 + 6274 + 8956

16.

$$\begin{array}{r} 91,426 \\ +\ 7,495 \\ \hline \end{array}$$

Write a subtraction sentence that corresponds to the situation. [1.3b], [1.3c]

17. By exercising daily, you lose 12 lb in one month. If you weighed 151 lb at the beginning of the month, what is your weight now?

18. Natosha has $196 and wants to buy a fax machine for $340. How much more does she need?

19. Write a related addition sentence: [1.3b]

10 − 6 = 4.

20. Write two related subtraction sentences: [1.3b]

8 + 3 = 11.

Subtract. [1.3d]

21. 8045 − 2897

22. 8465 − 7312

23. 6003 − 3729

24.

$$\begin{array}{r} 37,405 \\ -\ 19,648 \\ \hline \end{array}$$

Round 345,759 to the nearest: [1.4a]

25. Hundred.

26. Ten.

27. Thousand.

Estimate the sum, difference, or product by first rounding to the nearest hundred. Show your work. [1.4b], [1.5c]

28. 41,348 + 19,749

29. 38,652 − 24,549

30. 396 · 748

Use < or > for ▢ to write a true sentence. [1.4c]

31. 67 ▢ 56

32. 1 ▢ 23

33. Write a multiplication sentence that corresponds to the situation. [1.5a]

A farmer plants apple trees in a rectangular array. He plants 15 rows with 32 trees in each row. How many apple trees does he have altogether?

34. Find the area of the rectangle in Exercise 12.

Multiply. [1.5b]

35. $700 \cdot 600$

36. $7846 \cdot 800$

37. $726 \cdot 698$

38. $587 \cdot 47$

39.
$$\begin{array}{r} 8\,3\,0\,5 \\ \times \quad 6\,4\,2 \\ \hline \end{array}$$

Write a division sentence that corresponds to the situation. [1.6a]

40. A cheese factory made 176 lb of Monterey Jack cheese. The cheese was placed in 4-lb boxes. How many boxes were filled?

41. A beverage company packed 222 cans of soda into 6-can cartons. How many cartons did they fill?

42. Write a related multiplication sentence: [1.6b]

$56 \div 8 = 7.$

43. Write two related division sentences: [1.6b]

$13 \cdot 4 = 52.$

Divide. [1.6c]

44. $63 \div 5$

45. $80 \div 16$

46. $7\overline{)6\,3\,9\,4}$

47. $3073 \div 8$

48. $6\,0\overline{)2\,8\,6}$

49. $4266 \div 79$

50. $3\,8\overline{)1\,7{,}1\,7\,6}$

51. $1\,4\overline{)7\,0{,}1\,1\,2}$

52. $52{,}668 \div 12$

Solve. [1.7b]

53. $46 \cdot n = 368$

54. $47 + x = 92$

55. $x = 782 - 236$

Solve. [1.8a]

56. An apartment builder bought 3 electric ranges at $299 each and 4 dishwashers at $379 each. What was the total cost?

57. *Lincoln-Head Pennies.* In 1909, the first Lincoln-head pennies were minted. Seventy-three years later, these pennies were first minted with a decreased copper content. In what year was the copper content reduced?

58. A family budgets $4950 for food and clothing and $3585 for entertainment. The yearly income of the family was $28,283. How much of this income remained after these two allotments?

59. A chemist has 2753 mL of alcohol. How many 20-mL beakers can be filled? How much will be left over?

60. Write exponential notation: $4 \cdot 4 \cdot 4.$ [1.9a]

Evaluate. [1.9b]

61. 10^4

62. 6^2

Simplify. [1.9c, d]

63. $8 \cdot 6 + 17$

64. $10 \cdot 24 - (18 + 2) \div 4 - (9 - 7)$

65. $7 + (4 + 3)^2$

66. $7 + 4^2 + 3^2$

67. $(80 \div 16) \times [(20 - 56 \div 8) + (8 \cdot 8 - 5 \cdot 5)]$

68. Find the average of 157, 170, and 168.

Synthesis

69. Write a problem for a classmate to solve. Design the problem so that the solution is "Each of the 144 bottles will contain 8 oz of hot sauce." [1.8a]

70. Is subtraction associative? Why or why not? [1.2b], [1.3d]

71. Determine the missing digit d. [1.5b]

$$\begin{array}{r} 9\,d \\ \times \quad d\,2 \\ \hline 8\,0\,3\,6 \end{array}$$

72. Determine the missing digits a and b. [1.6c]

$$\begin{array}{r} 9\,a\,1 \\ 2\,b\,1\overline{)2\,3\,6{,}4\,2\,1} \end{array}$$

73. A mining company estimates that a crew must tunnel 2100 ft into a mountain to reach a deposit of copper ore. Each day the crew tunnels about 500 ft. Each night about 200 ft of loose rocks roll back into the tunnel. How many days will it take the mining company to reach the copper deposit? [1.8a]

Test: Chapter 1

1. Write expanded notation: 8843.

2. Write a word name: 38,403,277.

3. In the number 546,789, which digit tells the number of hundred thousands?

Add.

4. $\begin{array}{r} 6811 \\ +\ 3178 \\ \hline \end{array}$

5. $\begin{array}{r} 45{,}889 \\ +\ 17{,}902 \\ \hline \end{array}$

6. $\begin{array}{r} 12 \\ 8 \\ 3 \\ 7 \\ +\ 4 \\ \hline \end{array}$

7. $\begin{array}{r} 6203 \\ +\ 4312 \\ \hline \end{array}$

Subtract.

8. $\begin{array}{r} 7983 \\ -\ 4353 \\ \hline \end{array}$

9. $\begin{array}{r} 2974 \\ -\ 1935 \\ \hline \end{array}$

10. $\begin{array}{r} 8907 \\ -\ 2059 \\ \hline \end{array}$

11. $\begin{array}{r} 23{,}067 \\ -\ 17{,}892 \\ \hline \end{array}$

Multiply.

12. $\begin{array}{r} 4568 \\ \times\ 9 \\ \hline \end{array}$

13. $\begin{array}{r} 8876 \\ \times\ 600 \\ \hline \end{array}$

14. $\begin{array}{r} 65 \\ \times\ 37 \\ \hline \end{array}$

15. $\begin{array}{r} 678 \\ \times\ 788 \\ \hline \end{array}$

Divide.

16. $15 \div 4$

17. $420 \div 6$

18. $89 \overline{)8633}$

19. $44 \overline{)35{,}428}$

Solve.

20. *James Dean.* James Dean was 24 yr old when he died. He was born in 1931. In what year did he die?

21. A beverage company produces 739 cans of soda. How many 8-can packages can be filled? How many cans will be left over?

22. *Area of New England.* Listed below are the areas, in square miles, of the New England states (***Source:*** U.S. Bureau of the Census). What is the total area of New England?

Maine	30,865
Massachusetts	7,838
New Hampshire	8,969
Vermont	9,249
Connecticut	4,845
Rhode Island	1,045

23. A rectangular lot measures 200 m by 600 m. What is the area of the lot? What is the perimeter of the lot?

24. A sack of oranges weighs 27 lb. A sack of apples weighs 32 lb. Find the total weight of 16 bags of oranges and 43 bags of apples.

25. A box contains 5000 staples. How many staplers can be filled from the box if each stapler holds 250 staples?

Answers

1. ______
2. ______
3. ______
4. ______
5. ______
6. ______
7. ______
8. ______
9. ______
10. ______
11. ______
12. ______
13. ______
14. ______
15. ______
16. ______
17. ______
18. ______
19. ______
20. ______
21. ______
22. ______
23. ______
24. ______
25. ______

Answers

26. ______

27. ______

28. ______

29. ______

30. ______

31. ______

32. ______

33. ______

34. ______

35. ______

36. ______

37. ______

38. ______

39. ______

40. ______

41. ______

42. ______

43. ______

44. ______

45. ______

46. ______

47. ______

48. ______

49. ______

Solve.

26. $28 + x = 74$ **27.** $169 \div 13 = n$ **28.** $38 \cdot y = 532$

Round 34,578 to the nearest:

29. Thousand. **30.** Ten. **31.** Hundred.

Estimate the sum, difference, or product by first rounding to the nearest hundred. Show your work.

32. $\begin{array}{r} 23,649 \\ +\ 54,746 \\ \hline \end{array}$ **33.** $\begin{array}{r} 54,751 \\ -\ 23,649 \\ \hline \end{array}$ **34.** $\begin{array}{r} 824 \\ \times\ 489 \\ \hline \end{array}$

Use $<$ or $>$ for ▒ to write a true sentence.

35. 34 ▒ 17 **36.** 117 ▒ 157

37. Write exponential notation: $12 \cdot 12 \cdot 12 \cdot 12$.

Evaluate.

38. 7^3 **39.** 2^3

Simplify.

40. $(10 - 2)^2$ **41.** $10^2 - 2^2$ **42.** $(25 - 15) \div 5$

43. $8 \times \{(20 - 11) \cdot [(12 + 48) \div 6 - (9 - 2)]\}$ **44.** $2^4 + 24 \div 12$

45. Find the average of 97, 98, 87, and 86.

Synthesis

46. An open cardboard shoe box is 8 in. wide, 12 in. long, and 6 in. high. How many square inches of cardboard are used?

47. Cara spends $229 a month to repay her student loan. If she has already paid $9160 on the 10-yr loan, how many payments remain?

48. Jennie scores three 90's, four 80's, and a 74 on her eight quizzes. Find her average.

49. Use trials to find the single digit number a for which

$$359 - 46 + a \div 3 \times 25 - 7^2 = 339.$$

2

Fractional Notation: Multiplication and Division

An Application

Liz and her road crew paint the lines in the middle and on the sides of a highway. They average about $\frac{5}{16}$ mi each hour. How long will it take them to paint the lines on 70 mi of highway?

This problem appears as Exercise 53 in the Summary and Review Exercises.

The Mathematics

We let $t =$ the time it takes to paint 70 mi of highway. The problem then translates to this equation:

$$\underbrace{t = 70 \div \frac{5}{16}}.$$

Here is how division using fractional notation occurs in applied problems.

For more information, visit us at www.mathmax.com

Introduction

We consider multiplication and division using fractional notation in this chapter. To aid our study, the chapter begins with factorizations and rules for divisibility. After multiplication and division are discussed, those skills are used to solve equations and problems.

Pretest: Chapter 2

1. Determine whether 59 is prime, composite, or neither.

2. Find the prime factorization of 140.

3. Determine whether 1503 is divisible by 9.

4. Determine whether 788 is divisible by 8.

Simplify.

5. $\frac{57}{57}$

6. $\frac{68}{1}$

7. $\frac{0}{50}$

8. $\frac{8}{32}$

Multiply and simplify.

9. $\frac{1}{3} \cdot \frac{18}{5}$

10. $\frac{5}{6} \cdot 24$

11. $\frac{2}{5} \cdot \frac{25}{8}$

Find the reciprocal.

12. $\frac{7}{8}$

13. 11

Divide and simplify.

14. $15 \div \frac{5}{8}$

15. $\frac{2}{3} \div \frac{8}{9}$

16. Solve:

$$\frac{7}{10} \cdot x = 21.$$

17. Use $=$ or $\neq$ for ■ to write a true sentence:

$$\frac{5}{11} \ ■ \ \frac{1}{2}.$$

Solve.

18. A person earns $48 for working a full day. How much is earned for working $\frac{3}{4}$ of a day?

19. A piece of tubing $\frac{5}{8}$ m long is to be cut into 15 pieces of the same length. What is the length of each piece?

Objectives for Retesting

The objectives to be tested in addition to the material in this chapter are as follows.

[1.3d] Subtract whole numbers.

[1.6c] Divide whole numbers.

[1.7b] Solve equations like $t + 28 = 54$, $28 \cdot x = 268$, and $98 \div 2 = y$.

[1.8a] Solve applied problems involving addition, subtraction, multiplication, or division of whole numbers.

2.1 Factorizations

In this chapter, we begin our work with fractions. Certain skills make such work easier. For example, in order to simplify

$$\frac{12}{32},$$

it is important that we be able to *factor* the 12 and the 32, as follows:

$$\frac{12}{32} = \frac{4 \cdot 3}{4 \cdot 8}.$$

Then we "remove" a factor of 1:

$$\frac{4 \cdot 3}{4 \cdot 8} = \frac{4}{4} \cdot \frac{3}{8} = 1 \cdot \frac{3}{8} = \frac{3}{8}.$$

Thus factoring is an important skill in working with fractions.

Objectives

- a Find the factors of a number.
- b Find some multiples of a number, and determine whether a number is divisible by another.
- c Given a number from 1 to 100, tell whether it is prime, composite, or neither.
- d Find the prime factorization of a composite number.

For Extra Help

TAPE 3

TAPE 3A

MAC WIN

CD-ROM

a Factors and Factorization

In Sections 2.1 and 2.2, we consider only the **natural numbers** 1, 2, 3, and so on.

Let's look at the product $3 \cdot 4 = 12$. We say that 3 and 4 are **factors** of 12. Since $12 = 12 \cdot 1$, we also know that 12 and 1 are factors of 12.

> A **factor** of a given number is a number multiplied in a product.
>
> A **factorization** of a number is an expression for the number that shows it as a product of natural numbers.

For example, each of the following gives a factorization of 12.

$12 = 4 \cdot 3$ — **This factorization shows that 4 and 3 are factors of 12.**

$12 = 12 \cdot 1$ — **This factorization shows that 12 and 1 are factors of 12.**

$12 = 6 \cdot 2$ — **This factorization shows that 6 and 2 are factors of 12.**

$12 = 2 \cdot 3 \cdot 2$ — **This factorization shows that 2 and 3 are factors of 12.**

Since $n = n \cdot 1$, every number has a factorization, and every number has factors even if its only factors are itself and 1.

Example 1 Find all the factors of 24.

We first find some factorizations.

$24 = 1 \cdot 24$ $\quad 24 = 3 \cdot 8$

$24 = 2 \cdot 12$ $\quad 24 = 4 \cdot 6$

Note that all but one of the factors of a natural number are *less* than the number.

Factors: 1, 2, 3, 4, 6, 8, 12, 24.

Do Exercises 1–4.

Find all the factors of the number. (*Hint*: Find some factorizations of the number.)

1. 6

2. 8

3. 10

4. 32

Answers on page A-4

5. Show that each of the numbers 5, 45, and 100 is a multiple of 5.

6. Show that each of the numbers 10, 60, and 110 is a multiple of 10.

7. Multiply by 1, 2, 3, and so on, to find ten multiples of 5.

Answers on page A-4

b Multiples and Divisibility

A **multiple** of a natural number is a product of it and some natural number. For example, some multiples of 2 are:

$$\left.\begin{array}{ll} 2 & (\text{because } 2 = 1 \cdot 2); \\ 4 & (\text{because } 4 = 2 \cdot 2); \\ 6 & (\text{because } 6 = 3 \cdot 2); \\ 8 & (\text{because } 8 = 4 \cdot 2); \\ 10 & (\text{because } 10 = 5 \cdot 2). \end{array}\right\}$$

Note that all but one of the multiples of a number are *larger* than the number.

We find multiples of 2 by counting by twos: 2, 4, 6, 8, and so on. We can find multiples of 3 by counting by threes: 3, 6, 9, 12, and so on.

Example 2 Show that each of the numbers 3, 6, 9, and 15 is a multiple of 3.

$3 = 1 \cdot 3 \qquad 6 = 2 \cdot 3 \qquad 9 = 3 \cdot 3 \qquad 15 = 5 \cdot 3$

Do Exercises 5 and 6.

Example 3 Multiply by 1, 2, 3, and so on, to find ten multiples of 7.

$1 \cdot 7 = 7$	$6 \cdot 7 = 42$
$2 \cdot 7 = 14$	$7 \cdot 7 = 49$
$3 \cdot 7 = 21$	$8 \cdot 7 = 56$
$4 \cdot 7 = 28$	$9 \cdot 7 = 63$
$5 \cdot 7 = 35$	$10 \cdot 7 = 70$

Do Exercise 7.

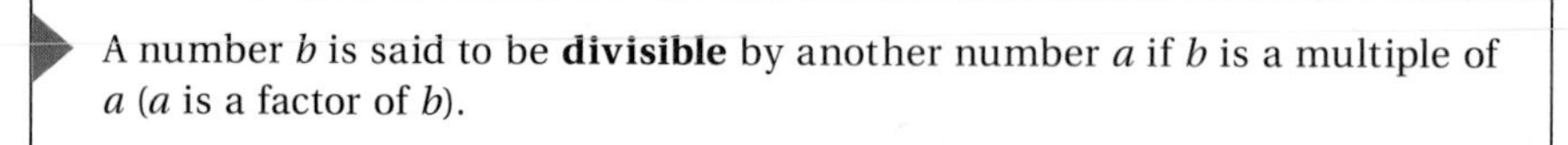

A number b is said to be **divisible** by another number a if b is a multiple of a (a is a factor of b).

Thus,

4 is divisible by 2 because 4 is a multiple of 2 ($4 = 2 \cdot 2$);
27 is divisible by 3 because 27 is a multiple of 3 ($27 = 9 \cdot 3$);
100 is divisible by 25 because 100 is a multiple of 25 ($100 = 4 \cdot 25$).

A number b is **divisible** by another number a if division of b by a results in a remainder of zero. We sometimes say that a divides b "evenly."

Example 4 Determine whether 24 is divisible by 3.

We divide 24 by 3:

$$\begin{array}{r} 8 \\ 3\overline{)24} \\ \underline{24} \\ 0 \end{array}$$

Because the remainder is 0, 24 is divisible by 3.

Example 5 Determine whether 98 is divisible by 4.

We divide 98 by 4:

$$\begin{array}{r} 24 \\ 4\overline{)98} \\ \underline{80} \\ 18 \\ \underline{16} \\ 2 \end{array} \longleftarrow \text{Not 0!}$$

Since the remainder is not 0, 98 is *not* divisible by 4.

Do Exercises 8–10.

Calculator Spotlight

Rather than list remainders, most calculators display quotients using decimal notation. Although decimal notation is not studied until Chapter 4, it is still possible for us to now check for divisibility using a calculator.

To determine whether a number, like 551, is divisible by another number, like 19, we simply press [5] [5] [1] [÷] [1] [9] [=]. If the resulting quotient contains no digits to the right of the decimal point, the first number is divisible by the second. Thus, since $551 \div 19 = 29$, we know that 551 is divisible by 19. On the other hand, since $551 \div 20 = 27.55$, we know that 551 is not divisible by 20.

Exercises

For each pair of numbers, determine whether the first number is divisible by the second number.

1. 731, 17

2. 1502, 79

3. 1053, 36

4. 4183, 47

c Prime and Composite Numbers

A natural number that has exactly two *different* factors, only itself and 1, is called a **prime number**.

Example 6 Tell whether the numbers 2, 3, 5, 7, and 11 are prime.

The number 2 is prime. It has only the factors 1 and 2.
The number 5 is prime. It has only the factors 1 and 5.
The numbers 3, 7, and 11 are also prime.

Some natural numbers are not prime.

8. Determine whether 16 is divisible by 2.

9. Determine whether 125 is divisible by 5.

10. Determine whether 125 is divisible by 6.

Answers on page A-4

11. Tell whether each number is prime, composite, or neither.

1, 4, 6, 8, 13, 19, 41

Answer on page A-4

Example 7 Tell whether the numbers 4, 6, 8, 10, 63, and 1 are prime.

The number 4 is not prime. It has the factors 1, 2, and 4.

The numbers 6, 8, 10, and 63 are not prime. Each has more than two different factors.

The number 1 is not prime. It does not have two *different* factors.

A natural number, other than 1, that is not prime is called **composite**.

In other words, if a number can be factored into a product of natural numbers, some of which are not the number itself or 1, it is composite. Thus,

2, 3, 5, 7, and 11 are prime;

4, 6, 8, 10, and 63 are composite;

1 is neither prime nor composite.

The number 0 is also neither prime nor composite, but 0 is *not* a natural number and thus is not considered here. We are considering only natural numbers.

Do Exercise 11.

The following is a table of the prime numbers from 2 to 157. There are more extensive tables, but these prime numbers will be the most helpful to you in this text.

A TABLE OF PRIMES

2, 3, 5, 7, 11, 13, 17, 19, 23, 29, 31, 37, 41, 43, 47, 53, 59, 61, 67, 71, 73, 79, 83, 89, 97, 101, 103, 107, 109, 113, 127, 131, 137, 139, 149, 151, 157

d Prime Factorizations

To factor a composite number into a product of primes is to find a **prime factorization** of the number. To do this, we consider the primes

2, 3, 5, 7, 11, 13, 17, 19, 23, and so on,

and determine whether a given number is divisible by the primes.

Example 8 Find the prime factorization of 39.

a) We divide by the first prime, 2.

$$\begin{array}{r} 19 \quad \text{R} = 1 \\ 2\overline{)39} \\ \underline{38} \\ 1 \end{array}$$

Because the remainder is not 0, 2 is not a factor of 39, and 39 is not divisible by 2.

b) We divide by the next prime, 3.

$$\begin{array}{r} 13 \quad \text{R} = 0 \\ 3\overline{)39} \end{array}$$

Because 13 is a prime, we are finished. The prime factorization is

$$39 = 3 \cdot 13.$$

Example 9 Find the prime factorization of 76.

a) We divide by the first prime, 2.

$$2\overline{)76}\quad 38 \quad R = 0$$

b) Because 38 is composite, we start with 2 again:

$$2\overline{)38}\quad 19 \quad R = 0$$

Because 19 is a prime, we are finished. The prime factorization is

$76 = 2 \cdot 2 \cdot 19.$

We abbreviate our procedure as follows.

```
     19
   2)38
 2)76
```

$76 = 2 \cdot 2 \cdot 19$

Multiplication is commutative so a factorization such as $2 \cdot 2 \cdot 19$ could also be expressed as $2 \cdot 19 \cdot 2$ or $19 \cdot 2 \cdot 2$ (or in exponential notation, as $2^2 \cdot 19$ or $19 \cdot 2^2$), but the prime factors are still the same. For this reason, we agree that any of these is "the" prime factorization of 76.

> Every number has just one (unique) prime factorization.

Example 10 Find the prime factorization of 72.

We can do divisions "up" as follows.

```
         3
       3)9
     2)18
   2)36
 2)72 ←——— Begin here.
```

$72 = 2 \cdot 2 \cdot 2 \cdot 3 \cdot 3$

Or, we can also do divisions "down":

```
2)72 ← Begin here.
2)36
2)18
3) 9
   3
```

Some other ways to find a prime factorization using **factor trees** are as follows:

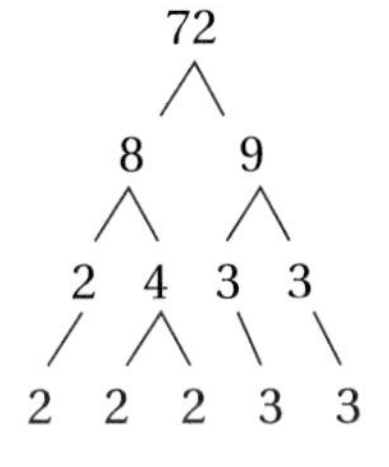

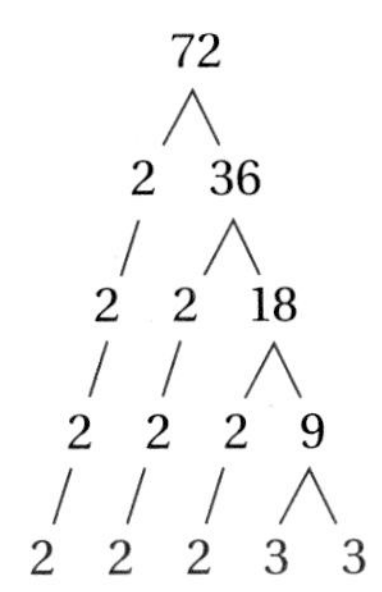

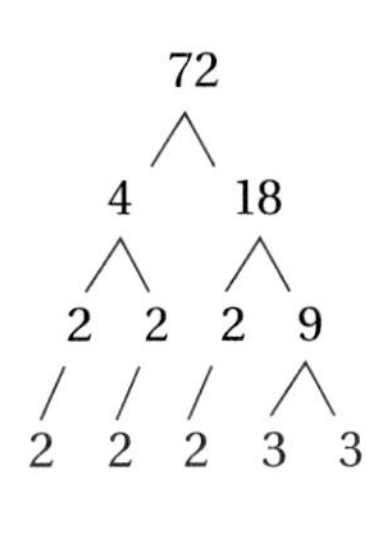

Find the prime factorization of the number.

12. 6

13. 12

14. 45

15. 98

16. 126

17. 144

To the student and the instructor: Recall that the Skill Maintenance exercises, which occur at the end of the exercise sets, review any skill that has been studied before in the text. Beginning with this chapter, however, certain objectives from four particular sections, along with the material of this chapter, will be tested on the chapter test. For this chapter, the objectives to be retested are Sections [1.3d], [1.6c], [1.7b], and [1.8a].

Answers on page A-4

Example 11 Find the prime factorization of 189.

We can use a string of successive divisions.

$$3\overline{)189} \to 3\overline{)63} \to 3\overline{)21} \to 7$$

189 is not divisible by 2. We move to 3.
63 is not divisible by 2. We move to 3.
21 is not divisible by 2. We move to 3.

$$189 = 3 \cdot 3 \cdot 3 \cdot 7$$

We can also use a factor tree.

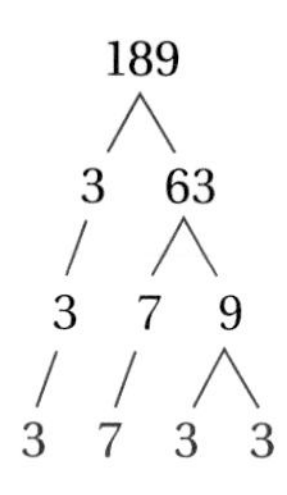

Example 12 Find the prime factorization of 65.

We can use a string of successive divisions.

$$5\overline{)65} \to 13$$

65 is not divisible by 2 or 3. We move to 5.

$$65 = 5 \cdot 13$$

We can also use a factor tree.

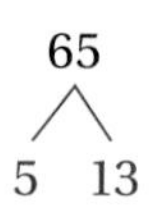

Do Exercises 12–17.

Calculator Spotlight

Exercises

For each pair of numbers, determine whether the second number is a factor of the first number.

1. 1502; 79 **2.** 4183; 47 **3.** 6888; 56

4. 864; 64 **5.** 32,768; 256 **6.** 32,768; 864

Exercise Set 2.1

a Find all the factors of the number.

1. 18

2. 16

3. 54

4. 48

5. 4

6. 9

7. 7

8. 11

9. 1

10. 3

11. 98

12. 100

b Multiply by 1, 2, 3, and so on, to find ten multiples of the number.

13. 4

14. 11

15. 20

16. 50

17. 3

18. 5

19. 12

20. 13

21. 10

22. 6

23. 9

24. 14

25. Determine whether 26 is divisible by 6.

26. Determine whether 29 is divisible by 9.

27. Determine whether 1880 is divisible by 8.

28. Determine whether 4227 is divisible by 3.

29. Determine whether 256 is divisible by 16.

30. Determine whether 102 is divisible by 4.

31. Determine whether 4227 is divisible by 9.

32. Determine whether 200 is divisible by 25.

33. Determine whether 8650 is divisible by 16.

34. Determine whether 4143 is divisible by 7.

c Determine whether the number is prime, composite, or neither.

35. 1

36. 2

37. 9

38. 19

39. 11

40. 27

41. 29

42. 49

d Find the prime factorization of the number.

43. 8

44. 16

45. 14

46. 15

47. 42

48. 32

49. 25

50. 40

51. 50

52. 62

53. 169

54. 140

55. 100

56. 110

57. 35

58. 70

59. 72

60. 86

61. 77

62. 99

63. 2884

64. 484

65. 51

66. 91

Skill Maintenance

Multiply. [1.5b]

67. $2 \cdot 13$

68. $8 \cdot 32$

69. $17 \cdot 25$

70. $25 \cdot 168$

Divide. [1.6c]

71. $0 \div 22$

72. $22 \div 1$

73. $22 \div 22$

74. $66 \div 22$

Solve. [1.8a]

75. Find the total cost of 7 shirts at $48 each and 4 pairs of pants at $69 each.

76. Sandy can type 62 words per minute. How long will it take her to type 12,462 words?

Synthesis

77. ◈ Explain a method for finding a composite number that contains exactly two factors other than itself and 1.

78. ◈ Is every natural number a multiple of 1? Why or why not?

79. *Factors and Sums.* To *factor* a number is to express it as a product. Since $15 = 5 \cdot 3$, we say that 15 is *factored* and that 5 and 3 are *factors* of 15. In the table below, the top number in each column has been factored in such a way that the sum of the factors is the bottom number in the column. For example, in the first column, 56 has been factored as $7 \cdot 8$, and $7 + 8 = 15$, the bottom number. Such thinking will be important in understanding the meaning of a factor and in algebra.

Product	56	63	36	72	140	96		168	110			
Factor	7									9	24	3
Factor	8						8	8		10	18	
Sum	15	16	20	38	24	20	14		21			24

Find the missing numbers in the table.

Collaborative Learning Manual

Find all the prime numbers less than 200, using the Sieve of Eratosthenes.

2.2 Divisibility

Suppose you are asked to find the simplest fractional notation for

$$\frac{117}{225}.$$

Since the numbers are quite large, you might feel that the task is difficult. However, both the numerator and the denominator have 9 as a factor. If you knew this, you could factor and simplify quickly as follows:

$$\frac{117}{225} = \frac{9 \cdot 13}{9 \cdot 25} = \frac{9}{9} \cdot \frac{13}{25} = 1 \cdot \frac{13}{25} = \frac{13}{25}.$$

How did we know that both numbers have 9 as a factor? There are fast tests for such determinations. If the sum of the digits of a number is divisible by 9, then the number is divisible by 9; that is, it has 9 as a factor. Since $1 + 1 + 7 = 9$ and $2 + 2 + 5 = 9$, both numbers have 9 as a factor.

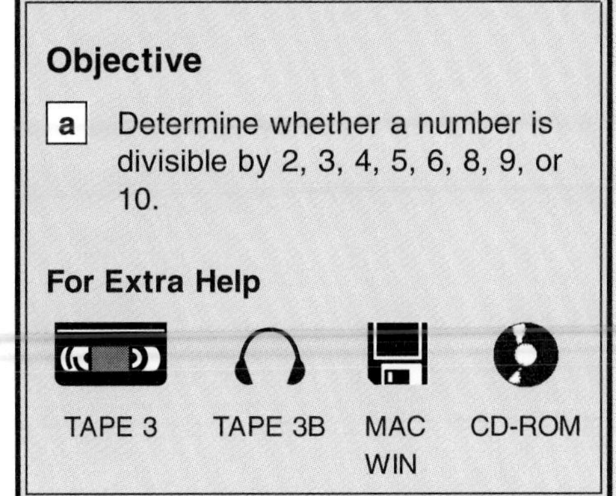

a Rules for Divisibility

In this section, we learn fast ways of determining whether numbers are divisible by 2, 3, 4, 5, 6, 8, 9, and 10. This will make simplifying fractional notation much easier.

Divisibility by 2

You may already know the test for divisibility by 2.

> A number is divisible by 2 (is *even*) if it has a ones digit of 0, 2, 4, 6, or 8 (that is, it has an even ones digit).

Let's see why. Consider 354, which is

3 hundreds + 5 tens + 4.

Hundreds and tens are both multiples of 2. If the last digit is a multiple of 2, then the entire number is a multiple of 2.

Examples Determine whether the number is divisible by 2.

1. 355 is not a multiple of 2; 5 is *not* even.

2. 4786 is a multiple of 2; 6 is even.

3. 8990 is a multiple of 2; 0 is even.

4. 4261 is not a multiple of 2; 1 is *not* even.

Do Exercises 1–4.

Determine whether the number is divisible by 2.

1. 84

2. 59

3. 998

4. 2225

Answers on page A-4

Divisibility by 3

> A number is divisible by 3 if the sum of its digits is divisible by 3.

Examples Determine whether the number is divisible by 3.

5. 18 $1 + 8 = 9$

6. 93 $9 + 3 = 12$

7. 201 $2 + 0 + 1 = 3$

All divisible by 3 because the sums of their digits are divisible by 3.

8. 256 $2 + 5 + 6 = 13$

The sum, 13, is not divisible by 3, so 256 is not divisible by 3.

Do Exercises 5–8.

Divisibility by 6

A number divisible by 6 is a multiple of 6. But $6 = 2 \cdot 3$, so the number is also a multiple of 2 and 3. Thus we have the following.

> A number is divisible by 6 if its ones digit is 0, 2, 4, 6, or 8 (is even) and the sum of its digits is divisible by 3.

Examples Determine whether the number is divisible by 6.

9. 720

Because 720 is even, it is divisible by 2. Also, $7 + 2 + 0 = 9$, so 720 is divisible by 3. Thus, 720 is divisible by 6.

720 $\quad 7 + 2 + 0 = 9$

↑ Even $\quad$ ↑ Divisible by 3

10. 73

73 is *not* divisible by 6 because it is *not* even.

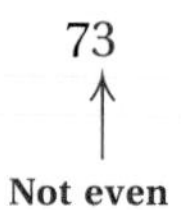

11. 256

256 is *not* divisible by 6 because the sum of its digits is *not* divisible by 3.

$2 + 5 + 6 = 13$

↑ Not divisible by 3

Do Exercises 9–12.

Determine whether the number is divisible by 3.

5. 111

6. 1111

7. 309

8. 17,216

Determine whether the number is divisible by 6.

9. 420

10. 106

11. 321

12. 444

Answers on page A-4

Divisibility by 9

The test for divisibility by 9 is similar to the test for divisibility by 3.

> A number is divisible by 9 if the sum of its digits is divisible by 9.

Example 12 The number 6984 is divisible by 9 because

$$6 + 9 + 8 + 4 = 27$$

and 27 is divisible by 9.

Example 13 The number 322 is *not* divisible by 9 because

$$3 + 2 + 2 = 7$$

and 7 is not divisible by 9.

Do Exercises 13–16.

Divisibility by 10

> A number is divisible by 10 if its ones digit is 0.

We know that this test works because the product of 10 and *any* number has a ones digit of 0.

Examples Determine whether the number is divisible by 10.

14. 3440 is divisible by 10 because the ones digit is 0.

15. 3447 is *not* divisible by 10 because the ones digit is not 0.

Do Exercises 17–20.

Divisibility by 5

> A number is divisible by 5 if its ones digit is 0 or 5.

Examples Determine whether the number is divisible by 5.

16. 220 is divisible by 5 because the ones digit is 0.

17. 475 is divisible by 5 because the ones digit is 5.

18. 6514 is *not* divisible by 5 because the ones digit is neither a 0 nor a 5.

Do Exercises 21–24.

Let's see why the test for 5 works. Consider 7830:

$$7830 = 10 \cdot 783 = 5 \cdot 2 \cdot 783.$$

Since 7830 is divisible by 10 and 5 is a factor of 10, 7830 is divisible by 5.

Determine whether the number is divisible by 9.

13. 16

14. 117

15. 930

16. 29,223

Determine whether the number is divisible by 10.

17. 305

18. 300

19. 847

20. 8760

Determine whether the number is divisible by 5.

21. 5780

22. 3427

23. 34,678

24. 7775

Answers on page A-4

Consider 6734:

6734 = 673 tens + 4.

Tens are multiples of 5, so the only number that must be checked is the ones digit. If the last digit is a multiple of 5, the entire number is. In this case, 4 is not a multiple of 5, so 6734 is not divisible by 5.

Divisibility by 4

The test for divisibility by 4 is similar to the test for divisibility by 2.

> A number is divisible by 4 if the number named by its last *two* digits is divisible by 4.

Examples Determine whether the number is divisible by 4.

19. 8212 is divisible by 4 because 12 is divisible by 4.

20. 5216 is divisible by 4 because 16 is divisible by 4.

21. 8211 is *not* divisible by 4 because 11 is *not* divisible by 4.

22. 7515 is *not* divisible by 4 because 15 is *not* divisible by 4.

Do Exercises 25–28.

To see why the test for divisibility by 4 works, consider 516:

516 = 5 hundreds + 16.

Hundreds are multiples of 4. If the number named by the last two digits is a multiple of 4, then the entire number is a multiple of 4.

Divisibility by 8

The test for divisibility by 8 is an extension of the tests for divisibility by 2 and 4.

> A number is divisible by 8 if the number named by its last *three* digits is divisible by 8.

Examples Determine whether the number is divisible by 8.

23. 5648 is divisible by 8 because 648 is divisible by 8.

24. 96,088 is divisible by 8 because 88 is divisible by 8.

25. 7324 is *not* divisible by 8 because 324 is *not* divisible by 8.

26. 13,420 is *not* divisible by 8 because 420 is *not* divisible by 8.

Do Exercises 29–32.

A Note About Divisibility by 7

There are several tests for divisibility by 7, but all of them are more complicated than simply dividing by 7. So if you want to test for divisibility by 7, simply divide by 7.

Determine whether the number is divisible by 4.

25. 216

26. 217

27. 5865

28. 23,524

Determine whether the number is divisible by 8.

29. 7564

30. 7864

31. 17,560

32. 25,716

Answers on page A-4

Exercise Set 2.2

a To answer Exercises 1–8, consider the following numbers.

46	300	85	256
224	36	711	8064
19	45,270	13,251	1867
555	4444	254,765	21,568

1. Which of the above are divisible by 2?

2. Which of the above are divisible by 3?

3. Which of the above are divisible by 4?

4. Which of the above are divisible by 5?

5. Which of the above are divisible by 6?

6. Which of the above are divisible by 8?

7. Which of the above are divisible by 9?

8. Which of the above are divisible by 10?

To answer Exercises 9–16, consider the following numbers.

56	200	75	35
324	42	812	402
784	501	2345	111,111
55,555	3009	2001	1005

9. Which of the above are divisible by 3?

10. Which of the above are divisible by 2?

11. Which of the above are divisible by 5?

12. Which of the above are divisible by 4?

13. Which of the above are divisible by 9?

14. Which of the above are divisible by 6?

15. Which of the above are divisible by 10?

16. Which of the above are divisible by 8?

Skill Maintenance

Solve. [1.7b]

17. $56 + x = 194$

18. $y + 124 = 263$

19. $18 \cdot t = 1008$

20. $24 \cdot m = 624$

Divide. [1.6c]

21. $2106 \div 9$

22. $45 \overline{)180{,}135}$

Solve. [1.8a]

23. An automobile with a 5-speed transmission gets 33 mpg in city driving. How many gallons of gas will it use to travel 1485 mi?

24. There are 60 min in 1 hr. How many minutes are there in 72 hr?

Synthesis

25. ◆ Are the divisibility tests useful for finding prime factorizations? Why or why not?

26. ◆ Which of the years from 1990 to 2010, if any, also happen to be prime numbers? Explain at least two ways in which you might go about solving this problem.

Find the prime factorization of the number.

27. 7800

28. 2520

29. 2772

30. 1998

31. Using each of the digits 1, 2, 3, . . . , 7, find the largest seven-digit number for which the sum of any two consecutive digits is a prime number.

32. A passenger in a taxicab asks for the driver's company number. The driver says abruptly, "Sure—you can have my number. Work it out: If you divide it by 2, 3, 4, 5, or 6, you will get a remainder of 1. If you divide it by 11, the remainder will be 0 and no driver has a company number that is smaller than this one." Determine the number.

Use the divisibility rules and properties of numbers to discover an unknown number.

2.3 Fractions

The study of arithmetic begins with the set of whole numbers

0, 1, 2, 3, 4, 5, 6, 7, 8, 9, 10, 11, and so on.

The need soon arises for fractional parts of numbers such as halves, thirds, fourths, and so on. Here are some examples:

$\frac{1}{25}$ of the parking spaces in a commercial area in the state of Indiana are to be marked for the handicapped.

$\frac{1}{11}$ of all women develop breast cancer.

$\frac{1}{4}$ of the minimum daily requirement of calcium is provided by a cup of frozen yogurt.

$\frac{43}{100}$ of all corporate travel money is spent on airfares.

Objectives

a Identify the numerator and the denominator of a fraction.

b Write fractional notation for part of an object or part of a set of objects.

c Simplify fractional notation like n/n to 1, $0/n$ to 0, and $n/1$ to n.

For Extra Help

TAPE 3 | TAPE 3B | MAC WIN | CD-ROM

a Identifying Numerators and Denominators

The following are some additional examples of fractions:

$$\frac{1}{2}, \quad \frac{3}{4}, \quad \frac{8}{5}, \quad \frac{11}{23}.$$

This way of writing number names is called **fractional notation.** The top number is called the **numerator** and the bottom number is called the **denominator**.

Example 1 Identify the numerator and the denominator.

$\frac{7}{8}$ ← Numerator (7) ← Denominator (8)

Do Exercises 1–3.

b Writing Fractional Notation

Example 2 What part is shaded?

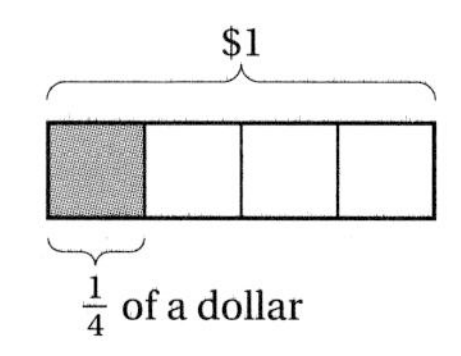

When an object is divided into 4 parts of the same size, each of these parts is $\frac{1}{4}$ of the object. Thus, $\frac{1}{4}$ (*one-fourth*) is shaded.

Do Exercises 4–7.

Identify the numerator and the denominator.

1. $\frac{1}{6}$ **2.** $\frac{5}{7}$ **3.** $\frac{22}{3}$

What part is shaded?

4.

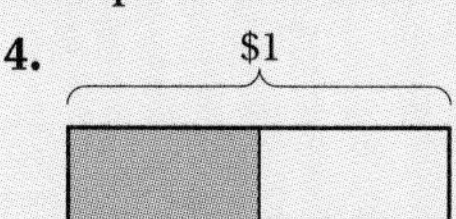

5.

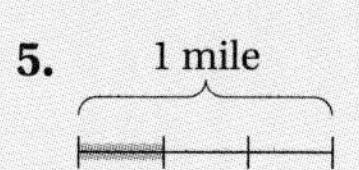

6.

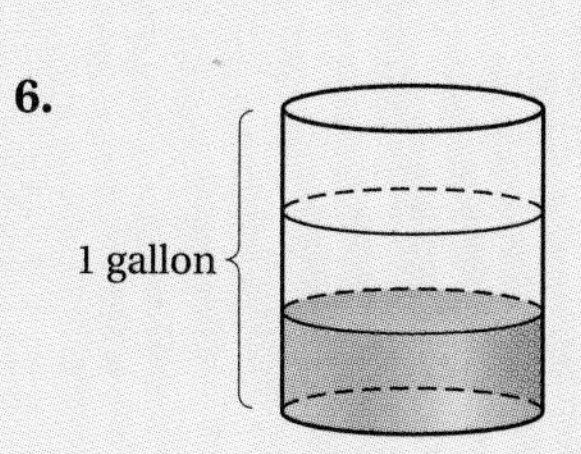

7.

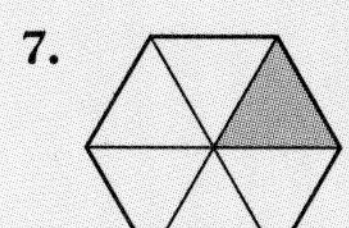

Answers on page A-4

What part is shaded?

8.

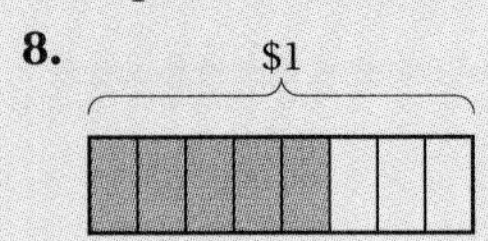

9.

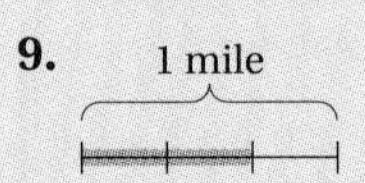

10.

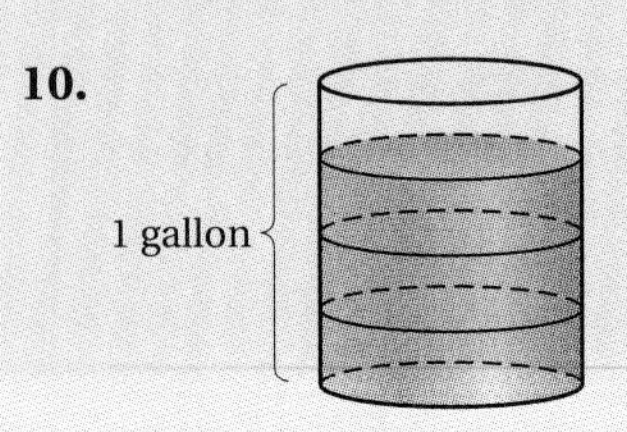

11.

What part is shaded?

12.

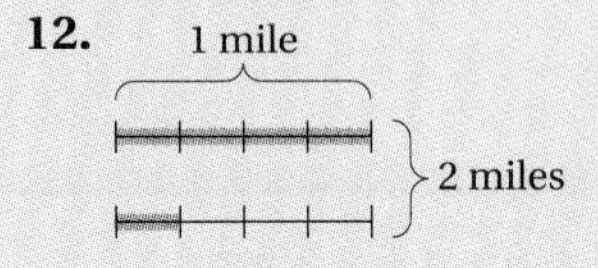

13.

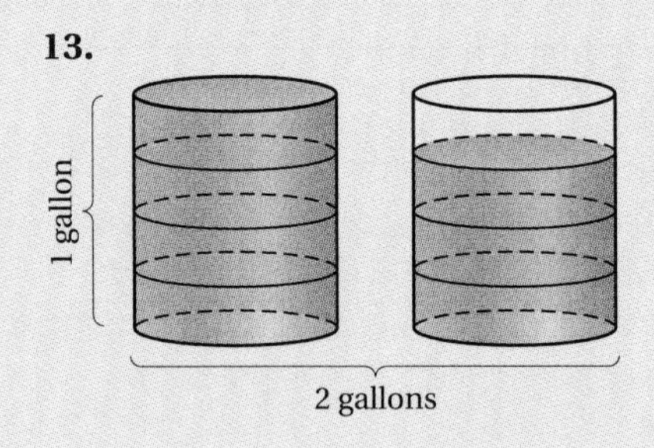

14. What part of the set of tools in Example 5 are hammers?

15. What part of this set is shaded?

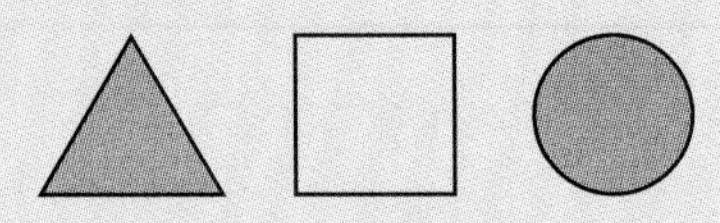

16. What part of this set are or were United States presidents? are recording stars?

Abraham Lincoln
Whitney Houston
Garth Brooks
Bill Clinton
Sheryl Crow
Gloria Estefan

Answers on page A-4

Example 3
What part is shaded?

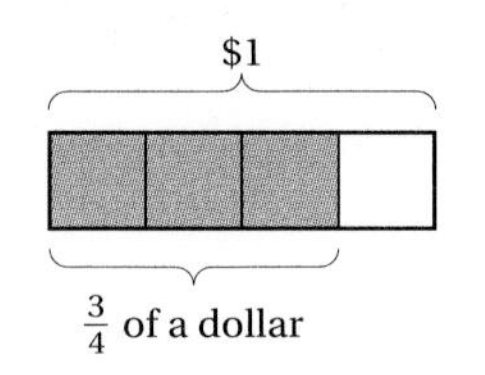

The object is divided into 4 parts of the same size, and 3 of them are shaded. This is $3 \cdot \frac{1}{4}$, or $\frac{3}{4}$. Thus, $\frac{3}{4}$ (*three-fourths*) of the object is shaded.

Do Exercises 8–11.

Fractions greater than 1 correspond to situations like the following.

Example 4
What part is shaded?

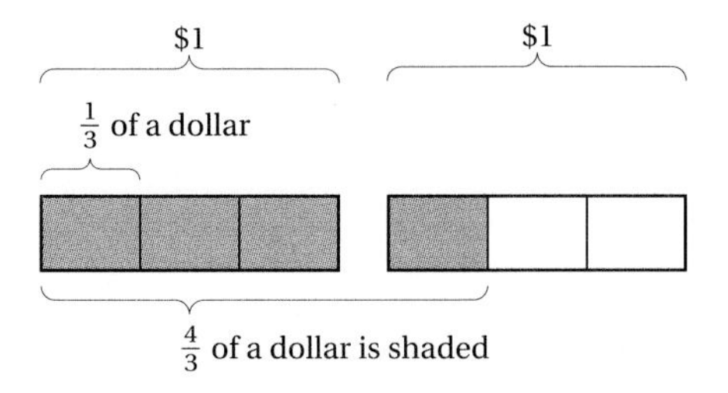

We divide the two objects into 3 parts each and take 4 of those parts. We have more than one whole object. In this case, it is $4 \cdot \frac{1}{3}$, or $\frac{4}{3}$.

Do Exercises 12 and 13.

Fractional notation also corresponds to situations involving part of a set.

Example 5 What part of this set, or collection, of tools are wrenches?

There are 5 tools, and 3 are wrenches. We say that three-fifths of the tools are wrenches; that is, $\frac{3}{5}$ of the set consists of wrenches.

Do Exercises 14–16.

Circle graphs, or pie charts, are often used to illustrate the relationships of fractional parts of a whole. The following graph shows color preferences of bicycles.

Bicycle Color Preferences

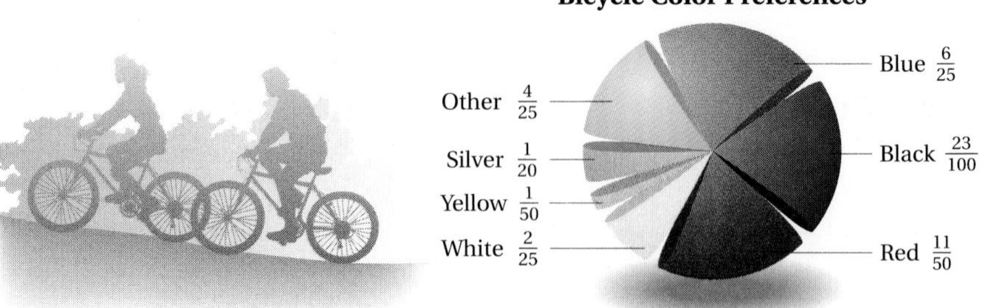

Source: Bicycle Market Research Institute

c Some Fractional Notation for Whole Numbers

Fractional Notation for 1

The number 1 corresponds to situations like those shown here.

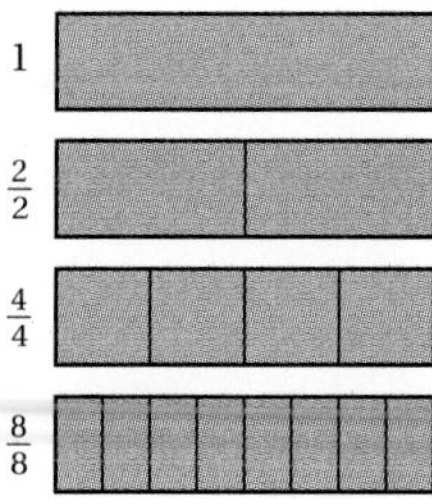

If we divide an object into n parts and take n of them, we get all of the object (1 whole object).

> $\frac{n}{n} = 1$, for any whole number n that is not 0.

Examples Simplify.

6. $\frac{5}{5} = 1$ **7.** $\frac{9}{9} = 1$ **8.** $\frac{23}{23} = 1$

Do Exercises 17–22.

Fractional Notation for 0

Consider $\frac{0}{4}$. This corresponds to dividing an object into 4 parts and taking none of them. We get 0.

> $\frac{0}{n} = 0$, for any whole number n that is not 0.

Examples Simplify.

9. $\frac{0}{1} = 0$ **10.** $\frac{0}{9} = 0$ **11.** $\frac{0}{23} = 0$

Fractional notation with a denominator of 0, such as $n/0$, is meaningless because we cannot speak of an object divided into *zero* parts. (If it is not divided at all, then we say that it is undivided and remains in one part.)

> $\frac{n}{0}$ is not defined for any whole number n.

Do Exercises 23–28.

Other Whole Numbers

Consider $\frac{4}{1}$. This corresponds to taking 4 objects and dividing each into 1 part. (We do not divide them.) We have 4 objects.

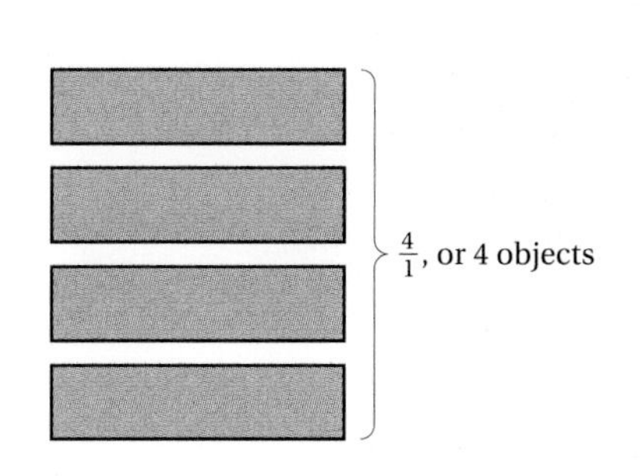

Simplify.

17. $\frac{1}{1}$ **18.** $\frac{4}{4}$

19. $\frac{34}{34}$ **20.** $\frac{100}{100}$

21. $\frac{2347}{2347}$ **22.** $\frac{103}{103}$

Simplify, if possible.

23. $\frac{0}{1}$ **24.** $\frac{0}{8}$

25. $\frac{0}{107}$ **26.** $\frac{4-4}{567}$

27. $\frac{15}{0}$ **28.** $\frac{0}{3-3}$

Answers on page A-4

Simplify.

29. $\frac{8}{1}$

30. $\frac{10}{1}$

31. $\frac{346}{1}$

32. $\frac{24 - 1}{23 - 22}$

Any whole number divided by 1 is the whole number. That is,

$$\frac{n}{1} = n, \quad \text{for any whole number } n.$$

Examples Simplify.

12. $\frac{2}{1} = 2$

13. $\frac{9}{1} = 9$

14. $\frac{34}{1} = 34$

Do Exercises 29–32.

Improving Your Math Study Skills

Studying for Tests and Making the Most of Tutoring Sessions

This math study skill feature focuses on the very important task of test preparation.

Test-Taking Tips

- **Make up your own test questions as you study.** After you have done your homework over a particular objective, write one or two questions on your own that you think might be on a test. You will be amazed at the insight this will provide. You are actually carrying out a task similar to what a teacher does in preparing an exam.
- **Do an overall review of the chapter, focusing on the objectives and the examples.** This should be accompanied by a study of any class notes you may have taken.
- **Do the review exercises at the end of the chapter.** Check your answers at the back of the book. If you have trouble with an exercise, use the objective symbol as a guide to go back and do further study of that objective.
- **Do the chapter test at the end of the chapter.** Check the answers and objective symbols at the back of the book.
- **Ask former students for old exams.** Working such exams can be very helpful and allows you to see what various professors think is important.
- **When taking a test, read each question carefully and try to do all the questions the first time through, but pace yourself.** Answer all the questions, and mark those to recheck if you have time at the end. Very often, your first hunch will be correct.
- **Try to write your test in a neat and orderly manner.** Very often, your instructor tries to give you partial credit when grading an exam. If your test paper is sloppy and disorderly, it is difficult to verify the partial credit. Doing your work neatly can ease such a task for the instructor.

Making the Most of Tutoring and Help Sessions

- **Work on the topics before you go to the help or tutoring session. Do not go to such sessions viewing yourself as an empty cup and the tutor as a magician who will pour in the learning.** We have seen so many students over the years go to help or tutoring sessions with no advanced preparation. You are often wasting your time and perhaps your money if you are paying for such sessions. Go to class, study the textbook, and mark trouble spots. Then use the help and tutoring sessions to deal with these difficulties most efficiently.
- **Do not be afraid to ask questions in these sessions!** The more you talk to your tutor, the more the tutor can help you with your difficulties.
- **Try being a "tutor" yourself.** Explaining a topic to someone else—a classmate, your instructor—is often the best way to learn it.

Answers on page A-4

Exercise Set 2.3

a Identify the numerator and the denominator.

1. $\frac{3}{4}$ **2.** $\frac{9}{10}$ **3.** $\frac{11}{20}$ **4.** $\frac{18}{5}$

b What part of the object or set of objects is shaded?

5.
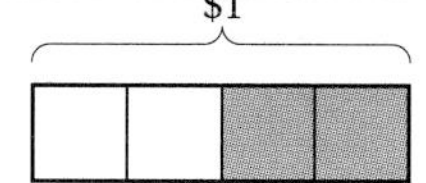

6.
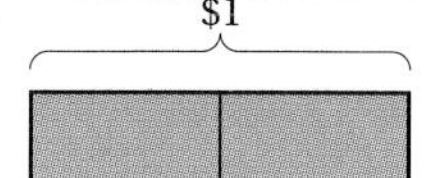

7.

1 yard

8.
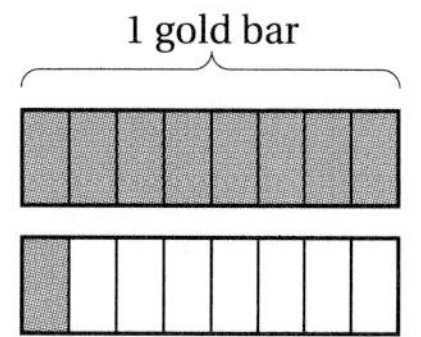

9.
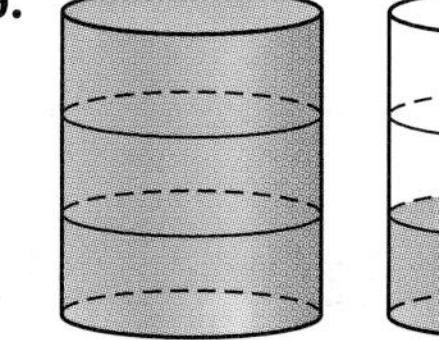
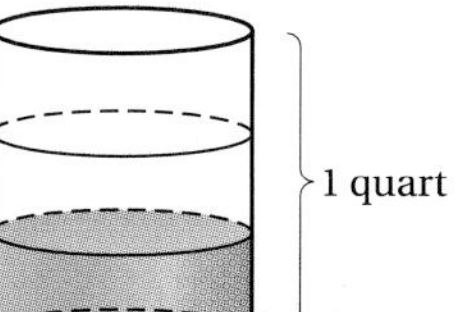

10.
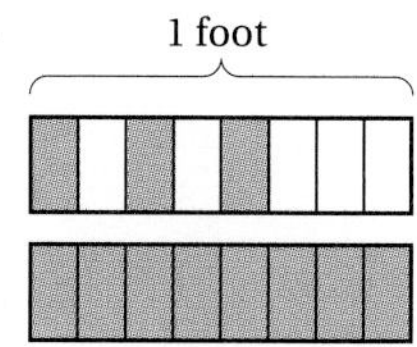

11.
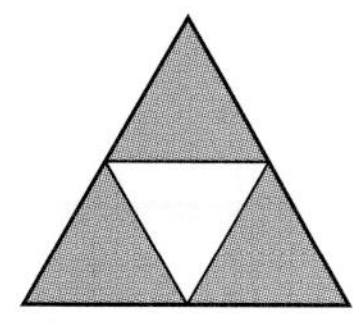

12.
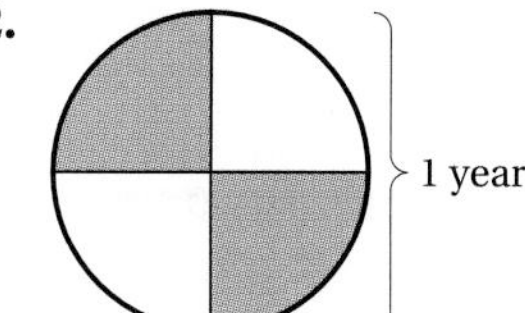

13.
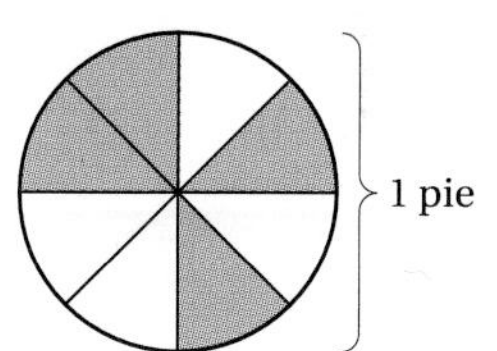

14.
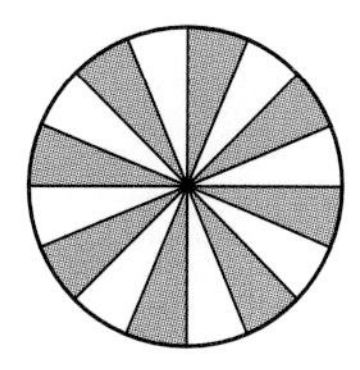

15.
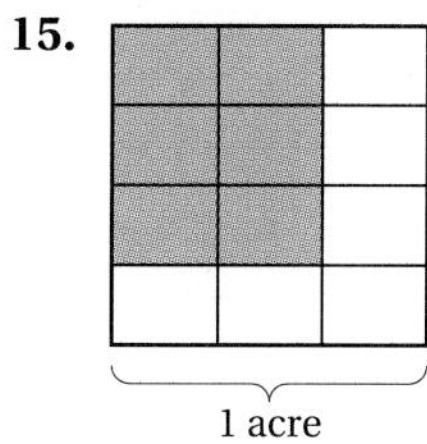

16.
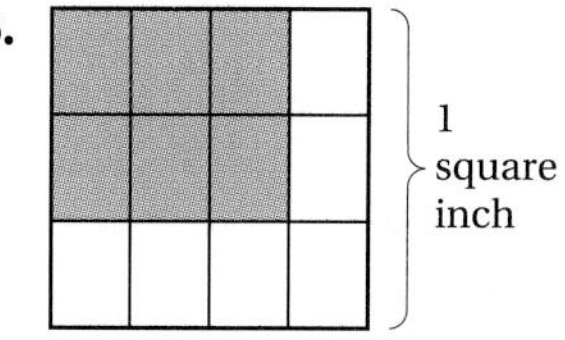

17.
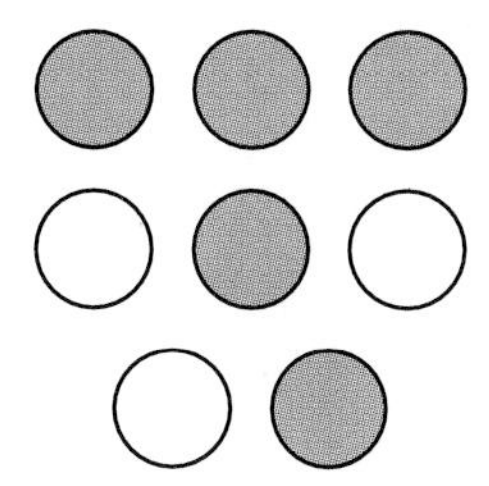

18.
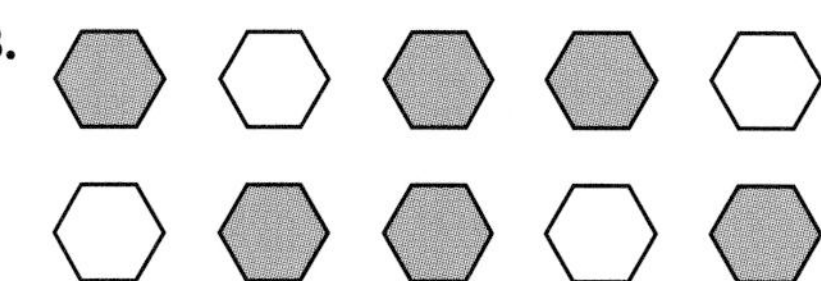

19.
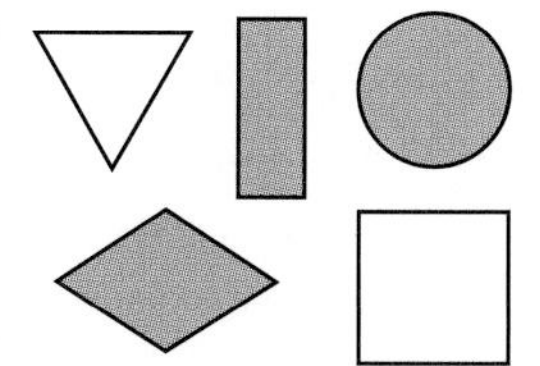

20.
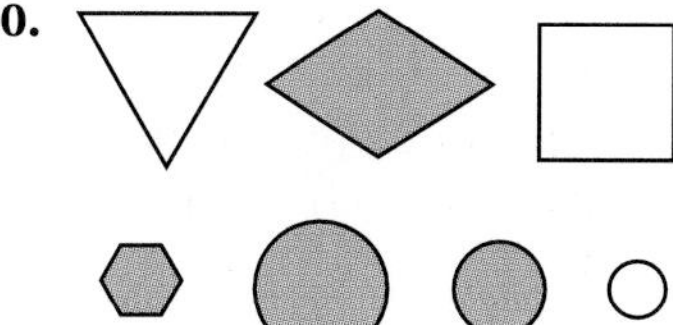

c Simplify.

21. $\frac{0}{8}$

22. $\frac{8}{8}$

23. $\frac{8-1}{9-8}$

24. $\frac{16}{1}$

25. $\frac{20}{20}$

26. $\frac{20}{1}$

27. $\frac{45}{45}$

28. $\frac{11-1}{10-9}$

29. $\frac{0}{238}$

30. $\frac{238}{1}$

31. $\frac{238}{238}$

32. $\frac{0}{16}$

33. $\frac{3}{3}$

34. $\frac{56}{56}$

35. $\frac{87}{87}$

36. $\frac{98}{98}$

37. $\frac{18}{18}$

38. $\frac{0}{18}$

39. $\frac{18}{1}$

40. $\frac{8-8}{1247}$

41. $\frac{729}{0}$

42. $\frac{1317}{0}$

43. $\frac{5}{6-6}$

44. $\frac{13}{10-10}$

Skill Maintenance

Round 34,562 to the nearest: [1.4a]

45. Ten.

46. Hundred.

47. Thousand.

Solve. [1.8a]

48. *Annual Income.* The average annual income of people living in Alaska is $24,182 per person. In Colorado, the average annual income is $23,449. (***Source*:** U.S. Bureau of Economic Analysis) How much more does a person in Alaska make, on average, than does a person living in Colorado?

49. A pet-care service cut 29,824 lb of hair in 8 yr of operation. On the average, how many pounds did it cut each year?

Subtract. [1.3d]

50. 2037 − 1189

51. 9001 − 6798

52. 67,113 − 29,874

Synthesis

53. ◆ Explain in your own words why $0/n = 0$, for any natural number n.

54. ◆ Explain in your own words why $n/n = 1$, for any natural number n.

55. The surface of the earth is 3 parts water and 1 part land. What fractional part of the earth is water? land?

56. A couple had 3 boys, each of whom had 3 daughters. If each daughter gave birth to 3 sons, what fractional part of the couple's descendants is female?

2.4 Multiplication and Applications

Objectives

[a] Multiply a whole number and a fraction.

[b] Multiply using fractional notation.

[c] Solve applied problems involving multiplication of fractions.

For Extra Help

TAPE 3 TAPE 4A MAC WIN CD-ROM

a Multiplication by a Whole Number

We can find $3 \cdot \frac{1}{4}$ by thinking of repeated addition. We add three $\frac{1}{4}$'s.

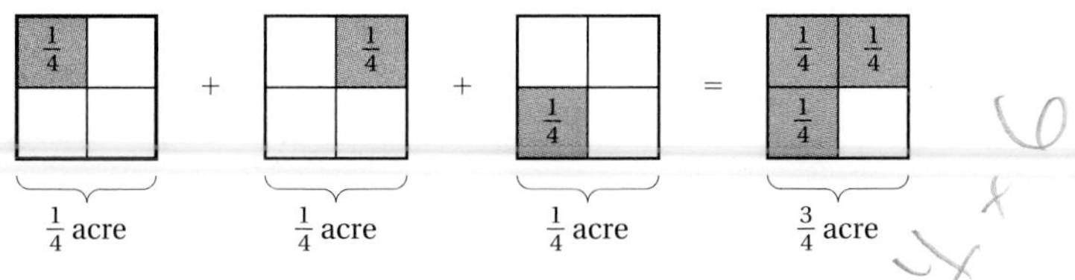

We see that $3 \cdot \frac{1}{4}$ is $\frac{3}{4}$.

Do Exercises 1 and 2.

To multiply a fraction by a whole number,

a) multiply the top number (the numerator) by the whole number, and

b) keep the same denominator.

$$6 \cdot \frac{4}{5} = \frac{6 \cdot 4}{5} = \frac{24}{5}$$

Examples Multiply.

1. $5 \times \frac{3}{8} = \frac{5 \times 3}{8} = \frac{15}{8}$

Skip this step whenever you can.

2. $\frac{2}{5} \cdot 13 = \frac{2 \cdot 13}{5} = \frac{26}{5}$

3. $10 \cdot \frac{1}{3} = \frac{10}{3}$

Do Exercises 3–5.

b Multiplication Using Fractional Notation

We find a product such as $\frac{9}{7} \cdot \frac{3}{4}$ as follows.

To multiply a fraction by a fraction,

a) multiply the numerators to get the new numerator, and

b) multiply the denominators to get the new denominator.

$$\frac{9}{7} \cdot \frac{3}{4} = \frac{9 \cdot 3}{7 \cdot 4} = \frac{27}{28}$$

1. Find $2 \cdot \frac{1}{3}$.

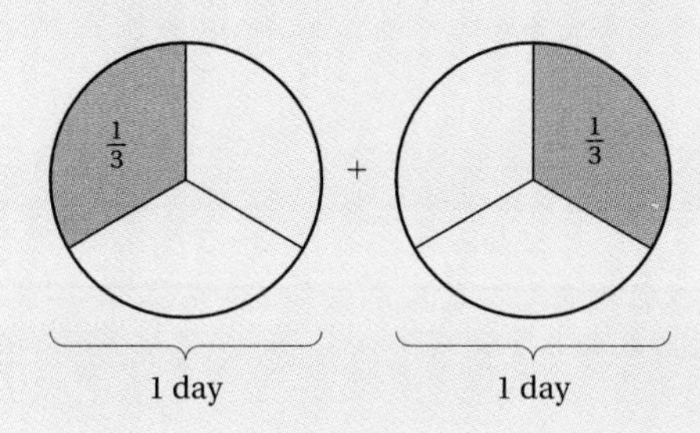

2. Find $5 \cdot \frac{1}{8}$.

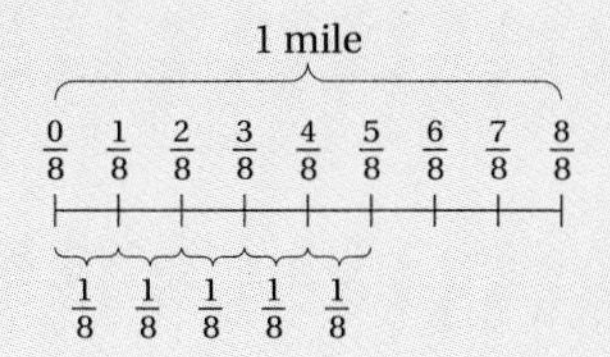

Multiply.

3. $5 \times \frac{2}{3}$

4. $11 \times \frac{3}{8}$

5. $23 \cdot \frac{2}{5}$

Answers on page A-4

Examples Multiply.

4. $\dfrac{5}{6} \times \dfrac{7}{4} = \dfrac{5 \times 7}{6 \times 4} = \dfrac{35}{24}$

Skip this step whenever you can.

5. $\dfrac{3}{5} \cdot \dfrac{7}{8} = \dfrac{3 \cdot 7}{5 \cdot 8} = \dfrac{21}{40}$

6. $\dfrac{3}{5} \cdot \dfrac{3}{4} = \dfrac{9}{20}$

7. $\dfrac{1}{4} \cdot \dfrac{1}{3} = \dfrac{1}{12}$

8. $6 \cdot \dfrac{4}{5} = \dfrac{6}{1} \cdot \dfrac{4}{5} = \dfrac{24}{5}$

Do Exercises 6–9.

Unless one of the factors is a whole number, multiplication using fractional notation does not correspond to repeated addition. Let's see how multiplication of fractions corresponds to situations in the real world. We consider the multiplication

$$\frac{3}{5} \cdot \frac{3}{4}.$$

We first consider some object and take $\frac{3}{4}$ of it. We divide it into 4 vertical parts, or columns of the same area, and take 3 of them. That is shown in the shading below.

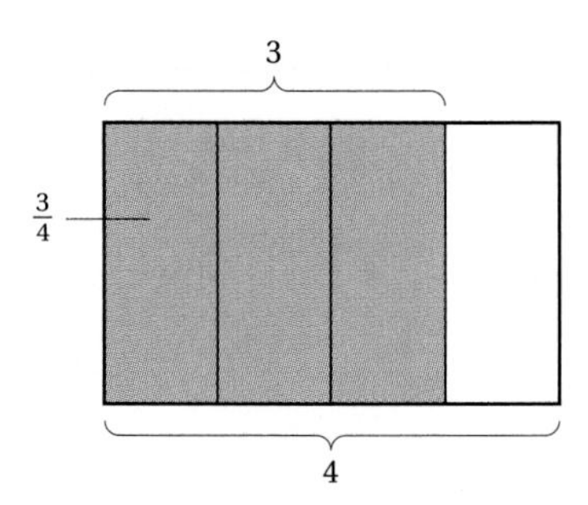

Next, we take $\frac{3}{5}$ of the result. We divide the shaded part into 5 horizontal parts, or rows of the same area, and take 3 of them. That is shown below.

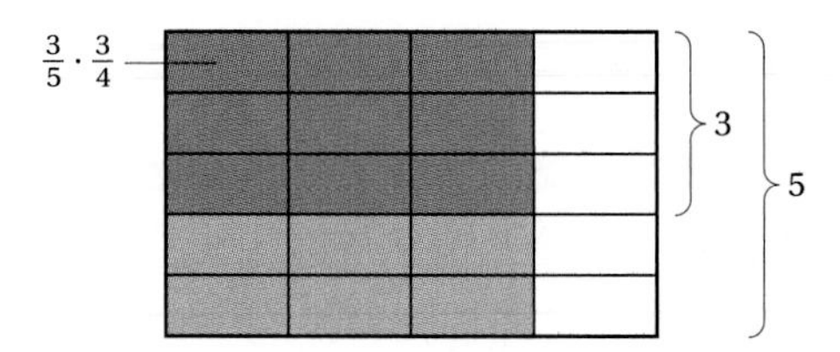

The entire object has been divided into 20 parts, and we have shaded 9 of them for a second time:

$$\frac{3}{5} \cdot \frac{3}{4} = \frac{3 \cdot 3}{5 \cdot 4} = \frac{9}{20}.$$

The figure above shows a rectangular array inside a rectangular array. The number of pieces in the entire array is $5 \cdot 4$ (the product of the denominators). The number of pieces shaded a second time is $3 \cdot 3$ (the product of the numerators). For the answer, we take 9 pieces out of a set of 20 to get $\frac{9}{20}$.

Do Exercise 10.

Multiply.

6. $\dfrac{3}{8} \cdot \dfrac{5}{7}$

7. $\dfrac{4}{3} \times \dfrac{8}{5}$

8. $\dfrac{3}{10} \cdot \dfrac{1}{10}$

9. $7 \cdot \dfrac{2}{3}$

10. Draw diagrams like those in the text to show how the multiplication $\frac{1}{3} \cdot \frac{4}{5}$ corresponds to a real-world situation.

Answers on page A-4

c Applications and Problem Solving

Most problems that can be solved by multiplying fractions can be thought of in terms of rectangular arrays.

Example 9 A rancher owns a square mile of land. He gives $\frac{4}{5}$ of it to his daughter and she gives $\frac{2}{3}$ of her share to her son. How much land goes to the son?

1. **Familiarize.** We first make a drawing to help solve the problem. The land may not be square. It could be in a shape like A or B below, or it could even be in more than one piece. But to think out the problem, we can think of it as a square, as shown by shape C.

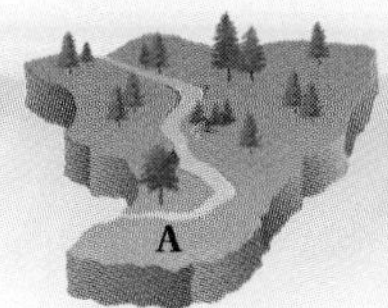

1 square mile

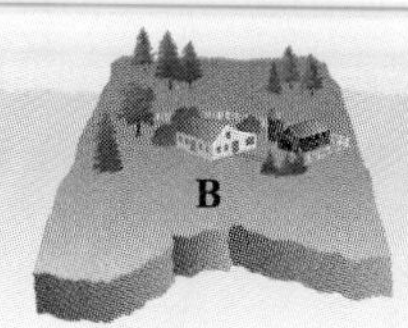

1 square mile

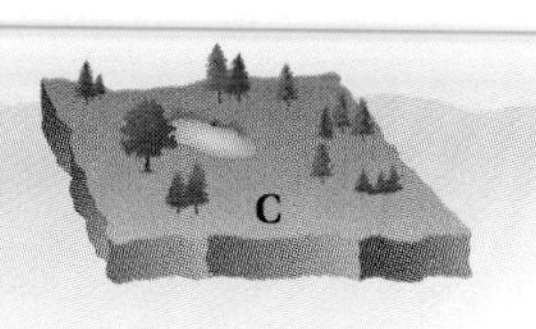

1 square mile

The daughter gets $\frac{4}{5}$ of the land. We shade $\frac{4}{5}$.

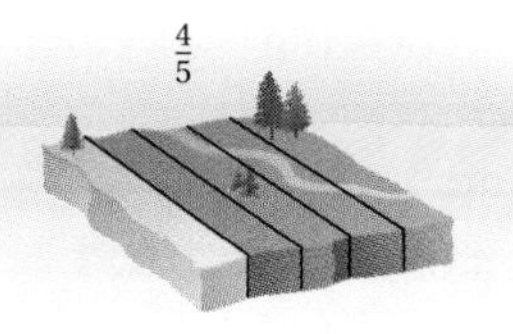

Her son gets $\frac{2}{3}$ of her part. We shade that.

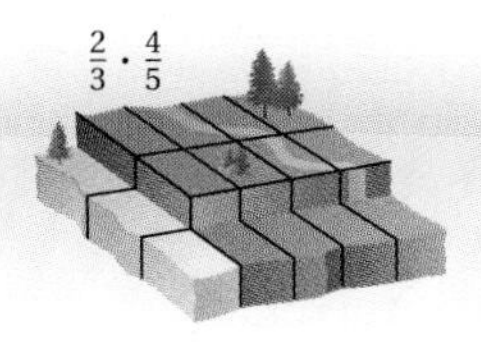

2. **Translate.** We let $n =$ the part of the land that goes to the son. We are taking "two-thirds of four-fifths." The word "of" corresponds to multiplication. Thus the following multiplication sentence corresponds to the situation:

$$\frac{2}{3} \cdot \frac{4}{5} = n.$$

3. **Solve.** The number sentence tells us what to do. We multiply:

$$\frac{2}{3} \cdot \frac{4}{5} = \frac{8}{15}.$$

4. **Check.** We can check partially by noting that the answer is smaller than the original area, 1, which we expect since the rancher is giving parts of the land away. Thus, $\frac{8}{15}$ is a reasonable answer. We can also check this in the figure above, where we see that 8 of 15 parts have been shaded a second time.

5. **State.** The son gets $\frac{8}{15}$ of a square mile of land.

Do Exercise 11.

11. A resort hotel uses $\frac{3}{4}$ of its extra land for recreational purposes. Of that, $\frac{1}{2}$ is used for swimming pools. What part of the land is used for swimming pools?

Answer on page A-4

12. *Area of Fax Key.* The length of a button on a fax machine is $\frac{9}{10}$ cm. The width is $\frac{7}{10}$ cm. What is the area?

13. Of the students at Overton Junior College, $\frac{1}{8}$ participate in sports and $\frac{3}{5}$ of these play football. What fractional part of the students play football?

Answers on page A-4

Example 9 and the preceding discussion indicate that the area of a rectangular region can be found by multiplying length by width. That is true whether length and width are whole numbers or not. Remember, the area of a rectangular region is given by the formula

$$A = l \cdot w.$$

Example 10 *Area of Calculator Key.* The length of a rectangular key on a calculator is $\frac{7}{10}$ cm. The width is $\frac{3}{10}$ cm. What is the area?

1. **Familiarize.** Recall that area is length times width. We draw a picture, letting A = the area of the calculator key.

2. **Translate.** Then we translate.

Area	is	Length	times	Width
↓	↓	↓	↓	↓
A	$=$	$\frac{7}{10}$	$\times$	$\frac{3}{10}$

3. **Solve.** The sentence tells us what to do. We multiply:

$$\frac{7}{10} \cdot \frac{3}{10} = \frac{7 \cdot 3}{10 \cdot 10} = \frac{21}{100}.$$

4. **Check.** We check by repeating the calculation. This is left to the student.

5. **State.** The area is $\frac{21}{100}$ cm^2.

Do Exercise 12.

Example 11 A recipe calls for $\frac{3}{4}$ cup of cornmeal. A chef is making $\frac{1}{2}$ of the recipe. How much cornmeal should the chef use?

1. **Familiarize.** We first make a drawing or at least visualize the situation. We let n = the amount of cornmeal the chef should use.

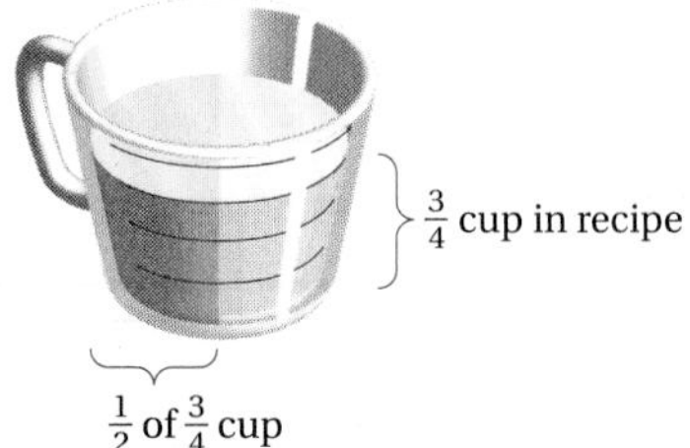

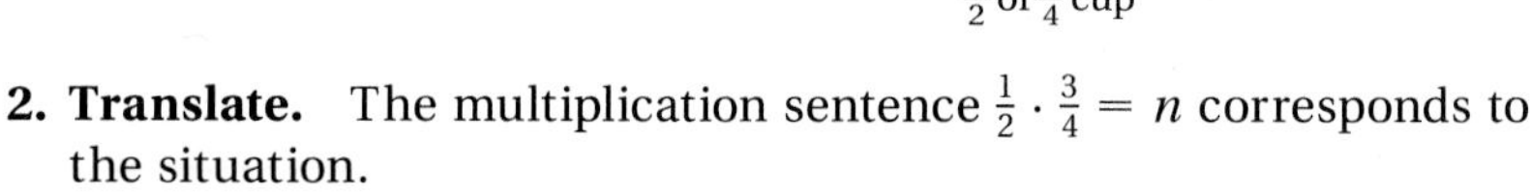

2. **Translate.** The multiplication sentence $\frac{1}{2} \cdot \frac{3}{4} = n$ corresponds to the situation.

3. **Solve.** We carry out the multiplication:

$$\frac{1}{2} \cdot \frac{3}{4} = \frac{1 \cdot 3}{2 \cdot 4} = \frac{3}{8}.$$

4. **Check.** We check by repeating the calculation. This is left to the student.

5. **State.** The chef should use $\frac{3}{8}$ cup of cornmeal.

Do Exercise 13.

Exercise Set 2.4

a Multiply.

1. $3 \cdot \frac{1}{5}$

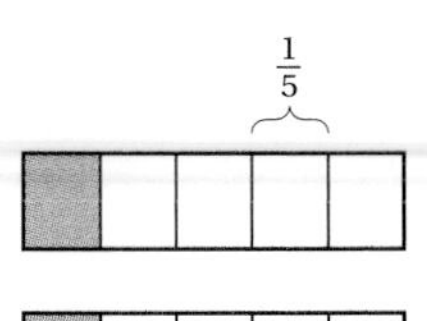

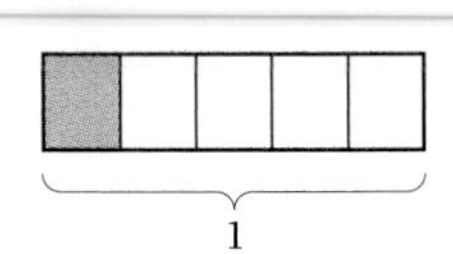

2. $2 \cdot \frac{1}{3}$

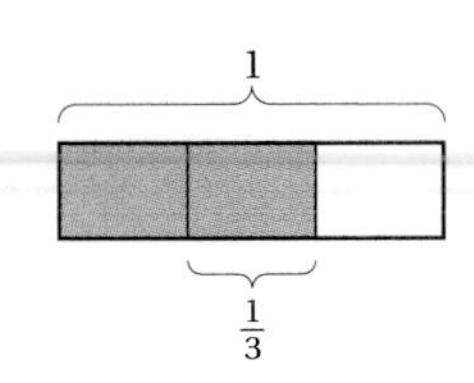

3. $5 \times \frac{1}{8}$

4. $4 \times \frac{1}{5}$

5. $\frac{2}{11} \cdot 4$

6. $\frac{2}{5} \cdot 3$

7. $10 \cdot \frac{7}{9}$

8. $9 \cdot \frac{5}{8}$

9. $\frac{2}{5} \cdot 1$

10. $\frac{3}{8} \cdot 1$

11. $\frac{2}{5} \cdot 3$

12. $\frac{3}{5} \cdot 4$

13. $7 \cdot \frac{3}{4}$

14. $7 \cdot \frac{2}{5}$

15. $17 \times \frac{5}{6}$

16. $\frac{3}{7} \cdot 40$

b Multiply.

17. $\frac{1}{2} \cdot \frac{1}{3}$

18. $\frac{1}{6} \cdot \frac{1}{4}$

19. $\frac{1}{4} \times \frac{1}{10}$

20. $\frac{1}{3} \times \frac{1}{10}$

21. $\frac{2}{3} \times \frac{1}{5}$

22. $\frac{3}{5} \times \frac{1}{5}$

23. $\frac{2}{5} \cdot \frac{2}{3}$

24. $\frac{3}{4} \cdot \frac{3}{5}$

25. $\frac{3}{4} \cdot \frac{3}{4}$

26. $\frac{3}{7} \cdot \frac{4}{5}$

27. $\frac{2}{3} \cdot \frac{7}{13}$

28. $\frac{3}{11} \cdot \frac{4}{5}$

29. $\frac{1}{10} \cdot \frac{7}{10}$

30. $\frac{3}{10} \cdot \frac{3}{10}$

31. $\frac{7}{8} \cdot \frac{7}{8}$

32. $\frac{4}{5} \cdot \frac{4}{5}$

33. $\frac{1}{10} \cdot \frac{1}{100}$

34. $\frac{3}{10} \cdot \frac{7}{100}$

35. $\frac{14}{15} \cdot \frac{13}{19}$

36. $\frac{12}{13} \cdot \frac{12}{13}$

c Solve.

37. A rectangular table top measures $\frac{4}{5}$ m long by $\frac{3}{5}$ m wide. What is its area?

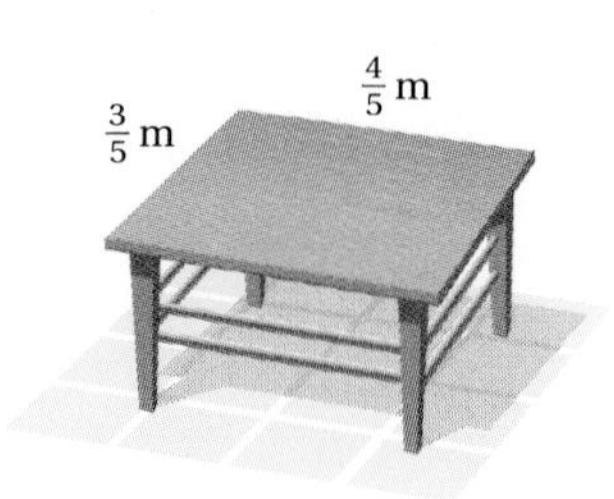

38. If each piece of pie is $\frac{1}{6}$ of a pie, how much of the pie is $\frac{1}{2}$ of a piece?

39. One of 39 high school football players plays college football. One of 39 college players plays professional football. What fractional part of high school players play professional football?

40. A gasoline can holds $\frac{7}{8}$ liter (L). How much will the can hold when it is $\frac{1}{2}$ full?

41. A cereal recipe calls for $\frac{3}{4}$ cup of granola. How much is needed to make $\frac{1}{2}$ of a recipe?

42. It takes $\frac{2}{3}$ yd of ribbon to make a bow. How much ribbon is needed to make 5 bows?

43. *Orange Juice Purchased.* Of every 100 containers of juice bought in grocery stores, 56 are orange juice. What fractional part of juice purchased is orange juice?

44. *Moviegoers.* Of every 1000 people who attend movies, 340 are in the 18–24 age group (**Source:** *American Demographics*). What fractional part of all moviegoers are in the 18–24 age group?

Skill Maintenance

Divide. [1.6c]

45. 7140 ÷ 35

46. 32,200 ÷ 46

47. $9\overline{)27{,}009}$

48. $35\overline{)7148}$

What does the digit 8 mean in each number? [1.1e]

49. 4,678,952

50. 8,473,901

51. 7148

52. 23,803

Simplify. [1.9c]

53. $12 - 3^2$

54. $(12 - 3)^2$

55. $8 \cdot 12 - (63 \div 9 + 13 \cdot 3)$

56. $(10 - 3)^4 + 10^3 \cdot 4 - 10 \div 5$

Synthesis

57. ◆ Following Example 8, we explained, using words and pictures, why $\frac{3}{5} \cdot \frac{3}{4}$ equals $\frac{9}{20}$. Present a similar explanation of why $\frac{2}{3} \cdot \frac{4}{7}$ equals $\frac{8}{21}$.

58. ◆ Write a problem for a classmate to solve. Design the problem so that the solution is "About $\frac{1}{30}$ of the students are left-handed women."

Multiply. Write the answer using fractional notation.

59. ▦ $\frac{341}{517} \cdot \frac{209}{349}$

60. ▦ $\left(\frac{57}{61}\right)^3$

61. $\left(\frac{2}{5}\right)^3\left(\frac{7}{9}\right)$

62. $\left(\frac{1}{2}\right)^5\left(\frac{3}{5}\right)$

2.5 Simplifying

Objectives

a Find another name for a number, but having a new denominator. Use multiplying by 1.

b Simplify fractional notation.

c Test fractions for equality.

For Extra Help

TAPE 4 | TAPE 4A | MAC WIN | CD-ROM

a Multiplying by 1

Recall the following:

$$1 = \frac{1}{1} = \frac{2}{2} = \frac{3}{3} = \frac{4}{4} = \frac{10}{10} = \frac{45}{45} = \frac{100}{100} = \frac{n}{n}.$$

Any nonzero number divided by itself is 1.

> When we multiply a number by 1, we get the same number.
>
> $$\frac{3}{5} = \frac{3}{5} \cdot 1 = \frac{3}{5} \cdot \frac{4}{4} = \frac{12}{20}$$

Since $\frac{3}{5} \cdot 1 = \frac{12}{20}$, we know that $\frac{3}{5}$ and $\frac{12}{20}$ are two names for the same number. We also say that $\frac{3}{5}$ and $\frac{12}{20}$ are **equivalent**.

Do Exercises 1–4.

Suppose we want to find a name for $\frac{2}{3}$, but one that has a denominator of 9. We can multiply by 1 to find equivalent fractions:

$$\frac{2}{3} = \frac{2}{3} \cdot \frac{3}{3} = \frac{2 \cdot 3}{3 \cdot 3} = \frac{6}{9}.$$

We chose $\frac{3}{3}$ for 1 in order to get a denominator of 9.

Example 1 Find a name for $\frac{1}{4}$ with a denominator of 24.

Since $4 \cdot 6 = 24$, we multiply by $\frac{6}{6}$:

$$\frac{1}{4} = \frac{1}{4} \cdot \frac{6}{6} = \frac{1 \cdot 6}{4 \cdot 6} = \frac{6}{24}.$$

Example 2 Find a name for $\frac{2}{5}$ with a denominator of 35.

Since $5 \cdot 7 = 35$, we multiply by $\frac{7}{7}$:

$$\frac{2}{5} = \frac{2}{5} \cdot \frac{7}{7} = \frac{2 \cdot 7}{5 \cdot 7} = \frac{14}{35}.$$

Do Exercises 5–9.

b Simplifying

All of the following are names for three-fourths:

$$\frac{3}{4}, \frac{6}{8}, \frac{9}{12}, \frac{12}{16}, \frac{15}{20}.$$

We say that $\frac{3}{4}$ is **simplest** because it has the smallest numerator and the smallest denominator. That is, the numerator and the denominator have no common factor other than 1.

Multiply.

1. $\frac{1}{2} \cdot \frac{8}{8}$

2. $\frac{3}{5} \cdot \frac{10}{10}$

3. $\frac{13}{25} \cdot \frac{4}{4}$

4. $\frac{8}{3} \cdot \frac{25}{25}$

Find another name for the number, but with the denominator indicated. Use multiplying by 1.

5. $\frac{4}{3} = \frac{?}{9}$

6. $\frac{3}{4} = \frac{?}{24}$

7. $\frac{9}{10} = \frac{?}{100}$

8. $\frac{3}{15} = \frac{?}{45}$

9. $\frac{8}{7} = \frac{?}{49}$

Answers on page A-5

Simplify.

10. $\frac{2}{8}$

11. $\frac{10}{12}$

12. $\frac{40}{8}$

13. $\frac{24}{18}$

Answers on page A-5

To simplify, we reverse the process of multiplying by 1.

$$\frac{12}{18} = \frac{2 \cdot 6}{3 \cdot 6} \quad \begin{array}{l}\leftarrow \text{Factoring the numerator} \\ \leftarrow \text{Factoring the denominator}\end{array}$$

$$= \frac{2}{3} \cdot \frac{6}{6} \qquad \text{Factoring the fraction}$$

$$= \frac{2}{3} \cdot 1 \qquad \frac{6}{6} = 1$$

$$= \frac{2}{3} \qquad \text{Removing a factor of 1: } \frac{2}{3} \cdot 1 = \frac{2}{3}$$

Examples Simplify.

3. $\frac{8}{20} = \frac{2 \cdot 4}{5 \cdot 4} = \frac{2}{5} \cdot \frac{4}{4} = \frac{2}{5}$

4. $\frac{2}{6} = \frac{1 \cdot 2}{3 \cdot 2} = \frac{1}{3} \cdot \frac{2}{2} = \frac{1}{3}$

The number 1 allows for pairing of factors in the numerator and the denominator.

5. $\frac{30}{6} = \frac{5 \cdot 6}{1 \cdot 6} = \frac{5}{1} \cdot \frac{6}{6} = \frac{5}{1} = 5$

We could also simplify $\frac{30}{6}$ by doing the division $30 \div 6$. That is, $\frac{30}{6} = 30 \div 6 = 5$.

Do Exercises 10–13.

The use of prime factorizations can be helpful for simplifying when numerators and/or denominators are larger numbers.

Example 6 Simplify: $\frac{90}{84}$.

$$\frac{90}{84} = \frac{2 \cdot 3 \cdot 3 \cdot 5}{2 \cdot 2 \cdot 3 \cdot 7} \qquad \text{Factoring the numerator and the denominator into primes}$$

$$= \frac{2 \cdot 3 \cdot 3 \cdot 5}{2 \cdot 3 \cdot 2 \cdot 7} \qquad \text{Changing the order so that like primes are above and below each other}$$

$$= \frac{2}{2} \cdot \frac{3}{3} \cdot \frac{3 \cdot 5}{2 \cdot 7} \qquad \text{Factoring the fraction}$$

$$= 1 \cdot 1 \cdot \frac{3 \cdot 5}{2 \cdot 7}$$

$$= \frac{3 \cdot 5}{2 \cdot 7} \qquad \text{Removing factors of 1}$$

$$= \frac{15}{14}$$

We could have shortened the preceding example had we recalled our tests for divisibility (Section 2.2) and noted that 6 is a factor of both the numerator and the denominator. Then

$$\frac{90}{84} = \frac{6 \cdot 15}{6 \cdot 14} = \frac{6}{6} \cdot \frac{15}{14} = \frac{15}{14}.$$

The tests for divisibility are very helpful in simplifying.

Example 7 Simplify: $\frac{603}{207}$.

At first glance this looks difficult. But note, using the test for divisibility by 9 (sum of digits divisible by 9), that both the numerator and the denominator are divisible by 9. Thus we can factor 9 from both numbers:

$$\frac{603}{207} = \frac{9 \cdot 67}{9 \cdot 23} = \frac{9}{9} \cdot \frac{67}{23} = \frac{67}{23}.$$

Do Exercises 14–18.

Simplify.

14. $\frac{35}{40}$

15. $\frac{801}{702}$

16. $\frac{24}{21}$

17. $\frac{75}{300}$

18. Simplify each fraction in this circle graph.

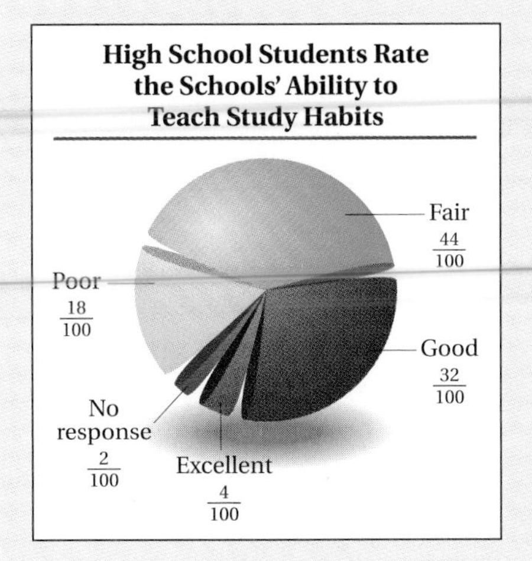

CANCELING Canceling is a shortcut that you may have used for removing a factor of 1 when working with fractional notation. With *great* concern, we mention it as a possibility for speeding up your work. Canceling may be done only when removing common factors in numerators and denominators. Each common factor allows us to remove a factor of 1 in a product.

Our concern is that canceling be done with care and understanding. In effect, slashes are used to indicate factors of 1 that have been removed. For instance, Example 6 might have been done faster as follows:

$$\frac{90}{84} = \frac{2 \cdot 3 \cdot 3 \cdot 5}{2 \cdot 2 \cdot 3 \cdot 7}$$ **Factoring the numerator and the denominator**

$$= \frac{\not{2} \cdot \not{3} \cdot 3 \cdot 5}{2 \cdot \not{2} \cdot \not{3} \cdot 7}$$ **When a factor of 1 is noted, it is "canceled" as shown: $\frac{2 \cdot 3}{2 \cdot 3} = 1$.**

$$= \frac{3 \cdot 5}{2 \cdot 7} = \frac{15}{14}.$$

CAUTION! The difficulty with canceling is that it is often applied incorrectly in situations like the following:

$\frac{\not{2} + 3}{\not{2}} = 3;$ Wrong! $\frac{\not{4} + 1}{\not{4} + 2} = \frac{1}{2};$ Wrong! $\frac{1\not{5}}{\not{5}4} = \frac{1}{4}.$ Wrong!

The correct answers are

$$\frac{2 + 3}{2} = \frac{5}{2}; \quad \frac{4 + 1}{4 + 2} = \frac{5}{6}; \quad \frac{15}{54} = \frac{5}{18}.$$

In each situation, the number canceled was not a factor of 1. Factors are parts of products. For example, in $2 \cdot 3$, 2 and 3 are factors, but in $2 + 3$, 2 and 3 are *not* factors. Canceling may not be done when sums or differences are in numerators or denominators, as shown here.

If you cannot factor, do not cancel! If in doubt, do not cancel!

c A Test for Equality

When denominators are the same, we say that fractions have a **common denominator.** Suppose we want to compare $\frac{2}{4}$ and $\frac{3}{6}$. We find a common denominator and compare numerators. To do this, we multiply by 1 using symbols for 1 formed by looking at contrasting denominators.

$$\frac{3}{6} = \frac{3}{6} \cdot \frac{4}{4} = \frac{3 \cdot 4}{6 \cdot 4} = \frac{12}{24}$$

$$\frac{2}{4} = \frac{2}{4} \cdot \frac{6}{6} = \frac{2 \cdot 6}{4 \cdot 6} = \frac{12}{24}$$

We see that $\frac{3}{6} = \frac{2}{4}$.

Calculator Spotlight

Fraction calculators are equipped with a key, often labeled [$a\,^b\!/_c$], that allows for simplification with fractional notation. To simplify

$$\frac{208}{256}$$

with such a fraction calculator, the following keystrokes can be used.

[2] [0] [8] [$a\,^b\!/_c$] [2] [5] [6] [=].

The display that appears

13 ⌟ 16.

represents simplified fractional notation $\frac{13}{16}$.

Exercises

Use a fraction calculator to simplify each of the following.

1. $\frac{84}{90}$ 2. $\frac{35}{40}$ 3. $\frac{690}{835}$ 4. $\frac{42}{150}$

Answers on page A-5

Use $=$ or $\neq$ for ▢ to write a true sentence.

19. $\frac{2}{6}$ ▢ $\frac{3}{9}$

20. $\frac{2}{3}$ ▢ $\frac{14}{20}$

Answers on page A-5

Note in the preceding that if

$$\frac{3}{6} = \frac{2}{4}, \quad \text{then} \quad 3 \cdot 4 = 6 \cdot 2.$$

We need only check the products $3 \cdot 4$ and $6 \cdot 2$ to compare the fractions.

A TEST FOR EQUALITY

We multiply these two numbers: $3 \cdot 4$.

We multiply these two numbers: $6 \cdot 2$.

$$\frac{3}{6} \quad \frac{2}{4}$$

We call $3 \cdot 4$ and $6 \cdot 2$ **cross products**. Since the cross products are the same, that is, $3 \cdot 4 = 6 \cdot 2$, we know that

$$\frac{3}{6} = \frac{2}{4}.$$

If a sentence $a = b$ is true, it means that a and b name the same number. If a sentence $a \neq b$ is true, it means that a and b do *not* name the same number.

Example 8 Use $=$ or $\neq$ for ▢ to write a true sentence:

$$\frac{6}{7} \ ▢ \ \frac{7}{8}.$$

We multiply these two numbers: $6 \cdot 8 = 48$.

We multiply these two numbers: $7 \cdot 7 = 49$.

$$\frac{6}{7} \quad \frac{7}{8}$$

Because $48 \neq 49$ (read "48 is not equal to 49"), $\frac{6}{7} = \frac{7}{8}$ is not a true sentence. Thus,

$$\frac{6}{7} \neq \frac{7}{8}.$$

Example 9 Use $=$ or $\neq$ for ▢ to write a true sentence:

$$\frac{6}{10} \ ▢ \ \frac{3}{5}.$$

We multiply these two numbers: $6 \cdot 5 = 30$.

We multiply these two numbers: $10 \cdot 3 = 30$.

$$\frac{6}{10} \quad \frac{3}{5}.$$

Because the cross products are the same, we have

$$\frac{6}{10} = \frac{3}{5}.$$

Do Exercises 19 and 20.

Exercise Set 2.5

a Find another name for the given number, but with the denominator indicated. Use multiplying by 1.

1. $\frac{1}{2} = \frac{?}{10}$
2. $\frac{1}{6} = \frac{?}{18}$
3. $\frac{5}{8} = \frac{?}{32}$
4. $\frac{2}{9} = \frac{?}{18}$
5. $\frac{9}{10} = \frac{?}{30}$
6. $\frac{5}{6} = \frac{?}{48}$
7. $\frac{7}{8} = \frac{?}{32}$
8. $\frac{2}{5} = \frac{?}{25}$
9. $\frac{5}{12} = \frac{?}{48}$
10. $\frac{3}{8} = \frac{?}{56}$
11. $\frac{17}{18} = \frac{?}{54}$
12. $\frac{11}{16} = \frac{?}{256}$
13. $\frac{5}{3} = \frac{?}{45}$
14. $\frac{11}{5} = \frac{?}{30}$
15. $\frac{7}{22} = \frac{?}{132}$
16. $\frac{10}{21} = \frac{?}{126}$

b Simplify.

17. $\frac{2}{4}$
18. $\frac{4}{8}$
19. $\frac{6}{8}$
20. $\frac{8}{12}$
21. $\frac{3}{15}$
22. $\frac{8}{10}$
23. $\frac{24}{8}$
24. $\frac{36}{9}$
25. $\frac{18}{24}$
26. $\frac{42}{48}$
27. $\frac{14}{16}$
28. $\frac{15}{25}$
29. $\frac{12}{10}$
30. $\frac{16}{14}$
31. $\frac{16}{48}$
32. $\frac{100}{20}$
33. $\frac{150}{25}$
34. $\frac{19}{76}$
35. $\frac{17}{51}$
36. $\frac{425}{525}$

c Use $=$ or $\neq$ for ▒ to write a true sentence.

37. $\frac{3}{4}$ ▒ $\frac{9}{12}$
38. $\frac{4}{8}$ ▒ $\frac{3}{6}$
39. $\frac{1}{5}$ ▒ $\frac{2}{9}$
40. $\frac{1}{4}$ ▒ $\frac{2}{9}$

41. $\frac{3}{8}$ ▒ $\frac{6}{16}$

42. $\frac{2}{6}$ ▒ $\frac{6}{18}$

43. $\frac{2}{5}$ ▒ $\frac{3}{7}$

44. $\frac{1}{3}$ ▒ $\frac{1}{4}$

45. $\frac{12}{9}$ ▒ $\frac{8}{6}$

46. $\frac{16}{14}$ ▒ $\frac{8}{7}$

47. $\frac{5}{2}$ ▒ $\frac{17}{7}$

48. $\frac{3}{10}$ ▒ $\frac{7}{24}$

49. $\frac{3}{10}$ ▒ $\frac{30}{100}$

50. $\frac{700}{1000}$ ▒ $\frac{70}{100}$

51. $\frac{5}{10}$ ▒ $\frac{520}{1000}$

52. $\frac{49}{100}$ ▒ $\frac{50}{1000}$

Skill Maintenance

Solve. [1.8a]

53. A playing field is 78 ft long and 64 ft wide. What is its area?

54. A landscaper buys 13 small maple trees and 17 small oak trees for a project. A maple costs $23 and an oak costs $37. How much is spent altogether for the trees?

Subtract. [1.3d]

55. $34 - 23$

56. $50 - 18$

57. $803 - 617$

58. $8344 - 5607$

Solve. [1.7b]

59. $30 \cdot x = 150$

60. $5280 = 1760 + t$

Synthesis

61. ◆ Explain in your own words when it *is* possible to "cancel" and when it *is not* possible to "cancel."

62. ◆ Can fractional notation be simplified if its numerator and its denominator are two different prime numbers? Why or why not?

Simplify. Use the list of prime numbers on p. 88.

63. ▦ $\frac{2603}{2831}$

64. ▦ $\frac{3197}{3473}$

65. *Shy People.* Sociologists have found that 4 out of 10 people are shy. Write fractional notation for the part of the population that is shy; the part that is not shy. Simplify.

66. *Left-handed People.* Sociologists estimate that 3 out of 20 people are left-handed. In a crowd of 460 people, how many would you expect to be left-handed?

67. ▦ *Batting Averages.* Bernie Williams of the New York Yankees got 168 hits in 551 times at bat. Ken Griffey, Jr., of the Seattle Mariners got 165 hits in 545 times at bat. (***Source:*** Major League Baseball) Did they have the same fraction of hits (batting average)? Why or why not?

68. ▦ On a test of 82 questions, a student got 63 correct. On another test of 100 questions, the student got 77 correct. Did the student get the same portion of each test correct? Why or why not?

Collaborative Learning Manual: Use fraction bars to represent equivalent fractions.

2.6 Multiplying, Simplifying, and Applications

Objectives

a. Multiply and simplify using fractional notation.

b. Solve applied problems involving multiplication.

For Extra Help

TAPE 4

TAPE 4B

MAC WIN

CD-ROM

a Simplifying After Multiplying

We usually simplify after we multiply. To make such simplifying easier, it is generally best not to carry out the products in the numerator and the denominator, but to factor and simplify before multiplying. Consider the product

$$\frac{3}{8} \cdot \frac{4}{9}.$$

We proceed as follows:

$$\frac{3}{8} \cdot \frac{4}{9} = \frac{3 \cdot 4}{8 \cdot 9}$$ We write the products in the numerator and the denominator, but we do not carry them out.

$$= \frac{3 \cdot 2 \cdot 2}{2 \cdot 2 \cdot 2 \cdot 3 \cdot 3}$$ Factoring the numerator and the denominator

$$= \frac{3 \cdot 2 \cdot 2}{3 \cdot 2 \cdot 2} \cdot \frac{1}{2 \cdot 3}$$ Factoring the fraction

$$= 1 \cdot \frac{1}{2 \cdot 3}$$

$$= \frac{1}{2 \cdot 3}$$ Removing a factor of 1

$$= \frac{1}{6}.$$

The procedure could have been shortened had we noticed that 4 is a factor of the 8 in the denominator:

$$\frac{3}{8} \cdot \frac{4}{9} = \frac{3 \cdot 4}{8 \cdot 9} = \frac{3 \cdot 4}{4 \cdot 2 \cdot 3 \cdot 3} = \frac{3 \cdot 4}{3 \cdot 4} \cdot \frac{1}{2 \cdot 3} = 1 \cdot \frac{1}{2 \cdot 3} = \frac{1}{2 \cdot 3} = \frac{1}{6}.$$

To multiply and simplify:

a) Write the products in the numerator and the denominator, but do not carry out the products.

b) Factor the numerator and the denominator.

c) Factor the fraction to remove factors of 1.

d) Carry out the remaining products.

Examples Multiply and simplify.

1. $\frac{2}{3} \cdot \frac{9}{4} = \frac{2 \cdot 9}{3 \cdot 4} = \frac{2 \cdot 3 \cdot 3}{3 \cdot 2 \cdot 2} = \frac{2 \cdot 3}{2 \cdot 3} \cdot \frac{3}{2} = 1 \cdot \frac{3}{2} = \frac{3}{2}$

2. $\frac{6}{7} \cdot \frac{5}{3} = \frac{6 \cdot 5}{7 \cdot 3} = \frac{3 \cdot 2 \cdot 5}{7 \cdot 3} = \frac{3}{3} \cdot \frac{2 \cdot 5}{7} = 1 \cdot \frac{2 \cdot 5}{7} = \frac{2 \cdot 5}{7} = \frac{10}{7}$

3. $40 \cdot \frac{7}{8} = \frac{40 \cdot 7}{8} = \frac{8 \cdot 5 \cdot 7}{8 \cdot 1} = \frac{8}{8} \cdot \frac{5 \cdot 7}{1} = 1 \cdot \frac{5 \cdot 7}{1} = \frac{5 \cdot 7}{1} = 35$

Multiply and simplify.

1. $\frac{2}{3} \cdot \frac{7}{8}$

2. $\frac{4}{5} \cdot \frac{5}{12}$

3. $16 \cdot \frac{3}{8}$

4. $\frac{5}{8} \cdot 4$

5. A landscaper uses $\frac{2}{3}$ lb of peat moss for a rosebush. How much will be needed for 21 rosebushes?

Answers on page A-5

CAUTION! Canceling can be used as follows for these examples.

1. $\frac{2}{3} \cdot \frac{9}{4} = \frac{2 \cdot 9}{3 \cdot 4} = \frac{\cancel{2} \cdot \cancel{3} \cdot 3}{\cancel{3} \cdot \cancel{2} \cdot 2} = \frac{3}{2}$ **Removing a factor of 1:** $\frac{2 \cdot 3}{2 \cdot 3} = 1$

2. $\frac{6}{7} \cdot \frac{5}{3} = \frac{6 \cdot 5}{7 \cdot 3} = \frac{\cancel{3} \cdot 2 \cdot 5}{7 \cdot \cancel{3}} = \frac{2 \cdot 5}{7} = \frac{10}{7}$ **Removing a factor of 1:** $\frac{3}{3} = 1$

3. $40 \cdot \frac{7}{8} = \frac{40 \cdot 7}{8} = \frac{\cancel{8} \cdot 5 \cdot 7}{\cancel{8} \cdot 1} = \frac{5 \cdot 7}{1} = 35$ **Removing a factor of 1:** $\frac{8}{8} = 1$

Remember, if you can't factor, you can't cancel!

Do Exercises 1–4.

b Applications and Problem Solving

Example 4 How much salmon will be needed to serve 30 people if each person gets $\frac{2}{5}$ lb?

1. **Familiarize.** We first make a drawing or at least visualize the situation. Repeated addition will work here.

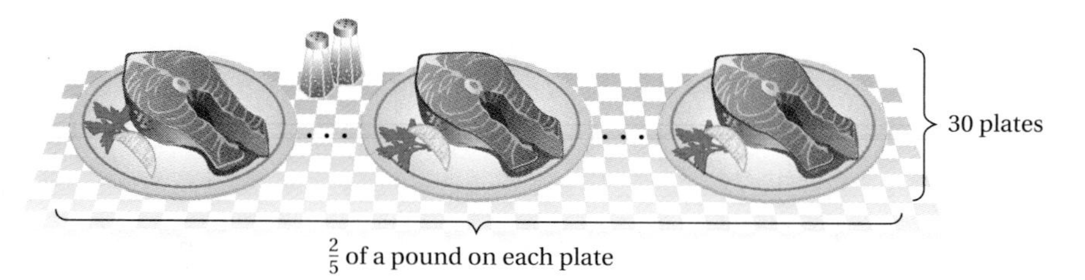

We let n = the number of pounds of salmon needed.

2. **Translate.** The problem translates to the following equation:

$$n = 30 \cdot \frac{2}{5}.$$

3. **Solve.** To solve the equation, we carry out the multiplication:

$$\begin{aligned} n = 30 \cdot \frac{2}{5} &= \frac{30 \cdot 2}{5} && \textbf{Multiplying} \\ &= \frac{5 \cdot 6 \cdot 2}{5 \cdot 1} \\ &= \frac{5}{5} \cdot \frac{6 \cdot 2}{1} \\ &= 12. && \textbf{Simplifying} \end{aligned}$$

4. **Check.** We check by repeating the calculation. (We leave the check to the student.) We can also think about the reasonableness of the answer. We are multiplying 30 by a number less than 1, so the product will be less than 30. Since 12 is less than 30, we have a partial check of the reasonableness of the answer. The number 12 checks.

5. **State.** Thus, 12 lb of salmon will be needed.

Do Exercise 5.

Exercise Set 2.6

a Multiply and simplify.

Don't forget to simplify!

1. $\frac{2}{3} \cdot \frac{1}{2}$

2. $\frac{3}{8} \cdot \frac{1}{3}$

3. $\frac{7}{8} \cdot \frac{1}{7}$

4. $\frac{4}{9} \cdot \frac{1}{4}$

5. $\frac{1}{8} \cdot \frac{4}{5}$

6. $\frac{2}{5} \cdot \frac{1}{6}$

7. $\frac{1}{4} \cdot \frac{2}{3}$

8. $\frac{4}{6} \cdot \frac{1}{6}$

9. $\frac{12}{5} \cdot \frac{9}{8}$

10. $\frac{16}{15} \cdot \frac{5}{4}$

11. $\frac{10}{9} \cdot \frac{7}{5}$

12. $\frac{25}{12} \cdot \frac{4}{3}$

13. $9 \cdot \frac{1}{9}$

14. $4 \cdot \frac{1}{4}$

15. $\frac{1}{3} \cdot 3$

16. $\frac{1}{6} \cdot 6$

17. $\frac{7}{10} \cdot \frac{10}{7}$

18. $\frac{8}{9} \cdot \frac{9}{8}$

19. $\frac{7}{5} \cdot \frac{5}{7}$

20. $\frac{2}{11} \cdot \frac{11}{2}$

21. $\frac{1}{4} \cdot 8$

22. $\frac{1}{3} \cdot 18$

23. $24 \cdot \frac{1}{6}$

24. $16 \cdot \frac{1}{2}$

25. $12 \cdot \frac{3}{4}$

26. $18 \cdot \frac{5}{6}$

27. $\frac{3}{8} \cdot 24$

28. $\frac{2}{9} \cdot 36$

29. $13 \cdot \frac{2}{5}$

30. $15 \cdot \frac{1}{6}$

31. $\frac{7}{10} \cdot 28$

32. $\frac{5}{8} \cdot 34$

33. $\frac{1}{6} \cdot 360$

34. $\frac{1}{3} \cdot 120$

35. $240 \cdot \frac{1}{8}$

36. $150 \cdot \frac{1}{5}$

37. $\frac{4}{10} \cdot \frac{5}{10}$

38. $\frac{7}{10} \cdot \frac{34}{150}$

39. $\frac{8}{10} \cdot \frac{45}{100}$

40. $\frac{3}{10} \cdot \frac{8}{10}$

41. $\frac{11}{24} \cdot \frac{3}{5}$

42. $\frac{15}{22} \cdot \frac{4}{7}$

43. $\frac{10}{21} \cdot \frac{3}{4}$

44. $\frac{17}{18} \cdot \frac{3}{5}$

b Solve.

45. Anna receives $36 for working a full day doing inventory at a hardware store. How much will she receive for working $\frac{3}{4}$ of the day?

46. After Jack completes 60 hr of teacher training in college, he can earn $45 for working a full day as a substitute teacher. How much will he receive for working $\frac{1}{5}$ of a day?

47. *Mailing-List Addresses.* Business people have determined that $\frac{1}{4}$ of the addresses on a mailing list will change in one year. A business has a mailing list of 2500 people. After one year, how many addresses on that list will be incorrect?

48. *Shy People.* Sociologists have determined that $\frac{2}{5}$ of the people in the world are shy. A sales manager is interviewing 650 people for an aggressive sales position. How many of these people might be shy?

49. A recipe calls for $\frac{2}{3}$ cup of flour. A chef is making $\frac{1}{2}$ of the recipe. How much flour should the chef use?

50. Of the students in the freshman class, $\frac{2}{5}$ have cameras; $\frac{1}{4}$ of these students also join the college photography club. What fraction of the students in the freshman class join the photography club?

51. A house worth $124,000 was assessed for $\frac{3}{4}$ of its value. What is the assessed value of the house?

52. A student's tuition was $2800. A loan was obtained for $\frac{3}{4}$ of the tuition. How much was the loan?

53. *Map Scaling.* On a map, 1 in. represents 240 mi. How much does $\frac{2}{3}$ in. represent?

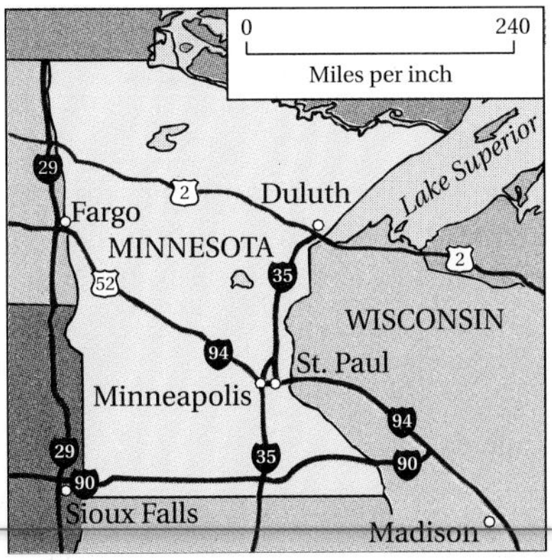

54. *Map Scaling.* On a map, 1 in. represents 120 mi. How much does $\frac{3}{4}$ in. represent?

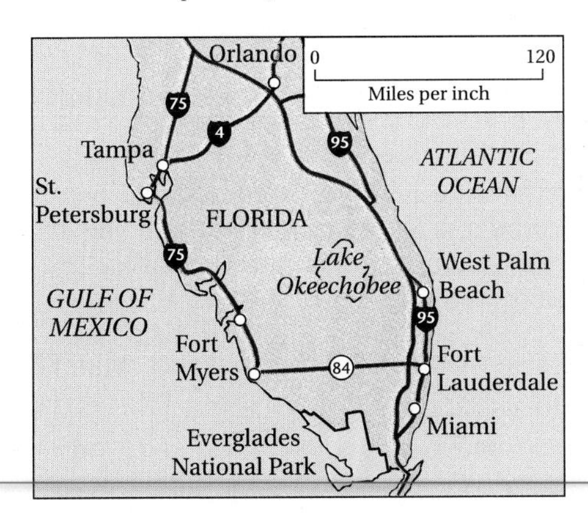

55. *Family Spending.* A family has an annual income of \$27,000. Of this, $\frac{1}{4}$ is spent for food, $\frac{1}{5}$ for housing, $\frac{1}{10}$ for clothing, $\frac{1}{9}$ for savings, $\frac{1}{4}$ for taxes, and the rest for other expenses. How much is spent for each?

56. *Family Spending.* A family has an annual income of \$25,200. Of this, $\frac{1}{4}$ is spent for food, $\frac{1}{5}$ for housing, $\frac{1}{10}$ for clothing, $\frac{1}{9}$ for savings, $\frac{1}{4}$ for taxes, and the rest for other expenses. How much is spent for each?

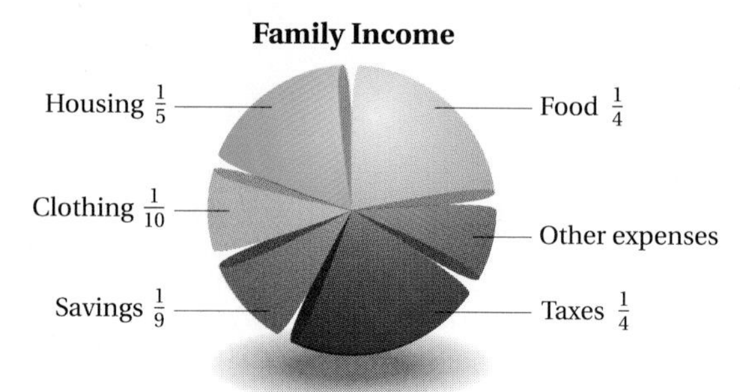

Source: U.S. Census Bureau

Skill Maintenance

Solve. [1.7b]

57. $48 \cdot t = 1680$

58. $74 \cdot x = 6290$

59. $t + 28 = 5017$

60. $456 + x = 9002$

Subtract. [1.3d]

61. $\begin{array}{r} 9060 \\ -4387 \\ \hline \end{array}$

62. $\begin{array}{r} 7800 \\ -2462 \\ \hline \end{array}$

Synthesis

63. ◆ When multiplying using fractional notation, we form products in the numerator and the denominator, but do not immediately calculate the products. Why?

64. ◆ If a fraction's numerator and denominator have no factors (other than 1) in common, can the fraction be simplified? Why or why not?

Simplify. Use the list of prime numbers on p. 88 or a fraction calculator.

65. 🖩 $\frac{201}{535} \cdot \frac{4601}{6499}$

66. 🖩 $\frac{5767}{3763} \cdot \frac{159}{395}$

67. *College Profile.* Of students entering a college, $\frac{7}{8}$ have completed high school and $\frac{2}{3}$ are older than 20. If $\frac{1}{7}$ of all students are left-handed, what fraction of students entering the college are left-handed high school graduates over the age of 20?

68. *College Profile.* Refer to the information in Exercise 67. If 480 students are entering the college, how many of them are left-handed high school graduates 20 yr old or younger?

69. *College Profile.* Refer to Exercise 67. What fraction of students entering the college did not graduate high school, are 20 yr old or younger, and are left-handed?

2.7 Division and Applications

a Reciprocals

Look at these products:

$$8 \cdot \frac{1}{8} = \frac{8 \cdot 1}{8} = \frac{8}{8} = 1; \qquad \frac{2}{3} \cdot \frac{3}{2} = \frac{2 \cdot 3}{3 \cdot 2} = \frac{6}{6} = 1.$$

> If the product of two numbers is 1, we say that they are **reciprocals** of each other. To find a reciprocal, interchange the numerator and the denominator.
>
> Number $\longrightarrow \frac{3}{4} \quad \frac{4}{3} \longleftarrow$ Reciprocal

Examples Find the reciprocal.

1. The reciprocal of $\frac{4}{5}$ is $\frac{5}{4}$. $\quad \frac{4}{5} \cdot \frac{5}{4} = \frac{20}{20} = 1$

2. The reciprocal of $\frac{8}{7}$ is $\frac{7}{8}$. $\quad \frac{8}{7} \cdot \frac{7}{8} = \frac{56}{56} = 1$

3. The reciprocal of 8 is $\frac{1}{8}$. $\quad$ Think of 8 as $\frac{8}{1}$: $\frac{8}{1} \cdot \frac{1}{8} = \frac{8}{8} = 1.$

4. The reciprocal of $\frac{1}{3}$ is 3. $\quad \frac{1}{3} \cdot 3 = \frac{3}{3} = 1$

Do Exercises 1–4.

Does 0 have a reciprocal? If it did, it would have to be a number x such that

$$0 \cdot x = 1.$$

But 0 times any number is 0. Thus,

> The number 0, or $\frac{0}{n}$, has no reciprocal. $\left(\text{Recall that } \frac{n}{0} \text{ is not defined.}\right)$

b Division

Recall that $a \div b$ is the number that when multiplied by b gives a. Consider the division $\frac{3}{4} \div \frac{1}{8}$. We are asking how many $\frac{1}{8}$'s are in $\frac{3}{4}$. We can answer this by looking at the figure below.

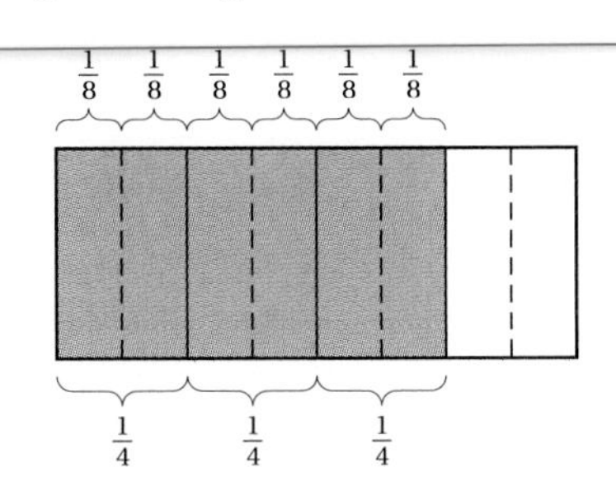

Objectives

a Find the reciprocal of a number.

b Divide and simplify using fractional notation.

c Solve equations of the type $a \cdot x = b$ and $x \cdot a = b$, where a and b may be fractions.

d Solve applied problems involving division.

For Extra Help

TAPE 4 | TAPE 4B | MAC WIN | CD-ROM

Find the reciprocal.

1. $\frac{2}{5}$

2. $\frac{10}{7}$

3. 9

4. $\frac{1}{5}$

Answers on page A-5

Divide and simplify.

5. $\frac{6}{7} \div \frac{3}{4}$

6. $\frac{2}{3} \div \frac{1}{4}$

7. $\frac{4}{5} \div 8$

8. $60 \div \frac{3}{5}$

9. $\frac{3}{5} \div \frac{3}{5}$

Answers on page A-5

We see that there are six $\frac{1}{8}$'s in $\frac{3}{4}$. Thus,

$$\frac{3}{4} \div \frac{1}{8} = 6.$$

We can check this by multiplying:

$$6 \cdot \frac{1}{8} = \frac{6}{8} = \frac{3}{4}.$$

Here is a faster way to divide.

To divide fractions, multiply the dividend by the reciprocal of the divisor:

$$\frac{2}{5} \div \frac{3}{4} = \frac{2}{5} \cdot \frac{4}{3} = \frac{2 \cdot 4}{5 \cdot 3} = \frac{8}{15}.$$

Multiply by the reciprocal of the divisor.

Examples Divide and simplify.

5. $\frac{5}{6} \div \frac{2}{3} = \frac{5}{6} \cdot \frac{3}{2} = \frac{5 \cdot 3}{6 \cdot 2} = \frac{5 \cdot 3}{3 \cdot 2 \cdot 2} = \frac{3}{3} \cdot \frac{5}{2 \cdot 2} = \frac{5}{2 \cdot 2} = \frac{5}{4}$

6. $\frac{3}{4} \div \frac{1}{8} = \frac{3}{4} \cdot 8 = \frac{3 \cdot 8}{4} = \frac{3 \cdot 4 \cdot 2}{4 \cdot 1} = \frac{4}{4} \cdot \frac{3 \cdot 2}{1} = \frac{3 \cdot 2}{1} = 6$

7. $\frac{2}{5} \div 6 = \frac{2}{5} \cdot \frac{1}{6} = \frac{2 \cdot 1}{5 \cdot 6} = \frac{2 \cdot 1}{5 \cdot 2 \cdot 3} = \frac{2}{2} \cdot \frac{1}{5 \cdot 3} = \frac{1}{5 \cdot 3} = \frac{1}{15}$

8. $\frac{3}{5} \div \frac{1}{2} = \frac{3}{5} \cdot 2 = \frac{3 \cdot 2}{5} = \frac{6}{5}$

CAUTION! Canceling can be used as follows for Examples 5–7.

5. $\frac{5}{6} \div \frac{2}{3} = \frac{5}{6} \cdot \frac{3}{2} = \frac{5 \cdot 3}{6 \cdot 2} = \frac{5 \cdot \cancel{3}}{\cancel{3} \cdot 2 \cdot 2} = \frac{5}{2 \cdot 2} = \frac{5}{4}$ **Removing a factor of 1:** $\frac{3}{3} = 1$

6. $\frac{3}{4} \div \frac{1}{8} = \frac{3}{4} \cdot 8 = \frac{3 \cdot 8}{4} = \frac{3 \cdot \cancel{4} \cdot 2}{\cancel{4} \cdot 1} = \frac{3 \cdot 2}{1} = 6$ **Removing a factor of 1:** $\frac{4}{4} = 1$

7. $\frac{2}{5} \div 6 = \frac{2}{5} \cdot \frac{1}{6} = \frac{2 \cdot 1}{5 \cdot 6} = \frac{\cancel{2} \cdot 1}{5 \cdot \cancel{2} \cdot 3} = \frac{1}{5 \cdot 3} = \frac{1}{15}$ **Removing a factor of 1:** $\frac{2}{2} = 1$

Remember, if you can't factor, you can't cancel!

Do Exercises 5–9.

Why do we multiply by a reciprocal when dividing? To see this, let's consider $\frac{2}{3} \div \frac{7}{5}$. We will multiply by 1. The name for 1 that we will use is (5/7)/(5/7); it comes from the reciprocal of $\frac{7}{5}$.

$$\frac{2}{3} \div \frac{7}{5} = \frac{\frac{2}{3}}{\frac{7}{5}}.$$ **Writing fractional notation for the division**

Then

$$= \frac{\frac{2}{3}}{\frac{7}{5}} \cdot 1 \qquad \text{Multiplying by 1}$$

$$= \frac{\frac{2}{3}}{\frac{7}{5}} \cdot \frac{\frac{5}{7}}{\frac{5}{7}} \qquad \text{Multiplying by 1; } \tfrac{5}{7} \text{ is the reciprocal of } \tfrac{7}{5} \text{ and } \frac{\frac{5}{7}}{\frac{5}{7}} = 1$$

$$= \frac{\frac{2}{3} \cdot \frac{5}{7}}{\frac{7}{5} \cdot \frac{5}{7}} \qquad \text{Multiplying the numerators and the denominators}$$

$$= \frac{\frac{2}{3} \cdot \frac{5}{7}}{1} = \frac{2}{3} \cdot \frac{5}{7} = \frac{10}{21}.$$

After we multiplied, we got 1 for the denominator. The numerator (in color) shows the multiplication by the reciprocal.

Do Exercise 10.

10. Divide by multiplying by 1:

$$\frac{\frac{4}{5}}{\frac{6}{7}}.$$

c Solving Equations

Now let's solve equations $a \cdot x = b$ and $x \cdot a = b$, where a and b may be fractions. We proceed as we did with equations involving whole numbers. We divide by a on both sides.

Example 9 Solve: $\frac{4}{3} \cdot x = \frac{6}{7}$.

$$\frac{4}{3} \cdot x = \frac{6}{7}$$

$$x = \frac{6}{7} \div \frac{4}{3} \qquad \text{Dividing by } \tfrac{4}{3} \text{ on both sides}$$

$$= \frac{6}{7} \cdot \frac{3}{4} \qquad \text{Multiplying by the reciprocal}$$

$$= \frac{2 \cdot 3 \cdot 3}{7 \cdot 2 \cdot 2} = \frac{2}{2} \cdot \frac{3 \cdot 3}{7 \cdot 2} = \frac{3 \cdot 3}{7 \cdot 2} = \frac{9}{14}$$

The solution is $\frac{9}{14}$.

Example 10 Solve: $t \cdot \frac{4}{5} = 80$.

Dividing by $\frac{4}{5}$ on both sides, we get

$$t = 80 \div \frac{4}{5} = 80 \cdot \frac{5}{4} = \frac{80 \cdot 5}{4} = \frac{4 \cdot 20 \cdot 5}{4 \cdot 1} = \frac{4}{4} \cdot \frac{20 \cdot 5}{1} = \frac{20 \cdot 5}{1} = 100.$$

The solution is 100.

Do Exercises 11 and 12.

Solve.

11. $\frac{5}{6} \cdot y = \frac{2}{3}$

12. $\frac{3}{4} \cdot n = 24$

d Applications and Problem Solving

Example 11 *Test Tubes.* How many test tubes, each containing $\frac{3}{5}$ mL, can a nursing student fill from a container of 60 mL?

Answers on page A-5

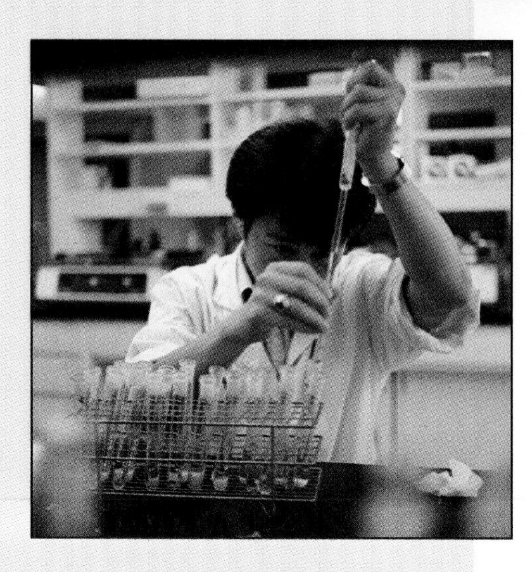

13. Each loop in a spring uses $\frac{3}{8}$ in. of wire. How many loops can be made from 120 in. of wire?

14. A service station tank had 175 gal of oil when it was $\frac{7}{8}$ full. How much could the tank hold altogether?

Answers on page A-5

1. **Familiarize.** Repeated addition will apply here. We let $n =$ the number of test tubes in all. We make a drawing.

$\frac{3}{5}$ of a milliliter in each test tube

n test tubes in all

2. **Translate.** The equation that corresponds to the situation is

$$n \cdot \frac{3}{5} = 60.$$

3. **Solve.** We solve the equation by dividing by $\frac{3}{5}$ on both sides and carrying out the division:

$$n = 60 \div \frac{3}{5} = 60 \cdot \frac{5}{3} = \frac{60 \cdot 5}{3} = \frac{3 \cdot 20 \cdot 5}{3 \cdot 1}$$
$$= \frac{3}{3} \cdot \frac{20 \cdot 5}{1} = 100.$$

4. **Check.** We check by repeating the calculation.
5. **State.** Thus, 100 test tubes can be filled.

Do Exercise 13.

Example 12 After driving 210 mi, $\frac{5}{6}$ of a sales trip was completed. How long was the total trip?

1. **Familiarize.** We make a drawing or at least visualize the situation. We let $n =$ the length of the trip.

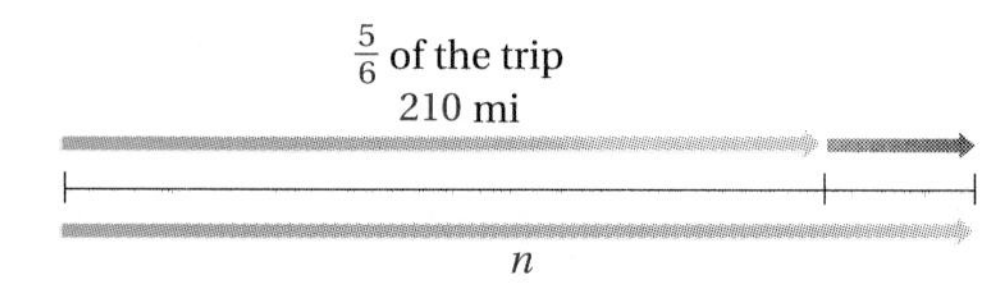

2. **Translate.** We translate to an equation.

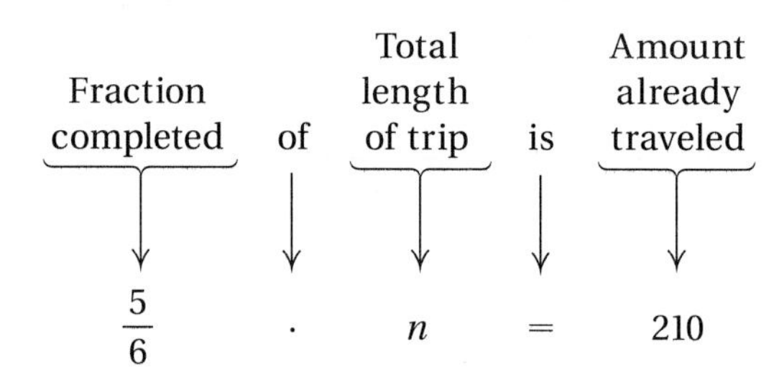

$$\frac{5}{6} \cdot n = 210$$

3. **Solve.** The equation that corresponds to the situation is $\frac{5}{6} \cdot n = 210$. We divide by $\frac{5}{6}$ on both sides and carry out the division:

$$n = 210 \div \frac{5}{6} = 210 \cdot \frac{6}{5} = \frac{210 \cdot 6}{5} = \frac{5 \cdot 42 \cdot 6}{5 \cdot 1} = \frac{5}{5} \cdot \frac{42 \cdot 6}{1} = 252.$$

4. **Check.** We check by repeating the calculation.
5. **State.** The total trip was 252 mi.

Do Exercise 14.

Exercise Set 2.7

a Find the reciprocal.

1. $\frac{5}{6}$
2. $\frac{7}{8}$
3. 6
4. 4
5. $\frac{1}{6}$
6. $\frac{1}{4}$
7. $\frac{10}{3}$
8. $\frac{17}{4}$

b Divide and simplify.

Don't forget to simplify!

9. $\frac{3}{5} \div \frac{3}{4}$
10. $\frac{2}{3} \div \frac{3}{4}$
11. $\frac{3}{5} \div \frac{9}{4}$
12. $\frac{6}{7} \div \frac{3}{5}$
13. $\frac{4}{3} \div \frac{1}{3}$
14. $\frac{10}{9} \div \frac{1}{3}$
15. $\frac{1}{3} \div \frac{1}{6}$
16. $\frac{1}{4} \div \frac{1}{5}$
17. $\frac{3}{8} \div 3$
18. $\frac{5}{6} \div 5$
19. $\frac{12}{7} \div 4$
20. $\frac{18}{5} \div 2$
21. $12 \div \frac{3}{2}$
22. $24 \div \frac{3}{8}$
23. $28 \div \frac{4}{5}$
24. $40 \div \frac{2}{3}$
25. $\frac{5}{8} \div \frac{5}{8}$
26. $\frac{2}{5} \div \frac{2}{5}$
27. $\frac{8}{15} \div \frac{4}{5}$
28. $\frac{6}{13} \div \frac{3}{26}$
29. $\frac{9}{5} \div \frac{4}{5}$
30. $\frac{5}{12} \div \frac{25}{36}$
31. $120 \div \frac{5}{6}$
32. $360 \div \frac{8}{7}$

c Solve.

33. $\frac{4}{5} \cdot x = 60$
34. $\frac{3}{2} \cdot t = 90$
35. $\frac{5}{3} \cdot y = \frac{10}{3}$
36. $\frac{4}{9} \cdot m = \frac{8}{3}$
37. $x \cdot \frac{25}{36} = \frac{5}{12}$
38. $p \cdot \frac{4}{5} = \frac{8}{15}$
39. $n \cdot \frac{8}{7} = 360$
40. $y \cdot \frac{5}{6} = 120$

d Solve.

41. Benny uses $\frac{2}{5}$ gram (g) of toothpaste each time he brushes his teeth. If Benny buys a 30-g tube, how many times will he be able to brush his teeth?

42. A piece of coaxial cable $\frac{4}{5}$ meter (m) long is to be cut into 8 pieces of the same length. What is the length of each piece?

43. A pair of basketball shorts requires $\frac{3}{4}$ yd of nylon. How many pairs of shorts can be made from 24 yd of nylon?

44. A child's baseball shirt requires $\frac{5}{6}$ yd of a certain fabric. How many shirts can be made from 25 yd of the fabric?

45. How many $\frac{2}{3}$-cup sugar bowls can be filled from 16 cups of sugar?

46. How many $\frac{2}{3}$-cup cereal bowls can be filled from 10 cups of cornflakes?

47. A bucket had 12 L of water in it when it was $\frac{3}{4}$ full. How much could it hold altogether?

48. A tank had 20 L of gasoline in it when it was $\frac{4}{5}$ full. How much could it hold altogether?

49. After driving 180 kilometers (km), $\frac{5}{8}$ of a trip is completed. How long is the total trip? How many kilometers are left to drive?

50. A road crew repaves $\frac{1}{12}$ mi of road each day. How long will it take the crew to repave a $\frac{3}{4}$-mi stretch of road?

Skill Maintenance

Divide. [1.6c]

51. $268 \div 4$

52. $268 \div 8$

53. $6842 \div 24$

54. $8765 \div 85$

Solve. [1.7b]

55. $4 \cdot x = 268$

56. $4 + x = 268$

57. $y + 502 = 9001$

58. $56 \cdot 78 = T$

Synthesis

59. ◆ Without performing the division, explain why $5 \div \frac{1}{7}$ is a greater number than $5 \div \frac{2}{3}$.

60. ◆ A student incorrectly insists that $\frac{2}{5} \div \frac{3}{4}$ is $\frac{15}{8}$. What mistake is the student probably making?

Simplify. Use the list of prime numbers on p. 88.

61. 🖩 $\frac{711}{1957} \div \frac{10{,}033}{13{,}081}$

62. 🖩 $\frac{8633}{7387} \div \frac{485}{581}$

63. $\left(\frac{9}{10} \div \frac{2}{5} \div \frac{3}{8}\right)^2$

64. $\dfrac{\left(\frac{3}{7}\right)^2 \div \frac{12}{5}}{\left(\frac{2}{9}\right)\left(\frac{9}{2}\right)}$

65. If $\frac{1}{3}$ of a number is $\frac{1}{4}$, what is $\frac{1}{2}$ of the number?

Summary and Review Exercises: Chapter 2

Beginning with this chapter, certain objectives, from four particular sections of preceding chapters, will be retested on the chapter test. The objectives to be tested in addition to the material in this chapter are [1.3d], [1.6c], [1.7b], and [1.8a].

Find the prime factorization of the number. [2.1d]

1. 70

2. 30

3. 45

4. 150

Determine whether: [2.2a]

5. 2432 is divisible by 6.

6. 182 is divisible by 4.

7. 4344 is divisible by 9.

8. 4344 is divisible by 8.

9. Determine whether 37 is prime, composite, or neither. [2.1c]

10. Identify the numerator and the denominator of $\frac{2}{7}$. [2.3a]

11. What fractional part is shaded? [2.3b]

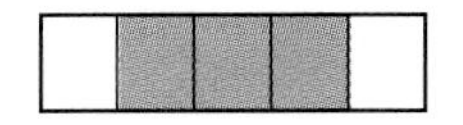

12. Simplify, if possible, the fractions on this circle graph. [2.5b]

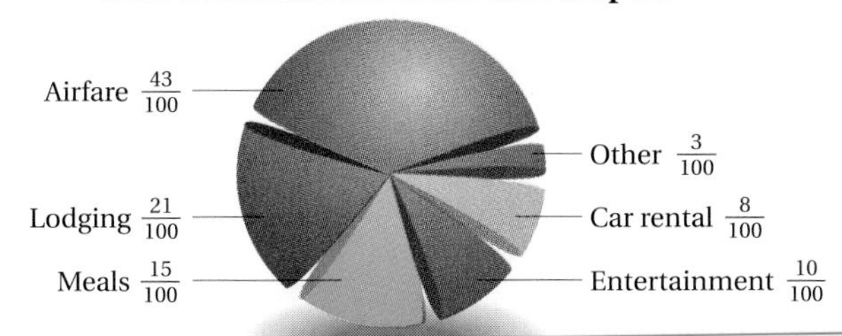

Simplify. [2.3c], [2.5b]

13. $\frac{0}{4}$

14. $\frac{23}{23}$

15. $\frac{48}{1}$

16. $\frac{48}{8}$

17. $\frac{10}{15}$

18. $\frac{7}{28}$

19. $\frac{21}{21}$

20. $\frac{0}{25}$

21. $\frac{12}{30}$

22. $\frac{18}{1}$

23. $\frac{32}{8}$

24. $\frac{9}{27}$

25. $\frac{18}{0}$

26. $\frac{5}{8-8}$

Use $=$ or $\neq$ for ▢ to write a true sentence. [2.5c]

27. $\frac{3}{5}$ ▢ $\frac{4}{6}$

28. $\frac{4}{7}$ ▢ $\frac{8}{14}$

29. $\frac{4}{5}$ ▢ $\frac{5}{6}$

30. $\frac{4}{3}$ ▢ $\frac{28}{21}$

Multiply and simplify. [2.6a]

31. $4 \cdot \frac{3}{8}$

32. $\frac{7}{3} \cdot 24$

33. $9 \cdot \frac{5}{18}$

34. $\frac{6}{5} \cdot 20$

35. $\frac{3}{4} \cdot \frac{8}{9}$

36. $\frac{5}{7} \cdot \frac{1}{10}$

37. $\frac{3}{7} \cdot \frac{14}{9}$

38. $\frac{1}{4} \cdot \frac{2}{11}$

Find the reciprocal. [2.7a]

39. $\frac{4}{5}$

40. 3

41. $\frac{1}{9}$

42. $\frac{47}{36}$

Divide and simplify. [2.7b]

43. $6 \div \frac{4}{3}$

44. $\frac{5}{9} \div \frac{5}{18}$

45. $\frac{1}{6} \div \frac{1}{11}$

46. $\frac{3}{14} \div \frac{6}{7}$

47. $\frac{1}{4} \div \frac{1}{9}$

48. $180 \div \frac{3}{5}$

49. $\frac{23}{25} \div \frac{23}{25}$

50. $\frac{2}{3} \div \frac{3}{2}$

Solve. [2.7c]

51. $\frac{5}{4} \cdot t = \frac{3}{8}$

52. $x \cdot \frac{2}{3} = 160$

Solve. [2.6b], [2.7d]

53. Liz and her road crew paint the lines in the middle and on the sides of a highway. They average about $\frac{5}{16}$ of a mile each day. How long will it take to paint the lines on 70 mi of highway?

54. After driving 60 km, $\frac{3}{8}$ of a vacation is complete. How long is the total trip?

55. A recipe calls for $\frac{2}{3}$ cup of diced bell peppers. In making $\frac{1}{2}$ of this recipe, how much diced pepper should be used?

56. Bernardo usually earns \$42 for working a full day. How much does he receive for working $\frac{1}{7}$ of a day?

Skill Maintenance

Solve. [1.7b]

57. $17 \cdot x = 408$

58. $765 + t = 1234$

Solve. [1.8a]

59. The balance in your checking account is \$789. After writing checks for \$78, \$97, and \$102 and making a deposit of \$400, what is your new balance?

60. An economy car gets 43 mpg on the highway. How far can the car be driven on a full tank of 18 gal of gasoline?

61. Divide: [1.6c]

$$36\overline{)14{,}697}$$

62. Subtract: [1.3d]

$$\begin{array}{r} 5604 \\ -\,1997 \\ \hline \end{array}$$

Synthesis

63. ◆ Write, in your own words, a series of steps that can be used when simplifying fractional notation. [2.5b]

64. ◆ A student claims that "taking $\frac{1}{2}$ of a number is the same as dividing by $\frac{1}{2}$." Explain the error in this reasoning. [2.7b]

65. 🖩 In the division below, find a and b. [2.7b]

$$\frac{19}{24} \div \frac{a}{b} = \frac{187{,}853}{268{,}224}$$

66. A prime number that becomes a prime number when its digits are reversed is called a **palindrome prime**. For example, 17 is a palindrome prime because both 17 and 71 are primes. Which of the following numbers are palindrome primes? [2.1c]

13, 91, 16, 11, 15, 24, 29, 101, 201

Test: Chapter 2

Find the prime factorization of the number.

1. 18

2. 60

3. Determine whether 1784 is divisible by 8.

4. Determine whether 784 is divisible by 9.

5. Identify the numerator and the denominator of $\frac{4}{9}$.

6. What part is shaded?

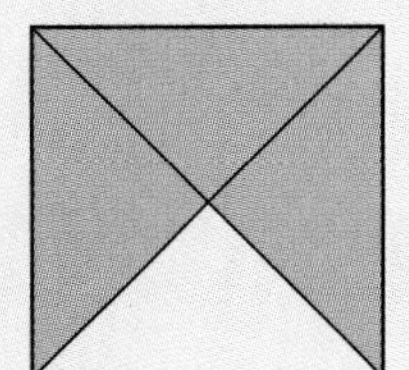

Simplify.

7. $\frac{26}{1}$

8. $\frac{12}{12}$

9. $\frac{0}{16}$

10. $\frac{12}{24}$

11. $\frac{42}{7}$

12. $\frac{2}{28}$

13. $\frac{9}{0}$

14. $\frac{7}{2-2}$

Use $=$ or $\neq$ for ▒ to write a true sentence.

15. $\frac{3}{4}$ ▒ $\frac{6}{8}$

16. $\frac{5}{4}$ ▒ $\frac{9}{7}$

Multiply and simplify.

17. $\frac{4}{3} \cdot 24$

18. $5 \cdot \frac{3}{10}$

19. $\frac{2}{3} \cdot \frac{15}{4}$

20. $\frac{3}{5} \cdot \frac{1}{6}$

Find the reciprocal.

21. $\frac{5}{8}$

22. $\frac{1}{4}$

23. 18

Answers

1. ______
2. ______
3. ______
4. ______
5. ______
6. ______
7. ______
8. ______
9. ______
10. ______
11. ______
12. ______
13. ______
14. ______
15. ______
16. ______
17. ______
18. ______
19. ______
20. ______
21. ______
22. ______
23. ______

Answers

24. ______

25. ______

26. ______

27. ______

28. ______

29. ______

30. ______

31. ______

32. ______

33. ______

34. ______

35. ______

36. ______

37. ______

38. ______

39. ______

Divide and simplify.

24. $\frac{3}{8} \div \frac{5}{4}$

25. $\frac{1}{5} \div \frac{1}{8}$

26. $12 \div \frac{2}{3}$

Solve.

27. $\frac{7}{8} \cdot x = 56$

28. $\frac{2}{5} \cdot t = \frac{7}{10}$

Solve.

29. It takes $\frac{7}{8}$ lb of salt to use in the ice of one batch of homemade ice cream. How much salt is required for 32 batches?

30. A strip of taffy $\frac{9}{10}$ m long is cut into 12 equal pieces. What is the length of each piece?

Skill Maintenance

Solve.

31. $x + 198 = 2003$

32. $47 \cdot t = 4747$

33. It is 2060 mi from San Francisco to Winnipeg, Canada. It is 1575 mi from Winnipeg to Atlanta. What is the total length of a route from San Francisco to Winnipeg to Atlanta?

34. Divide: $24\overline{)9127}$

35. Subtract: $\begin{array}{r} 8001 \\ -3567 \\ \hline \end{array}$

Synthesis

36. A recipe for a batch of buttermilk pancakes calls for $\frac{3}{4}$ teaspoon (tsp) of salt. Jacqueline plans to cut the amount of salt in half for each of 5 batches of pancakes. How much salt will she need?

37. Grandma Phyllis left $\frac{2}{3}$ of her $\frac{7}{8}$-acre tree farm to Karl. Karl gave $\frac{1}{4}$ of his share to his oldest daughter, Irene. How much land did Irene receive?

38. Simplify: $\left(\frac{3}{8}\right)^2 \div \frac{6}{7} \cdot \frac{2}{9} \div 5.$

39. Solve: $\frac{33}{38} \cdot \frac{34}{55} = \frac{17}{35} \cdot \frac{15}{19}x.$

Cumulative Review: Chapters 1–2

1. Write standard notation for the number in the following sentence:

   The earth travels five hundred eighty-four million, seventeen thousand, eight hundred miles around the sun.

2. Write a word name: 5,380,621.

3. In the number 2,751,043, which digit tells the number of hundreds?

Add.

4. $\begin{array}{r} 14{,}862 \\ +\ \ 2{,}935 \\ \hline \end{array}$

5. $\begin{array}{r} 7989 \\ 798 \\ +\ \ \ 79 \\ \hline \end{array}$

Subtract.

6. $\begin{array}{r} 5376 \\ -\ \ 430 \\ \hline \end{array}$

7. $\begin{array}{r} 2004 \\ -\ \ 579 \\ \hline \end{array}$

Multiply and simplify.

8. $\begin{array}{r} 621 \\ \times\ \ 27 \\ \hline \end{array}$

9. $\begin{array}{r} 2505 \\ \times 3300 \\ \hline \end{array}$

10. $5 \times \frac{3}{100}$

11. $\frac{4}{9} \cdot \frac{3}{8}$

Divide and simplify.

12. $19 \overline{)4580}$

13. $62 \overline{)3844}$

14. $\frac{3}{10} \div 5$

15. $\frac{8}{9} \div \frac{15}{6}$

16. Round 427,931 to the nearest thousand.

17. Round 5309 to the nearest hundred.

Estimate the sum or product by rounding to the nearest hundred. Show your work.

18. $\begin{array}{r} 749{,}559 \\ +\,301{,}362 \\ \hline \end{array}$

19. $\begin{array}{r} 749 \\ \times 531 \\ \hline \end{array}$

20. Use $<$ or $>$ for ▒ to write a true sentence:

    26 ▒ 17.

21. Use $=$ or $\neq$ for ▒ to write a true sentence:

    $\frac{7}{10}$ ▒ $\frac{5}{7}$.

22. Evaluate: 3^4.

Simplify.

23. $35 - 25 \div 5 + 2 \times 3$

24. $\{17 - [8 - (5 - 2 \times 2)]\} \div (3 + 12 \div 6)$

25. Find all the factors of 28.

26. Find the prime factorization of 28.

27. Determine whether 39 is prime, composite, or neither.

28. Determine whether 32,712 is divisible by 3.

29. Determine whether 32,712 is divisible by 5.

Simplify.

30. $\frac{35}{1}$

31. $\frac{77}{11}$

32. $\frac{28}{98}$

33. $\frac{0}{47}$

Solve.

34. $x + 13 = 50$

35. $\frac{1}{5} \cdot t = \frac{3}{10}$

36. $13 \cdot y = 39$

37. $384 \div 16 = n$

Solve.

38. *Baseball Salaries.* The average salary of a major-league baseball player was $1,176,967 in 1996 and $512,804 in 1989 (***Source***: Major League Baseball). How much more was the average salary in 1996 than in 1989?

39. *Hotel Rooms in Las Vegas.* Four of the largest hotels in the United States are in Las Vegas. One has 3174 rooms, the second has 2920 rooms, the third 2832 rooms, and the fourth 2793 rooms. What is the total number of rooms in these four hotels?

40. A student is offered a part-time job paying $3900 a year. How much is each weekly paycheck?

41. One $\frac{3}{4}$-cup serving of macaroni and cheese contains 290 calories. A box makes 4 servings. How many cups of macaroni and cheese does the box make?

42. It takes 6 hr to paint the trim on a certain house. If the painter can work only $\frac{3}{4}$ hr per day, how many days will it take to finish the job?

43. Eastside Appliance sells a refrigerator for $600 and $30 tax with no delivery charge. Westside Appliance sells the same model for $560 and $28 tax plus a $25 delivery charge. Which is the better buy?

Synthesis

44. A student works 35 hr a week and earns $8 an hour. The employer withholds $\frac{1}{4}$ of the total salary for taxes. Room and board expenses are $85 a week and tuition is $800 for a 16-week semester. Books cost $250 for the semester. Will the student make enough to cover the listed expenses? If so, how much will be left over at the end of the semester?

45. A can of mixed nuts is 1 part cashews, 1 part almonds, 1 part pecans, and 3 parts peanuts. What part of the mixture is peanuts?

3

Fractional Notation and Mixed Numerals

An Application

The tape in an audio cassette is played at a rate of $1\frac{7}{8}$ in. per second. A defective tape player has destroyed 30 in. of tape. How many seconds of music have been lost?

This problem appears as Example 9 in Section 3.6.

The Mathematics

We let $t =$ the number of seconds of music lost. The problem then translates to the equation

$$\underbrace{t = 30 \div 1\frac{7}{8}}.$$

Division using mixed numerals occurs often in applications and problem solving.

World Wide Web For more information, visit us at www.mathmax.com

Introduction

In this chapter, we consider addition and subtraction using fractional notation. Also discussed are addition, subtraction, multiplication, and division using mixed numerals. We then apply all these operations to rules for order of operations, estimating, and applied problems.

Pretest: Chapter 3

1. Find the LCM of 15 and 24.

2. Use < or > for ■ to write a true sentence:

$$\frac{7}{9} \; ■ \; \frac{4}{5}.$$

3. Convert to fractional notation: $7\frac{5}{8}$.

4. Convert to a mixed numeral: $\frac{11}{2}$.

5. Divide. Write a mixed numeral for the answer.

$$12 \overline{)\,4789}$$

6. Add. Write a mixed numeral for the answer.

$$\begin{array}{r} 8\frac{11}{12} \\ +\,2\frac{3}{5} \\ \hline \end{array}$$

7. Subtract. Write a mixed numeral for the answer.

$$\begin{array}{r} 14 \\ -\;7\frac{5}{6} \\ \hline \end{array}$$

8. Multiply. Write a mixed numeral for the answer.

$$3 \cdot 4\frac{8}{15}$$

9. Multiply. Write a mixed numeral for the answer.

$$6\frac{2}{3} \cdot 3\frac{1}{4}$$

10. Divide. Write a mixed numeral for the answer.

$$35 \div 5\frac{5}{6}$$

11. Divide. Write a mixed numeral for the answer.

$$5\frac{5}{12} \div 3\frac{1}{4}$$

12. Solve:

$$\frac{2}{3} + x = \frac{8}{9}.$$

13. At a summer camp, the cook bought 100 lb of potatoes and used $78\frac{3}{4}$ lb. How many pounds were left?

14. *Weight of Water.* The weight of water is $62\frac{1}{2}$ lb per cubic foot. How many cubic feet would be occupied by $265\frac{5}{8}$ lb of water?

15. A courier drove $214\frac{3}{10}$ mi one day and $136\frac{9}{10}$ mi the next. How far did she travel in all?

16. A cake recipe calls for $3\frac{3}{4}$ cups of flour. How much flour would be used to make 6 cakes?

Estimate each of the following as 0, $\frac{1}{2}$, or 1.

17. $\frac{29}{30}$

18. $\frac{2}{41}$

19. $\frac{15}{29}$

Estimate each of the following as a whole number or as a mixed numeral where the fractional part is $\frac{1}{2}$.

20. $10\frac{2}{17}$

21. $\frac{1}{10} + \frac{7}{8} + \frac{41}{39}$

22. $33\frac{14}{15} + 28\frac{3}{4} - 4\frac{25}{28} \div \frac{75}{76}$

Objectives for Retesting

The objectives to be tested in addition to the material in this chapter are as follows.

[1.5b] Multiply whole numbers.
[1.8a] Solve applied problems involving addition, subtraction, multiplication, or division of whole numbers.
[2.6a] Multiply and simplify using fractional notation.
[2.7b] Divide and simplify using fractional notation.

3.1 Least Common Multiples

In this chapter, we study addition and subtraction using fractional notation. Suppose we want to add $\frac{2}{3}$ and $\frac{1}{2}$. To do so, we find the least common multiple of the denominators: $\frac{2}{3} + \frac{1}{2} = \frac{4}{6} + \frac{3}{6}$. Then we add the numerators and keep the common denominator, 6. Before we do this, though, we study finding the **least common denominator (LCD),** or **least common multiple (LCM),** of the denominators.

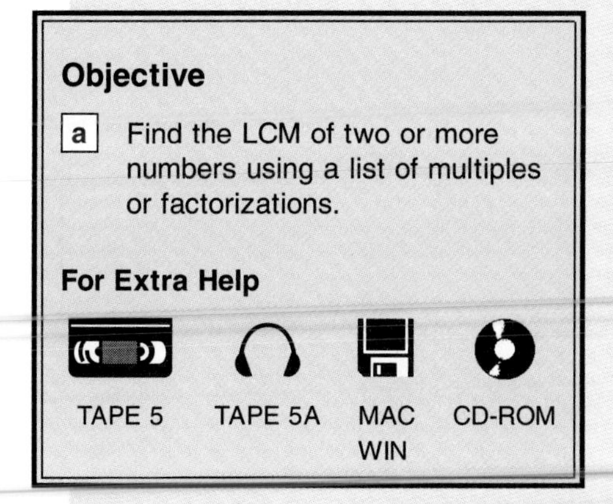

a Finding Least Common Multiples

The **least common multiple,** or LCM, of two natural numbers is the smallest number that is a multiple of both.

Example 1 Find the LCM of 20 and 30.

a) First list some multiples of 20 by multiplying 20 by 1, 2, 3, and so on:

20, 40, 60, 80, 100, 120, 140, 160, 180, 200, 220, 240, ...

b) Then list some multiples of 30 by multiplying 30 by 1, 2, 3, and so on:

30, 60, 90, 120, 150, 180, 210, 240, ...

c) Now list the numbers *common* to both lists, the common multiples:

60, 120, 180, 240, ...

d) These are the common multiples of 20 and 30. Which is the smallest? The LCM of 20 and 30 is 60.

Do Exercise 1.

1. By examining lists of multiples, find the LCM of 9 and 15.

Next we develop two methods that are more efficient for finding LCMs. You may choose to learn either method (consult with your instructor), or both, but if you are going on to a study of algebra, you should definitely learn method 2.

Method 1: Finding LCMs Using One List of Multiples

Method 1. To find the LCM of a set of numbers (9, 12):

a) Determine whether the largest number is a multiple of the others. If it is, it is the LCM. That is, if the largest number has the others as factors, the LCM is that number.

(12 is not a multiple of 9)

b) If not, check multiples of the largest number until you get one that is a multiple of the others.

($2 \cdot 12 = 24$, not a multiple of 9)

($3 \cdot 12 = 36$, a multiple of 9)

c) That number is the LCM.

LCM = 36

Answer on page A-6

2. By examining lists of multiples, find the LCM of 8 and 10.

Find the LCM.

3. 10, 15

4. 6, 8

Find the LCM.

5. 5, 10

6. 20, 40, 80

Answers on page A-6

Example 2 Find the LCM of 12 and 15.

a) 15 is not a multiple of 12.

b) Check multiples:

$2 \cdot 15 = 30,$ Not a multiple of 12

$3 \cdot 15 = 45,$ Not a multiple of 12

$4 \cdot 15 = 60.$ A multiple of 12

c) The LCM = 60.

Do Exercise 2.

Example 3 Find the LCM of 4 and 14.

a) 14 is not a multiple of 4.

b) Check multiples:

$2 \cdot 14 = 28.$ A multiple of 4

c) The LCM = 28.

Do Exercises 3 and 4.

Example 4 Find the LCM of 8 and 32.

a) 32 is a multiple of 8, so it is the LCM.

c) The LCM = 32.

Example 5 Find the LCM of 10, 100, and 1000.

a) 1000 is a multiple of 10 and 100, so it is the LCM.

c) The LCM = 1000.

Do Exercises 5 and 6.

Method 2: Finding LCMs Using Factorizations

A second method for finding LCMs uses prime factorizations. Consider again 20 and 30. Their prime factorizations are

$$20 = 2 \cdot 2 \cdot 5 \quad \text{and} \quad 30 = 2 \cdot 3 \cdot 5.$$

Let's look at these prime factorizations in order to find the LCM. Any multiple of 20 will have to have *two* 2's as factors and *one* 5 as a factor. Any multiple of 30 will have to have *one* 2, *one* 3, and *one* 5 as factors. The smallest number satisfying these conditions is

$$2 \cdot 2 \cdot 3 \cdot 5.$$

Two 2's, one 5; makes 20 a factor (first 2, second 2, and 5)

One 2, one 3, one 5; makes 30 a factor (second 2, 3, and 5)

The LCM must have all the factors of 20 and all the factors of 30, but the factors need not be repeated when they are common to both numbers.

The greatest number of times a 2 occurs as a factor of either 20 or 30 is two, and the LCM has 2 as a factor twice. The greatest number of times a 3 occurs as a factor of either 20 or 30 is one, and the LCM has 3 as a factor once. The greatest number of times that 5 occurs as a factor of either 20 or 30 is one, and the LCM has 5 as a factor once. The LCM is the product $2 \cdot 2 \cdot 3 \cdot 5$, or 60.

Method 2. To find the LCM of a set of numbers using prime factorizations:

a) Find the prime factorization of each number.

b) Create a product of factors, using each factor the greatest number of times that it occurs in any one factorization.

Example 6 Find the LCM of 6 and 8.

a) Find the prime factorization of each number.

$$6 = 2 \cdot 3, \qquad 8 = 2 \cdot 2 \cdot 2$$

b) Create a product by writing factors, using each the greatest number of times that it occurs in any one factorization.

Consider the factor 2. The greatest number of times that 2 occurs in any one factorization is three. We write 2 as a factor three times.

$$2 \cdot 2 \cdot 2 \cdot ?$$

Consider the factor 3. The greatest number of times that 3 occurs in any one factorization is one. We write 3 as a factor one time.

$$2 \cdot 2 \cdot 2 \cdot 3 \cdot ?$$

Since there are no other prime factors in either factorization, the

LCM is $2 \cdot 2 \cdot 2 \cdot 3$, or 24.

Example 7 Find the LCM of 24 and 36.

a) Find the prime factorization of each number.

$$24 = 2 \cdot 2 \cdot 2 \cdot 3, \qquad 36 = 2 \cdot 2 \cdot 3 \cdot 3$$

b) Create a product by writing factors, using each the greatest number of times that it occurs in any one factorization.

Consider the factor 2. The greatest number of times that 2 occurs in any one factorization is three. We write 2 as a factor three times:

$$2 \cdot 2 \cdot 2 \cdot ?$$

Consider the factor 3. The greatest number of times that 3 occurs in any one factorization is two. We write 3 as a factor two times:

$$2 \cdot 2 \cdot 2 \cdot 3 \cdot 3 \cdot ?$$

Since there are no other prime factors in either factorization, the

LCM is $2 \cdot 2 \cdot 2 \cdot 3 \cdot 3$, or 72.

Do Exercises 7–9.

Use prime factorizations to find the LCM.

7. 8, 10

8. 18, 40

9. 32, 54

Answers on page A-6

10. Find the LCM of 24, 35, and 45.

Answer on page A-6

Example 8 Find the LCM of 27, 90, and 84.

a) Find the prime factorization of each number.

$27 = 3 \cdot 3 \cdot 3, \qquad 90 = 2 \cdot 3 \cdot 3 \cdot 5, \qquad 84 = 2 \cdot 2 \cdot 3 \cdot 7$

b) Create a product by writing factors, using each the greatest number of times that it occurs in any one factorization.

Consider the factor 2. The greatest number of times that 2 occurs in any one factorization is two. We write 2 as a factor two times:

$2 \cdot 2 \cdot ?$

Consider the factor 3. The greatest number of times that 3 occurs in any one factorization is three. We write 3 as a factor three times:

$2 \cdot 2 \cdot 3 \cdot 3 \cdot 3 \cdot ?$

Consider the factor 5. The greatest number of times that 5 occurs in any one factorization is one. We write 5 as a factor one time:

$2 \cdot 2 \cdot 3 \cdot 3 \cdot 3 \cdot 5 \cdot ?$

Consider the factor 7. The greatest number of times that 7 occurs in any one factorization is one. We write 7 as a factor one time:

$2 \cdot 2 \cdot 3 \cdot 3 \cdot 3 \cdot 5 \cdot 7 \cdot ?$

Since no other prime factors are possible in any of the factorizations, the

LCM is $2 \cdot 2 \cdot 3 \cdot 3 \cdot 3 \cdot 5 \cdot 7$, or 3780.

The use of exponents might be helpful to you as an extension of the factorization method. Let's reconsider Example 8. We want to find the LCM of 27, 90, and 84. We factor and then convert to exponential notation:

$27 = 3 \cdot 3 \cdot 3 = 3^3,$

$90 = 2 \cdot 3 \cdot 3 \cdot 5 = 2^1 \cdot 3^2 \cdot 5^1,$ and

$84 = 2 \cdot 2 \cdot 3 \cdot 7 = 2^2 \cdot 3^1 \cdot 7^1.$

Thus the

LCM is $2^2 \cdot 3^3 \cdot 5^1 \cdot 7^1$, or 3780.

Note that in 84, the 2 in 2^2 is the largest exponent of 2 in any of the factorizations. It is also the exponent of 2 in the LCM. It indicates the greatest number of times that 2 occurs as a factor of any of the numbers. Similarly in 27, the 3 in 3^3 is the largest exponent of 3 in any of the factorizations. It is also the exponent of 3 in the LCM. Likewise, the 1's in 5^1 and 7^1 tell us the exponents of 5 and 7 in the LCM. They indicate the greatest number of times that 5 and 7 occur as factors.

Do Exercise 10.

Example 9 Find the LCM of 7 and 21.

a) Find the prime factorization of each number. Because 7 is prime, it has no prime factorization. We think of $7 = 7$ as a "factorization" in order to carry out our procedure.

$7 = 7, \qquad 21 = 3 \cdot 7$

b) Create a product by writing factors, using each the greatest number of times that it occurs in any one factorization.

Consider the factor 7. The greatest number of times that 7 occurs in any one factorization is one. We write 7 as a factor one time:

$7 \cdot ?$

Consider the factor 3. The greatest number of times that 3 occurs in any one factorization is one. We write 3 as a factor one time:

$7 \cdot 3 \cdot ?$

Since no other prime factors are possible in any of the factorizations, the

LCM is $7 \cdot 3$, or 21.

Note in Example 9 that 7 is a factor of 21. We stated earlier that if one number is a factor of another, the LCM is the larger of the numbers. Thus, if you note this at the outset, you can find the LCM quickly without using factorizations.

Do Exercises 11 and 12.

Example 10 Find the LCM of 8 and 9.

a) Find the prime factorization of each number.

$8 = 2 \cdot 2 \cdot 2, \qquad 9 = 3 \cdot 3$

b) Create a product by writing factors, using each the greatest number of times that it occurs in any one factorization.

Consider the factor 2. The greatest number of times that 2 occurs in any one factorization is three. We write 2 as a factor three times.

$2 \cdot 2 \cdot 2 \cdot ?$

Consider the factor 3. The greatest number of times that 3 occurs in any one factorization is two. We write 3 as a factor two times.

$2 \cdot 2 \cdot 2 \cdot 3 \cdot 3 \cdot ?$

Since no other prime factors are possible in any of the factorizations, the

LCM is $2 \cdot 2 \cdot 2 \cdot 3 \cdot 3$, or 72.

Note in Example 10 that the two numbers, 8 and 9, have no common prime factor. When this happens, the LCM is just the product of the two numbers. Thus, when you note this at the outset, you can find the LCM quickly by multiplying the two numbers.

Do Exercises 13 and 14.

Find the LCM.

11. 3, 18

12. 12, 24

Find the LCM.

13. 4, 9

14. 5, 6, 7

Answers on page A-6

Find the LCM using the optional method.

15. 24, 35, 45

16. 27, 90, 84

Let's compare the two methods considered for finding LCMs: the multiples method and the factorization method.

Method 1, the **multiples method,** can be longer than the factorization method when the LCM is large or when there are more than two numbers. But this method is faster and easier to use mentally for two numbers.

Method 2, the **factorization method,** works well for several numbers. It is just like a method used in algebra. If you are going to study algebra, you should definitely learn the factorization method.

Method 3: A Third Method for Finding LCMs (Optional)

Here is another method for finding LCMs that may work well for you. Suppose you want to find the LCM of 48, 72, and 80. If possible, find a prime number that divides any two of these numbers with no remainder. Do the division and bring the third number down, unless the third number is divisible by the prime also. Repeat the process until you can divide no more. Multiply, as shown at the right, all the numbers at the side by all the numbers at the bottom. The LCM is

2	48	72	80
3	24	36	40
2	8	12	40
2	4	6	20
2	2	3	10
	1	3	5

$$2 \cdot 3 \cdot 2 \cdot 2 \cdot 2 \cdot 1 \cdot 3 \cdot 5, \text{ or } 720.$$

Do Exercises 15 and 16.

Improving Your Math Study Skills

Better Test Taking

How often do you make the following statement after taking a test: "I was able to do the homework, but I froze during the test"? Instructors have heard this comment for years, and in most cases, it is merely a coverup for a lack of proper study habits. Here are two related tips, however, to help you with this difficulty. Both are intended to make test taking less stressful by getting you to practice good test-taking habits on a daily basis.

- **Treat *every* homework exercise as if it were a test question.** If you had to work a problem at your job with no backup answer provided, what would you do? You would probably work it very deliberately, checking and rechecking every step. You might work it more than one time, or you might try to work it another way to check the result. Try to use this approach when doing your homework. Treat every exercise as though it were a test question and no answer were provided at the back of the book.
- **Be sure that you do questions without answers as part of every homework assignment whether or not the instructor has assigned them!** One reason a test may seem such a different task is that questions on a test lack answers. That is the reason for taking a test: to see if you can do the questions without assistance. As part of your test preparation, be sure you do some exercises for which you do not have the answers. Thus when you take a test, you are doing a more familiar task.

The purpose of doing your homework using these approaches is to give you more test-taking practice beforehand. Let's make a sports analogy here. At a basketball game, the players take lots of practice shots before the game. They play the first half, go to the locker room, and come out for the second half. What do they do before the second half, even though they have just played 20 minutes of basketball? They shoot baskets again! We suggest the same approach here. Create more and more situations in which you practice taking test questions by treating each homework exercise like a test question and by doing exercises for which you have no answers. Good luck! Please send me an e-mail (exponent@aol.com) and let me know how it works for you.

Answers on page A-6

Exercise Set 3.1

a Find the LCM of the set of numbers.

1. 2, 4

2. 3, 15

3. 10, 25

4. 10, 15

5. 20, 40

6. 8, 12

7. 18, 27

8. 9, 11

9. 30, 50

10. 24, 36

11. 30, 40

12. 21, 27

13. 18, 24

14. 12, 18

15. 60, 70

16. 35, 45

17. 16, 36

18. 18, 20

19. 32, 36

20. 36, 48

21. 2, 3, 5

22. 5, 18, 3

23. 3, 5, 7

24. 6, 12, 18

25. 24, 36, 12

26. 8, 16, 22

27. 5, 12, 15

28. 12, 18, 40

29. 9, 12, 6

30. 8, 16, 12

31. 180, 100, 450

32. 18, 30, 50, 48

33. 8, 48

34. 16, 32

35. 5, 50

36. 12, 72

37. 11, 13

38. 13, 14

39. 12, 35

40. 23, 25

41. 54, 63

42. 56, 72

43. 81, 90

44. 75, 100

Applications of LCMs: Planet Orbits. The earth, Jupiter, Saturn, and Uranus all revolve around the sun. The earth takes 1 yr, Jupiter 12 yr, Saturn 30 yr, and Uranus 84 yr to make a complete revolution. On a certain night, you look at those three distant planets and wonder how many years it will take before they have the same position again. (*Hint*: To find out, you find the LCM of 12, 30, and 84. It will be that number of years.)

45. How often will Jupiter and Saturn appear in the same direction in the night sky as seen from the earth?

46. How often will Jupiter, Saturn, and Uranus appear in the same direction in the night sky as seen from the earth?

Skill Maintenance

47. Joy uses $\frac{1}{2}$ yd of dental floss each day. How long will a 45-yd container of dental floss last for Joy? [2.7d]

48. A performing arts center was sold out for a musical. Its seats sell for $13 each. Total receipts were $3250. How many seats does this auditorium contain? [1.8a]

49. Multiply: $23 \cdot 345$. [1.5b]

50. Multiply and simplify: $\frac{4}{5} \cdot \frac{10}{12}$. [2.6a]

51. Divide and simplify: $\frac{4}{5} \div \frac{7}{10}$. [2.7b]

52. Subtract: $10{,}007 - 3068$. [1.3d]

Synthesis

53. ◆ Is the LCM of two prime numbers always their product? Why or why not?

54. ◆ Is the LCM of two numbers always at least twice as large as the larger of the two numbers? Why or why not?

Use a calculator and the multiples method to find the LCM of each pair of numbers.

55. 288, 324

56. 2700, 7800

57. Find the LCM of 27, 90, 84, 210, 108, and 50.

58. Find the LCM of 18, 21, 24, 36, 63, 56, and 20.

59. A pencil company uses two sizes of boxes, 5 in. by 6 in. and 5 in. by 8 in. These boxes are packed in bigger cartons for shipping. Find the width and the length of the smallest carton that will accommodate boxes of either size without any room left over. (Each carton can contain only one type of box and all boxes must point in the same direction.)

60. Consider 8 and 12. Determine whether each of the following is the LCM of 8 and 12. Tell why or why not.

a) $2 \cdot 2 \cdot 3 \cdot 3$
b) $2 \cdot 2 \cdot 3$
c) $2 \cdot 3 \cdot 3$
d) $2 \cdot 2 \cdot 2 \cdot 3$

Collaborative Learning Manual: Find the least common multiple of two or more numbers using shaped markers.

3.2 Addition and Applications

a Like Denominators

Addition using fractional notation corresponds to combining or putting like things together, just as addition with whole numbers does. For example,

We combine two sets, each of which consists of fractional parts of one object that are the same size.

This is the resulting set.

$$\frac{2}{8} \quad + \quad \frac{3}{8} \quad = \quad \frac{5}{8}$$

2 eighths + 3 eighths = 5 eighths,

or $\quad 2 \cdot \frac{1}{8} + 3 \cdot \frac{1}{8} = 5 \cdot \frac{1}{8}, \qquad$ or $\quad \frac{2}{8} + \frac{3}{8} = \frac{5}{8}$.

We see that to add when denominators are the same, we add the numerators, keep the denominator, and simplify, if possible.

Do Exercise 1.

> To add when denominators are the same,
>
> a) add the numerators,
>
> b) keep the denominator, and
>
> c) simplify, if possible.
>
> $$\frac{2}{6} + \frac{5}{6} = \frac{2+5}{6} = \frac{7}{6}$$

Examples Add and simplify.

1. $\frac{2}{4} + \frac{1}{4} = \frac{2+1}{4} = \frac{3}{4}$ **No simplifying is possible.**

2. $\frac{11}{6} + \frac{3}{6} = \frac{11+3}{6} = \frac{14}{6} = \frac{2 \cdot 7}{2 \cdot 3} = \frac{2}{2} \cdot \frac{7}{3} = 1 \cdot \frac{7}{3} = \frac{7}{3}$ **Here we simplified.**

3. $\frac{3}{12} + \frac{5}{12} = \frac{3+5}{12} = \frac{8}{12} = \frac{4 \cdot 2}{4 \cdot 3} = \frac{4}{4} \cdot \frac{2}{3} = 1 \cdot \frac{2}{3} = \frac{2}{3}$

Do Exercises 2–4.

b Addition Using the LCD: Different Denominators

What do we do when denominators are different? We try to find a common denominator. We can do this by multiplying by 1. Consider adding $\frac{1}{6}$ and $\frac{3}{4}$. There are several common denominators that can be obtained. Let's look at two possibilities.

Objectives

a Add using fractional notation when denominators are the same.

b Add using fractional notation when denominators are different, by multiplying by 1 to find the least common denominator.

c Solve applied problems involving addition with fractional notation.

For Extra Help

TAPE 5

TAPE 5A

MAC WIN

CD-ROM

1. Find $\frac{1}{5} + \frac{3}{5}$.

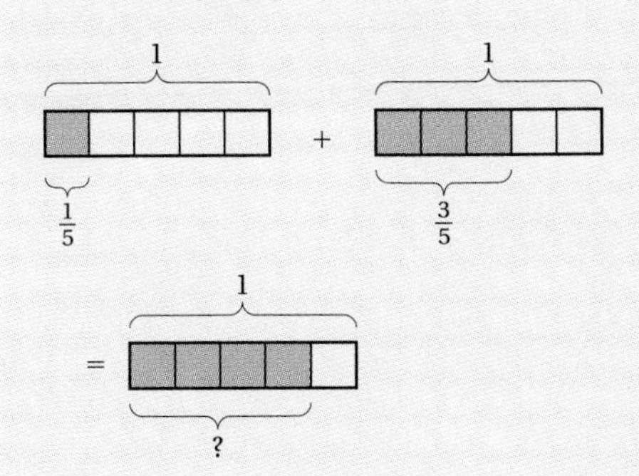

Add and simplify.

2. $\frac{1}{3} + \frac{2}{3}$

3. $\frac{5}{12} + \frac{1}{12}$

4. $\frac{9}{16} + \frac{3}{16}$

Answers on page A-6

5. Add. (Find the least common denominator.)

$$\frac{2}{3} + \frac{1}{6}$$

6. Add: $\frac{3}{8} + \frac{5}{6}$.

A. $\frac{1}{6} + \frac{3}{4} = \frac{1}{6} \cdot 1 + \frac{3}{4} \cdot 1$

$= \frac{1}{6} \cdot \frac{4}{4} + \frac{3}{4} \cdot \frac{6}{6}$

$= \frac{4}{24} + \frac{18}{24}$

$= \frac{22}{24}$

$= \frac{11}{12}$

B. $\frac{1}{6} + \frac{3}{4} = \frac{1}{6} \cdot 1 + \frac{3}{4} \cdot 1$

$= \frac{1}{6} \cdot \frac{2}{2} + \frac{3}{4} \cdot \frac{3}{3}$

$= \frac{2}{12} + \frac{9}{12}$

$= \frac{11}{12}$

We had to simplify in (A). We didn't have to simplify in (B). In (B), we used the least common multiple of the denominators, 12. That number is called the **least common denominator,** or **LCD.**

> To add when denominators are different:
>
> a) Find the least common multiple of the denominators. That number is the least common denominator, LCD.
>
> b) Multiply by 1, using an appropriate notation, n/n, to express each number in terms of the LCD.
>
> c) Add the numerators, keeping the same denominator.
>
> d) Simplify, if possible.

Example 4 Add: $\frac{3}{4} + \frac{1}{8}$.

The LCD is 8. **4 is a factor of 8 so the LCM of 4 and 8 is 8.**

$\frac{3}{4} + \frac{1}{8} = \frac{3}{4} \cdot 1 + \frac{1}{8}$ ← **This fraction already has the LCD as its denominator.**

$= \frac{3}{4} \cdot \frac{2}{2} + \frac{1}{8}$ — ***Think:*** **$4 \times \blacksquare = 8$. The answer is 2, so we multiply by 1, using $\frac{2}{2}$.**

$= \frac{6}{8} + \frac{1}{8}$

$= \frac{7}{8}$ **No simplification is necessary.**

Do Exercise 5.

Example 5 Add: $\frac{1}{9} + \frac{5}{6}$.

The LCD is 18. **$9 = 3 \cdot 3$ and $6 = 2 \cdot 3$, so the LCM of 9 and 6 is $2 \cdot 3 \cdot 3$, or 18.**

$\frac{1}{9} + \frac{5}{6} = \frac{1}{9} \cdot 1 + \frac{5}{6} \cdot 1 = \frac{1}{9} \cdot \frac{2}{2} + \frac{5}{6} \cdot \frac{3}{3}$

Think: **$6 \times \blacksquare = 18$. The answer is 3, so we multiply by 1 using $\frac{3}{3}$.**

Think: **$9 \times \blacksquare = 18$. The answer is 2, so we multiply by 1, using $\frac{2}{2}$.**

$= \frac{2}{18} + \frac{15}{18} = \frac{17}{18}$

Do Exercise 6.

Answers on page A-6

Example 6 Add: $\frac{5}{9} + \frac{11}{18}$.

The LCD is 18.

$$\frac{5}{9} + \frac{11}{18} = \frac{5}{9} \cdot \frac{2}{2} + \frac{11}{18}$$
$$= \frac{10}{18} + \frac{11}{18}$$
$$= \frac{21}{18}$$
$$= \frac{7}{6}$$

We may still have to simplify, but it is usually easier if we have used the LCD.

Do Exercise 7.

Example 7 Add: $\frac{1}{10} + \frac{3}{100} + \frac{7}{1000}$.

Since 10 and 100 are factors of 1000, the LCD is 1000. Then

$$\frac{1}{10} + \frac{3}{100} + \frac{7}{1000} = \frac{1}{10} \cdot \frac{100}{100} + \frac{3}{100} \cdot \frac{10}{10} + \frac{7}{1000}$$
$$= \frac{100}{1000} + \frac{30}{1000} + \frac{7}{1000} = \frac{137}{1000}.$$

Look back over this example. Try to think it out so that you can do it mentally.

Example 8 Add: $\frac{13}{70} + \frac{11}{21} + \frac{6}{15}$.

We have

$$\frac{13}{70} + \frac{11}{21} + \frac{6}{15} = \frac{13}{2 \cdot 5 \cdot 7} + \frac{11}{3 \cdot 7} + \frac{6}{3 \cdot 5}.$$ **Factoring denominators**

The LCD is $2 \cdot 3 \cdot 5 \cdot 7$, or 210. Then

$$\frac{13}{70} + \frac{11}{21} + \frac{6}{15} = \frac{13}{2 \cdot 5 \cdot 7} \cdot \frac{3}{3} + \frac{11}{3 \cdot 7} \cdot \frac{2 \cdot 5}{2 \cdot 5} + \frac{6}{3 \cdot 5} \cdot \frac{7 \cdot 2}{7 \cdot 2}$$
$$= \frac{13 \cdot 3}{2 \cdot 5 \cdot 7 \cdot 3} + \frac{11 \cdot 2 \cdot 5}{3 \cdot 7 \cdot 2 \cdot 5} + \frac{6 \cdot 7 \cdot 2}{3 \cdot 5 \cdot 7 \cdot 2}$$
$$= \frac{39}{3 \cdot 5 \cdot 7 \cdot 2} + \frac{110}{3 \cdot 5 \cdot 7 \cdot 2} + \frac{84}{3 \cdot 5 \cdot 7 \cdot 2}$$
$$= \frac{233}{3 \cdot 5 \cdot 7 \cdot 2}$$
$$= \frac{233}{210}.$$

We left 210 factored until we knew we could not simplify.

The LCD of 70, 21, and 15 is $2 \cdot 3 \cdot 5 \cdot 7$. In each case, we multiply by 1 to obtain the LCD.

Do Exercises 8–10.

7. Add: $\frac{1}{6} + \frac{7}{18}$.

Add.

8. $\frac{4}{10} + \frac{1}{100} + \frac{3}{1000}$

9. $\frac{7}{10} + \frac{5}{100} + \frac{9}{1000}$
(Try to do this one mentally.)

10. $\frac{7}{10} + \frac{2}{21} + \frac{1}{7}$

Answers on page A-6

11. A consumer bought $\frac{1}{2}$ lb of peanuts and $\frac{3}{5}$ lb of cashews. How many pounds of nuts were bought altogether?

Answer on page A-6

c Applications and Problem Solving

Example 9 A jogger ran $\frac{4}{5}$ mi, rested, and then ran another $\frac{1}{10}$ mi. How far did she run in all?

1. **Familiarize.** We first make a drawing. We let D = the distance run in all.

2. **Translate.** The problem can be translated to an equation as follows.

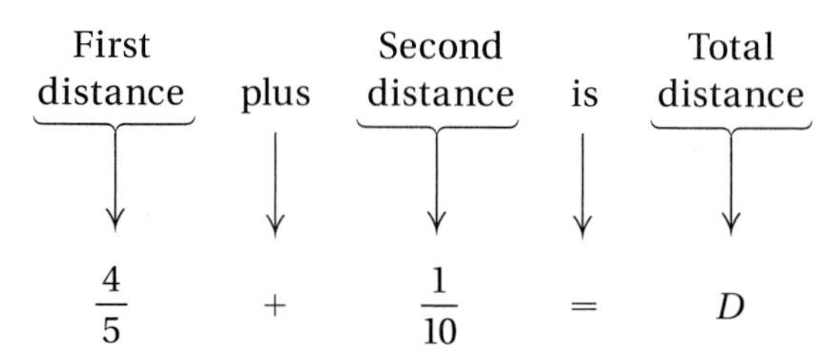

$$\frac{4}{5} + \frac{1}{10} = D$$

3. **Solve.** To solve the equation, we carry out the addition. The LCM of the denominators is 10 because 5 is a factor of 10. We multiply by 1 in order to obtain the LCD:

$$\frac{4}{5} \cdot \frac{2}{2} + \frac{1}{10} = D$$

$$\frac{8}{10} + \frac{1}{10} = D$$

$$\frac{9}{10} = D.$$

4. **Check.** We check by repeating the calculation. We also note that the sum should be larger than either of the individual distances, which it is. This gives us a partial check on the reasonableness of the answer.

5. **State.** In all, the jogger ran $\frac{9}{10}$ mi.

Do Exercise 11.

Exercise Set 3.2

a, **b** Add and simplify.

1. $\frac{7}{8} + \frac{1}{8}$

2. $\frac{2}{5} + \frac{3}{5}$

3. $\frac{1}{8} + \frac{5}{8}$

4. $\frac{3}{10} + \frac{3}{10}$

5. $\frac{2}{3} + \frac{5}{6}$

6. $\frac{5}{6} + \frac{1}{9}$

7. $\frac{1}{8} + \frac{1}{6}$

8. $\frac{1}{6} + \frac{3}{4}$

9. $\frac{4}{5} + \frac{7}{10}$

10. $\frac{3}{4} + \frac{1}{12}$

11. $\frac{5}{12} + \frac{3}{8}$

12. $\frac{7}{8} + \frac{1}{16}$

13. $\frac{3}{20} + \frac{3}{4}$

14. $\frac{2}{15} + \frac{2}{5}$

15. $\frac{5}{6} + \frac{7}{9}$

16. $\frac{5}{8} + \frac{5}{6}$

17. $\frac{3}{10} + \frac{1}{100}$

18. $\frac{9}{10} + \frac{3}{100}$

19. $\frac{5}{12} + \frac{4}{15}$

20. $\frac{3}{16} + \frac{1}{12}$

21. $\frac{9}{10} + \frac{99}{100}$

22. $\frac{3}{10} + \frac{27}{100}$

23. $\frac{7}{8} + \frac{0}{1}$

24. $\frac{0}{1} + \frac{5}{6}$

25. $\frac{3}{8} + \frac{1}{6}$

26. $\frac{5}{8} + \frac{1}{6}$

27. $\frac{5}{12} + \frac{7}{24}$

28. $\frac{1}{18} + \frac{7}{12}$

29. $\frac{3}{16} + \frac{5}{16} + \frac{4}{16}$

30. $\frac{3}{8} + \frac{1}{8} + \frac{2}{8}$

31. $\frac{8}{10} + \frac{7}{100} + \frac{4}{1000}$

32. $\frac{1}{10} + \frac{2}{100} + \frac{3}{1000}$

33. $\frac{3}{8} + \frac{5}{12} + \frac{8}{15}$

34. $\frac{1}{2} + \frac{3}{8} + \frac{1}{4}$

35. $\frac{15}{24} + \frac{7}{36} + \frac{91}{48}$

36. $\frac{5}{7} + \frac{25}{52} + \frac{7}{4}$

c Solve.

37. Rene bought $\frac{1}{3}$ lb of orange pekoe tea and $\frac{1}{2}$ lb of English cinnamon tea. How many pounds of tea did he buy?

38. Spike bought $\frac{1}{4}$ lb of gumdrops and $\frac{1}{2}$ lb of caramels. How many pounds of candy did he buy?

39. Russ walked $\frac{7}{6}$ mi to a friend's dormitory, and then $\frac{3}{4}$ mi to class. How far did he walk?

40. Elaine walked $\frac{7}{8}$ mi to the student union, and then $\frac{2}{5}$ mi to class. How far did she walk?

41. *Concrete Mix.* A cubic meter of concrete mix contains 420 kilograms (kg) of cement, 150 kg of stone, and 120 kg of sand. What is the total weight of the cubic meter of concrete mix? What part is cement? stone? sand? Add these amounts. What is the result?

42. *Punch Recipe.* A recipe for strawberry punch calls for $\frac{1}{5}$ quart (qt) of ginger ale and $\frac{3}{5}$ qt of strawberry soda. How much liquid is needed? If the recipe is doubled, how much liquid is needed? If the recipe is halved, how much liquid is needed?

43. A tile $\frac{5}{8}$ in. thick is glued to a board $\frac{7}{8}$ in. thick. The glue is $\frac{3}{32}$ in. thick. How thick is the result?

44. A baker used $\frac{1}{2}$ lb of flour for rolls, $\frac{1}{4}$ lb for donuts, and $\frac{1}{3}$ lb for cookies. How much flour was used?

Skill Maintenance

Multiply. [1.5b]

45. $408 \cdot 516$

46. $1125 \cdot 3728$

47. $423 \cdot 8009$

48. $2025 \cdot 174$

Solve.

49. A shopper has $3458 in a checking account and writes checks for $329 and $52. How much is left in the account? [1.8a]

50. What is the total cost of 5 sweaters at $89 each and 6 shirts at $49 each? [1.8a]

51. Elsa has $9 to spend on ride tickets at the fair. If the tickets cost 75¢, or $\$\frac{3}{4}$, each, how many tickets can she purchase? [2.7d]

52. The Bingham community garden is to be split into 16 equally sized plots. If the garden occupies $\frac{3}{4}$ acre of land, how large will each plot be? [2.7d]

Synthesis

53. ◆ Explain the role of multiplication when adding using fractional notation with different denominators.

54. ◆ To add numbers with different denominators, a student consistently uses the product of the denominators as a common denominator. Is this correct? Why or why not?

55. A guitarist's band is booked for Friday and Saturday nights at a local club. The guitarist is part of a trio on Friday and part of a quintet on Saturday. Thus the guitarist is paid one-third of one-half the weekend's pay for Friday and one-fifth of one-half the weekend's pay for Saturday. What fractional part of the band's pay did the guitarist receive for the weekend's work? If the band was paid $1200, how much did the guitarist receive?

3.3 Subtraction, Order, and Applications

a Subtraction

Like Denominators

We can consider the difference $\frac{4}{8} - \frac{3}{8}$ as we did before, as either "take away" or "how much more." Let's consider "take away."

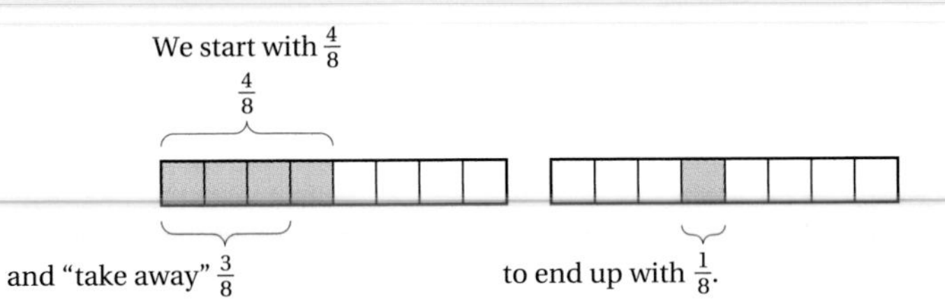

We start with 4 eighths and take away 3 eighths:

4 eighths − 3 eighths = 1 eighth,

or $\quad 4 \cdot \frac{1}{8} - 3 \cdot \frac{1}{8} = \frac{1}{8}, \qquad$ or $\quad \frac{4}{8} - \frac{3}{8} = \frac{1}{8}.$

To subtract when denominators are the same,

a) subtract the numerators,

b) keep the denominator, and

c) simplify, if possible.

$$\frac{7}{10} - \frac{4}{10} = \frac{7-4}{10} = \frac{3}{10}$$

Examples Subtract and simplify.

1. $\frac{7}{10} - \frac{3}{10} = \frac{7-3}{10} = \frac{4}{10} = \frac{2 \cdot 2}{5 \cdot 2} = \frac{2}{5} \cdot \frac{2}{2} = \frac{2}{5} \cdot 1 = \frac{2}{5}$

2. $\frac{8}{9} - \frac{2}{9} = \frac{8-2}{9} = \frac{6}{9} = \frac{2 \cdot 3}{3 \cdot 3} = \frac{2}{3} \cdot \frac{3}{3} = \frac{2}{3} \cdot 1 = \frac{2}{3}$

3. $\frac{32}{12} - \frac{25}{12} = \frac{32-25}{12} = \frac{7}{12}$

Do Exercises 1–3.

Different Denominators

To subtract when denominators are different:

a) Find the least common multiple of the denominators. That number is the least common denominator, LCD.

b) Multiply by 1, using an appropriate notation, n/n, to express each number in terms of the LCD.

c) Subtract the numerators, keeping the same denominator.

d) Simplify, if possible.

Objectives

a Subtract using fractional notation.

b Use < or > with fractional notation to write a true sentence.

c Solve equations of the type $x + a = b$ and $a + x = b$, where a and b may be fractions.

d Solve applied problems involving subtraction with fractional notation.

For Extra Help

TAPE 5 | TAPE 5B | MAC WIN | CD-ROM

Subtract and simplify.

1. $\frac{7}{8} - \frac{3}{8}$

2. $\frac{10}{16} - \frac{4}{16}$

3. $\frac{8}{10} - \frac{3}{10}$

Answers on page A-6

4. Subtract: $\frac{3}{4} - \frac{2}{3}$.

Example 4 Subtract: $\frac{2}{5} - \frac{3}{8}$.

The LCM of 5 and 8 is 40. The LCD is 40.

$$\frac{2}{5} - \frac{3}{8} = \frac{2}{5} \cdot \frac{8}{8} - \frac{3}{8} \cdot \frac{5}{5}$$

Think: 8 × ■ = 40. The answer is 5, so we multiply by 1, using $\frac{5}{5}$.
Think: 5 × ■ = 40. The answer is 8, so we multiply by 1, using $\frac{8}{8}$.

$$= \frac{16}{40} - \frac{15}{40}$$

$$= \frac{16 - 15}{40} = \frac{1}{40}$$

Do Exercise 4.

Subtract.

5. $\frac{5}{6} - \frac{1}{9}$

6. $\frac{4}{5} - \frac{3}{10}$

Example 5 Subtract: $\frac{5}{6} - \frac{7}{12}$.

Since 6 is a factor of 12, the LCM of 6 and 12 is 12. The LCD is 12.

$$\frac{5}{6} - \frac{7}{12} = \frac{5}{6} \cdot \frac{2}{2} - \frac{7}{12}$$

$$= \frac{10}{12} - \frac{7}{12}$$

$$= \frac{10 - 7}{12} = \frac{3}{12}$$

$$= \frac{3 \cdot 1}{3 \cdot 4} = \frac{3}{3} \cdot \frac{1}{4}$$

$$= \frac{1}{4}$$

Do Exercises 5 and 6.

7. Subtract: $\frac{11}{28} - \frac{5}{16}$.

Example 6 Subtract: $\frac{17}{24} - \frac{4}{15}$.

We have

$$\frac{17}{24} - \frac{4}{15} = \frac{17}{3 \cdot 2 \cdot 2 \cdot 2} - \frac{4}{5 \cdot 3}.$$

The LCD is $3 \cdot 2 \cdot 2 \cdot 2 \cdot 5$, or 120. Then

$$\frac{17}{24} - \frac{4}{15} = \frac{17}{3 \cdot 2 \cdot 2 \cdot 2} \cdot \frac{5}{5} - \frac{4}{5 \cdot 3} \cdot \frac{2 \cdot 2 \cdot 2}{2 \cdot 2 \cdot 2}$$

$$= \frac{17 \cdot 5}{3 \cdot 2 \cdot 2 \cdot 2 \cdot 5} - \frac{4 \cdot 2 \cdot 2 \cdot 2}{5 \cdot 3 \cdot 2 \cdot 2 \cdot 2}$$

$$= \frac{85}{120} - \frac{32}{120} = \frac{53}{120}.$$

The LCD of 24 and 15 is $2 \cdot 2 \cdot 2 \cdot 3 \cdot 5$. In each case, we multiply by 1 to obtain the LCD.

Do Exercise 7.

Answers on page A-6

b Order

We see from this figure that $\frac{4}{5} > \frac{3}{5}$. That is, $\frac{4}{5}$ is greater than $\frac{3}{5}$.

$\frac{4}{5}$ 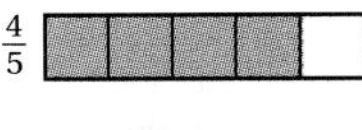

$\frac{3}{5}$ 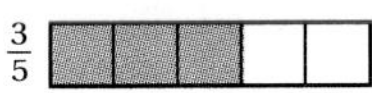

> To determine which of two numbers is greater when there is a common denominator, compare the numerators:
>
> $$\frac{4}{5}, \frac{3}{5}, \qquad 4 > 3 \qquad \frac{4}{5} > \frac{3}{5}.$$

Do Exercises 8 and 9.

When denominators are different, we cannot compare numerators. We multiply by 1 to make the denominators the same.

Example 7 Use < or > for ▒ to write a true sentence:

$$\frac{2}{5} \;▒\; \frac{3}{4}.$$

We have

$$\frac{2}{5} \cdot \frac{4}{4} = \frac{8}{20};$$ **We multiply by 1 using $\frac{4}{4}$ to get the LCD.**

$$\frac{3}{4} \cdot \frac{5}{5} = \frac{15}{20}.$$ **We multiply by 1 using $\frac{5}{5}$ to get the LCD.**

Now that the denominators are the same, 20, we can compare the numerators. Since $8 < 15$, it follows that $\frac{8}{20} < \frac{15}{20}$, so

$$\frac{2}{5} < \frac{3}{4}.$$

Example 8 Use < or > for ▒ to write a true sentence:

$$\frac{9}{10} \;▒\; \frac{89}{100}.$$

The LCD is 100.

$$\frac{9}{10} \cdot \frac{10}{10} = \frac{90}{100}$$ **We multiply by $\frac{10}{10}$ to get the LCD.**

Since $90 > 89$, it follows that $\frac{90}{100} > \frac{89}{100}$, so

$$\frac{9}{10} > \frac{89}{100}.$$

Do Exercises 10–12.

c Solving Equations

Now let's solve equations of the form $x + a = b$ or $a + x = b$, where a and b may be fractions. Proceeding as we have before, we subtract a on both sides of the equation.

8. Use < or > for ▒ to write a true sentence:

$$\frac{3}{8} \;▒\; \frac{5}{8}.$$

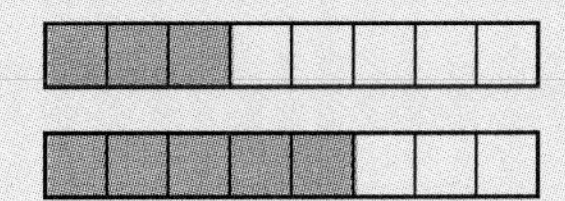

9. Use < or > for ▒ to write a true sentence:

$$\frac{7}{10} \;▒\; \frac{6}{10}.$$

Use < or > for ▒ to write a true sentence.

10. $\frac{2}{3} \;▒\; \frac{5}{8}$

11. $\frac{3}{4} \;▒\; \frac{8}{12}$

12. $\frac{5}{6} \;▒\; \frac{7}{8}$

Answers on page A-6

Solve.

13. $x + \frac{2}{3} = \frac{5}{6}$

14. $\frac{3}{5} + t = \frac{7}{8}$

15. There is $\frac{1}{4}$ cup of olive oil in a measuring cup. How much oil must be added to make a total of $\frac{4}{5}$ cup of oil in the measuring cup?

Answers on page A-6

Example 9 Solve: $x + \frac{1}{4} = \frac{3}{5}$.

$x + \frac{1}{4} - \frac{1}{4} = \frac{3}{5} - \frac{1}{4}$ — **Subtracting $\frac{1}{4}$ on both sides**

$x + 0 = \frac{3}{5} \cdot \frac{4}{4} - \frac{1}{4} \cdot \frac{5}{5}$ — **The LCD is 20. We multiply by 1 to get the LCD.**

$x = \frac{12}{20} - \frac{5}{20} = \frac{7}{20}$

Do Exercises 13 and 14.

d Applications and Problem Solving

Example 10 A jogger has run $\frac{2}{3}$ mi and will stop running when she has run $\frac{7}{8}$ mi. How much farther does the jogger have to go?

1. **Familiarize.** We first make a drawing or at least visualize the situation. We let d = the distance to go.

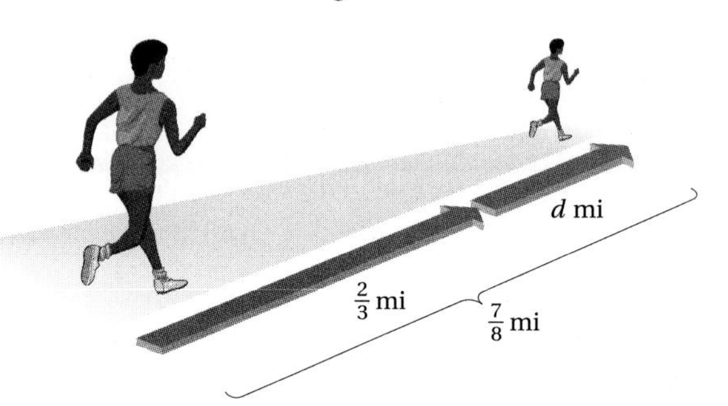

2. **Translate.** We see that this is a "how much more" situation. Now we translate to an equation.

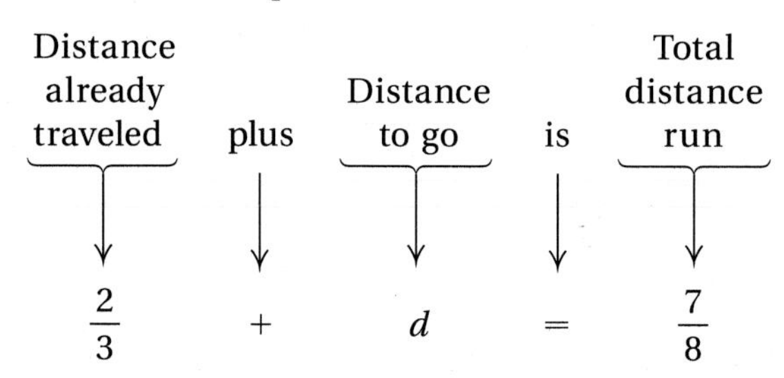

$$\frac{2}{3} + d = \frac{7}{8}$$

3. **Solve.** To solve the equation, we subtract $\frac{2}{3}$ on both sides:

$\frac{2}{3} + d - \frac{2}{3} = \frac{7}{8} - \frac{2}{3}$ — **Subtracting $\frac{2}{3}$ on both sides**

$d + 0 = \frac{7}{8} \cdot \frac{3}{3} - \frac{2}{3} \cdot \frac{8}{8}$ — **The LCD is 24. We multiply by 1 to obtain the LCD.**

$d = \frac{21}{24} - \frac{16}{24} = \frac{5}{24}$.

4. **Check.** To check, we return to the original problem and add:

$$\frac{2}{3} + \frac{5}{24} = \frac{2}{3} \cdot \frac{8}{8} + \frac{5}{24} = \frac{16}{24} + \frac{5}{24} = \frac{21}{24} = \frac{7}{8} \cdot \frac{3}{3} = \frac{7}{8}.$$

5. **State.** The jogger has $\frac{5}{24}$ mi to go.

Do Exercise 15.

Exercise Set 3.3

a Subtract and simplify.

1. $\frac{5}{6} - \frac{1}{6}$
2. $\frac{5}{8} - \frac{3}{8}$
3. $\frac{11}{12} - \frac{2}{12}$
4. $\frac{17}{18} - \frac{11}{18}$
5. $\frac{3}{4} - \frac{1}{8}$
6. $\frac{2}{3} - \frac{1}{9}$
7. $\frac{1}{8} - \frac{1}{12}$
8. $\frac{1}{6} - \frac{1}{8}$
9. $\frac{4}{3} - \frac{5}{6}$
10. $\frac{7}{8} - \frac{1}{16}$
11. $\frac{3}{4} - \frac{3}{28}$
12. $\frac{2}{5} - \frac{2}{15}$
13. $\frac{3}{4} - \frac{3}{20}$
14. $\frac{5}{6} - \frac{1}{2}$
15. $\frac{3}{4} - \frac{1}{20}$
16. $\frac{3}{4} - \frac{4}{16}$
17. $\frac{5}{12} - \frac{2}{15}$
18. $\frac{9}{10} - \frac{11}{16}$
19. $\frac{6}{10} - \frac{7}{100}$
20. $\frac{9}{10} - \frac{3}{100}$
21. $\frac{7}{15} - \frac{3}{25}$
22. $\frac{18}{25} - \frac{4}{35}$
23. $\frac{99}{100} - \frac{9}{10}$
24. $\frac{78}{100} - \frac{11}{20}$
25. $\frac{2}{3} - \frac{1}{8}$
26. $\frac{3}{4} - \frac{1}{2}$
27. $\frac{3}{5} - \frac{1}{2}$
28. $\frac{5}{6} - \frac{2}{3}$
29. $\frac{5}{12} - \frac{3}{8}$
30. $\frac{7}{12} - \frac{2}{9}$
31. $\frac{7}{8} - \frac{1}{16}$
32. $\frac{5}{12} - \frac{5}{16}$
33. $\frac{17}{25} - \frac{4}{15}$
34. $\frac{11}{18} - \frac{7}{24}$
35. $\frac{23}{25} - \frac{112}{150}$
36. $\frac{89}{90} - \frac{53}{120}$

b Use $<$ or $>$ for ▒ to write a true sentence.

37. $\frac{5}{8}$ ▒ $\frac{6}{8}$
38. $\frac{7}{9}$ ▒ $\frac{5}{9}$
39. $\frac{1}{3}$ ▒ $\frac{1}{4}$
40. $\frac{1}{8}$ ▒ $\frac{1}{6}$
41. $\frac{2}{3}$ ▒ $\frac{5}{7}$
42. $\frac{3}{5}$ ▒ $\frac{4}{7}$
43. $\frac{4}{5}$ ▒ $\frac{5}{6}$
44. $\frac{3}{2}$ ▒ $\frac{7}{5}$
45. $\frac{19}{20}$ ▒ $\frac{4}{5}$
46. $\frac{5}{6}$ ▒ $\frac{13}{16}$
47. $\frac{19}{20}$ ▒ $\frac{9}{10}$
48. $\frac{3}{4}$ ▒ $\frac{11}{15}$
49. $\frac{31}{21}$ ▒ $\frac{41}{13}$
50. $\frac{12}{7}$ ▒ $\frac{132}{49}$

c Solve.

51. $x + \frac{1}{30} = \frac{1}{10}$
52. $y + \frac{9}{12} = \frac{11}{12}$
53. $\frac{2}{3} + t = \frac{4}{5}$
54. $\frac{2}{3} + p = \frac{7}{8}$
55. $m + \frac{5}{6} = \frac{9}{10}$
56. $x + \frac{1}{3} = \frac{5}{6}$

c Solve.

57. Monica spent $\frac{3}{4}$ hr listening to tapes of Beethoven and Brahms. She spent $\frac{1}{3}$ hr listening to Beethoven. How many hours were spent listening to Brahms?

58. From a $\frac{4}{5}$-lb wheel of cheese, a $\frac{1}{4}$-lb piece was served. How much cheese remained on the wheel?

59. *Tire Tread.* A new long-life tire has a tread depth of $\frac{3}{8}$ in. instead of the more typical $\frac{11}{32}$ in. (***Source:*** *Popular Science*). How much deeper is the new tread depth?

60. As part of a fitness program, Deb swims $\frac{1}{2}$ mi every day. She has already swum $\frac{1}{5}$ mi. How much farther should Deb swim?

61. A hamburger franchise was owned by three people. One owned $\frac{7}{12}$ of the business and the second owned $\frac{1}{6}$. How much did the third person own?

62. An estate was left to four children. One received $\frac{1}{4}$ of the estate, the second $\frac{1}{16}$, and the third $\frac{3}{8}$. How much did the fourth receive?

Skill Maintenance

Divide and simplify. [2.7b]

63. $\frac{3}{7} \div \frac{9}{4}$

64. $\frac{9}{10} \div \frac{3}{5}$

65. $7 \div \frac{1}{3}$

66. $\frac{1}{4} \div 8$

Solve. [2.6b]

67. A small box of cornflakes weighs $\frac{3}{4}$ lb. How much does 8 small boxes of cornflakes weigh?

68. A batch of fudge requires $\frac{3}{4}$ cup of sugar. How much sugar is needed to make 12 batches?

Synthesis

69. ◆ A fellow student made the following error:

$$\frac{8}{5} - \frac{8}{2} = \frac{8}{3}.$$

Find at least two ways to convince the student of the mistake.

70. ◆ Explain how one could use pictures to convince someone that $\frac{7}{29}$ is larger than $\frac{13}{57}$.

Solve.

71. 🖩 $x + \frac{16}{323} = \frac{10}{187}$

72. 🖩 $x + \frac{7}{253} = \frac{12}{299}$

73. A mountain climber, beginning at sea level, climbs $\frac{3}{5}$ km, descends $\frac{1}{4}$ km, climbs $\frac{1}{3}$ km, and then descends $\frac{1}{7}$ km. At what elevation does the climber finish?

Simplify. Use the rules for order of operations given in Section 1.9.

74. $\frac{2}{5} + \frac{1}{6} \div 3$

75. $\frac{7}{8} - \frac{1}{10} \times \frac{5}{6}$

76. $5 \times \frac{3}{7} - \frac{1}{7} \times \frac{4}{5}$

77. $\left(\frac{2}{3}\right)^2 + \left(\frac{3}{4}\right)^2$

78. A VCR can record up to 6 hr on one tape. It can also fill that same tape in either 4 hr or 2 hr when running at faster speeds. A tape is placed in the machine, which records for $\frac{1}{2}$ hr at the 4-hr speed and $\frac{3}{4}$ hr at the 2-hr speed. How much time is left on the tape to record at the 6-hr speed?

79. As part of a rehabilitation program, an athlete must swim and then walk a total of $\frac{9}{10}$ km each day. If one lap in the swimming pool is $\frac{3}{80}$ km, how far must the athlete walk after swimming 10 laps?

Use <, >, or = for ▢ to write a true sentence.

80. 🖩 $\frac{12}{97} + \frac{67}{139}$ ▢ $\frac{8167}{13,289}$

81. 🖩 $\frac{37}{157} + \frac{19}{107}$ ▢ $\frac{6941}{16,799}$

Collaborative Learning Manual

Arrange sockets and drill bits in fractional sizes from smallest to largest.

3.4 Mixed Numerals

a What Is a Mixed Numeral?

A symbol like $2\frac{3}{4}$ is called a **mixed numeral.**

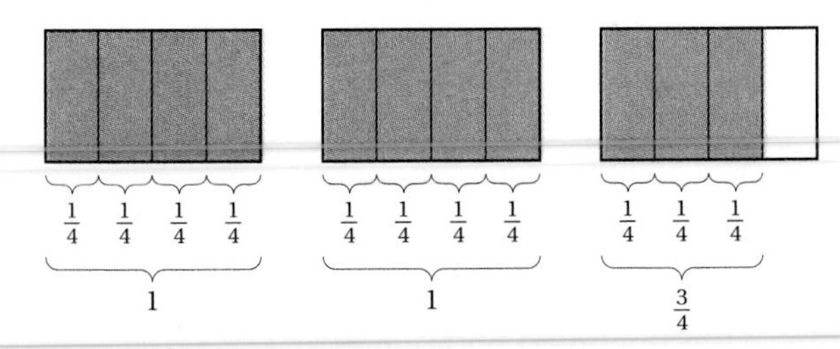

$$2\frac{3}{4} \quad \text{means} \quad 2 + \frac{3}{4}$$

This is a whole number. This is a fraction less than 1.

Examples Convert to a mixed numeral.

1. $7 + \frac{2}{5} = 7\frac{2}{5}$

2. $4 + \frac{3}{10} = 4\frac{3}{10}$

Do Exercises 1–3.

The notation $2\frac{3}{4}$ has a plus sign left out. To aid in understanding, we sometimes write the missing plus sign.

Examples Convert to fractional notation.

3. $2\frac{3}{4} = 2 + \frac{3}{4}$ **Inserting the missing plus sign**

$= \frac{2}{1} + \frac{3}{4}$ $\quad 2 = \frac{2}{1}$

$= \frac{2}{1} \cdot \frac{4}{4} + \frac{3}{4}$ **Finding a common denominator**

$= \frac{8}{4} + \frac{3}{4}$

$= \frac{11}{4}$ **Adding**

4. $4\frac{3}{10} = 4 + \frac{3}{10} = \frac{4}{1} + \frac{3}{10} = \frac{4}{1} \cdot \frac{10}{10} + \frac{3}{10} = \frac{40}{10} + \frac{3}{10} = \frac{43}{10}$

Do Exercises 4 and 5.

Objectives

a Convert from mixed numerals to fractional notation.

b Convert from fractional notation to mixed numerals.

c Divide, writing a mixed numeral for the quotient.

For Extra Help

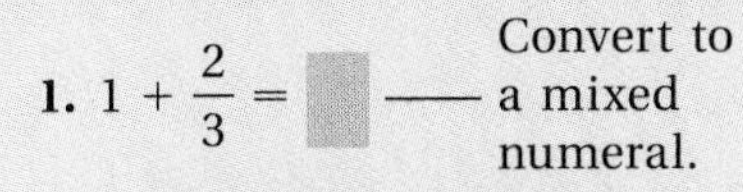

TAPE 5 | TAPE 5B | MAC WIN | CD-ROM

1. $1 + \frac{2}{3} = \square$ —— Convert to a mixed numeral.

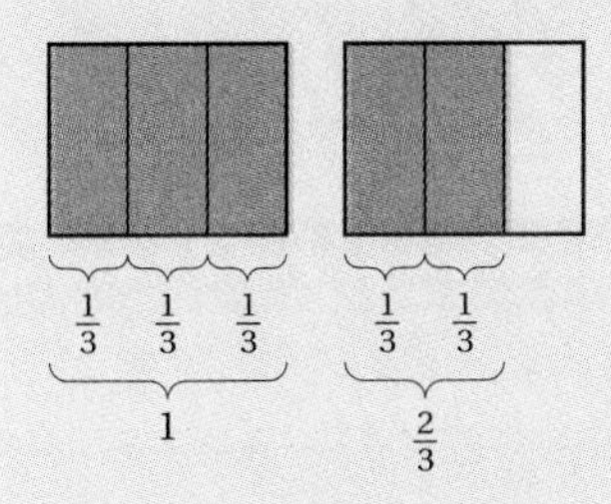

Convert to a mixed numeral.

2. $8 + \frac{3}{4}$

3. $12 + \frac{2}{3}$

Convert to fractional notation.

4. $4\frac{2}{5}$

5. $6\frac{1}{10}$

Answers on page A-6

Convert to fractional notation. Use the faster method.

6. $4\frac{5}{6}$

7. $9\frac{1}{4}$

8. $20\frac{2}{3}$

Answers on page A-6

Let's now consider a faster method for converting a mixed numeral to fractional notation.

> To convert from a mixed numeral to fractional notation:
>
> (a) Multiply the whole number by the denominator: $4 \cdot 10 = 40$.
>
> (b) Add the result to the numerator: $40 + 3 = 43$.
>
> (c) Keep the denominator.
>
> $$4\frac{3}{10} = \frac{43}{10}$$

Examples Convert to fractional notation.

5. $6\frac{2}{3} = \frac{20}{3}$ $\quad$ **$6 \cdot 3 = 18,\ 18 + 2 = 20$**

6. $8\frac{2}{9} = \frac{74}{9}$

7. $10\frac{7}{8} = \frac{87}{8}$

Do Exercises 6–8.

b Writing Mixed Numerals

We can find a mixed numeral for $\frac{5}{3}$ as follows:

$$\frac{5}{3} = \frac{3}{3} + \frac{2}{3} = 1 + \frac{2}{3} = 1\frac{2}{3}.$$

Fractional symbols like $\frac{5}{3}$ also indicate division. Let's divide the numerator by the denominator.

$$3\overline{)5} = 1 \text{ R } 2$$

$$2 \leftarrow 3\overline{)2} = \frac{2}{3} \quad \text{or} \quad 2 \div 3 = \frac{2}{3}$$

Thus, $\frac{5}{3} = 1\frac{2}{3}$.

In terms of objects, we can think of $\frac{5}{3}$ as $\frac{3}{3}$, or 1, plus $\frac{2}{3}$, as shown below.

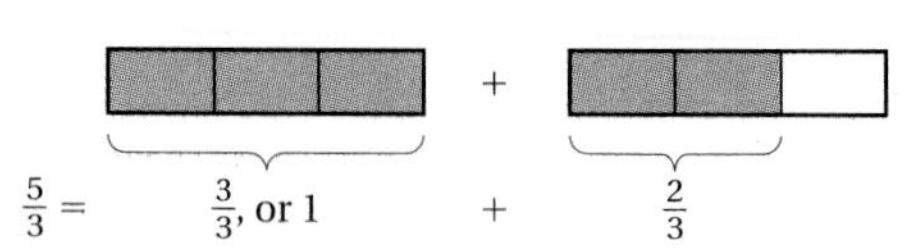

$\frac{5}{3} = \frac{3}{3}$, or 1 $+ \frac{2}{3}$

> To convert from fractional notation to a mixed numeral, divide.
>
> $\frac{13}{5}$ $\quad$ $5\overline{)13}$ gives quotient 2 and remainder 3 ($13 - 10 = 3$), with divisor 5, so $\frac{13}{5} = 2\frac{3}{5}$.
>
> The quotient (2) becomes the whole number; the remainder (3) becomes the numerator; the divisor (5) becomes the denominator.

Examples Convert to a mixed numeral.

8. $\frac{8}{5}$ $\qquad 5\overline{)8}$ with quotient 1, -5, remainder 3 $\qquad \frac{8}{5} = 1\frac{3}{5}$

> A fraction larger than 1, such as $\frac{8}{5}$, is sometimes referred to as an "improper" fraction. We have intentionally avoided such terminology. The use of such notation as $\frac{8}{5}$, $\frac{69}{10}$, and so on, is quite proper and very common in algebra.

9. $\frac{69}{10}$ $\qquad 10\overline{)69}$ with quotient 6, -60, remainder 9 $\qquad \frac{69}{10} = 6\frac{9}{10}$

10. $\frac{122}{8}$ $\qquad 8\overline{)122}$ with quotient 15, -80, 42, -40, remainder 2 $\qquad \frac{122}{8} = 15\frac{2}{8} = 15\frac{1}{4}$

Do Exercises 9–11.

Convert to a mixed numeral.

9. $\frac{7}{3}$

10. $\frac{11}{10}$

11. $\frac{110}{6}$

c Finding Mixed Numerals for Quotients

It is quite common when dividing whole numbers to write the quotient using a mixed numeral. The remainder is the numerator of the fractional part of the mixed numeral.

Example 11 Divide. Write a mixed numeral for the quotient.

$$7\overline{)6\ 3\ 4\ 1}$$

We first divide as usual.

$$\begin{array}{r} 9\ 0\ 5 \\ 7\overline{)6\ 3\ 4\ 1} \\ \underline{6\ 3\ 0\ 0} \\ 4\ 1 \\ \underline{3\ 5} \\ 6 \end{array}$$

The answer is 905 R 6. We write a mixed numeral for the answer as follows:

$$905\frac{6}{7}.$$

The division $6341 \div 7$ can be expressed using fractional notation or a mixed numeral as follows:

$$\frac{6341}{7} = 905\frac{6}{7}.$$

Answers on page A-6

Divide. Write a mixed numeral for the answer.

12. $6\overline{)4846}$

13. $45\overline{)6053}$

Answers on page A-6

Example 12 Divide. Write a mixed numeral for the answer.

$42\overline{)8915}$

We first divide as usual.

$$\begin{array}{r} 212 \\ 42\overline{)8915} \\ \underline{8400} \\ 515 \\ \underline{420} \\ 95 \\ \underline{84} \\ 11 \end{array} \qquad \frac{8915}{42} = 212\frac{11}{42}$$

The answer is $212\frac{11}{42}$.

Do Exercises 12 and 13.

Calculator Spotlight

Exercises

If your calculator has the capability to find whole-number quotients and remainders (see Section 1.6), use it to find mixed numerals for the answers to each of the following divisions.

1. $6\overline{)8857}$

2. $9\overline{)6088}$

3. $56\overline{)44,851}$

4. $18\overline{)234,567}$

5. $11\overline{)567,895}$

6. $32\overline{)234,567}$

7. $45\overline{)6033}$

8. $213\overline{)567,988}$

9. $112\overline{)400,003}$

10. $908\overline{)11,234}$

Exercise Set 3.4

a Convert to fractional notation.

1. $5\frac{2}{3}$

2. $3\frac{4}{5}$

3. $3\frac{1}{4}$

4. $6\frac{1}{2}$

5. $10\frac{1}{8}$

6. $20\frac{1}{5}$

7. $5\frac{1}{10}$

8. $9\frac{1}{10}$

9. $20\frac{3}{5}$

10. $30\frac{4}{5}$

11. $9\frac{5}{6}$

12. $8\frac{7}{8}$

13. $7\frac{3}{10}$

14. $6\frac{9}{10}$

15. $1\frac{5}{8}$

16. $1\frac{3}{5}$

17. $12\frac{3}{4}$

18. $15\frac{2}{3}$

19. $4\frac{3}{10}$

20. $5\frac{7}{10}$

21. $2\frac{3}{100}$

22. $5\frac{7}{100}$

23. $66\frac{2}{3}$

24. $33\frac{1}{3}$

25. $5\frac{29}{50}$

26. $84\frac{3}{8}$

b Convert to a mixed numeral.

27. $\frac{18}{5}$

28. $\frac{17}{4}$

29. $\frac{14}{3}$

30. $\frac{39}{8}$

31. $\frac{27}{6}$

32. $\frac{30}{9}$

33. $\frac{57}{10}$

34. $\frac{89}{10}$

35. $\frac{53}{7}$

36. $\frac{59}{8}$

37. $\frac{45}{6}$

38. $\frac{50}{8}$

39. $\frac{46}{4}$

40. $\frac{39}{9}$

41. $\frac{12}{8}$

42. $\frac{28}{6}$

43. $\frac{757}{100}$

44. $\frac{467}{100}$

45. $\frac{345}{8}$

46. $\frac{223}{4}$

c Divide. Write a mixed numeral for the answer.

47. $8\overline{)869}$

48. $3\overline{)2126}$

49. $5\overline{)3091}$

50. $9\overline{)9110}$

51. $21\overline{)852}$

52. $85\overline{)7672}$

53. $102\overline{)5612}$

54. $46\overline{)1081}$

Skill Maintenance

Multiply and simplify. [2.4a], [2.6a]

55. $\frac{6}{5} \cdot 15$

56. $\frac{5}{12} \cdot 6$

57. $\frac{7}{10} \cdot \frac{5}{14}$

58. $\frac{1}{10} \cdot \frac{20}{5}$

Divide and simplify. [2.7b]

59. $\frac{2}{3} \div \frac{1}{36}$

60. $28 \div \frac{4}{7}$

61. $200 \div \frac{15}{64}$

62. $\frac{3}{4} \div \frac{9}{16}$

Synthesis

63. ◆ Describe in your own words a method for rewriting a fraction as a mixed numeral.

64. ◆ Are the numbers $2\frac{1}{3}$ and $2 \cdot \frac{1}{3}$ equal? Why or why not?

Write a mixed numeral.

65. 🖩 $\frac{128{,}236}{541}$

66. 🖩 $\frac{103{,}676}{349}$

67. $\frac{56}{7} + \frac{2}{3}$

68. $\frac{72}{12} + \frac{5}{6}$

69. There are $\frac{366}{7}$ weeks in a leap year.

70. There are $\frac{365}{7}$ weeks in a year.

3.5 Addition and Subtraction Using Mixed Numerals; Applications

Objectives

a. Add using mixed numerals.

b. Subtract using mixed numerals.

c. Solve applied problems involving addition and subtraction with mixed numerals.

For Extra Help

TAPE 6 TAPE 6A MAC WIN CD-ROM

a Addition

To find the sum $1\frac{5}{8} + 3\frac{1}{8}$, we first add the fractions. Then we add the whole numbers.

$$\begin{array}{rl} 1\frac{5}{8} & = \\ +\ 3\frac{1}{8} & = \\ \hline \frac{6}{8} & \end{array} \qquad \begin{array}{r} 1\frac{5}{8} \\ +\ 3\frac{1}{8} \\ \hline 4\frac{6}{8} = 4\frac{3}{4} \end{array}$$

Simplifying

Add the fractions. Add the whole numbers.

Do Exercise 1.

Example 1 Add: $5\frac{2}{3} + 3\frac{5}{6}$. Write a mixed numeral for the answer.

The LCD is 6.

$$\begin{array}{rcl} 5\frac{2}{3}\cdot\frac{2}{2} & = & 5\frac{4}{6} \\ +\ 3\frac{5}{6} & = & +\ 3\frac{5}{6} \\ \hline & & 8\frac{9}{6} = 8 + \frac{9}{6} \\ & & = 8 + 1\frac{1}{2} \\ & & = 9\frac{1}{2} \end{array}$$

To find a mixed numeral for $\frac{9}{6}$, we divide:

$$6\overline{)9}\ \text{quotient } 1,\ \text{remainder } 3 \qquad \frac{9}{6} = 1\frac{3}{6} = 1\frac{1}{2}$$

$\frac{19}{2}$ is also a correct answer, but it is not a mixed numeral, which is what we are working with in Sections 3.4, 3.5, and 3.6.

Do Exercise 2.

Example 2 Add: $10\frac{5}{6} + 7\frac{3}{8}$.

The LCD is 24.

$$\begin{array}{rcl} 10\frac{5}{6}\cdot\frac{4}{4} & = & 10\frac{20}{24} \\ +\ 7\frac{3}{8}\cdot\frac{3}{3} & = & +\ 7\frac{9}{24} \\ \hline & & 17\frac{29}{24} = 18\frac{5}{24} \end{array}$$

Do Exercise 3.

1. Add.

$$\begin{array}{r} 2\frac{3}{10} \\ +\ 5\frac{1}{10} \\ \hline \end{array}$$

2. Add.

$$\begin{array}{r} 8\frac{2}{5} \\ +\ 3\frac{7}{10} \\ \hline \end{array}$$

3. Add.

$$\begin{array}{r} 9\frac{3}{4} \\ +\ 3\frac{5}{6} \\ \hline \end{array}$$

Answers on page A-6

Subtract.

4. $10\frac{7}{8}$
$- 9\frac{3}{8}$

5. $8\frac{2}{3}$
$- 5\frac{1}{2}$

6. Subtract.

$5\frac{1}{12}$
$- 1\frac{3}{4}$

7. Subtract.

5
$- 1\frac{1}{3}$

Answers on page A-6

b Subtraction

Example 3 Subtract: $7\frac{3}{4} - 2\frac{1}{4}$.

$$\begin{array}{rl} 7\frac{3}{4} & = \\ -\ 2\frac{1}{4} & = \\ \hline \frac{2}{4} & \end{array} \qquad \begin{array}{r} 7\frac{3}{4} \\ -\ 2\frac{1}{4} \\ \hline 5\frac{2}{4} = 5\frac{1}{2} \end{array}$$

Subtract the fractions. Subtract the whole numbers. Simplifying

Example 4 Subtract: $9\frac{4}{5} - 3\frac{1}{2}$.

The LCD is 10.

$$\begin{array}{rcr} 9\frac{4}{5}\cdot\frac{2}{2} & = & 9\frac{8}{10} \\ -\ 3\frac{1}{2}\cdot\frac{5}{5} & = & -\ 3\frac{5}{10} \\ \hline & & 6\frac{3}{10} \end{array}$$

Do Exercises 4 and 5.

Example 5 Subtract: $7\frac{1}{6} - 2\frac{1}{4}$.

The LCD is 12.

$$\begin{array}{rcr} 7\frac{1}{6}\cdot\frac{2}{2} & = & 7\frac{2}{12} \\ -\ 2\frac{1}{4}\cdot\frac{3}{3} & = & -\ 2\frac{3}{12} \\ \hline & & \end{array}$$

We cannot subtract $\frac{3}{12}$ from $\frac{2}{12}$. We borrow 1, or $\frac{12}{12}$, from 7: $7\frac{2}{12} = 6 + 1 + \frac{2}{12} = 6 + \frac{12}{12} + \frac{2}{12} = 6\frac{14}{12}$.

We can write this as

$$\begin{array}{rcr} 7\frac{2}{12} & = & 6\frac{14}{12} \\ -\ 2\frac{3}{12} & = & -\ 2\frac{3}{12} \\ \hline & & 4\frac{11}{12} \end{array}$$

Do Exercise 6.

Example 6 Subtract: $12 - 9\frac{3}{8}$.

$$\begin{array}{rcr} 12 & = & 11\frac{8}{8} \\ -\ 9\frac{3}{8} & = & -\ 9\frac{3}{8} \\ \hline & & 2\frac{5}{8} \end{array}$$

$12 = 11 + 1 = 11 + \frac{8}{8} = 11\frac{8}{8}$

Do Exercise 7.

c Applications and Problem Solving

Example 7 *Travel Distance.* On two business days, a salesperson drove $144\frac{9}{10}$ mi and $87\frac{1}{4}$ mi. What was the total distance driven?

1. **Familiarize.** We let d = the total distance driven.
2. **Translate.** We translate as follows.

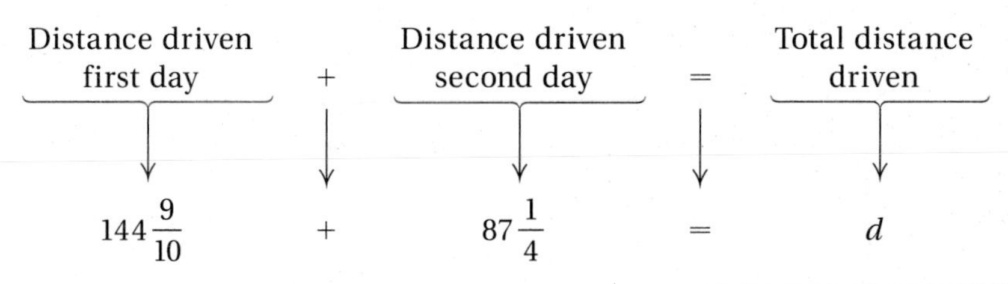

3. **Solve.** The sentence tells us what to do. We add. The LCD is 20.

$$\begin{array}{rcrcr} 144\frac{9}{10} & = & 144\left[\frac{9}{10}\cdot\frac{2}{2}\right] & = & 144\frac{18}{20} \\ +\ 87\frac{1}{4} & = & +\ 87\left[\frac{1}{4}\cdot\frac{5}{5}\right] & = & +\ 87\frac{5}{20} \\ \hline & & & & 231\frac{23}{20} = 232\frac{3}{20} \end{array}$$

Thus, $d = 232\frac{3}{20}$.

4. **Check.** We check by repeating the calculation. We also note that the answer is larger than either of the distances driven, which means that the answer is reasonable.
5. **State.** The total distance driven was $232\frac{3}{20}$ mi.

Do Exercise 8.

Example 8 *NCAA Football Goalposts.* Recently, in college football, the distance between goalposts was reduced from $23\frac{1}{3}$ ft to $18\frac{1}{2}$ ft. How much was it reduced?

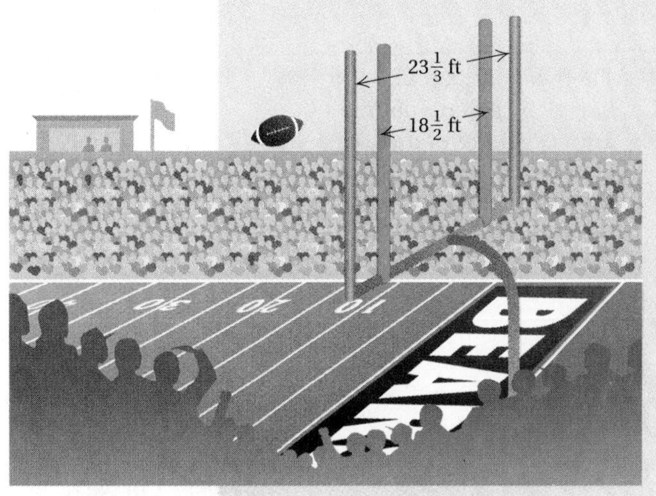

Source: NCAA

1. **Familiarize.** We let d = the amount of reduction and make a drawing to illustrate the situation.
2. **Translate.** We translate as follows.

Former distance	−	New distance	=	Amount of reduction
↓	↓	↓	↓	↓
$23\frac{1}{3}$	−	$18\frac{1}{2}$	=	d

3. **Solve.** To solve the equation, we carry out the subtraction. The LCD is 6.

$$\begin{array}{rcrcrcr} 23\frac{1}{3} & = & 23\left[\frac{1}{3}\cdot\frac{2}{2}\right] & = & 23\frac{2}{6} & = & 22\frac{8}{6} \\ -\ 18\frac{1}{2} & = & -\ 18\left[\frac{1}{2}\cdot\frac{3}{3}\right] & = & -18\frac{3}{6} & = & -\ 18\frac{3}{6} \\ \hline & & & & & & 4\frac{5}{6} \end{array}$$

Thus, $d = 4\frac{5}{6}$ ft.

8. A car-seat upholstering company sold two pieces of leather $6\frac{1}{4}$ yd and $10\frac{5}{6}$ yd long. What was the total length of the leather?

Answer on page A-6

4. Check. To check, we add the reduction to the new distance:

$$18\frac{1}{2} + 4\frac{5}{6} = 18\frac{3}{6} + 4\frac{5}{6} = 22\frac{8}{6} = 23\frac{2}{6} = 23\frac{1}{3}.$$

This checks.

5. State. The reduction in the goalpost distance was $4\frac{5}{6}$ ft.

Do Exercise 9.

Multistep Problems

Example 9 *Intel Stock Price.* One morning, the stock of Intel Corporation opened at a price of $\$100\frac{3}{8}$ per share. By noon, the price had risen $\$4\frac{7}{8}$. At the end of the day, it had fallen $\$10\frac{3}{4}$ from the price at noon. What was the closing price?

1. **Familiarize.** We first make a drawing or at least visualize the situation. We let p = the price at noon, after the rise, and c = the price at the close, after the drop.

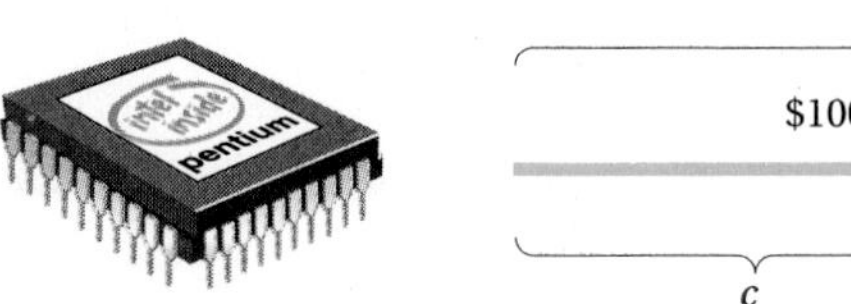

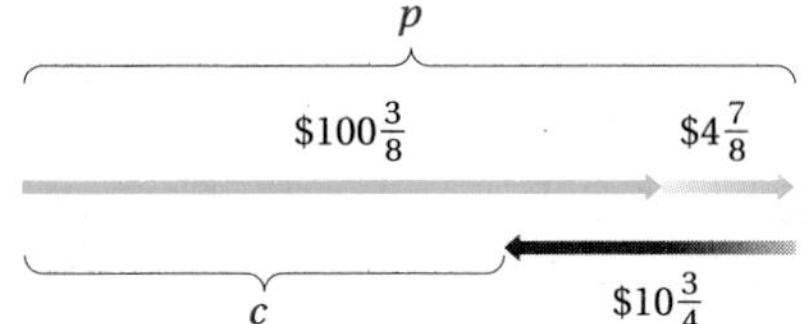

2. **Translate.** From the figure, we see that the price at the close is the price at noon minus the amount of the drop. Thus,

$$c = p - \$10\frac{3}{4} = \left(\$100\frac{3}{8} + \$4\frac{7}{8}\right) - \$10\frac{3}{4}.$$

3. **Solve.** This is a two-step problem.

a) We first add $\$4\frac{7}{8}$ to $\$100\frac{3}{8}$ to find the price p of the stock at noon.

$$\begin{array}{r} 100\frac{3}{8} \\ + \quad 4\frac{7}{8} \\ \hline 104\frac{10}{8} = 105\frac{1}{4} = p \end{array}$$

b) Next we subtract $\$10\frac{3}{4}$ from $\$105\frac{1}{4}$ to find the price c of the stock at closing.

$$\begin{array}{rcr} 105\frac{1}{4} & = & 104\frac{5}{4} \\ -\ 10\frac{3}{4} & = & -\ 10\frac{3}{4} \\ \hline & & 94\frac{2}{4} = 94\frac{1}{2} = c \end{array}$$

4. **Check.** We check by repeating the calculation.

5. **State.** The price of the stock at closing is $\$94\frac{1}{2}$.

Do Exercise 10.

9. *Damascus Blade.* The Damascus blade of a pearl-handled folding knife is $3\frac{3}{4}$ in. long. The same blade in an ATS-34 is $4\frac{1}{8}$ in. long (***Source:*** *Blade Magazine* 23, no. 10, October 1996: 26–27). How many inches longer is the ATS-34 blade?

10. There are $20\frac{1}{3}$ gal of water in a barrel; $5\frac{3}{4}$ gal are poured out and $8\frac{2}{3}$ gal are poured back in. How many gallons of water are then in the barrel?

Answers on page A-6

Exercise Set 3.5

a Add. Write a mixed numeral for the answer.

1. $2\frac{7}{8} + 3\frac{5}{8}$

2. $4\frac{5}{6} + 3\frac{5}{6}$

3. $1\frac{1}{4} + 1\frac{2}{3}$

4. $4\frac{1}{3} + 5\frac{2}{9}$

5. $8\frac{3}{4} + 5\frac{5}{6}$

6. $4\frac{3}{8} + 6\frac{5}{12}$

7. $3\frac{2}{5} + 8\frac{7}{10}$

8. $5\frac{1}{2} + 3\frac{7}{10}$

9. $5\frac{3}{8} + 10\frac{5}{6}$

10. $\frac{5}{8} + 1\frac{5}{6}$

11. $12\frac{4}{5} + 8\frac{7}{10}$

12. $15\frac{5}{8} + 11\frac{3}{4}$

13. $14\frac{5}{8} + 13\frac{1}{4}$

14. $16\frac{1}{4} + 15\frac{7}{8}$

15. $7\frac{1}{8} + 9\frac{2}{3} + 10\frac{3}{4}$

16. $45\frac{2}{3} + 31\frac{3}{5} + 12\frac{1}{4}$

b Subtract. Write a mixed numeral for the answer.

17. $4\frac{1}{5} - 2\frac{3}{5}$

18. $5\frac{1}{8} - 2\frac{3}{8}$

19. $6\frac{3}{5} - 2\frac{1}{2}$

20. $7\frac{2}{3} - 6\frac{1}{2}$

21. $34\frac{1}{3} - 12\frac{5}{8}$

22. $23\frac{5}{16} - 16\frac{3}{4}$

23. $21 - 8\frac{3}{4}$

24. $42 - 3\frac{7}{8}$

25. $\begin{array}{r} 34 \\ -\ 18\frac{5}{8} \\ \hline \end{array}$

26. $\begin{array}{r} 23 \\ -\ 19\frac{3}{4} \\ \hline \end{array}$

27. $\begin{array}{r} 21\frac{1}{6} \\ -\ 13\frac{3}{4} \\ \hline \end{array}$

28. $\begin{array}{r} 42\frac{1}{10} \\ -\ 23\frac{7}{12} \\ \hline \end{array}$

29. $\begin{array}{r} 14\frac{1}{8} \\ -\ \frac{3}{4} \\ \hline \end{array}$

30. $\begin{array}{r} 28\frac{1}{6} \\ -\ 5 \\ \hline \end{array}$

31. $\begin{array}{r} 25\frac{1}{9} \\ -\ 13\frac{5}{6} \\ \hline \end{array}$

32. $\begin{array}{r} 23\frac{5}{16} \\ -\ 14\frac{7}{12} \\ \hline \end{array}$

c Solve.

33. A butcher sold packages of hamburger weighing $1\frac{2}{3}$ lb and $5\frac{3}{4}$ lb. What was the total weight of the meat?

34. A butcher sold packages of sliced turkey breast weighing $1\frac{1}{3}$ lb and $4\frac{3}{5}$ lb. What was the total weight of the meat?

35. Tricia is 66 in. tall and her son is $59\frac{7}{12}$ in. tall. How much taller is Tricia?

36. Nicholas is $73\frac{2}{3}$ in. tall and his daughter is $71\frac{5}{16}$ in. tall. How much taller is Nicholas?

37. A plumber uses pipes of lengths $10\frac{5}{16}$ ft and $8\frac{3}{4}$ ft in the installation of a sink. How much pipe was used?

38. The standard pencil is $6\frac{7}{8}$ in. wood and $\frac{1}{2}$ in. eraser (***Source*:** Eberhard Faber American). What is the total length of the standard pencil?

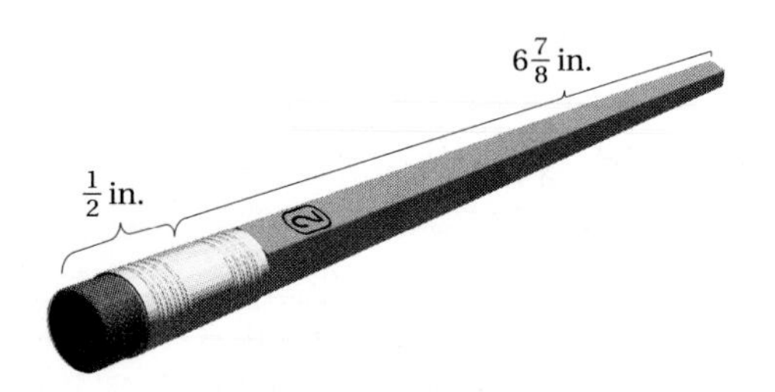

39. One day, a computer technician drove $180\frac{7}{10}$ mi away from Los Angeles for a service call. The next day, she drove $85\frac{1}{2}$ mi back toward Los Angeles for another service call. How far was the technician from Los Angeles?

40. A woman is $4\frac{1}{2}$ in. taller than her daughter. The daughter is $66\frac{2}{3}$ in. tall. How tall is the woman?

41. One standard book size is $8\frac{1}{2}$ in. by $9\frac{3}{4}$ in. What is the total distance around (perimeter of) the front cover of such a book?

42. A standard sheet of paper is $8\frac{1}{2}$ in. by 11 in. What is the total distance around (perimeter of) the paper?

43. *Toys "R" Us Stock.* During a recent year, the price of one share of stock in Toys "R" Us varied between a low of $\$20\frac{1}{2}$ and a high of $\$37\frac{5}{8}$ (***Source:*** Toys "R" Us annual report). What was the difference between the high and the low?

44. *Nike, Inc., Stock.* During a recent year, the lowest price of one share of Nike, Inc., stock was $\$31\frac{3}{4}$. Its highest price was $\$22\frac{1}{4}$ more than its lowest price (***Source:*** Nike, Inc., annual report). What was the highest price?

45. When cutting wood with a saw, a carpenter must take into account the thickness of the saw blade. Suppose that from a piece of wood 36 in. long, a carpenter cuts a $15\frac{3}{4}$-in. length with a saw blade that is $\frac{1}{8}$ in. in thickness. How long is the piece that remains?

46. When redecorating, a painter used $1\frac{3}{4}$ gal of paint for the living room and $1\frac{1}{3}$ gal for the family room. How much paint was used in all?

47. Rene is $5\frac{1}{4}$ in. taller than his son, who is $72\frac{5}{6}$ in. tall. How tall is Rene?

48. A plane flew 640 mi on a nonstop flight. On the return flight, it landed after having flown $320\frac{3}{10}$ mi. How far was the plane from its original point of departure?

49. Sue, an interior designer, worked $10\frac{1}{2}$ hr over a three-day period. If Sue worked $2\frac{1}{2}$ hr on the first day and $4\frac{1}{5}$ hr on the second, how many hours did Sue work on the third day?

50. A painter had $3\frac{1}{2}$ gal of paint. It took $2\frac{3}{4}$ gal for a family room. It was estimated that it would take $2\frac{1}{4}$ gal to paint the living room. How much more paint was needed?

Find the perimeter of (distance around) the figure.

51.

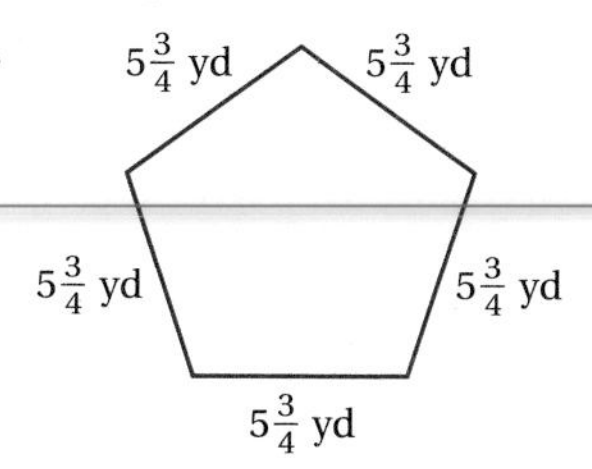

52.

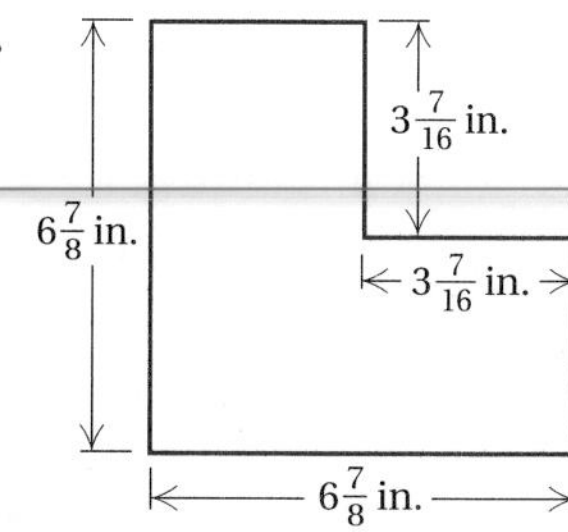

Find the length d in the figure.

53.

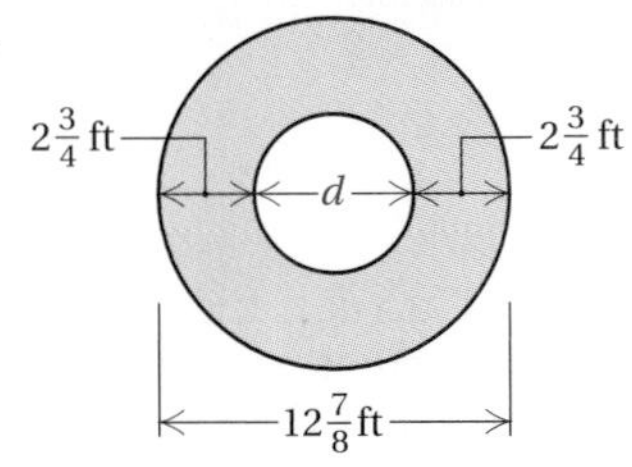

54.

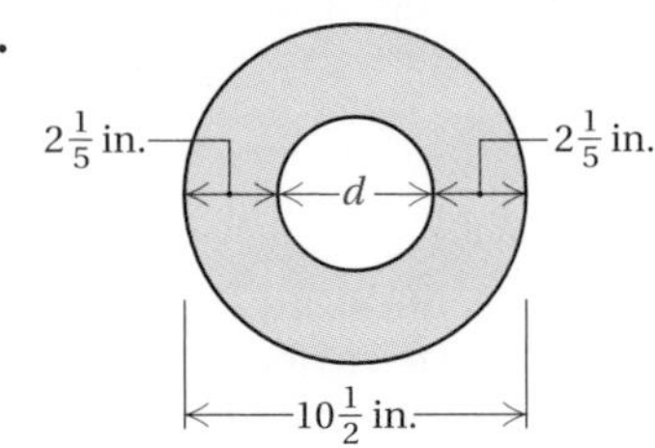

55. Find the smallest length of a bolt that will pass through a piece of tubing with an outside diameter of $\frac{1}{2}$ in., a washer $\frac{1}{16}$ in. thick, a piece of tubing with a $\frac{3}{4}$-in. outside diameter, another washer, and a nut $\frac{3}{16}$ in. thick.

Skill Maintenance

Solve.

56. A dairy produced 4578 oz of milk one week. How many 16-oz cartons were filled? How much milk was left over? [1.8a]

57. Rick's Market prepackages Swiss cheese in $\frac{3}{4}$-lb packages. How many packages can be made from a 12-lb slab of cheese? [2.7d]

58. Divide and simplify: $\frac{12}{25} \div \frac{24}{5}$. [2.7b]

59. Multiply and simplify: $\frac{15}{9} \cdot \frac{18}{39}$. [2.6a]

Synthesis

60. ◆ Is the sum of two mixed numerals always a mixed numeral? Why or why not?

61. ◆ Write a problem for a classmate to solve. Design the problem so the solution is "The larger package holds $4\frac{1}{2}$ oz more than the smaller package."

Calculate each of the following. Write the result as a mixed numeral.

62. 🖩 $5798\frac{17}{53} - 3909\frac{1957}{2279}$

63. 🖩 $3289\frac{1047}{1189} + 5278\frac{32}{41}$

64. Solve: $47\frac{2}{3} + n = 56\frac{1}{4}$.

65. A post for a pier is 29 ft long. Half of the post extends above the water's surface and $8\frac{3}{4}$ ft of the post is buried in mud. How deep is the water at that location?

Collaborative Learning Manual: Add and subtract mixed numerals using fraction bars.

3.6 Multiplication and Division Using Mixed Numerals; Applications

a Multiplication

Carrying out addition and subtraction with mixed numerals is easier if the numbers are left as mixed numerals. With multiplication and division, however, it is easier to convert the numbers first to fractional notation.

To multiply using mixed numerals, first convert to fractional notation. Then multiply with fractional notation and convert the answer back to a mixed numeral, if appropriate.

Example 1 Multiply: $6 \cdot 2\frac{1}{2}$.

$$6 \cdot 2\frac{1}{2} = \frac{6}{1} \cdot \frac{5}{2} = \frac{6 \cdot 5}{1 \cdot 2} = \frac{2 \cdot 3 \cdot 5}{2 \cdot 1} = \frac{2}{2} \cdot \frac{3 \cdot 5}{1} = 15$$

Here we write fractional notation.

Do Exercise 1.

Example 2 Multiply: $3\frac{1}{2} \cdot \frac{3}{4}$.

$$3\frac{1}{2} \cdot \frac{3}{4} = \frac{7}{2} \cdot \frac{3}{4} = \frac{21}{8} = 2\frac{5}{8}$$

Note that fractional notation is needed to carry out the multiplication.

Do Exercise 2.

Example 3 Multiply: $8 \cdot 4\frac{2}{3}$.

$$8 \cdot 4\frac{2}{3} = \frac{8}{1} \cdot \frac{14}{3} = \frac{112}{3} = 37\frac{1}{3}$$

Do Exercise 3.

Example 4 Multiply: $2\frac{1}{4} \cdot 3\frac{2}{5}$.

$$2\frac{1}{4} \cdot 3\frac{2}{5} = \frac{9}{4} \cdot \frac{17}{5} = \frac{153}{20} = 7\frac{13}{20}$$

CAUTION! $2\frac{1}{4} \cdot 3\frac{2}{5} \neq 6\frac{2}{20}$. A common error is to multiply the whole numbers and then the fractions. This does not give the correct answer, $7\frac{13}{20}$, which is found by converting first to fractional notation.

Do Exercise 4.

Objectives

a Multiply using mixed numerals.

b Divide using mixed numerals.

c Solve applied problems involving multiplication and division with mixed numerals.

For Extra Help

TAPE 6 | TAPE 6A | MAC WIN | CD-ROM

1. Multiply: $6 \cdot 3\frac{1}{3}$.

2. Multiply: $2\frac{1}{2} \cdot \frac{3}{4}$.

3. Multiply: $2 \cdot 6\frac{2}{5}$.

4. Multiply: $3\frac{1}{3} \cdot 2\frac{1}{2}$.

Answers on page A-7

5. Divide: $84 \div 5\frac{1}{4}$.

6. Divide: $26 \div 3\frac{1}{2}$.

Divide.

7. $2\frac{1}{4} \div 1\frac{1}{5}$

8. $1\frac{3}{4} \div 2\frac{1}{2}$

Answers on page A-7

b Division

The division $1\frac{1}{2} \div \frac{1}{6}$ is shown here.

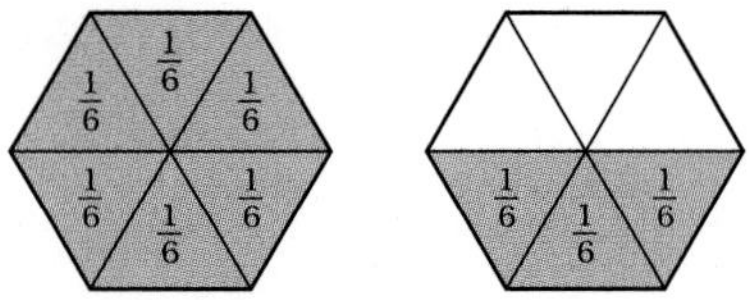

$$1\frac{1}{2} \div \frac{1}{6} = \frac{3}{2} \div \frac{1}{6}$$

$$= \frac{3}{2} \cdot 6 = \frac{3 \cdot 6}{2} = \frac{3 \cdot 3 \cdot 2}{2 \cdot 1} = \frac{3 \cdot 3}{1} \cdot \frac{2}{2} = \frac{3 \cdot 3}{1} \cdot 1 = 9$$

To divide using mixed numerals, first write fractional notation. Then divide with fractional notation and convert the answer back to a mixed numeral, if appropriate.

Example 5 Divide: $32 \div 3\frac{1}{5}$.

$$32 \div 3\frac{1}{5} = \frac{32}{1} \div \frac{16}{5}$$

$$= \frac{32}{1} \cdot \frac{5}{16} = \frac{32 \cdot 5}{1 \cdot 16} = \frac{2 \cdot 16 \cdot 5}{1 \cdot 16} = \frac{16}{16} \cdot \frac{2 \cdot 5}{1} = 10$$

Remember to multiply by the reciprocal.

Do Exercise 5.

Example 6 Divide: $35 \div 4\frac{1}{3}$.

$$35 \div 4\frac{1}{3} = \frac{35}{1} \div \frac{13}{3} = \frac{35}{1} \cdot \frac{3}{13} = \frac{105}{13} = 8\frac{1}{13}$$

Do Exercise 6.

Example 7 Divide: $2\frac{1}{3} \div 1\frac{3}{4}$.

$$2\frac{1}{3} \div 1\frac{3}{4} = \frac{7}{3} \div \frac{7}{4} = \frac{7}{3} \cdot \frac{4}{7} = \frac{7 \cdot 4}{7 \cdot 3} = \frac{7}{7} \cdot \frac{4}{3} = 1 \cdot \frac{4}{3} = \frac{4}{3} = 1\frac{1}{3}$$

Caution! The reciprocal of $1\frac{3}{4}$ is *not* $1\frac{4}{3}$!

Example 8 Divide: $1\frac{3}{5} \div 3\frac{1}{3}$.

$$1\frac{3}{5} \div 3\frac{1}{3} = \frac{8}{5} \div \frac{10}{3} = \frac{8}{5} \cdot \frac{3}{10} = \frac{2 \cdot 4 \cdot 3}{5 \cdot 2 \cdot 5} = \frac{2}{2} \cdot \frac{4 \cdot 3}{5 \cdot 5} = 1 \cdot \frac{4 \cdot 3}{5 \cdot 5} = \frac{12}{25} = \frac{12}{25}$$

Do Exercises 7 and 8.

C Applications and Problem Solving

Example 9 *Cassette Tape Music.* The tape in an audio cassette is played at a rate of $1\frac{7}{8}$ in. per second. A defective tape player has destroyed 30 in. of tape. How many seconds of music have been lost?

1. **Familiarize.** We can make a drawing.

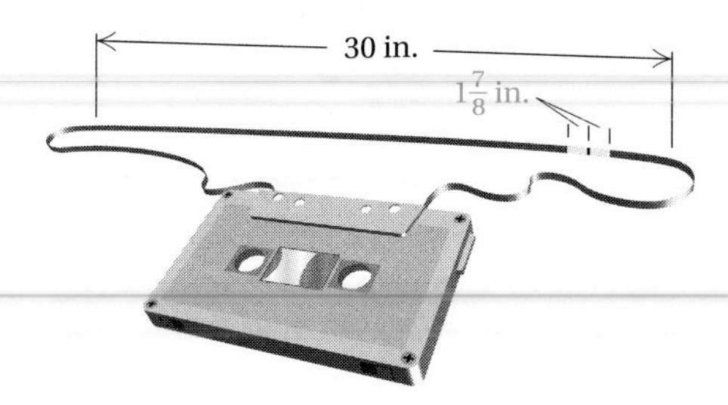

Since each $1\frac{7}{8}$ in. of tape represents 1 sec of lost music, the question can be regarded as asking how many times 30 can be divided by $1\frac{7}{8}$. We let t = the number of seconds of music lost.

2. **Translate.** The situation corresponds to a division sentence:

$$t = 30 \div 1\frac{7}{8}.$$

3. **Solve.** To solve the equation, we perform the division:

$$\begin{aligned} t &= 30 \div 1\frac{7}{8} \\ &= \frac{30}{1} \div \frac{15}{8} \\ &= \frac{30}{1} \cdot \frac{8}{15} \\ &= \frac{15 \cdot 2 \cdot 8}{1 \cdot 15} \\ &= \frac{15}{15} \cdot \frac{2 \cdot 8}{1} \\ &= 16. \end{aligned}$$

4. **Check.** We check by multiplying. If 16 sec of music were lost, then

$$\begin{aligned} 16 \cdot 1\frac{7}{8} &= \frac{16}{1} \cdot \frac{15}{8} \\ &= \frac{8 \cdot 2 \cdot 15}{1 \cdot 8} \\ &= \frac{8}{8} \cdot \frac{2 \cdot 15}{1} = 30 \text{ in.} \end{aligned}$$

of tape were destroyed. A quicker, but less precise, check can be made by noting that $1\frac{7}{8} \approx 2$. Then $16 \cdot 1\frac{7}{8} \approx 16 \cdot 2 = 32 \approx 30$. Our answer checks.

5. **State.** The cassette has lost 16 sec of music.

Do Exercises 9 and 10.

9. Kyle's pickup truck travels on an interstate highway at 65 mph for $3\frac{1}{2}$ hr. How far does it travel?

10. Holly's minivan travels 302 mi on $15\frac{1}{10}$ gal of gas. How many miles per gallon did it get?

Answers on page A-7

11. A room is $22\frac{1}{2}$ ft by $15\frac{1}{2}$ ft. A 9-ft by 12-ft Oriental rug is placed in the center of the room. How much area is not covered by the rug?

Example 10 An L-shaped room consists of a rectangle that is $8\frac{1}{2}$ by 11 ft and one that is $6\frac{1}{2}$ by $7\frac{1}{2}$ ft. What is the total area of a carpet that covers the floor?

1. **Familiarize.** We make a drawing of the situation. We let a = the total floor area.

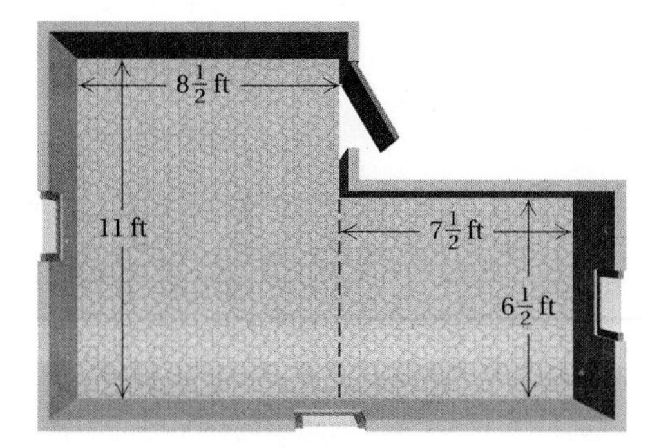

2. **Translate.** The total area is the sum of the areas of the two rectangles. This gives us the following equation:

$$a = 8\frac{1}{2} \cdot 11 + 7\frac{1}{2} \cdot 6\frac{1}{2}.$$

3. **Solve.** This is a multistep problem. We perform each multiplication and then add. This follows the rules for order of operations:

$$\begin{aligned} a &= 8\frac{1}{2} \cdot 11 + 7\frac{1}{2} \cdot 6\frac{1}{2} \\ &= \frac{17}{2} \cdot 11 + \frac{15}{2} \cdot \frac{13}{2} \\ &= \frac{17 \cdot 11}{2} + \frac{15 \cdot 13}{2 \cdot 2} \\ &= \frac{187}{2} + \frac{195}{4} \\ &= 93\frac{1}{2} + 48\frac{3}{4} \\ &= 93\frac{2}{4} + 48\frac{3}{4} \\ &= 141\frac{5}{4} \\ &= 142\frac{1}{4}. \end{aligned}$$

4. **Check.** We perform a partial check by estimating the total area as $11 \cdot 9 + 7 \cdot 7 = 99 + 49 = 148$ ft^2. Our answer, $142\frac{1}{4}$ ft^2, seems reasonable.

5. **State.** The total area of the carpet is $142\frac{1}{4}$ ft^2.

Do Exercise 11.

Answer on page A-7

Exercise Set 3.6

a Multiply. Write a mixed numeral for the answer.

1. $8 \cdot 2\frac{5}{6}$

2. $5 \cdot 3\frac{3}{4}$

3. $3\frac{5}{8} \cdot \frac{2}{3}$

4. $6\frac{2}{3} \cdot \frac{1}{4}$

5. $3\frac{1}{2} \cdot 2\frac{1}{3}$

6. $4\frac{1}{5} \cdot 5\frac{1}{4}$

7. $3\frac{2}{5} \cdot 2\frac{7}{8}$

8. $2\frac{3}{10} \cdot 4\frac{2}{5}$

9. $4\frac{7}{10} \cdot 5\frac{3}{10}$

10. $6\frac{3}{10} \cdot 5\frac{7}{10}$

11. $20\frac{1}{2} \cdot 10\frac{1}{5} \cdot 4\frac{2}{3}$

12. $21\frac{1}{3} \cdot 11\frac{1}{3} \cdot 3\frac{5}{8}$

b Divide. Write a mixed numeral for the answer.

13. $20 \div 3\frac{1}{5}$

14. $18 \div 2\frac{1}{4}$

15. $8\frac{2}{5} \div 7$

16. $3\frac{3}{8} \div 3$

17. $4\frac{3}{4} \div 1\frac{1}{3}$

18. $5\frac{4}{5} \div 2\frac{1}{2}$

19. $1\frac{7}{8} \div 1\frac{2}{3}$

20. $4\frac{3}{8} \div 2\frac{5}{6}$

21. $5\frac{1}{10} \div 4\frac{3}{10}$

22. $4\frac{1}{10} \div 2\frac{1}{10}$

23. $20\frac{1}{4} \div 90$

24. $12\frac{1}{2} \div 50$

c Solve.

25. Irene wants to build a bookcase to hold her collection of favorite videocassette movies. Each shelf in the bookcase will be 27 in. long and each videocassette is $1\frac{1}{8}$ in. thick. How many cassettes can she place on each shelf?

26. A bicycle wheel makes $66\frac{2}{3}$ revolutions per minute. If it rotates for 21 min, how many revolutions does it make?

27. One serving of meat is about $3\frac{1}{2}$ oz. Art eats 2 servings a day. How many ounces of meat is this?

28. *Sodium Consumption.* The average American woman consumes $1\frac{1}{3}$ tsp of sodium each day (***Source***: *Nutrition Action Health Letter,* March 1994, p. 6. 1875 Connecticut Ave., N.W., Washington, DC 20009-5728). How much sodium do 10 average American women consume in one day?

29. *Weight of Water.* The weight of water is $62\frac{1}{2}$ lb per cubic foot. What is the weight of $5\frac{1}{2}$ cubic feet of water?

30. *Weight of Water.* The weight of water is $62\frac{1}{2}$ lb per cubic foot. What is the weight of $2\frac{1}{4}$ cubic feet of water?

31. *Chicken à la King.* Listed below are the ingredients for a low-fat, heart-healthy dish called *Chicken à la King.* What are the ingredients for $\frac{1}{2}$ recipe? for 3 recipes?

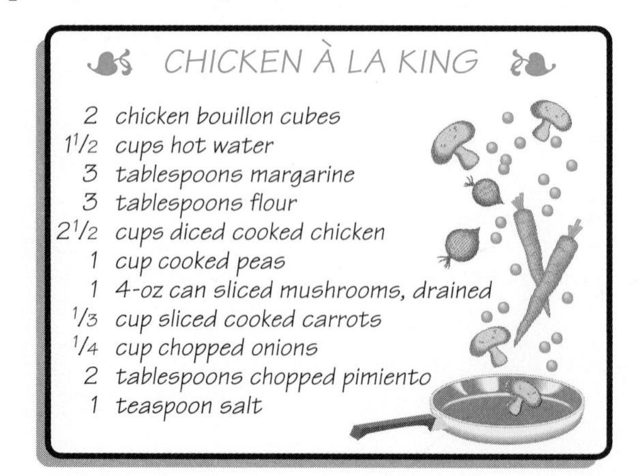

32. *Italian Stuffed Peppers.* Listed below are the ingredients for a low-fat, heart-healthy dish called *Italian Stuffed Peppers.* What are the ingredients for $\frac{1}{2}$ recipe? for 3 recipes?

33. *Temperatures.* Fahrenheit temperature can be obtained from Celsius (centigrade) temperature by multiplying by $1\frac{4}{5}$ and adding 32°. What Fahrenheit temperature corresponds to a Celsius temperature of 20°?

34. *Temperature.* Fahrenheit temperature can be obtained from Celsius (centigrade) temperature by multiplying by $1\frac{4}{5}$ and adding 32°. What Fahrenheit temperature corresponds to the Celsius temperature of boiling water, which is 100°?

35. The tape in a VCR operating in the short-play mode travels at a rate of $1\frac{3}{8}$ in. per second. How many inches of tape are used to record for 60 sec in the short-play mode?

36. The tape in an audio cassette is played at the rate of $1\frac{7}{8}$ in. per second. How many inches of tape are used when a cassette is played for $5\frac{1}{2}$ sec?

37. A car traveled 213 mi on $14\frac{2}{10}$ gal of gas. How many miles per gallon did it get?

38. A car traveled 385 mi on $15\frac{4}{10}$ gal of gas. How many miles per gallon did it get?

39. *Weight of Water.* The weight of water is $62\frac{1}{2}$ lb per cubic foot. How many cubic feet would be occupied by 250 lb of water?

40. *Weight of Water.* The weight of water is $62\frac{1}{2}$ lb per cubic foot. How many cubic feet would be occupied by 375 lb of water?

41. *Shuttle Orbits.* Most space shuttles orbit the earth once every $1\frac{1}{2}$ hr. How many orbits are made every 24 hr?

42. *Turkey Servings.* Turkey contains $1\frac{1}{3}$ servings per pound. How many pounds are needed for 32 servings?

Find the area of the shaded region.

43.

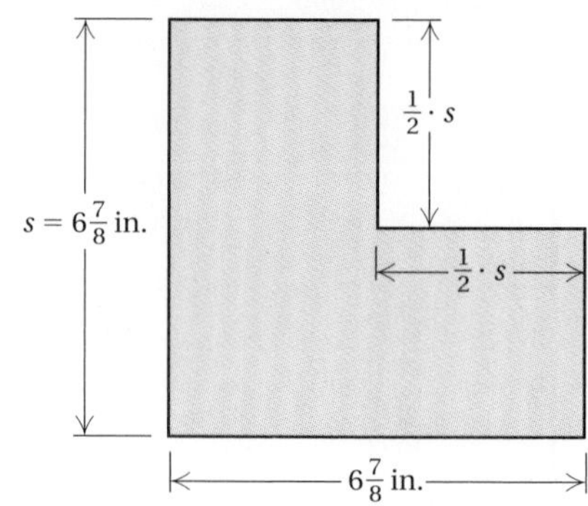

44.

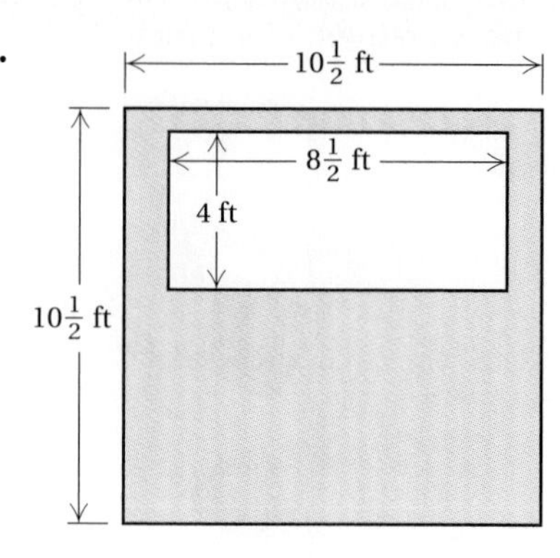

45. A rectangular lot has dimensions of $302\frac{1}{2}$ ft by $205\frac{1}{4}$ ft. A building with dimensions of 100 ft by $25\frac{1}{2}$ ft is built on the lot. How much area is left over?

46. *Word Processing.* Kelly wants to create a table using Microsoft® Word software for word processing. She needs to have two columns, each $1\frac{1}{2}$ in. wide, and five columns, each $\frac{3}{4}$ in. wide. Will this table fit on a piece of standard paper that is $8\frac{1}{2}$ in. wide? If so, how wide will each margin be if her margins on each side are to be of equal width?

Skill Maintenance

47. Multiply. [1.5b]

$$\begin{array}{r} 6\,7\,0\,9 \\ \times \quad 2\,1\,3 \\ \hline \end{array}$$

48. Round to the nearest hundred: 45,765. [1.4a]

49. Solve: $\frac{5}{7} \cdot t = 420$. [2.7c]

50. Divide and simplify: $\frac{4}{5} \div \frac{6}{5}$. [2.7b]

51. Multiply and simplify: $\frac{3}{8} \cdot \frac{4}{9}$. [2.6a]

52. Round to the nearest ten: 45,765. [1.4a]

Synthesis

53. ◆ Write a problem for a classmate to solve. Design the problem so that its solution is found by performing the multiplication $4\frac{1}{2} \cdot 33\frac{1}{3}$.

54. ◆ Under what circumstances is a pair of mixed numerals more easily added than multiplied?

Multiply. Write the answer as a mixed numeral whenever possible.

55. 🖩 $15\frac{2}{11} \cdot 23\frac{31}{43}$

56. 🖩 $17\frac{23}{31} \cdot 19\frac{13}{15}$

Simplify.

57. $8 \div \frac{1}{2} + \frac{3}{4} + \left(5 - \frac{5}{8}\right)^2$

58. $\left(\frac{5}{9} - \frac{1}{4}\right) \times 12 + \left(4 - \frac{3}{4}\right)^2$

59. $\frac{1}{3} \div \left(\frac{1}{2} - \frac{1}{5}\right) \times \frac{1}{4} + \frac{1}{6}$

60. $\frac{7}{8} - 1\frac{1}{8} \times \frac{2}{3} + \frac{9}{10} \div \frac{3}{5}$

61. $4\frac{1}{2} \div 2\frac{1}{2} + 8 - 4 \div \frac{1}{2}$

62. $6 - 2\frac{1}{3} \times \frac{3}{4} + \frac{5}{8} \div \frac{2}{3}$

Collaborative Learning Manual — Analyze stock market prices.

3.7 Order of Operations; Estimation

a Order of Operations: Fractional Notation and Mixed Numerals

The rules for order of operations that we use with whole numbers (see Section 1.9) apply when we are simplifying expressions involving fractional notation and mixed numerals. For review, these rules are listed below.

> **RULES FOR ORDER OF OPERATIONS**
>
> 1. Do all calculations within parentheses before operations outside.
> 2. Evaluate all exponential expressions.
> 3. Do all multiplications and divisions in order from left to right.
> 4. Do all additions and subtractions in order from left to right.

Example 1 Simplify: $\frac{2}{3} \div \frac{1}{2} \cdot \frac{5}{8} + \frac{1}{6}$.

$\frac{2}{3} \div \frac{1}{2} \cdot \frac{5}{8} + \frac{1}{6} = \frac{2}{3} \cdot \frac{2}{1} \cdot \frac{5}{8} + \frac{1}{6}$ Doing the division first by multiplying by the reciprocal of $\frac{1}{2}$

$= \frac{4}{3} \cdot \frac{5}{8} + \frac{1}{6}$

$= \frac{4 \cdot 5}{3 \cdot 8} + \frac{1}{6}$ Doing the multiplication

$= \frac{\not{4} \cdot 5}{3 \cdot \not{4} \cdot 2} + \frac{1}{6}$ Factoring in order to simplify

$= \frac{5}{3 \cdot 2} + \frac{1}{6}$ Removing a factor of 1: $\frac{4}{4} = 1$

$= \frac{5}{6} + \frac{1}{6}$

$= \frac{6}{6}$, or 1 Doing the addition

Do Exercises 1 and 2.

Example 2 Simplify: $\frac{2}{3} \cdot 24 - 11\frac{1}{2}$.

$\frac{2}{3} \cdot 24 - 11\frac{1}{2} = \frac{2 \cdot 24}{3} - 11\frac{1}{2}$ Doing the multiplication first

$= \frac{2 \cdot \not{3} \cdot 8}{\not{3}} - 11\frac{1}{2}$ Factoring the numerator

$= 2 \cdot 8 - 11\frac{1}{2}$ Removing a factor of 1: $\frac{3}{3} = 1$

$= 16 - 11\frac{1}{2}$ Completing the multiplication

$= 4\frac{1}{2}$, or $\frac{9}{2}$ Doing the subtraction

Do Exercise 3.

Objectives

a Simplify expressions using the rules for order of operations.

b Estimate with fractional and mixed-numeral notation.

For Extra Help

TAPE 6

TAPE 6B

MAC WIN

CD-ROM

Simplify.

1. $\frac{2}{5} \cdot \frac{5}{8} + \frac{1}{4}$

2. $\frac{1}{3} \cdot \frac{3}{4} \div \frac{5}{8} - \frac{1}{10}$

3. Simplify: $\frac{3}{4} \cdot 16 + 8\frac{2}{3}$.

Answers on page A-7

4. Find the average of $\frac{1}{2}$, $\frac{1}{3}$, and $\frac{5}{6}$.

5. Find the average of $\frac{3}{4}$ and $\frac{4}{5}$.

6. Simplify:

$$\left(\frac{2}{3} + \frac{3}{4}\right) \div 2\frac{1}{3} - \left(\frac{1}{2}\right)^3.$$

Answers on page A-7

Example 3 To find the **average** of a set of numbers, we add the numbers and then divide by the number of addends. Find the average of $\frac{1}{2}$, $\frac{3}{4}$, and $\frac{7}{8}$.

The average is given by

$$\left(\frac{1}{2} + \frac{3}{4} + \frac{7}{8}\right) \div 3.$$

To find the average, we carry out the computation using the rules for order of operations:

$$\left(\frac{1}{2} + \frac{3}{4} + \frac{7}{8}\right) \div 3 = \left(\frac{4}{8} + \frac{6}{8} + \frac{7}{8}\right) \div 3$$ **Doing the operations inside parentheses first: adding by finding a common denominator**

$$= \frac{17}{8} \div 3$$ **Adding**

$$= \frac{17}{8} \cdot \frac{1}{3}$$ **Dividing by multiplying by the reciprocal**

$$= \frac{17}{24}$$ **Multiplying**

The average is $\frac{17}{24}$.

Do Exercises 4 and 5.

Example 4 Simplify: $\left(\frac{7}{8} - \frac{1}{3}\right) \times 48 + \left(13 + \frac{4}{5}\right)^2$.

$$\left(\frac{7}{8} - \frac{1}{3}\right) \times 48 + \left(13 + \frac{4}{5}\right)^2$$

$$= \left(\frac{7}{8} \cdot \frac{3}{3} - \frac{1}{3} \cdot \frac{8}{8}\right) \times 48 + \left(13 \cdot \frac{5}{5} + \frac{4}{5}\right)^2$$ **Carrying out operations inside parentheses first. To do so, we first multiply by 1 to obtain the LCD.**

$$= \left(\frac{21}{24} - \frac{8}{24}\right) \times 48 + \left(\frac{65}{5} + \frac{4}{5}\right)^2$$

$$= \frac{13}{24} \times 48 + \left(\frac{69}{5}\right)^2$$ **Completing the operations within parentheses**

$$= \frac{13}{24} \times 48 + \frac{4761}{25}$$ **Evaluating exponential expressions next**

$$= 26 + \frac{4761}{25}$$ **Doing the multiplication**

$$= 26 + 190\frac{11}{25}$$ **Converting to a mixed numeral**

$$= 216\frac{11}{25}, \text{ or } \frac{5411}{25}$$ **Adding**

Answers can be given using either fractional notation or mixed numerals as desired. Consult with your instructor.

Do Exercise 6.

b Estimation with Fractional Notation and Mixed Numerals

We now estimate with fractional notation and mixed numerals.

Examples Estimate each of the following as 0, $\frac{1}{2}$, or 1.

5. $\frac{2}{17}$

A fraction is very close to 0 when the numerator is very small in comparison to the denominator. Thus, 0 is an estimate for $\frac{2}{17}$ because 2 is very small in comparison to 17. Thus, $\frac{2}{17} \approx 0$.

6. $\frac{11}{23}$

A fraction is very close to $\frac{1}{2}$ when the denominator is about twice the numerator. Thus, $\frac{1}{2}$ is an estimate for $\frac{11}{23}$ because $2 \cdot 11 = 22$ and 22 is close to 23. Thus, $\frac{11}{23} \approx \frac{1}{2}$.

7. $\frac{37}{38}$

A fraction is very close to 1 when the numerator is nearly equal to the denominator. Thus, 1 is an estimate for $\frac{37}{38}$ because 37 is nearly equal to 38. Thus, $\frac{37}{38} \approx 1$.

8. $\frac{43}{41}$

As in the preceding example, the numerator 43 is very close to the denominator 41. Thus, $\frac{43}{41} \approx 1$.

Do Exercises 7–10.

Example 9 Find a number for the blank so that $\frac{9}{\square}$ is close to but less than 1. Answers may vary.

If the number in the blank were 9, we would have 1, so we increase 9 to 10. The answer is 10; $\frac{9}{10}$ is close to 1. The number 11 would also be a correct answer; $\frac{9}{11}$ is close to 1.

Do Exercises 11 and 12.

Example 10 Estimate $16\frac{8}{9} + 11\frac{2}{13} - 4\frac{22}{43}$ as a whole number or as a mixed number where the fractional part is $\frac{1}{2}$.

We estimate each fraction as 0, $\frac{1}{2}$, or 1. Then we calculate:

$$16\frac{8}{9} + 11\frac{2}{13} - 4\frac{22}{43} \approx 17 + 11 - 4\frac{1}{2} = 23\frac{1}{2}.$$

Do Exercises 13–15.

Estimate each of the following as 0, $\frac{1}{2}$, or 1.

7. $\frac{3}{59}$

8. $\frac{61}{59}$

9. $\frac{29}{59}$

10. $\frac{57}{59}$

Find a number for the blank so that the fraction is close to but less than 1.

11. $\frac{11}{\square}$

12. $\frac{\square}{33}$

Estimate each of the following as a whole number or as a mixed numeral where the fractional part is $\frac{1}{2}$.

13. $5\frac{9}{10} + 26\frac{1}{2} - 10\frac{3}{29}$

14. $10\frac{7}{8} \cdot \left(25\frac{11}{13} - 14\frac{1}{9}\right)$

15. $\left(10\frac{4}{5} + 7\frac{5}{9}\right) \div \frac{17}{30}$

Answers on page A-7

Calculator Spotlight

Mixed Numerals on a Calculator. Fraction calculators are equipped with a key, often labeled [a b/c], that allows for computations with fractional notation and mixed numerals. To calculate

$$\frac{2}{3} + \frac{4}{5}$$

with such a fraction calculator, the following keystrokes can be used:

[2] [a b/c] [3] [+] [4] [a b/c] [5] [=].

The display that appears,

1 ⌋ 7 ⌋ 15 ,

represents the mixed numeral $1\frac{7}{15}$.

To express the answer in fractional notation, we use the following keystrokes:

[Shift] [d/c].

The display that appears,

22 ⌋ 15 ,

represents the fraction $\frac{22}{15}$.

To enter a mixed numeral like $3\frac{2}{5}$ on a fraction calculator equipped with an [a b/c] key, we press

[3] [a b/c] [2] [a b/c] [5].

The calculator's display is in the form

3 ⌋ 2 ⌋ 5 .

Some calculators are capable of displaying mixed numerals in the way in which we write them, as shown below.

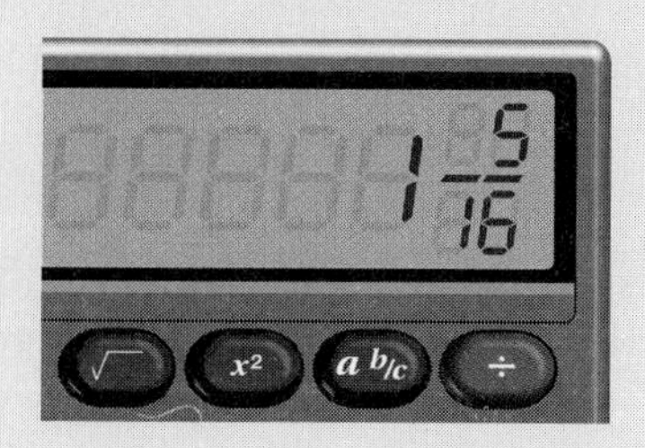

Exercises

Calculate using a fraction calculator. Give the answer in fractional notation.

1. $\frac{3}{8} + \frac{1}{4}$
2. $\frac{5}{12} + \frac{7}{10} - \frac{5}{12}$
3. $\frac{15}{7} \cdot \frac{1}{3}$
4. $\frac{19}{20} \div \frac{17}{35}$
5. $\frac{29}{30} - \frac{18}{25} \cdot \frac{2}{3}$
6. $\frac{1}{2} + \frac{13}{29} \cdot \frac{3}{4}$

Calculate using a fraction calculator. Give the answer in mixed numerals.

7. $4\frac{1}{2} \cdot 5\frac{3}{7}$
8. $7\frac{2}{3} \div 9\frac{4}{5}$
9. $8\frac{3}{7} + 5\frac{2}{9}$
10. $13\frac{4}{9} - 7\frac{5}{8}$
11. $13\frac{1}{4} - 2\frac{1}{5} \cdot 4\frac{3}{8}$
12. $2\frac{5}{6} + 5\frac{1}{6} \cdot 3\frac{1}{4}$

Exercise Set 3.7

a Simplify.

1. $\frac{1}{2} \cdot \frac{1}{3} \cdot \frac{1}{4}$

2. $\frac{1}{3} \cdot \frac{1}{4} \cdot \frac{1}{5}$

3. $6 \div 3 \div 5$

4. $12 \div 4 \div 8$

5. $\frac{2}{3} \div \frac{4}{3} \div \frac{7}{8}$

6. $\frac{5}{6} \div \frac{3}{4} \div \frac{2}{5}$

7. $\frac{5}{8} \div \frac{1}{4} - \frac{2}{3} \cdot \frac{4}{5}$

8. $\frac{4}{7} \cdot \frac{7}{15} + \frac{2}{3} \div 8$

9. $\frac{3}{4} - \frac{2}{3} \cdot \left(\frac{1}{2} + \frac{2}{5}\right)$

10. $\frac{3}{4} \div \frac{1}{2} \cdot \left(\frac{8}{9} - \frac{2}{3}\right)$

11. $28\frac{1}{8} - 5\frac{1}{4} + 3\frac{1}{2}$

12. $10\frac{3}{5} - 4\frac{1}{10} - 1\frac{1}{2}$

13. $\frac{7}{8} \div \frac{1}{2} \cdot \frac{1}{4}$

14. $\frac{7}{10} \cdot \frac{4}{5} \div \frac{2}{3}$

15. $\left(\frac{2}{3}\right)^2 - \frac{1}{3} \cdot 1\frac{1}{4}$

16. $\left(\frac{3}{4}\right)^2 + 3\frac{1}{2} \div 1\frac{1}{4}$

17. $\frac{1}{2} - \left(\frac{1}{2}\right)^2 + \left(\frac{1}{2}\right)^3$

18. $1 + \frac{1}{4} + \left(\frac{1}{4}\right)^2 - \left(\frac{1}{4}\right)^3$

19. Find the average of $\frac{2}{3}$ and $\frac{7}{8}$.

20. Find the average of $\frac{1}{4}$ and $\frac{1}{5}$.

21. Find the average of $\frac{1}{6}$, $\frac{1}{8}$, and $\frac{3}{4}$.

22. Find the average of $\frac{4}{5}$, $\frac{1}{2}$, and $\frac{1}{10}$.

23. Find the average of $3\frac{1}{2}$ and $9\frac{3}{8}$.

24. Find the average of $10\frac{2}{3}$ and $24\frac{5}{6}$.

Simplify.

25. $\left(\frac{2}{3} + \frac{3}{4}\right) \div \left(\frac{5}{6} - \frac{1}{3}\right)$

26. $\left(\frac{3}{5} - \frac{1}{2}\right) \div \left(\frac{3}{4} - \frac{3}{10}\right)$

27. $\left(\frac{1}{2} + \frac{1}{3}\right)^2 \cdot 144 - \frac{5}{8} \div 10\frac{1}{2}$

28. $\left(3\frac{1}{2} - 2\frac{1}{3}\right)^2 + 6 \cdot 2\frac{1}{2} \div 32$

b Estimate each of the following as 0, $\frac{1}{2}$, or 1.

29. $\frac{2}{47}$

30. $\frac{4}{5}$

31. $\frac{1}{13}$

32. $\frac{7}{8}$

33. $\frac{6}{11}$

34. $\frac{10}{13}$

35. $\frac{7}{15}$

36. $\frac{1}{16}$

37. $\frac{7}{100}$

38. $\frac{5}{9}$

39. $\frac{19}{20}$

40. $\frac{5}{12}$

Find a number for the blank so that the fraction is close to but greater than $\frac{1}{2}$. Answers may vary.

41. $\frac{\square}{11}$

42. $\frac{\square}{8}$

43. $\frac{\square}{23}$

44. $\frac{\square}{35}$

45. $\frac{10}{\square}$

46. $\frac{7}{\square}$

47. $\frac{8}{\square}$

48. $\frac{51}{\square}$

Find a number for the blank so that the fraction is close to but greater than 1. Answers may vary.

49. $\dfrac{7}{\blacksquare}$ **50.** $\dfrac{11}{\blacksquare}$ **51.** $\dfrac{13}{\blacksquare}$ **52.** $\dfrac{27}{\blacksquare}$

53. $\dfrac{\blacksquare}{15}$ **54.** $\dfrac{\blacksquare}{9}$ **55.** $\dfrac{\blacksquare}{18}$ **56.** $\dfrac{\blacksquare}{100}$

Estimate each part of the following as a whole number or as a mixed numeral where the fractional part is $\frac{1}{2}$.

57. $2\dfrac{7}{8}$ **58.** $1\dfrac{1}{3}$ **59.** $12\dfrac{5}{6}$

60. $26\dfrac{6}{13}$ **61.** $\dfrac{4}{5} + \dfrac{7}{8}$ **62.** $\dfrac{1}{12} \cdot \dfrac{7}{15}$

63. $\dfrac{2}{3} + \dfrac{7}{13} + \dfrac{5}{9}$ **64.** $\dfrac{8}{9} + \dfrac{4}{5} + \dfrac{11}{12}$ **65.** $\dfrac{43}{100} + \dfrac{1}{10} - \dfrac{11}{1000}$

66. $\dfrac{23}{24} + \dfrac{37}{39} + \dfrac{51}{50}$ **67.** $7\dfrac{29}{60} + 10\dfrac{12}{13} \cdot 24\dfrac{2}{17}$ **68.** $5\dfrac{13}{14} - 1\dfrac{5}{8} + 1\dfrac{23}{28} \cdot 6\dfrac{35}{74}$

69. $24 \div 7\dfrac{8}{9}$ **70.** $43\dfrac{16}{17} \div 11\dfrac{2}{13}$ **71.** $76\dfrac{3}{14} + 23\dfrac{19}{20}$

72. $76\dfrac{13}{14} \cdot 23\dfrac{17}{20}$ **73.** $16\dfrac{1}{5} \div 2\dfrac{1}{11} + 25\dfrac{9}{10} - 4\dfrac{11}{23}$ **74.** $96\dfrac{2}{13} \div 5\dfrac{19}{20} + 3\dfrac{1}{7} \cdot 5\dfrac{18}{21}$

Skill Maintenance

75. Multiply: $27 \cdot 126$. [1.5b]

76. Multiply: $132 \cdot 7865$. [1.5b]

77. Divide: $7865 \div 132$. [1.6c]

78. Multiply: $\dfrac{2}{3} \cdot 63$. [2.4a]

79. Divide and simplify: $\frac{4}{5} \div \frac{3}{10}$. [2.7b]

80. Tell whether the given number is prime, composite, or neither. [2.1c]

1, 5, 7, 9, 14, 23, 43

Solve.

81. Ian purchased 6 lb of cold cuts for a luncheon. If Ian is to allow $\frac{3}{8}$ lb per person, how many people can he invite to the luncheon? [2.7d]

82. A 3-oz serving of crabmeat contains 85 milligrams (mg) of cholesterol. A 3-oz serving of shrimp contains 128 mg of cholesterol. How much more cholesterol is in the shrimp? [1.8a]

Synthesis

83. ◆ A student insists that $3\frac{2}{5} \cdot 1\frac{3}{7} = 3\frac{6}{35}$. What mistake is the student making and how should he have proceeded?

84. ◆ A student insists that $5 \cdot 3\frac{2}{7} = (5 \cdot 3) \cdot \left(5 \cdot \frac{2}{7}\right)$. What mistake is the student making and how should she have proceeded?

85. a) Find an expression for the sum of the areas of the two rectangles shown here.
b) Simplify the expression.
c) How is the computation in part (b) related to the rules for order of operations?

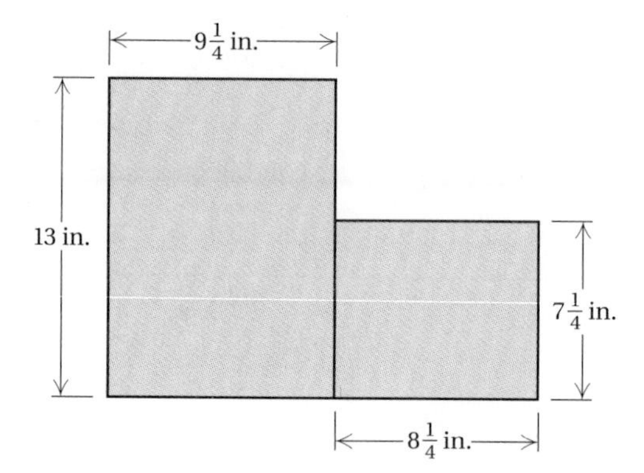

86. Find r if

$$\frac{1}{r} = \frac{1}{100} + \frac{1}{150} + \frac{1}{200}.$$

87. ▦ In the sum below, a and b are digits. Find a and b.

$$\frac{a}{17} + \frac{1b}{23} = \frac{35a}{391}$$

88. ▦ Consider only the numbers 3, 4, 5, and 6. Assume each can be placed in a blank in the following.

$$\square + \frac{\square}{\square} \cdot \square = ?$$

What placement of the numbers in the blanks yields the largest number?

89. ▦ Consider only the numbers 2, 3, 4, and 5. Assume each is placed in a blank in the following.

$$\frac{\square}{\square} + \frac{\square}{\square} = ?$$

What placement of the numbers in the blanks yields the largest sum?

90. ▦ Use a standard calculator. Arrange the following in order from smallest to largest.

$$\frac{3}{4}, \frac{17}{21}, \frac{13}{15}, \frac{7}{9}, \frac{15}{17}, \frac{13}{12}, \frac{19}{22}$$

Summary and Review Exercises: Chapter 3

The objectives to be tested in addition to the material in this chapter are [1.5b], [1.8a], [2.6a], and [2.7b].

Find the LCM. [3.1a]

1. 12 and 18

2. 18 and 45

3. 3, 6, and 30

4. 26, 36, and 54

Add and simplify. [3.2b]

5. $\frac{6}{5} + \frac{3}{8}$

6. $\frac{5}{16} + \frac{1}{12}$

7. $\frac{6}{5} + \frac{11}{15}$

8. $\frac{5}{16} + \frac{1}{8}$

Subtract and simplify. [3.3a]

9. $\frac{5}{9} - \frac{2}{9}$

10. $\frac{7}{8} - \frac{3}{4}$

11. $\frac{11}{27} - \frac{2}{9}$

12. $\frac{5}{6} - \frac{2}{9}$

Use < or > for ▒ to write a true sentence. [3.3b]

13. $\frac{4}{7}$ ▒ $\frac{5}{9}$

14. $\frac{8}{9}$ ▒ $\frac{11}{13}$

Solve. [3.3c]

15. $x + \frac{2}{5} = \frac{7}{8}$

16. $\frac{1}{2} + y = \frac{9}{10}$

Convert to fractional notation. [3.4a]

17. $7\frac{1}{2}$

18. $8\frac{3}{8}$

19. $4\frac{1}{3}$

20. $10\frac{5}{7}$

Convert to a mixed numeral. [3.4b]

21. $\frac{7}{3}$

22. $\frac{27}{4}$

23. $\frac{63}{5}$

24. $\frac{7}{2}$

Divide. Write a mixed numeral for the answer. [3.4c]

25. $9\,\overline{)\,7896}$

26. $23\,\overline{)\,10{,}493}$

Add. Write a mixed numeral for the answer. [3.5a]

27. $5\frac{3}{5} + 4\frac{4}{5}$

28. $8\frac{1}{3} + 3\frac{2}{5}$

29. $5\frac{5}{6} + 4\frac{5}{6}$

30. $2\frac{3}{4} + 5\frac{1}{2}$

Subtract. Write a mixed numeral for the answer. [3.5b]

31. $12 - 4\frac{2}{9}$

32. $9\frac{3}{5} - 4\frac{13}{15}$

33. $10\frac{1}{4} - 6\frac{1}{10}$

34. $24 - 10\frac{5}{8}$

Multiply. Write a mixed numeral for the answer. [3.6a]

35. $6 \cdot 2\frac{2}{3}$

36. $5\frac{1}{4} \cdot \frac{2}{3}$

37. $2\frac{1}{5} \cdot 1\frac{1}{10}$

38. $2\frac{2}{5} \cdot 2\frac{1}{2}$

Divide. Write a mixed numeral for the answer. [3.6b]

39. $27 \div 2\frac{1}{4}$

40. $2\frac{2}{5} \div 1\frac{7}{10}$

41. $3\frac{1}{4} \div 26$

42. $4\frac{1}{5} \div 4\frac{2}{3}$

Solve. [3.5c], [3.6c]

43. A curtain requires $2\frac{3}{5}$ yd of material. How many curtains can be made from 39 yd of material?

44. *Alcoa Stock Price.* On the first day of trading on the stock market, stock in Alcoa opened at $\$67\frac{3}{4}$ and rose by $\$2\frac{5}{8}$ at the close of trading. What was the stock's closing price?

45. A board $\frac{9}{10}$ in. thick is glued to a board $\frac{8}{10}$ in. thick. The glue is $\frac{3}{100}$ in. thick. How thick is the result?

46. What is the sum of the areas in the figure below?

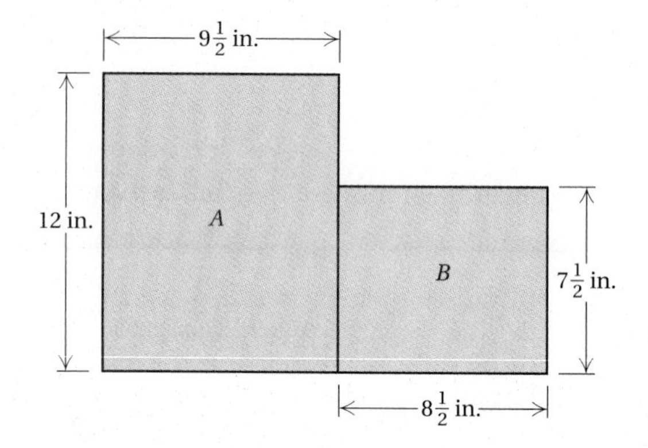

47. In the figure above, how much larger is the area of rectangle A than the area of rectangle B?

48. A wedding-cake recipe requires 12 cups of shortening. Being calorie-conscious, the wedding couple decides to reduce the shortening by $3\frac{5}{8}$ cups and replace it with prune purée. How many cups of shortening are used in their new recipe?

49. Simplify this expression using the rules for order of operations: [3.7a]

$$\frac{1}{8} \div \frac{1}{4} + \frac{1}{2}.$$

50. Find the average of $\frac{1}{2}$, $\frac{1}{4}$, $\frac{1}{3}$, and $\frac{1}{5}$. [3.7a]

Estimate each of the following as 0, $\frac{1}{2}$, or 1. [3.7b]

51. $\frac{29}{59}$

52. $\frac{2}{59}$

53. $\frac{61}{59}$

Estimate each of the following as a whole number or as a mixed numeral where the fractional part is $\frac{1}{2}$. [3.7b]

54. $6\frac{7}{8}$

55. $10\frac{2}{17}$

56. $\frac{3}{10} + \frac{5}{6} + \frac{31}{29}$

57. $32\frac{14}{15} + 27\frac{3}{4} - 4\frac{25}{28} \cdot 6\frac{37}{76}$

Skill Maintenance

58. Multiply and simplify: $\frac{9}{10} \cdot \frac{4}{3}$. [2.6a]

59. Divide and simplify: $\frac{5}{4} \div \frac{5}{6}$. [2.7b]

60. Multiply: $176 \cdot 4023$. [1.5b]

61. A factory produces 85 digital alarm clocks per day. How long will it take to fill an order for 1445 clocks? [1.8a]

Synthesis

62. ◆ Discuss the role of least common multiples in adding and subtracting with fractional notation. [3.2b], [3.3a]

63. ◆ Find a real-world situation that fits this equation: [3.5c], [3.6c]

$$2 \cdot 15\frac{3}{4} + 2 \cdot 28\frac{5}{8} = 88\frac{3}{4}.$$

64. *Orangutan Circus Act.* Yuri and Olga are orangutans who perform in a circus by riding bicycles around a circular track. It takes Yuri 6 min and Olga 4 min to make one trip around the track. Suppose they start at the same point and then complete their act when they again reach the same point. How long is their act? [3.1a]

Test: Chapter 3

1. Find the LCM of 12 and 16.

Add and simplify.

2. $\frac{1}{2} + \frac{5}{2}$

3. $\frac{7}{8} + \frac{2}{3}$

4. $\frac{7}{10} + \frac{9}{100}$

Subtract and simplify.

5. $\frac{5}{6} - \frac{3}{6}$

6. $\frac{5}{6} - \frac{3}{4}$

7. $\frac{17}{24} - \frac{5}{8}$

8. Use < or > for ▒ to write a true sentence:

$\frac{6}{7}$ ▒ $\frac{21}{25}$.

9. Solve: $x + \frac{2}{3} = \frac{11}{12}$.

Convert to fractional notation.

10. $3\frac{1}{2}$

11. $9\frac{7}{8}$

Convert to a mixed numeral.

12. $\frac{9}{2}$

13. $\frac{74}{9}$

Divide. Write a mixed numeral for the answer.

14. $11\overline{)1789}$

Add. Write a mixed numeral for the answer.

15. $6\frac{2}{5} + 7\frac{4}{5}$

16. $9\frac{1}{4} + 5\frac{1}{6}$

Subtract. Write a mixed numeral for the answer.

17. $10\frac{1}{6} - 5\frac{7}{8}$

18. $14 - 7\frac{5}{6}$

Multiply. Write a mixed numeral for the answer.

19. $9 \cdot 4\frac{1}{3}$

20. $6\frac{3}{4} \cdot \frac{2}{3}$

21. $3\frac{1}{3} \cdot 1\frac{3}{4}$

Answers

1. ______
2. ______
3. ______
4. ______
5. ______
6. ______
7. ______
8. ______
9. ______
10. ______
11. ______
12. ______
13. ______
14. ______
15. ______
16. ______
17. ______
18. ______
19. ______
20. ______
21. ______

Answers

22. ____

23. ____

24. ____

25. ____

26. ____

27. ____

28. ____

29. ____

30. ____

31. ____

32. ____

33. ____

34. ____

35. ____

36. ____

37. ____

38. ____

39. ____

40. ____

41. ____

42. a) ____

b) ____

43. ____

Divide. Write a mixed numeral for the answer.

22. $33 \div 5\frac{1}{2}$

23. $2\frac{1}{3} \div 1\frac{1}{6}$

24. $2\frac{1}{12} \div 75$

Solve.

25. *Turkey Loaf.* A low-cholesterol turkey loaf recipe calls for $3\frac{1}{2}$ cups of turkey breast. How much turkey is needed for 5 recipes?

26. An order of books for a math course weighs 220 lb. Each book weighs $2\frac{3}{4}$ lb. How many books are in the order?

27. The weights of two students are $183\frac{2}{3}$ lb and $176\frac{3}{4}$ lb. What is their total weight?

28. A standard piece of paper is $8\frac{1}{2}$ in. by 11 in. By how much does the length exceed the width?

29. Simplify: $\frac{2}{3} + 1\frac{1}{3} \cdot 2\frac{1}{8}$.

30. Find the average of $\frac{2}{5}$, $\frac{3}{4}$, and $\frac{1}{2}$.

Estimate each of the following as 0, $\frac{1}{2}$, or 1.

31. $\frac{44}{89}$

32. $\frac{3}{82}$

33. $\frac{93}{91}$

Estimate each of the following as a whole number or as a mixed numeral where the fractional part is $\frac{1}{2}$.

34. $3\frac{8}{9}$

35. $18\frac{9}{17}$

36. $256 \div 15\frac{19}{21}$

37. $43\frac{15}{31} \cdot 27\frac{3}{4} - 9\frac{15}{28} + 6\frac{5}{76}$

Skill Maintenance

38. Multiply:
$$\begin{array}{r} 4\,5\,6\,1 \\ \times \quad 7\,6 \\ \hline \end{array}$$

39. Divide and simplify: $\frac{4}{3} \div \frac{5}{6}$.

40. Multiply and simplify: $\frac{4}{3} \cdot \frac{5}{6}$.

41. A container has 8570 oz of beverage with which to fill 16-oz bottles. How many of these bottles can be filled? How much beverage will be left over?

Synthesis

42. The students in a math class can be organized into study groups of 8 each such that no students are left out. The same class of students can also be organized into groups of 6 such that no students are left out.

a) Find some class sizes for which this will work.

b) Find the smallest such class size.

43. Dolores runs 17 laps at her health club. Terence runs 17 laps at his health club. If the track at Dolores's health club is $\frac{1}{7}$ mi long, and the track at Terence's is $\frac{1}{8}$ mi long, who runs farther? How much farther?

Cumulative Review: Chapters 1–3

1. In the number 2753, what digit names tens?

2. Write expanded notation for 6075.

3. Write a word name for the number in the following sentence: The diameter of Uranus is 29,500 miles.

Add and simplify.

4. $\begin{array}{r} 628 \\ +\ 271 \\ \hline \end{array}$

5. $\begin{array}{r} 3704 \\ +\ 5278 \\ \hline \end{array}$

6. $\frac{3}{8} + \frac{1}{24}$

7. $\begin{array}{r} 2\frac{3}{4} \\ +\ 5\frac{1}{2} \\ \hline \end{array}$

Subtract and simplify.

8. $\begin{array}{r} 7469 \\ -\ 2345 \\ \hline \end{array}$

9. $\begin{array}{r} 7605 \\ -\ 3087 \\ \hline \end{array}$

10. $\frac{3}{4} - \frac{1}{3}$

11. $\begin{array}{r} 2\frac{1}{3} \\ -\ 1\frac{1}{6} \\ \hline \end{array}$

Multiply and simplify.

12. $\begin{array}{r} 278 \\ \times\ 18 \\ \hline \end{array}$

13. $\begin{array}{r} 894 \\ \times\ 328 \\ \hline \end{array}$

14. $\frac{9}{10} \cdot \frac{5}{3}$

15. $18 \cdot \frac{5}{6}$

16. $2\frac{1}{3} \cdot 3\frac{1}{7}$

Divide. Write the answer with the remainder in the form 34 R 7.

17. $6\overline{)4290}$

18. $45\overline{)2531}$

19. In Question 18, write a mixed numeral for the answer.

Divide and simplify, where appropriate.

20. $\frac{2}{5} \div \frac{7}{10}$

21. $2\frac{1}{5} \div \frac{3}{10}$

22. Round 38,478 to the nearest hundred.

23. Find the LCM of 18 and 24.

24. Determine whether 3718 is divisible by 8.

25. Find all factors of 16.

26. What part is shaded?

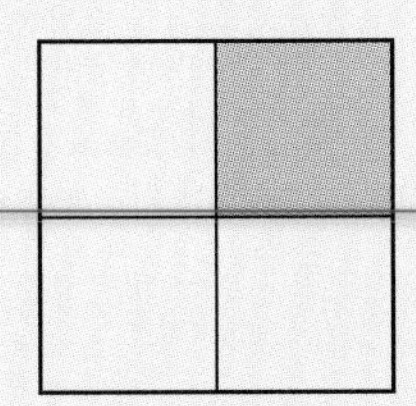

Use $<$, $>$, or $=$ for ■ to write a true sentence.

27. $\frac{4}{5}$ ■ $\frac{4}{6}$

28. $\frac{5}{12}$ ■ $\frac{3}{7}$

Simplify.

29. $\frac{36}{45}$

30. $\frac{320}{10}$

31. Convert to fractional notation: $4\frac{5}{8}$.

32. Convert to a mixed numeral: $\frac{17}{3}$.

Solve.

33. $x + 24 = 117$

34. $x + \frac{7}{9} = \frac{4}{3}$

35. $\frac{7}{9} \cdot t = \frac{4}{3}$

36. $y = 32{,}580 \div 36$

Solve.

37. A jacket costs \$87 and a coat costs \$148. How much does it cost to buy both?

38. An emergency food pantry fund contains \$423. From this fund, \$148 and \$167 are withdrawn for expenses. How much is left in the fund?

39. A lot measures 27 ft by 11 ft. What is its area?

40. How many people can get equal \$16 shares from a total of \$496?

41. A recipe calls for $\frac{4}{5}$ tsp of salt. How much salt should be used in $\frac{1}{2}$ recipe?

42. A book weighs $2\frac{3}{5}$ lb. How much do 15 books weigh?

43. How many pieces, each $2\frac{3}{8}$ ft long, can be cut from a piece of wire 38 ft long?

44. In a walkathon, one person walked $\frac{9}{10}$ mi and another walked $\frac{75}{100}$ mi. What was the total distance walked?

Estimate each of the following as 0, $\frac{1}{2}$, or 1.

45. $\frac{29}{30}$

46. $\frac{15}{29}$

47. $\frac{2}{43}$

Estimate each of the following as a whole number or as a mixed numeral where the fractional part is $\frac{1}{2}$.

48. $30\frac{4}{53}$

49. $\frac{9}{10} - \frac{7}{8} + \frac{41}{39}$

50. $78\frac{14}{15} - 28\frac{3}{4} - 7\frac{25}{28} \div \frac{65}{66}$

Synthesis

51. a) Simplify each of the following, using fractional notation for your answers.

$$\frac{1}{1 \cdot 2}$$

$$\frac{1}{1 \cdot 2} + \frac{1}{2 \cdot 3}$$

$$\frac{1}{1 \cdot 2} + \frac{1}{2 \cdot 3} + \frac{1}{3 \cdot 4}$$

$$\frac{1}{1 \cdot 2} + \frac{1}{2 \cdot 3} + \frac{1}{3 \cdot 4} + \frac{1}{4 \cdot 5}$$

b) Look for a pattern in your answers to part (a). Then find the following without carrying out the computations.

$$\frac{1}{1 \cdot 2} + \frac{1}{2 \cdot 3} + \frac{1}{3 \cdot 4} + \frac{1}{4 \cdot 5} + \frac{1}{5 \cdot 6} + \frac{1}{6 \cdot 7} + \frac{1}{7 \cdot 8} + \frac{1}{8 \cdot 9} + \frac{1}{9 \cdot 10}$$

4

Decimal Notation

An Application

As life becomes busier, Americans are eating many more meals outside the home. In 1995, the average check for a casual meal eaten out was $28.90 and in 1996, it was $39.51 (***Source:*** Sandelman and Associates, Brea CA). How much more is the average check for 1996 than that for 1995?

This problem appears as Example 1 in Section 4.7.

The Mathematics

We let c = the additional amount spent in 1996. The problem then translates to the equation

$$\underbrace{28.90 + c = 39.51.}$$

This equation involves decimal notation, which arises often in applied problems.

World Wide Web For more information, visit us at www.mathmax.com

Introduction

In this chapter, we consider the operations of addition, subtraction, multiplication, and division with decimal notation. This will allow us to solve applied problems like the one below. We will also study estimating sums, differences, products, and quotients. Conversion between fractional and decimal notation in which the decimal notation may be repeating will also be discussed.

Pretest: Chapter 4

1. Write a word name: 2.347.

2. Write a word name, as on a check, for $3264.78.

Write fractional notation.

3. 0.21

4. 5.408

Write decimal notation.

5. $\frac{379}{1000}$

6. $28\frac{439}{1000}$

Which number is larger?

7. 3.2, 0.321

8. 0.099, 0.091

Round 21.0448 to the nearest:

9. Tenth.

10. Thousandth.

11. Add:
$$\begin{array}{r} 601.3 \\ 5.81 \\ +\ 0.109 \\ \hline \end{array}$$

12. Subtract:
$$\begin{array}{r} 40.0 \\ -\ 0.9099 \\ \hline \end{array}$$

Multiply.

13.
$$\begin{array}{r} 0.835 \\ \times\ 0.74 \\ \hline \end{array}$$

14. 0.001×324.56

Divide.

15. $6.6\,\overline{)\,200.64}$

16. $\frac{576.98}{1000}$

Solve.

17. $9.6 \cdot y = 808.896$

18. $54.96 + q = 6400.117$

Solve.

19. On a three-day trip, a traveler drove these distances: 432.6 mi, 179.2 mi, and 469.8 mi. What is the total number of miles driven?

20. A checking account contained $434.19. After a $148.24 check was drawn, how much was left in the account?

21. What is the cost of 6 videotapes at $14.95 each?

22. A developer paid $47,567.89 for 14 acres of land. How much was paid for 1 acre? Round to the nearest cent.

23. Estimate the product 6.92×32.458 by rounding to the nearest one.

Find decimal notation. Use multiplying by 1.

24. $\frac{7}{5}$

25. $\frac{37}{40}$

Find decimal notation. Use division.

26. $\frac{11}{4}$

27. $\frac{29}{7}$

Round the answer to Exercise 27 to the nearest:

28. Tenth.

29. Hundredth.

30. Thousandth.

31. Convert from cents to dollars: 949 cents.

32. Convert to standard notation: 490 trillion.

Calculate.

33. $(1 - 0.06)^2 + 8[5(12.1 - 7.8) + 20(17.3 - 8.7)]$

34. $\frac{2}{3} \times 89.95 - \frac{5}{9} \times 3.234$

The objectives to be tested in addition to the material in this chapter are as follows.

[2.1d] Find the prime factorization of a composite number.
[2.5b] Simplify fractional notation.
[3.5a, b] Add and subtract using mixed numerals.
[3.6a, b] Multiply and divide using mixed numerals.

4.1 Decimal Notation, Order, and Rounding

Objectives

- a Given decimal notation, write a word name, and write a word name for an amount of money.
- b Convert from decimal notation to fractional notation.
- c Convert from fractional notation and mixed numerals to decimal notation.
- d Given a pair of numbers in decimal notation, tell which is larger.
- e Round to the nearest thousandth, hundredth, tenth, one, ten, hundred, or thousand.

For Extra Help

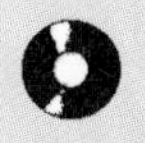

TAPE 7 TAPE 6B MAC WIN CD-ROM

The set of **arithmetic numbers,** or **nonnegative rational numbers,** consists of the whole numbers 0, 1, 2, 3, 4, 5, 6, 7, 8, 9, 10, and so on, and fractions like $\frac{1}{2}$, $\frac{2}{3}$, $\frac{7}{8}$, $\frac{17}{10}$, and so on. We studied the use of fractional notation for arithmetic numbers in Chapters 2 and 3. In Chapter 4, we will study the use of *decimal notation.* Although we are using different notation, we are still considering the same set of numbers. For example, instead of using fractional notation for $\frac{7}{8}$, we use decimal notation, 0.875.

a Decimal Notation and Word Names

Decimal notation for the women's shotput record is 74.249 ft. To understand what 74.249 means, we use a **place-value chart.** The value of each place is $\frac{1}{10}$ as large as the one to its left.

PLACE-VALUE CHART							
Hundreds	Tens	Ones	Ten*ths*	Hundred*ths*	Thousand*ths*	Ten-Thousand*ths*	Hundred-Thousand*ths*
100	10	1	$\frac{1}{10}$	$\frac{1}{100}$	$\frac{1}{1000}$	$\frac{1}{10,000}$	$\frac{1}{100,000}$
	7	4 .	2	4	9		

The decimal notation 74.249 means

$$7 \text{ tens} + 4 \text{ ones} + 2 \text{ tenths} + 4 \text{ hundredths} + 9 \text{ thousandths},$$

or $$7 \cdot 10 + 4 \cdot 1 + 2 \cdot \frac{1}{10} + 4 \cdot \frac{1}{100} + 9 \cdot \frac{1}{1000},$$

or $$70 + 4 + \frac{2}{10} + \frac{4}{100} + \frac{9}{1000}.$$

A mixed numeral for 74.249 is $74\frac{249}{1000}$. We read 74.249 as "seventy-four and two hundred forty-nine thousandths." When we come to the decimal point, we read "and." We can also read 74.249 as "seven four *point* two four nine."

To write a word name from decimal notation,

a) write a word name for the whole number (the number named to the left of the decimal point),

397.685 → Three hundred ninety-seven

b) write the word "and" for the decimal point, and

397.685 → Three hundred ninety-seven and

c) write a word name for the number named to the right of the decimal point, followed by the place value of the last digit.

397.685 → Three hundred ninety-seven and six hundred eighty-five *thousandths*

Write a word name for the number.

1. Each person in this country consumes an average of 21.1 gallons of coffee per year (***Source:*** Department of Agriculture).

2. The racehorse *Swale* won the Belmont Stakes in a time of 2.4533 minutes.

3. 245.89

4. 31,079.764

Write a word name as on a check.

5. $4217.56

6. $13.98

Answers on page A-8

Example 1 Write a word name for the number in this sentence: Each person consumes an average of 41.2 gallons of water per year.

Forty-one and two tenths

Example 2 Write a word name for 410.87.

Four hundred ten and eighty-seven hundredths

Example 3 Write a word name for the number in this sentence: The world record in the men's marathon is 2.1833 hours.

Two and one thousand eight hundred thirty-three ten-thousandths

Example 4 Write a word name for 1788.405.

One thousand, seven hundred eighty-eight and four hundred five thousandths

Do Exercises 1–4.

Decimal notation is also used with money. It is common on a check to write "and ninety-five cents" as "and $\frac{95}{100}$ dollars."

Example 5 Write a word name for the amount on the check, $5876.95.

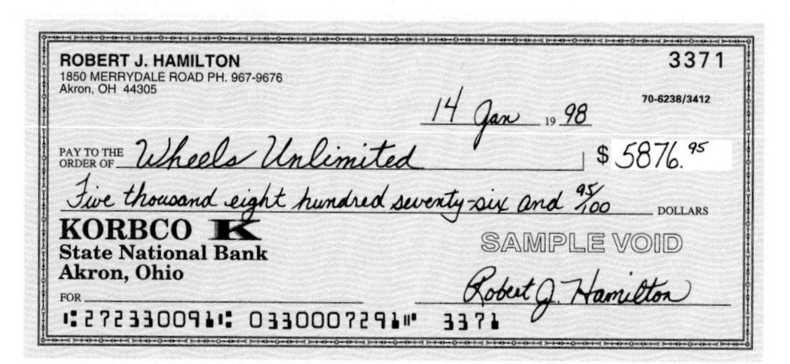

Five thousand, eight hundred seventy-six and $\frac{95}{100}$ dollars

Do Exercises 5 and 6.

b Converting from Decimal Notation to Fractional Notation

We can find fractional notation as follows:

$$9.875 = 9 + \frac{8}{10} + \frac{7}{100} + \frac{5}{1000}$$

$$= 9 \cdot \frac{1000}{1000} + \frac{8}{10} \cdot \frac{100}{100} + \frac{7}{100} \cdot \frac{10}{10} + \frac{5}{1000}$$

$$= \frac{9000}{1000} + \frac{800}{1000} + \frac{70}{1000} + \frac{5}{1000} = \frac{9875}{1000}.$$

Decimal notation → 9.875 (3 decimal places)

Fractional notation → $\frac{9875}{1000}$ (3 zeros)

To convert from decimal to fractional notation,

a) count the number of decimal places, — 4.98 (2 places)

b) move the decimal point that many places to the right, and — 4.98. Move 2 places.

c) write the answer over a denominator with a 1 followed by that number of zeros. — $\frac{498}{100}$ 2 zeros

Example 6 Write fractional notation for 0.876. Do not simplify.

0.876   0.876. (3 places) $0.876 = \frac{876}{1000}$ (3 zeros)

For a number like 0.876, we generally write a 0 before the decimal point to avoid forgetting or omitting it.

Example 7 Write fractional notation for 56.23. Do not simplify.

56.23   56.23. (2 places) $56.23 = \frac{5623}{100}$ (2 zeros)

Example 8 Write fractional notation for 1.5018. Do not simplify.

1.5018   1.5018. (4 places) $1.5018 = \frac{15{,}018}{10{,}000}$ (4 zeros)

Do Exercises 7–10.

c Converting from Fractional Notation and Mixed Numerals to Decimal Notation

If fractional notation has a denominator that is a power of ten, such as 10, 100, 1000, and so on, we reverse the procedure we used before.

To convert from fractional notation to decimal notation when the denominator is 10, 100, 1000, and so on,

a) count the number of zeros, and — $\frac{8679}{1000}$ (3 zeros)

b) move the decimal point that number of places to the left. Leave off the denominator. — 8.679. Move 3 places. $\frac{8679}{1000} = 8.679$

Write fractional notation.

7. 0.896

8. 23.78

9. 5.6789

10. 1.9

Answers on page A-8

Write decimal notation.

11. $\frac{743}{100}$

12. $\frac{406}{1000}$

13. $\frac{67,089}{10,000}$

14. $\frac{9}{10}$

15. $\frac{57}{1000}$

16. $\frac{830}{10,000}$

Write decimal notation.

17. $4\frac{3}{10}$

18. $283\frac{71}{100}$

19. $456\frac{13}{1000}$

Answers on page A-8

Example 9 Write decimal notation for $\frac{47}{10}$.

$\frac{47}{10}$ ← 1 zero   4.7. 1 place $\frac{47}{10} = 4.7$

Example 10 Write decimal notation for $\frac{123,067}{10,000}$.

$\frac{123,067}{10,000}$ ← 4 zeros   12.3067. 4 places $\frac{123,067}{10,000} = 12.3067$

Example 11 Write decimal notation for $\frac{13}{1000}$.

$\frac{13}{1000}$ ← 3 zeros   0.013. 3 places $\frac{13}{1000} = 0.013$

Example 12 Write decimal notation for $\frac{570}{100,000}$.

$\frac{570}{100,000}$ ← 5 zeros   0.00570. 5 places $\frac{570}{100,000} = 0.0057$

Do Exercises 11–16.

When denominators are numbers other than 10, 100, and so on, we will use another method for conversion. It will be considered in Section 4.5.

If a mixed numeral has a fractional part with a denominator that is a power of ten, such as 10, 100, or 1000, and so on, we first write the mixed numeral as a sum of a whole number and a fraction. Then we convert to decimal notation.

Example 13 Write decimal notation for $23\frac{59}{100}$.

$$23\frac{59}{100} = 23 + \frac{59}{100} = 23 \text{ and } \frac{59}{100} = 23.59$$

Example 14 Write decimal notation for $772\frac{129}{10,000}$.

$$772\frac{129}{10,000} = 772 + \frac{129}{10,000} = 772 \text{ and } \frac{129}{10,000} = 772.0129$$

Do Exercises 17–19.

d Order

To understand how to compare numbers in decimal notation, consider 0.85 and 0.9. First note that 0.9 = 0.90 because $\frac{9}{10} = \frac{90}{100}$. Then $0.85 = \frac{85}{100}$ and $0.90 = \frac{90}{100}$. Since $\frac{85}{100} < \frac{90}{100}$, it follows that $0.85 < 0.90$. This leads us to a quick way to compare two numbers in decimal notation.

> To compare two numbers in decimal notation, start at the left and compare corresponding digits moving from left to right. If two digits differ, the number with the larger digit is the larger of the two numbers. To ease the comparison, extra zeros can be written to the right of the last decimal place.

Example 15 Which of 2.109 and 2.1 is larger?

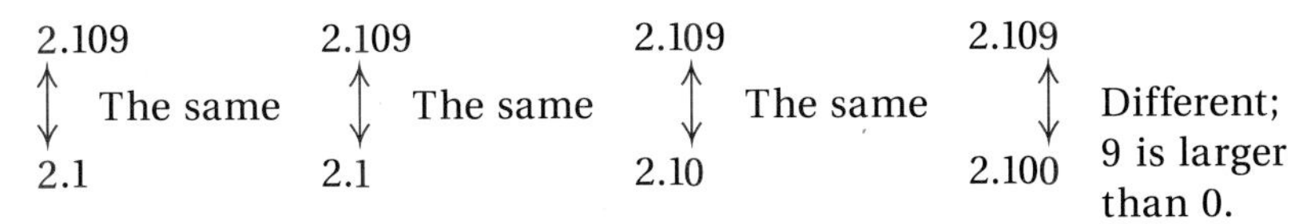

Thus, 2.109 is larger.

Example 16 Which of 0.09 and 0.108 is larger?

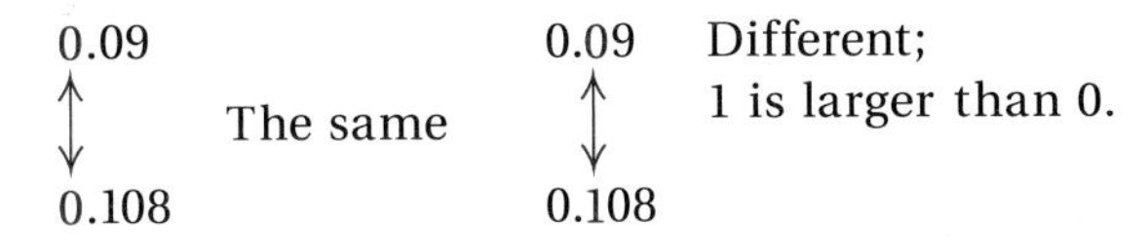

Thus, 0.108 is larger.

Do Exercises 20–25.

e Rounding

Rounding is done as for whole numbers. To understand, we first consider an example using a number line. It might help to review Section 1.4.

Example 17 Round 0.37 to the nearest tenth.

Here is part of a number line.

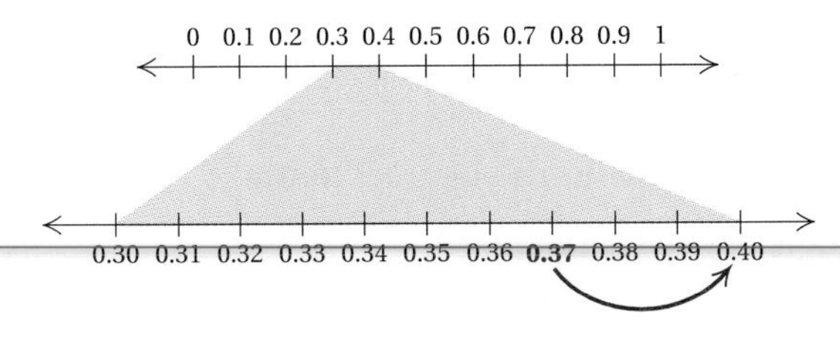

We see that 0.37 is closer to 0.40 than to 0.30. Thus, 0.37 rounded to the nearest tenth is 0.4.

Which number is larger?

20. 2.04, 2.039

21. 0.06, 0.008

22. 0.5, 0.58

23. 1, 0.9999

24. 0.8989, 0.09898

25. 21.006, 21.05

Answers on page A-8

Round to the nearest tenth.

26. 2.76 **27.** 13.85

28. 234.448 **29.** 7.009

Round to the nearest hundredth.

30. 0.636 **31.** 7.834

32. 34.675 **33.** 0.025

Round to the nearest thousandth.

34. 0.9434 **35.** 8.0038

36. 43.1119 **37.** 37.4005

Round 7459.3548 to the nearest:

38. Thousandth.

39. Hundredth.

40. Tenth.

41. One.

42. Ten. (*Caution*: "Tens" are not "tenths.")

43. Hundred.

44. Thousand.

Answers on page A-8

To round to a certain place:

a) Locate the digit in that place.

b) Consider the next digit to the right.

c) If the digit to the right is 5 or higher, round up; if the digit to the right is 4 or lower, round down.

Example 18 Round 3872.2459 to the nearest tenth.

a) Locate the digit in the tenths place.

3 8 7 2.2 4 5 9
↑

b) Consider the next digit to the right.

3 8 7 2.2 4 5 9
↑

c) Since that digit, 4, is less than 5, round down.

3 8 7 2.2 ← **This is the answer.**

CAUTION! 3872.3 is not a correct answer to Example 18. It is incorrect to round from the ten-thousandths digit over to the tenths digit, as follows:

3872.246, 3872.25, 3872.3.

Example 19 Round 3872.2459 to the nearest thousandth, hundredth, tenth, one, ten, hundred, and thousand.

Thousandth:	3872.246	Ten:	3870
Hundredth:	3872.25	Hundred:	3900
Tenth:	3872.2	Thousand:	4000
One:	3872		

Example 20 Round 14.8973 to the nearest hundredth.

a) Locate the digit in the hundredths place. 1 4.8 9 7 3 (arrow under 9)

b) Consider the next digit to the right. 1 4.8 9 7 3 (arrow under 9)

c) Since that digit, 7, is 5 or higher, round up. When we make the hundredths digit a 10, we carry 1 to the tenths place.

The answer is 14.90. Note that the 0 in 14.90 indicates that the answer is correct to the nearest hundredth.

Example 21 Round 0.008 to the nearest tenth.

a) Locate the digit in the tenths place. 0.0 0 8 (arrow under first 0 after point)

b) Consider the next digit to the right. 0.0 0 8 (arrow under 0)

c) Since that digit, 0, is less than 5, round down.

The answer is 0.0.

Do Exercises 26–44.

Exercise Set 4.1

a Write a word name for the number in the sentence.

1. The largest pumpkin ever grown weighed 449.06 kilograms (***Source***: *Guinness Book of Records,* 1997).

2. The average loss of daylight in October in Anchorage, Alaska, is 5.63 min per day.

3. Recently, one British pound was worth about $1.5599 in U.S. currency.

4. The cost of a fast modem for a computer is about $289.95.

Write a word name.

5. 34.891

6. 27.1245

Write a word name as on a check.

7. $326.48

8. $125.99

9. $36.72

10. $0.67

b Write fractional notation. Do not simplify.

11. 8.3

12. 0.17

13. 3.56

14. 203.6

15. 46.03

16. 1.509

17. 0.00013

18. 0.0109

19. 1.0008

20. 2.0114

21. 20.003

22. 4567.2

c Write decimal notation.

23. $\frac{8}{10}$

24. $\frac{51}{10}$

25. $\frac{889}{100}$

26. $\frac{92}{100}$

27. $\frac{3798}{1000}$

28. $\frac{780}{1000}$

29. $\frac{78}{10,000}$

30. $\frac{56,788}{100,000}$

31. $\frac{19}{100,000}$

32. $\frac{2173}{100}$

33. $\frac{376,193}{1,000,000}$

34. $\frac{8,953,074}{1,000,000}$

35. $99\frac{44}{100}$

36. $4\frac{909}{1000}$

37. $3\frac{798}{1000}$

38. $67\frac{83}{100}$

39. $2\frac{1739}{10,000}$

40. $9243\frac{1}{10}$

41. $8\frac{953,073}{1,000,000}$

42. $2256\frac{3059}{10,000}$

d Which number is larger?

43. 0.06, 0.58

44. 0.008, 0.8

45. 0.905, 0.91

46. 42.06, 42.1

47. 0.0009, 0.001

48. 7.067, 7.054

49. 234.07, 235.07

50. 0.99999, 1

51. 0.004, $\frac{4}{100}$

52. $\frac{73}{10}$, 0.73

53. 0.432, 0.4325

54. 0.8437, 0.84384

e Round to the nearest tenth.

55. 0.11

56. 0.85

57. 0.49

58. 0.5794

59. 2.7449

60. 4.78

61. 123.65

62. 36.049

Round to the nearest hundredth.

63. 0.893

64. 0.675

65. 0.6666

66. 6.529

67. 0.995

68. 207.9976

69. 0.094

70. 11.4246

Round to the nearest thousandth.

71. 0.3246

72. 0.6666

73. 17.0015

74. 123.4562

75. 10.1011

76. 0.1161

77. 9.9989

78. 67.100602

Round 809.4732 to the nearest:

79. Hundred.

80. Tenth.

81. Thousandth.

82. Hundredth.

83. One.

84. Ten.

Round 34.54389 to the nearest:

85. Ten-thousandth.

86. Thousandth.

87. Hundredth.

88. Tenth.

89. One.

90. Ten.

Skill Maintenance

Round 6172 to the nearest: [1.4a]

91. Ten.

92. Hundred.

93. Thousand.

Add or subtract.

94. $\begin{array}{r} 681 \\ +\ 149 \\ \hline \end{array}$ [1.2b]

95. $\frac{681}{1000} + \frac{149}{1000}$ [3.2a]

96. $\begin{array}{r} 267 \\ -\ 85 \\ \hline \end{array}$ [1.3d]

97. $\frac{267}{100} - \frac{85}{100}$ [3.3a]

Synthesis

98. ◆ Describe in your own words a procedure for converting from decimal notation to fractional notation.

99. ◆ A fellow student rounds 236.448 to the nearest one and gets 237. Explain the possible error.

There are other methods of rounding decimal notation. A computer often uses a method called **truncating**. To round using truncating, drop off all decimal places past the rounding place, which is the same as changing all digits to the right to zeros. For example, rounding 6.78093456285102 to the ninth decimal place, using truncating, gives us 6.780934562. Use truncating to round each of the following to the fifth decimal place, that is, the nearest hundred thousandth.

100. 6.78346123

101. 6.783461902

102. 99.999999999

103. 0.030303030303

4.2 Addition and Subtraction with Decimal Notation

Objectives

a Add using decimal notation.

b Subtract using decimal notation.

c Solve equations of the type $x + a = b$ and $a + x = b$, where a and b may be in decimal notation.

For Extra Help

TAPE 7 | TAPE 7A | MAC WIN | CD-ROM

a Addition

Adding with decimal notation is similar to adding whole numbers. First we line up the decimal points so that we can add corresponding place-value digits. Then we add digits from the right. For example, we add the thousandths, then the hundredths, and so on, carrying if necessary. If desired, we can write extra zeros to the right of the decimal point so that the number of places is the same.

Example 1 Add: 56.314 + 17.78.

```
   5 6 . 3 1 4      Lining up the decimal points in order to add
 + 1 7 . 7 8 0      Writing an extra zero to the right
 -------------      of the decimal point

   5 6 . 3 1 4      Adding thousandths
 + 1 7 . 7 8 0
 -------------
             4

   5 6 . 3 1 4      Adding hundredths
 + 1 7 . 7 8 0
 -------------
           9 4

     1
   5 6 . 3 1 4      Adding tenths
 + 1 7 . 7 8 0      Write a decimal point in the answer.
 -------------
       . 0 9 4      We get 10 tenths = 1 one + 0 tenths,
                    so we carry the 1 to the ones column.

   1 1
   5 6 . 3 1 4      Adding ones
 + 1 7 . 7 8 0
 -------------
     4 . 0 9 4      We get 14 ones = 1 ten + 4 ones,
                    so we carry the 1 to the tens column.

   1 1
   5 6 . 3 1 4      Adding tens
 + 1 7 . 7 8 0
 -------------
   7 4 . 0 9 4
```

Do Exercises 1 and 2.

Add.

1.
```
     0.8 4 7
  + 1 0.0 7
```

2.
```
      2.1
      0.7 3 9
  + 3 1.3 6 8 9
```

Remember, we can write extra zeros to the right of the decimal point to get the same number of decimal places.

Example 2 Add: 3.42 + 0.237 + 14.1.

```
     3.4 2 0     Lining up the decimal points
     0.2 3 7     and writing extra zeros
  + 1 4.1 0 0
  -----------
    1 7.7 5 7    Adding
```

Do Exercises 3–5.

Add.

3. 0.02 + 4.3 + 0.649

4. 0.12 + 3.006 + 0.4357

5. 0.4591 + 0.2374 + 8.70894

Answers on page A-8

Add.

6. 789 + 123.67

Consider the addition 3456 + 19.347. Keep in mind that a whole number, such as 3456, has an "unwritten" decimal point at the right, with 0 fractional parts. When adding, we can always write in that decimal point and extra zeros if desired.

Example 3 Add: 3456 + 19.347.

$$\begin{array}{rl} \scriptstyle 1 & \\ 3\,4\,5\,6.0\,0\,0 & \textbf{Writing in the decimal point and extra zeros} \\ +\quad 1\,9.3\,4\,7 & \textbf{Lining up the decimal points} \\ \hline 3\,4\,7\,5.3\,4\,7 & \textbf{Adding} \end{array}$$

Do Exercises 6 and 7.

7. 45.78 + 2467 + 1.993

b Subtraction

Subtracting with decimal notation is similar to subtracting whole numbers. First we line up the decimal points so that we can subtract corresponding place-value digits. Then we subtract digits from the right. For example, we subtract the thousandths, then the hundredths, the tenths, and so on, borrowing if necessary.

Example 4 Subtract: 56.314 − 17.78.

$$\begin{array}{rl} 5\,6.3\,1\,4 & \textbf{Lining up the decimal points in order to subtract} \\ -\,1\,7.7\,8\,0 & \textbf{Writing an extra 0} \end{array}$$

$$\begin{array}{rl} 5\,6.3\,1\,4 & \textbf{Subtracting thousandths} \\ -\,1\,7.7\,8\,0 & \\ \hline 4 & \end{array}$$

$$\begin{array}{rl} \scriptstyle 2\;11 & \\ 5\,6.\cancel{3}\,\cancel{1}\,4 & \textbf{Borrowing tenths to subtract hundredths} \\ -\,1\,7.7\,8\,0 & \\ \hline 3\,4 & \end{array}$$

$$\begin{array}{rl} \scriptstyle 12 & \\ \scriptstyle 5\;\cancel{2}\;11 & \\ 5\,\cancel{6}.\cancel{3}\,\cancel{1}\,4 & \textbf{Borrowing ones to subtract tenths} \\ -\,1\,7.7\,8\,0 & \\ \hline .5\,3\,4 & \textbf{Writing a decimal point} \end{array}$$

$$\begin{array}{rl} \scriptstyle 15\;12 & \\ \scriptstyle 4\;\cancel{5}\;\cancel{2}\;11 & \\ \cancel{5}\,\cancel{6}.\cancel{3}\,\cancel{1}\,4 & \textbf{Borrowing tens to subtract ones} \\ -\,1\,7.7\,8\,0 & \\ \hline 8.5\,3\,4 & \end{array}$$

$$\begin{array}{rl} \scriptstyle 15\;12 & \\ \scriptstyle 4\;\cancel{5}\;\cancel{2}\;11 & \\ \cancel{5}\,\cancel{6}.\cancel{3}\,\cancel{1}\,4 & \textbf{Subtracting tens} \\ -\,1\,7.7\,8\,0 & \\ \hline 3\,8.5\,3\,4 & \end{array}$$

CHECK:

$$\begin{array}{r} \scriptstyle 1\;\;1\;\;1 \\ 3\,8.5\,3\,4 \\ +\,1\,7.7\,8\,0 \\ \hline 5\,6.3\,1\,4 \end{array}$$

Do Exercises 8 and 9.

Subtract.

8. 37.428 − 26.674

9.

$$\begin{array}{r} 0.3\,4\,7 \\ -\,0.0\,0\,8 \\ \hline \end{array}$$

Answers on page A-8

Example 5 Subtract: 13.07 − 9.205.

```
    12
    2̸ 10 6 10
  1̸ 3̸.0̸ 7̸ 0̸     Writing an extra zero
−    9.2 0 5
  ---------
     3.8 6 5     Subtracting
```

Example 6 Subtract: 23.08 − 5.0053.

```
  1 13    7 9 10
  2̸ 3̸.0 8̸ 0̸ 0̸     Writing two extra zeros
−    5.0 0 5 3
  -------------
   1 8.0 7 4 7     Subtracting
```

Do Exercises 10–12.

When subtraction involves a whole number, again keep in mind that there is an "unwritten" decimal point that can be written in if desired. Extra zeros can also be written in to the right of the decimal point.

Example 7 Subtract: 456 − 2.467.

```
        5 9 9 10
  4 5 6̸.0̸ 0̸ 0̸     Writing in the decimal point and extra zeros
−       2.4 6 7
  -------------
  4 5 3.5 3 3     Subtracting
```

Do Exercises 13 and 14.

c Solving Equations

Now let's solve equations $x + a = b$ and $a + x = b$, where a and b may be in decimal notation. Proceeding as we have before, we subtract a on both sides.

Example 8 Solve: $x + 28.89 = 74.567$.

We have

$$x + 28.89 - 28.89 = 74.567 - 28.89 \quad \text{Subtracting 28.89 on both sides}$$
$$x = 45.677.$$

```
  6 13 14 16
  7̸ 4̸.5̸ 6̸ 7
− 2 8.8 9 0
  ---------
  4 5.6 7 7
```

The solution is 45.677.

Subtract.

10. 1.2345 − 0.7

11. 0.9564 − 0.4392

12. 7.37 − 0.00008

Subtract.

13. 1277 − 82.78

14. 5 − 0.0089

Answers on page A-8

Solve.

15. $x + 17.78 = 56.314$

16. $8.906 + t = 23.07$

17. Solve: $241 + y = 2374.5$.

Answers on page A-8

Example 9 Solve: $0.8879 + y = 9.0026$.

We have

$$0.8879 + y - 0.8879 = 9.0026 - 0.8879 \quad \text{Subtracting 0.8879 on both sides}$$

$$y = 8.1147.$$

$$\begin{array}{r} \overset{8}{\cancel{9}}.\overset{9}{\cancel{0}}\overset{9}{\cancel{0}}\overset{11}{\cancel{2}}\overset{16}{\cancel{6}} \\ -\ 0.8879 \\ \hline 8.1147 \end{array}$$

The solution is 8.1147.

Do Exercises 15 and 16.

Example 10 Solve: $120 + x = 4380.6$.

We have

$$120 + x - 120 = 4380.6 - 120 \quad \text{Subtracting 120 on both sides}$$

$$x = 4260.6$$

$$\begin{array}{r} 4380.6 \\ -\ 120.0 \\ \hline 4260.6 \end{array}$$

The solution is 4260.6.

Do Exercise 17.

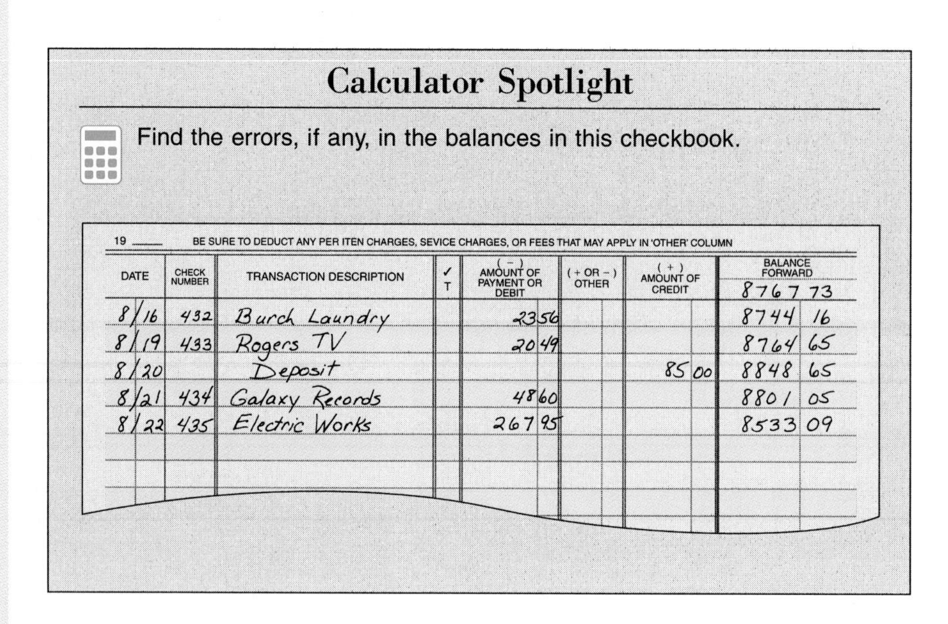

Calculator Spotlight

Find the errors, if any, in the balances in this checkbook.

19 ____ BE SURE TO DEDUCT ANY PER ITEN CHARGES, SEVICE CHARGES, OR FEES THAT MAY APPLY IN 'OTHER' COLUMN

DATE	CHECK NUMBER	TRANSACTION DESCRIPTION	✓ T	(−) AMOUNT OF PAYMENT OR DEBIT	(+ OR −) OTHER	(+) AMOUNT OF CREDIT	BALANCE FORWARD
							8767 73
8/16	432	Burch Laundry		23 56			8744 16
8/19	433	Rogers TV		20 49			8764 65
8/20		Deposit				85 00	8848 65
8/21	434	Galaxy Records		48 60			8801 05
8/22	435	Electric Works		267 95			8533 09

Exercise Set 4.2

a Add.

1. $\begin{array}{r} 316.25 \\ +\ 18.12 \\ \hline \end{array}$

2. $\begin{array}{r} 641.803 \\ +\ 14.935 \\ \hline \end{array}$

3. $\begin{array}{r} 659.403 \\ +916.812 \\ \hline \end{array}$

4. $\begin{array}{r} 4203.28 \\ +\ 3.39 \\ \hline \end{array}$

5. $\begin{array}{r} 9.104 \\ +123.456 \\ \hline \end{array}$

6. $\begin{array}{r} 6.1528 \\ +5.2777 \\ \hline \end{array}$

7. $\begin{array}{r} 81.008 \\ +\ 3.409 \\ \hline \end{array}$

8. 0.8096 + 0.7856

9. 20.0124 + 30.0124

10. 0.687 + 0.9

11. 39 + 1.007

12. 0.845 + 10.02

13. 0.34 + 3.5 + 0.127 + 768

14. 2.3 + 0.729 + 23

15. 17 + 3.24 + 0.256 + 0.3689

16. $\begin{array}{r} 47.8 \\ 219.852 \\ 43.59 \\ +666.713 \\ \hline \end{array}$

17. $\begin{array}{r} 2.703 \\ 78.33 \\ 28.0009 \\ +118.4341 \\ \hline \end{array}$

18. $\begin{array}{r} 13.72 \\ 9.112 \\ 6542.7908 \\ +\ 23.901 \\ \hline \end{array}$

19. 99.6001 + 7285.18 + 500.042 + 870

20. 65.987 + 9.4703 + 6744.02 + 1.0003 + 200.895

b Subtract.

21. $\begin{array}{r} 5.2 \\ -\ 3.9 \\ \hline \end{array}$

22. $\begin{array}{r} 44.345 \\ -\ 3.105 \\ \hline \end{array}$

23. $\begin{array}{r} 51.31 \\ -\ 2.29 \\ \hline \end{array}$

24. $\begin{array}{r} 87.46 \\ -\ 6.32 \\ \hline \end{array}$

25. $\begin{array}{r} 48.76 \\ -\ 3.15 \\ \hline \end{array}$

26. $\begin{array}{r} 97.01 \\ -\ 3.15 \\ \hline \end{array}$

27. $\begin{array}{r} 92.341 \\ -\ 6.42 \\ \hline \end{array}$

28. $\begin{array}{r} 0.8468 \\ -\ 0.034 \\ \hline \end{array}$

29. $\begin{array}{l} \ \ 2.5 \\ -\ 0.0025 \\ \hline \end{array}$

30. $\begin{array}{r} 39.0 \\ -\ 0.28 \\ \hline \end{array}$

31. $\begin{array}{l} \ \ 3.4 \\ -\ 0.003 \\ \hline \end{array}$

32. $\begin{array}{l} \ \ 2.8 \\ -\ 2.08 \\ \hline \end{array}$

33. 28.2 − 19.35

34. 100.16 − 0.118

35. 34.07 − 30.7

36. 36.2 − 16.28

37. 8.45 − 7.405

38. 3.801 − 2.81

39. 6.003 − 2.3

40. 9.087 − 8.807

41. 1 − 0.0098

42. 2 − 1.0908

43. 100 − 0.34

44. 624 − 18.79

45. 7.48 − 2.6

46. $18.4 - 5.92$

47. $3 - 2.006$

48. $263.7 - 102.08$

49. $19 - 1.198$

50. $2548.98 - 2.007$

51. $65 - 13.87$

52. $45 - 0.999$

53. $3.907 - 1.416$

54. $70.0009 - 23.0567$

55.
$$\begin{array}{r} 32.7978 \\ -\ 0.0592 \\ \hline \end{array}$$

56.
$$\begin{array}{r} 0.49634 \\ -\ 0.12678 \\ \hline \end{array}$$

57.
$$\begin{array}{r} 3.0074 \\ -\ 1.3408 \\ \hline \end{array}$$

58.
$$\begin{array}{r} 6.07 \\ -\ 2.0078 \\ \hline \end{array}$$

59.
$$\begin{array}{r} 2345.9078\,6 \\ -\ 0.999 \\ \hline \end{array}$$

60.
$$\begin{array}{r} 1.0 \\ -\ 0.9999 \\ \hline \end{array}$$

c Solve.

61. $x + 17.5 = 29.15$

62. $t + 50.7 = 54.07$

63. $3.205 + m = 22.456$

64. $4.26 + q = 58.32$

65. $17.95 + p = 402.63$

66. $w + 1.3004 = 47.8$

67. $13{,}083.3 = x + 12{,}500.33$

68. $100.23 = 67.8 + z$

69. $x + 2349 = 17{,}684.3$

70. $1830.4 + t = 23{,}067$

Skill Maintenance

71. Round 34,567 to the nearest thousand. [1.4a]

72. Round 34,496 to the nearest thousand. [1.4a]

Subtract.

73. $\frac{13}{24} - \frac{3}{8}$ [3.3a]

74. $\frac{8}{9} - \frac{2}{15}$ [3.3a]

75. $8805 - 2639$ [1.3d]

76. $8005 - 2639$ [1.3d]

Solve.

77. A serving of filleted fish is generally considered to be about $\frac{1}{3}$ lb. How many servings can be prepared from $5\frac{1}{2}$ lb of flounder fillet? [3.6c]

78. A photocopier technician drove $125\frac{7}{10}$ mi away from Scottsdale for a repair call. The next day he drove $65\frac{1}{2}$ mi back toward Scottsdale for another service call. How far was the technician from Scottsdale? [3.5c]

Synthesis

79. ◆ Explain the error in the following:

Add.

$$\begin{array}{r} 1\,3.0\,7 \\ +\quad 9.2\,0\,5 \\ \hline 1\,0.5\,1\,2 \end{array}$$

80. ◆ Explain the error in the following:

Subtract.

$$\begin{array}{r} 7\,3.0\,8\,9 \\ -\,5.0\,0\,6\,1 \\ \hline 2.3\,0\,2\,8 \end{array}$$

81. A student presses the wrong button when using a calculator and adds 235.7 instead of subtracting it. The incorrect answer is 817.2. What is the correct answer?

4.3 Multiplication with Decimal Notation

Objectives

a Multiply using decimal notation.

b Convert from notation like 45.7 million to standard notation, and from dollars to cents and cents to dollars.

For Extra Help

TAPE 7

TAPE 7A

MAC WIN

CD-ROM

a Multiplication

Let's find the product

$2.3 \times 1.12.$

To understand how we find such a product, we first convert each factor to fractional notation. Next, we multiply the whole numbers 23 and 112, and then divide by 1000.

$$2.3 \times 1.12 = \frac{23}{10} \times \frac{112}{100} = \frac{23 \times 112}{10 \times 100} = \frac{2576}{1000} = 2.576$$

Note the number of decimal places.

$$\begin{array}{rl} 1.1\,2 & \text{(2 decimal places)} \\ \times \quad 2.3 & \text{(1 decimal place)} \\ \hline 2.5\,7\,6 & \text{(3 decimal places)} \end{array}$$

Now consider

$$0.011 \times 15.0002 = \frac{11}{1000} \times \frac{150{,}002}{10{,}000} = \frac{1{,}650{,}022}{10{,}000{,}000} = 0.1650022.$$

Note the number of decimal places.

$$\begin{array}{rl} 1\,5.0\,0\,0\,2 & \text{(4 decimal places)} \\ \times \quad 0.0\,1\,1 & \text{(3 decimal places)} \\ \hline 0.1\,6\,5\,0\,0\,2\,2 & \text{(7 decimal places)} \end{array}$$

To multiply using decimals:

a) Ignore the decimal points and multiply as though both factors were whole numbers.

b) Then place the decimal point in the result. The number of decimal places in the product is the sum of the numbers of places in the factors (count places from the right).

0.8×0.43

$$\begin{array}{rl} \overset{2}{0.4\,3} & \\ \times \quad 0.8 & \text{Ignore the decimal} \\ \hline 3\,4\,4 & \text{points for now.} \end{array}$$

$$\begin{array}{rl} 0.4\,3 & \text{(2 decimal places)} \\ \times \quad 0.8 & \text{(1 decimal place)} \\ \hline 0.3\,4\,4 & \text{(3 decimal places)} \end{array}$$

1. Multiply.

$$\begin{array}{r} 85.4 \\ \times \quad 6.2 \\ \hline \end{array}$$

Multiply.

2. $$\begin{array}{r} 1234 \\ \times 0.0041 \\ \hline \end{array}$$

3. $$\begin{array}{r} 42.65 \\ \times 0.804 \\ \hline \end{array}$$

Answers on page A-8

Example 1 Multiply: 8.3 × 74.6.

a) Ignore the decimal points and multiply as though factors were whole numbers:

$$\begin{array}{r} {}^{3\ 4} \\ {}^{1\ 1} \\ 74.6 \\ \times \quad 8.3 \\ \hline 2238 \\ 59680 \\ \hline 61918 \end{array}$$

b) Place the decimal point in the result. The number of decimal places in the product is the sum, 1 + 1, of the number of places in the factors.

$$\begin{array}{rl} 74.6 & \text{(1 decimal place)} \\ \times \quad 8.3 & \text{(1 decimal place)} \\ \hline 2238 & \downarrow \\ 59680 & \\ \hline 619.18 & \text{(2 decimal places)} \end{array}$$

Do Exercise 1.

Example 2 Multiply: 0.0032 × 2148.

As we catch on to the skill, we can combine the two steps.

$$\begin{array}{rl} 2148 & \text{(0 decimal places)} \\ \times 0.0032 & \text{(4 decimal places)} \\ \hline 4296 & \downarrow \\ 64440 & \\ \hline 6.8736 & \text{(4 decimal places)} \end{array}$$

Example 3 Multiply: 0.14 × 0.867.

$$\begin{array}{rl} 0.867 & \text{(3 decimal places)} \\ \times \quad 0.14 & \text{(2 decimal places)} \\ \hline 3468 & \downarrow \\ 8670 & \\ \hline 0.12138 & \text{(5 decimal places)} \end{array}$$

Do Exercises 2 and 3.

Now let's consider some special kinds of products. The first involves multiplying by a tenth, hundredth, thousandth, or ten-thousandth. Let's look at those products.

$$0.1 \times 38 = \frac{1}{10} \times 38 = \frac{38}{10} = 3.8$$

$$0.01 \times 38 = \frac{1}{100} \times 38 = \frac{38}{100} = 0.38$$

$$0.001 \times 38 = \frac{1}{1000} \times 38 = \frac{38}{1000} = 0.038$$

$$0.0001 \times 38 = \frac{1}{10{,}000} \times 38 = \frac{38}{10{,}000} = 0.0038$$

Note in each case that the product is *smaller* than 38.

Multiply.

4. 0.1×3.48

5. 0.01×3.48

6. 0.001×3.48

7. 0.0001×3.48

To multiply any number by a tenth, hundredth, or thousandth,

a) count the number of decimal places in the tenth, hundredth, or thousandth, and

0.001×34.45678 → 3 places

b) move the decimal point that many places to the left.

$0.001 \times 34.45678 = 0.034.45678$

Move 3 places to the left.

$0.001 \times 34.45678 = 0.03445678$

Examples Multiply.

4. $0.1 \times 14.605 = 1.4605$ 1.4.605

5. $0.01 \times 14.605 = 0.14605$

6. $0.001 \times 14.605 = 0.014605$ ← We write an extra zero.

7. $0.0001 \times 14.605 = 0.0014605$ ← We write two extra zeros.

Do Exercises 4–7.

Now let's consider multiplying by a power of ten, such as 10, 100, 1000, and so on. Let's look at those products.

$$10 \times 97.34 = 973.4$$

$$100 \times 97.34 = 9734.$$

$$1000 \times 97.34 = 97{,}340$$

$$10{,}000 \times 97.34 = 973{,}400$$

Note in each case that the product is *larger* than 97.34.

Answers on page A-8

Multiply.

8. 10×3.48

9. 100×3.48

10. 1000×3.48

11. $10{,}000 \times 3.48$

Answers on page A-8

To multiply any number by a power of ten, such as 10, 100, 1000, and so on,

a) count the number of zeros, and

1000×34.45678 → 3 zeros

b) move the decimal point that many places to the right.

$1000 \times 34.45678 = 34.456.78$

Move 3 places to the right.

$1000 \times 34.45678 = 34{,}456.78$

Examples Multiply.

8. $10 \times 14.605 = 146.05$ 14.6.05

9. $100 \times 14.605 = 1460.5$

10. $1000 \times 14.605 = 14{,}605$

11. $10{,}000 \times 14.605 = 146{,}050$ 14.6050.

Do Exercises 8–11.

b Applications Using Multiplication with Decimal Notation

Naming Large Numbers

We often see notation like the following in newspapers and magazines and on television.

> O'Hare International Airport handles 67.3 million passengers per year.
>
> Americans eat $1.1 billion worth of lettuce each year.
>
> The population of the world is 6.6 billion.

To understand such notation, it helps to consider the following table.

1 hundred = $100 = 10^2$ → 2 zeros

1 thousand = $1000 = 10^3$ → 3 zeros

1 million = $1{,}000{,}000 = 10^6$ → 6 zeros

1 billion = $1{,}000{,}000{,}000 = 10^9$ → 9 zeros

1 trillion = $1{,}000{,}000{,}000{,}000 = 10^{12}$ → 12 zeros

To convert to standard notation, we proceed as follows.

Example 12 Convert the number in this sentence to standard notation: O'Hare's International Airport handles 67.3 million passengers per year.

$$\begin{aligned} 67.3 \text{ million} &= 67.3 \times 1 \text{ million} \\ &= 67.3 \times 1{,}000{,}000 \\ &= 67{,}300{,}000 \end{aligned}$$

Do Exercises 12 and 13.

Convert the number in the sentence to standard notation.

12. In a recent year, the total payroll of major league baseball was $938 million.

13. In a recent year, the U.S. trade deficit with Japan was $44.1 billion.

Money Conversion

Converting from dollars to cents is like multiplying by 100. To see why, consider $19.43.

$19.43 = 19.43 × $1	**We think of $19.43 as 19.43 × 1 dollar, or 19.43 × $1.**
= 19.43 × 100¢	**Substituting 100¢ for $1: $1 = 100¢**
= 1943¢	**Multiplying**

> To convert from dollars to cents, move the decimal point two places to the right and change from the $ sign in front to the ¢ sign at the end.

Examples Convert from dollars to cents.

13. $189.64 = 18,964¢

14. $0.75 = 75¢

Do Exercises 14 and 15.

Convert from dollars to cents.

14. $15.69

15. $0.17

Converting from cents to dollars is like multiplying by 0.01. To see why, consider 65¢.

65¢ = 65 × 1¢	**We think of 65¢ as 65 × 1 cent, or 65 × 1¢.**
= 65 × $0.01	**Substituting $0.01 for 1¢: 1¢ = $0.01**
= $0.65	**Multiplying**

> To convert from cents to dollars, move the decimal point two places to the left and change from the ¢ sign at the end to the $ sign in front.

Examples Convert from cents to dollars.

15. 395¢ = $3.95

16. 8503¢ = $85.03

Do Exercises 16 and 17.

Convert from cents to dollars.

16. 35¢

17. 577¢

Answers on page A-8

Improving Your Math Study Skills

Learning Resources and Time Management

Two other topics to consider in enhancing your math study skills are learning resources and time management.

Learning Resources

- **Textbook supplements.** Are you aware of all the supplements that exist for this textbook? Many details are given in the preface. Now that you are more familiar with the book, let's discuss them.
  1. The *Student's Solutions Manual* contains worked-out solutions to the odd-numbered exercises in the exercise sets. Consider obtaining a copy if you are having trouble. It should be your first choice if you can make an additional purchase.
  2. An extensive set of *videotapes* supplement this text. These may be available to you on your campus at a learning center or math lab. Check with your instructor.
  3. *Tutorial software* also accompanies the text. If not available in the campus learning center, you might order it by calling the number 1-800-322-1377.
- **The Internet.** Our on-line World Wide Web supplement provides additional practice resources. If you have internet access, you can reach this site through the address:

  http://www.mathmax.com

  It contains many helpful ideas as well as many links to other resources for learning mathematics.
- **Your college or university.** Your own college or university probably has resources to enhance your math learning.
  1. For example, is there a learning lab or tutoring center for drop-in tutoring?
  2. Are there special lab classes or formal tutoring sessions tailored for the specific course you are taking?
  3. Perhaps there is a bulletin board or network where you can locate the names of experienced private tutors.
- **Your instructor.** Although it may seem obvious, students neglect to consider the most underused resource available to them: their instructor. Find out your instructor's office hours and make it a point to visit when you need additional help.

Time Management

- **Juggling time.** Have reasonable expectations about the time you need to study math. Unreasonable expectations may lead to lower grades and frustrations. Working 40 hours per week and taking 12 hours of credit is equivalent to working two full-time jobs. Can you handle such a load? As a rule of thumb, your ratio of work hours to credit load should be about 40/3, 30/6, 20/9, 10/12, and 5/14. Budget about 2–3 hours of homework and studying per hour of class.
- **Daily schedule.** Make an hour-by-hour schedule of your typical week. Include work, college, home, personal, sleep, study, and leisure times. Be realistic about the amount of time needed for sleep and home duties. If possible, try to schedule time for study when you are most alert.

Exercise Set 4.3

a Multiply.

1. $\begin{array}{r} 8.6 \\ \times\ 7 \\ \hline \end{array}$ **2.** $\begin{array}{r} 5.7 \\ \times\ 0.8 \\ \hline \end{array}$ **3.** $\begin{array}{r} 0.84 \\ \times\ 8 \\ \hline \end{array}$ **4.** $\begin{array}{r} 9.4 \\ \times\ 0.6 \\ \hline \end{array}$

5. $\begin{array}{r} 6.3 \\ \times\ 0.04 \\ \hline \end{array}$ **6.** $\begin{array}{r} 9.8 \\ \times\ 0.08 \\ \hline \end{array}$ **7.** $\begin{array}{r} 87 \\ \times\ 0.006 \\ \hline \end{array}$ **8.** $\begin{array}{r} 18.4 \\ \times\ 0.07 \\ \hline \end{array}$

9. 10×23.76

10. 100×3.8798

11. 1000×583.686852

12. 0.34×1000

13. 7.8×100

14. 0.00238×10

15. 0.1×89.23

16. 0.01×789.235

17. 0.001×97.68

18. 8976.23×0.001

19. 78.2×0.01

20. 0.0235×0.1

21. $\begin{array}{r} 32.6 \\ \times\ 16 \\ \hline \end{array}$ **22.** $\begin{array}{r} 9.28 \\ \times\ 8.6 \\ \hline \end{array}$ **23.** $\begin{array}{r} 0.984 \\ \times\ 3.3 \\ \hline \end{array}$ **24.** $\begin{array}{r} 8.489 \\ \times\ 7.4 \\ \hline \end{array}$

25. $\begin{array}{r} 374 \\ \times\ 2.4 \\ \hline \end{array}$ **26.** $\begin{array}{r} 865 \\ \times\ 1.08 \\ \hline \end{array}$ **27.** $\begin{array}{r} 749 \\ \times\ 0.43 \\ \hline \end{array}$ **28.** $\begin{array}{r} 978 \\ \times\ 20.5 \\ \hline \end{array}$

29. $\begin{array}{r} 0.87 \\ \times\ 64 \\ \hline \end{array}$ **30.** $\begin{array}{r} 7.25 \\ \times\ 60 \\ \hline \end{array}$ **31.** $\begin{array}{r} 46.50 \\ \times\ 75 \\ \hline \end{array}$ **32.** $\begin{array}{r} 8.24 \\ \times\ 703 \\ \hline \end{array}$

33. $\begin{array}{r} 81.7 \\ \times\ 0.612 \\ \hline \end{array}$ **34.** $\begin{array}{r} 31.82 \\ \times\ 7.15 \\ \hline \end{array}$ **35.** $\begin{array}{r} 10.105 \\ \times\ 11.324 \\ \hline \end{array}$ **36.** $\begin{array}{r} 151.2 \\ \times\ 4.555 \\ \hline \end{array}$

37. $\begin{array}{r} 12.3 \\ \times\ 1.08 \\ \hline \end{array}$ **38.** $\begin{array}{r} 7.82 \\ \times\ 0.024 \\ \hline \end{array}$ **39.** $\begin{array}{r} 32.4 \\ \times\ 2.8 \\ \hline \end{array}$ **40.** $\begin{array}{r} 8.09 \\ \times\ 0.0075 \\ \hline \end{array}$

41. $\begin{array}{r} 0.0\,0\,3\,4\,2 \\ \times \quad 0.8\,4 \\ \hline \end{array}$

42. $\begin{array}{r} 2.0\,0\,5\,6 \\ \times \quad 3.8 \\ \hline \end{array}$

43. $\begin{array}{r} 0.3\,4\,7 \\ \times \quad 2.0\,9 \\ \hline \end{array}$

44. $\begin{array}{r} 2.5\,3\,2 \\ \times 1.0\,6\,7 \\ \hline \end{array}$

45. $\begin{array}{r} 3.0\,0\,5 \\ \times 0.6\,2\,3 \\ \hline \end{array}$

46. $\begin{array}{r} 1\,6.3\,4 \\ \times 0.0\,0\,0\,5\,1\,2 \\ \hline \end{array}$

47. 1000×45.678

48. 0.001×45.678

b Convert from dollars to cents.

49. $28.88

50. $67.43

51. $0.66

52. $1.78

Convert from cents to dollars.

53. 34¢

54. 95¢

55. 3445¢

56. 933¢

Convert the number in the sentence to standard notation.

57. In a recent year, the net sales of Morton International, Inc., were $3.6 billion.

58. Annual production of sugarcane is 1.075 billion tons.

59. The total surface area of the earth is 196.8 million square miles.

60. Annual sales of *Sports Illustrated* magazine is 3.2 million copies per year.

Skill Maintenance

61. Multiply: $2\frac{1}{3} \cdot 4\frac{4}{5}$. [3.6a]

62. Divide: $2\frac{1}{3} \div 4\frac{4}{5}$. [3.6b]

Divide. [1.6c]

63. $24 \overline{)\,8208}$

64. $4 \overline{)\,348}$

65. $7 \overline{)\,31{,}962}$

66. $18 \overline{)\,22{,}626}$

Synthesis

67. ◆ If two rectangles have the same perimeter, will they also have the same area? Why?

68. ◆ A student insists that 346.708×0.1 is 3467.08. How could you convince the student that a mistake has been made?

Express as a power of 10.

69. (1 trillion) · (1 billion)

70. (1 million) · (1 billion)

4.4 Division with Decimal Notation

a Division

Whole-Number Divisors

Compare these divisions by a whole number.

$$\frac{588}{7} = 84$$
$$\frac{58.8}{7} = 8.4$$
$$\frac{5.88}{7} = 0.84$$
$$\frac{0.588}{7} = 0.084$$

When we are dividing by a whole number, the number of decimal places in the *quotient* is the same as the number of decimal places in the *dividend.*

These examples lead us to the following method for dividing by a whole number.

To divide by a whole number,

a) place the decimal point directly above the decimal point in the dividend, and

b) divide as though dividing whole numbers.

```
      0.8 4  <--- Quotient
7 ) 5.8 8
    5 6 0
      2 8
      2 8
        0  <--- Remainder
```

Example 1 Divide: 379.2 ÷ 8.

```
      4 7.4      <- Place the decimal point.
8 ) 3 7 9.2
    3 2 0 0
      5 9 2
      5 6 0      Divide as though dividing whole numbers.
        3 2
        3 2
          0
```

Example 2 Divide: 82.08 ÷ 24.

```
         3.4 2     <- Place the decimal point.
2 4 ) 8 2.0 8
      7 2 0 0
      1 0 0 8
        9 6 0      Divide as though dividing whole numbers.
          4 8
          4 8
            0
```

Do Exercises 1–3.

Objectives

a Divide using decimal notation.

b Solve equations of the type $a \cdot x = b$, where a and b may be in decimal notation.

c Simplify expressions using the rules for order of operations.

For Extra Help

TAPE 7

TAPE 7B

MAC WIN

CD-ROM

Divide.

1. $9\overline{)5.4}$

2. $15\overline{)22.5}$

3. $82\overline{)38.54}$

Answers on page A-9

Divide.

4. $25\overline{)8}$

5. $4\overline{)15}$

6. $86\overline{)21.5}$

Answers on page A-9

Sometimes it helps to write some extra zeros to the right of the decimal point. They don't change the number.

Example 3 Divide: $30 \div 8$.

```
     3.
8 ) 3 0.
    2 4
      6
```
Place the decimal point and divide to find how many ones.

```
     3.
8 ) 3 0.0
    2 4 ↓
      6 0
```
Write an extra zero.

```
     3.7
8 ) 3 0.0
    2 4
      6 0
      5 6
        4
```
Divide to find how many tenths.

```
     3.7
8 ) 3 0.0 0
    2 4
      6 0
      5 6 ↓
        4 0
```
Write an extra zero.

```
     3.7 5
8 ) 3 0.0 0
    2 4
      6 0
      5 6
        4 0
        4 0
          0
```
Divide to find how many hundredths.

Example 4 Divide: $4 \div 25$.

```
       0.1 6
2 5 ) 4.0 0
      2 5
      1 5 0
      1 5 0
          0
```

Do Exercises 4–6.

Divisors That Are Not Whole Numbers

Consider the division

$0.24 \overline{)8.208}$

We write the division as $\frac{8.208}{0.24}$. Then we multiply by 1 to change to a whole-number divisor:

$$\frac{8.208}{0.24} = \frac{8.208}{0.24} \times \frac{100}{100} = \frac{820.8}{24}.$$

The divisor is now a whole number. The division

$0.24 \overline{)8.208}$

is the same as

$24 \overline{)820.8}$

To divide when the divisor is not a whole number,

a) move the decimal point (multiply by 10, 100, and so on) to make the divisor a whole number;

$0.24 \overline{)8.208}$

Move 2 places to the right.

b) move the decimal point (multiply the same way) in the dividend the same number of places; and

$0.24 \overline{)8.208}$

Move 2 places to the right.

c) place the decimal point directly above the new decimal point in the dividend and divide as though dividing whole numbers.

```
           3 4.2
0.2 4 ) 8.2 0^8
        7 2 0 0
        1 0 0 8
          9 6 0
            4 8
            4 8
              0
```

(The new decimal point in the dividend is indicated by a caret.)

Example 5 Divide: 5.848 ÷ 8.6.

$8.6 \overline{)5.848}$

Multiply the divisor by 10 (move the decimal point 1 place). Multiply the same way in the dividend (move 1 place).

```
        0.6 8
8.6 ) 5.8^4 8
      5 1 6 0
        6 8 8
        6 8 8
            0
```

Then divide.

Note: $\frac{5.848}{8.6} = \frac{5.848}{8.6} \cdot \frac{10}{10} = \frac{58.48}{86}$.

Do Exercises 7–9.

7. a) Complete.

$$\frac{3.75}{0.25} = \frac{3.75}{0.25} \times \frac{100}{100} = \frac{(\quad)}{25}$$

b) Divide.

$0.25 \overline{)3.75}$

Divide.

8. $0.83 \overline{)4.067}$

9. $3.5 \overline{)44.8}$

Answers on page A-9

10. Divide.

$1.6\,\overline{)\,25}$

Answer on page A-9

Example 6 Divide: $12 \div 0.64$.

$0.6\,4\,\overline{)\,1\,2.}$ Put a decimal point at the end of the whole number.

$0.6\,4\,\overline{)\,1\,2.0\,0}$ Multiply the divisor by 100 (move the decimal point 2 places). Multiply the same way in the dividend (move 2 places).

$$
\begin{array}{r}
18.75 \\
0.64_{\wedge}\,\overline{)\,12.00_{\wedge}00} \\
\underline{640} \\
560 \\
\underline{512} \\
480 \\
\underline{448} \\
320 \\
\underline{320} \\
0
\end{array}
$$

Then divide.

Do Exercise 10.

It is often helpful to be able to divide quickly by a ten, hundred, or thousand, or by a tenth, hundredth, or thousandth. The procedure we use is based on multiplying by 1. Consider the following examples:

$$\frac{23.789}{1000} = \frac{23.789}{1000} \cdot \frac{1000}{1000} = \frac{23{,}789}{1{,}000{,}000} = 0.023789.$$

We are dividing by a number greater than 1: The result is *smaller* than 23.789.

$$\frac{23.789}{0.01} = \frac{23.789}{0.01} \cdot \frac{100}{100} = \frac{2378.9}{1} = 2378.9.$$

We are dividing by a number less than 1: The result is *larger* than 23.789.
We use the following procedure.

To divide by a power of ten, such as 10, 100, or 1000, and so on,

a) count the number of zeros in the divisor, and $\frac{713.49}{100}$ → 2 zeros

b) move the decimal point that number of places to the left.

$\frac{713.49}{100}$, 7.13.49 (2 places to the left) $\frac{713.49}{100} = 7.1349$

To divide by a tenth, hundredth, or thousandth,

a) count the number of decimal places in the divisor, and $\frac{713.49}{0.001}$ → 3 places

b) move the decimal point that number of places to the right.

$\frac{713.49}{0.001}$, 713.490. (3 places to the right) $\frac{713.49}{0.001} = 713{,}490$

Example 7 Divide: $\dfrac{0.0104}{10}$.

$\dfrac{0.0104}{10}$, 0.0.0104, $\dfrac{0.0104}{10} = 0.00104$

1 zero — 1 place to the left to change 10 to 1

Example 8 Divide: $\dfrac{23.738}{0.001}$.

$\dfrac{23.738}{0.001}$, 23.738. $\dfrac{23.738}{0.001} = 23{,}738$

3 places — 3 places to the right to change 0.001 to 1

Do Exercises 11–14.

b Solving Equations

Now let's solve equations of the type $a \cdot x = b$, where a and b may be in decimal notation. Proceeding as before, we divide by a on both sides.

Example 9 Solve: $8 \cdot x = 27.2$.

We have

$$\frac{8 \cdot x}{8} = \frac{27.2}{8} \quad \text{Dividing by 8 on both sides}$$

$x = 3.4$.

```
     3.4
8 ) 2 7.2
    2 4 0
      3 2
      3 2
        0
```

The solution is 3.4.

Example 10 Solve: $2.9 \cdot t = 0.14616$.

We have

$$\frac{2.9 \cdot t}{2.9} = \frac{0.14616}{2.9} \quad \text{Dividing by 2.9 on both sides}$$

$t = 0.0504$.

```
         0.0 5 0 4
2.9 ) 0.1^4 6 1 6
        1 4 5 0 0
            1 1 6
            1 1 6
                0
```

The solution is 0.0504.

Do Exercises 15 and 16.

Divide.

11. $\dfrac{0.1278}{0.01}$

12. $\dfrac{0.1278}{100}$

13. $\dfrac{98.47}{1000}$

14. $\dfrac{6.7832}{0.1}$

Solve.

15. $100 \cdot x = 78.314$

16. $0.25 \cdot y = 276.4$

Answers on page A-9

Simplify.

17. $0.25 \cdot (1 + 0.08) - 0.0274$

18. $20^2 - 3.4^2 + \{2.5[20(9.2 - 5.6)] + 5(10 - 5)\}$

19. *Tickets Sold at the Movies.* The number of tickets sold at the movies in each of the four years from 1993 to 1996 is shown in the bar graph below. Find the average number of tickets sold.

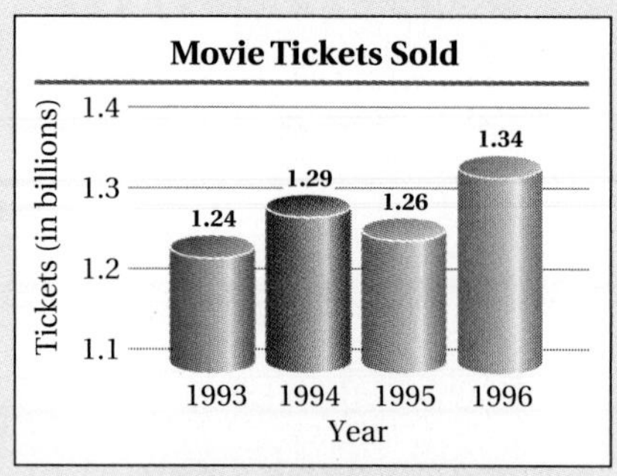

Source: Motion Picture Association of America

Answers on page A-9

c Order of Operations: Decimal Notation

The same rules for order of operations used with whole numbers and fractional notation apply when simplifying expressions with decimal notation.

RULES FOR ORDER OF OPERATIONS

1. Do all calculations within parentheses before operations outside.
2. Evaluate all exponential expressions.
3. Do all multiplications and divisions in order from left to right.
4. Do all additions and subtractions in order from left to right.

Example 11 Simplify: $(5 - 0.06) \div 2 + 3.42 \times 0.1$.

$(5 - 0.06) \div 2 + 3.42 \times 0.1 = 4.94 \div 2 + 3.42 \times 0.1$ **Carrying out operations inside parentheses**

$= 2.47 + 0.342$ **Doing all multiplications and divisions in order from left to right**

$= 2.812$

Example 12 Simplify: $10^2 \times \{[(3 - 0.24) \div 2.4] - (0.21 - 0.092)\}$.

$10^2 \times \{[(3 - 0.24) \div 2.4] - (0.21 - 0.092)\}$

$= 10^2 \times \{[2.76 \div 2.4] - 0.118\}$ **Doing the calculations in the innermost parentheses first**

$= 10^2 \times \{1.15 - 0.118\}$ **Again, doing the calculations in the innermost parentheses**

$= 10^2 \times 1.032$ **Subtracting inside the parentheses**

$= 100 \times 1.032$ **Evaluating the exponential expression**

$= 103.2$

Do Exercises 17 and 18.

Example 13 *Average Movie Revenue.* The bar graph shows movie box-office revenue (money taken in), in billions, in each of the four years from 1993 to 1996. Find the average revenue.

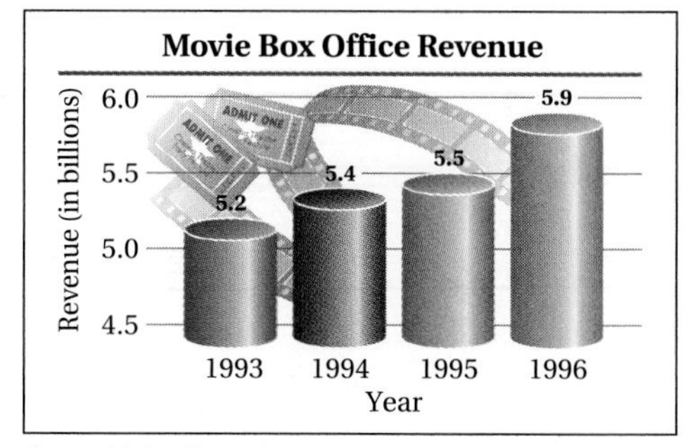

Source: Motion Picture Association of America

To find the average of a set of numbers, we add them. Then we divide by the number of addends. In this case, we are finding the average of 5.2, 5.4, 5.5, and 5.9. The average is given by

$$(5.2 + 5.4 + 5.5 + 5.9) \div 4.$$

Thus,

$$(5.2 + 5.4 + 5.5 + 5.9) \div 4 = 22 \div 4 = 5.5.$$

The average box-office revenue was \$5.5 billion.

Do Exercise 19.

Exercise Set 4.4

a Divide.

1. $2\overline{)5.98}$
2. $5\overline{)18}$
3. $4\overline{)95.12}$
4. $8\overline{)25.92}$
5. $12\overline{)89.76}$
6. $23\overline{)25.07}$
7. $33\overline{)237.6}$
8. $12.4 \div 4$
9. $9.144 \div 8$
10. $4.5 \div 9$
11. $12.123 \div 3$
12. $7\overline{)5.6}$
13. $5\overline{)0.35}$
14. $0.04\overline{)1.68}$
15. $0.12\overline{)8.4}$
16. $0.36\overline{)2.88}$
17. $3.4\overline{)68}$
18. $0.25\overline{)5}$
19. $15\overline{)6}$
20. $12\overline{)1.8}$
21. $36\overline{)14.76}$
22. $52\overline{)119.6}$
23. $3.2\overline{)27.2}$
24. $8.5\overline{)27.2}$
25. $4.2\overline{)39.06}$
26. $4.8\overline{)0.1104}$
27. $8\overline{)5}$
28. $8\overline{)3}$

29. $0.47\overline{)0.1222}$

30. $1.08\overline{)0.54}$

31. $4.8\overline{)75}$

32. $0.28\overline{)63}$

33. $0.032\overline{)0.07488}$

34. $0.017\overline{)1.581}$

35. $82\overline{)38.54}$

36. $34\overline{)0.1462}$

37. $\dfrac{213.4567}{1000}$

38. $\dfrac{213.4567}{100}$

39. $\dfrac{213.4567}{10}$

40. $\dfrac{100.7604}{0.1}$

41. $\dfrac{1.0237}{0.001}$

42. $\dfrac{1.0237}{0.01}$

b Solve.

43. $4.2 \cdot x = 39.06$

44. $36 \cdot y = 14.76$

45. $1000 \cdot y = 9.0678$

46. $789.23 = 0.25 \cdot q$

47. $1048.8 = 23 \cdot t$

48. $28.2 \cdot x = 423$

c Simplify.

49. $14 \times (82.6 + 67.9)$

50. $(26.2 - 14.8) \times 12$

51. $0.003 + 3.03 \div 0.01$

52. $9.94 + 4.26 \div (6.02 - 4.6) - 0.9$

53. $42 \times (10.6 + 0.024)$

54. $(18.6 - 4.9) \times 13$

55. $4.2 \times 5.7 + 0.7 \div 3.5$

56. $123.3 - 4.24 \times 1.01$

57. $9.0072 + 0.04 \div 0.1^2$

58. $12 \div 0.03 - 12 \times 0.03^2$

59. $(8 - 0.04)^2 \div 4 + 8.7 \times 0.4$

60. $(5 - 2.5)^2 \div 100 + 0.1 \times 6.5$

61. $86.7 + 4.22 \times (9.6 - 0.03)^2$

62. $2.48 \div (1 - 0.504) + 24.3 - 11 \times 2$

63. $4 \div 0.4 + 0.1 \times 5 - 0.1^2$

64. $6 \times 0.9 + 0.1 \div 4 - 0.2^3$

65. $5.5^2 \times [(6 - 4.2) \div 0.06 + 0.12]$

66. $12^2 \div (12 + 2.4) - [(2 - 1.6) \div 0.8]$

67. $200 \times \{[(4 - 0.25) \div 2.5] - (4.5 - 4.025)\}$

68. $0.03 \times \{1 \times 50.2 - [(8 - 7.5) \div 0.05]\}$

69. Find the average of \$1276.59, \$1350.49, \$1123.78, and \$1402.56.

70. Find the average weight of two wrestlers who weigh 308 lb and 296.4 lb.

Global Warming. The following table lists the global average temperature for the years 1984 through 1994. Use the table for Exercises 71 and 72.

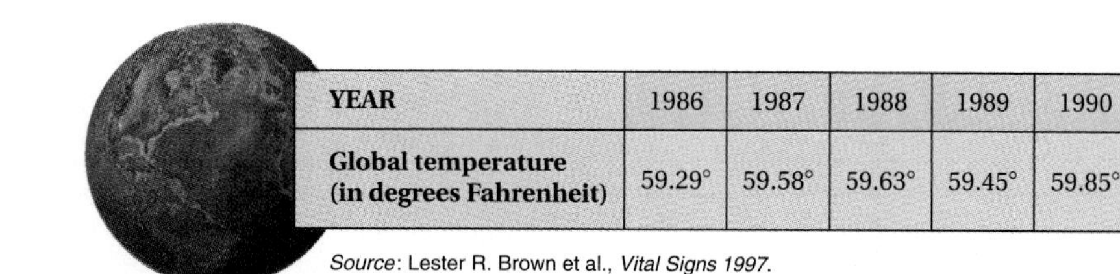

YEAR	1986	1987	1988	1989	1990	1991	1992	1993	1994	1995	1996
Global temperature (in degrees Fahrenheit)	59.29°	59.58°	59.63°	59.45°	59.85°	59.74°	59.23°	59.36°	59.56°	59.72°	59.58°

Source: Lester R. Brown et al., *Vital Signs 1997*.

71. Find the average temperature for the years 1992 through 1996.

72. Find the average temperature for the years 1987 through 1991.

Skill Maintenance

73. Add: $10\frac{1}{2} + 4\frac{5}{8}$. [3.5a]

74. Subtract: $10\frac{1}{2} - 4\frac{5}{8}$. [3.5b]

75. Simplify: $\frac{36}{42}$. [2.5b]

76. Find the prime factorization of 162. [2.1d]

77. Find the prime factorization of 684. [2.1d]

78. Simplify: $\frac{56}{64}$. [2.5b]

Synthesis

79. ◆ A student insists that $0.247 \div 0.1$ is 0.0247. How could you convince this student that a mistake has been made?

80. ◆ A student insists that $0.247 \div 10$ is 2.47. How could you convince this student that a mistake has been made?

Simplify.

81. 🖩 $9.0534 - 2.041^2 \times 0.731 \div 1.043^2$

82. 🖩 $23.042(7 - 4.037 \times 1.46 - 0.932^2)$

In Exercises 83–86, find the missing value.

83. $439.57 \times 0.01 \div 1000 \times \square = 4.3957$

84. $5.2738 \div 0.01 \times 1000 \div \square = 52.738$

85. $0.0329 \div 0.001 \times 10^4 \div \square = 3290$

86. $0.0047 \times 0.01 \div 10^4 \times \square = 4.7$

4.5 Converting from Fractional Notation to Decimal Notation

Objectives

a Convert from fractional notation to decimal notation.

b Round numbers named by repeating decimals.

c Calculate using fractional and decimal notation together.

For Extra Help

TAPE 8 TAPE 7B MAC WIN CD-ROM

a Fractional Notation to Decimal Notation

When a denominator has no prime factors other than 2's and 5's, we can find decimal notation by multiplying by 1. We multiply to get a denominator that is a power of ten, like 10, 100, or 1000.

Example 1 Find decimal notation for $\frac{3}{5}$.

$$\frac{3}{5} = \frac{3}{5} \cdot \frac{2}{2} = \frac{6}{10} = 0.6$$

We use $\frac{2}{2}$ for 1 to get a denominator of 10.

Example 2 Find decimal notation for $\frac{7}{20}$.

$$\frac{7}{20} = \frac{7}{20} \cdot \frac{5}{5} = \frac{35}{100} = 0.35$$

We use $\frac{5}{5}$ for 1 to get a denominator of 100.

Example 3 Find decimal notation for $\frac{9}{40}$.

$$\frac{9}{40} = \frac{9}{40} \cdot \frac{25}{25} = \frac{225}{1000} = 0.225$$

We use $\frac{25}{25}$ for 1 to get a denominator of 1000.

Example 4 Find decimal notation for $\frac{87}{25}$.

$$\frac{87}{25} = \frac{87}{25} \cdot \frac{4}{4} = \frac{348}{100} = 3.48$$

We use $\frac{4}{4}$ for 1 to get a denominator of 100.

Do Exercises 1–4.

We can also divide to find decimal notation.

Example 5 Find decimal notation for $\frac{3}{5}$.

$$\frac{3}{5} = 3 \div 5 \qquad 5\overline{)3.0} = 0.6 \qquad \frac{3}{5} = 0.6$$

```
    0.6
5 ) 3.0
    3 0
    ---
      0
```

Example 6 Find decimal notation for $\frac{7}{8}$.

$$\frac{7}{8} = 7 \div 8 \qquad \frac{7}{8} = 0.875$$

```
    0.8 7 5
8 ) 7.0 0 0
    6 4
    ---
      6 0
      5 6
      ---
        4 0
        4 0
        ---
          0
```

Do Exercises 5 and 6.

Find decimal notation. Use multiplying by 1.

1. $\frac{4}{5}$

2. $\frac{9}{20}$

3. $\frac{11}{40}$

4. $\frac{33}{25}$

Find decimal notation.

5. $\frac{2}{5}$

6. $\frac{3}{8}$

Answers on page A-9

Find decimal notation.

7. $\dfrac{1}{6}$

8. $\dfrac{2}{3}$

Answers on page A-9

In Examples 5 and 6, the division *terminated,* meaning that eventually we got a remainder of 0. A **terminating decimal** occurs when the denominator has only 2's or 5's, or both, as factors, as in $\frac{17}{25}$, $\frac{5}{8}$, or $\frac{83}{100}$. This assumes that the fractional notation has been simplified.

Consider a different situation:

$$\frac{5}{6}, \quad \text{or} \quad \frac{5}{2 \cdot 3}.$$

Since 6 has a 3 as a factor, the division will not terminate. Although we can still use division to get decimal notation, the answer will be a **repeating decimal,** as follows.

Example 7 Find decimal notation for $\frac{5}{6}$.

We have

$$\frac{5}{6} = 5 \div 6 \qquad \begin{array}{r} 0.8\,3\,3 \\ 6\,\overline{)\,5.0\,0\,0} \\ \underline{4\,8} \\ 2\,0 \\ \underline{1\,8} \\ 2\,0 \\ \underline{1\,8} \\ 2 \end{array}$$

Since 2 keeps reappearing as a remainder, the digits repeat and will continue to do so; therefore,

$$\frac{5}{6} = 0.83333\ldots.$$

The dots indicate an endless sequence of digits in the quotient. When there is a repeating pattern, the dots are often replaced by a bar to indicate the repeating part—in this case, only the 3:

$$\frac{5}{6} = 0.8\overline{3}.$$

Do Exercises 7 and 8.

Example 8 Find decimal notation for $\frac{4}{11}$.

$$\frac{4}{11} = 4 \div 11 \qquad \begin{array}{r} 0.3\,6\,3\,6 \\ 11\,\overline{)\,4.0\,0\,0\,0} \\ \underline{3\,3} \\ 7\,0 \\ \underline{6\,6} \\ 4\,0 \\ \underline{3\,3} \\ 7\,0 \\ \underline{6\,6} \\ 4 \end{array}$$

Since 7 and 4 keep reappearing as remainders, the sequence of digits "36" repeats in the quotient, and

$$\frac{4}{11} = 0.363636\ldots, \quad \text{or} \quad 0.\overline{36}.$$

Do Exercises 9 and 10.

Example 9 Find decimal notation for $\frac{5}{7}$.

We have

$$\begin{array}{r} 0.714285 \\ 7\overline{)5.000000} \\ \underline{4\,9} \\ 10 \\ \underline{7} \\ 30 \\ \underline{28} \\ 20 \\ \underline{14} \\ 60 \\ \underline{56} \\ 40 \\ \underline{35} \\ 5 \end{array}$$

Since 5 appears as a remainder, the sequence of digits "714285" repeats in the quotient, and

$$\frac{5}{7} = 0.714285714285\ldots, \quad \text{or} \quad 0.\overline{714285}.$$

The length of a repeating part can be very long—too long to find on a calculator. An example is $\frac{5}{97}$, which has a repeating part of 96 digits.

Do Exercise 11.

b Rounding in Problem Solving

In applied problems, repeating decimals are rounded to get approximate answers.

Examples Round each to the nearest tenth, hundredth, and thousandth.

	Nearest tenth	*Nearest hundredth*	*Nearest thousandth*
10. $0.8\overline{3} = 0.83333\ldots$	0.8	0.83	0.833
11. $0.\overline{09} = 0.090909\ldots$	0.1	0.09	0.091
12. $0.\overline{714285} = 0.714285714285\ldots$	0.7	0.71	0.714

Do Exercises 12–14.

Find decimal notation.

9. $\frac{5}{11}$

10. $\frac{12}{11}$

11. Find decimal notation for $\frac{3}{7}$.

Round each to the nearest tenth, hundredth, and thousandth.

12. $0.\overline{6}$

13. $0.\overline{808}$

14. $6.\overline{245}$

Answers on page A-9

15. Calculate: $\frac{5}{6} \times 0.864$.

Calculate.

16. $\frac{1}{3} \times 0.384 + \frac{5}{8} \times 0.6784$

17. $\frac{5}{6} \times 0.864 + 14.3 \div \frac{8}{5}$

Answers on page A-9

C Calculations with Fractional and Decimal Notation Together

In certain kinds of calculations, fractional and decimal notation might occur together. In such cases, there are at least three ways in which we might proceed.

Example 13 Calculate: $\frac{2}{3} \times 0.576$.

METHOD 1. One way to do this calculation is to convert the decimal notation to fractional notation so that both numbers are in fractional notation. The answer can be left in fractional notation and simplified, or we can convert back to decimal notation and round, if appropriate.

$$\begin{aligned}\frac{2}{3} \times 0.576 &= \frac{2}{3} \cdot \frac{576}{1000} = \frac{2 \cdot 576}{3 \cdot 1000} \\ &= \frac{2 \cdot 2 \cdot 2 \cdot 2 \cdot 2 \cdot 2 \cdot 2 \cdot 3 \cdot 3}{2 \cdot 2 \cdot 2 \cdot 3 \cdot 5 \cdot 5 \cdot 5} \\ &= \frac{2 \cdot 2 \cdot 2 \cdot 3}{2 \cdot 2 \cdot 2 \cdot 3} \cdot \frac{2 \cdot 2 \cdot 2 \cdot 2 \cdot 3}{5 \cdot 5 \cdot 5} \\ &= 1 \cdot \frac{2 \cdot 2 \cdot 2 \cdot 2 \cdot 3}{5 \cdot 5 \cdot 5} \\ &= \frac{2 \cdot 2 \cdot 2 \cdot 2 \cdot 3}{5 \cdot 5 \cdot 5} = \frac{48}{125}, \text{ or } 0.384\end{aligned}$$

METHOD 2. A second way to do this calculation is to convert the fractional notation to decimal notation so that both numbers are in decimal notation. Since $\frac{2}{3}$ converts to repeating decimal notation, it is first rounded to some chosen decimal place. We choose three decimal places. Then, using decimal notation, we multiply. Note that the answer is not as accurate as that found by method 1, due to the rounding.

$$\frac{2}{3} \times 0.576 = 0.\overline{6} \times 0.576 \approx 0.667 \times 0.576 = 0.384192$$

METHOD 3. A third way to do this calculation is to treat 0.576 as $\frac{0.576}{1}$. Then we multiply 0.576 by 2, and divide the result by 3.

$$\frac{2}{3} \times 0.576 = \frac{2}{3} \times \frac{0.576}{1} = \frac{2 \times 0.576}{3} = \frac{1.152}{3} = 0.384$$

Do Exercise 15.

Example 14 Calculate: $\frac{2}{3} \times 0.576 + 3.287 \div \frac{4}{5}$.

We use the rules for order of operations, doing first the multiplication and then the division. Then we add.

$$\begin{aligned}\frac{2}{3} \times 0.576 + 3.287 \div \frac{4}{5} &= 0.384 + 3.287 \cdot \frac{5}{4} \\ &= 0.384 + 4.10875 \\ &= 4.49275\end{aligned}$$

Do Exercises 16 and 17.

Exercise Set 4.5

a Find decimal notation.

1. $\frac{3}{5}$ **2.** $\frac{19}{20}$ **3.** $\frac{13}{40}$ **4.** $\frac{3}{16}$ **5.** $\frac{1}{5}$ **6.** $\frac{3}{20}$

7. $\frac{17}{20}$ **8.** $\frac{9}{40}$ **9.** $\frac{19}{40}$ **10.** $\frac{81}{40}$ **11.** $\frac{39}{40}$ **12.** $\frac{31}{40}$

13. $\frac{13}{25}$ **14.** $\frac{61}{125}$ **15.** $\frac{2502}{125}$ **16.** $\frac{181}{200}$ **17.** $\frac{1}{4}$ **18.** $\frac{1}{2}$

19. $\frac{23}{40}$ **20.** $\frac{11}{20}$ **21.** $\frac{18}{25}$ **22.** $\frac{37}{25}$ **23.** $\frac{19}{16}$ **24.** $\frac{5}{8}$

25. $\frac{4}{15}$ **26.** $\frac{7}{9}$ **27.** $\frac{1}{3}$ **28.** $\frac{1}{9}$ **29.** $\frac{4}{3}$ **30.** $\frac{8}{9}$

31. $\frac{7}{6}$ **32.** $\frac{7}{11}$ **33.** $\frac{4}{7}$ **34.** $\frac{14}{11}$ **35.** $\frac{11}{12}$ **36.** $\frac{5}{12}$

b

37.–47. Round each answer of the odd-numbered Exercises 25–35 to the nearest tenth, hundredth, and thousandth.

38.–48. Round each answer of the even-numbered Exercises 26–36 to the nearest tenth, hundredth, and thousandth.

Round each to the nearest tenth, hundredth, and thousandth.

49. $0.\overline{18}$ **50.** $0.\overline{83}$ **51.** $0.2\overline{7}$ **52.** $3.5\overline{4}$

c Calculate.

53. $\frac{7}{8} \times 12.64$ **54.** $\frac{4}{5} \times 384.8$ **55.** $2\frac{3}{4} + 5.65$ **56.** $4\frac{4}{5} + 3.25$

57. $\frac{47}{9} \times 79.95$ **58.** $\frac{7}{11} \times 2.7873$ **59.** $\frac{1}{2} - 0.5$ **60.** $3\frac{1}{8} - 2.75$

61. $4.875 - 2\frac{1}{16}$

62. $55\frac{3}{5} - 12.22$

63. $\frac{5}{6} \times 0.0765 + \frac{5}{4} \times 0.1124$

64. $\frac{3}{5} \times 6384.1 - \frac{3}{8} \times 156.56$

65. $\frac{4}{5} \times 384.8 + 24.8 \div \frac{8}{3}$

66. $102.4 \div \frac{2}{5} - 12 \times \frac{5}{6}$

67. $\frac{7}{8} \times 0.86 - 0.76 \times \frac{3}{4}$

68. $17.95 \div \frac{5}{8} + \frac{3}{4} \times 16.2$

69. $3.375 \times 5\frac{1}{3}$

70. $2.5 \times 3\frac{5}{8}$

71. $6.84 \div 2\frac{1}{2}$

72. $8\frac{1}{2} \div 2.125$

Skill Maintenance

73. Multiply: $9 \cdot 2\frac{1}{3}$. [3.6a]

74. Divide: $84 \div 8\frac{2}{5}$. [3.6b]

75. Subtract: $20 - 16\frac{3}{5}$. [3.5b]

76. Add: $14\frac{3}{5} + 16\frac{1}{10}$. [3.5a]

Solve. [3.2c]

77. A recipe for bread calls for $\frac{2}{3}$ cup of water, $\frac{1}{4}$ cup of milk, and $\frac{1}{8}$ cup of oil. How many cups of liquid ingredients does the recipe call for?

78. A board $\frac{9}{10}$ in. thick is glued to a board $\frac{8}{10}$ in. thick. The glue is $\frac{3}{100}$ in. thick. How thick is the result?

Synthesis

79. ◆ When is long division *not* the fastest way of converting a fraction to decimal notation?

80. ◆ Examine Example 13 of this section. How could the problem be changed so that method 2 would give a result that is completely accurate?

Find decimal notation.

81. $\frac{1}{7}$

82. $\frac{2}{7}$

83. $\frac{3}{7}$

84. $\frac{4}{7}$

85. $\frac{5}{7}$

86. From the pattern of Exercises 81–85, guess the decimal notation for $\frac{6}{7}$. Check on your calculator.

Find decimal notation.

87. $\frac{1}{9}$

88. $\frac{1}{99}$

89. $\frac{1}{999}$

90. From the pattern of Exercises 87–89, guess the decimal notation for $\frac{1}{9999}$. Check on your calculator.

4.6 Estimating

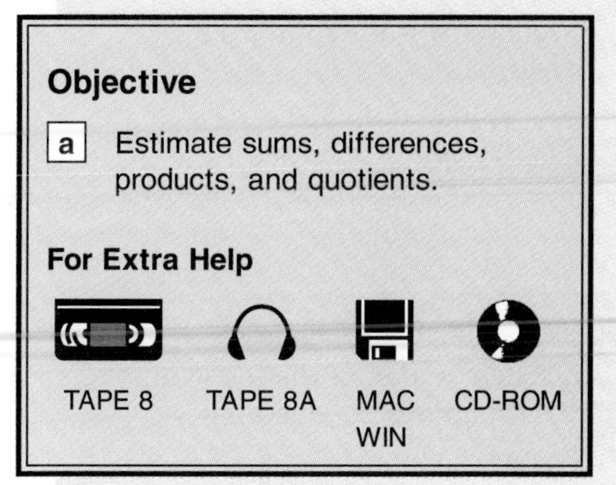

a Estimating Sums, Differences, Products and Quotients

Estimating has many uses. It can be done before a problem is even attempted in order to get an idea of the answer. It can be done afterward as a check, even when we are using a calculator. In many situations, an estimate is all we need. We usually estimate by rounding the numbers so that there are one or two nonzero digits. Consider the following advertisements for Examples 1–4.

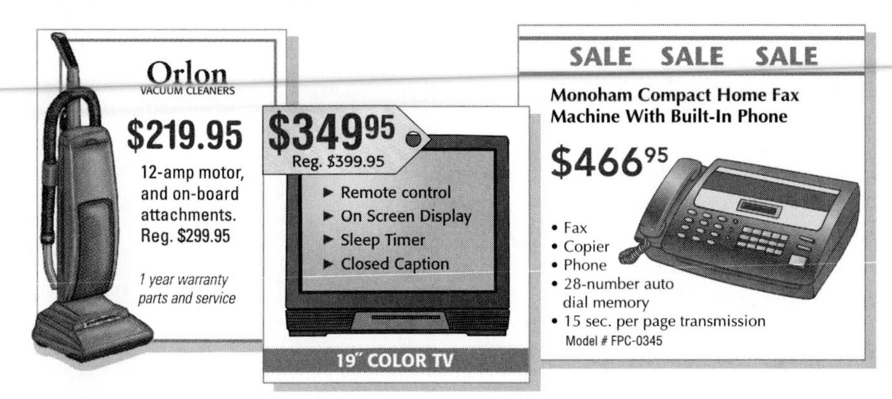

Example 1 Estimate to the nearest ten the total cost of one fax machine and one TV.

We are estimating the sum

$466.95 + $349.95 = Total cost.

The estimate to the nearest ten is

$470 + $350 = $820. (Estimated total cost)

We rounded $466.95 to the nearest ten and $349.95 to the nearest ten. The estimated sum is $820.

Do Exercise 1.

1. Estimate to the nearest ten the total cost of one TV and one vacuum cleaner. Which of the following is an appropriate estimate?

a) $5700 b) $570
c) $790 d) $57

Example 2 About how much more does the fax machine cost than the TV? Estimate to the nearest ten.

We are estimating the difference

$466.95 − $349.95 = Price difference.

The estimate to the nearest ten is

$470 − $350 = $120. (Estimated price difference)

Do Exercise 2.

2. About how much more does the TV cost than the vacuum cleaner? Estimate to the nearest ten. Which of the following is an appropriate estimate?

a) $130 b) $1300
c) $580 d) $13

Answers on page A-9

Example 3 Estimate the total cost of 4 vacuum cleaners.

We are estimating the product

$$4 \times \$219.95 = \text{Total cost.}$$

The estimate is found by rounding $219.95 to the nearest ten:

$$4 \times \$220 = \$880.$$

Do Exercise 3.

Example 4 About how many fax machines can be bought for $1580?

We estimate the quotient

$$\$1580 \div \$466.95.$$

Since we want a whole-number estimate, we choose our rounding appropriately. Rounding $466.95 to the nearest hundred, we get $500. Since $1580 is close to $1500, which is a multiple of 500, we estimate

$$\$1500 \div \$500,$$

so the answer is 3.

Do Exercise 4.

Example 5 Estimate: 4.8×52. Do not find the actual product. Which of the following is an appropriate estimate?

a) 25 b) 250 c) 2500 d) 360

We have

$$5 \times 50 = 250. \quad \text{(Estimated product)}$$

We rounded 4.8 to the nearest one and 52 to the nearest ten. Thus an appropriate estimate is (b).

Compare these estimates for the product 4.94×38:

$$5 \times 40 = 200, \quad 5 \times 38 = 190, \quad 4.9 \times 40 = 196.$$

The first estimate was the easiest. You could probably do it mentally. The others had more nonzero digits.

Do Exercises 5–10.

3. Estimate the total cost of 6 fax machines. Which of the following is an appropriate estimate?

a) $4400 b) $300
c) $30,000 d) $3000

4. About how many vacuum cleaners can be bought for $1100? Which of the following is an appropriate estimate?

a) 8 b) 5
c) 11 d) 124

Estimate the product. Do not find the actual product. Which of the following is an appropriate estimate?

5. 2.4×8

a) 16 b) 34
c) 125 d) 5

6. 24×0.6

a) 200 b) 5
c) 110 d) 20

7. 0.86×0.432

a) 0.04 b) 0.4
c) 1.1 d) 4

8. 0.82×0.1

a) 800 b) 8
c) 0.08 d) 80

9. 0.12×18.248

a) 180 b) 1.8
c) 0.018 d) 18

10. 24.234×5.2

a) 200 b) 125
c) 12.5 d) 234

Answers on page A-9

Example 6 Estimate: 82.08 ÷ 24. Which of the following is an appropriate estimate?

a) 400 b) 16 c) 40 d) 4

This is about 80 ÷ 20, so the answer is about 4. Thus an appropriate estimate is (d).

Example 7 Estimate: 94.18 ÷ 3.2. Which of the following is an appropriate estimate?

a) 30 b) 300 c) 3 d) 60

This is about 90 ÷ 3, so the answer is about 30. Thus an appropriate estimate is (a).

Example 8 Estimate: 0.0156 ÷ 1.3. Which of the following is an appropriate estimate?

a) 0.2 b) 0.002 c) 0.02 d) 20

This is about 0.02 ÷ 1, so the answer is about 0.02. Thus an appropriate estimate is (c).

Do Exercises 11–13.

In some cases, it is easier to estimate a quotient directly rather than by rounding the divisor and the dividend.

Example 9 Estimate: 0.0074 ÷ 0.23. Which of the following is an appropriate estimate?

a) 0.3 b) 0.03 c) 300 d) 3

We estimate 3 for a quotient. We check by multiplying.

$0.23 \times 3 = 0.69$

We make the estimate smaller. We estimate 0.3 and check by multiplying.

$0.23 \times 0.3 = 0.069$

We make the estimate smaller. We estimate 0.03 and check by multiplying.

$0.23 \times 0.03 = 0.0069$

This is about 0.0074, so the quotient is about 0.03. Thus an appropriate estimate is (b).

Do Exercise 14.

Estimate the quotient. Which of the following is an appropriate estimate?

11. 59.78 ÷ 29.1

a) 200 b) 20
c) 2 d) 0.2

12. 82.08 ÷ 2.4

a) 40 b) 4.0
c) 400 d) 0.4

13. 0.1768 ÷ 0.08

a) 8 b) 10
c) 2 d) 20

14. Estimate: 0.0069 ÷ 0.15. Which of the following is an appropriate estimate?

a) 0.5 b) 50
c) 0.05 d) 23.4

Answers on page A-9

Calculator Spotlight

Calculators can perform calculations so quickly that repeated experimental trials are not particularly time-consuming.

1. Use one of $+$, $-$, $\times$, and $\div$ in each blank to make a true sentence.

a) (0.37 ___ 18.78) ___ 2^{13} = 156,876.8

b) 2.56 ___ 6.4 ___ 51.2 ___ 17.4 = 312.84

2. In the subtraction below, *a* and *b* are digits. Find *a* and *b*.

$$\begin{array}{r} b876.a4321 \\ -1234.a678b \\ \hline 8641.b7a32 \end{array}$$

3. Look for a pattern in the following list, and find the missing numbers.

$22.22, $33.34, $44.46, $55.58, ____, ____, ____, ____, ____, ____, ____.

4. Look for a pattern in the following list, and find the missing numbers.

$2344.78, $2266, $2187.22, $2108.44, ____, ____, ____, ____.

Each of the following is called a *magic square.* The sum along each row, column, or diagonal is the same. Find the missing numbers.

5.

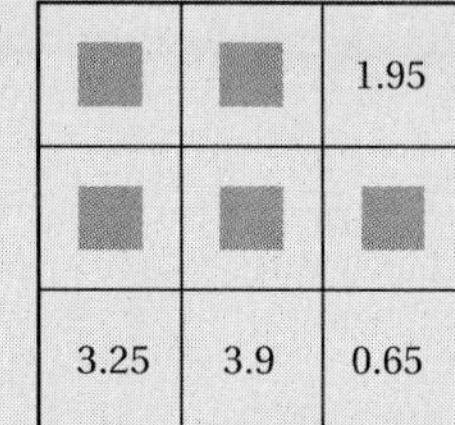

		1.95
3.25	3.9	0.65

Magic sum = ___

6.

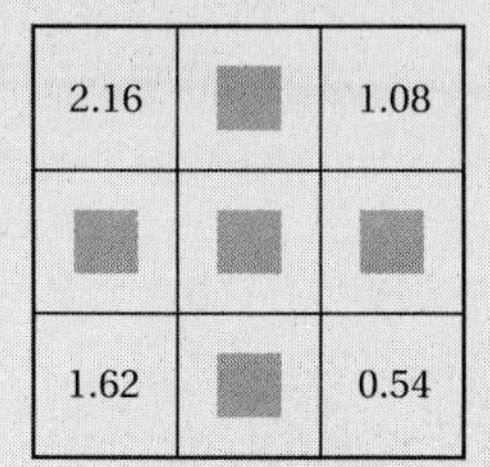

2.16		1.08
1.62		0.54

Magic sum = 4.05

7.

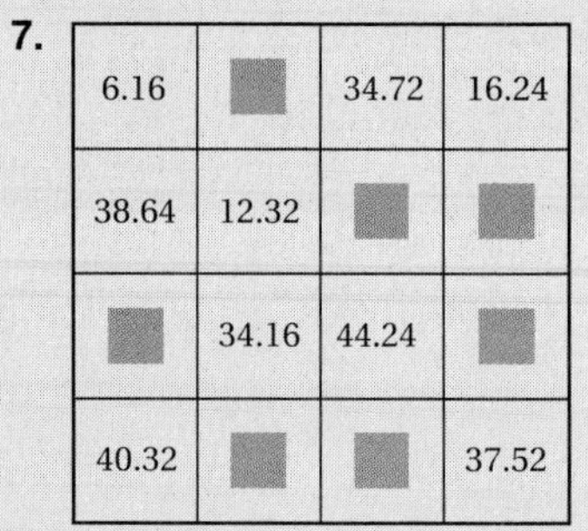

6.16		34.72	16.24
38.64	12.32		
	34.16	44.24	
40.32			37.52

Magic sum = 100.24

Exercise Set 4.6

a Consider the following advertisements for Exercises 1–8. Estimate the sums, differences, products, or quotients involved in these problems. Indicate which of the choices is an appropriate estimate.

1. Estimate the total cost of one entertainment center and one sound system.

a) $36 b) $72 c) $3.60 d) $360

2. Estimate the total cost of one entertainment center and one TV.

a) $410 b) $820 c) $41 d) $4.10

3. About how much more does the TV cost than the sound system?

a) $500 b) $80 c) $50 d) $5

4. About how much more does the TV cost than the entertainment center?

a) $100 b) $190 c) $250 d) $150

5. Estimate the total cost of 9 TVs.

a) $2700 b) $27 c) $270 d) $540

6. Estimate the total cost of 16 sound systems.

a) $5010 b) $4000 c) $40 d) $410

7. About how many TVs can be bought for $1700?

a) 600 b) 72 c) 6 d) 60

8. About how many sound systems can be bought for $1300?

a) 10 b) 5 c) 50 d) 500

Estimate by rounding as directed.

9. $0.02 + 1.31 + 0.34$; nearest tenth

10. $0.88 + 2.07 + 1.54$; nearest one

11. $6.03 + 0.007 + 0.214$; nearest one

12. $1.11 + 8.888 + 99.94$; nearest one

13. $52.367 + 1.307 + 7.324$; nearest one

14. $12.9882 + 1.0115$; nearest tenth

15. $2.678 - 0.445$; nearest tenth

16. $12.9882 - 1.0115$; nearest one

17. $198.67432 - 24.5007$; nearest ten

Estimate. Choose a rounding digit that gives one or two nonzero digits. Indicate which of the choices is an appropriate estimate.

18. 234.12321 − 200.3223
a) 600 b) 60
c) 300 d) 30

19. 49 × 7.89
a) 400 b) 40
c) 4 d) 0.4

20. 7.4 × 8.9
a) 95 b) 63
c) 124 d) 6

21. 98.4 × 0.083
a) 80 b) 12
c) 8 d) 0.8

22. 78 × 5.3
a) 400 b) 800
c) 40 d) 8

23. 3.6 ÷ 4
a) 10 b) 1
c) 0.1 d) 0.01

24. 0.0713 ÷ 1.94
a) 4 b) 0.4
c) 0.04 d) 40

25. 74.68 ÷ 24.7
a) 9 b) 3
c) 12 d) 120

26. 914 ÷ 0.921
a) 9 b) 90
c) 900 d) 0.9

27. *Movie Revenue.* Total summer box-office revenue (money taken in) for the movie *Eraser* was $53.6 million (***Source*:** *Hollywood Reporter Magazine*). Each theater showing the movie averaged $6716 in revenue. Estimate how many screens were showing this movie.

28. *Nintendo and the Sears Tower.* The Nintendo Game Boy portable video game is 4.5 in. (0.375 ft) tall (***Source*:** Nintendo of America). Estimate how many game units it would take to reach the top of the Sears Tower, which is 1454 ft tall. Round to the nearest one.

Skill Maintenance

Find the prime factorization. [2.1d]

29. 108 **30.** 400 **31.** 325 **32.** 666

Simplify. [2.5b]

33. $\frac{125}{400}$ **34.** $\frac{3225}{6275}$ **35.** $\frac{72}{81}$ **36.** $\frac{325}{625}$

Synthesis

37. ◆ A roll of fiberglass insulation costs $21.95. Describe two situations involving estimating and the cost of fiberglass insulation. Devise one situation so that $21.95 is rounded to $22. Devise the other situation so that $21.95 is rounded to $20.

38. ◆ Describe a situation in which an estimation is made by rounding to the nearest 10,000 and then multiplying.

The following were done on a calculator. Estimate to see if the decimal point was placed correctly.

39. 178.9462 × 61.78 = 11,055.29624

40. 14,973.35 ÷ 298.75 = 501.2

41. 19.7236 − 1.4738 × 4.1097 = 1.366672414

42. 28.46901 ÷ 4.9187 − 2.5081 = 3.279813473

Estimate the cost of food for a catered party.

4.7 Applications and Problem Solving

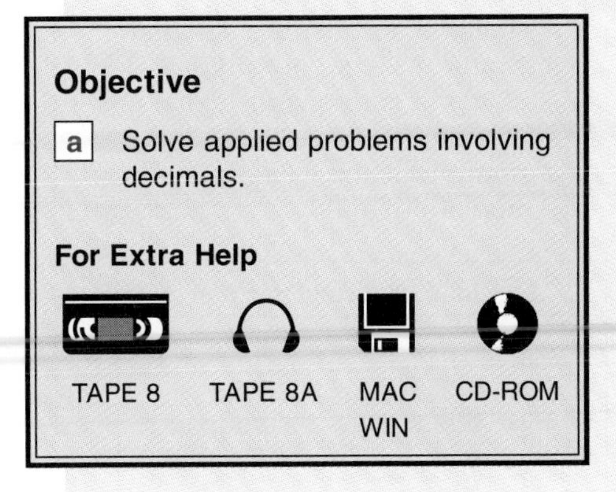

a Solving applied problems with decimals is like solving applied problems with whole numbers. We translate first to an equation that corresponds to the situation. Then we solve the equation.

Example 1 *Eating Out.* More and more Americans are eating meals outside the home. The following graph compares the average check for meals of various types for the years 1995 and 1996. How much more is the average check for casual dining in 1996 than in 1995?

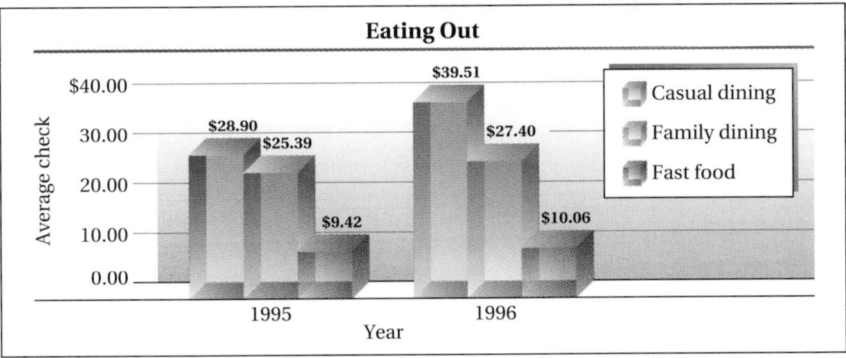

Source: Sandelman and Associates, Brea, California

1. **Familiarize.** We use the bar graph to visualize the situation and to obtain the appropriate data. We let c = the additional amount spent in 1996.
2. **Translate.** This is a "how-much-more" situation. We translate as follows, using the data from the bar graph.

Average check in 1995	plus	Additional amount	is	Average check in 1996
↓	↓	↓	↓	↓
$28.90	$+$	c	$=$	$39.51

3. **Solve.** We solve the equation, first subtracting 28.90 from both sides:

$$28.90 + c - 28.90 = 39.51 - 28.90$$
$$c = 10.61.$$

$$\begin{array}{r} \overset{8\ 15}{3\ \not{9}.\not{5}\ 1} \\ -\ 2\ 8.9\ 0 \\ \hline 1\ 0.6\ 1 \end{array}$$

4. **Check.** We can check by adding 10.61 to 28.90 to get 39.51.
5. **State.** The average check for casual dining in 1996 was $10.61 more than in 1995.

Do Exercise 1.

1. *Body Temperature.* Normal body temperature is 98.6°F. When fevered, most people will die if their bodies reach 107°F. This is a rise of how many degrees?

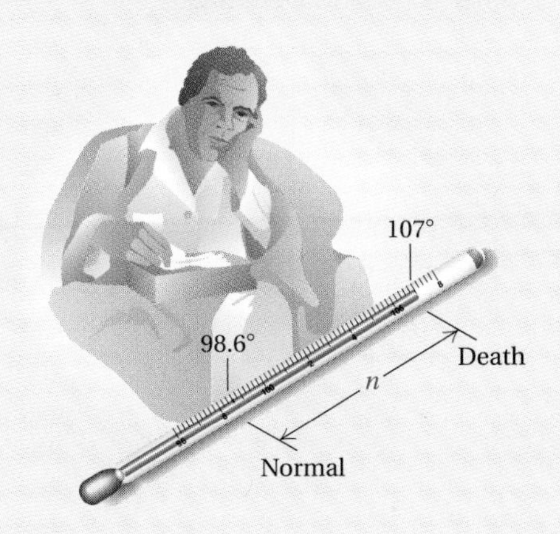

Answer on page A-9

2. Each year, the average American drinks about 49.0 gal of soft drinks, 41.2 gal of water, 25.3 gal of milk, 24.8 gal of coffee, and 7.8 gal of fruit juice. What is the total amount that the average American drinks?

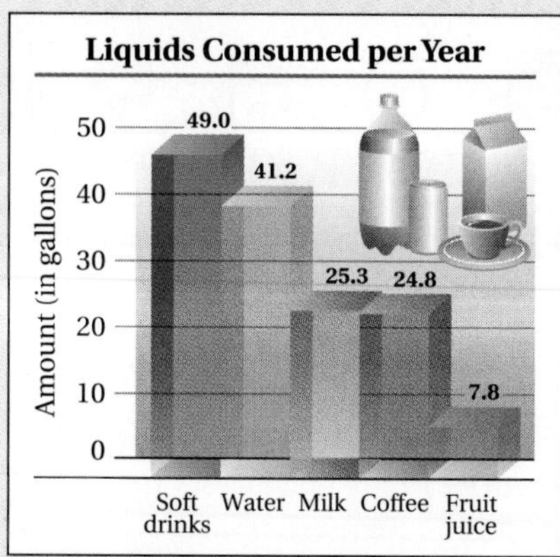

Source: U.S. Department of Agriculture

Answer on page A-9

Example 2 *Injections of Medication.* A patient was given injections of 2.8 mL, 1.35 mL, 2.0 mL, and 1.88 mL over a 24-hr period. What was the total amount of the injections?

1. **Familiarize.** We make a drawing or at least visualize the situation. We let t = the amount of the injections.

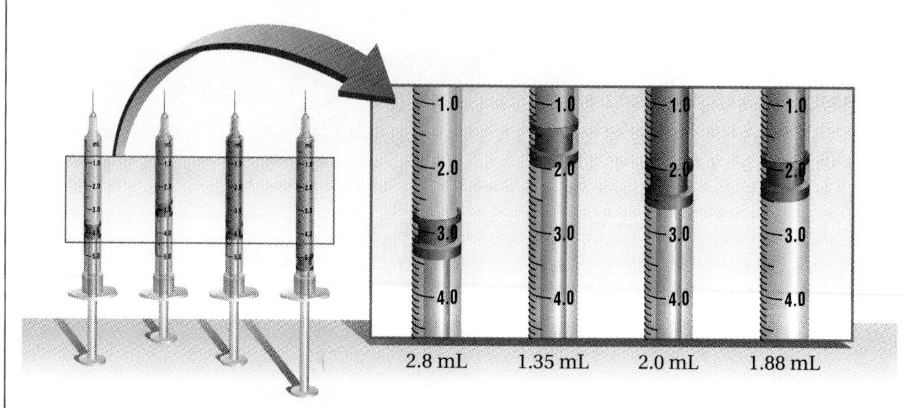

2. **Translate.** Amounts are being combined. We translate to an equation:

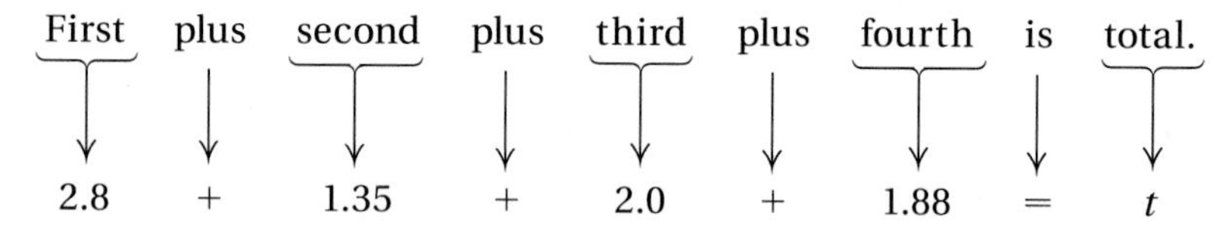

3. **Solve.** To solve, we carry out the addition.

$$\begin{array}{r} {}^{2\ 1} \\ 2.8\ 0 \\ 1.3\ 5 \\ 2.0\ 0 \\ +\ 1.8\ 8 \\ \hline 8.0\ 3 \end{array}$$

Thus, $t = 8.03$.

4. **Check.** We can check by repeating our addition. We can also see whether our answer is reasonable by first noting that it is indeed larger than any of the numbers being added. We can also check by rounding:

$$\begin{aligned} 2.8 + 1.35 + 2.0 + 1.88 &\approx 3 + 1 + 2 + 2 \\ &= 8 \approx 8.03. \end{aligned}$$

If we had gotten an answer like 80.3 or 0.803, then our estimate, 8, would have told us that we did something wrong, like not lining up the decimal points.

5. **State.** The total amount of the injections was 8.03 mL.

Do Exercise 2.

Example 3 *IRS Driving Allowance.* In a recent year, the Internal Revenue Service allowed a tax deduction of 31¢ per mile for mileage driven for business purposes. What deduction, in dollars, would be allowed for driving 127 mi?

1. **Familiarize.** We first make a drawing or at least visualize the situation. Repeated addition fits this situation. We let d = the deduction, in dollars, allowed for driving 127 mi.

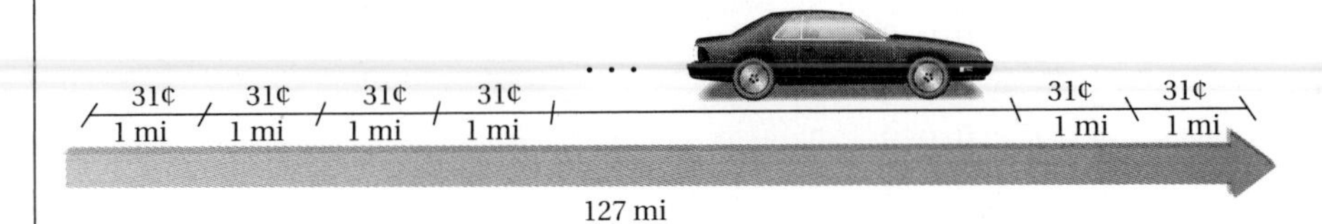

2. **Translate.** We translate as follows.

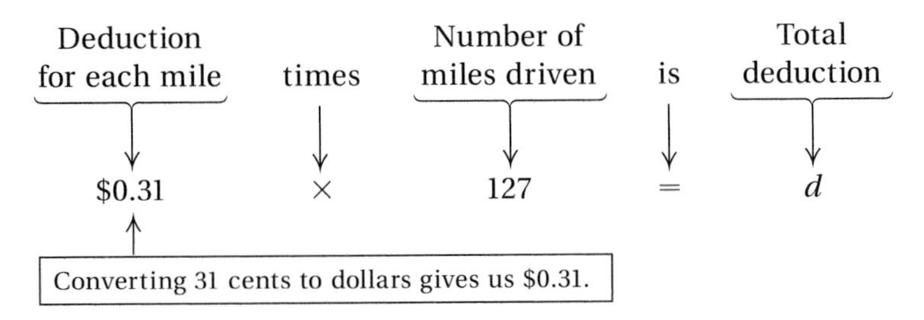

3. **Solve.** To solve the equation, we carry out the multiplication.

$$\begin{array}{r} 1\,2\,7 \\ \times \quad 0.3\,1 \\ \hline 1\,2\,7 \\ 3\,8\,1\,0 \\ \hline 3\,9.3\,7 \end{array}$$

Thus, $d = 39.37$.

4. **Check.** We can obtain a partial check by rounding and estimating:

$$127 \times 0.31 \approx 130 \times 0.3$$
$$= 39 \approx 39.37.$$

5. **State.** The total allowable deduction would be $39.37.

Do Exercise 3.

Example 4 *Loan Payments.* A car loan of $7382.52 is to be paid off in 36 monthly payments. How much is each payment?

1. **Familiarize.** We first make a drawing. We let n = the amount of each payment.

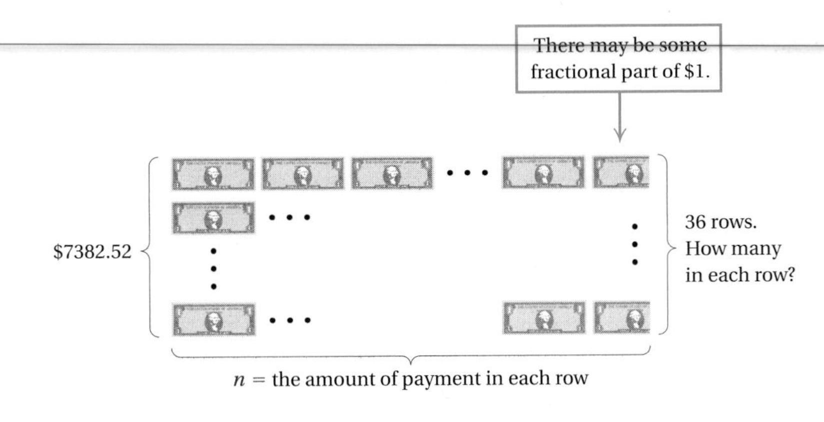

3. *Printing Costs.* At a printing company, the cost of copying is 12 cents per page. How much, in dollars, would it cost to make 466 copies?

Answer on page A-9

4. *Loan Payments.* A loan of \$4425 is to be paid off in 12 monthly payments. How much is each payment?

Answer on page A-9

2. Translate. The problem can be translated to the following equation, thinking that

$$(\text{Total loan}) \div (\text{Number of payments}) = \text{Amount of each payment}$$
$$\$7382.52 \div 36 = n.$$

3. Solve. To solve the equation, we carry out the division.

$$\begin{array}{r} 205.07 \\ 36\overline{)7382.52} \\ \underline{720000} \\ 18252 \\ \underline{18000} \\ 252 \\ \underline{252} \\ 0 \end{array}$$

Thus, $n = 205.07$.

4. Check. A partial check can be obtained by estimating the quotient: $\$7382.56 \div 36 \approx 8000 \div 40 = 200 \approx 205.07$. The estimate checks.

5. State. Each payment is \$205.07.

Do Exercise 4.

The area of a rectangular region is given by the formula *Area = Length · Width,* or $A = l \cdot w$. We can use this formula with decimal notation.

Example 5 *Poster Area.* A rectangular poster measures 73.2 cm by 61.8 cm. Find the area.

1. Familiarize. We first make a drawing, letting A = the area.

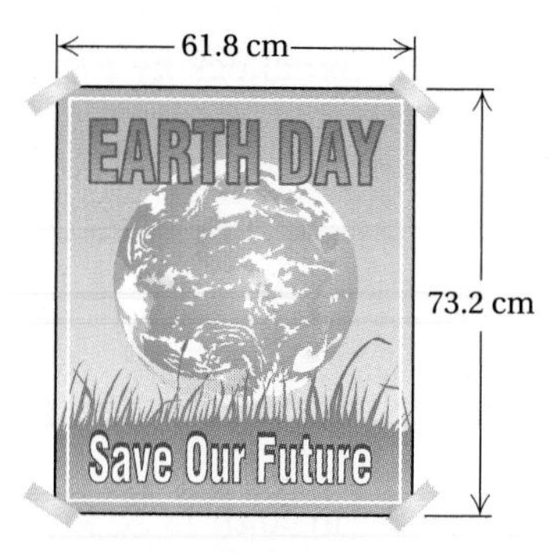

2. Translate. Then we use the formula $A = l \cdot w$ and translate:

$A = 73.2 \times 61.8.$

3. Solve. We solve by carrying out the multiplication.

$$\begin{array}{r} 73.2 \\ \times\quad 61.8 \\ \hline 5856 \\ 7320 \\ \underline{439200} \\ 4523.76 \end{array}$$

Thus, $A = 4523.76$.

4. **Check.** We obtain a partial check by estimating the product:

$$73.2 \times 61.8 \approx 70 \times 60 = 4200 \approx 4523.76.$$

Since this estimate is not too close, we might repeat our calculation or change our estimate to be more certain. We leave this to the student. We see that 4523.76 checks.

5. **State.** The area is 4523.76 cm^2.

Do Exercise 5.

5. A standard-size index card measures 12.7 cm by 7.6 cm. Find its area.

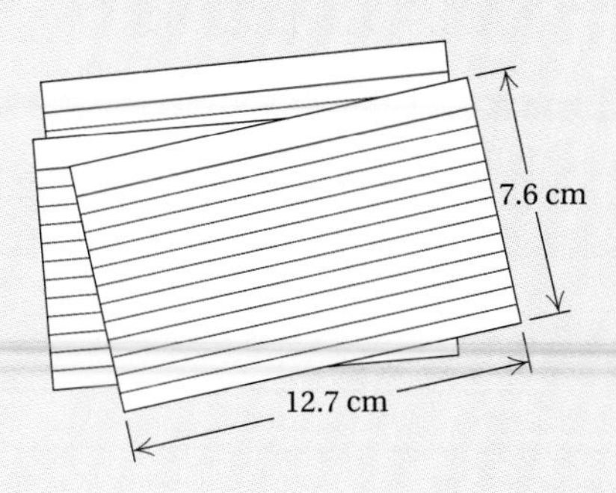

Example 6 *Cost of Crabmeat.* One pound of crabmeat makes 3 servings at the Key West Seafood Restaurant. It costs \$14.98 per pound. What is the cost per serving? Round to the nearest cent.

1. **Familiarize.** We let c = the cost per serving.

2. **Translate.** We translate as follows.

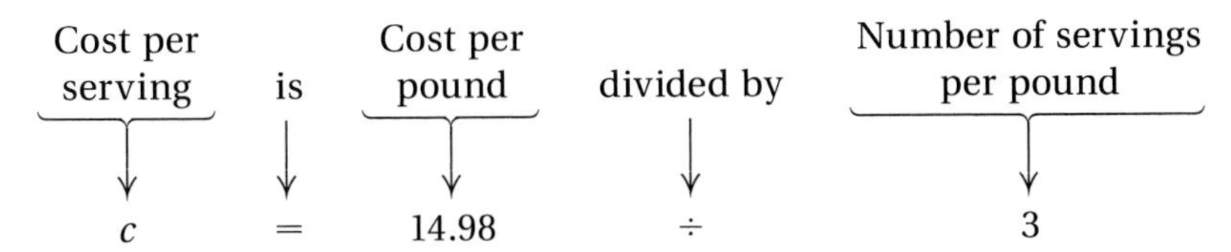

3. **Solve.** To solve, we carry out the division.

```
      4.9 9 3 3
3 ) 1 4.9 8 0 0      c = 4.993̅
    1 2
      2 9
      2 7
        2 8
        2 7
          1 0
            9
          1 0
```

4. **Check.** We check by estimating the quotient:

$$14.98 \div 3 \approx 15 \div 3 = 5 \approx 4.99\overline{3}.$$

In this case, our check provides a good estimate.

5. **State.** We round $4.99\overline{3}$ and find the cost per serving to be about \$4.99.

Do Exercise 6.

6. One pound of lean boneless ham contains 4.5 servings. It costs \$3.99 per pound. What is the cost per serving? Round to the nearest cent.

Answers on page A-9

7. *Gas Mileage.* A driver filled the gasoline tank and noted that the odometer read 38,320.8. After the next filling, the odometer read 38,735.5. It took 14.5 gal to fill the tank. How many miles per gallon did the driver get?

Answer on page A-9

Multistep Problems

Example 7 *Gas Mileage.* A driver filled the gasoline tank and noted that the odometer read 67,507.8. After the next filling, the odometer read 68,006.1. It took 16.5 gal to fill the tank. How many miles per gallon did the driver get?

1. Familiarize. We first make a drawing.

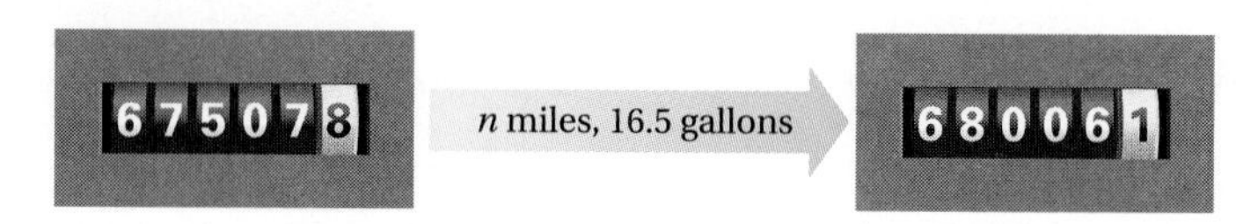

This is a two-step problem. First, we find the number of miles that have been driven between fillups. We let $n =$ the number of miles driven.

2., 3. Translate and **Solve.** This is a "how-much-more" situation. We translate and solve as follows.

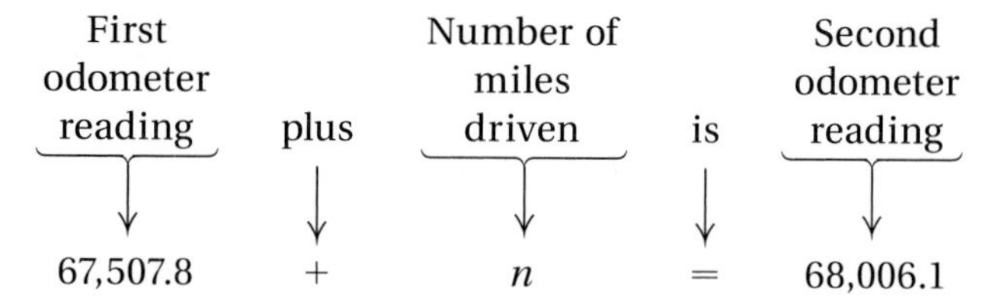

To solve the equation, we subtract 67,507.8 on both sides:

$$n = 68{,}006.1 - 67{,}507.8$$
$$= 498.3.$$

```
  6 8,0 0 6.1
- 6 7,5 0 7.8
-------------
      4 9 8.3
```

Second, we divide the total number of miles driven by the number of gallons. This gives us $m =$ the number of miles per gallon—that is, the mileage. The division that corresponds to the situation is

$$498.3 \div 16.5 = m.$$

To find the number m, we divide.

```
             3 0.2
1 6.5 ) 4 9 8.3 0
        4 9 5 0
        -------
            3 3 0
            3 3 0
            -----
                0
```

Thus, $m = 30.2$.

4. Check. To check, we first multiply the number of miles per gallon times the number of gallons:

$$16.5 \times 30.2 = 498.3.$$

Then we add 498.3 to 67,507.8:

$$67{,}507.8 + 498.3 = 68{,}006.1.$$

The mileage 30.2 checks.

5. State. The driver got 30.2 miles per gallon.

Do Exercise 7.

Example 8 *Home-Cost Comparison.* Suppose you own a home like the one shown here and it is valued at $250,000 in Indianapolis, Indiana. What would it cost to buy a similar (replacement) home in Beverly Hills, California? To find out, we can use an index table prepared by Coldwell Banker Real Estate Corporation (***Source*:** Coldwell Banker Real Estate Corporation. For a complete index table, contact your local representative.). We use the following formula:

$$\begin{pmatrix}\text{Cost of your}\\ \text{home in new city}\end{pmatrix} = \begin{pmatrix}\text{Value of}\\ \text{your home}\end{pmatrix} \div \begin{pmatrix}\text{Index of}\\ \text{your city}\end{pmatrix} \times \begin{pmatrix}\text{Index of}\\ \text{new city}\end{pmatrix}$$

Find the cost of your Indianapolis home in Beverly Hills.

State	City	Index
California	San Francisco Beverly Hills Fresno	286 376 82
Indiana	Indianapolis Fort Wayne	79 69
Arizona	Phoenix Tucson	90 79
Illinois	Chicago Naperville	214 101
Texas	Austin Dallas Houston	89 70 61
Florida	Miami Orlando Tampa	85 76 72
Massachusetts	Wellesley Cape Cod	231 84
Georgia	Atlanta	81
New York	Queens Albany	179 89

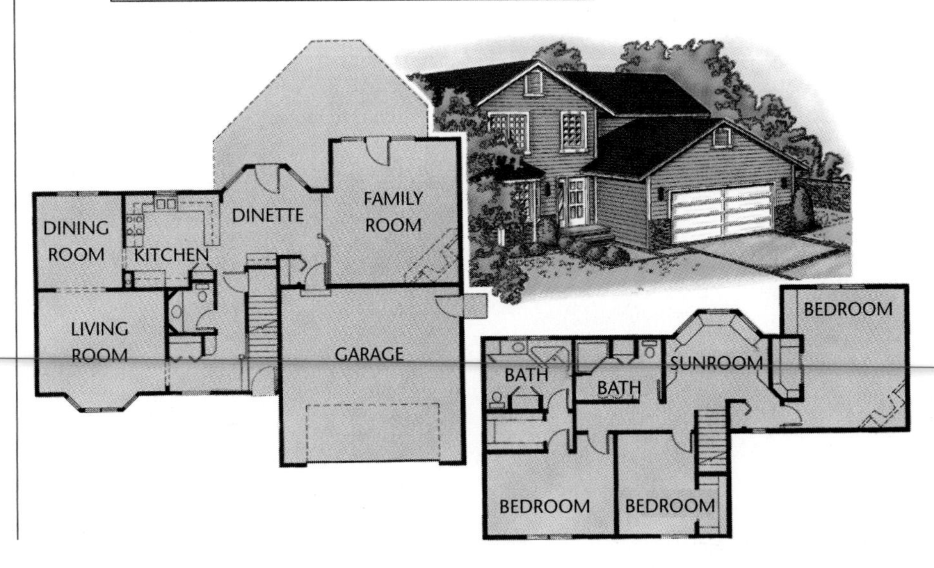

8. *Home-cost Comparison.* Find the replacement cost of a $250,000 home in Indianapolis if you were to try to replace it when moving to Dallas.

9. Find the replacement cost of a $250,000 home in Phoenix if you were to try to replace it when moving to Chicago.

Answers on page A-9

1. **Familiarize.** We let C = the cost of the home in Beverly Hills. We use the table and look up the indexes of the city in which you now live and the city to which you are moving.
2. **Translate.** Using the formula, we translate to the following equation:

   $C = \$250{,}000 \div 79 \times 376.$

3. **Solve.** To solve, we carry out the computations using the rules for order of operations (see Section 4.4):

   $C = \$250{,}000 \div 79 \times 376$

   $= \$3164.557 \times 376$ — Carrying out the division first

   $\approx \$1{,}189{,}873.$ — Carrying out the multiplication and rounding to the nearest one

   On a calculator, the computation could be done in one step.

4. **Check.** We can repeat our computations or round and estimate as we learned in Section 4.6:

   $C = \$250{,}000 \div 79 \times 376$

   $\approx \$250{,}000 \div 100 \times 400$

   $= \$1{,}000{,}000.$

   Since $\$1{,}000{,}000 \approx \$1{,}189{,}873$, we have a partial check.

5. **State.** A home selling for $250,000 in Indianapolis would cost about $1,189,873 in Beverly Hills.

Do Exercises 8 and 9.

Exercise Set 4.7

a Solve.

1. What is the cost of 8 pairs of socks at $4.95 per pair?

2. What is the cost of 7 shirts at $32.98 each?

3. *Gasoline Cost.* What is the cost, in dollars, of 17.7 gal of gasoline at 119.9 cents per gallon? (119.9 cents = $1.199) Round the answer to the nearest cent.

4. *Gasoline Cost.* What is the cost, in dollars, of 20.4 gal of gasoline at 149.9 cents per gallon? Round the answer to the nearest cent.

5. Roberto bought a CD for $16.99 and paid with a $20 bill. How much change did he receive?

6. Madeleine buys a book for $44.68 and pays with a $50 bill. How much change does she receive?

7. *Body Temperature.* Normal body temperature is 98.6°F. During an illness, a patient's temperature rose 4.2°. What was the new temperature?

8. *Blood Test.* A medical assistant draws 9.85 mL of blood and uses 4.68 mL in a blood test. How much is left?

9. *Lottery Winnings.* In Texas, one of the state lotteries is called "Cash 5." In a recent weekly game, the lottery prize of $127,315 was shared equally by 6 winners. How much was each winner's share? Round to the nearest cent.

10. A group of 4 students pays $40.76 for lunch. What is each person's share?

11. A rectangular parking lot measures 800.4 ft by 312.6 ft. What is its area?

12. A rectangular fenced yard measures 40.3 yd by 65.7 yd. What is its area?

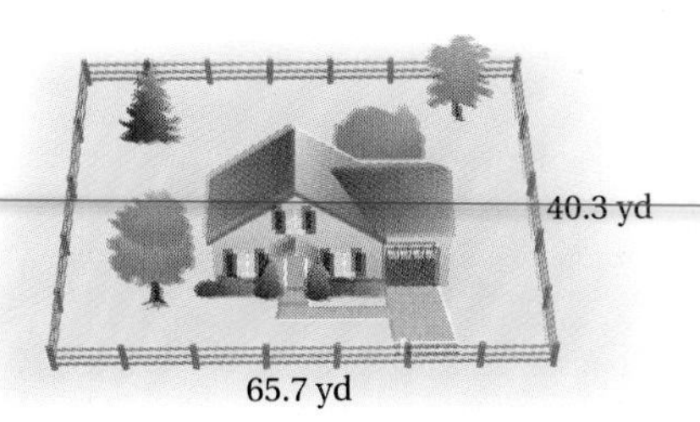

13. *Odometer Reading.* A family checked the odometer before starting a trip. It read 22,456.8 and they know that they will be driving 234.7 mi. What will the odometer read at the end of the trip?

14. *Miles Driven.* Petra bought gasoline when the odometer read 14,296.3. At the next gasoline purchase, the odometer read 14,515.8. How many miles had been driven?

15. *Eating Habits.* Each year, Americans eat 24.8 billion hamburgers and 15.9 billion hot dogs. How many more hamburgers than hot dogs do Americans eat?

16. *Gas Mileage.* A driver wants to estimate gas mileage per gallon. At 36,057.1 mi, the tank is filled with 10.7 gal. At 36,217.6 mi, the tank is filled with 11.1 gal. Find the mileage per gallon. Round to the nearest tenth.

17. *Jet-Powered Car.* A jet-powered car was measured on a computer to go from a speed of mach 0.85 to mach 1.15 (mach 1.0 is the speed of sound). What was the difference in these speeds?

18. *Fat Content.* There is 0.8 g of fat in one serving $\left(3\frac{1}{2}\text{ oz}\right)$ of raw scallops. In one serving of oysters, there is 2.5 g of fat. How much more fat is in one serving of oysters than in one serving of scallops?

19. *Gas Mileage.* Peggy filled her van's gas tank and noted that the odometer read 26,342.8. After the next filling, the odometer read 26,736.7. It took 19.5 gal to fill the tank. How many miles per gallon did the van get?

20. *Gas Mileage.* Peter filled his Honda's gas tank and noted that the odometer read 18,943.2. After the next filling, the odometer read 19,306.2. It took 13.2 gal to fill the tank. How many miles per gallon did the car get?

21. The water in a filled tank weighs 748.45 lb. One cubic foot of water weighs 62.5 lb. How many cubic feet of water does the tank hold?

22. *Highway Routes.* You can drive from home to work using either of two routes:

Route A: Via interstate highway, 7.6 mi, with a speed limit of 65 mph.

Route B: Via a country road, 5.6 mi, with a speed limit of 50 mph.

Assuming you drive at the posted speed limit, which route takes less time? (Use the formula *Distance* = *Speed* × *Time*.)

23. *Cost of Video Game.* The average video game costs 25 cents and runs for 1.5 min. Assuming a player does not win any free games and plays continuously, how much money, in dollars, does it cost to play a video game for 1 hr?

24. *Property Taxes.* The Colavitos own a house with an assessed value of $124,500. For every $1000 of assessed value, they pay $7.68 in taxes. How much do they pay in taxes?

Find the distance around (perimeter of) the figure.

25.

8.9 cm
23.8 cm
4.7 cm
18.6 cm
22.1 cm

26.

104.8 yd
111.9 yd
68.9 yd
56.7 yd
49.2 yd

Find the length d in the figure.

27.

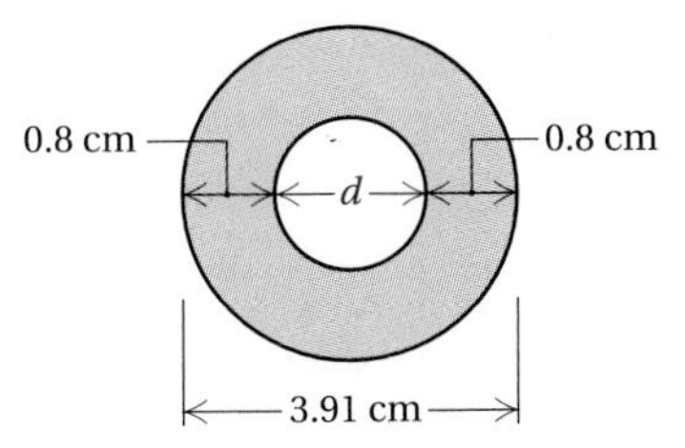

28.

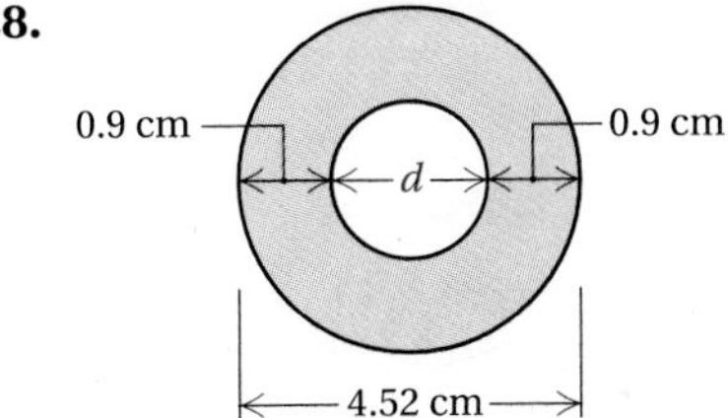

29. *Calories Burned Mowing.* A person weighing 150 lb burns 7.3 calories per minute while mowing a lawn with a power lawnmower (**Source:** *Hanely Science Answer Book*). How many calories would be burned in 2 hr of mowing?

30. Lot A measures 250.1 ft by 302.7 ft. Lot B measures 389.4 ft by 566.2 ft. What is the total area of the two lots?

31. Holly had $1123.56 in her checking account. She wrote checks of $23.82, $507.88, and $98.32 to pay some bills. She then deposited a bonus check of $678.20. How much is in her account after these changes?

32. Natalie Clad had $185.00 to spend for fall clothes: $44.95 was spent for shoes, $71.95 for a jacket, and $55.35 for pants. How much was left?

33. A rectangular yard is 20 ft by 15 ft. The yard is covered with grass except for an 8.5-ft square flower garden. How much grass is in the yard?

34. Rita earns a gross paycheck (before deductions) of $495.72. Her deductions are $59.60 for federal income tax, $29.00 for FICA, and $29.00 for medical insurance. What is her take-home paycheck?

35. *Batting Average.* In a recent year, Bernie Williams of the New York Yankees got 168 hits in 551 times at bat. What part of his at-bats were hits? Give decimal notation to the nearest thousandth. (This is a player's *batting average.*)

36. *Batting Average.* In a recent year, Chipper Jones of the Atlanta Braves got 185 hits in 598 times at bat. What was his batting average? Give decimal notation to the nearest thousandth. (See Exercise 35.)

37. It costs $24.95 a day plus 27 cents per mile to rent a compact car at Shuttles Rent-a-Car. How much, in dollars, would it cost to drive the car 120 mi in 1 day?

38. Zachary worked 53 hr during a week one summer. He earned $6.50 per hour for the first 40 hr and $9.75 per hour for overtime (hours exceeding 40). How much did Zachary earn during the week?

39. A family of five can save $6.72 per week by eating cooked cereal instead of ready-to-eat cereal. How much will they save in 1 year? Use 52 weeks for 1 year.

40. A medical assistant prepares 200 injections, each with 2.5 mL of penicillin. How much penicillin is used in all?

41. A restaurant owner bought 20 dozen eggs for $13.80. Find the cost of each egg to the nearest tenth of a cent (thousandth of a dollar).

42. *Weight Loss.* A person weighing 170 lb burns 8.6 calories per minute while mowing a lawn. One must burn about 3500 calories in order to lose 1 lb. How many pounds would be lost by mowing for 2 hr? Round to the nearest tenth.

43. *Soccer Field.* The dimensions of a World Cup soccer field are 114.9 yd by 74.4 yd. The dimensions of a standard football field are 120 yd by 53.3 yd. How much greater is the area of a soccer field?

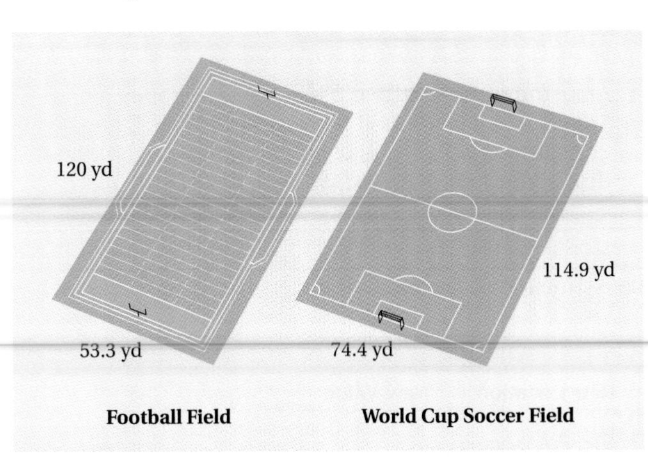

44. *Construction Pay.* A construction worker is paid \$13.50 per hour for the first 40 hr of work, and time and a half, or \$20.25 per hour, for any overtime exceeding 40 hr per week. One week she works 46 hr. How much is her pay?

45. *Loan Payment.* In order to make money on loans, financial institutions are paid back more money than they loan. You borrow \$120,000 to buy a house and agree to make monthly payments of \$880.52 for 30 yr. How much do you pay back altogether? How much more do you pay back than the amount of the loan?

46. *Car-Rental Cost.* Enterprise Rent-A-Car charges \$59.99 per day plus \$0.25 per mile for a luxury sedan (***Source:*** Enterprise Rent-A-Car). How much is the rental charge for a 4-day trip of 876 mi?

Airport Passengers. The following graph shows the number of passengers in a recent year who traveled through the country's busiest airports. (Use the graph for Exercises 47–50.)

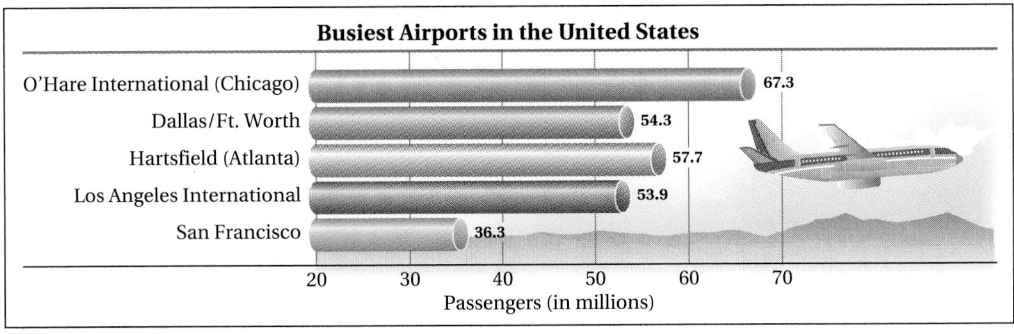

Source: Air Transport Association of America

47. How many more passengers does O'Hare handle than San Francisco?

48. How many more passengers does Hartsfield handle than Los Angeles?

49. How many passengers do Dallas/Ft. Worth and San Francisco handle altogether?

50. How many passengers do all these airports handle altogether?

51. *Body Temperature.* Normal body temperature is 98.6°F. A baby's bath water should be 100°F. How many degrees above normal body temperature is this?

52. *Body Temperature.* Normal body temperature is 98.6°F. The lowest temperature at which a patient has survived is 69°F. How many degrees below normal is this?

53. A student used a $20 bill to buy a poster for $10.75. The change was a five-dollar bill, three one-dollar bills, a dime, and two nickels. Was the change correct?

54. A customer bought two blank cassette tapes for $13.88. They were paid for with a $20 bill. The change was a five-dollar bill, a one-dollar bill, one dime, and two pennies. Was the change correct?

Home-Cost Comparison. Use the table and formula from Example 8. In each of the following cases, find the value of the house in the new location.

	Value	Present Location	New Location	New Value
55.	$125,000	Fresno	San Francisco	
56.	$180,000	Chicago	Beverly Hills	
57.	$96,000	Indianapolis	Tampa	
58.	$300,000	Miami	Queens	
59.	$240,000	San Francisco	Atlanta	
60.	$160,000	Cape Cod	Phoenix	

Skill Maintenance

Add.

61. 4569 + 1766 [1.2b]

62. $\frac{2}{3} + \frac{5}{8}$ [3.2b]

63. $4\frac{1}{3} + 2\frac{1}{2}$ [3.5a]

64. $\frac{5}{6} + \frac{7}{10}$ [3.2b]

65. 8099 + 5667 [1.2b]

Subtract.

66. 4569 − 1766 [1.3d]

67. $\frac{2}{3} - \frac{5}{8}$ [3.3a]

68. $4\frac{1}{3} - 2\frac{1}{2}$ [3.5b]

69. $\frac{5}{6} - \frac{7}{10}$ [3.3a]

70. 8099 − 5667 [1.3d]

Solve. [3.6c]

71. If a water wheel made 469 revolutions at a rate of $16\frac{3}{4}$ revolutions per minute, how long did it rotate?

72. If a bicycle wheel made 480 revolutions at a rate of $66\frac{2}{3}$ revolutions per minute, how long did it rotate?

Synthesis

73. ◆ Write a problem for a classmate to solve. Design the problem so that the solution is "Mona's Buick got 23.5 mpg."

74. ◆ Write a problem for a classmate to solve. Design the problem so that the solution is "The larger field is 200 m^2 bigger."

75. You buy a half-dozen packs of basketball cards with a dozen cards in each pack. The cost is a dozen cents for each half-dozen cards. How much do you pay for the cards?

Summary and Review Exercises: Chapter 4

The objectives to be tested in addition to the material in this chapter are [2.1d], [2.5b], [3.5a, b], and [3.6a, b].

Write a word name. [4.1a]

1. 3.47 **2.** 0.031

Write a word name as on a check. [4.1a]

3. $597.25 **4.** $0.96

Write fractional notation. [4.1b]

5. 0.09 **6.** 4.561

7. 0.089 **8.** 3.0227

Write decimal notation. [4.1c]

9. $\frac{34}{1000}$ **10.** $\frac{42,603}{10,000}$

11. $27\frac{91}{100}$ **12.** $867\frac{6}{1000}$

Which number is larger? [4.1d]

13. 0.034, 0.0185 **14.** 0.91, 0.19

15. 0.741, 0.6943 **16.** 1.038, 1.041

Round 17.4287 to the nearest: [4.1e]

17. Tenth. **18.** Hundredth.

19. Thousandth. **20.** One.

Add. [4.2a]

21. $\begin{array}{r} 2.048 \\ 65.371 \\ +\ 507.1 \\ \hline \end{array}$ **22.** $\begin{array}{r} 0.6 \\ 0.004 \\ 0.07 \\ +\ 0.0098 \\ \hline \end{array}$

23. 219.3 + 2.8 + 7

24. 0.41 + 4.1 + 41 + 0.041

Subtract. [4.2b]

25. $\begin{array}{r} 30.0 \\ -\ 0.7908 \\ \hline \end{array}$ **26.** $\begin{array}{r} 845.08 \\ -\ 54.79 \\ \hline \end{array}$

27. 37.645 − 8.497 **28.** 70.8 − 0.0109

Multiply. [4.3a]

29. $\begin{array}{r} 48 \\ \times\ 0.27 \\ \hline \end{array}$ **30.** $\begin{array}{r} 0.174 \\ \times\ 0.83 \\ \hline \end{array}$

31. 100 × 0.043 **32.** 0.001 × 24.68

Divide. [4.4a]

33. $8\overline{)60}$ **34.** $52\overline{)23.4}$

35. $2.6\overline{)117.52}$ **36.** $2.14\overline{)2.18708}$

37. $\frac{276.3}{1000}$ **38.** $\frac{13.892}{0.01}$

Solve. [4.2c], [4.4b]

39. $x + 51.748 = 548.0275$ **40.** $3 \cdot x = 20.85$

41. $10 \cdot y = 425.4$ **42.** $0.0089 + y = 5$

Solve. [4.7a]

43. Katrina earned $310.37 during a 40-hr week. What was her hourly wage? Round to the nearest cent.

44. Derek had $6274.53 in his checking account. He wrote a check for $385.79 to buy a fax-modem for his computer. How much was left in his account?

45. *Tea Consumption.* The average person drinks about 3.48 cups of tea per day. How many cups of tea does the average person drink in a week? in a 30-day month?

46. *Software Purchase.* Four software disks containing video instruction are purchased by a learning lab at a cost of $59.95 each. How much is spent altogether?

47. *Telephone Poles.* In the United States, there are 51.81 telephone poles for every 100 people. In Canada, there are 40.65. How many more telephone poles for every 100 people are there in the United States?

48. Antarctica Agricultural College has four subterranean cornfields. One year the harvest in each field was 1419.3 bushels, 1761.8 bushels, 1095.2 bushels, and 2088.8 bushels. What was the total harvest?

Estimate each of the following. [4.6a]

49. The product 7.82×34.487 by rounding to the nearest one

50. The difference $219.875 - 4.478$ by rounding to the nearest one

51. The quotient $82.304 \div 17.287$ by rounding to the nearest ten

52. The sum $45.78 + $78.99 by rounding to the nearest one

Find decimal notation. Use multiplying by 1. [4.5a]

53. $\frac{13}{5}$ 54. $\frac{32}{25}$ 55. $\frac{11}{4}$

Find decimal notation. Use division. [4.5a]

56. $\frac{13}{4}$ 57. $\frac{7}{6}$ 58. $\frac{17}{11}$

Round the answer to Exercise 58 to the nearest: [4.5b]

59. Tenth. 60. Hundredth. 61. Thousandth.

Convert from cents to dollars. [4.3b]

62. 8273 cents 63. 487 cents

Convert from dollars to cents. [4.3b]

64. $24.93 65. $9.86

Convert the number in the sentence to standard notation. [4.3b]

66. The statutory debt limit is $5.5 trillion.

67. Your blood travels 1.2 million miles in a week.

Calculate. [4.5c]

68. $(8 - 1.23) \div 4 + 5.6 \times 0.02$

69. $(1 + 0.07)^2 + 10^3 \div 10^2 + [4(10.1 - 5.6) + 8(11.3 - 7.8)]$

70. $\frac{3}{4} \times 20.85$

71. $\frac{1}{3} \times 123.7 + \frac{4}{9} \times 0.684$

Skill Maintenance

72. Multiply: $8\frac{1}{3} \cdot 5\frac{1}{4}$. [3.6a]

73. Divide: $20 \div 5\frac{1}{3}$. [3.6b]

74. Add: $12\frac{1}{2} + 7\frac{3}{10}$. [3.5a]

75. Subtract: $24 - 17\frac{2}{5}$. [3.5b]

76. Simplify: $\frac{28}{56}$. [2.5b]

77. Find the prime factorization of 192. [2.1d]

Synthesis

78. ◆ Consider finding decimal notation for $\frac{44}{125}$. Discuss as many ways as you can for finding such notation and give the answer. [4.5a]

79. ◆ Explain how we can use fractional notation to understand why we count decimal places when multiplying with decimal notation. [4.3a]

80. 🖩 In each of the following, use one of +, −, ×, and ÷ in each blank to make a true sentence. [4.4c]

a) $2.56 \;\square\; 6.4 \;\square\; 51.2 \;\square\; 17.4 \;\square\; 89.7 = 72.62$
b) $(0.37 \;\square\; 18.78) \;\square\; 2^{13} = 156{,}876.8$

81. Find repeating decimal notation for 1 and explain. Use the following hints. [4.5a]

$$\frac{1}{3} = 0.33333333\ldots,$$

$$\frac{2}{3} = 0.66666666\ldots$$

82. Find repeating decimal notation for 2. [4.5a]

Test: Chapter 4

1. Write a word name: 2.34.

2. Write a word name, as on a check, for \$1234.78.

Write fractional notation.

3. 0.91

4. 2.769

Write decimal notation.

5. $\frac{74}{1000}$

6. $\frac{37,047}{10,000}$

7. $756\frac{9}{100}$

8. $91\frac{703}{1000}$

Which number is larger?

9. 0.07, 0.162

10. 0.078, 0.06

11. 0.09, 0.9

Round 5.6783 to the nearest:

12. One.

13. Hundredth.

14. Thousandth.

15. Tenth.

Add.

16. $\begin{array}{r} 402.3 \\ 2.81 \\ +\quad 0.109 \\ \hline \end{array}$

17. $\begin{array}{r} 0.7 \\ 0.08 \\ 0.009 \\ +\,0.0012 \\ \hline \end{array}$

18. $102.4 + 6.1 + 78$

19. $0.93 + 9.3 + 93 + 930$

Subtract.

20. $\begin{array}{r} 52.678 \\ -\quad 4.321 \\ \hline \end{array}$

21. $\begin{array}{r} 20.0 \\ -\quad 0.9099 \\ \hline \end{array}$

22. $2 - 0.0054$

23. $234.6788 - 81.7854$

Multiply.

24. $\begin{array}{r} 0.125 \\ \times\quad 0.24 \\ \hline \end{array}$

25. $\begin{array}{r} 32 \\ \times\,0.25 \\ \hline \end{array}$

26. 0.001×213.45

27. 1000×73.962

Divide.

28. $4\,\overline{)\,19}$

29. $42\,\overline{)\,10.08}$

30. $3.3\,\overline{)\,100.32}$

31. $82\,\overline{)\,15.58}$

32. $\frac{346.89}{1000}$

33. $\frac{346.89}{0.01}$

Solve.

34. $4.8 \cdot y = 404.448$

35. $x + 0.018 = 9$

Answers

1. ______
2. ______
3. ______
4. ______
5. ______
6. ______
7. ______
8. ______
9. ______
10. ______
11. ______
12. ______
13. ______
14. ______
15. ______
16. ______
17. ______
18. ______
19. ______
20. ______
21. ______
22. ______
23. ______
24. ______
25. ______
26. ______
27. ______
28. ______
29. ______
30. ______
31. ______
32. ______
33. ______
34. ______
35. ______

Answers

36. ______
37. ______
38. ______
39. ______
40. ______
41. ______
42. ______
43. ______
44. ______
45. ______
46. ______
47. ______
48. ______
49. ______
50. ______
51. ______
52. ______
53. ______
54. ______
55. ______
56. ______
57. ______
58. ______
59. ______
60. ______
61. ______
62. ______
63. ______
64. ______

Solve.

36. A marathon runner ran 24.85 km in 5 hr. How far did she run in 1 hr?

37. Carla has a balance of $10,200 in her checking account before writing checks of $123.89, $56.68, and $3446.98. What was the balance after she had written the checks?

38. Ben Westlund paid $23,457 for 14 acres of land adjoining his ranch. How much did he pay for 1 acre? Round to the nearest cent.

39. A government agency bought 6 new flags at $79.95 each. How much was spent altogether?

Estimate each of the following.

40. The product 8.91×22.457 by rounding to the nearest one

41. The quotient $78.2209 \div 16.09$ by rounding to the nearest ten

Find decimal notation. Use multiplying by 1.

42. $\frac{8}{5}$

43. $\frac{22}{25}$

44. $\frac{21}{4}$

Find decimal notation. Use division.

45. $\frac{3}{4}$

46. $\frac{11}{9}$

47. $\frac{15}{7}$

Round the answer to Exercise 47 to the nearest:

48. Tenth.

49. Hundredth.

50. Thousandth.

51. Convert from cents to dollars: 949 cents.

52. Convert to standard notation: Procter & Gamble spent $2.8 billion on advertising in a recent year.

Calculate.

53. $256 \div 3.2 \div 2 - 1.56 + 78.325 \times 0.02$

54. $(1 - 0.08)^2 + 6[5(12.1 - 8.7) + 10(14.3 - 9.6)]$

55. $\frac{7}{8} \times 345.6$

56. $\frac{2}{3} \times 79.95 - \frac{7}{9} \times 1.235$

Skill Maintenance

57. Multiply: $2\frac{1}{10} \cdot 6\frac{2}{3}$.

58. Add: $2\frac{3}{16} + \frac{1}{2}$.

59. Subtract: $28\frac{2}{3} - 2\frac{1}{6}$.

60. Divide: $3\frac{3}{8} \div 3$.

61. Simplify: $\frac{33}{54}$.

62. Find the prime factorization of 360.

Synthesis

63. The Fit Fiddle health club generally charges a $79 membership fee and $42.50 a month. Allise has a coupon that will allow her to join the club for $299 for six months. How much will Allise save if she uses the coupon?

64. 🖩 Arrange from smallest to largest.

$\frac{2}{3}, \frac{15}{19}, \frac{11}{13}, \frac{5}{7}, \frac{13}{15}, \frac{17}{20}$

Cumulative Review: Chapters 1–4

Convert to fractional notation.

1. $2\frac{2}{9}$

2. 3.052

Find decimal notation.

3. $\frac{7}{5}$

4. $\frac{6}{11}$

5. Determine whether 43 is prime, composite, or neither.

6. Determine whether 2,053,752 is divisible by 4.

Calculate.

7. $48 + 12 \div 4 - 10 \times 2 + 6892 \div 4$

8. $4.7 - \{0.1[1.2(3.95 - 1.65) + 1.5 \div 2.5]\}$

Round to the nearest hundredth.

9. 584.903

10. $218.\overline{5}$

11. Estimate the product 16.392×9.715 by rounding to the nearest one.

12. Estimate by rounding to the nearest tenth:

$$2.714 + 4.562 - 3.31 - 0.0023.$$

13. Estimate the product 6418×1984 by rounding to the nearest hundred.

14. Estimate the quotient $717.832 \div 124.998$ by rounding to the nearest ten.

Add and simplify.

15. $$\begin{array}{r} 2\frac{1}{4} \\ +\ 3\frac{4}{5} \\ \hline \end{array}$$

16. $$\begin{array}{r} 34{,}921 \\ 93{,}092 \\ +\ 11{,}103 \\ \hline \end{array}$$

17. $\frac{1}{6} + \frac{2}{3} + \frac{8}{9}$

18. $143.9 + 2.053$

Subtract and simplify.

19. $723{,}041 - 12{,}904$

20. $19 - 5.903$

21. $5\frac{1}{7} - 4\frac{3}{7}$

22. $\frac{10}{11} - \frac{9}{10}$

Multiply and simplify.

23. $\frac{3}{8} \cdot \frac{4}{9}$

24. $$\begin{array}{r} 2532 \\ \times\ 2100 \\ \hline \end{array}$$

25. $$\begin{array}{r} 23.9 \\ \times\ 0.2 \\ \hline \end{array}$$

26. $$\begin{array}{r} 27.9431 \\ \times\ 0.001 \\ \hline \end{array}$$

Divide and simplify.

27. $16.5 \overline{)35.013}$

28. $26 \overline{)47{,}918}$

29. $13.8621 \div 0.001$

30. $\frac{4}{9} \div \frac{8}{15}$

Solve.

31. $8.32 + x = 9.1$

32. $75 \cdot x = 2100$

33. $y \cdot 9.47 = 81.6314$

34. $1062 + y = 368{,}313$

35. $t + \frac{5}{6} = \frac{8}{9}$

36. $\frac{7}{8} \cdot t = \frac{7}{16}$

Solve.

37. In a recent year, there were 1952 heart transplants, 9004 kidney transplants, 3229 liver transplants, and 89 pancreas transplants (***Source:*** U.S. Department of Health). How many transplants of these four organs were performed that year?

38. After making a $150 down payment on a sofa, $\frac{3}{10}$ of the total cost was paid. How much did the sofa cost?

39. There are 60 seconds in a minute and 60 minutes in an hour. How many seconds are in a day?

40. A student's tuition was $3600. A loan was obtained for $\frac{2}{3}$ of the tuition. How much was the loan?

41. The balance in a checking account is $314.79. After a check is written for $56.02, what is the balance in the account?

42. A clerk in a delicatessen sold $1\frac{1}{2}$ lb of ham, $2\frac{3}{4}$ lb of turkey, and $2\frac{1}{4}$ lb of roast beef. How many pounds of meat were sold?

43. A baker used $\frac{1}{2}$ lb of sugar for cookies, $\frac{2}{3}$ lb of sugar for pie, and $\frac{5}{6}$ lb of sugar for cake. How much sugar was used in all?

44. A rectangular family room measures 19.8 ft by 23.6 ft. Find its area.

45. Simplify:

$$\left(\frac{3}{4}\right)^2 - \frac{1}{8} \cdot \left(3 - 1\frac{1}{2}\right)^2.$$

Synthesis

46. A customer in a grocery store used a manufacturer's coupon to buy juice. With the coupon, if 5 cartons of juice were purchased, the sixth carton was free. The price of each carton was $1.09. What was the cost per carton with the coupon? Round to the nearest cent.

47. A box of gelatin mix packages weighs $15\frac{3}{4}$ lb. Each package weighs $1\frac{3}{4}$ oz. How many packages are in the box?

5

Ratio and Proportion

Introduction

The mathematics of the application below shows what is called a *proportion.* The expressions on either side of the equals sign are called *ratios.* In this chapter, we use ratios and proportions to solve problems such as this one. We will also study such topics as rates, unit pricing, and similar triangles.

An Application

To determine the number of fish in a lake, a conservationist catches 225 fish, tags them, and throws them back into the lake. Later, 108 fish are caught, and it is found that 15 of them are tagged. Estimate how many fish are in the lake.

This problem appears as Example 6 in Section 5.4.

The Mathematics

We let $F =$ the number of fish in the lake. Then we translate to a proportion.

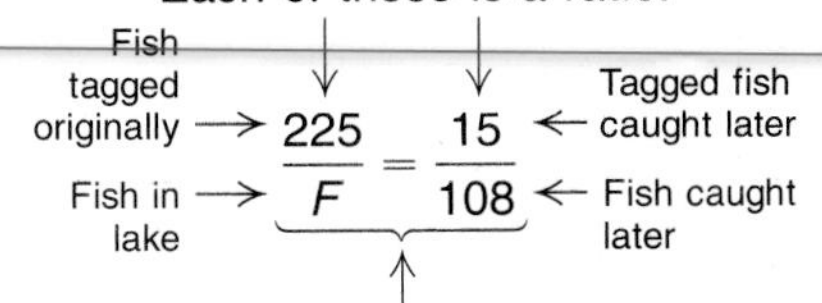

$$\frac{225}{F} = \frac{15}{108}$$

For more information, visit us at www.mathmax.com

Pretest: Chapter 5

Write fractional notation for the ratio.

1. 35 to 43

2. 0.079 to 1.043

Solve.

3. $\frac{5}{6} = \frac{x}{27}$

4. $\frac{y}{0.25} = \frac{0.3}{0.1}$

5. $\frac{3\frac{1}{2}}{4\frac{1}{3}} = \frac{6\frac{3}{4}}{x}$

6. What is the rate in miles per gallon?

408 miles, 16 gallons

7. A student picked 10 qt (quarts) of strawberries in 45 min. What is the rate in quarts per minute?

8. A 24-oz loaf of bread costs $1.39. Find the unit price in cents per ounce. Round to the nearest hundredth of a cent.

9. Which has the lower unit price?

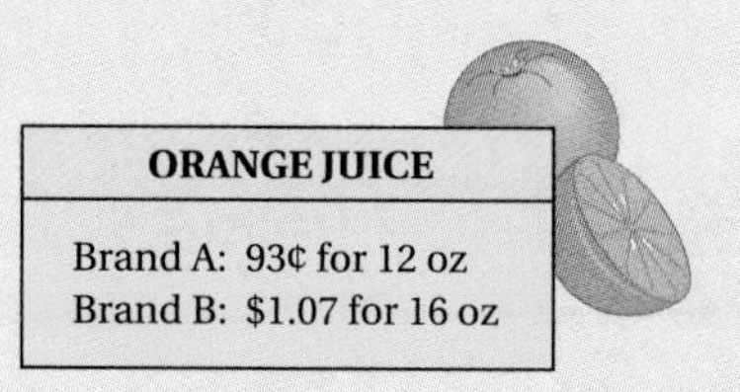

ORANGE JUICE
Brand A: 93¢ for 12 oz Brand B: $1.07 for 16 oz

Solve.

10. A man traveled 216 km in 6 hr. At this rate, how far will he travel in 54 hr?

11. If 4 packs of gum cost $5.16, how many packs of gum can you buy for $28.38?

12. Juan's digital car clock loses 5 min in 10 hr. At this rate, how much will it lose in 24 hr?

13. On a map, 4 in. represents 225 mi. If two cities are 7 in. apart on the map, how far are they apart in reality?

14. These triangles are similar. Find the missing lengths.

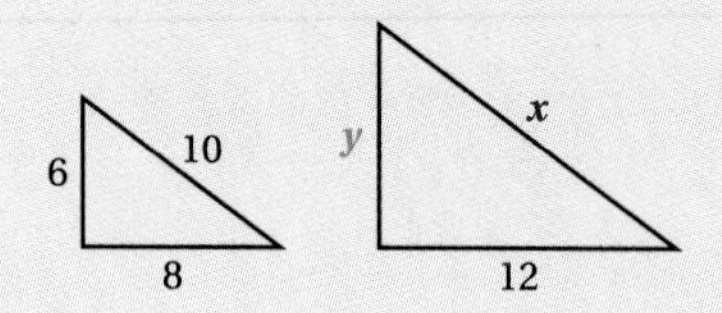

Objectives for Retesting

The objectives to be tested in addition to the material in this chapter are as follows.

[2.5c] Test fractions for equality.
[4.3a] Multiply using decimal notation.
[4.4a] Divide using decimal notation.
[4.7a] Solve applied problems involving decimals.

5.1 Introduction to Ratios

Objectives

a Find fractional notation for ratios.

b Simplify ratios.

For Extra Help

TAPE 9 TAPE 8B MAC WIN CD-ROM

a Ratios

> A **ratio** is the quotient of two quantities.

For example, each day in this country about 5200 people die. Of these, 1070 die of cancer. The *ratio* of those who die of cancer to those who die is shown by the fractional notation

$$\frac{1070}{5200} \text{ or by the notation } 1070:5200.$$

We read such notation as "the ratio of 1070 to 5200," listing the numerator first and the denominator second.

> The **ratio** of a to b is given by $\frac{a}{b}$, where a is the numerator and b is the denominator, or by $a:b$.

Example 1 Find the ratio of 7 to 8.

The ratio is $\frac{7}{8}$, or $7:8$.

Example 2 Find the ratio of 31.4 to 100.

The ratio is $\frac{31.4}{100}$, or $31.4:100$.

Example 3 Find the ratio of $4\frac{2}{3}$ to $5\frac{7}{8}$. You need not simplify.

The ratio is $\frac{4\frac{2}{3}}{5\frac{7}{8}}$, or $4\frac{2}{3}:5\frac{7}{8}$.

Do Exercises 1–3.

In most of our work, we will use fractional notation for ratios.

Example 4 Hank Aaron hit 755 home runs in 12,364 at-bats. Find the ratio of at-bats to home runs.

The ratio is $\frac{12,364}{755}$.

Example 5 A family earning $21,400 per year allots about $3210 for car expenses. Find the ratio of car expenses to yearly income.

The ratio is $\frac{3210}{21,400}$.

Do Exercises 4–6.

1. Find the ratio of 5 to 11.

2. Find the ratio of 57.3 to 86.1.

3. Find the ratio of $6\frac{3}{4}$ to $7\frac{2}{5}$.

4. The average American drinks 182.5 gal of liquid each year. Of this, 21.1 gal is milk. Find the ratio of milk drunk to total amount drunk.

5. *Fat Content.* In one serving $\left(3\frac{1}{2}\text{ oz}\right)$ of raw scallops, there is 0.8 g of fat. In one serving of oysters, there is 2.5 g of fat. Find the ratio of fat in one serving of oysters to that in one serving of scallops.

6. A pitcher gives up 4 earned runs in $7\frac{2}{3}$ innings of pitching. Find the ratio of earned runs to number of innings pitched.

Answers on page A-10

7. In the triangle below, what is the ratio of the length of the shortest side to the length of the longest side?

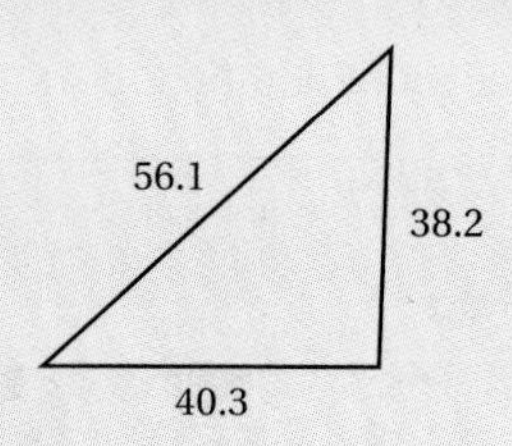

8. Find the ratio of 18 to 27. Then simplify and find two other numbers in the same ratio.

9. Find the ratio of 3.6 to 12. Then simplify and find two other numbers in the same ratio.

10. Find the ratio of 1.2 to 1.5. Then simplify and find two other numbers in the same ratio.

11. In Example 9, what is the ratio of the length of the shortest side of the television screen to the length of the longest side?

Answers on page A-10

Example 6 In the triangle at right:

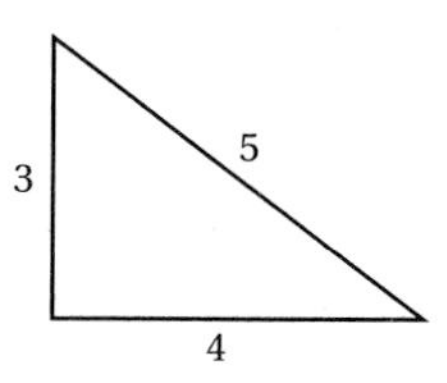

a) What is the ratio of the length of the longest side to the length of the shortest side?

$$\frac{5}{3}$$

b) What is the ratio of the length of the shortest side to the length of the longest side?

$$\frac{3}{5}$$

Do Exercise 7.

b Simplifying Notation for Ratios

Sometimes a ratio can be simplified. This provides a means of finding other numbers with the same ratio.

Example 7 Find the ratio of 6 to 8. Then simplify and find two other numbers in the same ratio.

We write the ratio in fractional notation and then simplify:

$$\frac{6}{8} = \frac{2 \cdot 3}{2 \cdot 4} = \frac{2}{2} \cdot \frac{3}{4} = 1 \cdot \frac{3}{4} = \frac{3}{4}.$$

Thus, 3 and 4 have the same ratio as 6 and 8. We can express this by saying "6 is to 8" as "3 is to 4."

Do Exercise 8.

Example 8 Find the ratio of 2.4 to 10. Then simplify and find two other numbers in the same ratio.

We first write the ratio. Next, we multiply by 1 to clear the decimal from the numerator. Then we simplify:

$$\frac{2.4}{10} = \frac{2.4}{10} \cdot \frac{10}{10} = \frac{24}{100} = \frac{4 \cdot 6}{4 \cdot 25} = \frac{4}{4} \cdot \frac{6}{25} = \frac{6}{25}.$$

Thus, 2.4 is to 10 as 6 is to 25.

Do Exercises 9 and 10.

Example 9 A standard television screen with a length of 16 in. has a width or height of 12 in. What is the ratio of length to width?

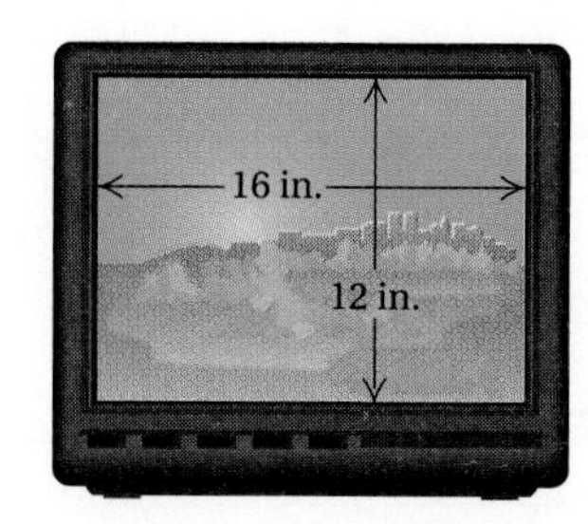

$$\text{The ratio is } \frac{16}{12} = \frac{4 \cdot 4}{4 \cdot 3} = \frac{4}{4} \cdot \frac{4}{3} = \frac{4}{3}.$$

Thus we can say that the ratio of length to width is 4 to 3.

Do Exercise 11.

Exercise Set 5.1

a Find fractional notation for the ratio. You need not simplify.

1. 4 to 5

2. 3 to 2

3. 178 to 572

4. 329 to 967

5. 0.4 to 12

6. 2.3 to 22

7. 3.8 to 7.4

8. 0.6 to 0.7

9. 56.78 to 98.35

10. 456.2 to 333.1

11. $8\frac{3}{4}$ to $9\frac{5}{6}$

12. $10\frac{1}{2}$ to $43\frac{1}{4}$

13. One person in four plays a musical instrument. In a typical group of people, what is the ratio of those who play an instrument to total number of people? What is the ratio of those who do not play an instrument to those who do?

14. Of the 365 days in each year, it takes 107 days of work for the average person to pay his or her taxes. What is the ratio of days worked for taxes to total number of days in a year?

15. *Corvette Accidents.* Of every 5 fatal accidents involving a Corvette, 4 do not involve another vehicle (***Source:*** *Harper's Magazine*). Find the ratio of fatal accidents involving just a Corvette to those involving a Corvette and at least one other vehicle.

16. *New York Commuters.* Of every 5 people who commute to work in New York City, 2 spend more than 90 min a day commuting (***Source:*** *The Amicus Journal*). Find the ratio of people whose daily commute to New York exceeds 90 min a day to those whose commute is 90 min or less.

17. In this rectangle, find the ratios of length to width and of width to length.

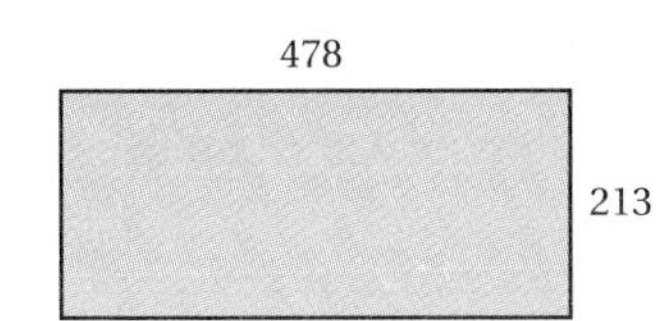

18. In this right triangle, find the ratios of shortest length to longest length and of longest length to shortest length.

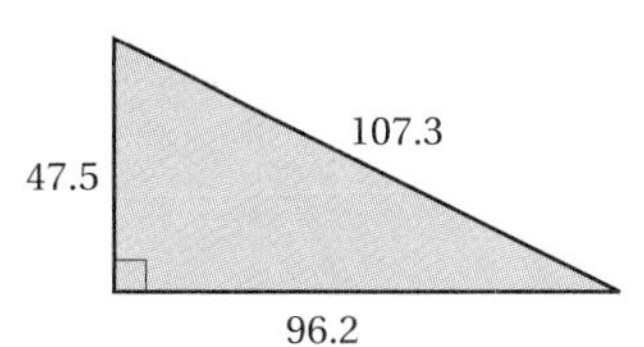

b Simplify the ratio.

19. 4 to 6

20. 6 to 10

21. 18 to 24

22. 28 to 36

23. 4.8 to 10

24. 5.6 to 10

25. 2.8 to 3.6

26. 4.8 to 6.4

27. 20 to 30

28. 40 to 60

29. 56 to 100

30. 42 to 100

31. 128 to 256

32. 232 to 116

33. 0.48 to 0.64

34. 0.32 to 0.96

35. The ratio of Americans aged 18–24 living with their parents is 54 to 100. Find the ratio and simplify.

36. Of every 100 hr, the average person spends 8 hr cooking. Find the ratio of hours spent cooking to total number of hours and simplify.

37. In this right triangle, find the ratio of shortest length to longest length and simplify.

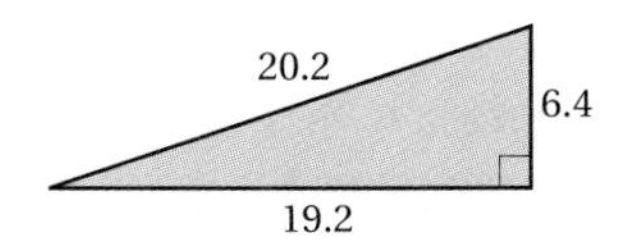

38. In this rectangle, find the ratio of width to length and simplify.

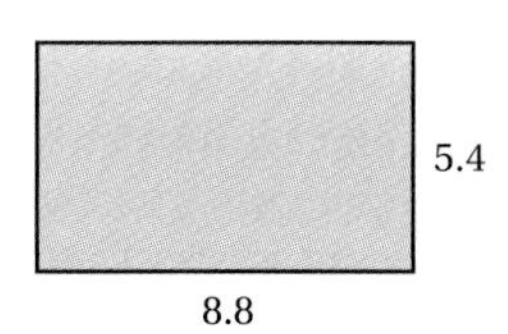

Skill Maintenance

Use = or ≠ for ▭ to write a true sentence. [2.5c]

39. $\frac{12}{8}$ ▭ $\frac{6}{4}$

40. $\frac{4}{7}$ ▭ $\frac{5}{9}$

Divide. Write decimal notation for the answer. [4.4a]

41. $200 \div 4$

42. $95 \div 10$

43. $232 \div 16$

44. $342 \div 2.25$

Solve. [3.5c]

45. Rocky is $187\frac{1}{10}$ cm tall and his daughter is $180\frac{3}{4}$ cm tall. How much taller is Rocky?

46. Aunt Louise is $168\frac{1}{4}$ cm tall and her son is $150\frac{7}{10}$ cm tall. How much taller is Aunt Louise?

Synthesis

47. ◆ Can every ratio be written as the ratio of some number to 1? Why or why not?

48. ◆ What can be concluded about a rectangle's width if the ratio of length to perimeter is 1 to 3? Make some sketches and explain your reasoning.

49. 🖩 In 1996, the total payroll of major league baseball teams was $937,905,284. The New York Yankees won the World Series that year. Their payroll was the highest at $61,511,870. Find the ratio in decimal notation of the Yankees payroll to the overall payroll.

50. 🖩 See Exercise 49. In 1995, the total payroll of major league baseball teams was $927,334,416. Find the ratio of the payroll in 1996 to the payroll in 1995.

51. Find the ratio of $3\frac{3}{4}$ to $5\frac{7}{8}$ and simplify.

Exercises 52 and 53 refer to a common fertilizer known as "5, 10, 15." This mixture contains 5 parts of potassium for every 10 parts of phosphorus and 15 parts of nitrogen (this is often denoted 5:10:15).

52. Find the ratio of potassium to nitrogen and of nitrogen to phosphorus.

53. Simplify the ratio 5:10:15.

Analyze the ratios of different-colored M&M candies.

5.2 Rates and Unit Prices

a When a ratio is used to compare two different kinds of measure, we call it a **rate.** Suppose that a car is driven 200 km in 4 hr. The ratio

$$\frac{200 \text{ km}}{4 \text{ hr}}, \quad \text{or } 50\frac{\text{km}}{\text{hr}}, \quad \text{or 50 kilometers per hour}, \quad \text{or 50 km/h}$$

Recall that "per" means "division," or "for each."

is the rate traveled in kilometers per hour, which is the division of the number of kilometers by the number of hours. A ratio of distance traveled to time is also called **speed.**

Example 1 A European driver travels 145 km on 2.5 L of gas. What is the rate in kilometers per liter?

$$\frac{145 \text{ km}}{2.5 \text{ L}}, \quad \text{or} \quad 58\frac{\text{km}}{\text{L}}$$

Example 2 It takes 60 oz of grass seed to seed 3000 sq ft of lawn. What is the rate in ounces per square foot?

$$\frac{60 \text{ oz}}{3000 \text{ sq ft}} = \frac{1}{50}\frac{\text{oz}}{\text{sq ft}}, \quad \text{or} \quad 0.02\frac{\text{oz}}{\text{sq ft}}$$

Example 3 A cook buys 10 lb of potatoes for \$3.69. What is the rate in cents per pound?

$$\frac{\$3.69}{10 \text{ lb}} = \frac{369 \text{ cents}}{10 \text{ lb}}, \quad \text{or} \quad 36.9\frac{\text{cents}}{\text{lb}}$$

Example 4 A student nurse working in a health center earned \$3690 for working 3 months one summer. What was the rate of pay per month?

The rate of pay is the ratio of money earned per length of time worked, or

$$\frac{\$3690}{3 \text{ mo}} = 1230\frac{\text{dollars}}{\text{month}}, \quad \text{or} \quad \$1230 \text{ per month.}$$

Example 5 *Strikeout-to-Home-Run Ratio.* One year Gary Sheffield of the Florida Marlins had the lowest strikeout-to-home-run ratio in the major leagues. He had 66 strikeouts and 42 home runs. What was his strikeout-to-home-run ratio?

$$\frac{66 \text{ strikeouts}}{42 \text{ home runs}} \approx 1.57\frac{\text{strikeouts}}{\text{home runs}}$$

Do Exercises 1–8.

Objectives

a Give the ratio of two different kinds of measure as a rate.

b Find unit prices and use them to determine which of two possible purchases has the lower unit price.

For Extra Help

TAPE 9 | TAPE 8B | MAC WIN | CD-ROM

What is the rate, or speed, in miles per hour?

1. 45 mi, 9 hr

2. 120 mi, 10 hr

3. 3 mi, 10 hr

What is the rate, or speed, in feet per second?

4. 2200 ft, 2 sec

5. 52 ft, 13 sec

6. 232 ft, 16 sec

7. A well-hit golf ball can travel 500 ft in 2 sec. What is the rate, or speed, of the golf ball in feet per second?

8. A leaky faucet can lose 14 gal of water in a week. What is the rate in gallons per day?

Answers on page A-11

9. A customer bought a 14-oz package of oat bran for $2.89. What is the unit price in cents per ounce? Round to the nearest hundredth of a cent.

10. Which has the lower unit price? [*Note*: 1 qt = 32 fl oz (fluid ounces).]

Answers on page A-11

b Unit Pricing

A **unit price** or **unit rate** is the ratio of price to the number of units.

By carrying out the division indicated by the ratio, we can find the price per unit.

Example 6 A customer bought a 20-lb box of powdered detergent for $19.47. What is the unit price in dollars per pound?

The unit price is the price in dollars for each pound.

$$\text{Unit price} = \frac{\text{Price}}{\text{Number of units}}$$
$$= \frac{\$19.47}{20 \text{ lb}} = \frac{19.47}{20} \cdot \frac{\$}{\text{lb}}$$
$$= 0.9735 \text{ dollars per pound}$$

Do Exercise 9.

For comparison shopping, it helps to find unit prices.

Example 7 Which has the lower unit price?

To find out, we compare the unit prices—in this case, the price per ounce.

For can A: $\dfrac{48 \text{ cents}}{14 \text{ oz}} \approx 3.429 \dfrac{\text{cents}}{\text{oz}}$.

For can B: We need to find the total number of ounces:

1 lb, 15 oz = 16 oz + 15 oz = 31 oz.

Then

$$\frac{99 \text{ cents}}{31 \text{ oz}} \approx 3.194 \frac{\text{cents}}{\text{oz}}.$$

Thus can B has the lower unit price.

In many stores, unit prices are now listed on the items or the shelves.

Do Exercise 10.

Exercise Set 5.2

a In Exercises 1–6, find the rate as a ratio of distance to time.

1. 120 km, 3 hr

2. 18 mi, 9 hr

3. 440 m, 40 sec

4. 200 mi, 25 sec

5. 342 yd, 2.25 days

6. 492 m, 60 sec

7. A car is driven 500 mi in 20 hr. What is the rate in miles per hour? in hours per mile?

8. A student eats 3 hamburgers in 15 min. What is the rate in hamburgers per minute? in minutes per hamburger?

9. A long-distance telephone call between two cities costs $5.75 for 10 min. What is the rate in cents per minute?

10. An 8-lb boneless ham contains 36 servings of meat. What is the ratio in servings per pound?

11. To water a lawn adequately requires 623 gal of water for every 1000 ft^2. What is the rate in gallons per square foot?

12. A car is driven 200 km on 40 L of gasoline. What is the rate in kilometers per liter?

13. Light travels 186,000 mi in 1 sec. What is its rate, or speed, in miles per second?

14. Sound travels 1100 ft in 1 sec. What is its rate, or speed, in feet per second?

15. Impulses in nerve fibers travel 310 km in 2.5 hr. What is the rate, or speed, in kilometers per hour?

16. A black racer snake can travel 4.6 km in 2 hr. What is its rate, or speed, in kilometers per hour?

17. A jet flew 2660 mi in 4.75 hr. What was its speed?

18. A turtle traveled 0.42 mi in 2.5 hr. What was its speed?

b

19. The fabric for a wedding gown costs $80.75 for 8.5 yd. Find the unit price.

20. An 8-oz bottle of shampoo costs $2.59. Find the unit price.

21. A 2-lb can of decaffeinated coffee costs $6.59. What is the unit price in cents per ounce? Round to the nearest hundredth of a cent.

22. A 24-can package of 12-oz cans of orange soda costs $6.99. What is the unit price in cents per ounce? Round to the nearest hundredth of a cent.

23. A $\frac{2}{3}$-lb package of Monterey Jack cheese costs $2.89. Find the unit price in dollars per pound. Round to the nearest hundredth of a dollar.

24. A $1\frac{1}{4}$-lb container of cottage cheese costs $1.62. Find the unit price in dollars per pound.

Which has the lower unit price?

25.

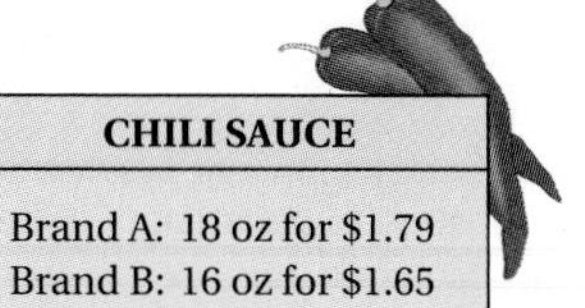

26.

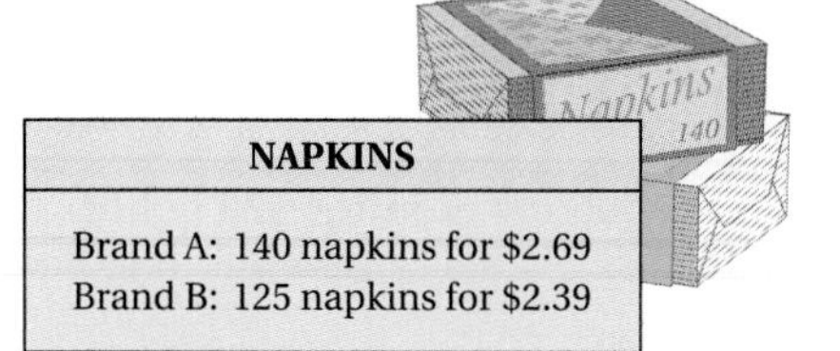

27.

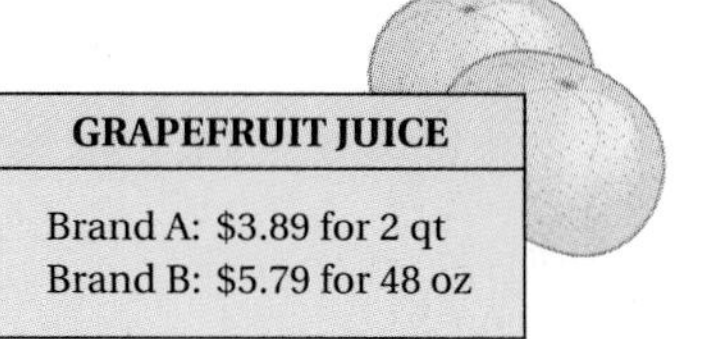

28.

29.

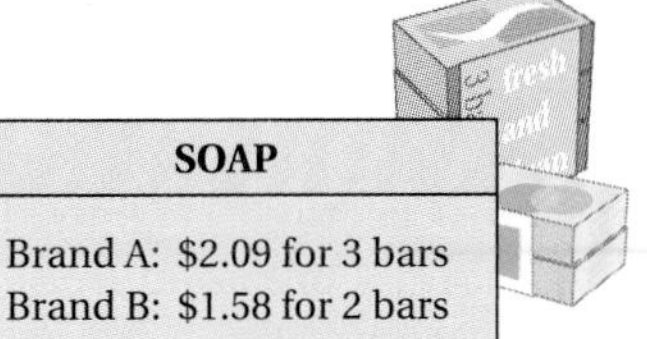

SOAP

Brand A: $2.09 for 3 bars
Brand B: $1.58 for 2 bars

30.

BROCCOLI SOUP

Brand A: 8.25 oz for 96 cents
Brand B: 10.75 oz for $1.11

31.

FANCY TUNA

Brand A: $1.19 for $6\frac{1}{8}$ oz
Brand B: $1.11 for 6 oz

32.

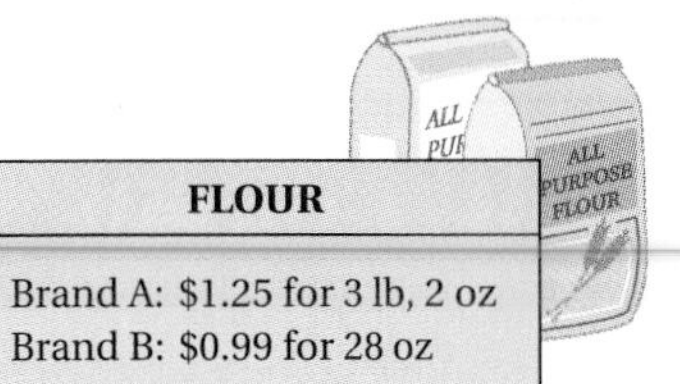

FLOUR

Brand A: $1.25 for 3 lb, 2 oz
Brand B: $0.99 for 28 oz

33.

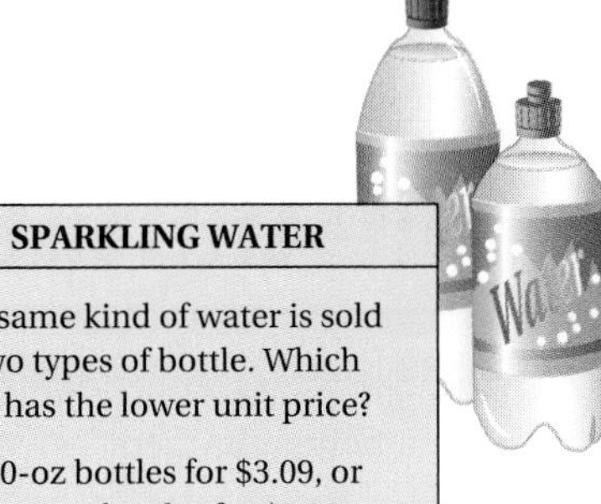

SPARKLING WATER

The same kind of water is sold in two types of bottle. Which type has the lower unit price?

Six 10-oz bottles for $3.09, or
Four 12-oz bottles for $2.39

34.

COLA

The same kind of cola is sold in two types of container. Which type has the lower unit price?

Six 12-oz cans for $2.19, or
One 30-oz bottle for 79¢

35.

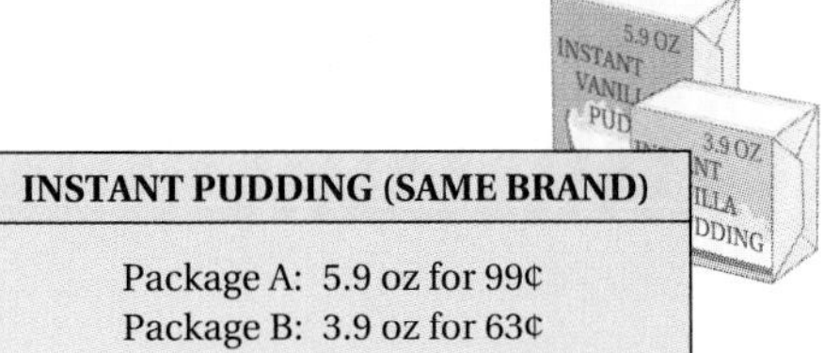

INSTANT PUDDING (SAME BRAND)

Package A: 5.9 oz for 99¢
Package B: 3.9 oz for 63¢

36.

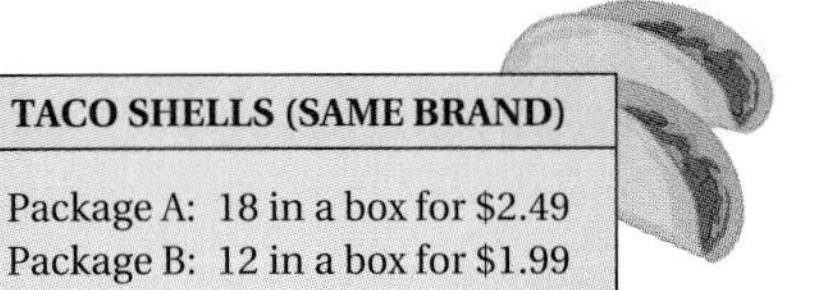

TACO SHELLS (SAME BRAND)

Package A: 18 in a box for $2.49
Package B: 12 in a box for $1.99

37.

BROCCOLI SOUP

Big Chunk: 10.5 oz, 2 for $1.59
Bert's: 11 oz for $0.82

38.

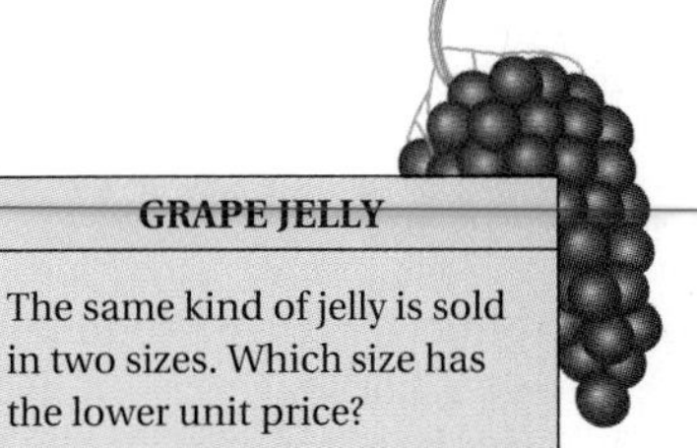

GRAPE JELLY

The same kind of jelly is sold in two sizes. Which size has the lower unit price?

18 oz for $1.59, or
32 oz for $3.59

Skill Maintenance

Solve.

39. There are 20.6 million people in this country who play the piano and 18.9 million who play the guitar. How many more play the piano than the guitar? [4.7a]

40. A serving of fish steak (cross section) is generally $\frac{1}{2}$ lb. How many servings can be prepared from a cleaned $18\frac{3}{4}$-lb tuna? [3.6c]

Multiply. [4.3a]

41. $\begin{array}{r} 45.67 \\ \times \quad 2.4 \\ \hline \end{array}$

42. $\begin{array}{r} 678.19 \\ \times \quad 100 \\ \hline \end{array}$

43. 84.3×69.2

44. 1002.56×465

Synthesis

45. ◆ The unit price of an item generally drops when larger packages of that item are purchased. Why?

46. ◆ Suppose that the same type of juice is available in two sizes and that the larger bottle has the lower unit price. If the larger bottle costs $3.79 and contains twice as much juice, what can you conclude about the price of the smaller bottle? Why?

47. Recently, certain manufacturers have been changing the size of their containers in such a way that the consumer thinks the price of a product has been lowered when in reality, a higher unit price is being charged.

a) Some aluminum juice cans are now concave (curved in) on the bottom. Suppose the volume of the can in the figure has been reduced from a fluid capacity of 6 oz to 5.5 oz, and the price of each can has been reduced from 65¢ to 60¢. Find the unit price of each container in cents per ounce.

b) Suppose that at one time the cost of a certain kind of paper towel was $0.89 for a roll containing 78 ft^2 of absorbent surface. Later the surface area was changed to 65 ft^2 and the price was decreased to $0.79. Find the unit price of each product in cents per square foot.

48. In 1994, Coca-Cola introduced a 20-oz soda bottle. At first it was sold for 64¢ a bottle, the same price as their 16-oz bottle. After about a month, the price of a 20-oz bottle rose to 80¢. How did the unit price change for a consumer who made the switch from the 16-oz to the 20-oz bottle?

49. Suppose that a pasta manufacturer shrinks the size of a box from 1 lb to 14 oz, but keeps the price at 85 cents a box. How does the unit price change?

5.3 Proportions

a Proportion

When two pairs of numbers (such as 3, 2 and 6, 4) have the same ratio, we say that they are **proportional**. The equation

$$\frac{3}{2} = \frac{6}{4}$$

states that the pairs 3, 2 and 6, 4 are proportional. Such an equation is called a **proportion**. We sometimes read $\frac{3}{2} = \frac{6}{4}$ as "3 is to 2 as 6 is to 4."

Since ratios are represented by fractional notation, we can test whether two ratios are the same by using the test for equality discussed in Section 2.5. (It is also a skill to review for the chapter test.)

Objectives

a Determine whether two pairs of numbers are proportional.

b Solve proportions.

For Extra Help

TAPE 9 TAPE 9A MAC WIN CD-ROM

Example 1 Determine whether 1, 2 and 3, 6 are proportional.

We can use cross products:

$$1 \cdot 6 = 6 \qquad \frac{1}{2} \quad \frac{3}{6} \qquad 2 \cdot 3 = 6.$$

Since the cross products are the same, 6 = 6, we know that $\frac{1}{2} = \frac{3}{6}$, so the numbers are proportional.

Example 2 Determine whether 2, 5 and 4, 7 are proportional.

We can use cross products:

$$2 \cdot 7 = 14 \qquad \frac{2}{5} \quad \frac{4}{7} \qquad 5 \cdot 4 = 20.$$

Since the cross products are not the same, $14 \neq 20$, we know that $\frac{2}{5} \neq \frac{4}{7}$, so the numbers are not proportional.

Do Exercises 1–3.

Example 3 Determine whether 3.2, 4.8 and 0.16, 0.24 are proportional.

We can use cross products:

$$3.2 \times 0.24 = 0.768 \qquad \frac{3.2}{4.8} \quad \frac{0.16}{0.24} \qquad 4.8 \times 0.16 = 0.768.$$

Since the cross products are the same, 0.768 = 0.768, we know that $\frac{3.2}{4.8} = \frac{0.16}{0.24}$, so the numbers are proportional.

Do Exercises 4 and 5.

Determine whether the two pairs of numbers are proportional.

1. 3, 4 and 6, 8

2. 1, 4 and 10, 39

3. 1, 2 and 20, 39

Determine whether the two pairs of numbers are proportional.

4. 6.4, 12.8 and 5.3, 10.6

5. 6.8, 7.4 and 3.4, 4.2

Answers on page A-11

6. Determine whether $4\frac{2}{3}$, $5\frac{1}{2}$ and 14, $16\frac{1}{2}$ are proportional.

7. Solve: $\dfrac{x}{63} = \dfrac{2}{9}$.

Answers on page A-11

Example 4 Determine whether $4\frac{2}{3}$, $5\frac{1}{2}$ and $8\frac{7}{8}$, $16\frac{1}{3}$ are proportional.

We can use cross products:

$$\frac{4\frac{2}{3}}{5\frac{1}{2}} \quad \frac{8\frac{7}{8}}{16\frac{1}{3}}$$

$$4\frac{2}{3} \cdot 16\frac{1}{3} = \frac{14}{3} \cdot \frac{49}{3} = \frac{686}{9} = 76\frac{2}{9}$$

$$5\frac{1}{2} \cdot 8\frac{7}{8} = \frac{11}{2} \cdot \frac{71}{8} = \frac{781}{16} = 48\frac{13}{16}.$$

Since the cross products are not the same, $76\frac{2}{9} \neq 48\frac{13}{16}$, we know that the numbers are not proportional.

Do Exercise 6.

b Solving Proportions

Let's now look at solving proportions. Consider the proportion

$$\frac{x}{3} = \frac{4}{6}.$$

One way to solve a proportion is to use cross products. Then we can divide on both sides to get the variable alone:

$x \cdot 6 = 3 \cdot 4$ **Equating cross products (finding cross products and setting them equal)**

$x = \dfrac{3 \cdot 4}{6}$ **Dividing by 6 on both sides**

$= \dfrac{12}{6}$ **Multiplying**

$= 2.$ **Dividing**

We can check that 2 is the solution by replacing x with 2 and using cross products:

$$2 \cdot 6 = 12 \quad \frac{2}{3} \quad \frac{4}{6} \quad 3 \cdot 4 = 12.$$

Since the cross products are the same, it follows that $\frac{2}{3} = \frac{4}{6}$ so the numbers 2, 3 and 4, 6 are proportional, and 2 is the solution of the equation.

> To solve $\dfrac{x}{a} = \dfrac{c}{d}$, equate cross products and divide on both sides to get x alone.

Do Exercise 7.

Example 5 Solve: $\frac{x}{7} = \frac{5}{3}$. Write a mixed numeral for the answer.

We have

$$\frac{x}{7} = \frac{5}{3}$$

$$3 \cdot x = 7 \cdot 5 \quad \text{Equating cross products}$$

$$x = \frac{7 \cdot 5}{3} \quad \text{Dividing by 3}$$

$$= \frac{35}{3}, \text{ or } 11\frac{2}{3}.$$

The solution is $11\frac{2}{3}$.

Do Exercise 8.

Example 6 Solve: $\frac{7.7}{15.4} = \frac{y}{2.2}$.

We have

$$\frac{7.7}{15.4} = \frac{y}{2.2}$$

$$7.7 \times 2.2 = 15.4 \times y \quad \text{Equating cross products}$$

$$\frac{7.7 \times 2.2}{15.4} = y \quad \text{Dividing by 15.4}$$

$$\frac{16.94}{15.4} = y \quad \text{Multiplying}$$

$$1.1 = y. \quad \text{Dividing:}$$

```
          1.1
1 5.4 ) 1 6.9 4
        1 5 4 0
          1 5 4
          1 5 4
              0
```

The solution is 1.1.

Do Exercise 9.

Example 7 Solve: $\frac{8}{x} = \frac{5}{3}$. Write decimal notation for the answer.

We have

$$\frac{8}{x} = \frac{5}{3}$$

$$8 \cdot 3 = x \cdot 5 \quad \text{Equating cross products}$$

$$\frac{8 \cdot 3}{5} = x \quad \text{Dividing by 5}$$

$$\frac{24}{5} = x \quad \text{Multiplying}$$

$$4.8 = x. \quad \text{Simplifying}$$

The solution is 4.8.

Do Exercise 10.

8. Solve: $\frac{x}{9} = \frac{5}{4}$.

9. Solve: $\frac{21}{5} = \frac{n}{2.5}$.

10. Solve: $\frac{6}{x} = \frac{25}{11}$.

Answers on page A-11

11. Solve: $\frac{0.4}{0.9} = \frac{4.8}{t}$.

12. Solve:

$$\frac{8\frac{1}{3}}{x} = \frac{10\frac{1}{2}}{3\frac{3}{4}}.$$

Calculator Spotlight

You may have noticed in Examples 5–9 that after equating cross products, we divided on both sides. Since this is always the case when solving proportions, calculators can be useful. For instance, to solve Example 8 with a calculator, we could press

[4] [.] [9] [3] [×] [1] [0] [÷] [3] [.] [4] [=].

Exercises

1. Use a calculator to solve Examples 5–8.
2. Use a calculator to check your answers to Margin Exercises 7–11.

Answers on page A-11

Example 8 Solve: $\frac{3.4}{4.93} = \frac{10}{n}$.

We have

$$\frac{3.4}{4.93} = \frac{10}{n}$$

$$3.4 \times n = 4.93 \times 10 \quad \text{Equating cross products}$$

$$n = \frac{4.93 \times 10}{3.4} \quad \text{Dividing by 3.4}$$

$$= \frac{49.3}{3.4} \quad \text{Multiplying}$$

$$= 14.5. \quad \text{Dividing: } 3.4\overline{)49.30} = 14.5$$

```
        1 4.5
3.4 ) 4 9.3,0
      3 4 0 0
      1 5 3 0
      1 3 6 0
        1 7 0
        1 7 0
            0
```

The solution is 14.5.

Do Exercise 11.

Example 9 Solve: $\frac{4\frac{2}{3}}{5\frac{1}{2}} = \frac{14}{x}$. Write a mixed numeral for the answer.

We have

$$\frac{4\frac{2}{3}}{5\frac{1}{2}} = \frac{14}{x}$$

$$4\frac{2}{3} \cdot x = 14 \cdot 5\frac{1}{2} \quad \text{Equating cross products}$$

$$\frac{14}{3} \cdot x = 14 \cdot \frac{11}{2} \quad \text{Converting to fractional notation}$$

$$x = 14 \cdot \frac{11}{2} \div \frac{14}{3} \quad \text{Dividing by } \frac{14}{3}$$

$$= \cancel{14} \cdot \frac{11}{2} \cdot \frac{3}{\cancel{14}} \quad \text{Multiplying by the reciprocal of the divisor}$$

$$= \frac{11 \cdot 3}{2} \quad \text{Simplifying by removing a factor of 1: } \frac{14}{14} = 1$$

$$= \frac{33}{2}, \text{ or } 16\frac{1}{2}.$$

The solution is $16\frac{1}{2}$.

Do Exercise 12.

Exercise Set 5.3

a Determine whether the two pairs of numbers are proportional.

1. 5, 6 and 7, 9

2. 7, 5 and 6, 4

3. 1, 2 and 10, 20

4. 7, 3 and 21, 9

5. 2.4, 3.6 and 1.8, 2.7

6. 4.5, 3.8 and 6.7, 5.2

7. $5\frac{1}{3}, 8\frac{1}{4}$ and $2\frac{1}{5}, 9\frac{1}{2}$

8. $2\frac{1}{3}, 3\frac{1}{2}$ and 14, 21

b Solve.

9. $\frac{18}{4} = \frac{x}{10}$

10. $\frac{x}{45} = \frac{20}{25}$

11. $\frac{x}{8} = \frac{9}{6}$

12. $\frac{8}{10} = \frac{n}{5}$

13. $\frac{t}{12} = \frac{5}{6}$

14. $\frac{12}{4} = \frac{x}{3}$

15. $\frac{2}{5} = \frac{8}{n}$

16. $\frac{10}{6} = \frac{5}{x}$

17. $\frac{n}{15} = \frac{10}{30}$

18. $\frac{2}{24} = \frac{x}{36}$

19. $\frac{16}{12} = \frac{24}{x}$

20. $\frac{7}{11} = \frac{2}{x}$

21. $\frac{6}{11} = \frac{12}{x}$

22. $\frac{8}{9} = \frac{32}{n}$

23. $\frac{20}{7} = \frac{80}{x}$

24. $\frac{5}{x} = \frac{4}{10}$

25. $\frac{12}{9} = \frac{x}{7}$

26. $\frac{x}{20} = \frac{16}{15}$

27. $\frac{x}{13} = \frac{2}{9}$

28. $\frac{1.2}{4} = \frac{x}{9}$

29. $\frac{t}{0.16} = \frac{0.15}{0.40}$

30. $\frac{x}{11} = \frac{7.1}{2}$

31. $\frac{100}{25} = \frac{20}{n}$

32. $\frac{35}{125} = \frac{7}{m}$

33. $\frac{7}{\frac{1}{4}} = \frac{28}{x}$

34. $\frac{x}{6} = \frac{1}{6}$

35. $\frac{\frac{1}{4}}{\frac{1}{2}} = \frac{\frac{1}{2}}{x}$

36. $\frac{1}{7} = \frac{x}{4\frac{1}{2}}$

37. $\frac{1}{2} = \frac{7}{x}$

38. $\frac{x}{3} = \frac{0}{9}$

39. $\frac{\frac{2}{7}}{\frac{3}{4}} = \frac{\frac{5}{6}}{y}$

40. $\frac{\frac{5}{4}}{\frac{5}{8}} = \frac{\frac{3}{2}}{Q}$

41. $\frac{2\frac{1}{2}}{3\frac{1}{3}} = \frac{x}{4\frac{1}{4}}$

42. $\frac{5\frac{1}{5}}{6\frac{1}{6}} = \frac{y}{3\frac{1}{2}}$

43. $\frac{1.28}{3.76} = \frac{4.28}{y}$

44. $\frac{10.4}{12.4} = \frac{6.76}{t}$

45. $\frac{10\frac{3}{8}}{12\frac{2}{3}} = \frac{5\frac{3}{4}}{y}$

46. $\frac{12\frac{7}{8}}{20\frac{3}{4}} = \frac{5\frac{2}{3}}{y}$

Skill Maintenance

Use $=$ or $\neq$ for ▒ to write a true sentence. [2.5c]

47. $\frac{3}{4}$ ▒ $\frac{5}{6}$

48. $\frac{18}{24}$ ▒ $\frac{36}{48}$

49. $\frac{7}{8}$ ▒ $\frac{7}{9}$

50. $\frac{19}{37}$ ▒ $\frac{15}{29}$

Divide. Write decimal notation for the answer. [4.4a]

51. $260 \div 4$

52. $395 \div 10$

53. $4648 \div 16$

54. $3427 \div 2.25$

Synthesis

55. ◆ Instead of equating cross products, a student solves $\frac{x}{7} = \frac{5}{3}$ (see Example 5) by multiplying on both sides by the least common denominator, 21. Is the student's approach a good one? Why or why not?

56. ◆ An instructor predicts that a student's test grade will be proportional to the amount of time the student spends studying. What is meant by this? Write an example of a proportion that involves the grades of two students and their study times.

▦ Solve.

57. $\frac{1728}{5643} = \frac{836.4}{x}$

58. $\frac{328.56}{627.48} = \frac{y}{127.66}$

5.4 Applications of Proportions

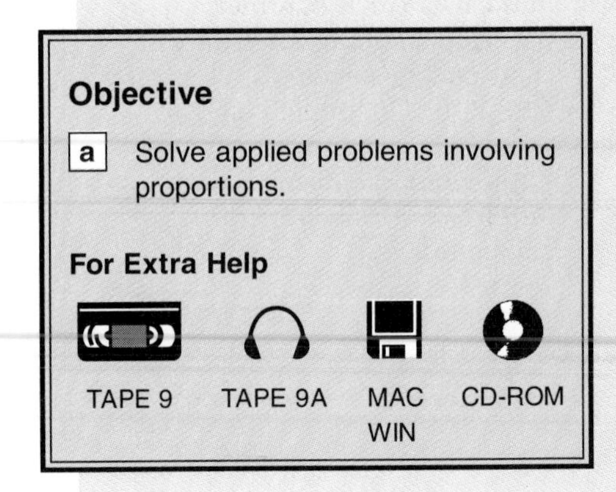

a Proportions have applications in such diverse fields as business, chemistry, health sciences, and home economics, as well as to many areas of daily life. Proportions are most useful in making predictions.

Example 1 *Predicting Total Distance.* Donna drives her delivery van 800 mi in 3 days. At this rate, how far will she drive in 15 days?

1. **Familiarize.** We let $d =$ the distance traveled in 15 days.
2. **Translate.** We translate to a proportion. We make each side the ratio of distance to time, with distance in the numerator and time in the denominator.

$$\begin{array}{r} \text{Distance in 15 days} \rightarrow \\ \text{Time} \rightarrow \end{array} \frac{d}{15} = \frac{800}{3} \begin{array}{l} \leftarrow \text{Distance in 3 days} \\ \leftarrow \text{Time} \end{array}$$

It may help to verbalize the proportion above as "the unknown distance d is to 15 days, as the known distance 800 miles is to 3 days."

3. **Solve.** Next, we solve the proportion:

$3 \cdot d = 15 \cdot 800$ **Equating cross products**

$d = \dfrac{15 \cdot 800}{3}$ **Dividing by 3 on both sides**

$= 4000.$ **Multiplying and dividing**

4. **Check.** We substitute into the proportion and check cross products:

$$\frac{4000}{15} = \frac{800}{3};$$

$4000 \cdot 3 = 12{,}000;\quad 15 \cdot 800 = 12{,}000.$

The cross products are the same.

5. **State.** Donna drives 4000 mi in 15 days.

1. *Calories Burned.* The author of this book generally exercises three times per week. The readout on a stairmaster machine tells him that if he exercises for 24 min, he will burn 356 calories. How many calories will he burn if he exercises for 30 min?

Do Exercise 1.

Proportion problems can be solved in more than one way. In Example 1, any one of the following is an appropriate translation:

$$\frac{800}{3} = \frac{x}{15}, \quad \frac{15}{x} = \frac{3}{800}, \quad \frac{15}{3} = \frac{x}{800}, \quad \frac{800}{x} = \frac{3}{15}.$$

Answer on page A-11

Example 2 *Predicting Medication.* To control a fever, a doctor suggests that a child who weighs 28 kg be given 420 mg of Tylenol. If the dosage is proportional to the child's weight, how much Tylenol is recommended for a child who weighs 35 kg?

1. **Familiarize.** We let $t =$ the number of milligrams of Tylenol.
2. **Translate.** We translate to a proportion, keeping the amount of Tylenol in the numerators.

$$\text{Tylenol suggested} \rightarrow \frac{420}{28} = \frac{t}{35} \leftarrow \text{Tylenol suggested}$$
$$\text{Child's weight} \rightarrow 28 \qquad 35 \leftarrow \text{Child's weight}$$

3. **Solve.** Next, we solve the proportion:

$420 \cdot 35 = 28 \cdot t$ **Equating cross products**

$\frac{420 \cdot 35}{28} = t$ **Dividing by 28 on both sides**

$525 = t.$ **Multiplying and dividing**

4. **Check.** We substitute into the proportion and check cross products:

$$\frac{420}{28} = \frac{525}{35};$$

$$420 \cdot 35 = 14{,}700; \quad 28 \cdot 525 = 14{,}700.$$

The cross products are the same.

5. **State.** The dosage for a child who weighs 35 kg is 525 mg.

Do Exercise 2.

2. *Predicting Paint Needs.* Lowell and Chris run a summer painting company to support their college expenses. They can paint 1700 ft^2 of clapboard with 4 gal of paint. How much paint would be needed for a building with 6000 ft^2 of clapboard?

Example 3 *Purchasing Tickets.* Carey bought 8 tickets to an international food festival for \$52. How many tickets could she purchase with \$90?

1. **Familiarize.** We let $n =$ the number of tickets that can be purchased with \$90.
2. **Translate.** We translate to a proportion, keeping the number of tickets in the numerators.

$$\text{Tickets} \rightarrow \frac{8}{52} = \frac{n}{90} \leftarrow \text{Tickets}$$
$$\text{Cost} \rightarrow 52 \qquad 90 \leftarrow \text{Cost}$$

3. **Solve.** Next, we solve the proportion:

$52 \cdot n = 8 \cdot 90$ **Equating cross products**

$n = \frac{8 \cdot 90}{52}$ **Dividing by 52 on both sides**

$\approx 13.8.$ **Multiplying and dividing**

Because it is impossible to buy a fractional part of a ticket, we must round our answer *down* to 13.

4. **Check.** As a check, we use a different approach: We find the cost per ticket and then divide \$90 by that price. Since

$$52 \div 8 = 6.50 \quad \text{and} \quad 90 \div 6.50 \approx 13.8,$$

we have a check.

5. **State.** Carey could purchase 13 tickets with \$90.

Do Exercise 3.

3. *Purchasing Shirts.* If 2 shirts can be bought for \$47, how many shirts can be bought with \$200?

Answers on page A-11

Example 4 *Women's Hip Measurements.* For improved health, it is recommended that a woman's waist-to-hip ratio be 0.85 (or lower) (***Source:*** David Schmidt, "Lifting Weight Myths," *Nutrition Action Newsletter* 20, no 4, October 1993). Marta's hip measurement is 40 in. To meet the recommendation, what should Marta's waist measurement be?

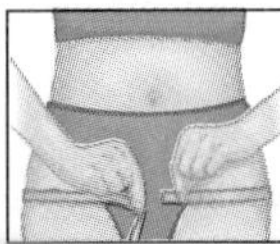
Hip measurement is the largest measurement around the widest part of the buttocks.

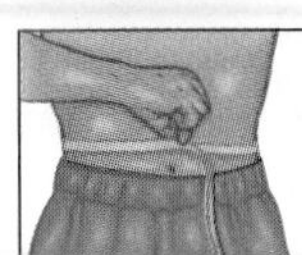
Waist measurement is the smallest measurement below the ribs but above the navel.

1. **Familiarize.** Note that $0.85 = \frac{85}{100}$. We let w = Marta's waist measurement
2. **Translate.** We translate to a proportion as follows:

$$\text{Waist measurement} \rightarrow \frac{w}{40} = \frac{85}{100} \leftarrow \text{Recommended waist-to-hip ratio}$$
(Hip measurement → 40)

3. **Solve.** Next, we solve the proportion:

$$100 \cdot w = 40 \cdot 85 \quad \textbf{Equating cross products}$$
$$w = \frac{40 \cdot 85}{100} \quad \textbf{Dividing by 100 on both sides}$$
$$= 34. \quad \textbf{Multiplying and dividing}$$

4. **Check.** As a check, we divide 34 by 40:

$34 \div 40 = 0.85.$

This is the desired ratio.

5. **State.** Marta's recommended waist measurement is 34 in. (or less).

Do Exercise 4.

4. *Men's Hip Measurements.* It is recommended that a man's waist-to-hip ratio be 0.95 (or lower). Malcolm's hip measurement is 40 in. To meet the recommendation, what should Malcolm's waist measurement be?

Answer on page A-11

5. *Construction Plans.* In Example 5, the length of the actual deck is 28.5 ft. What is the length of the deck on the blueprints?

Answer on page A-11

Example 5 *Construction Plans.* Architects make blueprints of projects being constructed. These are scale drawings in which lengths are in proportion to actual sizes. The Hennesseys are constructing a rectangular deck just outside their house. The architectural blueprints are rendered such that $\frac{3}{4}$ in. on the drawing is actually 2.25 ft on the deck. The width of the deck on the drawing is 4.3 in. How wide is the deck in reality?

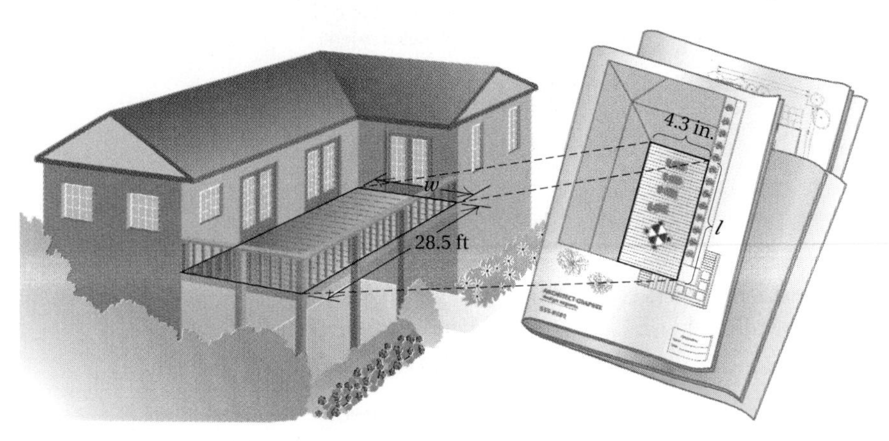

1. **Familiarize.** We let w = the width of the deck.
2. **Translate.** Then we translate to a proportion, using 0.75 for $\frac{3}{4}$ in.

$$\begin{array}{r} \text{Measure on drawing} \rightarrow \\ \text{Measure on deck} \rightarrow \end{array} \frac{0.75}{2.25} = \frac{4.3}{w} \begin{array}{l} \leftarrow \text{Width of drawing} \\ \leftarrow \text{Width of deck} \end{array}$$

3. **Solve.** Next, we solve the proportion:

$$0.75 \cdot w = 2.25 \cdot 4.3 \quad \text{Equating cross products}$$

$$w = \frac{2.25 \cdot 4.3}{0.75} \quad \text{Dividing by 0.75 on both sides}$$

$$= 12.9.$$

4. **Check.** We substitute into the proportion and check cross products:

$$\frac{0.75}{2.25} = \frac{4.3}{12.9};$$

$$0.75 \times 12.9 = 9.675; \quad 2.25 \times 4.3 = 9.675.$$

The cross products are the same.

5. **State.** The width of the deck is 12.9 ft.

Do Exercise 5.

Example 6 *Estimating a Wildlife Population.* To determine the number of fish in a lake, a conservationist catches 225 fish, tags them, and throws them back into the lake. Later, 108 fish are caught, and it is found that 15 of them are tagged. Estimate how many fish are in the lake.

1. **Familiarize.** We let $F =$ the number of fish in the lake.
2. **Translate.** We translate to a proportion as follows:

$$\text{Fish tagged originally} \rightarrow \frac{225}{F} = \frac{15}{108} \leftarrow \text{Tagged fish caught later}$$

(Fish in lake $\rightarrow F$; $108 \leftarrow$ Fish caught later)

3. **Solve.** Next, we solve the proportion:

$225 \cdot 108 = F \cdot 15$ Equating cross products

$\dfrac{225 \cdot 108}{15} = F$ Dividing by 15 on both sides

$1620 = F.$ Multiplying and dividing

4. **Check.** We substitute into the proportion and check cross products:

$\dfrac{225}{1620} = \dfrac{15}{108}$;

$225 \cdot 108 = 24{,}300$; $1620 \cdot 15 = 24{,}300.$

The cross products are the same.

5. **State.** We estimate that there are 1620 fish in the lake.

Do Exercise 6.

6. *Estimating a Deer Population.* To determine the number of deer in a forest, a conservationist catches 612 deer, tags them, and releases them. Later, 244 deer are caught, and it is found that 72 of them are tagged. Estimate how many deer are in the forest.

2074

Answer on page A-11

Improving Your Math Study Skills

Study Tips for Trouble Spots

By now you have probably encountered certain topics that gave you more difficulty than others. It is important to know that this happens to every person who studies mathematics. Unfortunately, frustration is often part of the learning process and it is important not to give up when difficulty arises.

One source of frustration for many students is not being able to set aside sufficient time for studying. Family commitments, work schedules, and athletics are just a few of the time demands that many students face. Couple these demands with a math lesson that seems to require a greater than usual amount of study time, and it is no wonder that many students often feel frustrated. Below are some study tips that might be useful if and when troubles arise.

- **Realize that everyone—even your instructor—has been stymied at times when studying math.** You are not the first person, nor will you be the last, to encounter a "roadblock."
- **Whether working alone or with a classmate, try to allow enough study time so that you won't need to constantly glance at a clock.** Difficult material is best mastered when your mind is completely focused on the subject matter. Thus, if you are tired, it is usually best to study early the next morning or to take a ten-minute "power-nap" in order to make the most productive use of your time.
- **Talk about your trouble spot with a classmate.** It is possible that she or he is also having difficulty with the same material. If that is the case, perhaps the majority of your class is confused and your instructor's coverage of the topic is not yet finished. If your classmate *does* understand the topic that is troubling you, patiently allow him or her to explain it to you. By verbalizing the math in question, your classmate may help clarify the material for both of you. Perhaps you will be able to return the favor for your classmate when he or she is struggling with a topic that you understand.
- **Try to study in a "controlled" environment.** What we mean by this is that you can often put yourself in a setting that will enable you to maximize your powers of concentration. For example, whereas some students may succeed in studying at home or in a dorm room, for many these settings are filled with distractions. Consider a trip to a library, classroom building, or perhaps the attic or basement if such a setting is more conducive to studying. If you plan on working with a classmate, try to find a location in which conversation will not be bothersome to others.
- **When working on difficult material, it is often helpful to first "back up" and review the most recent material that *did* make sense.** This can build your confidence and create a momentum that can often carry you through the roadblock. Sometimes a small piece of information that appeared in a previous section is all that is needed for your problem spot to disappear. When the difficult material is finally mastered, try to make use of what is fresh in your mind by taking a "sneak preview" of what your next topic for study will be.

Exercise Set 5.4

a Solve.

1. *Travel Distance.* Monica bicycled 234 mi in 14 days. At this rate, how far would Monica travel in 42 days?

2. *Gasoline Mileage.* Chuck's van traveled 84 mi on 6.5 gal of gasoline. At this rate, how many gallons would be needed to travel 126 mi?

3. If 2 tee shirts cost $18.80, how much would 9 tee shirts cost?

4. If 2 bars of soap cost $0.89, how many bars of soap can be purchased with $6.50?

5. In the rectangular paintings below, the ratio of length to height is the same. Find the height of the larger painting.

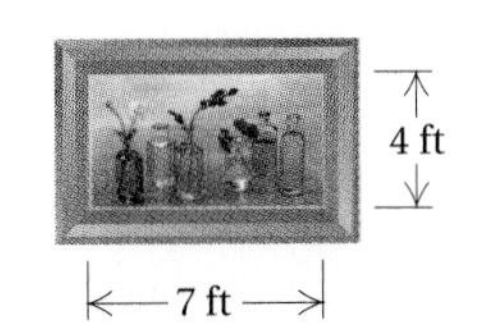

6. In the rectangles below, the ratio of length to width is the same. Find the width of the larger rectangle.

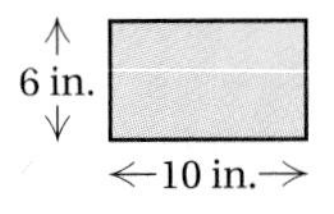

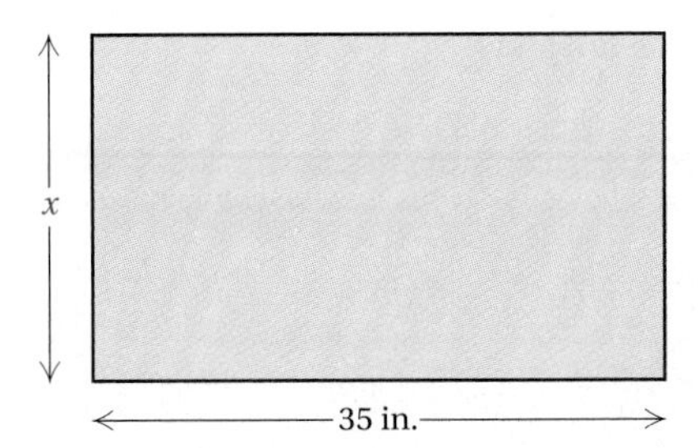

7. A bookstore manager knows that 24 books weigh 37 lb. How much do 40 books weigh?

8. *Turkey Servings.* An 8-lb turkey breast contains 36 servings of meat. How many pounds of turkey breast would be needed for 54 servings?

9. *Maple Syrup.* When 38 gal of maple sap are boiled down, the result is 2 gal of maple syrup. How much sap is needed to produce 9 gal of syrup?

10. In a class of 40 students, on average, 6 will be left-handed. If a class includes 9 "lefties," how many students would you estimate are in the class?

11. *Coffee.* Coffee beans from 14 trees are required to produce the 17 lb of coffee that the average person in the United States drinks each year. How many trees are required to produce 375 lb of coffee?

12. Jean bought a new car. In the first 8 months, it was driven 10,000 km. At this rate, how many kilometers will the car be driven in 1 yr?

13. A college advertises that its student-to-faculty ratio is 14 to 1. If 56 students register for Introductory Spanish, how many sections of the course would you expect to see offered?

14. In a metal alloy, the ratio of zinc to copper is 3 to 13. If there are 520 lb of copper, how many pounds of zinc are there?

15. *Deck Sealant.* Bonnie can waterproof 450 ft^2 of decking with 2 gal of sealant. How many gallons should Bonnie buy for a 1200-ft^2 deck?

16. *Paint Coverage.* Fred uses 3 gal of paint to cover 1275 ft^2 of siding. How much siding can Fred paint with 7 gal of paint?

17. *Estimating a Deer Population.* To determine the number of deer in a game preserve, a forest ranger catches 318 deer, tags them, and releases them. Later, 168 deer are caught, and it is found that 56 of them are tagged. Estimate how many deer are in the game preserve.

18. *Estimating a Trout Population.* To determine the number of trout in a lake, a conservationist catches 112 trout, tags them, and throws them back into the lake. Later, 82 trout are caught, and it is found that 32 of them are tagged. Estimate how many trout there are in the lake.

19. *Grass-Seed Coverage.* It takes 60 oz of grass seed to seed 3000 ft^2 of lawn. At this rate, how much would be needed for 5000 ft^2 of lawn?

20. *Grass-Seed Coverage.* In Exercise 19, how much seed would be needed for 7000 ft^2 of lawn?

21. *Quality Control.* A quality-control inspector examined 200 lightbulbs and found 18 of them to be defective. At this rate, how many defective bulbs will there be in a lot of 22,000?

22. A professor must grade 32 essays in a literature class. She can grade 5 essays in 40 min. At this rate, how long will it take her to grade all 32 essays?

23. *Map Scaling.* On a road atlas map, 1 in. represents 16.6 mi. If two cities are 3.5 in. apart on the map, how far apart are they in reality?

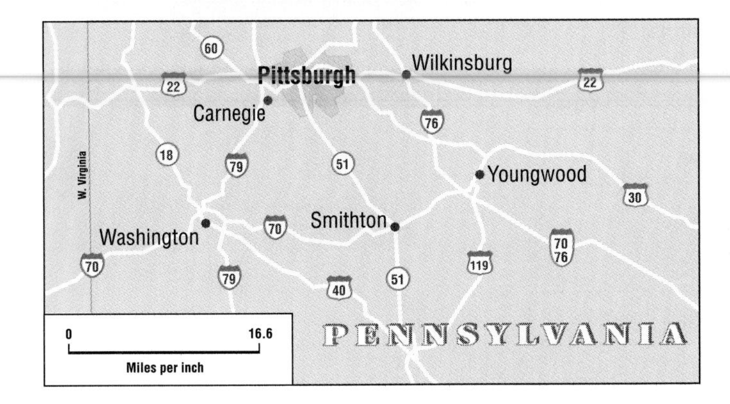

24. *Map Scaling.* On a map, $\frac{1}{4}$ in. represents 50 mi. If two cities are $3\frac{1}{4}$ in. apart on the map, how far apart are they in reality?

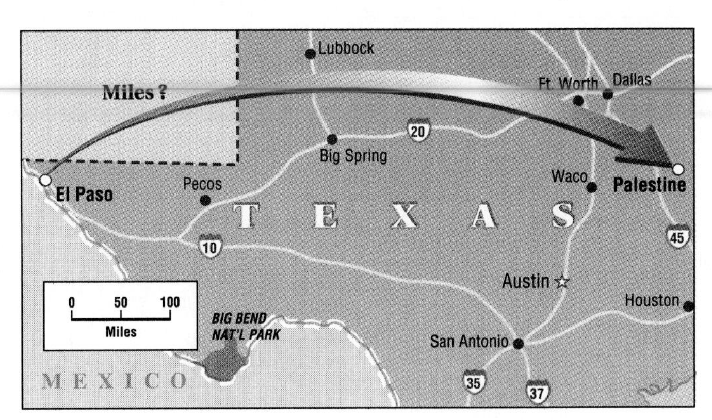

25. *Snow to Water.* Under typical conditions, $1\frac{1}{2}$ ft of snow will melt to 2 in. of water. To how many inches of water will $5\frac{1}{2}$ ft of snow melt?

26. *Tire Wear.* Tires are often priced according to the number of miles that they are expected to be driven. Suppose a tire priced at $59.76 is expected to be driven 35,000 mi. How much would you pay for a tire that is expected to be driven 40,000 mi?

27. *College Expenses.* A student attends a university whose academic year consists of two 16-week semesters. She budgets $800 for incidental expenses for the academic year. After 3 weeks, she has spent $80 for incidental expenses. Assuming the student continues to spend at the same rate, will the budget for incidental expenses be adequate? If not, when will the money be exhausted and how much more will be needed to complete the year?

28. *Sound-System Expense.* A basic sound system consists of a CD player, a receiver–amplifier, and two speakers. A rule of thumb used to estimate the relative investment in these components is 1:3:2. That is, the receiver–amplifier should cost three times the amount spent on the CD player and the speakers should cost twice as much as the amount spent on the CD player.

a) You have $1800 to spend. How should you allocate the funds if you use this rule of thumb?
b) How should you allocate a budget of $3000?

Earned Run Average. In baseball, the average number of earned runs given up by a pitcher in 9 innings is the pitcher's *earned run average,* or *ERA*. For example, John Smoltz of the Atlanta Braves gave up 83 earned runs in $253\frac{2}{3}$ innings during the year in which he earned the Cy Young award as the finest pitcher in the National League. His earned average is found by solving the following proportion:

$$\frac{\text{ERA}}{9} = \frac{83}{253\frac{2}{3}}$$

$$\text{ERA} = \frac{9 \cdot 83}{253\frac{2}{3}} \approx 2.94.$$

Complete the following table to find the ERA of each National League pitcher in the same year.

	Player	Team	Earned Runs	Innings Pitched	ERA
	John Smoltz	Atlanta Braves	83	$253\frac{2}{3}$	2.94
29.	Greg Maddux	Atlanta Braves	74	245	
30.	Jaime Navarro	Chicago Cubs	103	$236\frac{2}{3}$	
31.	Kevin Ritz	Colorado Rockies	125	213	
32.	Hideo Nomo	Los Angeles Dodgers	81	$228\frac{1}{3}$	

Skill Maintenance

Solve. [4.7a]

33. Dallas, Texas, receives an average of 31.1 in. (78.994 cm) of rain and 2.6 in. (6.604 cm) of snow each year (***Source*:** National Oceanic and Atmospheric Administration).

a) What is the total amount of precipitation in inches?

b) What is the total amount of precipitation in centimeters?

34. The distance, by air, from New York to St. Louis is 876 mi (1401.6 km) and from St. Louis to Los Angeles is 1562 mi (2499.2 km).

a) How far, in miles, is it from New York to Los Angeles?

b) How far, in kilometers, is it from New York to Los Angeles?

Synthesis

35. ◈ Polly solved Example 1 by forming the proportion $\frac{15}{3} = \frac{x}{800}$, whereas Rudy wrote $\frac{800}{15} = \frac{3}{x}$. Are both approaches valid? Why or why not?

36. ◈ Rob's waist and hips measure 35 in. and 33 in., respectively (see Margin Exercise 4). Suppose that Rob can either gain or lose 1 in. from one of his measurements. Where should the inch come from or go to? Why?

37. 🖩 Carney College is expanding from 850 to 1050 students. To avoid any rise in the student-to-faculty ratio, the faculty of 69 professors must also increase. How many new faculty positions should be created?

38. 🖩 In recognition of her outstanding work, Sheri's salary has been increased from $26,000 to $29,380. Tim is earning $23,000 and is requesting a proportional raise. How much more should he ask for?

39. Sue can paint 950 ft^2 with 2 gal of paint. How many gallons should Sue buy in order to paint a 30-ft by 100-ft wall?

40. Cy Young, one of the greatest baseball pitchers of all time, had an earned run average of 2.63. He pitched more innings, 7356, than anyone in the history of baseball. How many earned runs did he give up?

Use proportions to estimate your college's student population.

5.5 Similar Triangles

a Proportions and Similar Triangles

Look at the pair of triangles below. Note that they appear to have the same shape, but their sizes are different. These are examples of **similar triangles**. By using a magnifying glass, you could imagine enlarging the smaller triangle to get the larger. This process works because the corresponding sides of each triangle have the same ratio. That is, the following proportion is true.

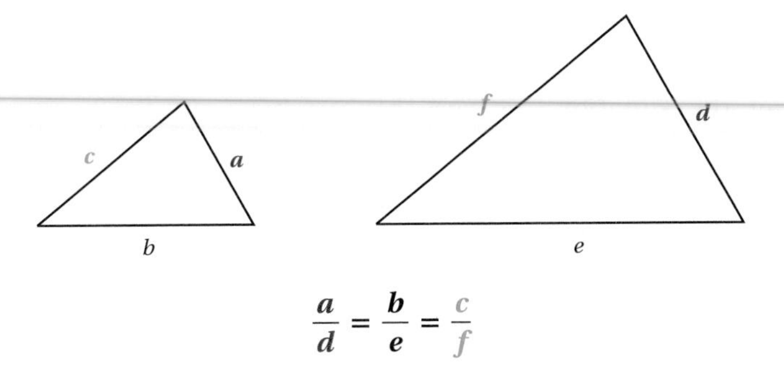

$$\frac{a}{d} = \frac{b}{e} = \frac{c}{f}$$

> **Similar triangles** have the same shape. Their corresponding sides have the same ratio—that is, they are proportional.

Example 1 The triangles at right are similar triangles. Find the missing length x.

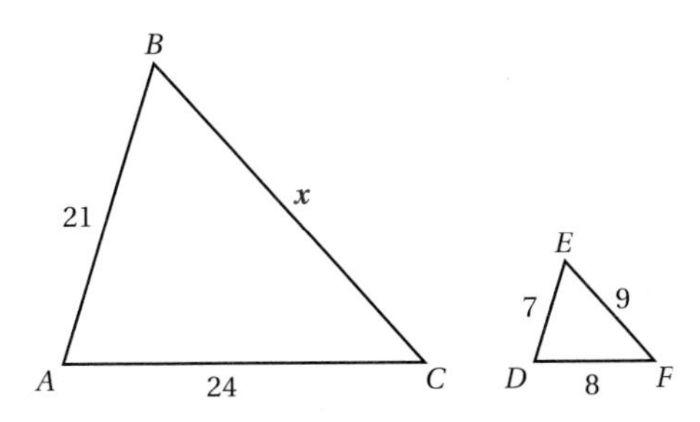

The ratio of x to 9 is the same as the ratio of 24 to 8. We get the proportion

$$\frac{x}{9} = \frac{24}{8}.$$

We solve the proportion:

$$8 \cdot x = 9 \cdot 24 \qquad \textbf{Equating cross products}$$

$$x = \frac{9 \cdot 24}{8} \qquad \textbf{Dividing by 8 on both sides}$$

$$= 27. \qquad \textbf{Simplifying}$$

The missing length x is 27. We could have also used $\frac{x}{9} = \frac{21}{7}$ to find x.

Do Exercise 1.

Similar triangles and proportions can often be used to find lengths that would ordinarily be difficult to measure. For example, we could find the height of a flagpole without climbing it or the distance across a river without crossing it.

Objective

a Find lengths of sides of similar triangles using proportions.

For Extra Help

TAPE 9 TAPE 9B MAC WIN CD-ROM

1. This pair of triangles is similar. Find the missing length x.

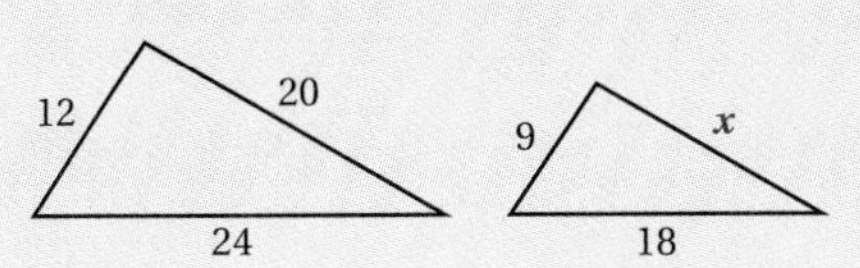

Answer on page A-11

2. How high is a flagpole that casts a 45-ft shadow at the same time that a 5.5-ft woman casts a 10-ft shadow?

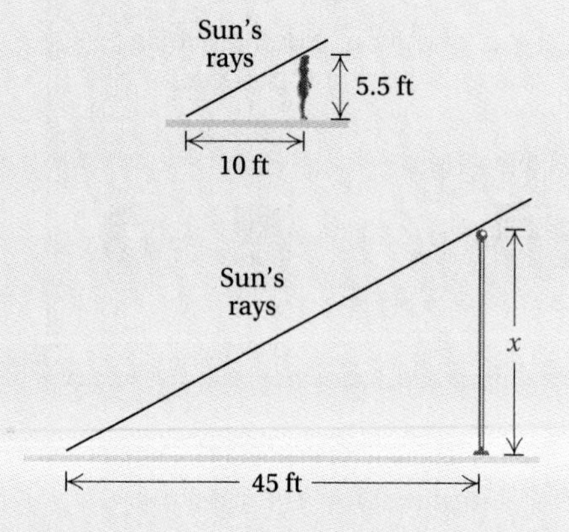

3. *F-106 Blueprint.* Referring to Example 3, find the length x of the wing.

Answers on page A-11

Example 2 How high is a flagpole that casts a 56-ft shadow at the same time that a 6-ft man casts a 5-ft shadow?

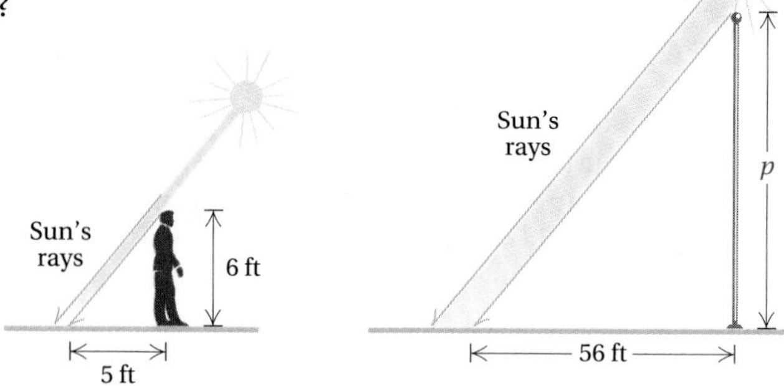

If we use the sun's rays to represent the third side of the triangle in our drawing of the situation, we see that we have similar triangles. Let p = the height of the flagpole. The ratio of 6 to p is the same as the ratio of 5 to 56. Thus we have the proportion

$$\text{Height of man} \rightarrow \frac{6}{p} = \frac{5}{56} \leftarrow \text{Length of shadow of man}$$
$$\text{Height of pole} \rightarrow \qquad\qquad \leftarrow \text{Length of shadow of pole}$$

Solve: $6 \cdot 56 = 5 \cdot p$ **Equating cross products**

$\frac{6 \cdot 56}{5} = p$ **Dividing by 5 on both sides**

$67.2 = p$ **Simplifying**

The height of the flagpole is 67.2 ft.

Do Exercise 2.

Example 3 *F-106 Blueprint.* A blueprint for an F-106 Delta Dart fighter plane is a scale drawing. Each wing of the plane has a triangular shape. The blueprint shows similar triangles. Find the length of side a of the wing.

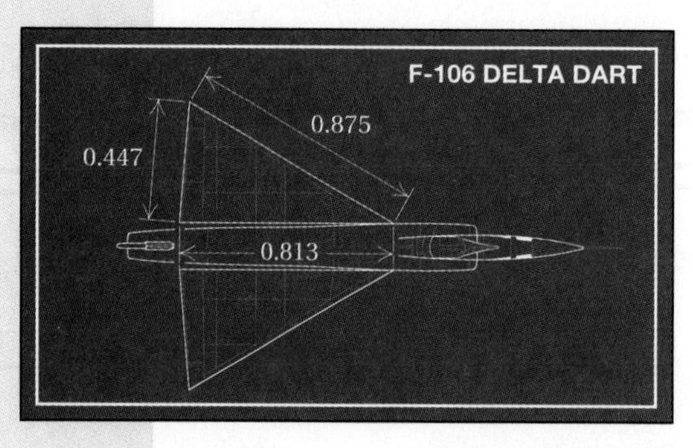

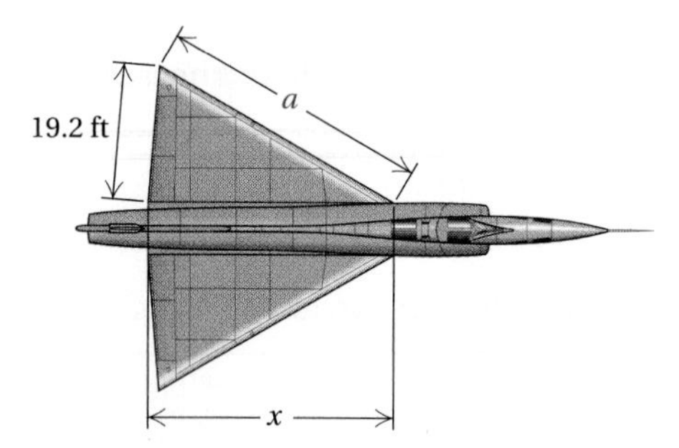

We let a = the length of the wing. Thus we have the proportion

$$\text{Length on the blueprint} \rightarrow \frac{0.447}{19.2} = \frac{0.875}{a} \leftarrow \text{Length on the blueprint}$$
$$\text{Length of the wing} \rightarrow \qquad\qquad \leftarrow \text{Length of the wing}$$

Solve: $0.447 \cdot a = 19.2 \cdot 0.875$ **Equating cross products**

$a = \frac{19.2 \cdot 0.875}{0.447}$ **Dividing by 0.447 on both sides**

≈ 37.6 ft

The length of side a of the wing is about 37.6 ft.

Do Exercise 3.

Exercise Set 5.5

a The triangles in each exercise are similar. Find the missing lengths.

1.

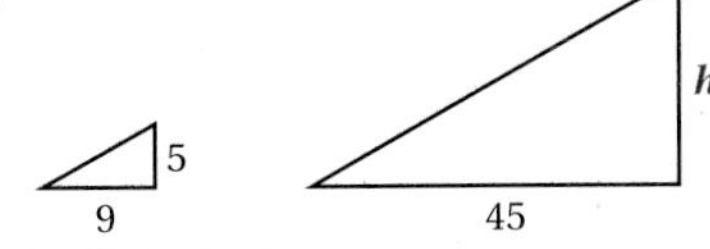

2.

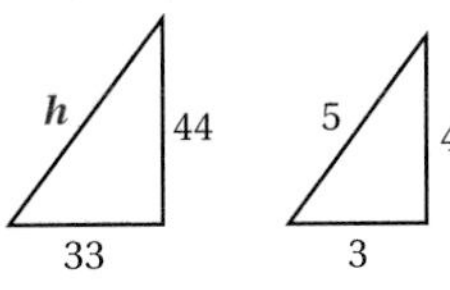

3.

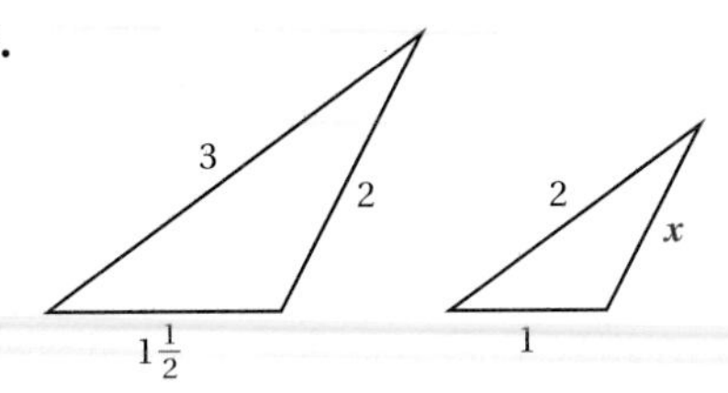

4.

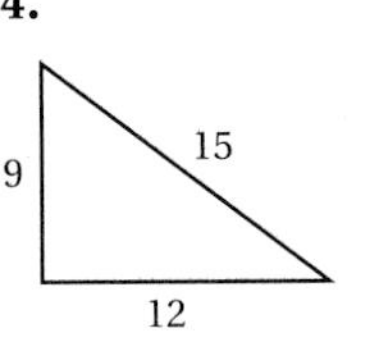

5.

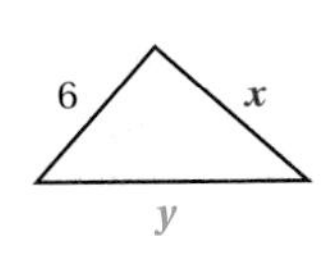

6.

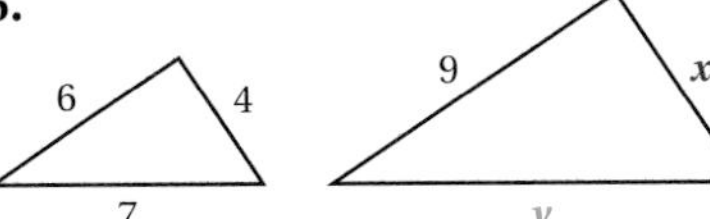

7.

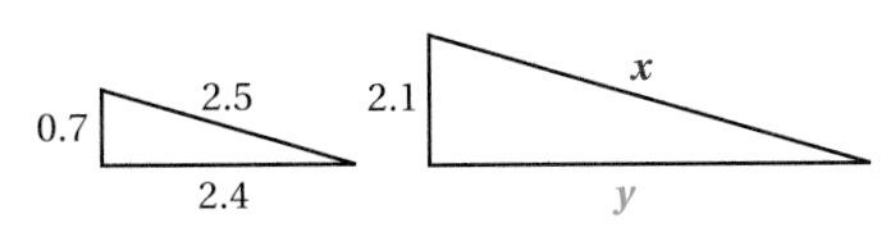

8.

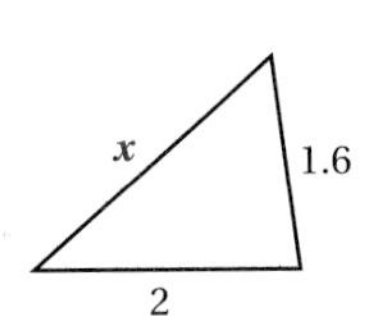

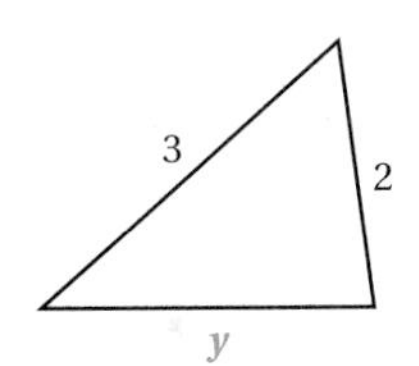

9. When a tree 8 m high casts a shadow 5 m long, how long a shadow is cast by a person 2 m tall?

10. How high is a flagpole that casts a 42-ft shadow at the same time that a $5\frac{1}{2}$-ft woman casts a 7-ft shadow?

11. How high is a tree that casts a 27-ft shadow at the same time that a 4-ft fence post casts a 3-ft shadow?

12. How high is a tree that casts a 32-ft shadow at the same time that an 8-ft light pole casts a 9-ft shadow?

13. Find the height h of the wall.

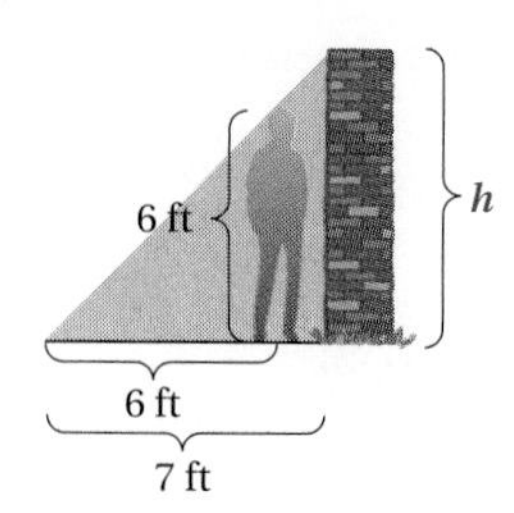

14. Find the length L of the lake.

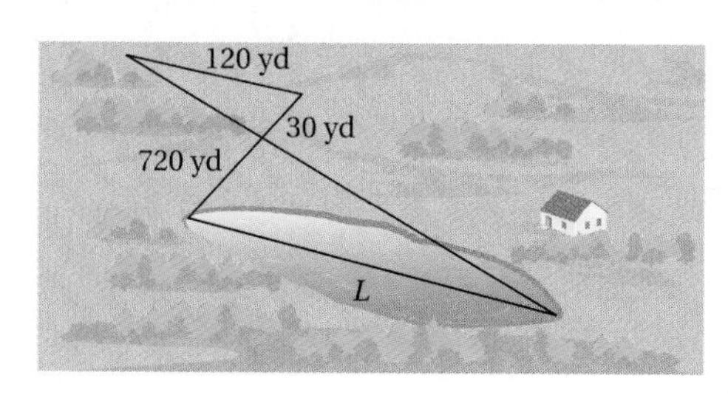

15. Find the distance across the river. Assume that the ratio of d to 25 ft is the same as the ratio of 40 ft to 10 ft.

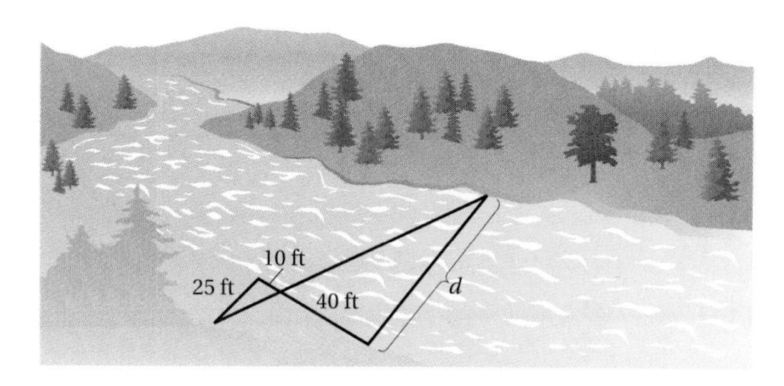

17. To measure the height of a hill, a string is drawn tight from level ground to the top of the hill. A 3-ft yardstick is placed under the string, touching it at point P, a distance of 5 ft from point G, where the string touches the ground. The string is then detached and found to be 120 ft long. How high is the hill?

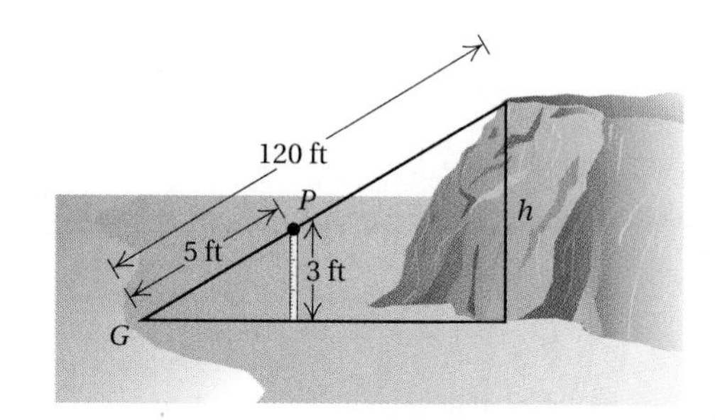

The sides in each pair of figures are proportional. Find the missing lengths.

17.

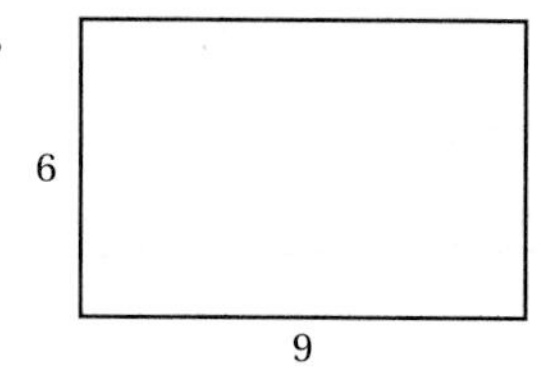

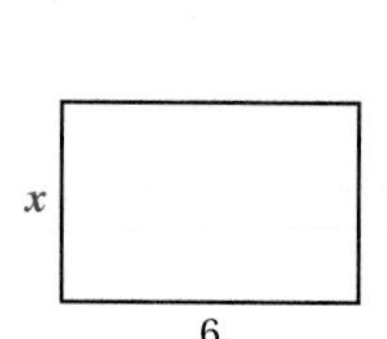

18.

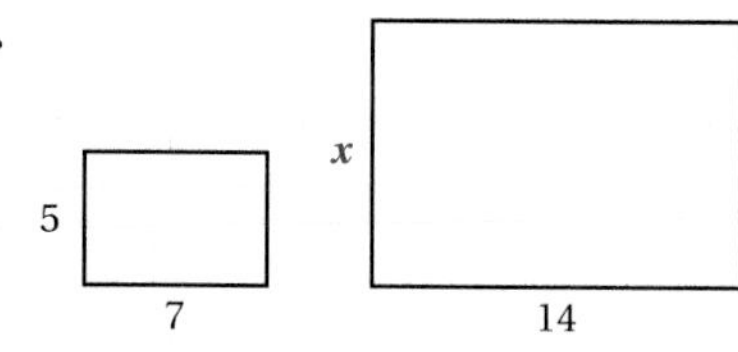

19.

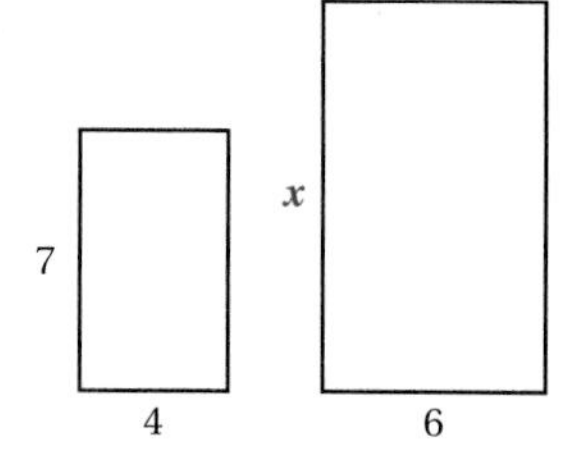

20.

11

x

4

3

21.

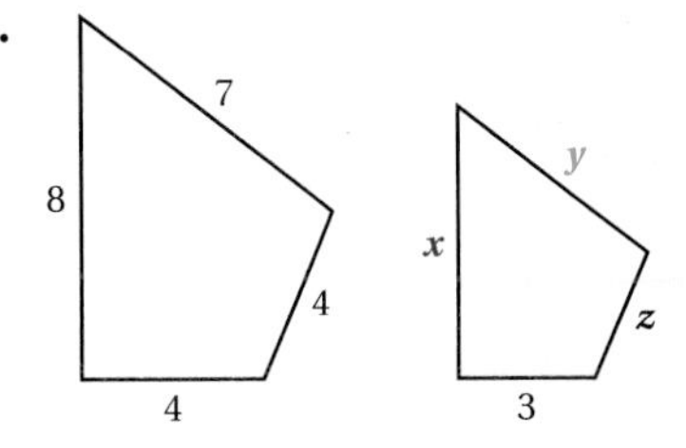

22.

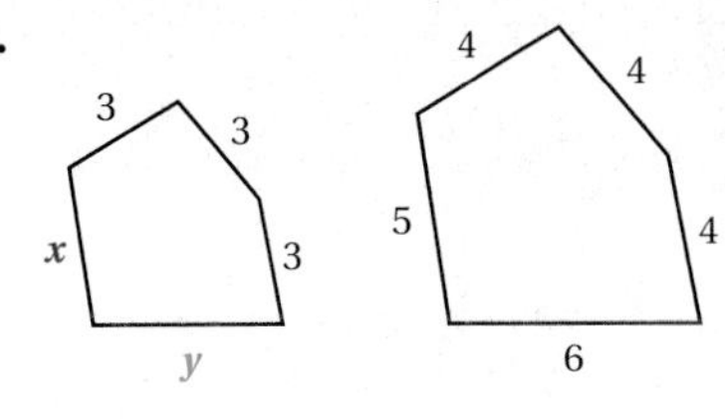

23.

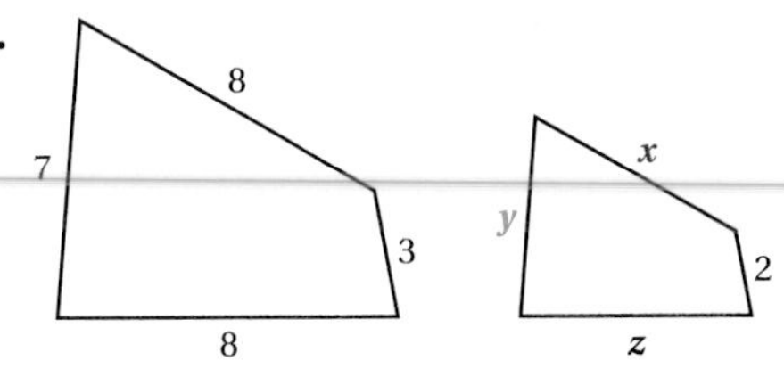

24.

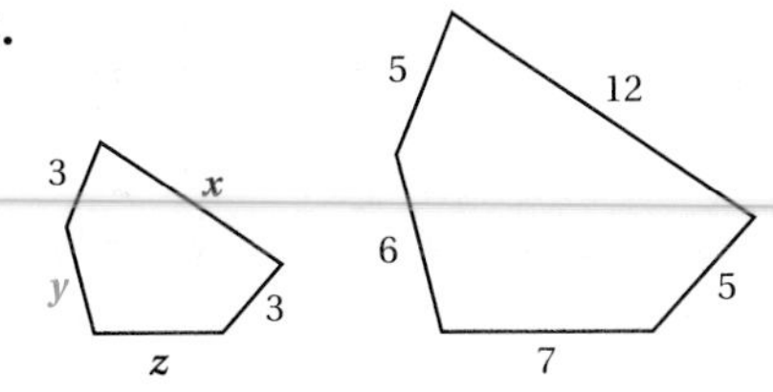

25.

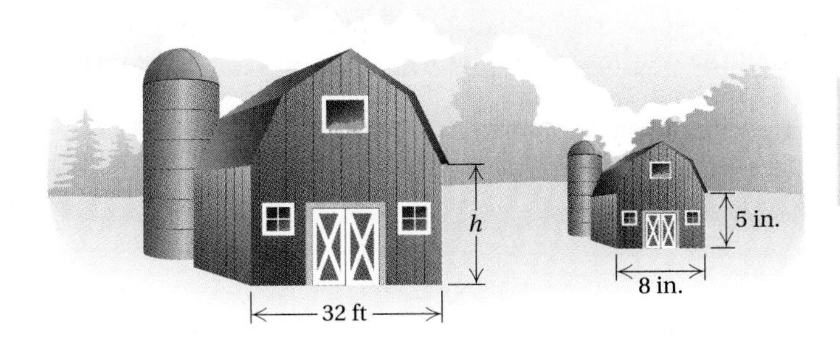

26.

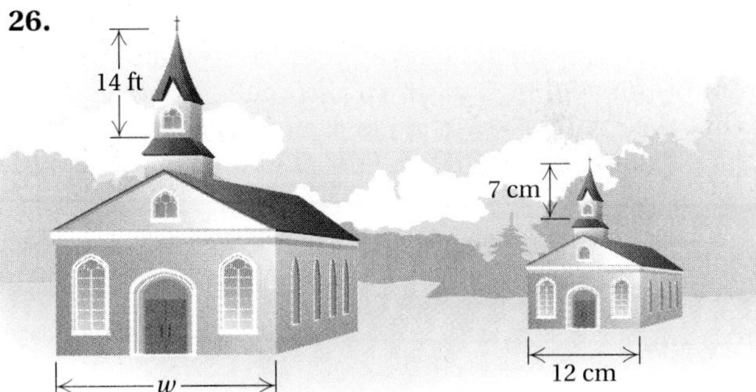

27. A scale model of an addition to an athletic facility is 12 cm wide at the base and rises to a height of 15 cm. If the actual base is to be 116 ft, what will be the height of the addition?

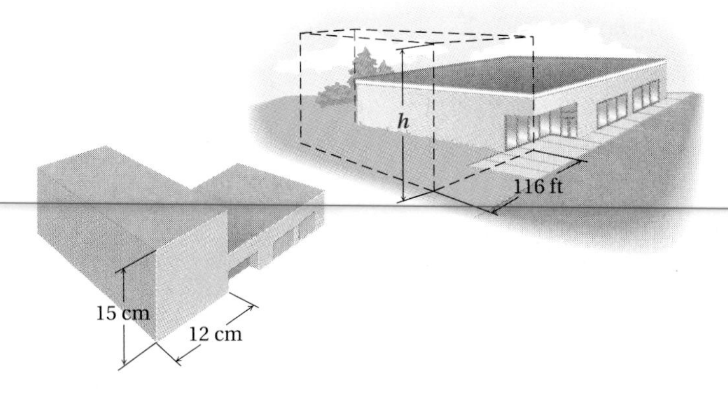

28. Refer to the figures in Exercise 27. If a model skylight is 3 cm wide, how wide will the actual skylight be?

Skill Maintenance

29. A student has $34.97 to spend for a book at $49.95, a CD at $14.88, and a sweatshirt at $29.95. How much more money does the student need to make these purchases? [4.7a]

30. Divide: $80.892 \div 8.4$. [4.4a]

Multiply. [4.3a]

31. 8.4×80.892

32. 0.01×274.568

33. 100×274.568

34. 0.002×274.568

Synthesis

35. ◈ Is it possible for two triangles to have two pairs of sides that are proportional without the triangles being similar? Why or why not?

36. ◈ Design for a classmate a problem involving similar triangles for which

$$\frac{18}{128.95} = \frac{x}{789.89}.$$

Hockey Goals. An official hockey goal is 6 ft wide. To make scoring more difficult, goalies often locate themselves far in front of the goal to "cut down the angle." In Exercises 37 and 38, suppose that a slapshot from point A is attempted and that the goalie is 2.7 ft wide. Determine how far from the goal the goalie should be located if point A is the given distance from the goal. (*Hint*: First find how far the goalie should be from point A.)

37. ▦ 25 ft

38. ▦ 35 ft

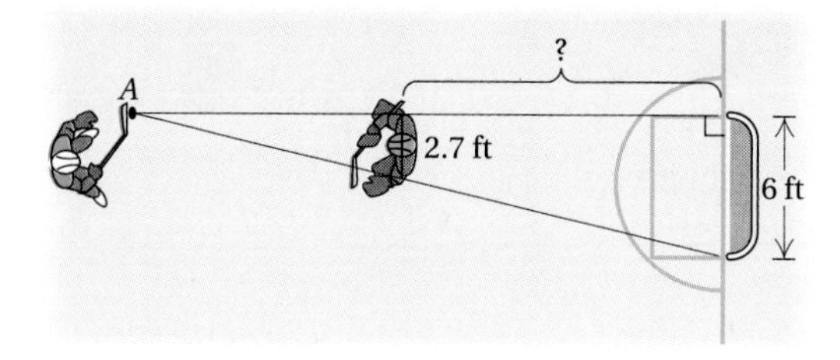

39. ▦ A miniature basketball hoop is built for the model referred to in Exercise 27. An actual hoop is 10 ft high. How high should the model hoop be? Round to the nearest thousandth of a centimeter.

▦ Solve. Round the answer to the nearest thousandth.

40. $\dfrac{8664.3}{10{,}344.8} = \dfrac{x}{9776.2}$

41. $\dfrac{12.0078}{56.0115} = \dfrac{789.23}{y}$

▦ The triangles in each exercise are similar triangles. Find the lengths not given.

42.

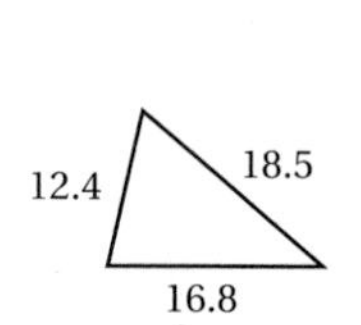

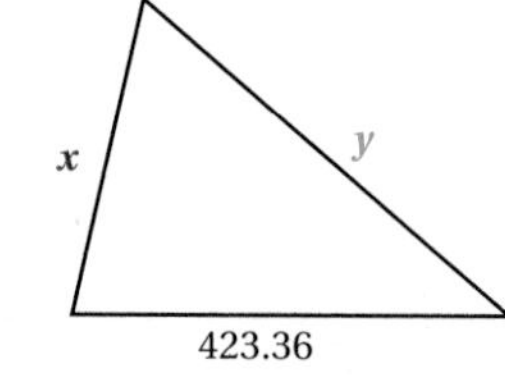

43.

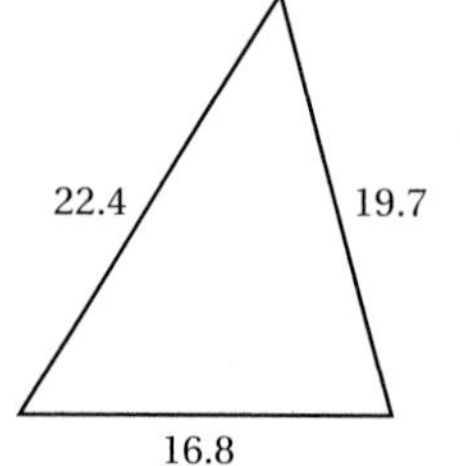

Summary and Review Exercises: Chapter 5

The objectives to be tested in addition to the material in this chapter are [2.5c], [4.3a], [4.4a], and [4.7a].

Write fractional notation for the ratio. Do not simplify. [5.1a]

1. 47 to 84

2. 46 to 1.27

3. 83 to 100

4. 0.72 to 197

5. In a recent year in the United States, 2,312,203 people died. Of these, 537,969 died of cancer. (***Source*:** U.S. National Center for Health Statistics) Write fractional notation for the ratio of the number of people who die of cancer to the number of people who die. [5.1a]

Simplify the ratio. [5.1b]

6. 9 to 12

7. 3.6 to 6.4

8. What is the rate in miles per hour? [5.2a]

117.7 miles, 5 hours

9. A lawn requires 319 gal of water for every 500 ft^2. What is the rate in gallons per square foot? [5.2a]

10. What is the rate in dollars per kilogram? [5.2a]

$355.04, 14 kilograms

11. *Turkey Servings.* A 25-lb turkey serves 18 people. What is the rate in servings per pound? [5.2a]

12. A 1-lb, 7-oz package of flour costs $1.30. Find the unit price in cents per ounce. Round to the nearest tenth of a cent. [5.2b]

13. *Unit Pricing.* It costs 79 cents for a $14\frac{1}{2}$-oz can of tomatoes. Find the unit price in cents per ounce. Round to the nearest hundredth of a cent. [5.2b]

Which has the lower unit price? [5.2b]

14.

WHITE BREAD
Brand A: 16 oz for 89 cents Brand B: 12 oz for 65 cents

15.

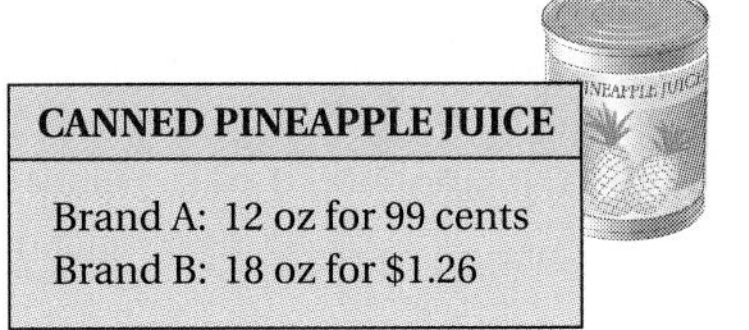

CANNED PINEAPPLE JUICE
Brand A: 12 oz for 99 cents Brand B: 18 oz for $1.26

Determine whether the two pairs of numbers are proportional. [5.3a]

16. 9, 15 and 36, 59

17. 24, 37 and 40, 46.25

Solve. [5.3b]

18. $\frac{8}{9} = \frac{x}{36}$

19. $\frac{6}{x} = \frac{48}{56}$

20. $\frac{120}{\frac{3}{7}} = \frac{7}{x}$

21. $\frac{4.5}{120} = \frac{0.9}{x}$

Solve. [5.4a]

22. If 3 dozen eggs cost $2.67, how much will 5 dozen eggs cost?

23. *Quality Control.* A factory manufacturing computer circuits found 39 defective circuits in a lot of 65 circuits. At this rate, how many defective circuits can be expected in a lot of 585 circuits?

24. A train travels 448 mi in 7 hr. At this rate, how far will it travel in 13 hr?

25. Fifteen acres are required to produce 54 bushels of tomatoes. At this rate, how many acres are required to produce 97.2 bushels of tomatoes?

26. *Garbage Production.* It is known that 5 people produce 13 kg of garbage in one day. San Diego, California, has 1,150,000 people. How many kilograms of garbage are produced in San Diego in one day?

27. *Snow to Water.* Under typical conditions, $1\frac{1}{2}$ ft of snow will melt to 2 in. of water. To how many inches of water will $4\frac{1}{2}$ ft of snow melt?

28. *Lawyers in Michigan.* In Michigan, there are 2.3 lawyers for every 1000 people. The population of Detroit is 4,307,000. (***Source:*** U.S. Bureau of the Census) How many lawyers would you expect there to be in Detroit?

29. The following triangles are similar. Find x and y. [5.5a]

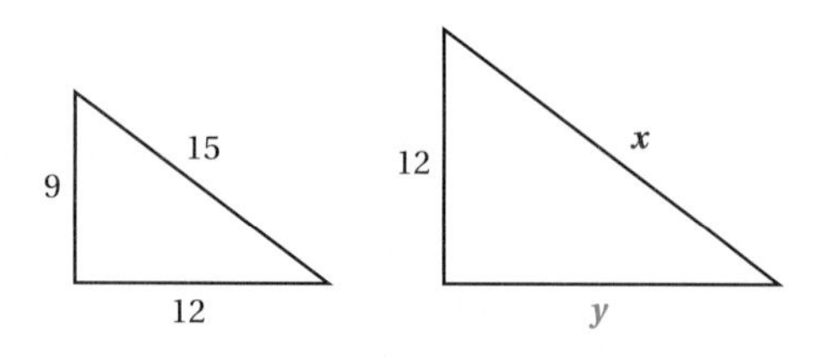

Skill Maintenance

Solve. [4.7a]

30. A family has $2347.89 in its checking account. Someone writes checks for $678.95 and $38.54. How much is left in the checking account?

31. What is the total cost of 8 CD players at $349.95 each?

Use = or ≠ for ▒ to write a true sentence. [2.5c]

32. $\frac{5}{2}$ ▒ $\frac{10}{4}$

33. $\frac{4}{6}$ ▒ $\frac{8}{10}$

34. Multiply. [4.3a]

$$\begin{array}{r} 456.1 \\ \times\ 23.4 \\ \hline \end{array}$$

35. Divide. Write decimal notation for the answer. [4.4a]

$$5.6\overline{)254.8}$$

Synthesis

36. ◈ If you were a college president, which would you prefer: a low or high faculty-to-student ratio? Why? What about the student-to-faculty ratio? Why? [5.1a]

37. ◈ Write a proportion problem for a classmate to solve. Design the problem so that the solution is "Leslie would need 16 gal of gasoline in order to travel 368 mi." [5.4a]

38. It takes Yancy Martinez 10 min to type two-thirds of a page of his term paper. At this rate, how long will it take him to type a 7-page term paper? [5.4a]

39. ▦ The following triangles are similar. Find the missing lengths. [5.5a]

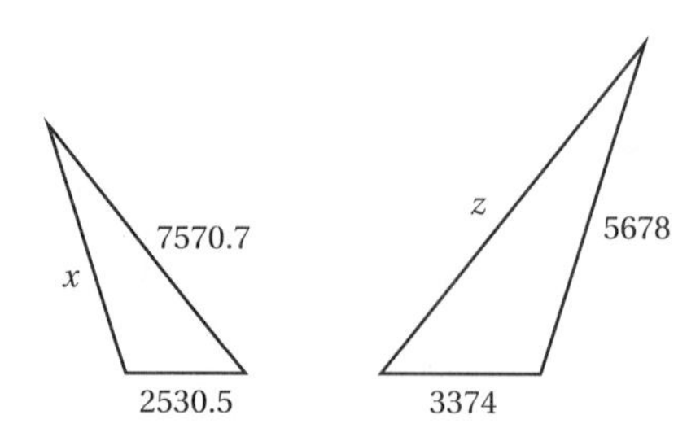

Test: Chapter 5

Answers

Write fractional notation for the ratio. Do not simplify.

1. 85 to 97

2. 0.34 to 124

Simplify the ratio.

3. 18 to 20

4. 0.75 to 0.96

5. What is the rate in feet per second?

10 feet, 16 seconds

6. *Ham Servings.* A 12-lb shankless ham contains 16 servings. What is the rate in servings per pound?

7. A 1-lb, 2-oz package of Mahi Mahi fish costs \$3.49. Find the unit price in cents per ounce. Round to the nearest hundredth of a cent.

8. Which has the lower unit price?

ORANGE JUICE
Brand A: \$1.19 for 12 oz Brand B: \$1.33 for 16 oz

Determine whether the two pairs of numbers are proportional.

9. 7, 8 and 63, 72

10. 1.3, 3.4 and 5.6, 15.2

Solve.

11. $\frac{9}{4} = \frac{27}{x}$

12. $\frac{150}{2.5} = \frac{x}{6}$

1. ______
2. ______
3. ______
4. ______
5. ______
6. ______
7. ______
8. ______
9. ______
10. ______
11. ______
12. ______

Answers

13. ______

14. ______

15. ______

16. ______

17. ______

18. ______

19. ______

20. ______

21. ______

22. ______

23. ______

Solve.

13. A woman traveled 432 km in 12 hr. At this rate, how far would she travel in 42 hr?

14. If 2 cans of apricots cost $3.39, how many cans of apricots can you buy for $74.58?

15. *Time Loss.* A watch loses 2 min in 10 hr. At this rate, how much will it lose in 24 hr?

16. *Map Scaling.* On a map, 3 in. represents 225 mi. If two cities are 7 in. apart on the map, how far apart are they in reality?

17. *Tower Height.* A birdhouse built on a pole that is 3 m high casts a shadow 5 m long. At the same time, the shadow of a tower is 110 m long. How high is the tower?

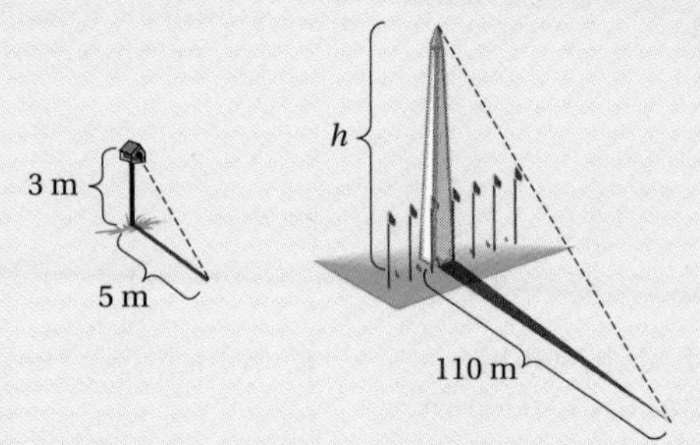

Skill Maintenance

18. In a recent year, Kellogg sold 146.2 million lb of Corn Flakes and 120.4 million lb of Frosted Flakes. How many more pounds of Corn Flakes did they sell than Frosted Flakes?

19. Use = or ≠ for ■ to write a true sentence:

$$\frac{6}{5} \;\blacksquare\; \frac{11}{9}.$$

20. Multiply:

$$\begin{array}{r} 2\,3\,4.1\,1 \\ \times \quad\quad 7\,4 \\ \hline \end{array}$$

21. Divide: $\dfrac{99.44}{100}$.

Synthesis

22. *NBA Score.* The score at the end of the first quarter of an NBA basketball game is the New York Knickerbockers 28, the Portland Trail Blazers 23. Assuming the teams continue to score at this rate, what will be the final score?

23. Nancy Morano-Smith wants to win a season football ticket from the local bookstore. Her goal is to guess the number of marbles in an 8-gal jar. She knows that there are 128 oz in a gallon. She goes home and fills an 8-oz jar with 46 marbles. How many marbles should she guess are in the jar?

Cumulative Review: Chapters 1–5

Add and simplify.

1.
$$\begin{array}{r} 27.68 \\ 3.019 \\ +\ 483.297 \\ \hline \end{array}$$

2.
$$\begin{array}{r} 2\frac{1}{3} \\ +\ 4\frac{5}{12} \\ \hline \end{array}$$

3. $\dfrac{6}{35} + \dfrac{5}{28}$

Subtract and simplify.

4.
$$\begin{array}{r} 40.2 \\ -\ 9.709 \\ \hline \end{array}$$

5. $73.82 - 0.908$

6. $\dfrac{4}{15} - \dfrac{3}{20}$

Multiply and simplify.

7.
$$\begin{array}{r} 37.64 \\ \times\ 5.9 \\ \hline \end{array}$$

8. 5.678×100

9. $2\dfrac{1}{3} \cdot 1\dfrac{2}{7}$

Divide and simplify.

10. $2.3\overline{)\,98.9}$

11. $54\overline{)\,48{,}546}$

12. $\dfrac{7}{11} \div \dfrac{14}{33}$

13. Write expanded notation: 30,074.

14. Write a word name for 120.07.

Which number is larger?

15. 0.7, 0.698

16. 0.799, 0.8

17. Find the prime factorization of 144.

18. Find the LCM of 28 and 35.

19. What part is shaded?

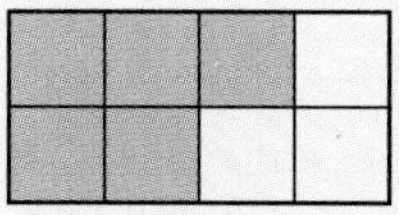

20. Simplify: $\dfrac{90}{144}$.

Calculate.

21. $\dfrac{3}{5} \times 9.53$

22. $\dfrac{1}{3} \times 0.645 - \dfrac{3}{4} \times 0.048$

23. Write fractional notation for the ratio 0.3 to 15.

24. Determine whether the pairs 3, 9 and 25, 75 are proportional.

25. What is the rate in meters per second?

660 meters, 12 seconds

26. A 14-oz jar of applesauce costs \$0.39. A 30-oz jar of applesauce costs \$0.99. Which has the lower unit price?

Solve.

27. $\frac{14}{25} = \frac{x}{54}$

28. $423 = 16 \cdot t$

29. $\frac{2}{3} \cdot y = \frac{16}{27}$

30. $\frac{7}{16} = \frac{56}{x}$

31. $34.56 + n = 67.9$

32. $t + \frac{7}{25} = \frac{5}{7}$

Solve.

33. A particular kind of fettucini alfredo has 520 calories in 1 cup. How many calories are there in $\frac{3}{4}$ cup?

34. *Hotel Rooms.* In a recent year, the total number of hotel rooms operated by the top 200 hotel companies in the world increased from 3,855,969 to 4,034,635 (***Source***: *Hotels,* July 1997). Find the increase in the number of rooms.

35. *Greyhound Mileage.* A Greyhound tour bus traveled 347.6 mi, 249.8 mi, and 379.5 mi on three separate trips. What was the total mileage of the bus?

36. A machine can stamp out 925 washers in 5 min. The company owning the machine needs 1295 washers by the end of the morning. How long will it take to stamp them out?

37. A 46-oz juice can contains $5\frac{3}{4}$ cups of juice. A recipe calls for $3\frac{1}{2}$ cups of juice. How many cups are left over?

38. It takes a carpenter $\frac{2}{3}$ hr to hang a door. How many doors can the carpenter hang in 8 hr?

39. A car travels 337.62 mi in 8 hr. How far does it travel in 1 hr?

40. *Shuttle Orbits.* A recent space shuttle made 16 orbits a day during an 8.25-day mission. How many orbits were made during the entire mission?

Synthesis

41. Find the ratio of 1.25 to $1\frac{3}{8}$ and simplify.

42. A 12-oz bag of shredded mozzarella cheese costs \$2.07. Blocks of mozzarella cheese are sold for \$2.79 per pound. Which is the better buy?

6

Percent Notation

Introduction

This chapter introduces a new kind of notation for numbers. We will see that $\frac{3}{8}$, 0.375, and 37.5% are all names for the same number. Then we will use percent notation and equations to solve applied problems.

An Application

In a treadmill test, a doctor's goal is to get the patient to reach his or her *maximum heart rate,* in beats per minute, which is found by subtracting the patient's age from 220 and taking 85% of the result. The author of this text took such a test at age 55. What was his maximum heart rate?

This problem appears as Exercise 11 in Section 6.5.

The Mathematics

We let $x =$ the maximum heart rate. This problem translates to the equation

$$x = \underbrace{85\%}_{\uparrow} \cdot (220 - 55).$$

This is percent notation.

For more information, visit us at www.mathmax.com

Pretest: Chapter 6

1. Find decimal notation for 87%.

2. Find percent notation for 0.537.

3. Find percent notation for $\frac{3}{4}$.

4. Find fractional notation for 37%.

5. Translate to an equation. Then solve.

   What is 60% of 75?

6. Translate to a proportion. Then solve.

   What percent of 50 is 35?

Solve.

7. *Weight of Muscles.* The weight of muscles in a human body is 40% of total body weight. A person weighs 225 lb. What do the muscles weigh?

8. The population of a town increased from 3000 to 3600. Find the percent of increase in population.

9. *Tax Rate in Massachusetts.* The sales tax rate in Massachusetts is 5%. How much tax is charged on a purchase of $286? What is the total price?

10. A salesperson's commission rate is 28%. What is the commission from the sale of $18,400 worth of merchandise?

11. The marked price of a stereo is $450. The stereo is on sale at Lowland Appliances for 25% off. What are the discount and the sale price?

12. What is the simple interest on $1200 principal at the interest rate of 8.3% for 1 year?

13. What is the simple interest on $500 at 8% for $\frac{1}{2}$ year?

14. Interest is compounded annually. Find the amount in an account if $6000 is invested at 9% for 2 years.

15. The Beechers invest $7500 in an investment account paying 6%, compounded monthly. How much is in the account after 5 months?

Objectives for Retesting

The objectives to be tested in addition to the material in this chapter are as follows.

[3.4b]	Convert from fractional notation to mixed numerals.
[4.4b]	Solve equations of the type $a \cdot x = b$, where a and b may be in decimal notation.
[4.1c], [4.5a]	Convert from fractional notation to decimal notation.
[5.3b]	Solve proportions.

6.1 Percent Notation

a Understanding Percent Notation

Of all wood harvested, 35% of it is used for paper production. What does this mean? It means that, on average, of every 100 tons of wood harvested, 35 tons is used to produce paper. Thus, 35% is a ratio of 35 to 100, or $\frac{35}{100}$.

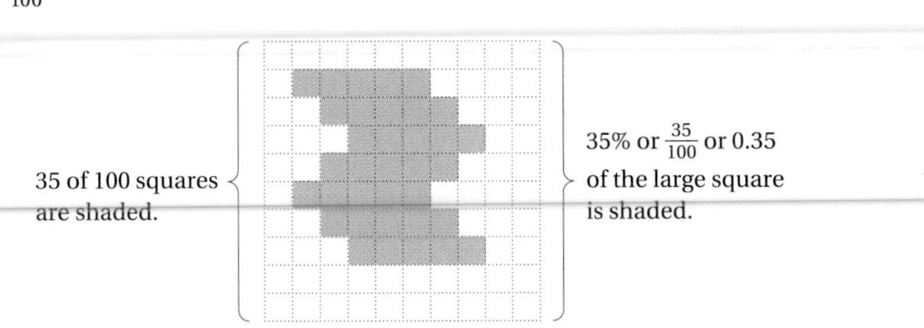

Percent notation is used extensively in our lives. Here are some examples:

Astronauts lose 1% of their bone mass for each month of weightlessness.

95% of hair spray is alcohol.

55% of all baseball merchandise sold is purchased by women.

62.4% of all aluminum cans were recycled in a recent year.

56% of all fruit juice purchased is orange juice.

45.8% of us sleep between 7 and 8 hours per night.

74% of the times a major-league baseball player strikes out swinging, the pitch was out of the strike zone.

Percent notation is often represented by pie charts to show how the parts of a quantity are related. For example, the chart below relates the amounts of different kinds of juices that are sold.

Juices Sold

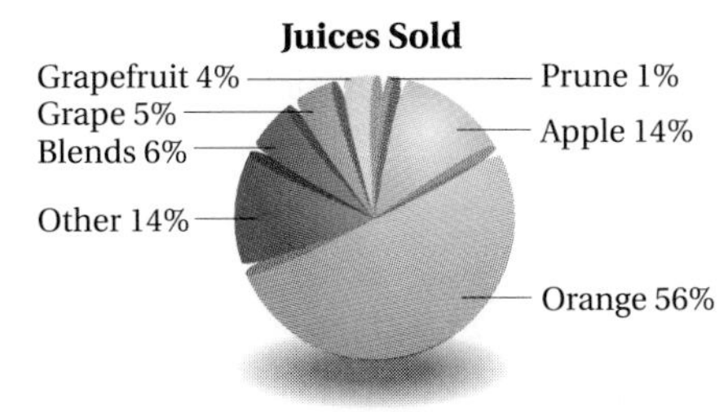

Source: Beverage Marketing Corporation

> The notation ***n*%** means "n per hundred."

Objectives

a. Write three kinds of notation for a percent.

b. Convert from percent notation to decimal notation.

c. Convert from decimal notation to percent notation.

For Extra Help

TAPE 10 | TAPE 9B | MAC WIN | CD-ROM

The number of jobs for professional chefs will increase by 43% from the year 1992 to the year 2005.

Write three kinds of notation as in Examples 1 and 2.

1. 70%

2. 23.4%

3. 100%

It is thought that the Roman Emperor Augustus began percent notation by taxing goods sold at a rate of $\frac{1}{100}$. In time, the symbol "%" evolved by interchanging the parts of the symbol "100" to "0/0" and then to "%".

Answers on page A-12

This definition leads us to the following equivalent ways of defining percent notation.

> **Percent notation, n%,** is defined using:
>
> ratio → $n\% = \text{the ratio of } n \text{ to } 100 = \frac{n}{100}$;
>
> fractional notation → $n\% = n \times \frac{1}{100}$;
>
> decimal notation → $n\% = n \times 0.01$.

Example 1 Write three kinds of notation for 35%.

Using ratio: $35\% = \frac{35}{100}$ **A ratio of 35 to 100**

Using fractional notation: $35\% = 35 \times \frac{1}{100}$ **Replacing % with $\times \frac{1}{100}$**

Using decimal notation: $35\% = 35 \times 0.01$ **Replacing % with × 0.01**

Example 2 Write three kinds of notation for 67.8%.

Using ratio: $67.8\% = \frac{67.8}{100}$ **A ratio of 67.8 to 100**

Using fractional notation: $67.8\% = 67.8 \times \frac{1}{100}$ **Replacing % with $\times \frac{1}{100}$**

Using decimal notation: $67.8\% = 67.8 \times 0.01$ **Replacing % with × 0.01**

Do Exercises 1–3.

b Converting from Percent Notation to Decimal Notation

Consider 78%. To convert to decimal notation, we can think of percent notation as a ratio and write

$78\% = \frac{78}{100}$ **Using the definition of percent as a ratio**

$= 0.78.$ **Converting to decimal notation**

Similarly,

$4.9\% = \frac{4.9}{100}$ **Using the definition of percent as a ratio**

$= 0.049.$ **Converting to decimal notation**

We could also convert 78% to decimal notation by replacing "%" with "× 0.01" and write

$78\% = 78 \times 0.01$ **Replacing % with × 0.01**

$= 0.78.$ **Multiplying**

Similarly,

$4.9\% = 4.9 \times 0.01$ **Replacing % with × 0.01**

$= 0.049.$ **Multiplying**

Dividing by 100 amounts to moving the decimal point two places to the left, which is the same as multiplying by 0.01. This leads us to a quick way to convert from percent notation to decimal notation: We drop the percent symbol and move the decimal point two places to the left.

To convert from percent notation to decimal notation,	36.5%	
a) replace the percent symbol % with $\times$ 0.01, and	36.5 $\times$ 0.01	
b) multiply by 0.01, which means move the decimal point two places to the left.	0.36.5	Move 2 places to the left.
	36.5% = 0.365	

Example 3 Find decimal notation for 99.44%.

a) Replace the percent symbol with $\times$ 0.01. 99.44 $\times$ 0.01

b) Move the decimal point two places to the left. 0.99.44

Thus, 99.44% = 0.9944.

Example 4 The interest rate on a $2\frac{1}{2}$-year certificate of deposit is 5.1%. Find decimal notation for 5.1%.

a) Replace the percent symbol with $\times$ 0.01. 5.1 $\times$ 0.01

b) Move the decimal point two places to the left. 0.05.1

Thus, 5.1% = 0.051.

Do Exercises 4–7.

c Converting from Decimal Notation to Percent Notation

To convert 0.38 to percent notation, we can first write fractional notation, as follows:

$0.38 = \frac{38}{100}$ **Converting to fractional notation**

$= 38\%.$ **Using the definition of percent as a ratio**

Note that 100% = 100 $\times$ 0.01 = 1. Thus to convert 0.38 to percent notation, we can multiply by 1, using 100% as a symbol for 1. Then

$0.38 = 0.38 \times 1$

$= 0.38 \times 100\%$

$= 0.38 \times 100 \times 0.01$

$= 38 \times 0.01$

$= 38\%.$ **Replacing "$\times$ 0.01" with the % symbol**

Even more quickly, since 0.38 = 0.38 $\times$ 100%, we can simply multiply 0.38 by 100 and write the % symbol.

Find decimal notation.

4. 34%

5. 78.9%

Find decimal notation for the percent notation in the sentence.

6. People forget 83% of all names they learn.

7. Soft drink sales in the United States have grown 4.2% annually over the past decade.

Answers on page A-12

Find percent notation.

8. 0.24

9. 3.47

10. 1

Find percent notation for the decimal notation in the sentence.

11. Blood is 0.9 water.

12. Of those accidents requiring medical attention, 0.108 of them occur on roads.

Answers on page A-12

To convert from decimal notation to percent notation, multiply by 100%—that is, move the decimal point two places to the right and write a percent symbol.

To convert from decimal notation to percent notation, multiply by 100%. That is,	0.675 = 0.675 × 100%	
a) move the decimal point two places to the right, and	0.67.5	Move 2 places to the right.
b) write a % symbol.	67.5%	
	0.675 = 67.5%	

Example 5 Find percent notation for 1.27.

a) Move the decimal point two places to the right. 1.27.

b) Write a % symbol. 127%

Thus, 1.27 = 127%.

Example 6 Television sets are on 0.25 of the time. Find percent notation for 0.25.

a) Move the decimal point two places to the right. 0.25.

b) Write a % symbol. 25%

Thus, 0.25 = 25%.

Example 7 Find percent notation for 5.6.

a) Move the decimal point two places to the right, adding an extra zero. 5.60.

b) Write a % symbol. 560%

Thus, 5.6 = 560%.

Do Exercises 8–12.

Exercise Set 6.1

a Write three kinds of notation as in Examples 1 and 2 on p. 304.

1. 90%
2. 58.7%
3. 12.5%
4. 130%

b Find decimal notation.

5. 67%
6. 17%
7. 45.6%
8. 76.3%
9. 59.01%
10. 30.02%
11. 10%
12. 40%
13. 1%
14. 100%
15. 200%
16. 300%
17. 0.1%
18. 0.4%
19. 0.09%
20. 0.12%
21. 0.18%
22. 5.5%
23. 23.19%
24. 87.99%

Find decimal notation for the percent notation in the sentence.

25. On average, about 40% of the body weight of an adult male is muscle.

26. On average, about 23% of the body weight of an adult female is muscle.

27. A person's brain is 2.5% of his or her body weight.

28. It is known that 16% of all dessert orders in restaurants is for pie.

29. It is known that 62.2% of us think Monday is the worst day of the week.

30. Of all 18-year-olds, 68.4% have a driver's license.

c Find percent notation.

31. 0.47 **32.** 0.87 **33.** 0.03 **34.** 0.01 **35.** 8.7

36. 4 **37.** 0.334 **38.** 0.889 **39.** 0.75 **40.** 0.99

41. 0.4 **42.** 0.5 **43.** 0.006 **44.** 0.008 **45.** 0.017

46. 0.024 **47.** 0.2718 **48.** 0.8911 **49.** 0.0239 **50.** 0.00073

Find percent notation for the decimal notation in the sentence.

51. Around the fourth of July, about 0.000104 of all children aged 15 to 19 suffer injuries from fireworks.

52. About 0.144 of all children are cared for by relatives.

53. It is known that 0.24 of us go to the movies once a month.

54. It is known that 0.458 of us sleep between 7 and 8 hours.

55. Of all CDs purchased, 0.581 of them are pop/rock.

56. About 0.026 of all college football players go on to play professional football.

Skill Maintenance

Convert to a mixed numeral. [3.4b]

57. $\frac{100}{3}$ **58.** $\frac{75}{2}$ **59.** $\frac{75}{8}$ **60.** $\frac{297}{16}$

Convert to decimal notation. [4.5a]

61. $\frac{2}{3}$ **62.** $\frac{1}{3}$ **63.** $\frac{5}{6}$ **64.** $\frac{17}{12}$

Synthesis

65. ◆ 🖩 What would you do to an entry on a calculator in order to get percent notation? Explain.

66. ◆ 🖩 What would you do to percent notation on a calculator in order to get decimal notation? Explain.

6.2 Percent Notation and Fractional Notation

Objectives

a Convert from fractional notation to percent notation.

b Convert from percent notation to fractional notation.

For Extra Help

TAPE 10

TAPE 10A

MAC WIN

CD-ROM

a Converting from Fractional Notation to Percent Notation

Consider the fractional notation $\frac{7}{8}$. To convert to percent notation, we use two skills we already have. We first find decimal notation by dividing:

$$\frac{7}{8} = 0.875 \qquad \begin{array}{r} 0.875 \\ 8\overline{)7.000} \\ \underline{6\,4} \\ 60 \\ \underline{56} \\ 40 \\ \underline{40} \\ 0 \end{array}$$

Then we convert the decimal notation to percent notation. We move the decimal point two places to the right

0.87.5

and write a % symbol:

$$\frac{7}{8} = 87.5\%, \text{ or } 87\frac{1}{2}\%.$$

To convert from fractional notation to percent notation,	$\frac{3}{5}$ Fractional notation
a) find decimal notation by division, and	$\begin{array}{r} 0.6 \\ 5\overline{)3.0} \\ \underline{3\,0} \\ 0 \end{array}$
b) convert the decimal notation to percent notation.	$0.6 = 0.60 = 60\%$ Percent notation $\frac{3}{5} = 60\%$

Example 1 Find percent notation for $\frac{3}{8}$.

a) Find decimal notation by division.

$$\begin{array}{r} 0.375 \\ 8\overline{)3.000} \\ \underline{2\,4} \\ 60 \\ \underline{56} \\ 40 \\ \underline{40} \\ 0 \end{array} \qquad \frac{3}{8} = 0.375$$

Calculator Spotlight

Conversion. Calculators are often used when we are converting fractional notation to percent notation. We simply perform the division on the calculator and then convert the decimal notation to percent notation. For example, percent notation for $\frac{17}{40}$ can be found by pressing

[1] [7] [÷] [4] [0] [=]

and then converting the result, 0.425, to percent notation, 42.5%.

Exercises

Find percent notation. Round to the nearest hundredth of a percent.

1. $\frac{13}{25}$

2. $\frac{5}{13}$

3. $\frac{42}{39}$

4. $\frac{12}{7}$

5. $\frac{217}{364}$

6. $\frac{2378}{8401}$

Find percent notation.

1. $\frac{1}{4}$

2. $\frac{5}{8}$

3. The human body is $\frac{2}{3}$ water. Find percent notation for $\frac{2}{3}$.

4. Find percent notation: $\frac{5}{6}$.

Find percent notation.

5. $\frac{57}{100}$

6. $\frac{19}{25}$

Answers on page A-12

b) Convert the decimal notation to percent notation. Move the decimal point two places to the right, and write a % symbol.

$$0.37.5$$

$$\frac{3}{8} = 37.5\%, \text{ or } 37\frac{1}{2}\%$$

Don't forget the % symbol.

Do Exercises 1 and 2.

Example 2 Of all meals, $\frac{1}{3}$ are eaten outside the home. Find percent notation for $\frac{1}{3}$.

a) Find decimal notation by division.

$$\begin{array}{r} 0.3\,3\,3 \\ 3\,)\overline{1.0\,0\,0} \\ \underline{9} \\ 1\,0 \\ \underline{9} \\ 1\,0 \\ \underline{9} \\ 1 \end{array}$$

We get a repeating decimal: $0.33\overline{3}$.

b) Convert the answer to percent notation.

$$0.33.\overline{3}$$

$$\frac{1}{3} = 33.\overline{3}\%, \text{ or } 33\frac{1}{3}\%$$

Do Exercises 3 and 4.

In some cases, division is not the fastest way to convert. The following are some optional ways in which conversion might be done.

Example 3 Find percent notation for $\frac{69}{100}$.

We use the definition of percent as a ratio.

$$\frac{69}{100} = 69\%$$

Example 4 Find percent notation for $\frac{17}{20}$.

We multiply by 1 to get 100 in the denominator. We think of what we have to multiply 20 by in order to get 100. That number is 5, so we multiply by 1 using $\frac{5}{5}$.

$$\frac{17}{20} \cdot \frac{5}{5} = \frac{85}{100} = 85\%$$

Note that this shortcut works only when the denominator is a factor of 100.

Do Exercises 5 and 6.

b Converting from Percent Notation to Fractional Notation

To convert from percent notation to fractional notation,	30%	Percent notation
a) use the definition of percent as a ratio, and	$\frac{30}{100}$	
b) simplify, if possible.	$\frac{3}{10}$	Fractional notation

Example 5 Find fractional notation for 75%.

$$75\% = \frac{75}{100} \quad \text{Using the definition of percent}$$

$$= \frac{3 \cdot 25}{4 \cdot 25} = \frac{3}{4} \cdot \frac{25}{25}$$

$$= \frac{3}{4} \quad \text{Simplifying}$$

Example 6 Find fractional notation for 62.5%.

$$62.5\% = \frac{62.5}{100} \quad \text{Using the definition of percent}$$

$$= \frac{62.5}{100} \times \frac{10}{10} \quad \text{Multiplying by 1 to eliminate the decimal point in the numerator}$$

$$= \frac{625}{1000}$$

$$= \frac{5 \cdot 125}{8 \cdot 125} = \frac{5}{8} \cdot \frac{125}{125}$$

$$= \frac{5}{8} \quad \text{Simplifying}$$

Example 7 Find fractional notation for $16\frac{2}{3}\%$.

$$16\frac{2}{3}\% = \frac{50}{3}\% \quad \text{Converting from the mixed numeral to fractional notation}$$

$$= \frac{50}{3} \times \frac{1}{100} \quad \text{Using the definition of percent}$$

$$= \frac{50 \cdot 1}{3 \cdot 50 \cdot 2} = \frac{1}{6} \cdot \frac{50}{50}$$

$$= \frac{1}{6} \quad \text{Simplifying}$$

Do Exercises 7–10.

The table on the inside front cover lists decimal, fractional, and percent equivalents used so often that it would speed up your work if you learned them. For example, $\frac{1}{3} = 0.\overline{3}$, so we say that the **decimal equivalent** of $\frac{1}{3}$ is $0.\overline{3}$, or that $0.\overline{3}$ has the **fractional equivalent** $\frac{1}{3}$.

Find fractional notation.

7. 60%

8. 3.25%

9. $66\frac{2}{3}\%$

10. Complete this table.

Fractional Notation	$\frac{1}{5}$		
Decimal Notation		$0.83\overline{3}$	
Percent Notation			$37\frac{1}{2}\%$

Answers on page A-12

Calculator Spotlight

Applications of Ratio and Percent: The Price–Earnings Ratio and Stock Yields

The Price–Earnings Ratio

If the total earnings of a company one year were $5,000,000 and 100,000 shares of stock were issued, the earnings per share was $50. At one time, the price per share of Coca-Cola was $\$48\frac{5}{8}$ and the earnings per share was $1.35. The **price–earnings ratio, *P*/*E*,** is the price of the stock divided by the earnings per share. For the Coca-Cola stock, the price–earnings ratio, *P*/*E*, is given by

$$\frac{P}{E} = \frac{48\frac{5}{8}}{1.35}$$

$$= \frac{48.625}{1.35} \quad \textbf{Converting to decimal notation}$$

$$\approx 36.02. \quad \textbf{Dividing, using a calculator, and rounding to the nearest tenth}$$

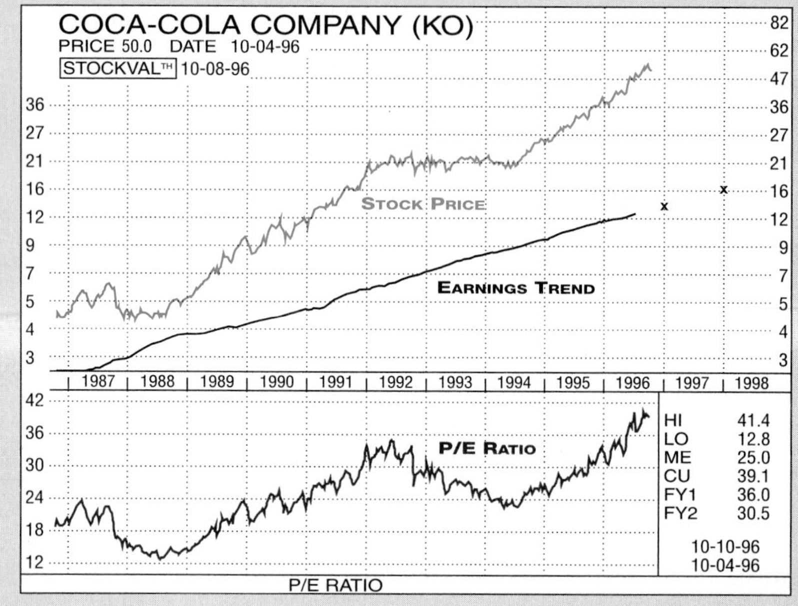

Source: C. H. Dean

Stock Yields

At one time, the price per share of Coca-Cola stock was $\$48\frac{5}{8}$ and the company was paying a yearly dividend of $0.50 per share. It is helpful to those interested in stocks to know what percent the dividend is of the price of the stock. The percent is called the **yield**. For the Coca-Cola stock, the yield is given by

$$\text{Yield} = \frac{\text{Dividend}}{\text{Price per share}}$$

$$= \frac{0.50}{48\frac{5}{8}}$$

$$= \frac{0.50}{48.625} \quad \textbf{Converting to decimal notation}$$

$$\approx 0.0103 \quad \textbf{Dividing and rounding to the nearest thousandth}$$

$$\approx 1.03\%. \quad \textbf{Converting to percent notation}$$

Exercises

Compute the price–earnings ratio and the yield for each stock.

Stock	*Price per Share*	*Earnings*	*Dividend*
1. Monsanto	$\$35\frac{3}{4}$	$1.41	$0.60
2. K-Mart	$10\frac{7}{8}$	0.91	0.00
3. Rubbermaid	24	0.42	0.60
4. AT&T	$38\frac{5}{8}$	0.98	1.32

Exercise Set 6.2

a Find percent notation.

1. $\frac{41}{100}$

2. $\frac{36}{100}$

3. $\frac{5}{100}$

4. $\frac{1}{100}$

5. $\frac{2}{10}$

6. $\frac{7}{10}$

7. $\frac{3}{10}$

8. $\frac{9}{10}$

9. $\frac{1}{2}$

10. $\frac{3}{4}$

11. $\frac{5}{8}$

12. $\frac{1}{8}$

13. $\frac{4}{5}$

14. $\frac{2}{5}$

15. $\frac{2}{3}$

16. $\frac{1}{3}$

17. $\frac{1}{6}$

18. $\frac{5}{6}$

19. $\frac{4}{25}$

20. $\frac{17}{25}$

21. $\frac{1}{20}$

22. $\frac{31}{50}$

23. $\frac{17}{50}$

24. $\frac{3}{20}$

Find percent notation for the fractional notation in the sentence.

25. Bread is $\frac{9}{25}$ water.

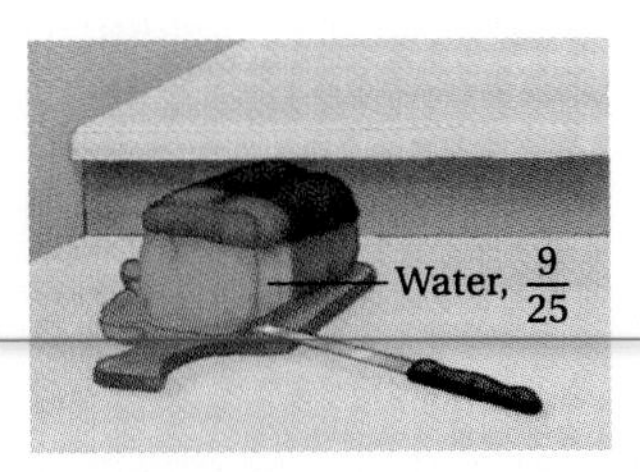

26. Milk is $\frac{7}{8}$ water.

Write percent notation for the fractions in this pie chart.

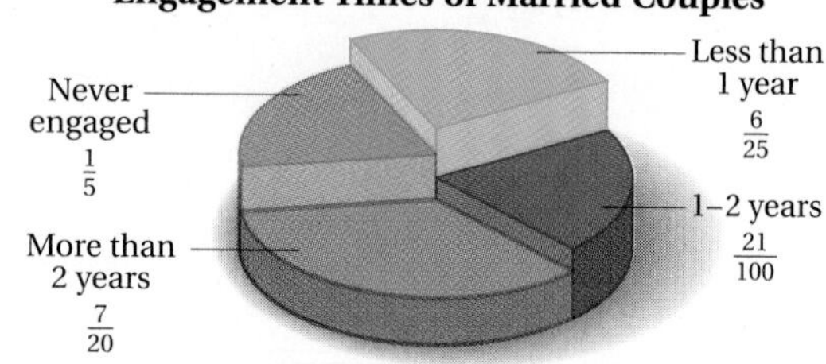

27. $\frac{21}{100}$

28. $\frac{1}{5}$

29. $\frac{6}{25}$

30. $\frac{7}{20}$

b Find fractional notation. Simplify.

31. 85%

32. 55%

33. 62.5%

34. 12.5%

35. $33\frac{1}{3}\%$

36. $83\frac{1}{3}\%$

37. $16.\overline{6}\%$

38. $66.\overline{6}\%$

39. 7.25%

40. 4.85%

41. 0.8%

42. 0.2%

43. $25\frac{3}{8}\%$

44. $48\frac{7}{8}\%$

45. $78\frac{2}{9}\%$

46. $16\frac{5}{9}\%$

47. $64\frac{7}{11}\%$

48. $73\frac{3}{11}\%$

49. 150%

50. 110%

51. 0.0325%

52. 0.419%

53. $33.\overline{3}\%$

54. $83.\overline{3}\%$

Find fractional notation for the percents in this bar graph.

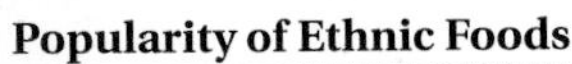

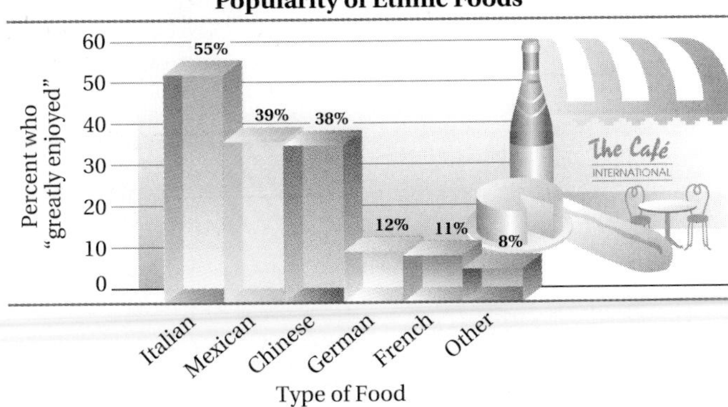

Note that there can be multiple responses.

55. 55%

56. 39%

57. 38%

58. 12%

59. 11%

60. 8%

Find fractional notation for the percent notation in the sentence.

61. A 30-g cup of Wheaties supplies 25% of the minimum daily requirement of vitamin A.

62. A 30-g cup of Wheaties supplies 45% of the minimum daily requirement of iron.

63. The interest rate on a 5-yr certificate of deposit was 5.69%.

64. One year the sales of *USA Today* increased 7.2%.

Complete the table.

65.

Fractional Notation	Decimal Notation	Percent Notation
$\frac{1}{8}$		$12\frac{1}{2}\%$, or 12.5%
$\frac{1}{6}$		
		20%
	0.25	
		$33\frac{1}{3}\%$, or $33.\overline{3}\%$
		$37\frac{1}{2}\%$, or 37.5%
		40%
$\frac{1}{2}$	0.5	50%

66.

Fractional Notation	Decimal Notation	Percent Notation
$\frac{3}{5}$		
	0.625	
$\frac{2}{3}$		
	0.75	75%
$\frac{4}{5}$		
$\frac{5}{6}$		$83\frac{1}{3}\%$, or $83.\overline{3}\%$
$\frac{7}{8}$		$87\frac{1}{2}\%$, or 87.5%
		100%

67.

Fractional Notation	Decimal Notation	Percent Notation
	0.5	
$\frac{1}{3}$		
		25%
		$16\frac{2}{3}\%$, or $16.\overline{6}\%$
	0.125	
$\frac{3}{4}$		
	$0.8\overline{3}$	
$\frac{3}{8}$		

68.

Fractional Notation	Decimal Notation	Percent Notation
		40%
		$62\frac{1}{2}\%$, or 62.5%
	0.875	
$\frac{1}{1}$		
	0.6	
	$0.\overline{6}$	
$\frac{1}{5}$		

Skill Maintenance

Solve.

69. $13 \cdot x = 910$ [1.7a]

70. $15 \cdot y = 75$ [1.7a]

71. $0.05 \times b = 20$ [4.4b]

72. $3 = 0.16 \times b$ [4.4b]

73. $\frac{24}{37} = \frac{15}{x}$ [5.3b]

74. $\frac{17}{18} = \frac{x}{27}$ [5.3b]

Convert to a mixed numeral. [3.4b]

75. $\frac{100}{3}$

76. $\frac{75}{2}$

77. $\frac{250}{3}$

78. $\frac{123}{6}$

79. $\frac{345}{8}$

80. $\frac{373}{6}$

81. $\frac{75}{4}$

82. $\frac{67}{9}$

Synthesis

83. ◆ Is it always best to convert from fractional notation to percent notation by first finding decimal notation? Why or why not?

84. ◆ Athletes sometimes speak of "giving 110%" effort. Does this make sense? Why or why not?

Find percent notation.

85. 🖩 $\frac{41}{369}$

86. 🖩 $\frac{54}{999}$

Find decimal notation.

87. $\frac{14}{9}\%$

88. $\frac{19}{12}\%$

Collaborative Learning Manual

Use percent squares to develop a number sense for percents.

6.3 Solving Percent Problems Using Equations

Objectives

a. Translate percent problems to equations.

b. Solve basic percent problems.

For Extra Help

TAPE 10

TAPE 10A

MAC WIN

CD-ROM

a Translating to Equations

To solve a problem involving percents, it is helpful to translate first to an equation.

Example 1 Translate:

23%	of	5	is	what?
↓	↓	↓	↓	↓
23%	$\cdot$	5	=	a

> "**Of**" translates to "$\cdot$", or "$\times$". "**Is**" translates to "=".
> "**What**" translates to any letter. **%** translates to "$\times \frac{1}{100}$" or "$\times$ 0.01".

Example 2 Translate:

What	is	11%	of	49?
↓	↓	↓	↓	↓
a	=	11%	$\cdot$	49

Any letter can be used.

Do Exercises 1 and 2.

Example 3 Translate:

3	is	10%	of	what?
↓	↓	↓	↓	↓
3	=	10%	$\cdot$	b

Example 4 Translate:

45%	of	what	is	23?
↓	↓	↓	↓	↓
45%	$\times$	b	=	23

Do Exercises 3 and 4.

Example 5 Translate:

10	is	what percent	of	20?
↓	↓	↓	↓	↓
10	=	n	$\times$	20

Example 6 Translate:

What percent	of	50	is	7?
↓	↓	↓	↓	↓
n	$\cdot$	50	=	7

Do Exercises 5 and 6.

Translate to an equation. Do not solve.

1. 12% of 50 is what?

2. What is 40% of 60?

Translate to an equation. Do not solve.

3. 45 is 20% of what?

4. 120% of what is 60?

Translate to an equation. Do not solve.

5. 16 is what percent of 40?

6. What percent of 84 is 10.5?

Answers on page A-13

7. Solve:

What is 12% of 50?

b Solving Percent Problems

In solving percent problems, we use the *Translate* and *Solve* steps in the problem-solving strategy used throughout this text.

Percent problems are actually of three different types. Although the method we present does *not* require that you be able to identify which type we are studying, it is helpful to know them.

We know that

15 is 25% of 60, or

$15 = 25\% \times 60.$

We can think of this as:

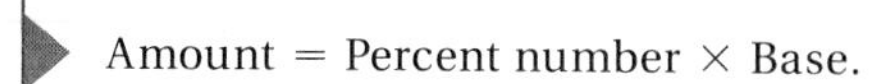

Amount = Percent number × Base.

Each of the three types of percent problems depends on which of the three pieces of information is missing.

1. Finding the *amount* (the result of taking the percent)

Example:	What	is	25%	of	60?
	↓	↓	↓	↓	↓
Translation:	y	=	25%	·	60

2. Finding the *base* (the number you are taking the percent of)

Example:	15	is	25%	of	what number?
	↓	↓	↓	↓	↓
Translation:	15	=	25%	·	y

3. Finding the *percent number* (the percent itself)

Example:	15	is	what percent	of	60?
	↓	↓	↓	↓	↓
Translation:	15	=	y	·	60

Finding the Amount

Example 7 What is 11% of 49?

Translate: $a = 11\% \times 49.$

Solve: The letter is by itself. To solve the equation, we just convert 11% to decimal notation and multiply.

$$\begin{array}{r} 4\,9 \\ \times\, 0.1\,1 \\ \hline 4\,9 \\ 4\,9\,0 \\ \hline a = 5.3\,9 \end{array} \qquad 11\% = 0.11$$

A way of checking answers is by estimating as follows:

$$11\% \times 49 \approx 10\% \times 50 = 0.10 \times 50 = 5.$$

Since 5 is close to 5.39, our answer is reasonable.

Thus, 5.39 is 11% of 49. The answer is 5.39.

Do Exercise 7.

Answer on page A-13

Example 8 120% of $42 is what?

Translate: $120\% \times 42 = a$.

Solve: The letter is by itself. To solve the equation, we carry out the calculation.

$$\begin{array}{r} 4\,2 \\ \times\ 1.2 \\ \hline 8\,4 \\ 4\,2\,0 \\ \hline a = 5\,0.4 \end{array} \qquad 120\% = 1.20 = 1.2$$

Thus, 120% of $42 is $50.40. The answer is $50.40.

Do Exercise 8.

Finding the Base

Example 9 5% of what is 20?

Translate: $5\% \times b = 20$.

Solve: This time the letter is *not* by itself. To solve the equation, we divide by 5% on both sides:

$$\frac{5\% \times b}{5\%} = \frac{20}{5\%} \qquad \text{Dividing by 5\% on both sides}$$

$$b = \frac{20}{0.05} \qquad 5\% = 0.05$$

$$= 400.$$

$$\begin{array}{r} 4\,0\,0. \\ 0.0\,5\,\overline{)\,2\,0.0\,0_{\wedge}} \\ 2\,0\,0\,0 \\ \hline 0 \end{array}$$

Thus, 5% of 400 is 20. The answer is 400.

Example 10 $3 is 16% of what?

Translate: $3 is 16% of what?

$$3 = 16\% \times b.$$

Solve: Again, the letter is not by itself. To solve the equation, we divide by 16% on both sides:

$$\frac{3}{16\%} = \frac{16\% \times b}{16\%} \qquad \text{Dividing by 16\% on both sides}$$

$$\frac{3}{0.16} = b \qquad 16\% = 0.16$$

$$18.75 = b.$$

$$\begin{array}{r} 1\,8.7\,5 \\ 0.1\,6\,\overline{)\,3.0\,0_{\wedge}0\,0} \\ 1\,6 \\ \hline 1\,4\,0 \\ 1\,2\,8 \\ \hline 1\,2\,0 \\ 1\,1\,2 \\ \hline 8\,0 \\ 8\,0 \\ \hline 0 \end{array}$$

Thus, $3 is 16% of $18.75. The answer is $18.75.

Do Exercises 9 and 10.

8. Solve:

64% of $55 is what?

Solve.

9. 20% of what is 45?

10. $60 is 120% of what?

Answers on page A-13

11. Solve:

16 is what percent of 40?

12. Solve:

What percent of $84 is $10.50?

Answers on page A-13

Finding the Percent Number

In solving these problems, you *must* remember to convert to percent notation after you have solved the equation.

Example 11 10 is what percent of 20?

Translate: 10 is what percent of 20?

$$10 = n \times 20.$$

Solve: To solve the equation, we divide by 20 on both sides and convert the result to percent notation:

$$n \cdot 20 = 10$$

$$\frac{n \cdot 20}{20} = \frac{10}{20} \qquad \text{Dividing by 20 on both sides}$$

$$n = 0.50 = 50\%. \qquad \text{Converting to percent notation}$$

Thus, 10 is 50% of 20. The answer is 50%.

Do Exercise 11.

Example 12 What percent of $50 is $16?

Translate: What percent of $50 is $16?

$$n \times 50 = 16.$$

Solve: To solve the equation, we divide by 50 on both sides and convert the answer to percent notation:

$$\frac{n \times 50}{50} = \frac{16}{50} \qquad \text{Dividing by 50 on both sides}$$

$$n = \frac{16}{50}$$

$$= \frac{16}{50} \cdot \frac{2}{2}$$

$$= \frac{32}{100}$$

$$= 32\%. \qquad \text{Converting to percent notation}$$

Thus, 32% of $50 is $16. The answer is 32%.

Do Exercise 12.

CAUTION! When a question asks "what percent?", be sure to give the answer in percent notation.

Exercise Set 6.3

a Translate to an equation. Do not solve.

1. What is 32% of 78?

2. 98% of 57 is what?

3. 89 is what percent of 99?

4. What percent of 25 is 8?

5. 13 is 25% of what?

6. 21.4% of what is 20?

b Solve.

7. What is 85% of 276?

8. What is 74% of 53?

9. 150% of 30 is what?

10. 100% of 13 is what?

11. What is 6% of $300?

12. What is 4% of $45?

13. 3.8% of 50 is what?

14. $33\frac{1}{3}\%$ of 480 is what? (*Hint:* $33\frac{1}{3}\% = \frac{1}{3}$.)

15. $39 is what percent of $50?

16. $16 is what percent of $90?

17. 20 is what percent of 10?

18. 60 is what percent of 20?

19. What percent of $300 is $150?

20. What percent of $50 is $40?

21. What percent of 80 is 100?

22. What percent of 60 is 15?

23. 20 is 50% of what?

24. 57 is 20% of what?

25. 40% of what is \$16?

26. 100% of what is \$74?

27. 56.32 is 64% of what?

28. 71.04 is 96% of what?

29. 70% of what is 14?

30. 70% of what is 35?

31. What is $62\frac{1}{2}\%$ of 10?

32. What is $35\frac{1}{4}\%$ of 1200?

33. What is 8.3% of \$10,200?

34. What is 9.2% of \$5600?

Skill Maintenance

Write fractional notation. [4.1b]

35. 0.09

36. 1.79

37. 0.875

38. 0.9375

Write decimal notation. [4.1c]

39. $\frac{89}{100}$

40. $\frac{7}{100}$

41. $\frac{3}{10}$

42. $\frac{17}{1000}$

Synthesis

43. ◆ Write a question that could be translated to the equation

$25 = 4\% \times b.$

44. ◆ To calculate a 15% tip on a \$24 bill, a customer adds \$2.40 and half of \$2.40, or \$1.20, to get \$3.60. Is this procedure valid? Why or why not?

Solve.

45. ▦ What is 7.75% of \$10,880?
Estimate ______________
Calculate ______________

46. ▦ 50,951.775 is what percent of 78,995?
Estimate ______________
Calculate ______________

47. *Recyclables.* It is estimated that 40% to 50% of all trash is recyclable. If a community produces 270 tons of trash, how much of their trash is recyclable?

48. 40% of $18\frac{3}{4}\%$ of \$25,000 is what?

6.4 Solving Percent Problems Using Proportions*

a Translating to Proportions

A percent is a ratio of some number to 100. For example, 75% is the ratio $\frac{75}{100}$. The numbers 3 and 4 have the same ratio as 75 and 100. Thus,

$$75\% = \frac{75}{100} = \frac{3}{4}.$$

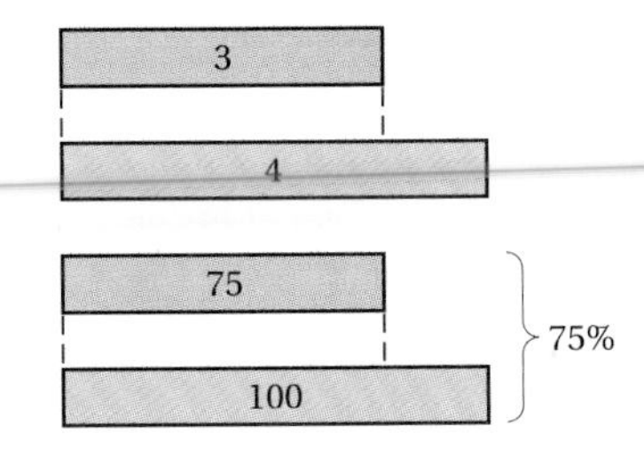

To solve a percent problem using a proportion, we translate as follows:

$$\begin{array}{l} \text{Number} \rightarrow \\ 100 \longrightarrow \end{array} \frac{N}{100} = \frac{a}{b} \begin{array}{l} \leftarrow \text{Amount} \\ \leftarrow \text{Base} \end{array}$$

You might find it helpful to read this as "part is to whole as part is to whole."

For example,

60% of 25 is 15

translates to

$$\frac{60}{100} = \frac{15}{25}. \begin{array}{l} \leftarrow \text{Amount} \\ \leftarrow \text{Base} \end{array}$$

A clue in translating is that the base, b, corresponds to 100 and usually follows the wording "percent of." Also, $N\%$ always translates to $N/100$. Another aid in translating is to make a comparison drawing. To do this, we start with the percent side and list 0% at the top and 100% near the bottom. Then we estimate where the specified percent—in this case, 60%—is located. The corresponding quantities are then filled in. The base—in this case, 25—always corresponds to 100% and the amount—in this case, 15—corresponds to the specified percent.

Percents	Quantities	Percents	Quantities	Percents	Quantities
0%	0	0%	0	0%	0
		60%		60%	15
100%		100%		100%	25

The proportion can then be read easily from the drawing.

Objectives

a Translate percent problems to proportions.

b Solve basic percent problems.

For Extra Help

TAPE 10 TAPE 10B MAC WIN CD-ROM

**Note*: This section presents an alternative method for solving basic percent problems. You can use either equations or proportions to solve percent problems, but you might prefer one method over the other, or your instructor may direct you to use one method over the other.*

Translate to a proportion. Do not solve.

1. 12% of 50 is what?

2. What is 40% of 60?

3. 130% of 72 is what?

Example 1 Translate to a proportion.

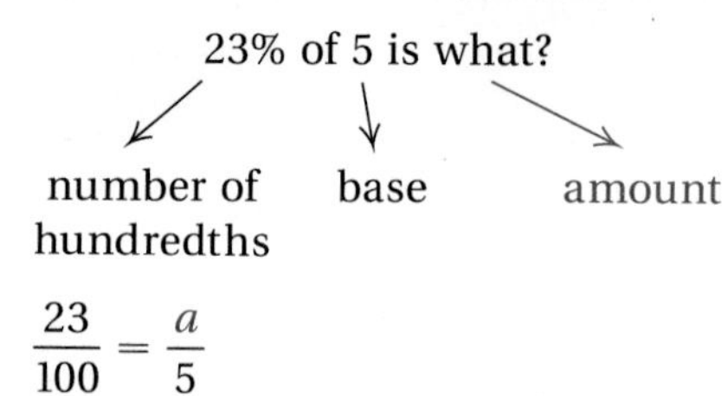

$$\frac{23}{100} = \frac{a}{5}$$

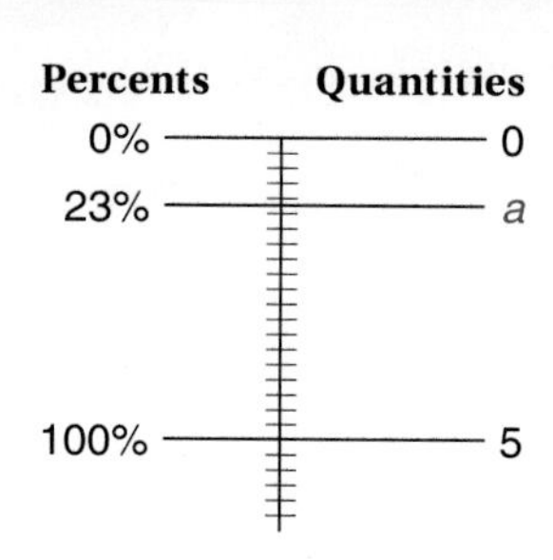

Example 2 Translate to a proportion.

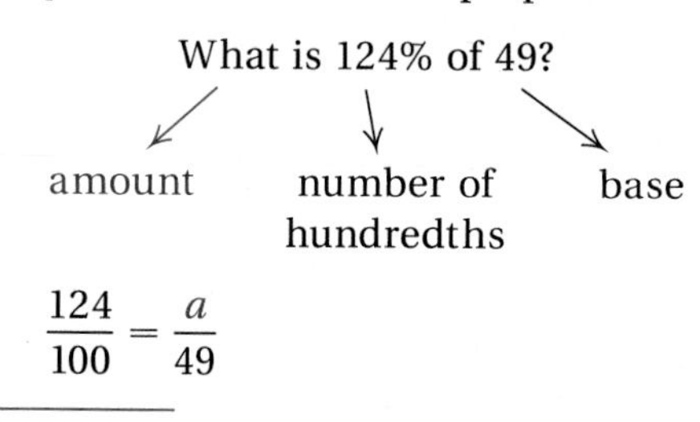

$$\frac{124}{100} = \frac{a}{49}$$

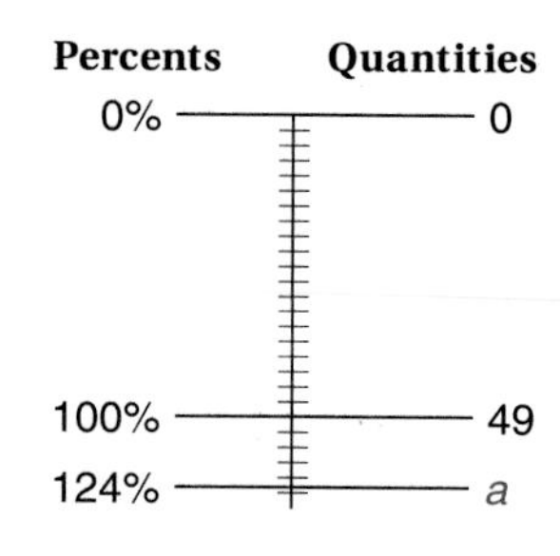

Do Exercises 1–3.

Example 3 Translate to a proportion.

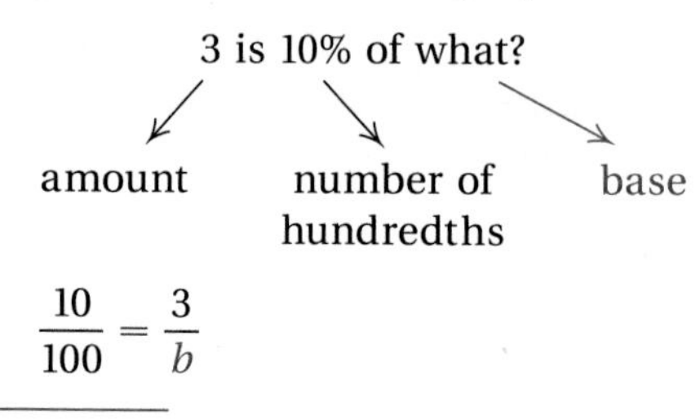

$$\frac{10}{100} = \frac{3}{b}$$

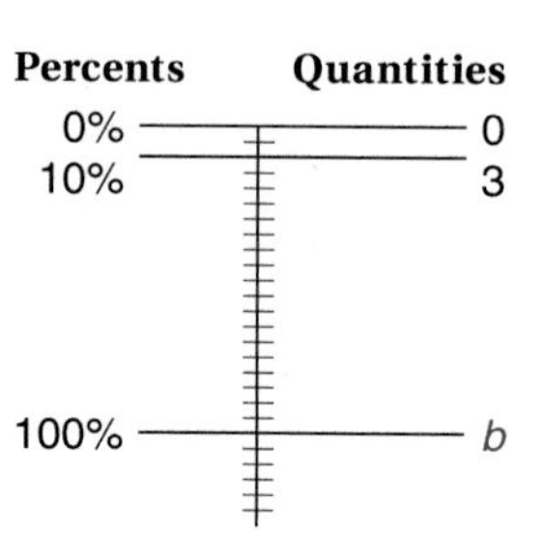

Translate to a proportion. Do not solve.

4. 45 is 20% of what?

5. 120% of what is 60?

Example 4 Translate to a proportion.

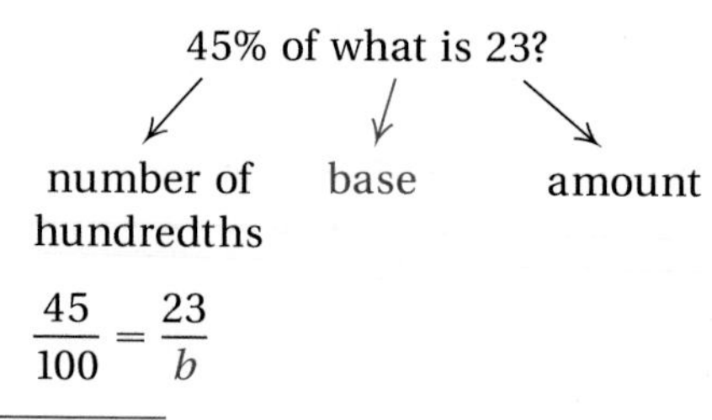

$$\frac{45}{100} = \frac{23}{b}$$

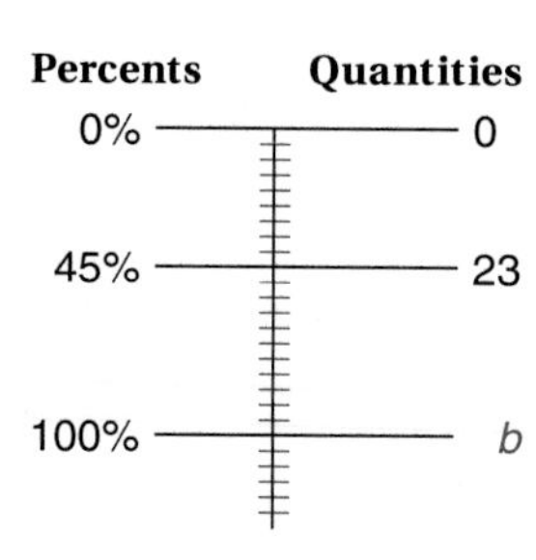

Do Exercises 4 and 5.

Example 5 Translate to a proportion.

10 is what percent of 20?

amount — number of hundredths — base

$$\frac{N}{100} = \frac{10}{20}$$

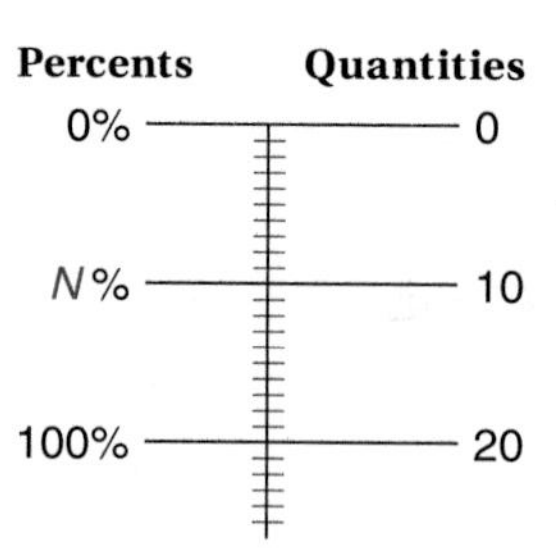

Answers on page A-13

Example 6 Translate to a proportion.

What percent of 50 is 7?

number of hundredths → What percent; base → 50; amount → 7

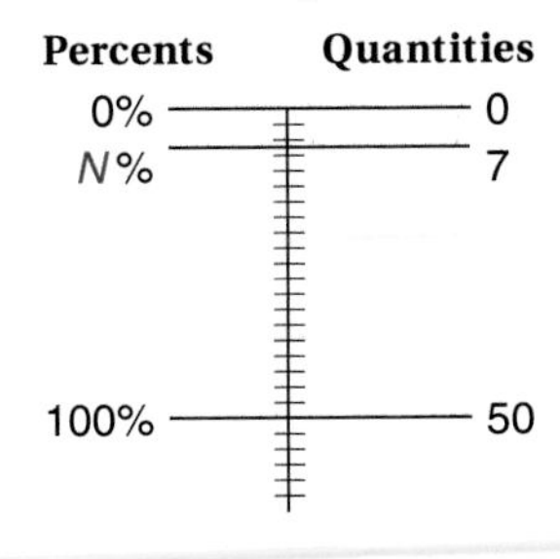

$$\frac{N}{100} = \frac{7}{50}$$

Do Exercises 6 and 7.

Translate to a proportion. Do not solve.

6. 16 is what percent of 40?

7. What percent of 84 is 10.5?

b Solving Percent Problems

After a percent problem has been translated to a proportion, we solve as in Section 5.3.

Example 7 5% of what is $20?

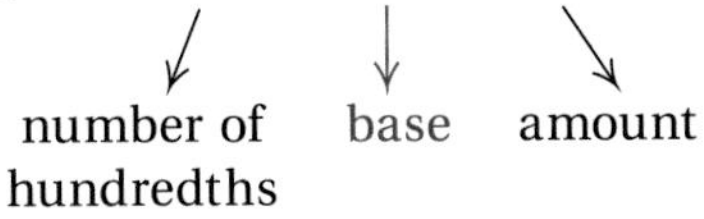

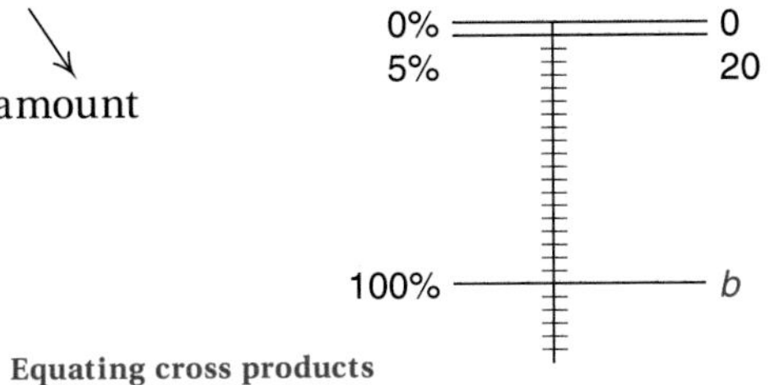

Translate: $\frac{5}{100} = \frac{20}{b}$

Solve: $5 \cdot b = 100 \cdot 20$ Equating cross products

$\frac{5 \cdot b}{5} = \frac{100 \cdot 20}{5}$ Dividing by 5

$b = \frac{5 \cdot 20 \cdot 20}{5}$ Factoring

$= 400$ Simplifying

Thus, 5% of $400 is $20. The answer is $400.

Do Exercise 8.

8. Solve:

20% of what is $45?

Example 8 120% of 42 is what?

number of hundredths → 120%; base → 42; amount → what

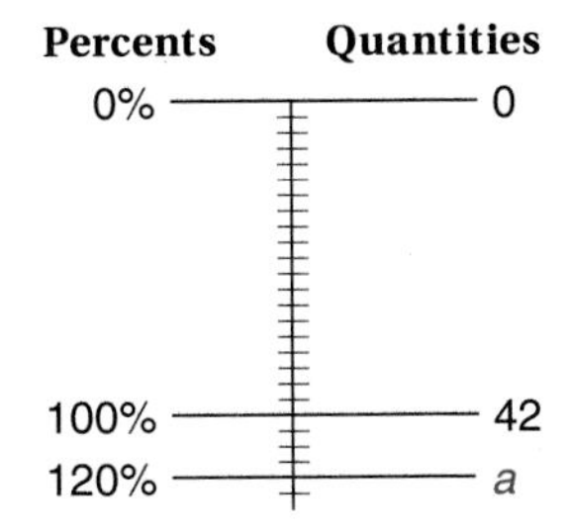

Translate: $\frac{120}{100} = \frac{a}{42}$

Solve: $120 \cdot 42 = 100 \cdot a$ Equating cross products

$\frac{120 \cdot 42}{100} = \frac{100 \cdot a}{100}$ Dividing by 100

$\frac{5040}{100} = a$

$50.4 = a$ Simplifying

Thus, 120% of 42 is 50.4. The answer is 50.4.

Do Exercises 9 and 10.

Solve.

9. 64% of 55 is what?

10. What is 12% of 50?

Answers on page A-13

11. Solve:

60 is 120% of what?

Example 9 3 is 16% of what?

3 → amount; 16% → number of hundredths; what → base

Percents	Quantities
0%	0
16%	3
100%	b

Translate: $\frac{3}{b} = \frac{16}{100}$

Solve: $3 \cdot 100 = b \cdot 16$ **Equating cross products**

$\frac{3 \cdot 100}{16} = \frac{b \cdot 16}{16}$ **Dividing by 16**

$\frac{3 \cdot 100}{16} = b$

$18.75 = b$ **Multiplying and dividing**

Thus, 3 is 16% of 18.75. The answer is 18.75.

Do Exercise 11.

12. Solve:

$12 is what percent of $40?

Example 10 $10 is what percent of $20?

$10 → amount; what percent → number of hundredths; $20 → base

Percents	Quantities
0%	0
N%	$10
100%	$20

Translate: $\frac{10}{20} = \frac{N}{100}$

Solve: $10 \cdot 100 = 20 \cdot N$ **Equating cross products**

$\frac{10 \cdot 100}{20} = \frac{20 \cdot N}{20}$ **Dividing by 20**

$\frac{10 \cdot 100}{20} = N$

$50 = N$ **Multiplying and dividing**

Thus, $10 is 50% of $20. The answer is 50%.

Do Exercise 12.

13. Solve:

What percent of 84 is 10.5?

Example 11 What percent of 50 is 16?

What percent → number of hundredths; 50 → base; 16 → amount

Percents	Quantities
0%	0
N%	16
100%	50

Translate: $\frac{N}{100} = \frac{16}{50}$

Solve: $50 \cdot N = 100 \cdot 16$ **Equating cross products**

$\frac{50 \cdot N}{50} = \frac{100 \cdot 16}{50}$ **Dividing by 50**

$N = \frac{100 \cdot 16}{50}$

$= 32$ **Multiplying and dividing**

Thus, 32% of 50 is 16. The answer is 32%.

Do Exercise 13.

Answers on page A-13

Exercise Set 6.4

a Translate to a proportion. Do not solve.

1. What is 37% of 74?

2. 66% of 74 is what?

3. 4.3 is what percent of 5.9?

4. What percent of 6.8 is 5.3?

5. 14 is 25% of what?

6. 133% of what is 40?

b Solve.

7. What is 76% of 90?

8. What is 32% of 70?

9. 70% of 660 is what?

10. 80% of 920 is what?

11. What is 4% of 1000?

12. What is 6% of 2000?

13. 4.8% of 60 is what?

14. 63.1% of 80 is what?

15. $24 is what percent of $96?

16. $14 is what percent of $70?

17. 102 is what percent of 100?

18. 103 is what percent of 100?

19. What percent of $480 is $120?

20. What percent of $80 is $60?

21. What percent of 160 is 150?

22. What percent of 33 is 11?

23. $18 is 25% of what?

24. $75 is 20% of what?

25. 60% of what is 54?

26. 80% of what is 96?

27. 65.12 is 74% of what?

28. 63.7 is 65% of what?

29. 80% of what is 16?

30. 80% of what is 10?

31. What is $62\frac{1}{2}\%$ of 40?

32. What is $43\frac{1}{4}\%$ of 2600?

33. What is 9.4% of $8300?

34. What is 8.7% of $76,000?

Skill Maintenance

Solve. [5.3b]

35. $\frac{x}{188} = \frac{2}{47}$

36. $\frac{15}{x} = \frac{3}{800}$

37. $\frac{4}{7} = \frac{x}{14}$

38. $\frac{612}{t} = \frac{72}{244}$

39. $\frac{5000}{t} = \frac{3000}{60}$

40. $\frac{75}{100} = \frac{n}{20}$

41. $\frac{x}{1.2} = \frac{36.2}{5.4}$

42. $\frac{y}{1\frac{1}{2}} = \frac{2\frac{3}{4}}{22}$

Solve.

43. A recipe for muffins calls for $\frac{1}{2}$ qt of buttermilk, $\frac{1}{3}$ qt of skim milk, and $\frac{1}{16}$ qt of oil. How many quarts of liquid ingredients does the recipe call for? [3.2c]

44. The Ferristown School District purchased $\frac{3}{4}$ ton (T) of clay. If the clay is to be shared equally among the district's 6 art departments, how much will each art department receive? [2.7d]

Synthesis

45. ◆ In your own words, list steps that a classmate could use to solve any percent problem in this section.

46. ◆ In solving Example 10, a student simplifies $\frac{10}{20}$ before solving. Is this a good idea? Why or why not?

Solve.

47. ▦ What is 8.85% of $12,640?
Estimate ____________
Calculate ____________

48. ▦ 78.8% of what is 9809.024?
Estimate ____________
Calculate ____________

6.5 Applications of Percent

a Applied Problems Involving Percent

Applied problems involving percent are not always stated in a manner easily translated to an equation. In such cases, it is helpful to rephrase the problem before translating. Sometimes it also helps to make a drawing.

Example 1 *Paper Recycling.* In a recent year, the United States generated 73.3 million tons of paper waste, of which 20.5 million tons were recycled (***Source***: Environmental Protection Agency). What percent of paper waste was recycled?

1. **Familiarize.** The question asks for a percent. We know that 10% of 73.3 is 7.33. Since $20.5 \approx 3 \times 7.33$, we expect the answer to be close to 30%. We let n = the percent of paper waste that was recycled.
2. **Translate.** We can rephrase the question and translate as follows:

20.5 million	is	what percent	of	73.3 million?
↓	↓	↓	↓	↓
20,500,000	$=$	n	$\times$	73,300,000

3. **Solve.** We solve as we did in Section 6.3:*

$$20{,}500{,}000 = n \times 73{,}300{,}000$$

$$\frac{20{,}500{,}000}{73{,}300{,}000} = \frac{n \times 73{,}300{,}000}{73{,}300{,}000} \quad \text{Dividing by 73,300,000 on both sides}$$

$$0.28 \approx n \quad \text{Rounding to the nearest hundredth}$$

$$28\% = n. \quad \text{Remember to find percent notation.}$$

4. **Check.** To check, we note that the answer, 28%, is close to 30%, as predicted in the *Familiarize* step.
5. **State.** About 28% of the paper waste was recycled.

Do Exercise 1.

*We can also use the proportion method of Section 6.4 and solve:

$$\frac{N}{100} = \frac{20{,}500{,}000}{73{,}300{,}000}.$$

Objectives

a Solve applied problems involving percent.

b Solve applied problems involving percent of increase or decrease.

For Extra Help

TAPE 11 TAPE 10B MAC WIN CD-ROM

1. *Desserts.* If a restaurant sells 250 desserts in an evening, it is typical that 40 of them will be pie. What percent of the desserts sold will be pie?

Answer on page A-13

2. *Desserts.* Of all desserts sold in restaurants, 20% of them are chocolate cake. One evening a restaurant sells 250 desserts. How many were chocolate cake?

Answer on page A-13

Example 2 *Junk Mail.* The U.S. Postal Service estimates that we read 78% of the junk mail we receive. Suppose that a business sends out 9500 advertising brochures. How many brochures can the business expect to be opened and read?

1. **Familiarize.** We can draw a pie chart to help familiarize ourselves with the problem. We let $a =$ the number of brochures that are opened and read.

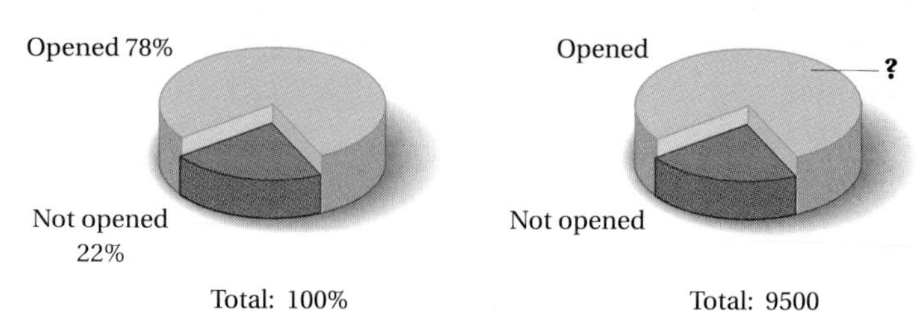

2. **Translate.** The question can be rephrased and translated as follows.

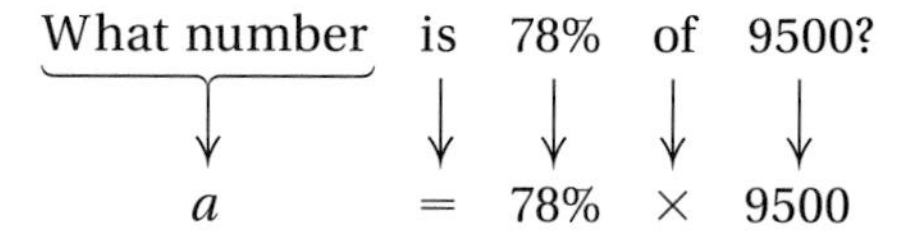

What number is 78% of 9500?

$$a = 78\% \times 9500$$

3. **Solve.** We convert 78% to decimal notation and multiply:*

$$a = 78\% \times 9500 = 0.78 \times 9500 = 7410.$$

4. **Check.** To check, we can repeat the calculation. We can also think about our answer. Since we are taking 78% of 9500, we would expect 7410 to be smaller than 9500 and about three-fourths of 9500, which it is.

5. **State.** The business can expect 7410 of its brochures to be opened and read.

Do Exercise 2.

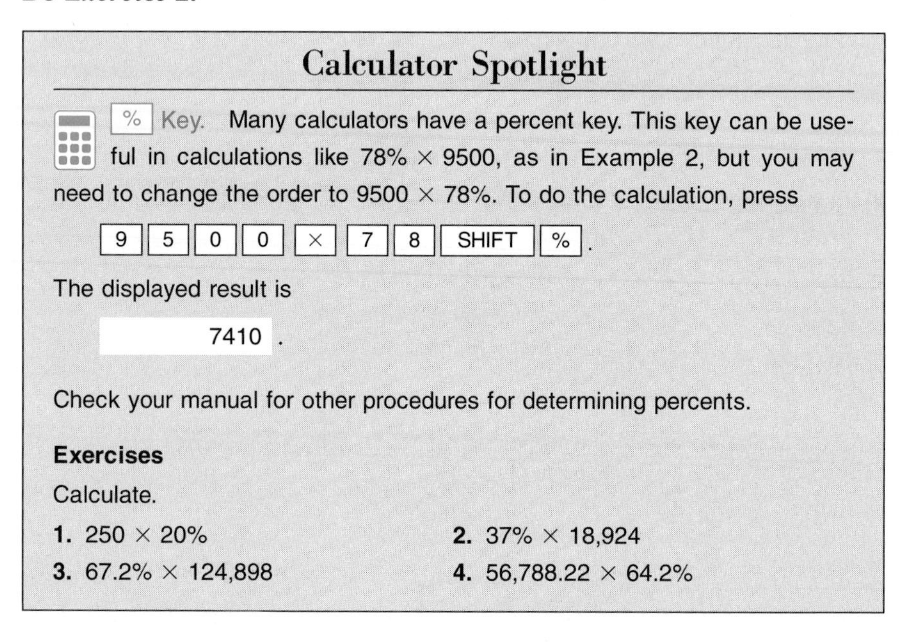

Calculator Spotlight

% Key. Many calculators have a percent key. This key can be useful in calculations like 78% × 9500, as in Example 2, but you may need to change the order to 9500 × 78%. To do the calculation, press

9 5 0 0 × 7 8 SHIFT %.

The displayed result is

7410.

Check your manual for other procedures for determining percents.

Exercises

Calculate.

1. 250 × 20%

2. 37% × 18,924

3. 67.2% × 124,898

4. 56,788.22 × 64.2%

*We can also use the proportion method of Section 6.4 and solve:

$$\frac{N}{9500} = \frac{78}{100}.$$

b Percent of Increase or Decrease

Percent is often used to state increases or decreases. For example, the average salary of an NBA basketball player increased from \$1.558 million in 1994 to \$1.867 million in 1995. To find the *percent of increase* in salary, we first subtract to find out how much more the salary was in 1995:

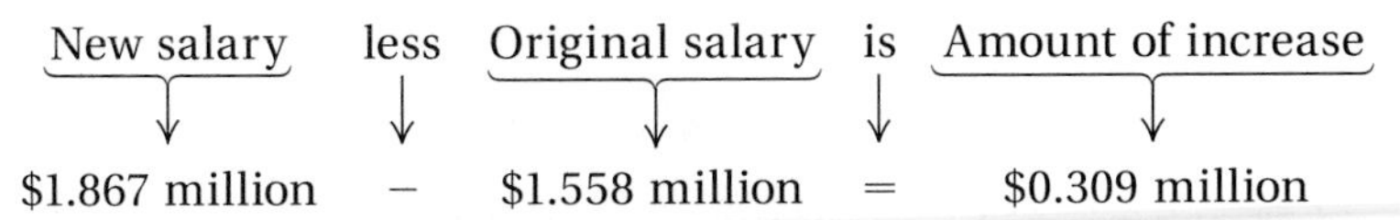

Let's first look at this with a drawing.

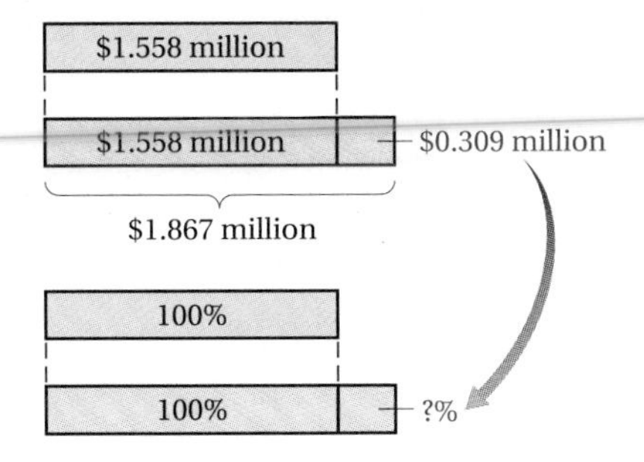

Then we determine what percent of the original amount the increase was. Since \$0.309 million = \$309,000, we are asking

\$309,000 is what percent of \$1,558,000?

This translates to the following:

$309{,}000 = x \cdot 1{,}558{,}000?$

This is an equation of the type studied in Sections 6.3 and 6.4. Solving the equation, we can confirm that \$0.309 million, or \$309,000, is about 19.8% of \$1.558 million. Thus the percent of increase in salary was 19.8%.

To find a percent of increase or decrease:

a) Find the amount of increase or decrease.

b) Then determine what percent this is of the original amount.

3. *Automobile Price.* The price of an automobile increased from \$15,800 to \$17,222. What was the percent of increase?

Example 3 *Digital-Camera Screen Size.* The diagonal of the display screen of a digital camera was recently increased from 1.8 in. to 2.5 in. What was the percent of increase in the diagonal?

1. **Familiarize.** We note that the increase in the diagonal was $2.5 - 1.8$, or 0.7 in. A drawing can help us to visualize the situation. We let $n =$ the percent of increase.

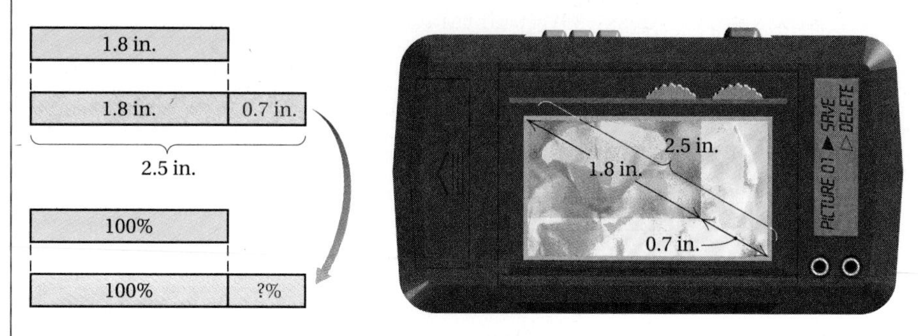

2. **Translate.** We rephrase the question and translate.

0.7 in. is what percent of 1.8 in.?

$$0.7 = n \times 1.8$$

3. **Solve.** To solve the equation, we divide by 1.8 on both sides:*

$$\frac{0.7}{1.8} = \frac{n \times 1.8}{1.8}$$

$0.389 \approx n$ Rounded to the nearest thousandth

$38.9\% \approx n.$ Remember to find percent notation.

4. **Check.** To check, we take 38.9% of 1.8:

$$38.9\% \times 1.8 = 0.389 \times 1.8 = 0.7002.$$

Since we rounded the percent, this approximation is close enough to 0.7 to be a good check.

5. **State.** The percent of increase of the screen diagonal is 38.9%.

Do Exercise 3.

What do we mean when we say that the price of Swiss cheese has decreased 8%? If the price was \$5.00 per pound and it went down to \$4.60 per pound, then the decrease is \$0.40, which is 8% of the original price. We can see this in the following figure.

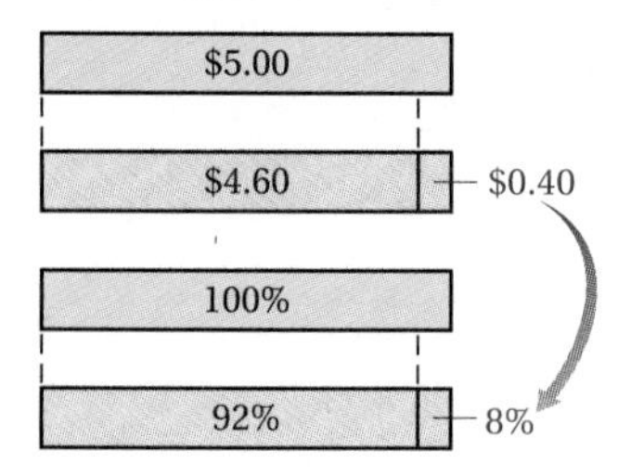

Answer on page A-13

*We can also use the proportion method of Section 6.4 and solve:

$$\frac{0.7}{1.8} = \frac{N}{100}.$$

Example 4 *Fuel Bill.* With proper furnace maintenance, a family that pays a monthly fuel bill of \$78.00 can reduce their bill to \$70.20. What is the percent of decrease?

1. **Familiarize.** We find the amount of decrease and then make a drawing.

7 8.0 0	Original bill
− 7 0.2 0	New bill
7.8 0	Decrease

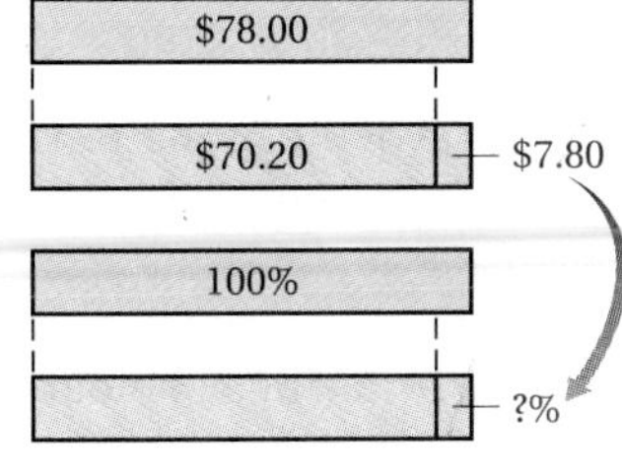

We let n = the percent of decrease.

2. **Translate.** We rephrase and translate as follows.*

$$\underbrace{7.80}_{7.80}\ \underbrace{\text{is}}_{=}\ \underbrace{\text{what percent}}_{n}\ \underbrace{\text{of}}_{\times}\ \underbrace{78.00?}_{78.00}$$

$$7.80 = n \times 78.00$$

3. **Solve.** To solve the equation, we divide by 78 on both sides:

$$\frac{7.80}{78.00} = \frac{n \times 78.00}{78.00}$$ Dividing by 78 on both sides

$0.1 = n$ You may have noticed earlier that 7.8 is 10% of 78.

$10\% = n.$ Changing from decimal to percent notation

4. **Check.** To check, we note that, with a 10% decrease, the reduced bill should be 90% of the original bill. Since 90% of 78 = 0.9 × 78 = 70.20, our answer checks.

5. **State.** The percent of decrease of the fuel bill is 10%.

Do Exercise 4.

Example 5 A part-time teacher's aide earns \$9700 one year and receives a 6% raise the next. What is the new salary?

1. **Familiarize.** We make a drawing.

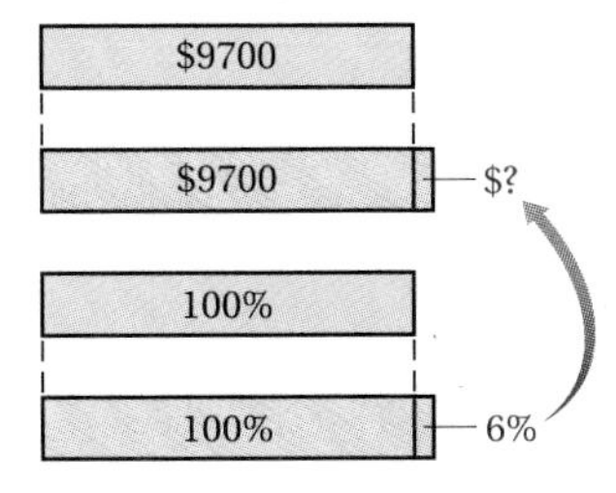

This is a two-step problem. First, we find the increase. We let a = the salary raise.

*We can also use the proportion method of Section 6.4 and solve:

$$\frac{7.8}{78} = \frac{N}{100}.$$

4. *Fuel Bill.* By using only cold water in the washing machine, a household with a monthly fuel bill of \$78.00 can reduce their bill to \$74.88. What is the percent of decrease?

Answer on page A-13

5. A part-time salesperson earns \$9800 one year and gets a 9% raise the next. What is the new salary?

Answer on page A-13

2. **Translate.** We rephrase the question and translate as follows.

$$\begin{array}{ccccc} \text{What} & \text{is} & 6\% & \text{of} & 9700? \\ \downarrow & \downarrow & \downarrow & \downarrow & \downarrow \\ a & = & 6\% & \times & 9700 \end{array}$$

3. **Solve.** We convert 6% to a decimal notation and multiply:

$$a = 0.06 \times 9700 = 582.$$

Next, we add \$582 to the old salary:

$$9700 + 582 = 10{,}282.$$

4. **Check.** To check, we can repeat the calculation. We can also check by estimating. The old salary, \$9700, is approximately \$10,000, and 6% of \$10,000 is 0.06 × 10,000, or \$600. The new salary would be about 9700 + 600, or \$10,300. Since \$10,282 is close to \$10,300, we have a partial check.
5. **State.** The new salary is \$10,282.

Do Exercise 5.

Calculator Spotlight

The % Key and Percent of Increase or Decrease. On a calculator with a percent key, there may be a fast way to find the result of adding or subtracting a percent from a number. In Example 5, the result of taking 6% of \$9700 and adding it to \$9700 might be found by pressing

The displayed result would be

10,282 .

If the salary had been reduced by 6%, the computation would be

The displayed result would be

9118 .

Check your manual for other procedures for determining percents.

Exercises

Use a calculator with a % key.

1. Find the result of Margin Exercise 5.
2. Find the result of Margin Exercise 5 if the salary were decreased by 9%.

Exercise Set 6.5

a Solve.

1. *Left-handed Professional Bowlers.* It has been determined by sociologists that 17% of the population is left-handed. Each tournament conducted by the Professional Bowlers Association has 120 entrants. How many would you expect to be left-handed? not left-handed? Round to the nearest one.

2. *Advertising Budget.* A common guideline for businesses is to use 5% of their operating budget for advertising. Ariel Electronics has an operating budget of $8000 per week. How much should it spend each week for advertising? for other expenses?

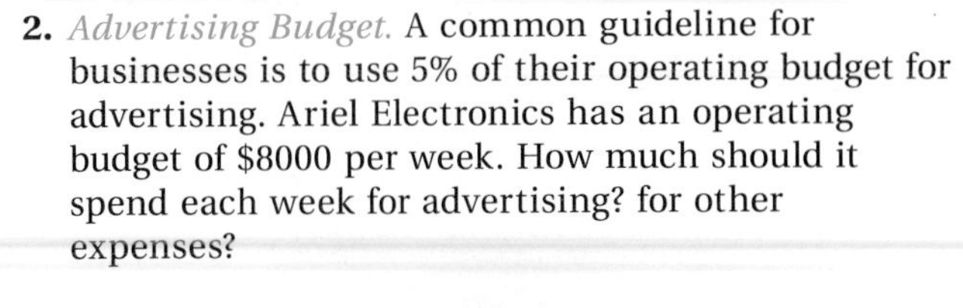

3. Of all moviegoers, 67% are in the 12–29 age group. A theater held 800 people for a showing of *Star Trek-18*. How many were in the 12–29 age group? not in this age group?

4. Deming, New Mexico, claims to have the purest drinking water in the world. It is 99.9% pure. If you had 240 L of water from Deming, how much of it, in liters, would be pure? impure?

5. A baseball player gets 13 hits in 40 at-bats. What percent are hits? not hits?

6. On a test of 80 items, Erika had 76 correct. What percent were correct? incorrect?

7. A lab technician has 680 mL of a solution of water and acid; 3% is acid. How many milliliters are acid? water?

8. A lab technician has 540 mL of a solution of alcohol and water; 8% is alcohol. How many milliliters are alcohol? water?

9. *TV Usage.* Of the 8760 hr in a year, most television sets are on for 2190 hr. What percent is this?

10. *Colds from Kissing.* In a medical study, it was determined that if 800 people kiss someone who has a cold, only 56 will actually catch a cold. What percent is this?

11. *Maximum Heart Rate.* Treadmill tests are often administered to diagnose heart ailments. A guideline in such a test is to try to get you to reach your *maximum heart rate,* in beats per minute. The maximum heart rate is found by subtracting your age from 220 and then multiplying by 85%. What is the maximum heart rate of someone whose age is 25? 36? 48? 55? 76? Round to the nearest one.

12. It costs an oil company $40,000 a day to operate two refineries. Refinery A accounts for 37.5% of the cost, and refinery B for the rest of the cost.

a) What percent of the cost does it take to run refinery B?

b) What is the cost of operating refinery A? refinery B?

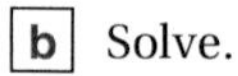

Solve.

13. The amount in a savings account increased from $200 to $216. What was the percent of increase?

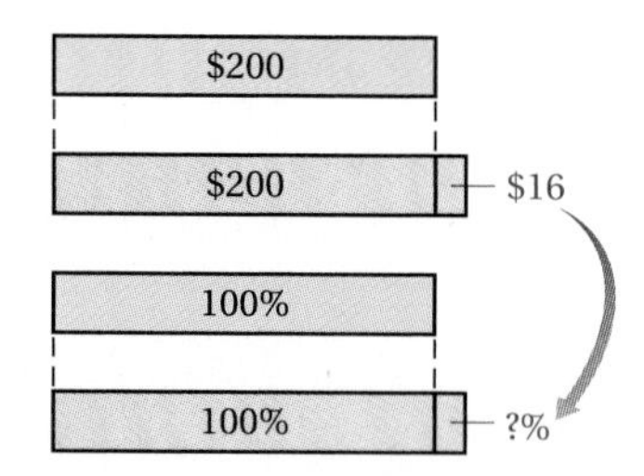

14. The population of a small mountain town increased from 840 to 882. What was the percent of increase?

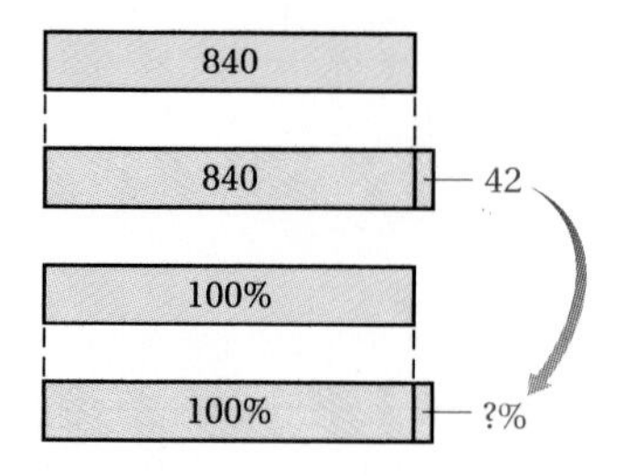

15. During a sale, a dress decreased in price from $90 to $72. What was the percent of decrease?

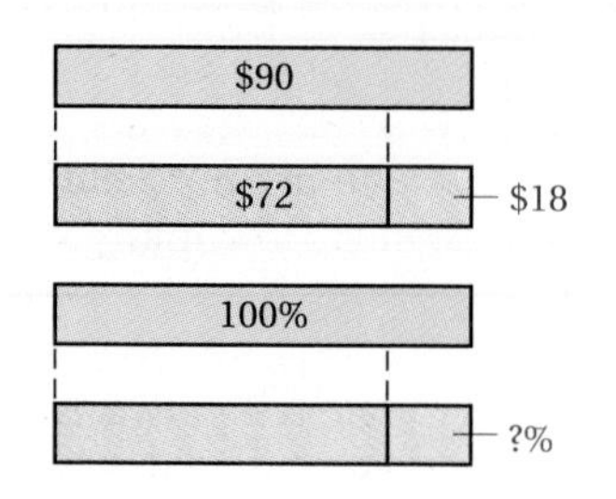

16. A person on a diet goes from a weight of 125 lb to a weight of 110 lb. What is the percent of decrease?

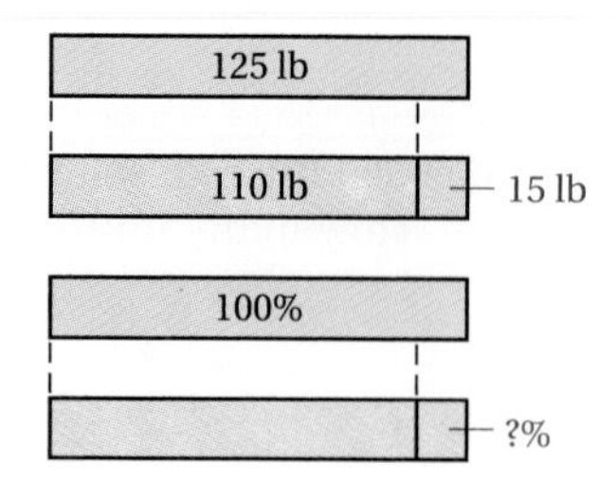

17. A person earns $28,600 one year and receives a 5% raise in salary. What is the new salary?

18. A person earns $20,400 one year and receives an 8% raise in salary. What is the new salary?

19. The value of a car typically decreases by 30% in the first year. A car is bought for $18,000. What is its value one year later?

20. One year the pilots of an airline shocked the business world by taking an 11% pay cut. The former salary was $55,000. What was the reduced salary?

21. *World Population.* World population is increasing by 1.6% each year. In 1999, it was 6.0 billion. How much will it be in 2000? 2001? 2002?

22. *Cooling Costs.* By increasing the thermostat from 72° to 78°, a family can reduce its cooling bill by 50%. If the cooling bill was $106.00, what would the new bill be? By what percent has the temperature been increased?

23. *Car Depreciation.* A car generally depreciates 30% of its original value in the first year. A car is worth $25,480 after the first year. What was its original cost?

24. *Car Depreciation.* Given normal use, an American-made car will depreciate 30% of its original cost the first year and 14% of its remaining value in the second year. What is the value of a car at the end of the second year if its original cost was $36,400? $28,400? $26,800?

25. *Tipping.* Diners frequently add a 15% tip when charging a meal to a credit card. What is the total amount charged if the cost of the meal, without tip, is $15? $34? $49?

26. *Two-by-Four.* A cross-section of a standard or nominal "two-by-four" board actually measures $1\frac{1}{2}$ in. by $3\frac{1}{2}$ in. The rough board is 2 in. by 4 in. but is planed and dried to the finished size. What percent of the wood is removed in planing and drying?

27. *MADD.* Despite efforts by groups such as MADD (Mothers Against Drunk Driving), the number of alcohol-related deaths is rising after many years of decline. The data in the table shows the number of deaths from 1986 to 1995.

a) What is the percent of increase in the number of alcohol-related deaths from 1994 to 1995?

b) What is the percent of decrease in the number of alcohol-related deaths from 1986 to 1994?

Alcohol-Related Traffic Deaths Back on the Increase!!

Year	Deaths
1986	24,045
1987	23,641
1988	23,626
1989	22,436
1990	22,084
1991	19,887
1992	17,859
1993	17,473
1994	16,589
1995	17,274

Source: National Highway Traffic Safety Administration

28. *Fetal Acoustic Stimulation.* Each year there are about 4 million births in the United States. Of these, about 120,000 births occur in breech position (delivery of a fetus with the buttocks or feet appearing first). A new technique, called *fetal acoustic stimulation (FAS),* uses sound directed through a mother's abdomen in order to stimulate movement of the fetus to a safer position. In a recent study of this low-risk and low-cost procedure, FAS enabled doctors to turn the baby in 34 of 38 cases (***Source*:** Johnson and Elliott, "Fetal Acoustic Stimulation, an Adjunct to External Cephalic Versions: A Blinded, Randomized Crossover Study," *American Journal of Obstetrics & Gynecology* **173**, no. 5 (1995): 1369–1372).

a) What percent of U.S. births are breech?

b) What percent (rounded to the nearest tenth) of cases showed success with FAS?

c) About how many breech babies yearly might be turned if FAS could be implemented in all births in the United States?

d) Breech position is one reason for performing Caesarean section (or C-section) birth surgery. Researchers expect that FAS alone can eliminate the need for about 2000 C-sections yearly in the United States. Given this information, how many yearly C-sections are due to breech position alone?

29. *Strike Zone.* In baseball, the *strike zone* is normally a 17-in. by 40-in. rectangle. Some batters give the pitcher an advantage by swinging at pitches thrown out of the strike zone. By what percent is the area of the strike zone increased if a 2-in. border is added to the outside?

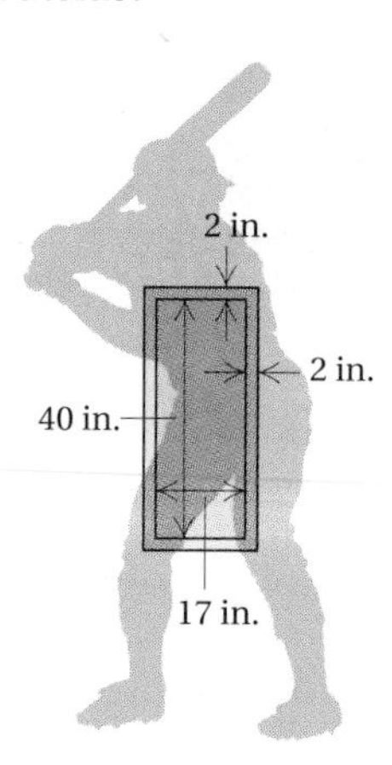

30. Tony is planting grass on a 24-ft by 36-ft area in his back yard. He installs a 6-ft by 8-ft garden. By what percent has he reduced the area he has to mow?

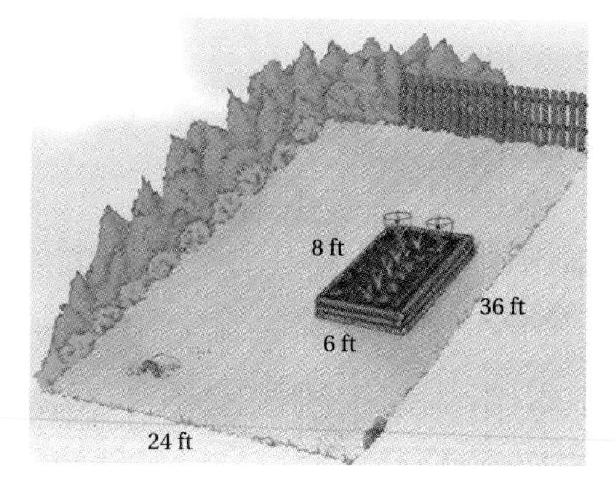

Skill Maintenance

Convert to decimal notation. [4.1c], [4.5a]

31. $\frac{25}{11}$

32. $\frac{11}{25}$

33. $\frac{27}{8}$

34. $\frac{43}{9}$

35. $\frac{23}{25}$

36. $\frac{20}{24}$

37. $\frac{14}{32}$

38. $\frac{2317}{1000}$

39. $\frac{34{,}809}{10{,}000}$

40. $\frac{27}{40}$

Synthesis

41. ◆ Which is better for a wage earner, and why: a 10% raise followed by a 5% raise a year later, or a 5% raise followed by a 10% raise a year later?

42. ◆ Write a problem for a classmate to solve. Design the problem so that the solution is "Jackie's raise was $7\frac{1}{2}$%."

43. 🖩 A worker receives raises of 3%, 6%, and then 9%. By what percent has the original salary increased?

44. ◆ 🖩 A workers' union is offered either a 5% "across-the-board" raise in which all salaries would increase 5%, or a flat $1650 raise for each worker. If the total payroll for the 123 workers is $4,213,365, which offer should the union select? Why?

45. *Adult Height.* It has been determined that at the age of 10, a girl has reached 84.4% of her final adult growth. Cynthia is 4 ft, 8 in. at the age of 10. What will be her final adult height?

46. *Adult Height.* It has been determined that at the age of 15, a boy has reached 96.1% of his final adult height. Claude is 6 ft, 4 in. at the age of 15. What will be his final adult height?

47. If p is 120% of q, then q is what percent of p?

48. A coupon allows a couple to have dinner and then have $10 subtracted from the bill. Before subtracting $10, however, the restaurant adds a tip of 15%. If the couple is presented with a bill for $44.05, how much would the dinner (without tip) have cost without the coupon?

6.6 Consumer Applications: Sales Tax, Commission, and Discount

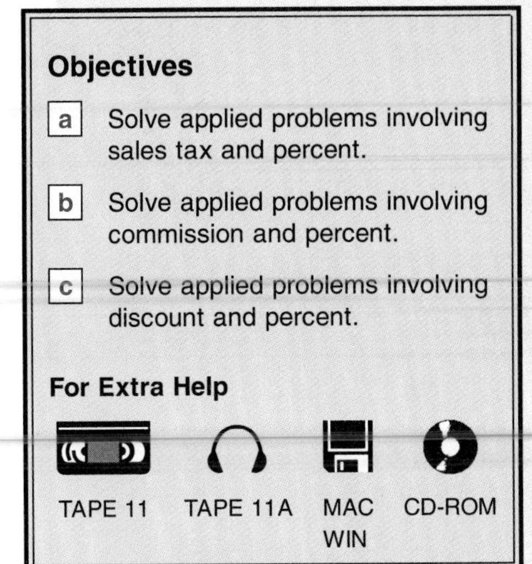

Objectives

[a] Solve applied problems involving sales tax and percent.

[b] Solve applied problems involving commission and percent.

[c] Solve applied problems involving discount and percent.

For Extra Help

TAPE 11 TAPE 11A MAC WIN CD-ROM

a Sales Tax

Sales tax computations represent a special type of percent of increase problem. The sales tax rate in Arkansas is 3%. This means that the tax is 3% of the purchase price. Suppose the purchase price on a coat is \$124.95. The sales tax is then

3% of \$124.95, or 0.03×124.95,

or

3.7485, or about \$3.75.

The total that you pay is the price plus the sales tax:

\$124.95 + \$3.75, or \$128.70.

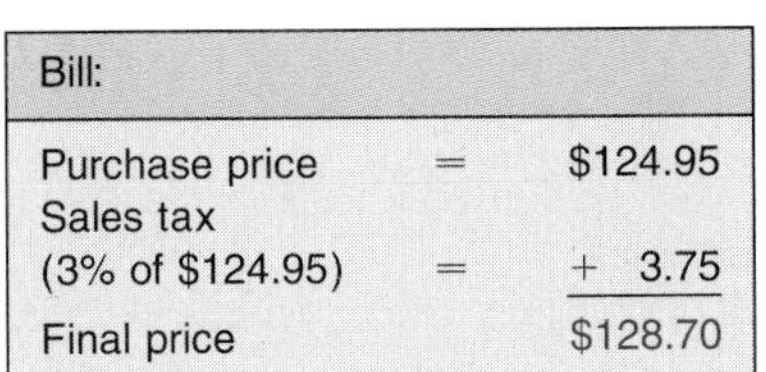

Bill:		
Purchase price	=	\$124.95
Sales tax (3% of \$124.95)	=	+ 3.75
Final price		\$128.70

Sales tax = Sales tax rate × Purchase price

Total price = Purchase price + Sales tax

Example 1 *Florida.* The sales tax rate in Florida is 6%. How much tax is charged on the purchase of 3 CDs at \$13.95 each? What is the total price?

a) We first find the cost of the CDs. It is

$3 \times \$13.95 = \$41.85$.

b) The sales tax on items costing \$41.85 is

$$\underbrace{\text{Sales tax rate}}_{\downarrow} \times \underbrace{\text{Purchase price}}_{\downarrow}$$
$$6\% \qquad \times \qquad \$41.85,$$

or 0.06×41.85, or 2.511. Thus the tax is \$2.51.

c) The total price is given by the purchase price plus the sales tax:

\$41.85 + \$2.51, or \$44.36.

To check, note that the total price is the purchase price plus 6% of the purchase price. Thus the total price is 106% of the purchase price. Since $1.06 \times 41.85 \approx 44.36$, we have a check. The total price is \$44.36.

Do Exercises 1 and 2.

1. *Connecticut.* The sales tax rate in Connecticut is 8%. How much tax is charged on the purchase of a refrigerator that sells for \$668.95? What is the total price?

2. *New Jersey.* Morris buys 5 blank audiocassettes in New Jersey, where the sales tax rate is 7%. If each tape costs \$2.95, how much tax will be charged? What is the total price?

Answers on page A-13

3. The sales tax is \$33 on the purchase of a \$550 washing machine. What is the sales tax rate?

Example 2 The sales tax is \$32 on the purchase of an \$800 sofa. What is the sales tax rate?

Rephrase: Sales tax is what percent of purchase price?

Translate: $32 = r \times 800$

To solve the equation, we divide by 800 on both sides:

$$\frac{32}{800} = \frac{r \times 800}{800}$$

$$0.04 = r$$

$$4\% = r.$$

The sales tax rate is 4%.

Do Exercise 3.

4. The sales tax on a television is \$25.20 and the sales tax rate is 6%. Find the purchase price (the price before taxes are added).

Example 3 The sales tax on a laser printer is \$31.74 and the sales tax rate is 5%. Find the purchase price (the price before taxes are added).

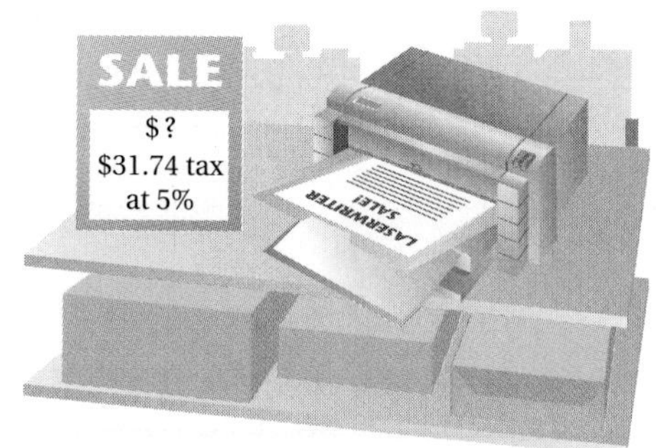

Rephrase: Sales tax is 5% of what?

Translate: $31.74 = 5\% \times b$, or $31.74 = 0.05 \times b$.

To solve, we divide by 0.05 on both sides:

$$\frac{31.74}{0.05} = \frac{0.05 \times b}{0.05}$$

$$634.8 = b.$$

$$\begin{array}{r} 634.8 \\ 0.05 \overline{)\,31.74_{\wedge}0} \\ \underline{3000} \\ 174 \\ \underline{150} \\ 24 \\ \underline{20} \\ 40 \\ \underline{40} \\ 0 \end{array}$$

The purchase price is \$634.80.

Do Exercise 4.

Answers on page A-13

b Commission

When you work for a **salary**, you receive the same amount of money each week or month. When you work for a **commission**, you are paid a percentage of the total sales for which you are responsible.

> **Commission** = Commission rate × Sales

Example 4 A salesperson's commission rate is 20%. What is the commission from the sale of $25,560 worth of stereophonic equipment?

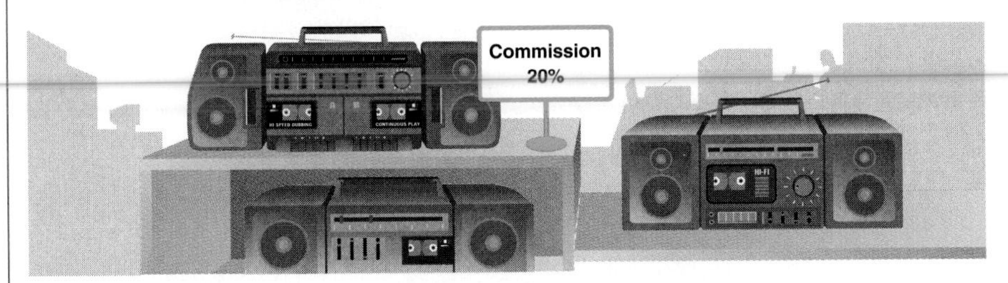

Commission	=	*Commission rate*	×	*Sales*
C	=	20%	×	25,560

This tells us what to do. We multiply.

$$\begin{array}{r} 25{,}560 \\ \times \quad 0.2 \\ \hline 5112.0 \end{array} \qquad 20\% = 0.20 = 0.2$$

The commission is $5112.

Do Exercise 5.

Example 5 Dawn earns a commission of $3000 selling $60,000 worth of farm machinery. What is the commission rate?

Commission	=	*Commission rate*	×	*Sales*
3000	=	r	×	60,000

To solve this equation, we divide by 60,000 on both sides:

$$\frac{3000}{60{,}000} = \frac{r \times 60{,}000}{60{,}000} = r.$$

5. Raul's commission rate is 30%. What is the commission from the sale of $18,760 worth of air conditioners?

Answer on page A-13

6. Liz earns a commission of $6000 selling $24,000 worth of refrigerators. What is the commission rate?

7. Ben's commission rate is 16%. He receives a commission of $268 from sales of clothing. How many dollars worth of clothing were sold?

Answers on page A-13

We can divide, but this time we simplify by removing a factor of 1:

$$r = \frac{3000}{60,000} = \frac{1}{20} \cdot \frac{3000}{3000} = \frac{1}{20} = 0.05 = 5\%.$$

The commission rate is 5%.

Do Exercise 6.

Example 6 Joyce's commission rate is 25%. She receives a commission of $425 on the sale of a motorbike. How much did the motorbike cost?

Commission	=	*Commission rate*	$\times$	*Sales*
425	=	25%	$\times$	S

To solve this equation, we divide by 0.25 on both sides:

$$\frac{425}{0.25} = \frac{0.25 \times S}{0.25}$$

$$425 \div 0.25 = S$$

$$1700 = S.$$

$$\begin{array}{r} 1700. \\ 0.25_{\wedge}\overline{)\,425.00_{\wedge}} \\ \underline{250} \\ 175 \\ \underline{175} \\ 0 \end{array}$$

The motorbike cost $1700.

Do Exercise 7.

c Discount

Suppose that the regular price of a rug is $60, and the rug is on sale at 25% off. Since 25% of $60 is $15, the sale price is $60 − $15, or $45. We call $60 the **original**, or **marked price,** 25% the **rate of discount,** $15 the **discount**, and $45 the **sale price.** Note that discount problems are a type of percent of decrease problem.

> **Discount** = Rate of discount × Original price
>
> **Sale price** = Original price − Discount

Example 7 A rug marked $240 is on sale at 25% off. What is the discount? the sale price?

a) *Discount* = *Rate of discount* × *Original price*

D = 25% × 240

This tells us what to do. We convert 25% to decimal notation and multiply.

$$\begin{array}{r} 240 \\ \times\ 0.25 \\ \hline 1200 \\ 4800 \\ \hline 60.00 \end{array} \qquad 25\% = 0.25$$

The discount is $60.

b) *Sale price* = *Marked price* − *Discount*

S = 240 − 60

This tells us what to do. We subtract.

$$\begin{array}{r} 240 \\ -\ 60 \\ \hline 180 \end{array}$$

To check, note that the sale price is 75% of the marked price: $0.75 \times 240 = 180$.

The sale price is $180.

Do Exercise 8.

8. A suit marked $140 is on sale at 24% off. What is the discount? the sale price?

Example 8 An antique table is marked down from $620 to $527. What is the rate of discount?

We first find the discount by subtracting the sale price from the original price:

$$\begin{array}{r} 620 \\ -\ 527 \\ \hline 93. \end{array}$$

The discount is $93.

Next, we use the equation for discount:

Discount = *Rate of discount* × *Original price*

93 = r × 620.

Answer on page A-13

9. A pair of hiking boots is reduced from \$75 to \$60. Find the rate of discount.

To solve, we divide by 620 on both sides:

$$\frac{93}{620} = \frac{r \times 620}{620}$$

$$0.15 = r$$

$$15\% = r.$$

$$\begin{array}{r} 0.15 \\ 620\overline{)\,93.00} \\ \underline{62\,0} \\ 31\,00 \\ \underline{31\,00} \\ 0 \end{array}$$

The discount rate is 15%.

To check, note that a 15% discount rate means that 85% of the original price is paid:
$0.85 \times 620 = 527.$

Do Exercise 9.

Improving Your Math Study Skills

A Checklist of Your Study Skills

You are now about halfway through this textbook as well as the course. How are you doing? If you are struggling, are you making full use of the study skills we have suggested to you in these inserts? To decide, go through the following checklist, marking the questions "Yes" or "No." This list is a review of many of the study skill suggestions considered so far in the text.

Study Skill Questions	Yes	No
1. Are you stopping to work the margin exercises when directed to do so?		
2. Are you doing your homework as soon as possible after class?		
3. Are you doing your homework at a specified time and in a quiet setting?		
4. Have you found a study group in which to work?		
5. Are you consistently trying to apply the five-step problem-solving strategy?		
6. If you are going to help or tutoring sessions, have you studied enough to generate questions to ask the tutor, rather than having the tutor teach you from scratch?		
7. Are you doing lots of even-numbered exercises for which answers are not available?		

Study Skill Questions	Yes	No
8. Are you keeping one section ahead on your syllabus?		
9. Are you using the book supplements, such as the *Student's Solutions Manual* and the *InterAct Math Tutorial Software*?		
10. When you study the book, are you marking the points you do not understand as a source for in-class questions?		
11. Are you reading and studying each step of each example?		
12. Are you noting and using the objective code symbols a, b, c, and so on, that appear at the beginning of each section, throughout the section, in the exercise sets, the summary–reviews, and the tests?		

If you have 7 or more "No" answers to these questions, and are struggling in the course, you now have many suggestions for improvement as you progress to the end of the course. A consultation with your instructor is strongly advised. Good luck!

Answer on page A-13

Exercise Set 6.6

a Solve.

1. *Indiana.* The sales tax rate in Indiana is 5%. How much tax is charged on a generator costing $586? What is the total price?

2. *New York City.* The sales tax rate in New York City is 8.25%. How much tax is charged on photo equipment costing $248? What is the total price?

3. *Illinois.* The sales tax rate in Illinois is 6.25%. How much tax is charged on a purchase of 5 telephones at $53 apiece? What is the total price?

4. *Kentucky.* The sales tax rate in Kentucky is 6%. How much tax is charged on a purchase of 5 teapots at $37.99 apiece? What is the total price?

5. The sales tax is $48 on the purchase of a dining room set that sells for $960. What is the sales tax rate?

6. The sales tax is $15 on the purchase of a diamond ring that sells for $500. What is the sales tax rate?

7. The sales tax is $35.80 on the purchase of a refrigerator–freezer that sells for $895. What is the sales tax rate?

8. The sales tax is $9.12 on the purchase of a patio set that sells for $456. What is the sales tax rate?

9. The sales tax on a used car is $100 and the sales tax rate is 5%. Find the purchase price (the price before taxes are added).

10. The sales tax on the purchase of a new boat is $112 and the sales tax rate is 2%. Find the purchase price.

11. The sales tax on a dining room set is $28 and the sales tax rate is 3.5%. Find the purchase price.

12. The sales tax on a stereo is $66 and the sales tax rate is 5.5%. Find the purchase price.

13. The sales tax rate in Dallas is 1% for the city and 6% for the state. Find the total amount paid for 2 shower units at $332.50 apiece.

14. The sales tax rate in Omaha is 1.5% for the city and 5% for the state. Find the total amount paid for 3 air conditioners at $260 apiece.

15. The sales tax is $1030.40 on an automobile purchase of $18,400. What is the sales tax rate?

16. The sales tax is $979.60 on an automobile purchase of $15,800. What is the sales tax rate?

b Solve.

17. Sondra's commission rate is 6%. What is the commission from the sale of $45,000 worth of furnaces?

18. Jose's commission rate is 32%. What is the commission from the sale of $12,500 worth of sailboards?

19. Vince earns $120 selling $2400 worth of television sets. What is the commission rate?

20. Donna earns $408 selling $3400 worth of shoes. What is the commission rate?

21. An art gallery's commission rate is 40%. They receive a commission of $392. How many dollars worth of artwork were sold?

22. A real estate agent's commission rate is 7%. She receives a commission of $5600 on the sale of a home. How much did the home sell for?

23. A real estate commission is 6%. What is the commission on the sale of a $98,000 home?

24. A real estate commission is 8%. What is the commission on the sale of a piece of land for $68,000?

25. Bonnie earns $280.80 selling $2340 worth of tee shirts. What is the commission rate?

26. Chuck earns $1147.50 selling $7650 worth of ski passes. What is the commission rate?

27. Miguel's commission is increased according to how much he sells. He receives a commission of 5% for the first $2000 and 8% on the amount over $2000. What is the total commission on sales of $6000?

28. Lucinda earns a salary of $500 a month, plus a 2% commission on sales. One month, she sold $990 worth of encyclopedias. What were her wages that month?

c Find what is missing.

29.

Marked Price	Rate of Discount	Discount	Sale Price
$300	10%		

30.

Marked Price	Rate of Discount	Discount	Sale Price
$2000	40%		

31.

Marked Price	Rate of Discount	Discount	Sale Price
$17.00	15%		

32.

Marked Price	Rate of Discount	Discount	Sale Price
$20.00	25%		

33.

Marked Price	Rate of Discount	Discount	Sale Price
	10%	$12.50	

34.

Marked Price	Rate of Discount	Discount	Sale Price
	15%	$65.70	

35.

Marked Price	Rate of Discount	Discount	Sale Price
$600		$240	

36.

Marked Price	Rate of Discount	Discount	Sale Price
$12,800		$1920	

37. Find the discount and the rate of discount for the ring in this ad.

1/2 CARAT T.W.
DIAMOND, 14K GOLD
LADY'S BRIDAL SET
was $1275.00
$888

38. Find the discount and the rate of discount for the calculator in this ad.

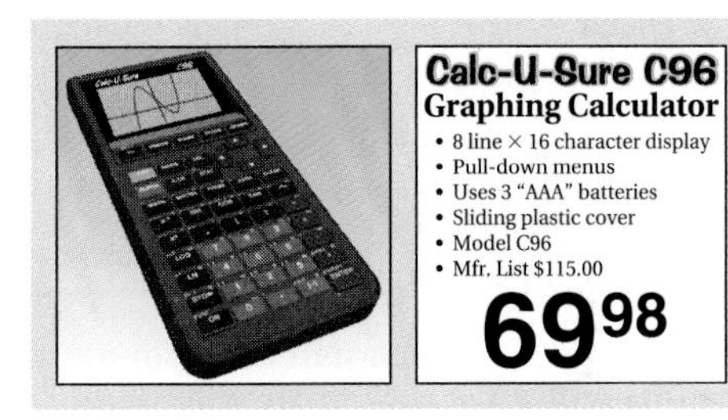

39. Find the marked price and the rate of discount for the camcorder in this ad.

REDUCED $83

Palmaster
VHS-C Camcorder
• Large Video Head Cylinder for Jitter-free, Crisp Pictures
• 12:1 Variable Speed Power Zoom
• Lens Cover Opens Automatically when Camera is Turned On
$377

40. Find the marked price and the rate of discount for the cedar chest in this ad.

Skill Maintenance

Solve. [5.3b]

41. $\frac{x}{12} = \frac{24}{16}$

42. $\frac{7}{2} = \frac{11}{x}$

Solve. [4.4b]

43. $0.64 \cdot x = 170$

44. $28.5 = 25.6 \times y$

Find decimal notation. [4.5a]

45. $\frac{5}{9}$

46. $\frac{23}{11}$

47. $\frac{11}{12}$

48. $\frac{13}{7}$

49. $\frac{15}{7}$

50. $\frac{19}{12}$

Convert to standard notation. [4.3b]

51. $4.03 trillion

52. 5.8 million

53. 42.7 million

54. 6.09 trillion

Synthesis

55. ◆ Is the following ad mathematically correct? Why or why not?

56. ◆ An item that is no longer on sale at "25% off" receives a price tag that is $33\frac{1}{3}\%$ more than the sale price. Has the item price been restored to its original price? Why or why not?

57. ◆ Which is better, a discount of 40% or a discount of 20% followed by another of 20%? Explain.

58. 🖩 A real estate commission rate is 7.5%. A house sells for $98,500. How much does the seller get for the house after paying the commission?

59. 🖩 *People Magazine.* In a recent subscription drive, *People* offered a subscription of 52 weekly issues for a price of $1.89 per issue. They advertised that this was a savings of 29.7% off the newsstand price. What was the newsstand price?

60. 🖩 Gordon receives a 10% commission on the first $5000 in sales and 15% on all sales beyond $5000. If Gordon receives a commission of $2405, how much did he sell? Use a calculator and trial and error if you wish.

61. Herb collects baseball memorabilia. He bought two autographed plaques, but became short of funds and had to sell them quickly for $200 each. On one, he made a 20% profit and on the other, he lost 20%. Did he make or lose money on the sale?

62. Tee shirts are being sold at the mall for $5 each, or 3 for $10. If you buy three tee shirts, what is the rate of discount?

Collaborative Learning Manual

Calculate the costs associated with the purchase of a car or truck.

6.7 Consumer Applications: Interest

a Simple Interest

Suppose you put $100 into an investment for 1 year. The $100 is called the **principal**. If the **interest rate** is 8%, in addition to the principal, you get back 8% of the principal, which is

8% of $100, or 0.08×100, or $8.00.

The $8.00 is called the **simple interest**. It is, in effect, the price that a financial institution pays for the use of the money over time.

> The **simple interest** I on principal P, invested for t years at interest rate r, is given by
>
> $$I = P \cdot r \cdot t.$$

Example 1 What is the interest on $2500 invested at an interest rate of 6% for 1 year?

We use the formula $I = P \cdot r \cdot t$:

$$I = P \cdot r \cdot t = \$2500 \times 6\% \times 1$$
$$= \$2500 \times 0.06$$
$$= \$150.$$

```
   2 5 0 0
×    0.0 6
 1 5 0.0 0
```

The interest for 1 year is $150.

Do Exercise 1.

Example 2 What is the interest on a principal of $2500 invested at an interest rate of 6% for $\frac{1}{4}$ year?

We use the formula $I = P \cdot r \cdot t$:

$$I = P \cdot r \cdot t = \$2500 \times 6\% \times \frac{1}{4}$$
$$= \frac{\$2500 \times 0.06}{4}$$
$$= \$37.50.$$

```
       3 7.5
4 ) 1 5 0.0
    1 2 0
      3 0
      2 8
        2 0
        2 0
          0
```

We could have instead found $\frac{1}{4}$ of 6% and then multiplied by 2500.

The interest for $\frac{1}{4}$ year is $37.50.

Do Exercise 2.

Objectives

a Solve applied problems involving simple interest and percent.

b Solve applied problems involving compound interest.

For Extra Help

TAPE 11 TAPE 11A MAC WIN CD-ROM

1. What is the interest on $4300 invested at an interest rate of 14% for 1 year?

2. What is the interest on a principal of $4300 invested at an interest rate of 14% for $\frac{3}{4}$ year?

Answers on page A-13

3. The Glass Nook borrows $4800 at 7% for 30 days. Find (a) the amount of simple interest due and (b) the total amount that must be paid after 30 days.

Answer on page A-13

When time is given in days, we usually divide it by 365 to express the time as a fractional part of a year.

Example 3 To pay for a shipment of tee shirts, New Wave Designs borrows $8000 at 9% for 60 days. Find (a) the amount of simple interest that is due and (b) the total amount that must be paid after 60 days.

a) We express 60 days as a fractional part of a year:

$$I = P \cdot r \cdot t = \$8000 \times 9\% \times \frac{60}{365}$$

$$= \$8000 \times 0.09 \times \frac{60}{365}$$

$$\approx \$118.36. \quad \text{Usng a calculator}$$

The interest due for 60 days is $118.36.

b) The total amount to be paid after 60 days is the principal plus the interest:

$$\$8000 + \$118.36 = \$8118.36.$$

The total amount due is $8118.36.

Do Exercise 3.

b Compound Interest

When interest is paid *on interest,* we call it **compound interest.** This is the type of interest usually paid on investments. Suppose you have $5000 in a savings account at 6%. In 1 year, the account will contain the original $5000 plus 6% of $5000. Thus the total in the account after 1 year will be

106% of $5000, or 1.06 × $5000, or $5300.

Now suppose that the total of $5300 remains in the account for another year. At the end of this second year, the account will contain the $5300 plus 6% of $5300. The total in the account would thus be

106% of $5300, or 1.06 × $5300, or $5618.

Note that in the second year, interest is earned on the first year's interest. When this happens, we say that interest is **compounded annually.**

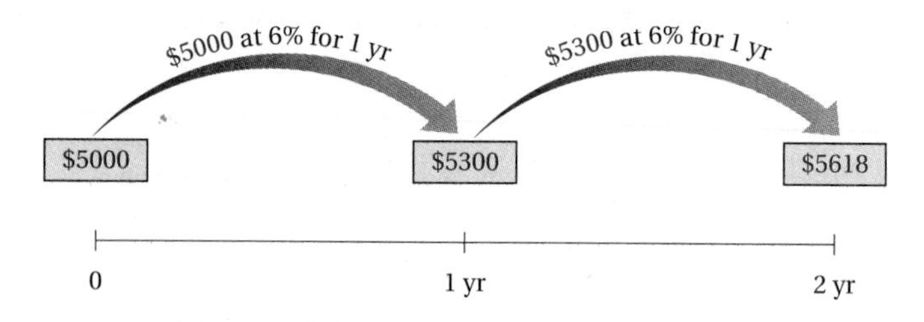

Example 4 Find the amount in an account if $2000 is invested at 8%, compounded annually, for 2 years.

a) After 1 year, the account will contain 108% of $2000:

$1.08 \times \$2000 = \$2160.$

$$\begin{array}{r} 2\,0\,0\,0 \\ \times \quad 1.0\,8 \\ \hline 1\,6\,0\,0\,0 \\ 0\,0\,0\,0 \\ 2\,0\,0\,0 \\ \hline 2\,1\,6\,0.0\,0 \end{array}$$

b) At the end of the second year, the account will contain 108% of $2160:

$1.08 \times \$2160 = \$2332.80.$

$$\begin{array}{r} 2\,1\,6\,0 \\ \times \quad 1.0\,8 \\ \hline 1\,7\,2\,8\,0 \\ 0\,0\,0\,0 \\ 2\,1\,6\,0 \\ \hline 2\,3\,3\,2.8\,0 \end{array}$$

The amount in the account after 2 years is $2332.80.

4. Find the amount in an account if $2000 is invested at 11%, compounded annually, for 2 years.

Do Exercise 4.

Suppose that the interest in Example 4 were **compounded semi-annually**—that is, every half year. Interest would then be calculated twice a year at a rate of 8% ÷ 2, or 4%, each time. The approach used in Example 4 can then be adapted, as follows.

After the first $\frac{1}{2}$ year, the account will contain 104% of $2000:

$1.04 \times \$2000 = \$2080.$ These calculations can be confirmed with a calculator.

After a second $\frac{1}{2}$ year (1 full year), the account will contain 104% of $2080:

$1.04 \times \$2080 = \$2163.20.$

After a third $\frac{1}{2}$ year $\left(1\frac{1}{2}\text{ full years}\right)$, the account will contain 104% of $2163.20:

$1.04 \times \$2163.20 = \$2249.728$

$\approx \$2249.73.$ Rounding to the nearest cent

Finally, after a fourth $\frac{1}{2}$ year (2 full years), the account will contain 104% of $2249.73:

$1.04 \times \$2249.73 = \$2339.7192$

$\approx \$2339.72.$ Rounding to the nearest cent

Note that each multiplication was by 1.04 and that

$\$2000 \times 1.04^4 = \$2339.72.$ Using a calculator and rounding to the nearest cent

Answer on page A-13

5. A couple invests \$7000 in an account paying 10%, compounded semiannually. Find the amount in the account after $1\frac{1}{2}$ years.

Answer on page A-13

We have illustrated the following result.

> If a principal P has been invested at interest rate r, compounded n times a year, in t years it will grow to an amount A given by
>
> $$A = P \cdot \left(1 + \frac{r}{n}\right)^{n \cdot t}.$$

Example 5 The Ibsens invest \$4000 in an account paying 8%, compounded quarterly. Find the amount in the account after $2\frac{1}{2}$ years.

We substitute \$4000 for P, 8% for r, 4 for n, and $2\frac{1}{2}$ fot t and solve for A:

$$\begin{aligned} A &= P \cdot \left(1 + \frac{r}{n}\right)^{n \cdot t} \\ &= 4000 \cdot \left(1 + \frac{0.08}{4}\right)^{4 \cdot (5/2)} && 8\% = 0.08;\ 2\tfrac{1}{2} = \tfrac{5}{2} \\ &= 4000 \cdot (1 + 0.02)^{10} \\ &= 4000 \cdot (1.02)^{10} \\ &\approx 4875.98. && \text{Using a calculator} \end{aligned}$$

The amount in the account after $2\frac{1}{2}$ years is \$4875.98.

Do Exercise 5.

Calculator Spotlight

When using a calculator for interest computations, it is important to remember the order in which operations are performed and to minimize "round-off error." For example, to find the amount due on a \$20,000 loan made for 25 days at 11%, compounded daily, we press the following sequence of keys:

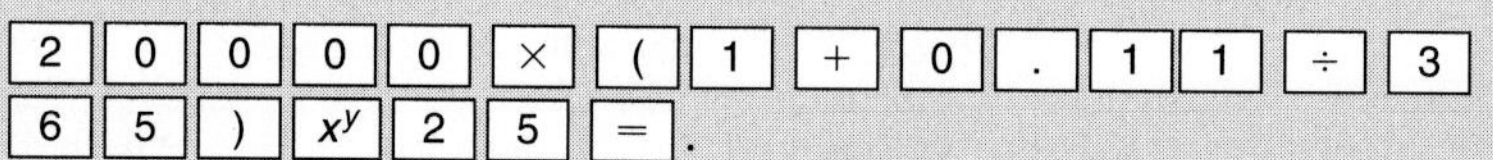

This key may appear differently. See p. 75.

Without parentheses keys, we would press

1 + 0 . 1 1 ÷ 3 6 5 = x^y 2 5 = × 2 0 0 0 0 = .

Note that in both sequences of keystrokes we raise $1 + \frac{0.11}{365}$, not just $\frac{0.11}{365}$, to the power.

Exercises

1. Find the amount due on a \$16,000 loan made for 62 days at 13%, compounded daily.
2. An investment of \$12,500 is made for 90 days at 8.5%, compounded daily. How much is the investment worth after 90 days?

Exercise Set 6.7

a Find the *simple* interest.

	Principal	*Rate of interest*	*Time*
1.	$200	13%	1 year
2.	$450	18%	1 year
3.	$2000	12.4%	$\frac{1}{2}$ year
4.	$200	7.7%	$\frac{1}{2}$ year
5.	$4300	14%	$\frac{1}{4}$ year
6.	$2000	15%	$\frac{1}{4}$ year

Solve. Assume that simple interest is being calculated in each case.

7. CopiPix, Inc., borrows $10,000 at 9% for 60 days. Find (a) the amount of interest due and (b) the total amount that must be paid after 60 days.

8. Sal's Laundry borrows $8000 at 10% for 90 days. Find (a) the amount of interest due and (b) the total amount that must be paid after 90 days.

9. Animal Instinct, a pet supply shop, borrows $6500 at 8% for 90 days. Find (a) the amount of interest due and (b) the total amount that must be paid after 90 days.

10. Andante's Cafe borrows $4500 at 9% for 60 days. Find (a) the amount of interest due and (b) the total amount that must be paid after 60 days.

11. Jean's Garage borrows $5600 at 10% for 30 days. Find (a) the amount of interest due and (b) the total amount that must be paid after 30 days.

12. Shear Delights, a hair salon, borrows $3600 at 8% for 30 days. Find (a) the amount of interest due and (b) the total amount that must be paid after 30 days.

b Interest is compounded annually. Find the amount in the account after the given length of time. Round to the nearest cent.

	Principal	*Rate of interest*	*Time*
13.	$400	10%	2 years
14.	$400	7.7%	2 years
15.	$200	8.8%	2 years
16.	$1000	15%	2 years

Interest is compounded semiannually. Find the amount in the account after the given length of time. Round to the nearest cent.

	Principal	*Rate of interest*	*Time*
17.	$4000	7%	1 year
18.	$1000	5%	1 year
19.	$2000	9%	3 years
20.	$5000	8%	30 months

Solve.

21. A family invests $4000 in an account paying 6%, compounded monthly. How much is in the account after 5 months?

22. A couple invests $2500 in an account paying 9%, compounded monthly. How much is in the account after 6 months?

23. A couple invests $1200 in an account paying 10%, compounded quarterly. How much is in the the account after 1 year?

24. The O'Hares invest $6000 in an account paying 8%, compounded quarterly. How much is in the account after 18 months?

Skill Maintenance

Solve. [5.3b]

25. $\frac{9}{10} = \frac{x}{5}$

26. $\frac{7}{x} = \frac{4}{5}$

27. $\frac{3}{4} = \frac{6}{x}$

28. $\frac{7}{8} = \frac{x}{100}$

Convert to a mixed numeral. [3.4b]

29. $\frac{100}{3}$

30. $\frac{64}{17}$

31. $\frac{38}{3}$

32. $\frac{38}{11}$

Convert from a mixed numeral to fractional notation. [3.4a]

33. $1\frac{1}{17}$

34. $20\frac{9}{10}$

35. $101\frac{1}{2}$

36. $32\frac{3}{8}$

Synthesis

37. ◆ Which is a better investment and why: $1000 invested at $14\frac{3}{4}\%$ simple interest for 1 year, or $1000 invested at 14% compounded monthly for 1 year?

38. ◆ A firm must choose between borrowing $5000 at 10% for 30 days and borrowing $10,000 at 8% for 60 days. Give arguments in favor of and against each option.

39. ▦ What is the simple interest on $24,680 at 7.75% for $\frac{3}{4}$ year?

40. ▦ Which gives the most interest, $\$1000 \times 8\% \times \frac{30}{360}$, or $\$1000 \times 8\% \times \frac{30}{365}$?

41. ▦ Interest is compounded semiannually. Find the value of the investment if $24,800 is invested at 6.4% for 5 years.

42. ▦ Interest is compounded quarterly. Find the value of the investment if $125,000 is invested at 9.2% for $2\frac{1}{2}$ years.

Effective Yield. The *effective yield* is the yearly rate of simple interest that corresponds to a rate for which interest is compounded two or more times a year. For example, if P is invested at 12%, compounded quarterly, we would multiply P by $(1 + 0.12/4)^4$, or 1.03^4. Since $1.03^4 \approx 1.126$, the 12% compounded quarterly corresponds to an effective yield of approximately 12.6%. In Exercises 38 and 39, find the effective yield for the indicated account.

43. ▦ The account pays 9% compounded monthly.

44. ▦ The account pays 10% compounded daily.

45. ▦ Rather than spend $20,000 on a new car that will lose 30% of its value in 1 year, the Coniglios invest the money at 9%, compounded daily. After 1 year, how much have the Coniglios saved by not buying the car?

Collaborative Learning Manual

Prepare an amortization table for a car loan.

Summary and Review Exercises: Chapter 6

Important Properties and Formulas

Commission = Commission rate × Sales
Sale price = Original price − Discount
Compounded Interest: $A = P \cdot \left(1 + \frac{r}{n}\right)^{n \cdot t}$
Discount = Rate of discount × Original price
Simple Interest: $I = P \cdot r \cdot t$

The objectives to be tested in addition to the material in this chapter are [3.4b], [4.4b], [4.1c], [4.5a], and [5.3b].

Find percent notation. [6.1c]

1. 0.483

2. 0.36

Find percent notation. [6.2a]

3. $\frac{3}{8}$

4. $\frac{1}{3}$

Find decimal notation. [6.1b]

5. 73.5%

6. $6\frac{1}{2}\%$

Find fractional notation. [6.2b]

7. 24%

8. 6.3%

Translate to an equation. Then solve. [6.3a, b]

9. 30.6 is what percent of 90?

10. 63 is 84 percent of what?

11. What is $38\frac{1}{2}\%$ of 168?

Translate to a proportion. Then solve. [6.4a, b]

12. 24 percent of what is 16.8?

13. 42 is what percent of 30?

14. What is 10.5% of 84?

Solve. [6.5a, b]

15. Food expenses account for 26% of the average family's budget. A family makes $2300 one month. How much do they spend for food?

16. The price of a television set was reduced from $350 to $308. Find the percent of decrease in price.

17. Jerome County has a population that is increasing 3% each year. This year the population is 80,000. What will it be next year?

18. The price of a box of cookies increased from $1.70 to $2.04. What was the percent of increase in the price?

19. Carney College has a student body of 960 students. Of these, 17.5% are seniors. How many students are seniors?

Solve. [6.6a, b, c]

20. A state charges a meals tax of $4\frac{1}{2}\%$. What is the meals tax charged on a dinner party costing $320?

21. In a certain state, a sales tax of $378 is collected on the purchase of a used car for $7560. What is the sales tax rate?

22. Kim earns $753.50 selling $6850 worth of televisions. What is the commission rate?

23. An air conditioner has a marked price of $350. It is placed on sale at 12% off. What are the discount and the sale price?

24. A fax machine priced at $305 is discounted at the rate of 14%. What are the discount and the sale price?

25. An insurance salesperson receives a 7% commission. If $42,000 worth of life insurance is sold, what is the commission?

Solve. [6.7a, b]

26. What is the simple interest on $1800 at 6% for $\frac{1}{3}$ year?

27. The Dress Shack borrows $24,000 at 10% simple interest for 60 days. Find (a) the amount of interest due and (b) the total amount that must be paid after 60 days.

28. What is the simple interest on $2200 principal at the interest rate of 5.5% for 1 year?

29. The Kleins invest $7500 in an investment account paying 12%, compounded monthly. How much is in the account after 3 months?

30. Find the amount in an investment account if $8000 is invested at 9%, compounded annually, for 2 years.

31. Find the rate of discount. [6.6c]

Skill Maintenance

Solve. [5.3c]

32. $\frac{3}{8} = \frac{7}{x}$

33. $\frac{1}{6} = \frac{7}{x}$

Solve. [4.4b]

34. $10.4 \times y = 665.6$

35. $100 \cdot x = 761.23$

Convert to decimal notation. [4.1c], [4.5a]

36. $\frac{11}{3}$

37. $\frac{11}{7}$

Convert to a mixed numeral. [3.4b]

38. $\frac{11}{3}$

39. $\frac{121}{7}$

Synthesis

40. ◆ Ollie buys a microwave oven during a 10%-off sale. The sale price that Ollie paid was $162. To find the original price, Ollie calculates 10% of $162 and adds that to $162. Is this correct? Why or why not? [6.6c]

41. ◆ Which is a better deal for a consumer and why: a discount of 40% or a discount of 20% followed by another of 22%? [6.6c]

42. ▦ *Land Area of the United States.* After Hawaii and Alaska became states, the total land area of the United States increased from 2,963,681 mi^2 to 3,540,939 mi^2. What was the percent of increase or decrease? [6.5b]

43. Rhonda's Dress Shop reduces the price of a dress by 40% during a sale. By what percent must the store increase the sale price, after the sale, to get back to the original price? [6.6c]

44. A $200 coat is marked up 20%. After 30 days, it is marked down 30% and sold. What was the final selling price of the coat? [6.6c]

Test: Chapter 6

1. Find decimal notation for 89%.

2. Find percent notation for 0.674.

3. Find percent notation for $\frac{11}{8}$.

4. Find fractional notation for 65%.

5. Translate to an equation. Then solve.

   What is 40% of 55?

6. Translate to a proportion. Then solve.

   What percent of 80 is 65?

Solve.

7. *Weight of Muscles.* The weight of muscles in a human body is 40% of total body weight. A person weighs 125 lb. What do the muscles weigh?

8. The population of Rippington increased from 1500 to 3600. Find the percent of increase in population.

9. *Arizona Tax Rate.* The sales tax rate in Arizona is 5%. How much tax is charged on a purchase of $324? What is the total price?

10. Gwen's commission rate is 15%. What is the commission from the sale of $4200 worth of merchandise?

11. The marked price of a CD player is $200 and the item is on sale at 20% off. What are the discount and the sale price?

12. What is the simple interest on a principal of $120 at the interest rate of 7.1% for 1 year?

13. The Burnham Parents–Teachers Association invests $5200 at 6% simple interest. How much is in the account after $\frac{1}{2}$ year?

14. Find the amount in an account if $1000 is invested at 5%, compounded annually, for 2 years.

Answers

1. ______
2. ______
3. ______
4. ______
5. ______
6. ______
7. ______
8. ______
9. ______
10. ______
11. ______
12. ______
13. ______
14. ______

Answers

15. ______

16. ______

17. ______

18. ______

19. ______

20. ______

21. ______

22. ______

23. a) ______

b) ______

c) ______

15. The Suarez family invests $10,000 at 9%, compounded monthly. How much is in the account after 3 months?

16. Find the discount and the discount rate of the bed in this ad.

Skill Maintenance

Solve.

17. $8.4 \times y = 1864.8$

18. $\frac{5}{8} = \frac{10}{x}$

19. Convert to decimal notation: $\frac{17}{12}$.

20. Convert to a mixed numeral: $\frac{153}{44}$.

Synthesis

21. By selling a home without using a realtor, Juan and Marie can avoid paying a 7.5% commission. They receive an offer of $109,000 from a potential buyer. In order to give a comparable offer, for what price would a realtor need to sell the house? Round to the nearest hundred.

22. Karen's commission rate is 16%. She invests her commission from the sale of $15,000 worth of merchandise at the interest rate of 12%, compounded quarterly. How much is Karen's investment worth after 6 months?

23. A housing development is constructed on a dead-end road along a river and ends in a cul-de-sac, as shown in the figure.

The property owners agree to share the cost of maintaining the road in the following manner. The first fifth of the road in front of lot 1 is to be shared equally among all five lot owners. The cost of the second fifth in front of lot 2 is to be shared equally among the owners of lots 2–5, and so on. Assume that all five sections of the road cost the same to maintain.

a) What fractional part of the cost is paid by each owner?

b) What percent of the cost is paid by each owner?

c) If lots 3, 4, and 5 were all owned by the same person, what percent of the cost of maintenance would this person pay?

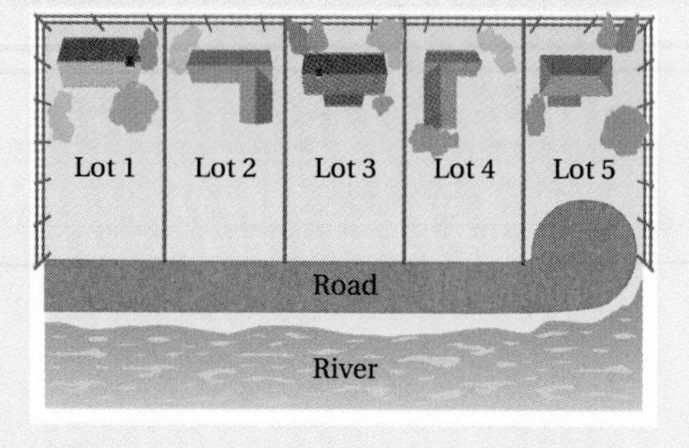

Cumulative Review: Chapters 1–6

1. Find fractional notation for 0.091.

2. Find decimal notation for $\frac{13}{6}$.

3. Find decimal notation for 3%.

4. Find percent notation for $\frac{9}{8}$.

5. Write fractional notation for the ratio 5 to 0.5.

6. Find the rate in kilometers per hour.

   350 km, 15 hr

Use $<$, $>$, or $=$ for $\blacksquare$ to write a true sentence.

7. $\frac{5}{7} \blacksquare \frac{6}{8}$

8. $\frac{6}{14} \blacksquare \frac{15}{25}$

Estimate the sum or difference by rounding to the nearest hundred.

9. $263{,}961 + 32{,}090 + 127.89$

10. $73{,}510 - 23{,}450$

Calculate.

11. $46 - [4(6 + 4 \div 2) + 2 \times 3 - 5]$

12. $[0.8(1.5 - 9.8 \div 49) + (1 + 0.1)^2] \div 1.5$

Add and simplify.

13. $\frac{6}{5} + 1\frac{5}{6}$

14. $46.9 + 2.84$

15. $$\begin{array}{r} 487{,}094 \\ 6{,}936 \\ +\ 21{,}120 \\ \hline \end{array}$$

Subtract and simplify.

16. $35 - 34.98$

17. $3\frac{1}{3} - 2\frac{2}{3}$

18. $\frac{8}{9} - \frac{6}{7}$

Multiply and simplify.

19. $\frac{7}{9} \cdot \frac{3}{14}$

20. $$\begin{array}{r} 236{,}984 \\ \times\ 3{,}600 \\ \hline \end{array}$$

21. $$\begin{array}{r} 46.012 \\ \times\ 0.03 \\ \hline \end{array}$$

Divide and simplify.

22. $6\frac{3}{5} \div 4\frac{2}{5}$

23. $431.2 \div 35.2$

24. $15\overline{)1850}$

Solve.

25. $36 \cdot x = 3420$

26. $y + 142.87 = 151$

27. $\frac{2}{15} \cdot t = \frac{6}{5}$

28. $\frac{3}{4} + x = \frac{5}{6}$

29. $\frac{y}{25} = \frac{24}{15}$

30. $\frac{16}{n} = \frac{21}{11}$

Solve.

31. On a checking account of $7428.63, a check was drawn for $549.79. What was left in the account?

32. A total of $57.50 was paid for 5 neckties. How much did each cost?

33. A 12-oz box of crackers costs $3.69. Find the unit price in cents per ounce.

34. A bus travels 456 km in 6 hr. At this rate, how far would the bus travel in 8 hr?

35. *Paper Disposal.* In a recent year, Americans threw away 50 million lb of paper. It is projected that this will increase to 65 million lb in the year 2000. Find the percent of increase.

36. The state of Utah has an area of 1,722,850 mi^2. Of this area, 60% is owned by the government. How many square miles are owned by the government?

37. *McDonald's.* In a recent year, there were 9145 McDonald's restaurants and 7258 Pizza Hut restaurants in the United States. How many more McDonald's restaurants were there?

38. How many pieces of ribbon $1\frac{4}{5}$ yd long can be cut from a length of ribbon 9 yd long?

39. A student walked $\frac{7}{10}$ mi to school and then $\frac{8}{10}$ mi to the library. How far did the student walk?

40. On a map, 1 in. represents 80 mi. How much does $\frac{3}{4}$ in. represent?

41. The Bakers invest $8500 in an investment account paying 8%, compounded monthly. How much is in the account after 5 months?

42. You manage to save $10,000 at 7% for 1 year, but your federal taxes take 20% of the interest earned. How much is in your account after 1 year? By what percent is the new amount higher than the old amount?

Synthesis

43. First National Bank offers 10% simple interest. Saturn Bank offers 9.75% interest compounded semiannually. Which bank offers the highest rate of return on the investment?

44. On a trip through the mountains, a car was driven 240 mi on $7\frac{1}{2}$ gal of gasoline. On a trip across the plains, the same car was driven 351 mi on $9\frac{3}{4}$ gal of gasoline. What was the percent of increase or decrease in miles per gallon?

7

Data Analysis, Graphs, and Statistics

An Application

Tricia adds one slice of chocolate cake with fudge frosting (560 calories) to her diet each day for one year (365 days) and makes no other changes in her eating or exercise habits. The consumption of 3500 extra calories will add about 1 lb to her body weight. How many pounds will she gain?

This problem appears as Exercise 12 in Exercise Set 7.3.

The Mathematics

We see that Tricia will consume

365×560, or 204,400 calories

in 1 year. Thus her weight gain is

$$\frac{204{,}400}{3500} \approx 58 \text{ lb.}$$

World Wide Web For more information, visit us at www.mathmax.com

Introduction

There are many ways in which we can analyze or describe data. One is to organize the data into a table or a graph. Another is to look at certain *statistics* related to the data. After such organization, many problems can be solved, including those of making predictions from the data. We will consider many kinds of graphs: pictographs, bar graphs, line graphs, and circle graphs. Statistics we will consider are *means, medians,* and *modes.*

Pretest: Chapter 7

In Questions 1–3, find (a) the average, (b) the median, and (c) the mode.

1. 46, 50, 53, 55

2. 5, 4, 3, 2, 1

3. 4, 17, 4, 18, 4, 17, 18, 20

4. A car was driven 660 mi in 12 hr. What was the average number of miles per hour?

5. To get a C in chemistry, Delia must average 70 on four tests. Scores on the first three tests were 68, 71, and 65. What is the lowest score that she can make on the last test and still get a C?

6. *Reasons for Exercising.* The following data show the percentage of women selecting a particular reason for exercising. Make a circle graph to show the data.

Health: 51%
Lose weight: 38%
Relieve stress: 11%

7. *Cost of Life Insurance.* The following table shows the comparison of the cost of a $100,000 life insurance policy for female smokers and nonsmokers at certain ages.

a) How much does it cost a female nonsmoker, age 32, for insurance?

b) How much more does it cost a female smoker, age 35, than a nonsmoker at the same age?

8. Using the data in Question 7, draw a vertical bar graph showing the cost of insurance for a female smoker at various ages. Use age on the horizontal scale and cost on the vertical scale.

9. Using the data in Question 7, draw a line graph showing the cost of insurance for a female smoker at various ages. Use age on the horizontal scale and cost on the vertical scale.

LIFE INSURANCE: FEMALE

Age	Cost (Smoker)	Cost (Nonsmoker)
31	$254	$201
32	273	208
33	294	221
34	319	236
35	341	249

Source: State Farm Insurance

Risk of Heart Disease. The line graph below shows the relationship between blood cholesterol level and risk of coronary heart disease.

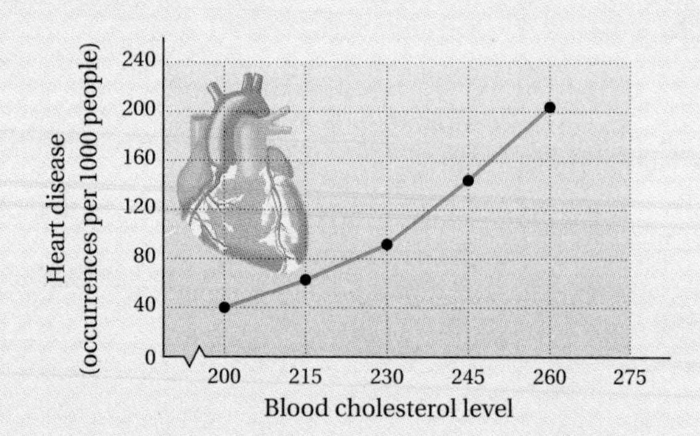

10. At what cholesterol level is the risk highest?

11. About how much higher is the risk at 260 than at 200?

12. *Study Time vs. Grades.* An English instructor asked his students to keep track of how much time each spent studying for a chapter test. He collected the information together with the test scores. The data are given in the table below.

Study Time (in hours)	Test Grade (in percent)
9	75
11	83
13	80
15	85
17	80
18	86
21	87
23	92
24	?

a) Draw a line graph of the data.

b) Estimate the missing data value.

Objectives for Retesting

The objectives to be tested in addition to the material in this chapter are as follows.

[2.7b]	Divide and simplify using fractional notation.
[5.4a]	Solve applied problems involving proportions.
[6.3b], [6.4b]	Solve basic percent problems.
[6.5a]	Solve applied problems involving percent.

7.1 Averages, Medians, and Modes

Data are often available regarding some kind of application involving mathematics. We can use tables and graphs of various kinds to show information about the data and to extract information from the data that can lead us to make analyses and predictions. Graphs allow us to communicate a message from the data.

For example, the following show data regarding credit-card spending between Thanksgiving and Christmas in recent years. Examine each method of presentation. Which method, if any, do you like the best and why? Which do you like the least and why?

Objectives

- **a** Find the average of a set of numbers and solve applied problems involving averages.
- **b** Find the median of a set of numbers and solve applied problems involving medians.
- **c** Find the mode of a set of numbers and solve applied problems involving modes.

For Extra Help

TAPE 12

TAPE 11B

MAC WIN

CD-ROM

PARAGRAPH FORM

The National Credit Counseling Services has recently released data regarding credit-card spending between Thanksgiving and Christmas for various years. In 1991, spending was $59.8 billion; in 1992, it was $66.8 billion; in 1993, it was $79.1 billion; in 1994, it was $96.9 billion; in 1995, it was $116.3 billion; and finally, in 1996, it was $131.4 billion.

TABLE

Year	Credit-Card Spending from Thanksgiving to Christmas (in billions)
1991	$ 59.8
1992	66.8
1993	79.1
1994	96.9
1995	116.3
1996	131.4

Source: RAM Research Group, National Credit Counseling Services

PICTOGRAPH

BAR GRAPH

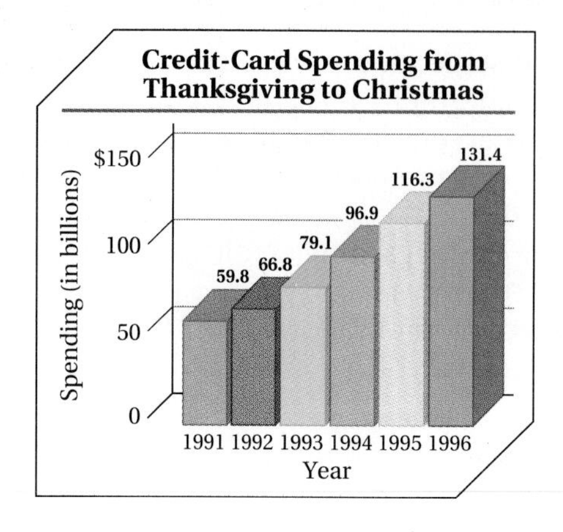

LINE GRAPH

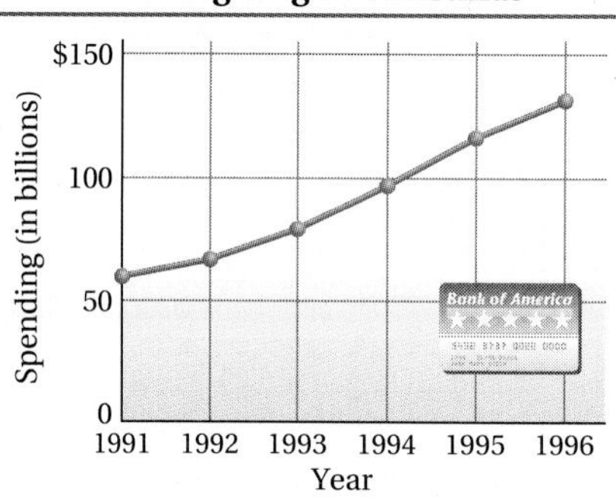

CIRCLE, OR PIE, GRAPH

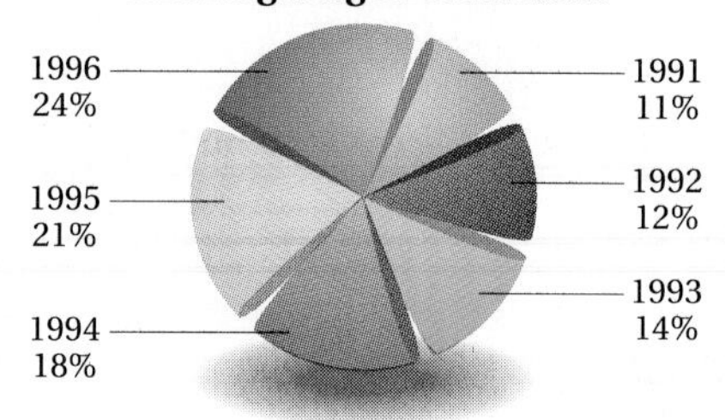

Most people would not find the paragraph method for displaying the data most useful. It takes time to read, and it is hard to look for a trend and make predictions. The circle, or pie, graph might be used to compare what part of the entire amount of spending over the six years each individual year represents, but that comparison is not the same comparison as those presented by the bar and line graphs. The bar and line graphs might be more worthwhile if we want to see the trend of increased spending and to make predictions about the years 1997 and beyond.

In this chapter, we will learn not only how to extract information from various kinds of tables and graphs, but also how to create various kinds of graphs.

a Averages

A **statistic** is a number describing a set of data. One statistic is a *center point* that characterizes the data. The most common kind of center point is the *mean,* or *average,* of a set of numbers. We first considered averages in Section 1.9.

Let's consider the data on credit-card spending (given in billions):

$59.8, $66.8, $79.1, $96.9, $116.3, $131.4.

What is the *average* of the numbers? First, we add the numbers:

$$59.8 + 66.8 + 79.1 + 96.9 + 116.3 + 131.4 = 550.3.$$

Next, we divide by the number of data items, 6:

$$\frac{550.3}{6} \approx \$91.7. \qquad \textbf{Rounding to the nearest tenth}$$

Note that

$$91.7 + 91.7 + 91.7 + 91.7 + 91.7 + 91.7 = 550.2 \approx 550.3.$$

The number 91.7 is called the **average** of the set of numbers. It is also called the **arithmetic** (pronounced ăr´ ĭth-mĕt´-ĭk) **mean** or simply the **mean**.

> To find the **average** of a set of numbers, add the numbers and then divide by the number of items of data.

Example 1 On a 4-day trip, a car was driven the following number of miles each day: 240, 302, 280, 320. What was the average number of miles per day?

$$\frac{240 + 302 + 280 + 320}{4} = \frac{1142}{4}, \quad \text{or} \quad 285.5$$

The car was driven an average of 285.5 mi per day. Had the car been driven exactly 285.5 mi each day, the same total distance (1142 mi) would have been traveled.

Do Exercises 1–4.

Example 2 *Food Waste.* Courtney is a typical American consumer. In the course of 1 yr, she discards 100 lb of food waste. What is the average number of pounds of food waste discarded each week? Round to the nearest tenth.

We already know the total amount of food waste for the year. Since there are 52 weeks in a year, we divide by 52 and round:

$$\frac{100}{52} \approx 1.9.$$

On average, Courtney discards 1.9 lb of food waste per week.

Do Exercise 5.

Find the average.

1. 14, 175, 36

2. 75, 36.8, 95.7, 12.1

3. A student scored the following on five tests: 68, 85, 82, 74, 96. What was the average score?

4. In the first five games, a basketball player scored points as follows: 26, 21, 13, 14, 23. Find the average number of points scored per game.

5. *Food Waste.* Courtney also composts (converts to dirt) 5 lb of food waste each year. How much, on average, does Courtney compost per month? Round to the nearest tenth.

Answers on page A-15

6. *Gas Mileage.* According to recent EPA estimates, a Toyota Camry LE can be expected to travel 209 mi (city) on 11 gal of gasoline (***Source:*** *Popular Science Magazine*). What is the average number of miles expected per gallon?

7. *GPA.* Jennifer earned the following grades one semester.

Grade	Number of Credit Hours in Course
B	3
C	4
C	4
A	2

What was Jennifer's grade point average? Assume that the grade point values are 4.0 for an A, 3.0 for a B, and so on. Round to the nearest tenth.

Answers on page A-15

Example 3 *Gas Mileage.* According to recent EPA estimates, an Oldsmobile Aurora can be expected to travel 204 mi (city) on 12 gal of gasoline. What is the average number of miles expected per gallon?

We divide the total number of miles, 204, by the number of gallons, 12:

$$\frac{204}{12} = 17 \text{ mpg.}$$

The Aurora's expected average is 17 miles per gallon.

Do Exercise 6.

Example 4 *GPA.* In most colleges, students are assigned grade point values for grades obtained. The **grade point average,** or **GPA,** is the average of the grade point values for each credit hour taken. At many colleges, grade point values are assigned as follows:

A: 4.0
B: 3.0
C: 2.0
D: 1.0
F: 0.0

Tom earned the following grades for one semester. What was his grade point average?

Course	Grade	Number of Credit Hours in Course
History	B	4
Basic mathematics	A	5
English	A	5
French	C	3
Physical education	F	1

To find the GPA, we first multiply the grade point value (in color below) by the number of credit hours in the course and then add, as follows:

History	$3.0 \cdot 4 = 12$
Basic mathematics	$4.0 \cdot 5 = 20$
English	$4.0 \cdot 5 = 20$
French	$2.0 \cdot 3 = 6$
Physical education	$0.0 \cdot 1 = 0$
	58 (Total)

The total number of credit hours taken is $4 + 5 + 5 + 3 + 1$, or 18. We divide 58 by 18 and round to the nearest tenth:

$$\text{GPA} = \frac{58}{18} \approx 3.2.$$

Tom's grade point average was 3.2.

Do Exercise 7.

Example 5 To get a B in math, Geraldo must score an average of 80 on the tests. On the first four tests, his scores were 79, 88, 64, and 78. What is the lowest score that Geraldo can get on the last test and still get a B?

We can find the total of the five scores needed as follows:

$$80 + 80 + 80 + 80 + 80 = 5 \cdot 80, \quad \text{or} \quad 400.$$

The total of the scores on the first four tests is

$$79 + 88 + 64 + 78 = 309.$$

Thus Geraldo needs to get at least

$$400 - 309, \quad \text{or} \quad 91$$

in order to get a B. We can check this as follows:

$$\frac{79 + 88 + 64 + 78 + 91}{5} = \frac{400}{5}, \quad \text{or} \quad 80.$$

Do Exercise 8.

8. To get an A in math, Rosa must score an average of 90 on the tests. On the first three tests, her scores were 80, 100, and 86. What is the lowest score that Rosa can get on the last test and still get an A?

b Medians

Another type of center-point statistic is the *median.* Medians are useful when we wish to de-emphasize unusually extreme scores. For example, suppose a small class scored as follows on an exam.

Phil:	78	Pat:	56
Jill:	81	Olga:	84
Matt:	82		

Let's first list the scores in order from smallest to largest:

56, 78, 81, 82, 84.

↑ Middle score (81)

The middle score—in this case, 81—is called the **median.** Note that because of the extremely low score of 56, the average of the scores is 76.2. In this example, the median may be a more appropriate center-point statistic.

Calculator Spotlight

Averages can be easily computed on a calculator if we remember the order in which operations are performed. For example, to calculate

$$\frac{85 + 92 + 79}{3}$$

on most calculators, we press

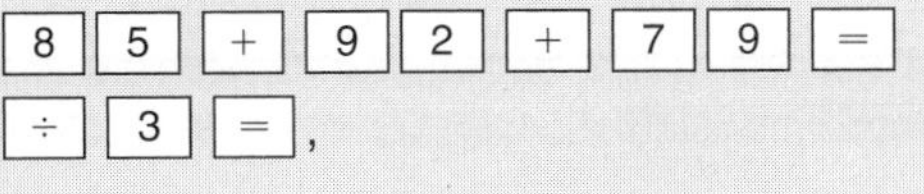

,

or

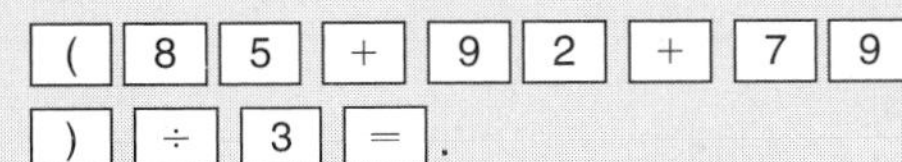

.

Exercises

1. What would the result have been if we had not used parentheses in the latter sequence of keystrokes?
2. Use a calculator to solve Examples 1–5.

Example 6 What is the median of this set of numbers?

99, 870, 91, 98, 106, 90, 98

We first rearrange the numbers in order from smallest to largest. Then we locate the middle number, 98.

90, 91, 98, 98, 99, 106, 870

↑ Middle number (98)

The median is 98.

Do Exercises 9–11.

Find the median.

9. 17, 13, 18, 14, 19

10. 20, 14, 13, 19, 16, 18, 17

11. 78, 81, 83, 91, 103, 102, 122, 119, 88

> Once a set of data is listed in order, from smallest to largest, the **median** is the middle number if there is an odd number of data items. If there is an even number of items, the median is the number that is the average of the two middle numbers.

Answers on page A-15

Find the median.

12. \$1300, \$2000, \$1900, \$1600, \$1800, \$1400

13. 68, 34, 67, 69, 34, 70

Find the modes of these data.

14. 23, 45, 45, 45, 78

15. 34, 34, 67, 67, 68, 70

16. 13, 24, 27, 28, 67, 89

17. In a lab, Gina determined the mass, in grams, of each of five eggs:

15 g, 19 g, 19 g, 14 g, 18 g.

a) What is the mean?

b) What is the median?

c) What is the mode?

Answers on page A-15

Example 7 What is the median of this set of numbers?

69, 80, 61, 63, 62, 65

We first rearrange the numbers in order from smallest to largest. There is an even number of numbers. We look for the middle two, which are 63 and 65. The median is halfway between 63 and 65, the number 64.

61, 62, 63, ↑ 65, 69, 80

The average of the middle numbers is $\frac{63 + 65}{2}$, or 64.

The median is 64.

Example 8 What is the median of this set of yearly salaries?

\$35,000, \$500,000, \$28,000, \$34,000

We rearrange the numbers in order from smallest to largest. The two middle numbers are \$34,000 and \$35,000. Thus the median is halfway between \$34,000 and \$35,000 (the average of \$34,000 and \$35,000):

\$28,000, \$34,000, \$35,000, \$500,000

$$\text{Median} = \frac{\$34{,}000 + \$35{,}000}{2} = \frac{\$69{,}000}{2} = \$34{,}500.$$

Do Exercises 12 and 13.

c Modes

The final type of center-point statistic is the **mode**.

> The **mode** of a set of data is the number or numbers that occur most often. If each number occurs the same number of times, there is *no* mode.

Example 9 Find the mode of these data.

13, 14, 17, 17, 18, 19

The number that occurs most often is 17. Thus the mode is 17.

A set of data has just one average (mean) and just one median, but it can have more than one mode. It is also possible for a set of data to have no mode—when all numbers are equally represented. For example, the set of data 5, 7, 11, 13, 19 has no mode.

Example 10 Find the modes of these data.

33, 34, 34, 34, 35, 36, 37, 37, 37, 38, 39, 40

There are two numbers that occur most often, 34 and 37. Thus the modes are 34 and 37.

Do Exercises 14–17.

Exercise Set 7.1

[a], [b], [c] For each set of numbers, find the average, the median, and any modes that exist.

1. 16, 18, 29, 14, 29, 19, 15

2. 72, 83, 85, 88, 92

3. 5, 30, 20, 20, 35, 5, 25

4. 13, 32, 25, 27, 13

5. 1.2, 4.3, 5.7, 7.4, 7.4

6. 13.4, 13.4, 12.6, 42.9

7. 234, 228, 234, 229, 234, 278

8. $29.95, $28.79, $30.95, $29.95

9. The following temperatures were recorded for seven days in Hartford:

   43°, 40°, 23°, 38°, 54°, 35°, 47°.

   What was the average temperature? the median? the mode?

10. Lauri Merten, a professional golfer, scored 71, 71, 70, and 68 to win the U.S. Women's Open in a recent year. What was the average score? the median? the mode?

11. *Gas Mileage.* According to recent EPA estimates, an Achieva can be expected to travel 297 mi (highway) on 9 gal of gasoline (***Source:*** *Motor Trend Magazine*). What is the average number of miles expected per gallon?

12. *Gas Mileage.* According to recent EPA estimates an Aurora can be expected to travel 192 mi (highway) on 8 gal of gasoline (***Source:*** *Motor Trend Magazine*). What is the average number of miles expected per gallon?

GPA. In Exercises 13 and 14 are the grades of a student for one semester. In each case, find the grade point average. Assume that the grade point values are 4.0 for an A, 3.0 for a B, and so on. Round to the nearest tenth.

13.

Grades	Number of Credit Hours in Course
B	4
B	5
B	3
C	4

14.

Grades	Number of Credit Hours in Course
A	5
B	4
B	3
C	5

15. The following prices per pound of Atlantic salmon were found at five fish markets:

    $7.99, $9.49, $9.99, $7.99, $10.49.

    What was the average price per pound? the median price? the mode?

16. The following prices per pound of Vermont cheddar cheese were found at five supermarkets:

    $4.99, $5.79, $4.99, $5.99, $5.79.

    What was the average price per pound? the median price? the mode?

17. To get a B in math, Rich must score an average of 80 on five tests. Scores on the first four tests were 80, 74, 81, and 75. What is the lowest score that Rich can get on the last test and still receive a B?

18. To get an A in math, Cybil must score an average of 90 on five tests. Scores on the first four tests were 90, 91, 81, and 92. What is the lowest score that Cybil can get on the last test and still receive an A?

19. Marta was pregnant 270 days, 259 days, and 272 days for her first three pregnancies. In order for Marta's average pregnancy to equal the worldwide average of 266 days, how long must her fourth pregnancy last? (***Source***: David Crystal (ed.), *The Cambridge Factfinder*. Cambridge CB2 1RP: Cambridge University Press, 1993, p. 84.)

20. Jason's brothers are 174 cm, 180 cm, 179 cm, and 172 cm tall. The average male is 176.5 cm tall. How tall is Jason if he and his brothers have an average height of 176.5 cm?

Skill Maintenance

Multiply.

21. $14 \cdot 14$ [1.5b]

22. $\frac{2}{3} \cdot \frac{2}{3}$ [2.6a]

23. 1.4×1.4 [4.3a]

24. 1.414×1.414 [4.3a]

Solve. [5.4a]

25. Four software CDs cost $239.80. How much would 19 comparable CDs cost?

26. A car is driven 700 mi in 5 days. At this rate, how far will it have been driven in 24 days?

Synthesis

27. ◆ You are applying for an entry-level job at a large firm. You can be informed of the mean, median, or mode salary. Which of the three figures would you request? Why?

28. ◆ Is it possible for a driver to average 20 mph on a 30-mi trip and still receive a ticket for driving 75 mph? Why or why not?

Bowling Averages. Bowling averages are always computed by rounding down to the nearest integer. For example, suppose a bowler gets a total of 599 for 3 games. To find the average, we divide 599 by 3 and drop the amount to the right of the decimal point:

$$\frac{599}{3} \approx 199.67. \quad \text{The bowler's average is 199.}$$

In each case, find the bowling average.

29. 547 in 3 games

30. 4621 in 27 games

31. *Hank Aaron.* Hank Aaron averaged $34\frac{7}{22}$ home runs per year over a 22-yr career. After 21 yr, Aaron had averaged $35\frac{10}{21}$ home runs per year. How many home runs did Aaron hit in his final year?

32. The ordered set of data 18, 21, 24, a, 36, 37, b has a median of 30 and an average of 32. Find a and b.

Perform a statistical analysis of pulse rates.

7.2 Tables and Pictographs

a Reading and Interpreting Tables

A **table** is often used to present data in rows and columns.

Example 1 *Cereal Data.* Let's assume that you generally have a 2-cup bowl of cereal each morning. The following table lists nutritional information for five name-brand cereals. (It does not consider the use of milk, sugar, or sweetener.) The data have been determined by doubling the information given for a 1-cup serving that is found in the Nutrition Facts panel on a box of cereal.

Cereal	Calories	Fat	Total Carbohydrate	Sodium
Ralston Rice Chex	240	0 g	54 g	460 mg
Kellogg's Complete Bran Flakes	240	1.3 g	64 g	613.3 mg
Kellogg's Special K	220	0 g	44 g	500 mg
Honey Nut Cheerios	240	3 g	48 g	540 mg
Wheaties	220	2 g	48 g	440 mg

a) Which cereal has the least amount of sodium per serving?

b) Which cereal has the greatest amount of fat?

c) Which cereal has the least amount of fat?

d) Find the average total carbohydrate in the cereals.

Careful examination of the table will give the answers.

a) To determine which cereal has the least amount of sodium, look down the column headed "Sodium" until you find the smallest number. That number is 440 mg. Then look across that row to find the brand of cereal, Wheaties.

b) To determine which cereal has the greatest amount of fat, look down the column headed "Fat" until you find the largest number. That number is 3 g. Then look across that row to find the cereal, Honey Nut Cheerios.

c) To determine which cereal has the least amount of fat, look down the column headed "Fat" until you find the smallest number. There are two listings of 0 g. Then look across those rows to find the cereals, Ralston Rice Chex and Kellogg's Special K.

d) Find the average of all the numbers in the column headed "Total Carbohydrate":

$$\frac{54 + 64 + 44 + 48 + 48}{5} = 51.6.$$

The average total carbohydrate content is 51.6 g.

Do Exercises 1–7. (Exercises 5–7 are on the following page.)

Objectives

a Extract and interpret data from tables.

b Extract and interpret data from pictographs.

c Draw simple pictographs.

For Extra Help

TAPE 12 | TAPE 11B | MAC WIN | CD-ROM

Use the table in Example 1 to answer each of the following.

1. Which cereal has the most total carbohydrate?

2. Which cereal has the least total carbohydrate?

3. Which cereal has the least number of calories?

4. Which cereal has the greatest number of calories?

Answers on page A-15

Example 2 *Wheaties Nutrition Facts.* Most foods are required by law to provide factual information regarding nutrition, as shown in the following table of Nutrition Facts from a box of Wheaties cereal. Although this can be very helpful to the consumer, one must be careful in interpreting the data. The % Daily Value figures shown here are based on a 2000-calorie diet. Your daily values may be higher or lower, depending on your calorie needs or intake.

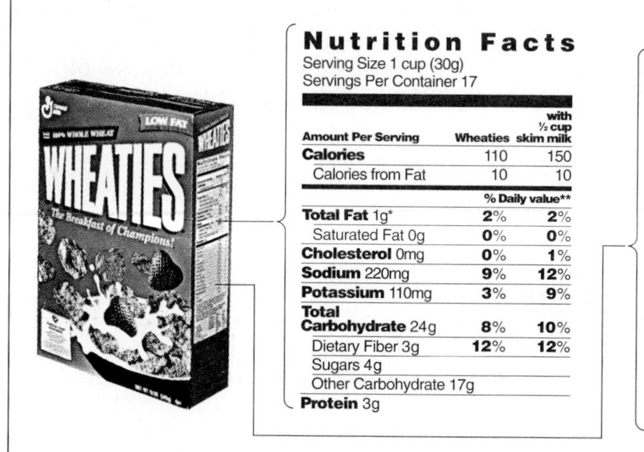

Nutrition Facts
Serving Size 1 cup (30g)
Servings Per Container 17

Amount Per Serving	Wheaties	with ½ cup skim milk
Calories	110	150
Calories from Fat	10	10
	% Daily value**	
Total Fat 1g*	**2**%	**2**%
Saturated Fat 0g	**0**%	**0**%
Cholesterol 0mg	**0**%	**1**%
Sodium 220mg	**9**%	**12**%
Potassium 110mg	**3**%	**9**%
Total Carbohydrate 24g	**8**%	**10**%
Dietary Fiber 3g	**12**%	**12**%
Sugars 4g		
Other Carbohydrate 17g		
Protein 3g		

Vitamin A	25%	30%
Vitamin C	25%	25%
Calcium	0%	15%
Iron	45%	45%
Vitamin D	10%	25%
Thiamin	25%	30%
Riboflavin	25%	35%
Niacin	25%	25%
Vitamin B_6	25%	25%
Folic Acid	25%	25%
Phosphorus	10%	20%
Magnesium	8%	10%
Zinc	4%	8%
Copper	4%	4%

*Amount in Cereal. A serving of cereal plus skim milk provides 1g fat, less than 5mg cholesterol, 280mg sodium, 310mg potassium, 30g carbohydrate (10g sugars), and 7g protein.

**Percent Daily Values are based on a 2,000 calorie diet. Your daily values may be higher or lower depending on your calorie needs.

Suppose your morning bowl of cereal consists of 2 cups of Wheaties together with 1 cup of skim milk, with artificial sweetener containing 0 calories.

a) How many calories have you consumed?

b) What percent of the daily value of total fat have you consumed?

c) A nutritionist recommends that you look for foods that provide 10% or more of the daily value for vitamin C. Do you get that with your bowl of Wheaties?

d) Suppose you are trying to limit your daily caloric intake to 2500 calories. How many bowls of cereal would it take to exceed the 2500 calories, even though you probably would not eat just cereal?

Careful examination of the table of nutrition facts will give the answers.

a) Look at the column marked "with ½ cup skim milk" and note that 1 cup of cereal with ½ cup skim milk contains 150 calories. Since you are having twice that amount, you are consuming

$$2 \times 150, \quad \text{or} \quad 300 \text{ calories.}$$

b) Read across from "Total Fat" and note that in 1 cup of cereal with ½ cup skim milk, you get 2% of the daily value of fat. Since you are doubling that, you get 4% of the daily value of fat.

c) Find the row labeled "Vitamin C" on the left and look under the column labeled "with ½ cup skim milk." Note that you get 25% of the daily value for "1 cup with ½ cup of skim milk," and since you are doubling that, you are more than satisfying the 10% requirement.

d) From part (a), we know that you are consuming 300 calories per bowl. Dividing 2500 by 300 gives $\frac{2500}{300} \approx 8.33$. Thus if you eat 9 bowls of cereal in this manner, you will exceed the 2500 calories.

Do Exercises 8–12.

5. Find the average amount of sodium in the cereals.

6. Find the median of the amount of sodium in the cereals.

7. Find the mean, the median, and the mode of the number of calories in the cereals.

Use the Nutrition Facts data from the Wheaties box and the bowl of cereal described in Example 2 to answer each of the following.

8. How many calories from fat are in your bowl of cereal?

9. A nutritionist recommends that you look for foods that provide 10% or more of the daily value for iron. Do you get that with your bowl of Wheaties?

10. How much sodium have you consumed?

11. What daily value of sodium have you consumed?

12. How much protein have you consumed?

Answers on page A-15

b Reading and Interpreting Pictographs

Pictographs (or *picture graphs*) are another way to show information. Instead of actually listing the amounts to be considered, a **pictograph** uses symbols to represent the amounts. In addition, a *key* is given telling what each symbol represents.

Example 3 *Elephant Population.* The following pictograph shows the elephant population of various countries in Africa. Located on the graph is a key that tells you that each symbol represents 10,000 elephants.

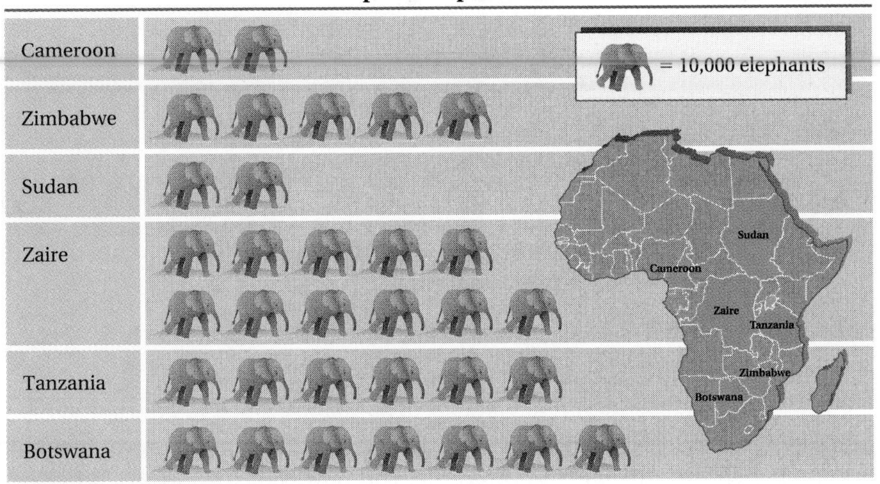

Source: National Geographic

a) Which country has the greatest number of elephants?

b) Which country has the least number of elephants?

c) How many more elephants are there in Zaire than in Botswana?

We can compute the answers by first reading the pictograph.

a) The country with the most symbols has the greatest number of elephants: Zaire, with $11 \times 10{,}000$, or 110,000 elephants.

b) The countries with the fewest symbols have the least number of elephants: Cameroon and Sudan, each with $2 \times 10{,}000$, or 20,000 elephants.

c) From part (a), we know that there are 110,000 elephants in Zaire. In Botswana there are $7 \times 10{,}000$, or 70,000 elephants. Thus there are $110{,}000 - 70{,}000$, or 40,000 more elephants in Zaire than in Botswana.

Do Exercises 13–15.

You have probably noticed that, although they seem to be very easy to read, pictographs are difficult to draw accurately because whole symbols reflect loose approximations due to significant rounding. In pictographs, you also need to use some mathematics to find the actual amounts.

Use the pictograph in Example 3 to answer each of the following.

13. How many elephants are there in Tanzania?

14. How does the elephant population of Zimbabwe compare to that of Cameroon?

15. What is the average number of elephants in these six countries?

Answers on page A-15

Use the pictograph in Example 4 to answer each of the following.

16. Determine the approximate coffee consumption per capita of France.

17. Determine the approximate coffee consumption per capita of Italy.

18. The approximate coffee consumption of Finland is about the same as the combined coffee consumptions of Switzerland and the United States. What is the approximate coffee consumption of Finland?

Answers on page A-15

Example 4 *Coffee Consumption.* For selected countries, the following pictograph shows approximately how many cups of coffee each person (per capita) drinks annually.

Coffee Consumption

Source: Beverage Marketing Corporation

a) Determine the approximate annual coffee consumption per capita of Germany.

b) Which two countries have the greatest difference in coffee consumption? Estimate that difference.

We use the data from the pictograph as follows.

a) Germany's consumption is represented by 11 whole symbols (1100 cups) and, though it is visually debatable, about $\frac{1}{8}$ of another symbol (about 13 cups), for a total of 1113 cups.

b) Visually, we see that Switzerland has the most consumption and that the United States has the least consumption. Switzerland's annual coffee consumption per capita is represented by 12 whole symbols (1200 cups) and about $\frac{1}{5}$ of another symbol (20 cups), for a total of 1220 cups. U.S. consumption is represented by 6 whole symbols (600 cups) and about $\frac{1}{10}$ of another symbol (10 cups), for a total of 610 cups. The difference between these amounts is $1220 - 610$, or 610 cups.

One advantage of pictographs is that the appropriate choice of a symbol will tell you, at a glance, the kind of measurement being made. Another advantage is that the comparison of amounts represented in the graph can be expressed more easily by just counting symbols. For instance, in Example 3, the ratio of elephants in Zaire to those in Cameroon is 11:2.

One disadvantage of pictographs is that, to make a pictograph easy to read, the amounts must be rounded significantly to the unit that a symbol represents. This makes it difficult to accurately represent an amount. Another problem is that it is difficult to determine very accurately how much a partial symbol represents. A third disadvantage is that you must use some mathematics to finally compute the amount represented, since there is usually no explicit statement of the amount.

Do Exercises 16–18.

c Drawing Pictographs

Example 5 *Concert Revenue.* The following list shows the top five concert acts in a recent year and their total gross revenue (money taken in) (***Source***: *Pollstar Magazine*). Draw a pictograph to represent the data. Let the symbol [dollar symbol] represent $10,000,000.

Kiss	$43,600,000
Garth Brooks	$34,500,000
Neil Diamond	$32,200,000
Rod Stewart	$29,100,000
Bob Seger	$26,300,000

Some computation is necessary before we can draw the pictograph.

Kiss: Note that $43{,}600{,}000 = 4.36 \times 10{,}000{,}000$. Thus we need 4 whole symbols and 0.36 of another symbol. Now 0.36 is hard to draw, but we estimate it to be about 33%, or $\frac{1}{3}$, of a symbol.

Garth Brooks: Note that $34{,}500{,}000 = 3.45 \times 10{,}000{,}000$. Thus we need 3 whole symbols and 0.45, or about half, of another symbol.

Neil Diamond: Note that $32{,}200{,}000 = 3.22 \times 10{,}000{,}000$. Thus we need 3 whole symbols and 0.22, or about 20% or $\frac{1}{5}$, of another symbol.

Rod Stewart: Note that $29{,}100{,}000 = 2.91 \times 10{,}000{,}000$. Thus we need 2 whole symbols and 0.91 of another, or about 3 whole symbols.

Bob Seger: Note that $26{,}300{,}000 = 2.63 \times 10{,}000{,}000$. Thus we need 2 whole symbols and 0.63, or about 60% or $\frac{3}{5}$, of another symbol.

The pictograph can now be drawn as follows. We list the concert act or performer in one column, draw the monetary amounts with their symbols, and title the overall graph "Total Gross Revenue."

Total Gross Revenue

Kiss	[4 1/3 symbols]
Garth Brooks	[3 1/2 symbols]
Neil Diamond	[3 1/5 symbols]
Rod Stewart	[3 symbols]
Bob Seger	[2 3/5 symbols]
	[symbol] = $10,000,000

Do Exercise 19.

19. *Concert Revenue.* The following is a list of the next five concert acts for the same year and their total gross revenues. Draw a pictograph to represent the data.

Jimmy Buffett	$26,200,000
Reba McEntire	$26,100,000
Alanis Morissette	$23,200,000
Hootie & the Blowfish	$21,400,000
Ozzy Osbourne	$21,300,000

Answer on page A-15

Improving Your Math Study Skills

Forming Math Study Groups, by James R. Norton

Dr. James Norton has taught at the University of Phoenix and Scottsdale Community College. He has extensive experience with the use of study groups to learn mathematics.

The use of math study groups for learning has become increasingly more common in recent years. Some instructors regard them as a primary source of learning, while others let students form groups on their own.

A study group generally consists of study partners who help each other learn the material and do the homework. You will probably meet outside of class at least once or twice a week. Here are some do's and don'ts to make your study group more valuable.

- DO make the group up of no more than four or five people. Research has shown clearly that this size works best.
- DO trade phone numbers so that you can get in touch with each other for help between team meetings.
- DO make sure that everyone in the group has a chance to contribute.
- DON'T let a group member copy from others without contributing. If this should happen, one member should speak with that student privately; if the situation continues, that student should be asked to leave the group.
- DON'T let the "A" students drop the ball. The group needs them! The benefits to even the best students are twofold: (1) Other students will benefit from their expertise and (2) the bright students will learn the material better by teaching it to someone else.
- DON'T let the slower students drop the ball either. *Everyone* can contribute something, and being in a group will actually improve their self-esteem as well as their performance.

How do you form study groups if the instructor has not already done so? A good place to begin is to get together with three or four friends and arrange a study time. If you don't know anyone, start getting acquainted with other people in the class during the first week of the semester.

What should you look for in a study partner?

- Do you live near each other to make it easy to get together?
- What are your class schedules like? Are you both on campus? Do you have free time?
- What about work schedules, athletic practice, and other out-of-school commitments that you might have to work around?

Making use of a study group is not a form of "cheating." You are merely helping each other learn. So long as everyone in the group is both contributing and doing the work, this method will bring you great success!

Exercise Set 7.2

a *Planets.* Use the following table, which lists information about the planets, for Exercises 1–10.

Planet	Average Distance from Sun (in miles)	Diameter (in miles)	Length of Planet's Day in Earth Time (in days)	Time of Revolution in Earth Time (in years)
Mercury	35,983,000	3,031	58.82	0.24
Venus	67,237,700	7,520	224.59	0.62
Earth	92,955,900	7,926	1.00	1.00
Mars	141,634,800	4,221	1.03	1.88
Jupiter	483,612,200	88,846	0.41	11.86
Saturn	888,184,000	74,898	0.43	29.46
Uranus	1,782,000,000	31,763	0.45	84.01
Neptune	2,794,000,000	31,329	0.66	164.78
Pluto	3,666,000,000	1,423	6.41	248.53

Source: *Handy Science Answer Book*, Gale Research, Inc.

1. Find the average distance from the sun to Jupiter.

2. How long is a day on Venus?

3. Which planet has a time of revolution of 164.78 yr?

4. Which planet has a diameter of 4221 mi?

5. Which planets have an average distance from the sun that is greater than 1,000,000 mi?

6. Which planets have a diameter that is less than 100,000 mi?

7. About how many earth diameters would it take to equal one Jupiter diameter?

8. How much longer is the longest time of revolution than the shortest?

9. What are the average, the median, and the mode of the diameters of the planets?

10. What are the average, the median, and the mode of the average distances from the sun of the planets?

Heat Index. In warm weather, a person can feel hotter due to reduced heat loss from the skin caused by higher humidity. The **temperature–humidity index,** or **apparent temperature,** is what the temperature would have to be with no humidity in order to give the same heat effect. The following table lists the apparent temperatures for various actual temperatures and relative humidities. Use this table for Exercises 11–22.

Actual Temperature (°F)	Relative Humidity									
	10%	20%	30%	40%	50%	60%	70%	80%	90%	100%
	Apparent Temperature (°F)									
75°	75	77	79	80	82	84	86	88	90	92
80°	80	82	85	87	90	92	94	97	99	102
85°	85	88	91	94	97	100	103	106	108	111
90°	90	93	97	100	104	107	111	114	118	121
95°	95	99	103	107	111	115	119	123	127	131
100°	100	105	109	114	118	123	127	132	137	141
105°	105	110	115	120	125	131	136	141	146	151

In Exercises 11–14, find the apparent temperature for the given actual temperature and humidity combinations.

11. 80°, 60%

12. 90°, 70%

13. 85°, 90%

14. 95°, 80%

15. How many temperature–humidity combinations give an apparent temperature of 100°?

16. How many temperature–humidity combinations give an apparent temperature of 111°?

17. At a relative humidity of 50%, what actual temperatures give an apparent temperature above 100°?

18. At a relative humidity of 90%, what actual temperatures give an apparent temperature above 100°?

19. At an actual temperature of 95°, what relative humidities give an apparent temperature above 100°?

20. At an actual temperature of 85°, what relative humidities give an apparent temperature above 100°?

21. At an actual temperature of 85°, by how much would the humidity have to increase in order to raise the apparent temperature from 97° to 111°?

22. At an actual temperature of 80°, by how much would the humidity have to increase in order to raise the apparent temperature from 87° to 102°?

Global Warming. Ecologists are increasingly concerned about global warming, that is, the trend of average global temperatures to rise over recent years. One possible effect is the melting of the polar icecaps. Use the following table for Exercises 23–26.

Year	Average Global Temperature (°F)
1986	59.29°
1987	59.58°
1988	59.63°
1989	59.45°
1990	59.85°
1991	59.74°
1992	59.23°
1993	59.36°
1994	59.56°
1995	59.72°
1996	59.58°

Source: *Vital Signs,* 1997

23. Find the average global temperatures in 1986 and 1987. What was the percent of increase in the temperature from 1986 to 1987?

24. Find the average global temperatures in 1992 and 1993. What was the percent of increase in the temperature from 1992 to 1993?

25. Find the average of the average global temperatures for the years 1986 to 1988. Find the average of the average global temperatures for the years 1994 to 1996. By how many degrees does the latter average exceed the former?

26. Find the average of the average global temperatures for the years 1994 to 1996. Find the ten-year average of the average global temperatures for the years 1987 to 1996. By how many degrees does the former average exceed the latter?

b *World Population Growth.* The following pictograph shows world population in various years. Use the pictograph for Exercises 27–34.

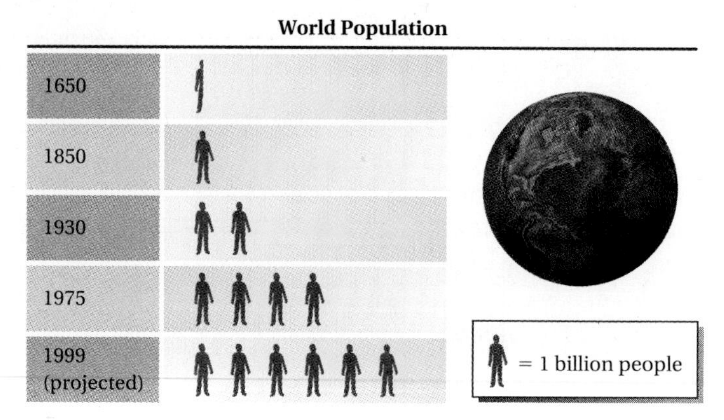

27. What was the world population in 1850?

28. What was the world population in 1975?

29. In which year will the population be the greatest?

30. In which year was the population the least?

31. Between which two years was the amount of growth the least?

32. Between which two years was the amount of growth the greatest?

33. How much greater will the world population in 1999 be than in 1975? What is the percent of increase?

34. How much greater will the world population be in 1999 than in 1930? What is the percent of increase?

Mountain Bikes. The following pictograph shows sales of mountain bikes for a bicycle company for six consecutive years. Use the pictograph for Exercises 35–42.

35. In which year was the greatest number of bikes sold?

36. Between which two consecutive years was there the greatest growth?

37. Between which two years did the least amount of positive growth occur?

38. How many sales does one bike symbol represent?

39. Approximately how many bikes were sold in 1996?

40. Approximately how many more bikes were sold in 1998 than in 1993?

41. In which year was there actually a decline in the number of bikes sold?

42. The sales for 1998 were how many times the sales for 1993?

c

43. *Lettuce Sales.* The sales of lettuce have experienced a tremendous increase in recent years due to the convenience of prepackaged, prewashed, and prechopped lettuce. Sales for recent years are listed below (***Source*:** International Fresh-Cut Produce Association). Draw a pictograph to represent lettuce sales for these years. Use the symbol to represent $100,000,000.

1992	$168,000,000
1993	$312,000,000
1994	$577,000,000
1995	$889,000,000
1996	$1,100,000,000

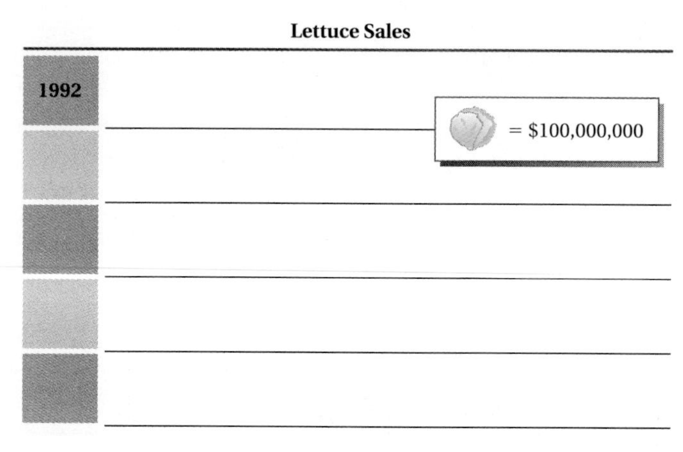

Skill Maintenance

Solve.

44. A football team has won 3 of its first 4 games. At this rate, how many games will it win in a 16-game season? [5.4a]

45. The state of Maine is 90% forest. The area of Maine is 30,955 mi^2. How many square miles of Maine are forest? [6.5a]

Find fractional notation for the percent notation in the sentence. [6.2b]

46. The United States uses 24% of the world's energy.

47. The United States has 4.8% of the world's population.

Synthesis

48. ◆ Loreena is drawing a pictograph in which dollar bills are used as symbols to represent the tuition at various private colleges. Should each dollar bill represent $8000, $4000, or $400? Why?

49. ◆ What advantage(s) does a table have over a pictograph?

50. Redraw the pictograph appearing in Example 4 as one in which each symbol represents 150 cups of coffee.

7.3 Bar Graphs and Line Graphs

A **bar graph** is convenient for showing comparisons because you can tell at a glance which amount represents the largest or smallest quantity. Of course, since a bar graph is a more abstract form of pictograph, this is true of pictographs as well. However, with bar graphs, a *second scale* is usually included so that a more accurate determination of the amount can be made.

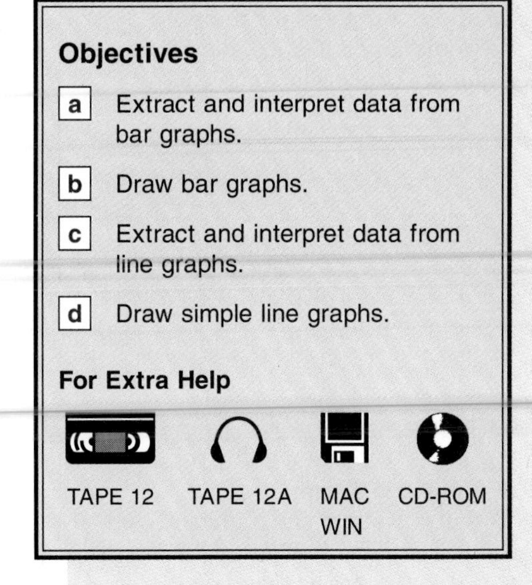

a Reading and Interpreting Bar Graphs

Example 1 *Fat Content in Fast Foods.* Wendy's Hamburgers is a national food franchise. The following bar graph shows the fat content of various sandwiches sold by Wendy's.

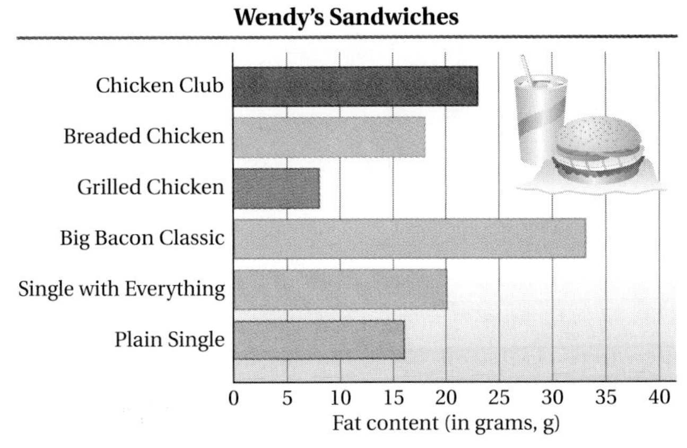

Source: Wendy's International

a) About how much fat is in a chicken club sandwich?

b) Which sandwich contains the least amount of fat?

c) Which sandwich contains about 20 g of fat?

We look at the graph to answer the questions.

a) We move to the right along the bar representing chicken club sandwiches. We can read, fairly accurately, that there is approximately 23 g of fat in the chicken club sandwich.

b) The shortest bar is for the grilled chicken sandwich. Thus that sandwich contains the least amount of fat.

c) We locate the line representing 20 g and then go up until we reach a bar that ends at approximately 20 g. We then go across to the left and read the name of the sandwich, which is the "Single with Everything."

Do Exercises 1–3.

Use the bar graph in Example 1 to answer each of the following.

1. About how much fat is in the plain single sandwich?

2. Which sandwich contains the greatest amount of fat?

3. Which sandwiches contain 20 g or more of fat?

Answers on page A-16

Use the bar graph in Example 2 to answer each of the following.

4. Approximately how many women, per 100,000, develop breast cancer between the ages of 35 and 39?

5. In what age group is the mortality rate the highest?

6. In what age group do about 350 out of every 100,000 women develop breast cancer?

7. Does the breast-cancer mortality rate seem to increase from the youngest to the oldest age group?

Answers on page A-16

Bar graphs are often drawn vertically and sometimes a double bar graph is used to make comparisons.

Example 2 *Breast Cancer.* The following graph indicates the incidence and mortality rates of breast cancer for women of various age groups.

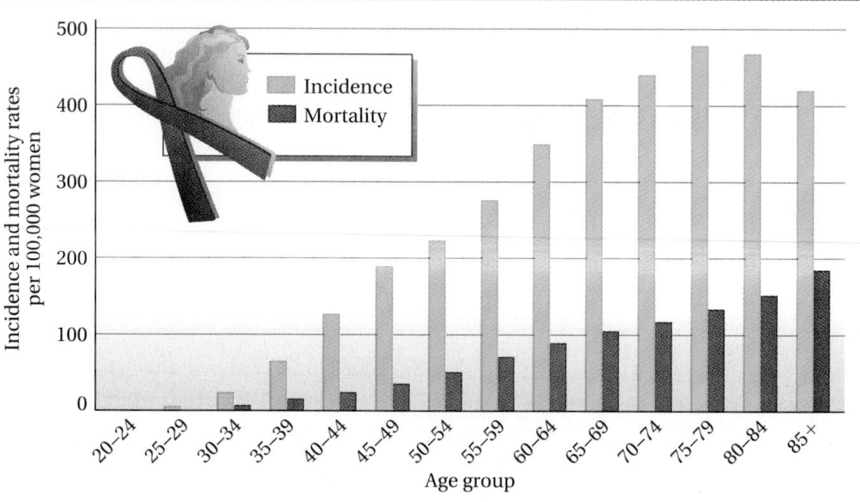

Source: National Cancer Institute

a) Approximately how many women, per 100,000, develop breast cancer between the ages of 40 and 44?

b) In what age range is the mortality rate for breast cancer approximately 100 for every 100,000 women?

c) In what age range is the incidence of breast cancer the highest?

d) Does the incidence of breast cancer seem to increase from the youngest to the oldest age group?

We look at the graph to answer the questions.

a) We go to the right, across the bottom, to the green bar above the age group 40–44. Next, we go up to the top of that bar and, from there, back to the left to read approximately 130 on the vertical scale. About 130 out of every 100,000 women develop breast cancer between the ages of 40 and 44.

b) We read up the vertical scale to the number 100. From there we move to the right until we come to the top of a red bar. Moving down that bar, we find that in the 65–69 age group, about 100 out of every 100,000 women die of breast cancer.

c) We look for the tallest green bar and read the age range below it. The incidence of breast cancer is highest for women in the 75–79 age group.

d) Looking at the heights of the bars, we see that the incidence of breast cancer increases to a high point in the 75–79 age group and then decreases.

Do Exercises 4–7.

b Drawing Bar Graphs

Example 3 *Heights of NBA Centers.* Listed below are the heights of some of the tallest centers in the NBA (***Source*:** National Basketball Association). Make a vertical bar graph of the data.

Shaquille O'Neal:	7′1″ (85 in.)
Shawn Bradley:	7′6″ (90 in.)
Rik Smits:	7′4″ (88 in.)
Gheorghe Muresan:	7′7″ (91 in.)
Arvydas Sabonis:	7′3″ (87 in.)
David Robinson:	7′1″ (85 in.)

First, we indicate on the base or horizontal scale in six equally spaced intervals the different names of the players and give the horizontal scale the title "Players." (See the figure on the left below.) Then we label the vertical scale with "Height (in inches)." We note that the largest number (in inches) is 91 and the smallest is 85. We could start the vertical scaling at 0, but then the bars would be very high. We decide to start at 83, using the jagged line to indicate the missing numbers. We label the marks by 1's from 83 to 91. Finally, we draw vertical bars to show the various heights (in inches), as shown in the figure on the right below. We give the graph the overall title "NBA Centers."

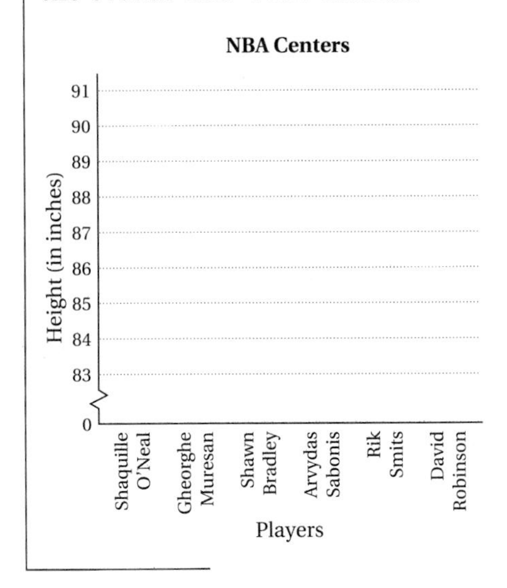

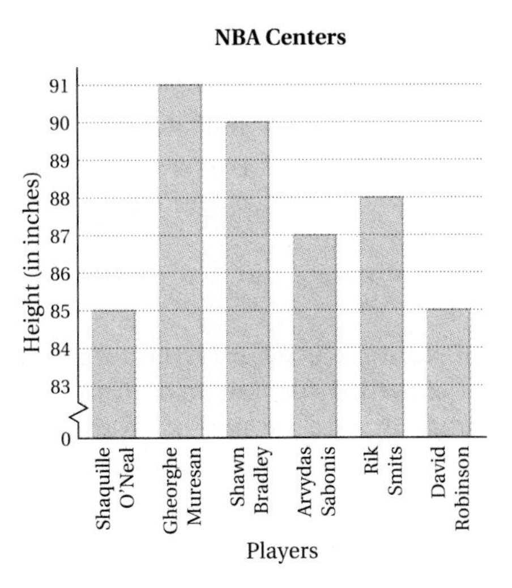

Do Exercise 8.

8. *Planetary Moons.* Make a horizontal bar graph to show the number of moons orbiting the various planets.

Planet	Number of Moons
Earth	1
Mars	2
Jupiter	16
Saturn	18
Uranus	15
Neptune	8
Pluto	1

Answer on page A-16

Use the line graph in Example 4 to answer each of the following.

9. For which month were new home sales lowest?

10. Between which months did new home sales decrease?

11. For which months were new home sales below 700 thousand?

Answers on page A-16

c Reading and Interpreting Line Graphs

Line graphs are often used to show a change over time as well as to indicate patterns or trends.

Example 4 *New Home Sales.* The following line graph shows the number of new home sales, in thousands, over a twelve-month period. The jagged line at the base of the vertical scale indicates an unnecessary portion of the scale. Note that the vertical scale differs from the horizontal scale so that the data can be shown reasonably.

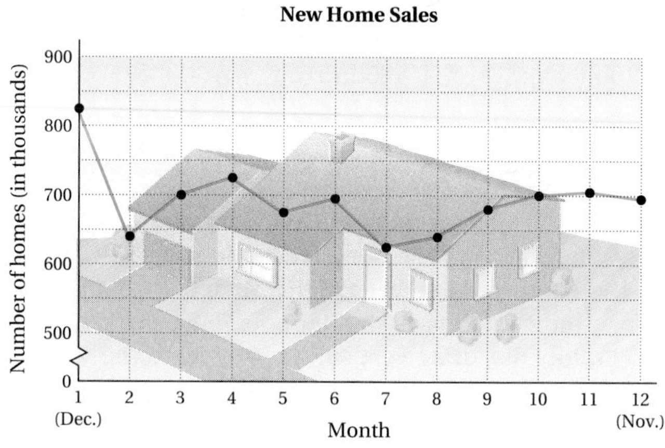

Source: U.S. Department of Commerce

a) For which month were new home sales the greatest?

b) Between which months did new home sales increase?

c) For which months were new home sales about 700 thousand?

We look at the graph to answer the questions.

a) The greatest number of new home sales was about 825 thousand in month 1.

b) Reading the graph from left to right, we see that new home sales increased from month 2 to month 3, from month 3 to month 4, from month 5 to month 6, from month 7 to month 8, from month 8 to month 9, from month 9 to month 10, and from month 10 to month 11.

c) We look from left to right along the line at 700.

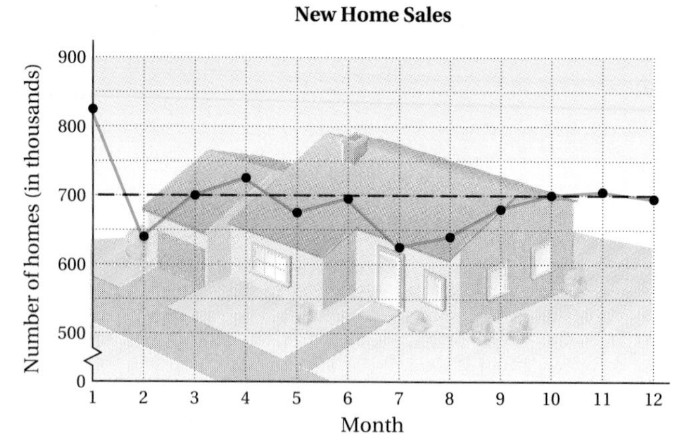

We see that points are closest to 700 thousand at months 3, 6, 10, 11, and 12.

Do Exercises 9–11.

Example 5 *Monthly Loan Payment.* Suppose that you borrow \$110,000 at an interest rate of 9% to buy a home. The following graph shows the monthly payment required to pay off the loan, depending on the length of the loan. (*Caution*: A low monthly payment means that you will pay more interest over the duration of the loan.)

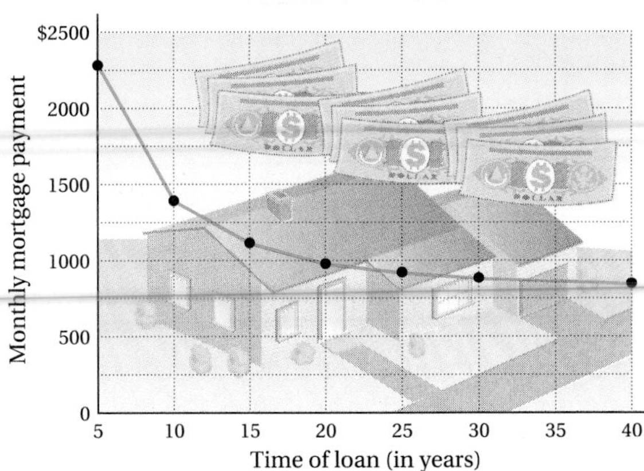

Note that there is no value for 35 yr. We will consider the reason for this in Section 7.5.

a) Estimate the monthly payment for a loan of 15 yr.

b) What time period corresponds to a monthly payment of about \$1400?

c) By how much does the monthly payment decrease when the loan period is increased from 10 yr to 20 yr?

We look at the graph to answer the questions.

a) We find the time period labeled "15" on the bottom scale and move up from that point to the line. We then go straight across to the left and find that the monthly payment is about \$1100.

b) We locate \$1400 on the vertical axis. Then we move to the right until we hit the line. The point \$1400 crosses the line at the 10-yr time period.

c) The graph shows that the monthly payment for 10 yr is about \$1400; for 20 yr, it is about \$990. Thus the monthly payment is decreased by \$1400 − \$990, or \$410. (It should be noted that you will pay back $\$990 \cdot 20 \cdot 12 - \$1400 \cdot 10 \cdot 12$, or \$69,600, more in interest for a 20-yr loan.)

Do Exercises 12–14.

Use the line graph in Example 5 to answer each of the following.

12. Estimate the monthly payment for a loan of 25 yr.

13. What time period corresponds to a monthly payment of about \$850?

14. By how much does the monthly payment decrease when the loan period is increased from 5 yr to 20 yr?

Answers on page A-16

d Drawing Line Graphs

Example 6 *Movie Releases.* Draw a line graph to show how the number of movies released each year has changed over a period of 6 yr. Use the following data (***Source*:** Motion Picture Association of America).

1991: 164 movies
1992: 150 movies
1993: 161 movies
1994: 184 movies
1995: 234 movies
1996: 260 movies

First, we indicate on the horizontal scale the different years and title it "Year." (See the graph below.) Then we mark the vertical scale appropriately by 50's to show the number of movies released and title it "Number per Year." We also give the overall title "Movies Released" to the graph.

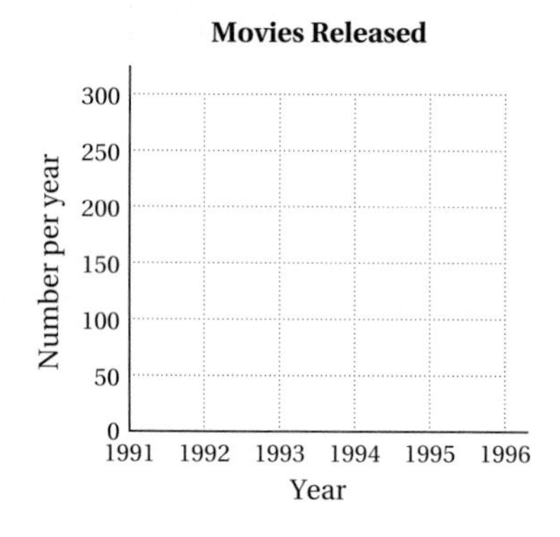

Next, we mark at the appropriate level above each year the points that indicate the number of movies released. Then we draw line segments connecting the points. The change over time can now be observed easily from the graph.

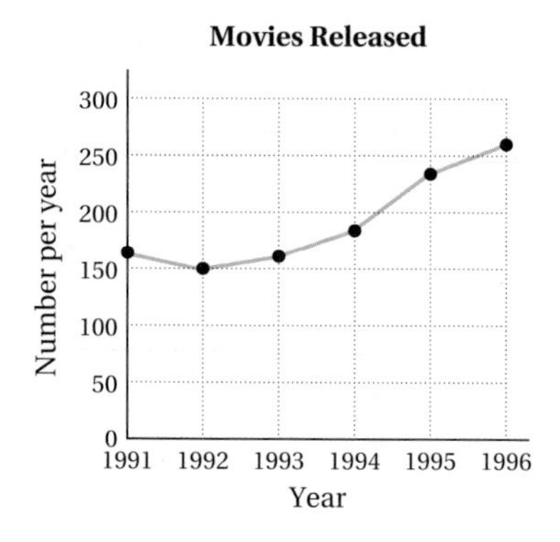

Do Exercise 15.

15. *SAT Scores.* Draw a line graph to show how the average combined verbal–math SAT score has changed over a period of 6 yr. Use the following data (***Source*:** The College Board).

1991: 999
1992: 1001
1993: 1003
1994: 1003
1995: 1010
1996: 1013

Answer on page A-16

Exercise Set 7.3

a *Chocolate Desserts.* The following horizontal bar graph shows the average caloric content of various kinds of chocolate desserts. Use the bar graph for Exercises 1–12.

Source: *Better Homes and Gardens*, December 1996

1. Estimate how many calories there are in 1 cup of hot cocoa with skim milk.

2. Estimate how many calories there are in 1 cup of premium chocolate ice cream.

3. Which dessert has the highest caloric content?

4. Which dessert has the lowest caloric content?

5. Which dessert contains about 460 calories?

6. Which desserts contain about 300 calories?

7. How many more calories are there in 1 cup of hot cocoa made with whole milk than in 1 cup of hot cocoa made with skim milk?

8. Fred generally drinks a 4-cup chocolate milkshake. How many calories does he consume?

9. Kristin likes to eat 2 cups of premium chocolate ice cream at bedtime. How many calories does she consume?

10. Barney likes to eat a 6-oz chocolate bar with peanuts for lunch. How many calories does he consume?

11. Paul adds a 2-oz chocolate bar with peanuts to his diet each day for 1 yr (365 days) and makes no other changes in his eating or exercise habits. Consumption of 3500 extra calories will add about 1 lb to his body weight. How many pounds will he gain?

12. Tricia adds one slice of chocolate cake with fudge frosting to her diet each day for 1 yr (365 days) and makes no other changes in her eating or exercise habits. Consumption of 3500 extra calories will add about 1 lb to her body weight. How many pounds will she gain?

Deforestation. The world is gradually losing its tropical forests. The following vertical triple bar graph shows the amount of forested land of three tropical regions in the years 1980 and 1990. Use the bar graph for Exercises 13–20.

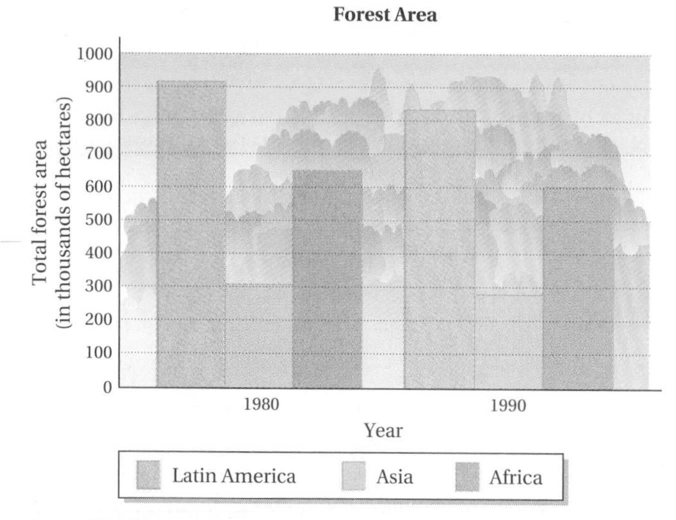

Source: World Resources Institute

13. What was the forest area of Latin America in 1980?

14. What was the forest area of Africa in 1990?

15. Which region experienced the greatest loss of forest area from 1980 to 1990?

16. Which region experienced the smallest loss of forest area from 1980 to 1990?

17. Which region had a forest area of about 600 thousand hectares in 1990?

18. Which region had a forest area of about 300 thousand hectares in 1980?

19. What was the average forest area in Latin America for the two years?

20. What was the average forest area in Asia for the two years?

b

21. *Commuting Time.* The following table lists the average commuting time in metropolitan areas with more than 1 million people. Make a vertical bar graph to illustrate the data.

City	Commuting Time (in minutes)
New York	30.6
Los Angeles	26.4
Phoenix	23.0
Dallas	24.1
Indianapolis	21.9
Orlando	22.9

Source: Census Bureau

Use the data and the bar graph in Exercise 21 to do Exercises 22–25.

22. Which city has the greatest commuting time?

23. Which city has the least commuting time?

24. What was the average commuting time for all six cities?

25. What was the median commuting time for the six cities?

26. *Deaths from Driving Incidents.* The following table lists for various years the number of driving incidents that resulted in death. Make a horizontal bar graph illustrating the data.

Year	Number of Incidents Causing Death
1990	1129
1991	1297
1992	1478
1993	1555
1994	1669
1995	1708

Source: AAA Foundation

Use the data and the bar graph in Exercise 26 to do Exercises 27–30.

27. Between which two years was the greatest increase in the number of incidents causing death?

28. Between which two years was the smallest increase in the number of incidents causing death?

29. What was the average number of incidents causing death over the 6-yr period?

30. What was the median number of incidents causing death over the 6-yr period?

c *Average Salary of Major-League Baseball Players.* The following graph shows the average salary of major-league baseball players over a recent 7-yr period. Use the graph for Exercises 31–36.

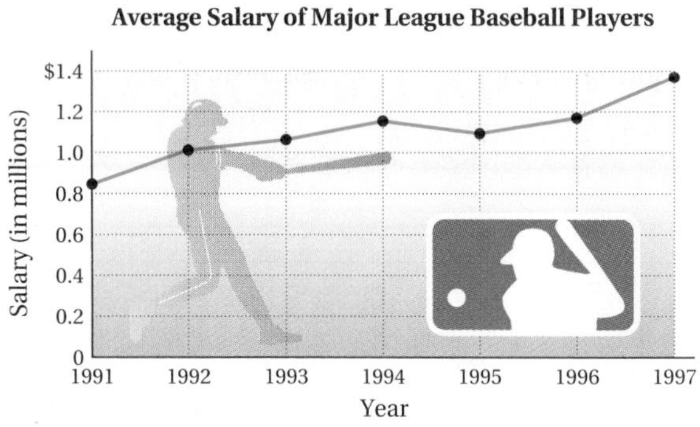

31. In which year was the average salary the highest?

32. In which year was the average salary the lowest?

33. What was the difference in salary between the highest and lowest salaries?

34. Between which two years was the increase in salary the greatest?

35. Between which two years did the salary decrease?

36. What was the percent of increase in salary between 1991 and 1996?

d Make a line graph of the data in the following tables (Exercises 37 and 42), using time on the horizontal scale.

37. *Ozone Layer.*

Year	Ozone Level (in parts per billion)
1991	2981
1992	3133
1993	3148
1994	3138
1995	3124

Source: National Oceanic and Atmospheric Administration

Use the data and the line graph in Exercise 37 to do Exercises 38–41.

38. Between which two years was the increase in the ozone level the greatest?

39. Between which two years was the decrease in the ozone level the greatest?

40. What was the average ozone level over the 5-yr period?

41. What was the median ozone level over the 5-yr period?

42. *Motion Picture Expense.*

Year	Average Expense per Picture (in millions)
1991	$38.2
1992	42.4
1993	44.0
1994	50.4
1995	54.1
1996	61.0

Source: Motion Picture Association of America

Use the data and the line graph in Exercise 42 to do Exercises 43–46.

43. Between which two years was the increase in motion-picture expense the greatest?

44. Between which two years was the increase in motion-picture expense the least?

45. What was the average motion-picture expense over the 6-yr period?

46. What was the median motion-picture expense over the 6-yr period?

47. What was the average motion-picture expense from 1991 through 1993?

48. What was the average motion-picture expense from 1994 through 1996?

Skill Maintenance

Solve.

49. A clock loses 3 min every 12 hr. At this rate, how much time will the clock lose in 72 hr? [5.4a]

50. It is known to operators of pizza restaurants that if 50 pizzas are ordered in an evening, people will request extra cheese on 9 of them. What percent of the pizzas sold are ordered with extra cheese? [6.5a]

51. 110% of 75 is what? [6.3b], [6.4b]

52. 34 is what percent of 51? [6.3b], [6.4b]

Synthesis

53. ◈ Compare bar graphs and line graphs. Discuss why you might use one over the other to graph a particular set of data.

54. ◈ Can bar graphs always, sometimes, or never be converted to line graphs? Why?

55. Referring to Exercise 42, what do you think was the average expense per picture in 1997? Justify your answer. How could you tell for sure?

Analyze class grades using line graphs and bar graphs.

7.4 Circle Graphs

We often use **circle graphs,** also called *pie charts,* to show the percent of a quantity used in different categories. Circle graphs can also be used very effectively to show visually the *ratio* of one category to another. In either case, it is quite often necessary to use mathematics to find the actual amounts represented for each specific category.

Objectives

a Extract and interpret data from circle graphs.

b Draw circle graphs.

For Extra Help

TAPE 12 TAPE 12A MAC WIN CD-ROM

a Reading and Interpreting Circle Graphs

Example 1 *Costs of Owning a Dog.* The following circle graph shows the relative costs of raising a dog from birth to death.

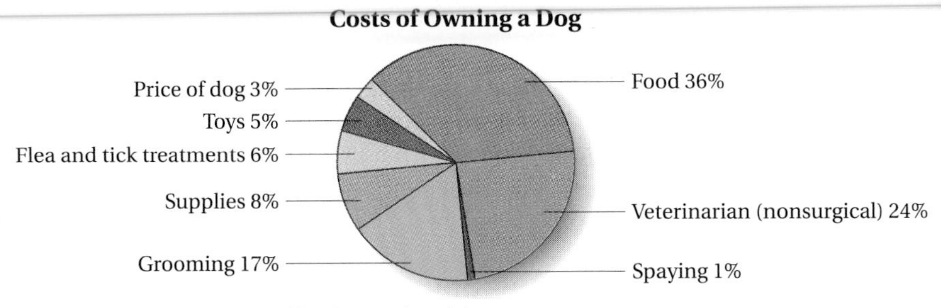

Source: The American Pet Products Manufacturers Association

a) Which item costs the most?

b) What percent of the total cost is spent on grooming?

c) Which item involves 24% of the cost?

d) The American Pet Products Manufacturers Association estimates that the total cost of owning a dog for its lifetime is $6600. How much of that amount is spent for food?

e) What percent of the expense is for grooming and flea and tick treatments?

We look at the sections of the graph to find the answers.

a) The largest section (or sector) of the graph, 36%, is for food.

b) We see that grooming is 17% of the cost.

c) Nonsurgical veterinarian bills account for 24% of the cost.

d) The section of the graph representing food costs is 36%; 36% of $6600 is $2376.

e) In a circle graph, we can add percents for questions like this. Therefore,

$$17\% \text{ (grooming)} + 6\% \text{ (flea and tick treatments)} = 23\%.$$

Do Exercises 1–4.

Use the circle graph in Example 1 to answer each of the following.

1. Which item costs the least?

2. What percent of the total cost is spent on toys?

3. How much of the $6600 lifetime cost of owning a dog is for grooming?

4. What part of the expense is for supplies and for buying the dog?

Answers on page A-17

5. *Lengths of Engagement of Married Couples.* The data below relate the percent of married couples who were engaged for a certain time period before marriage (***Source*:** Bruskin Goldring Research). Use this information to draw a circle graph.

Less than 1 yr:	24%
1–2 yr:	21%
More than 2 yr:	35%
Never engaged:	20%

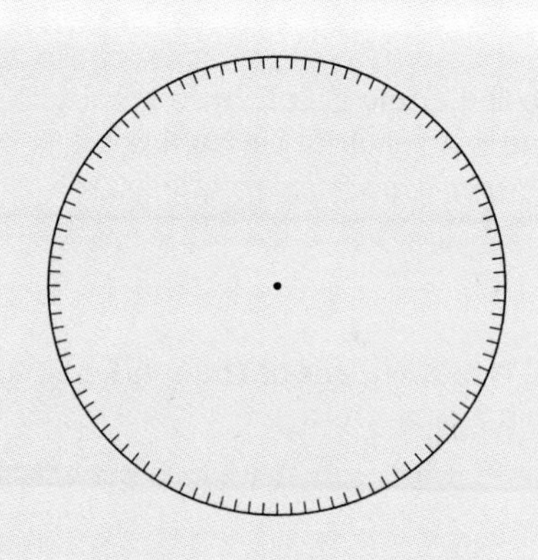

Answers on page A-17

b Drawing Circle Graphs

To draw a circle graph, or pie chart, like the one in Example 1, think of a pie cut into 100 equally sized pieces. We would then shade in a wedge equal in size to 36 of these pieces to represent 36% for food. We shade a wedge equal in size to 5 of these pieces to represent 5% for toys, and so on.

Example 2 *Fruit Juice Sales.* The percents of various kinds of fruit juice sold are given in the list at right (***Source*:** Beverage Marketing Corporation). Use this information to draw a circle graph.

Apple:	14%
Orange:	56%
Blends:	6%
Grape:	5%
Grapefruit:	4%
Prune:	1%
Other:	14%

Using a circle marked with 100 equally spaced ticks, we start with the 14% given for apple juice. We draw a line from the center to any one tick. Then we count off 14 ticks and draw another line. We shade the wedge with a color—in this case, red—and label the wedge as shown in the figure on the left below.

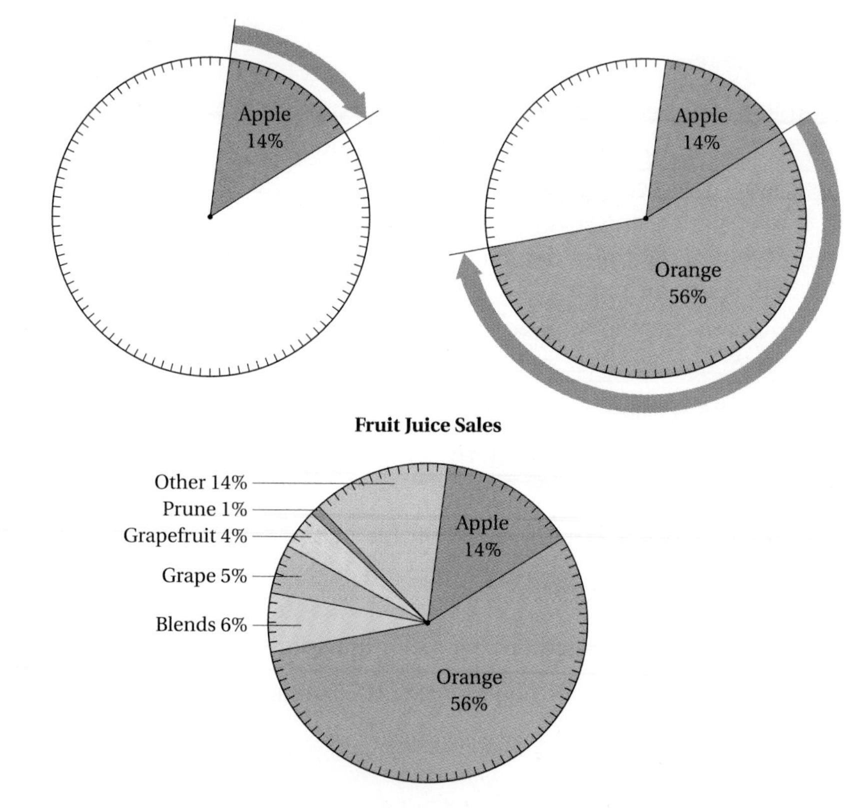

To shade a wedge for orange juice, at 56%, we count off 56 ticks and draw another line. We shade the wedge with a different color—in this case, orange—and label the wedge as shown in the figure on the right above. Continuing in this manner and choosing different colors, we obtain the graph shown above. Finally, we give the graph the overall title "Fruit Juice Sales."

Do Exercise 5.

Exercise Set 7.4

a *Musical Recordings.* This circle graph, in the shape of a CD, shows music preferences of customers on the basis of music store sales. Use the graph for Exercises 1–6.

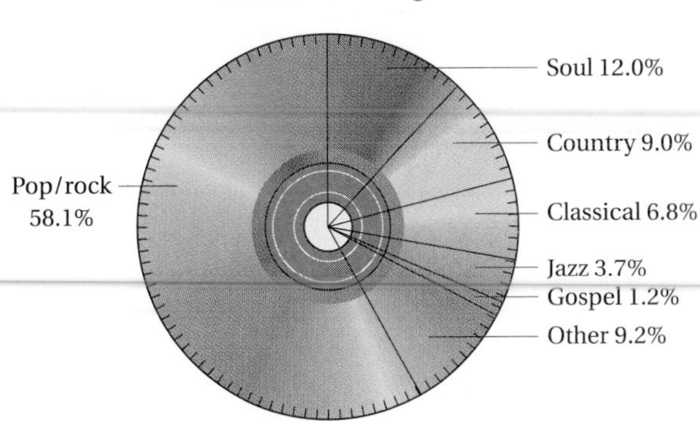

1. What percent of all recordings sold are jazz?

2. Together, what percent of all recordings sold are either soul or pop/rock?

3. Lou's Music Store sells 3000 recordings a month. How many are country?

4. Al's Music Store sells 2500 recordings a month. How many are gospel?

5. What percent of all recordings sold are classical?

6. Together, what percent of all recordings sold are either classical or jazz?

Family Expenses. This circle graph shows expenses as a percent of income in a family of four. (*Note*: Due to rounding, the sum of the percents is 101% instead of 100%.) Use the graph for Exercises 7–10.

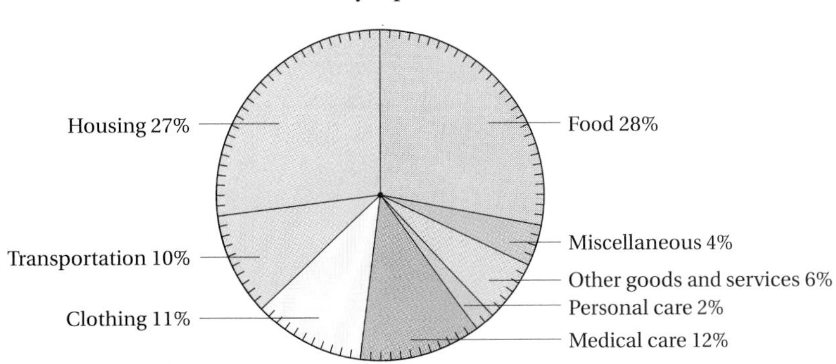

7. Which item accounts for the greatest expense?

8. In a family with a $2000 monthly income, how much is spent for transportation?

9. Some surveys combine medical care with personal care. What percent would be spent on those two items combined?

10. In a family with a $2000 monthly income, what is the ratio of the amount spent on medical care to the amount spent on personal care?

b Use the given information to complete a circle graph. Note that each circle is divided into 100 sections. (Circle graphs for Exercises 15 and 16 should be drawn on a separate sheet of paper.)

11. *Where Homebuyers Prefer to Live.*

City:	4%
Rural:	30%
Outlying suburbs:	34%
Nearby suburbs:	30%
Other:	2%

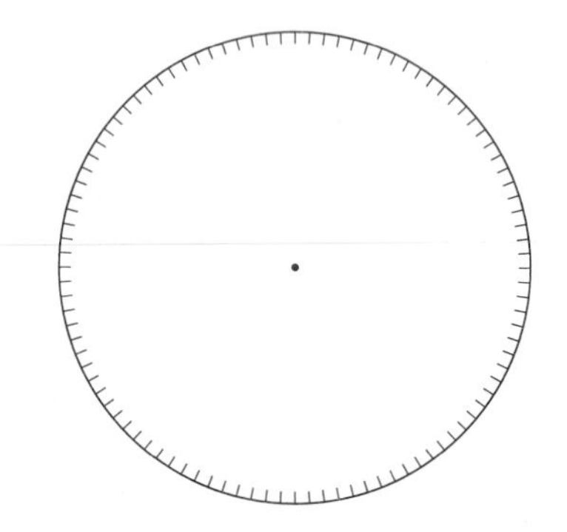

12. *How Vacation Money is Spent.*

Transportation:	15%
Meals:	20%
Lodging:	32%
Recreation:	18%
Other:	15%

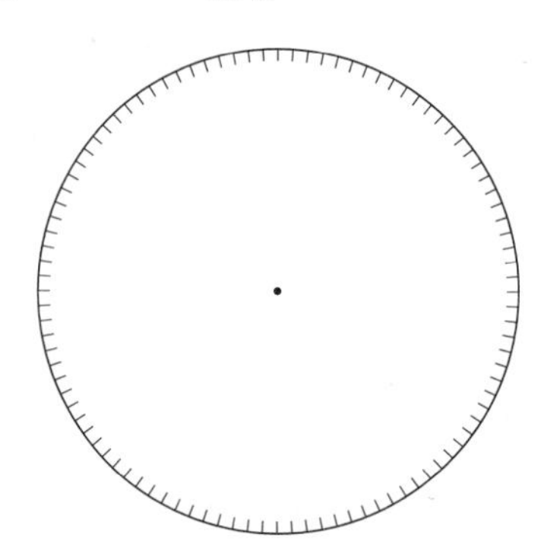

13. *Pilots' Ages.*

20–29:	6%
30–39:	32%
40–49:	36%
50–59:	26%

Source: Federal Aviation Administration

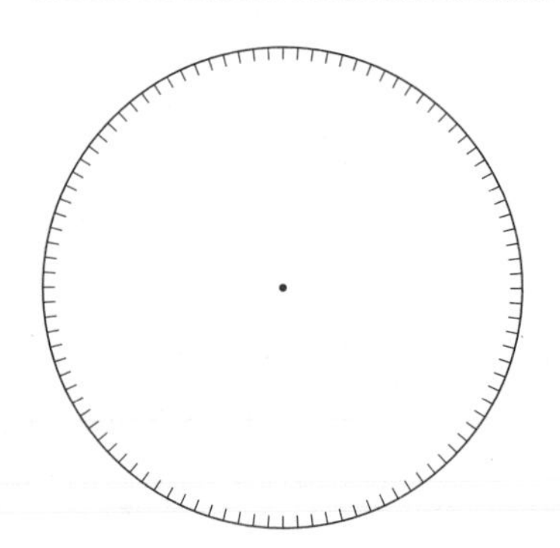

14. *Sources of Water.*

Drinking water:	29%
Tea and coffee:	24%
Vegetables:	9%
Milk/dairy products:	9%
Soft drinks:	8%
Other:	21%

Source: U.S. Department of Agriculture

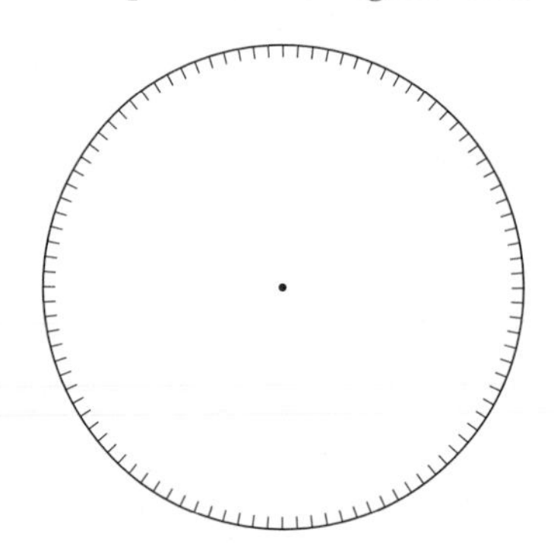

15. *Holiday Gift Giving by Men.*

More than 30 gifts:	14%
21–30 gifts:	13%
11–20 gifts:	32%
6–10 gifts:	24%
1–5 gifts:	13%
0 gifts:	4%

Source: Maritz AmeriPoll

16. *Reasons for Drinking Coffee.*

To get going in the morning:	32%
Like the taste:	33%
Not sure:	2%
To relax:	4%
As a pick-me-up:	10%
A habit:	19%

Source: LMK Associates survey for Au Bon Pain Co., Inc.

Skill Maintenance

Solve. [6.3b], [6.4b]

17. What is 45% of 668?

18. 16 is what percent of 64?

19. 23 is 20 percent of what?

Use a circle graph to show household expenses.

7.5 Data Analysis and Predictions

Objectives

a Compare two sets of data using their means.

b Make predictions from a set of data using interpolation or extrapolation.

For Extra Help

TAPE 12 TAPE 12B MAC WIN CD-ROM

a Comparing Two Sets of Data

We have seen how to organize, display, and interpret data using graphs and how to calculate averages, medians, and modes from data. Now we learn data analysis for the purpose of solving applied problems.

One goal of analyzing two sets of data is to make a determination about which of two groups is "better." One way to do so is by comparing the means.

Example 1 *Battery Testing.* An experiment is performed to compare battery quality. Two kinds of battery were tested to see how long, in hours, they kept a portable CD player running. On the basis of this test, which battery is better?

Battery A: EternReady Times (in hours)			Battery B: SturdyCell Times (in hours)		
27.9	28.3	27.4	28.3	27.6	27.8
27.6	27.9	28.0	27.4	27.6	27.9
26.8	27.7	28.1	26.9	27.8	28.1
28.2	26.9	27.4	27.9	28.7	27.6

Note that it is difficult to analyze the data at a glance because the numbers are close together. We need a way to compare the two groups. Let's compute the average of each set of data.

Battery A: Average

$$= \frac{27.9 + 28.3 + 27.4 + 27.6 + 27.9 + 28.0 + 26.8 + 27.7 + 28.1 + 28.2 + 26.9 + 27.4}{12}$$

$$= \frac{332.2}{12} \approx 27.68$$

Battery B: Average

$$= \frac{28.3 + 27.6 + 27.8 + 27.4 + 27.6 + 27.9 + 26.9 + 27.8 + 28.1 + 27.9 + 28.7 + 27.6}{12}$$

$$= \frac{333.6}{12} = 27.8$$

We see that the average of battery B is higher than that of battery A and thus conclude that battery B is "better." (It should be noted that statisticians might question whether these differences are what they call "significant." The answer to that question belongs to a later math course.)

Do Exercise 1.

1. *Growth of Wheat.* Rudy experiments to see which of two kinds of wheat is better. (In this situation, the shorter wheat is considered "better.") He grows both kinds under similar conditions and measures stalk heights, in inches, as follows. Which kind is better?

Wheat A Stalk Heights (in inches)			
16.2	42.3	19.5	25.7
25.6	18.0	15.6	41.7
22.6	26.4	18.4	12.6
41.5	13.7	42.0	21.6

Wheat B Stalk Heights (in inches)			
19.7	18.4	19.7	17.2
19.7	14.6	32.0	25.7
14.0	21.6	42.5	32.6
22.6	10.9	26.7	22.8

Answer on page A-17

2. *World Bicycle Production.* Use interpolation to estimate world bicycle production in 1994 from the information in the following table.

Year	World Bicycle Production (in millions)
1989	95
1990	90
1991	96
1992	103
1993	108
1994	?
1995	114

Source: United Nations Interbike Directory

Answer on page A-17

b Making Predictions

Sometimes we use data to make predictions or estimates of missing data points. One process for doing so is called **interpolation**. Let's return to some data first considered in Section 7.3.

Example 2 *Monthly Loan Payment.* The following table lists monthly repayments on a loan of $110,000 at 9% interest. Note that we have no data point for a 35-yr loan. Use interpolation to estimate its value.

Year	Monthly Payment
5	$2283.42
10	1393.43
15	1115.69
20	989.70
25	923.12
30	885.08
35	?
40	848.50

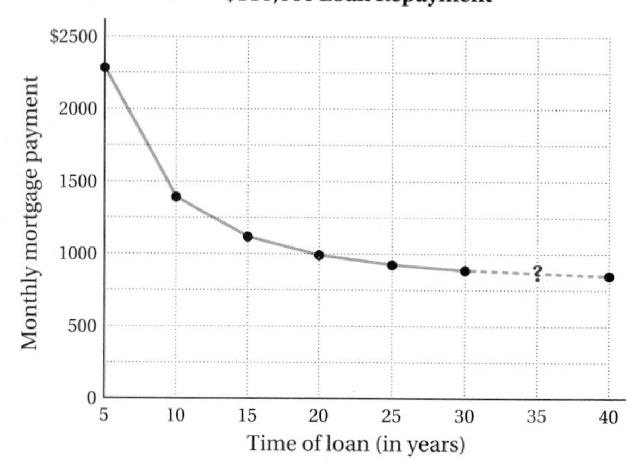

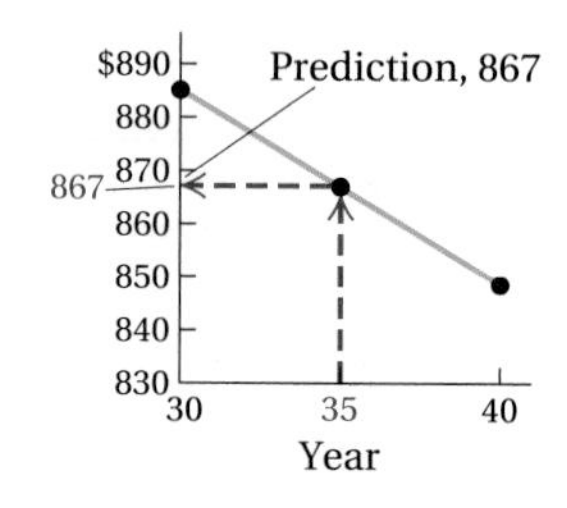

Refer to the figures shown above. First, we analyze the data and look for trends. Note that the monthly mortgage payments are decreasing and the data points resemble a straight line. It seems reasonable that we can draw a line between the points for 30 and 40. We draw a "zoomed-in" graph. Then we consider a vertical line up from the point for 35 and see where the vertical line crosses the line between the data points. We move to the left and read off a value—about $867. We can also estimate this value by taking the average of the data values $885.08 and $848.50:

$$\frac{\$885.08 + \$848.50}{2} = \$866.79.$$

When we estimate in this way to find an "in-between value," we are using a process called *interpolation*. Real-world information about the data might tell us that an estimate found in this way is unreliable. For example, data from the stock market might be very erratic.

Do Exercise 2.

We often analyze data with the view of going "beyond" the data. One process for doing so is called **extrapolation**.

Example 3 *Movies Released.* The data in the following table and graphs show the number of movie releases over a period of years. Use extrapolation to estimate the number of movies released in 1997.

Year	Movies Released
1991	164
1992	150
1993	161
1994	184
1995	234
1996	260
1997	?

Source: Motion Picture Association of America

Movies Released

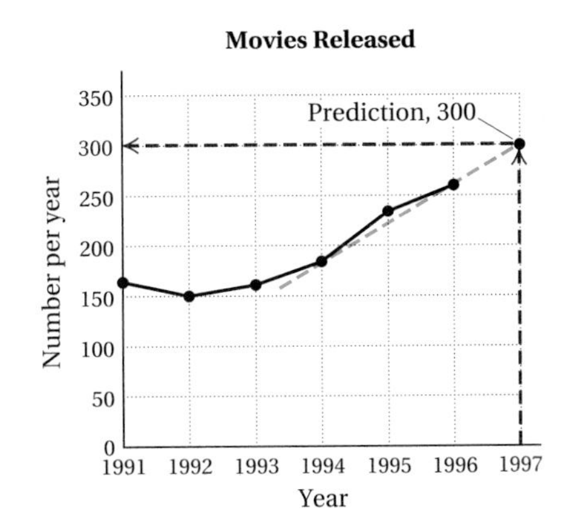

First, we analyze the data and note that they tend to follow a straight line past 1994. Keeping this trend in mind, we draw a "representative" line through the data and beyond. To estimate a value for 1997, we draw a vertical line up from 1997 until it hits the representative line. We go to the left and read off a value—about 300. When we estimate in this way to find a "go-beyond value," we are using a process called *extrapolation.* Answers found with this method can vary greatly depending on the points chosen to determine the "representative" line.

Do Exercise 3.

3. *Study Time and Test Scores.* A professor gathered the following data comparing study time and test scores. Use extrapolation to estimate the test score received when studying for 23 hr.

Study Time (in hours)	Test Grade (in percent)
19	83
20	85
21	88
22	91
23	?

Answer on page A-17

Improving Your Math Study Skills

How Many Women Have Won the Ultimate Math Contest?

Although this Study Skill feature does not contain specific tips on studying mathematics, we hope that you will find this article both challenging and encouraging.

Every year on college campuses across the United States and Canada, the most brilliant math students face the ultimate challenge. For six hours, they struggle with problems from the merely intractable to the seemingly impossible.

Every spring, five are chosen winners of the William Lowell Putnam Mathematical Competition, the Olympics of college mathematics. Every year for 56 years, all have been men.

Until this year.

This spring, Ioana Dumitriu (pronounced yo-AHN-na doo-mee-TREE-oo), 20, a New York University sophomore from Romania, became the first woman to win the award.

Ms. Dumitriu, the daughter of two electrical engineering professors in Romania, who as a girl solved math puzzles for fun, was identified as a math talent early in her schooling in Bucharest. At 11, Ms. Dumitriu was steered into years of math training camps as preparation for the Romanian entry in the International Mathematics Olympiad.

It was this training, and a handsome young coach, that led her to New York City. He was several years older. They fell in love. He chose N.Y.U. for its graduate school in mathematics, and at 19 she joined him in New York.

The test Ms. Dumitriu won is dauntingly difficult, even for math majors. About half of the 2,407 test-takers scored 2 or less of a possible 120, and a third scored 0. Some students simply walk out after staring at the questions for a while.

Ms. Dumitriu said that in the six hours allotted, she had time to do 8 of the 12 problems, each worth a maximum of 10 points. The last one she did in 10 minutes. This year, Ms. Dumitriu and her five co-winners (there was a tie for fifth place) scored between 76 and 98. She does not know her exact score or rank because the organizers do not announce them.

"I didn't ever tell myself that I was unlikely to win, that no woman before had ever won and therefore I couldn't," she said. "It is not that I forget that I'm a woman. It's just that I don't see it as an obstacle or a ——."

Her English is near-perfect, but she paused because she could not find the right word. "The mathematics community is made up of persons, and that is what I am primarily."

Prof. Joel Spencer, who was a Putnam winner himself, said her work for his class in problem solving last year was remarkable. "What really got me was her fearlessness," he said. "To be good at math, you have to go right at it and start playing around with it, and she had that from the start."

In the graduate lounge in the Courant Institute of Mathematical Sciences at N.Y.U., Ms. Dumitriu, a tall, striking redhead, stands out. Instead of jeans and T-shirts, she wears gray pin-striped slacks and a rust-colored turtleneck and vest.

"There is a social perception of women and math, a stereotype," Ms. Dumitriu said during an interview. "What's happening right now is that the stereotype is defied. It starts breaking."

Still, even as women began to flock to sciences, math has remained largely a male bastion.

"Math remains the bottom line of sex differences for many," said Sheila Tobias, author of "Overcoming Math Anxiety" (W.W. Norton & Company, 1994). "It's one thing for women to write books, negotiate bills through Congress, litigate, fire missiles; quite another for them to do math."

Besides collecting the $1,000 awarded to each Putnam fellow, Ms. Dumitriu also won the $500 Elizabeth Lowell Putman prize for the top woman finisher for the second year in a row, a prize created five years ago to encourage women to take the test. This year 414 did.

In her view, there are never too many problems, never too much practice.

Besides, each new problem holds its own allure: "When you have all the pieces and you put them together and you see the puzzle, that moment always amazes me."

Exercise Set 7.5

a

1. *Light-Bulb Testing.* An experiment is performed to compare the lives of two types of light bulb. Several bulbs of each type were tested and the results are listed in the following table. On the basis of this test, which bulb is better?

Bulb A: HotLight Times (in hours)			Bulb B: BrightBulb Times (in hours)		
983	964	1214	979	1083	1344
1417	1211	1521	984	1445	975
1084	1075	892	1492	1325	1283
1423	949	1322	1325	1352	1432

2. *Cola Testing.* An experiment is conducted to determine which of two colas tastes better. Students drank each cola and gave it a rating from 1 to 10. The results are given in the following table. On the basis of this test, which cola tastes better?

Cola A: Vervcola				Cola B: Cola-cola			
6	8	10	7	10	9	9	6
7	9	9	8	8	8	10	7
5	10	9	10	8	7	4	3
9	4	7	6	7	8	10	9

b Use interpolation or extrapolation to find the missing data values.

3. *Study Time vs. Grades.* A math instructor asked her students to keep track of how much time each spent studying the chapter on percent notation in her basic mathematics course. They collected the information together with test scores from that chapter's test. The data are given in the following table. Estimate the missing data value.

Study Time (in hours)	Test Grade (in percent)
9	75
11	93
13	80
15	85
16	85
17	80
18	?
21	86
23	91

4. *Maximum Heart Rate.* A person's maximum heart rate depends on his or her gender, age, and resting heart rate. The following table relates resting heart rate and maximum heart rate for a 20-yr-old man. Estimate the missing data value.

Resting Heart Rate (in beats per minute)	Maximum Heart Rate (in beats per minute)
50	166
60	168
65	?
70	170
80	172

Source: American Heart Association

Estimate the missing data value in each of the following tables.

5. *Ozone Layer.*

Year	Ozone Level (in parts per billion)
1991	2981
1992	3133
1993	3148
1994	3138
1995	3124
1996	?

Source: National Oceanic and Atmospheric Administration

6. *Motion Picture Expense.*

Year	Average Expense per Picture (in millions)
1991	$38.2
1992	42.4
1993	44.0
1994	50.4
1995	54.1
1996	61.0
1997	?

Source: Motion Picture Association of America

7. *Credit-Card Spending.*

Year	Credit-Card Spending from Thanksgiving to Christmas (in billions)
1991	$ 59.8
1992	66.8
1993	79.1
1994	96.9
1995	116.3
1996	131.4
1997	?

Source: RAM Research Group, National Credit Counseling Services

8. *U.S. Book-Buying Growth.*

Year	Book Sales (in billions)
1992	$21
1993	23
1994	24
1995	25
1996	26
1997	?

Source: Book Industry Trends 1995

Skill Maintenance

Solve. [5.4a]

9. The building costs on a 2200-ft^2 house are $118,000. Using this rate, find the building costs on a 2400-ft^2 house.

10. A glaucoma medication is mixed in the ratio of 25 parts of medicine to 400 parts of saline solution. How many cubic centimeters of medicine should be added to 10 mL of saline solution? (1 cubic centimeter = 1 milliliter)

11. Four software CDs cost $239.80. How much would 23 comparable CDs cost?

12. A car is driven 700 mi in 5 days. At this rate, how far will it have been driven in 36 days?

Divide. [2.7b]

13. $\frac{5}{6} \div \frac{7}{18}$

14. $256 \div \frac{6}{11}$

15. $\frac{17}{25} \div 1000$

16. $\frac{1}{12} \div \frac{1}{11}$

Synthesis

17. ◆ Discuss how you might test the estimates that you found in Exercises 3–8.

18. ◆ Compare and contrast the processes of interpolation and extrapolation.

Summary and Review Exercises: Chapter 7

The objectives to be tested in addition to the material in this chapter are [2.7b], [5.4a], [6.3b], [6.4b], and [6.5a].

FedEx Mailing Costs. Federal Express has three types of delivery service for packages of various weights, as shown in the following table. Use this table for Exercises 1–6. [7.2a]

Delivery by 10:00 a.m. next business day → FedEx Priority Overnight®

Delivery by 3:00 p.m. next business day → FedEx Standard Overnight®

Delivery by 4:30 p.m. second business day → FedEx 2Day®

		FedEx Priority Overnight®	FedEx Standard Overnight®	FedEx 2Day®
FedEx Letter up to 8 oz.		$ 13.25	$ 11.50	$ n/a
All other packaging/Weight in lbs.	1 lb.	$ 18.30	$ 16.00	$ 9.25
	2 lbs.	19.20	17.00	9.95
	3	21.00	18.00	11.00
	4	22.80	19.00	12.00
	5	24.90	20.00	13.00
	6	27.30	21.75	14.25
	7	29.70	23.50	15.25
	8	31.80	25.25	16.25
	9	34.20	27.00	17.25
	10	36.80	28.75	18.25
	11	37.80	30.75	19.25

Source: Federal Express Corporation

1. Find the cost of a 3-lb FedEx Priority Overnight delivery.

2. Find the cost of a 10-lb FedEx Standard Overnight delivery.

3. How much would you save by sending the package listed in Exercise 1 by FedEx 2Day delivery?

4. How much would you save by sending the package in Exercise 2 by FedEx 2Day delivery?

5. Is there any difference in price between sending a 5-oz package FedEx Priority Overnight and sending an 8-oz package in the same way?

6. An author has a 4-lb manuscript to send by FedEx Standard Overnight delivery to her publisher. She calls and the package is picked up. Later that day she completes work on another part of her manuscript that weighs 5 lb. She calls and sends it by FedEx Standard Overnight delivery to the same address. How much could she have saved if she had waited and sent both packages as one?

TV Sales. This pictograph shows the projected number of television sets to be sold by a company. Use it for Exercises 7–10. [7.2b]

7. About how many TV sets will be sold in 1999?

8. In which year does the company sell the least number of TV sets?

9. In which year does the company sell the greatest number of TV sets?

10. Estimate the average number of TV sets sold per year over the 4-yr period.

Find the average. [7.1a]

11. 26, 34, 43, 51

12. 7, 11, 14, 17, 18

13. 0.2, 1.7, 1.9, 2.4

14. 700, 900, 1900, 2700, 3000

15. \$2, \$14, \$17, \$17, \$21, \$29

16. 20, 190, 280, 470, 470, 500

Find the median. [7.1b]

17. 26, 34, 43, 51

18. 7, 11, 14, 17, 18

19. 0.2, 1.7, 1.9, 2.4

20. 700, 900, 1900, 2700, 3000

21. \$2, \$14, \$17, \$17, \$21, \$29

22. 20, 190, 280, 470, 470, 500

Find the mode. [7.1c]

23. 26, 34, 43, 26, 51

24. 7, 11, 11, 14, 17, 17, 18

25. 0.2, 0.2, 0.2, 1.7, 1.9, 2.4

26. 700, 700, 800, 2700, 800

27. \$2, \$14, \$17, \$17, \$21, \$29

28. 20, 20, 20, 20, 20, 500

29. One summer, a student earned the following amounts over a four-week period: \$102, \$112, \$130, and \$98. What was the average amount earned per week? the median? [7.1a, b]

30. The following temperatures were recorded in St. Louis every four hours on a certain day in June: 63°, 58°, 66°, 72°, 71°, 67°. What was the average temperature for that day? [7.1a]

31. To get an A in math, a student must score an average of 90 on four tests. Scores on the first three tests were 94, 78, and 92. What is the lowest score that the student can make on the last test and still get an A? [7.1a]

Calorie Content in Fast Foods. Wendy's Hamburgers is a national food franchise. The following bar graph shows the caloric content of various sandwiches sold by Wendy's. Use the bar graph for Exercises 32–39. [7.3a]

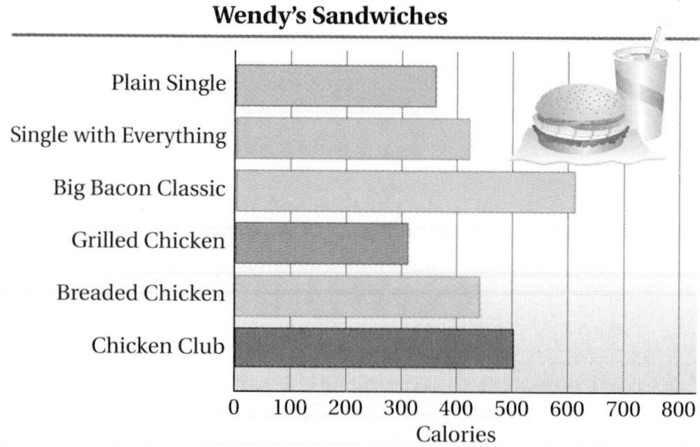

Source: Wendy's International

32. How many calories are in a Single with Everything?

33. How many calories are in a breaded chicken sandwich?

34. Which sandwich has the highest caloric content?

35. Which sandwich has the lowest caloric content?

36. Which sandwich contains about 360 calories?

37. Which sandwich contains about 500 calories?

38. How many more calories are in a chicken club than in a Single with Everything?

39. How many more calories are in a Big Bacon Classic than in a plain single?

Accidents of Drivers by Age. The following line graph shows the number of accidents per 100 drivers, by age. Use the graph for Exercises 40–45. [7.3c]

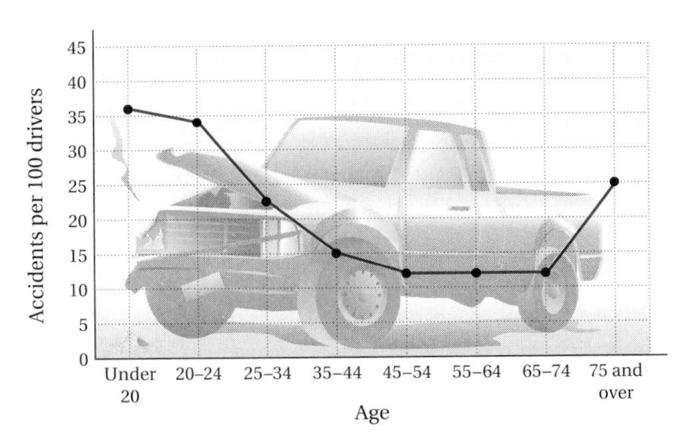

40. Which age group has the most accidents per 100 drivers?

41. What is the fewest number of accidents per 100 in any age group?

42. How many more accidents do people over 75 yr of age have than those in the age range of 65–74?

43. Between what ages does the number of accidents stay basically the same?

44. How many fewer accidents do people 25–34 yr of age have than those 20–24 yr of age?

45. Which age group has accidents more than three times as often as people 55–64 yr of age?

Hotel Preferences. This circle graph shows hotel preferences for travelers. Use the graph for Exercises 46–49. [7.4a]

Types of Hotels

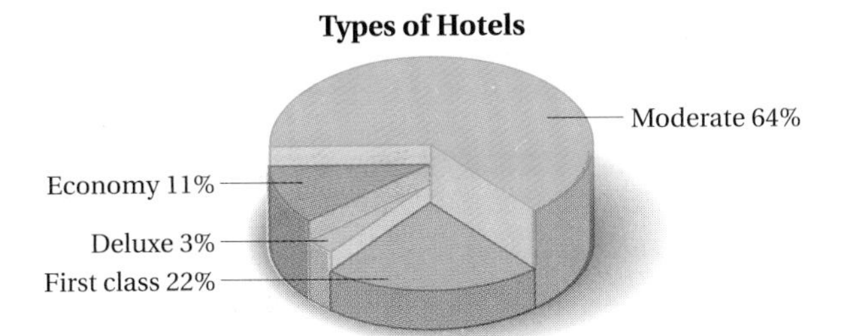

46. What percent of travelers prefer a first-class hotel?

47. What percent of travelers prefer an economy hotel?

48. Suppose 2500 travelers arrive in a city one day. How many of them might seek a moderate room?

49. What percent of travelers prefer either a first-class or a deluxe hotel?

Coca-Cola. The following table shows the total yearly revenue (money taken in), in millions, of the Coca-Cola Company from 1991 to 1996. Use the table for Exercises 50–53.

Year	Total Revenue (in millions)
1991	$465
1992	656
1993	687
1994	724
1995	762
1996	?

Source: Coca-Cola Bottling Consolidated Annual Report

50. Make a vertical bar graph of the data. [7.3b]

51. Make a line graph of the data. [7.3d]

52. Use extrapolation to find the missing data value for 1996. [7.5b]

53. Suppose you know that the revenue in 1997 was $850 million. Use this information, the other data in the table, and interpolation to estimate the missing data value for 1996. [7.5b]

54. *Battery Testing.* An experiment is performed to compare battery quality. Two kinds of battery were tested to see how long, in hours, they kept a hand radio running. On the basis of this test, which battery is better? [7.5a]

Battery A: Times (in hours)		
38.9	39.3	40.4
53.1	41.7	38.0
36.8	47.7	48.1
38.2	46.9	47.4

Battery B: Times (in hours)		
39.3	38.6	38.8
37.4	47.6	37.9
46.9	37.8	38.1
47.9	50.1	38.2

Skill Maintenance

Solve.

55. A company car was driven 4200 mi in the first 4 months of a year. At this rate, how far will it be driven in 12 months? [5.4a]

56. 92% of the world population does not have a telephone. The population is about 5.9 billion. How many do not have a telephone? [6.5a]

57. 789 is what percent of 355.05? [6.3b], [6.4b]

58. What percent of 98 is 49? [6.3b], [6.4b]

Divide and simplify. [2.7b]

59. $\frac{3}{4} \div \frac{5}{6}$

60. $\frac{5}{8} \div \frac{3}{2}$

Synthesis

61. ◈ Compare and contrast averages, medians, and modes. Discuss why you might use one over the others to analyze a set of data. [7.1a, b, c]

62. ◈ Find a real-world situation that fits this equation: [7.1a]

$$\frac{(20{,}500 + 22{,}800 + 23{,}400 + 26{,}000)}{4}.$$

63. The ordered set of data 298, 301, 305, a, 323, b, 390 has a median of 316 and an average of 326. Find a and b. [7.1a, b]

Test: Chapter 7

Retirement Savings. The following table lists estimates of the type of retirement savings a person should have, based on his or her household yearly income, age, gender, and marital status. Use the table for Exercises 1–6.

Household Yearly Income	Age 35	45	55	65
Couple				
$ 50,000	$ 2,756	$ 34,443	$117,739	$187,593
$100,000	$28,850	$101,462	$261,139	$474,590
$150,000	$60,538	$200,825	$468,837	$820,215
Single Male				
$ 50,000	$ 2,558	$ 38,939	$125,420	$180,953
$100,000	$26,345	$115,816	$275,744	$472,326
$150,000	$53,519	$209,960	$468,259	$779,456
Single Female				
$ 50,000	$ 35,158	$ 69,391	$121,242	$181,577
$100,000	$ 90,601	$193,985	$341,413	$504,500
$150,000	$152,725	$326,846	$565,817	$831,025

Source: Merrill Lynch

1. What is the recommended retirement savings for a 55-year-old single female with an annual income of $100,000?
2. What is the recommended retirement savings for a 35-year-old single male with an annual income of $50,000?
3. What type of person(s) needs a retirement savings of $474,590?
4. What type of person(s) needs a retirement savings of $326,846?
5. How much more retirement savings does a 45-year-old single female with an income of $100,000 need than a comparable single male?
6. How much more retirement savings does a 65-year-old couple with an income of $100,000 need than a comparable single male?

Shampoo Sales. The following pictograph shows projected sales of shampoo for a soap company for six consecutive years. Use the pictograph for Exercises 7–10.

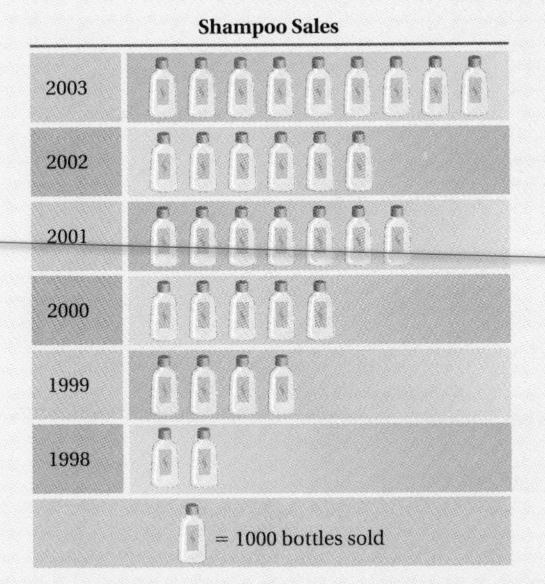

7. In which year will the greatest number of bottles be sold?
8. Between which two consecutive years will there be the greatest growth?
9. How many more bottles will be sold in 2003 than in 1998?
10. In which year will there actually be a decline in the number of bottles sold?

Answers

1. ______
2. ______
3. ______
4. ______
5. ______
6. ______
7. ______
8. ______
9. ______
10. ______

Answers

Find the average.

11. 45, 49, 52, 54

12. 1, 2, 3, 4, 5

13. 3, 17, 17, 18, 18, 20

Find the median and the mode.

14. 45, 49, 52, 54

15. 1, 2, 3, 4, 5

16. 3, 17, 17, 18, 18, 20

17. A car is driven 754 km in 13 hr. What is the average number of kilometers per hour?

18. To get a C in chemistry, a student must score an average of 70 on four tests. Scores on the first three tests were 68, 71, and 65. What is the lowest score that the student can make on the last test and still get a C?

Nike, Inc. The following line graph shows the revenues of Nike, Inc. Use the graph for Exercises 19–24.

Revenue of Nike, Inc.

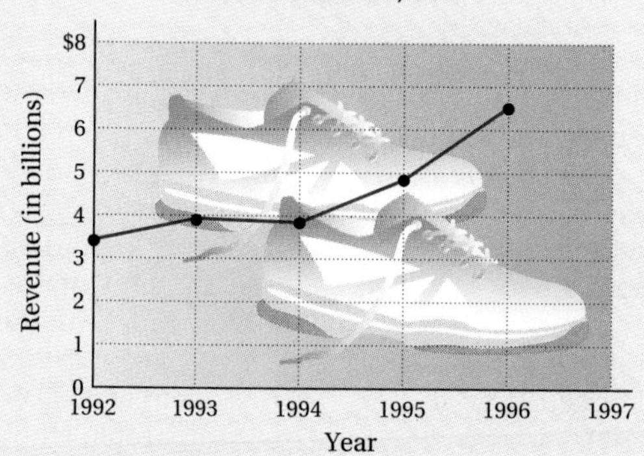

Source: Nike, Inc., annual report

19. How much revenue was earned in 1996?

20. What was the average of the revenues for the five years?

21. How much more revenue was earned in 1996 than in 1992?

22. In which year was the increase in revenue the greatest?

23. What was the median of the revenues for the five years?

24. Use extrapolation to estimate the revenue in 1997.

11. ______
12. ______
13. ______
14. ______
15. ______
16. ______
17. ______
18. ______
19. ______
20. ______
21. ______
22. ______
23. ______
24. ______

25. *Animal Speeds.* The following table lists maximum running speeds for various animals, in miles per hour, compared to the speed of the fastest human. Make a vertical bar graph of the data.

Animal	Speed (in miles per hour)
Antelope	61
Bear	30
Cheetah	70
Fastest Human	28
Greyhound	39
Lion	50
Zebra	40

Refer to the table and graph in Question 25 for Questions 26–29.

26. By how much does the fastest speed exceed the slowest speed?

27. Does a human have a chance of outrunning a lion? Explain.

28. Find the average of all the speeds.

29. Find the median of all the speeds.

30. *Popularity of Sport Utility Vehicles.* Use the following information to make a circle graph showing the reasons people buy sport utility vehicles.

Sporty look:	11%
Status symbol:	27%
Drives well in bad weather:	32%
Hauling or carrying capacity:	21%
Off-road capability:	6%
Other:	3%

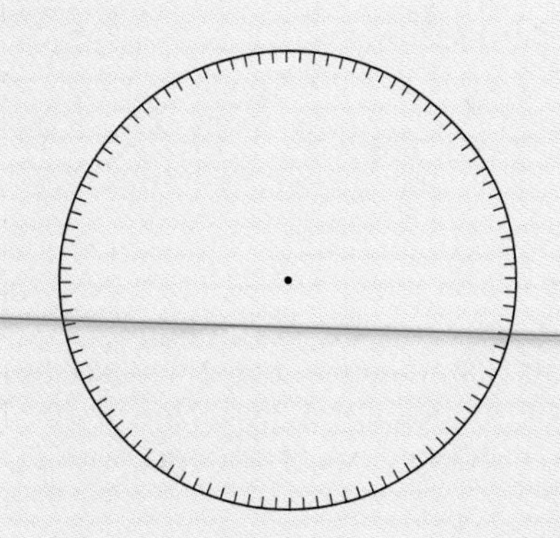

Answers

25. ______

26. ______

27. ______

28. ______

29. ______

30. ______

Answers

31. ____________

32. ____________

33. ____________

34. ____________

35. ____________

36. ____________

37. ____________

38. ____________

Pepsico. Pepsico makes soft drinks and snack foods and operates restaurants such as Taco Bell, Pizza Hut, and KFC. The following table shows the total revenue, in billions, of Pepsico for various years. Use the table for Questions 31 and 32.

Year	Total Revenue (in billions)
1992	$22.0
1993	25.0
1994	28.5
1995	30.4
1996	31.6
1997	?

Source: Pepsico Annual Report

31. Make a line graph of the data.

32. Suppose you know that the revenue in 1998 was $37.2 billion. Use this information, the other data in the table, and interpolation to estimate the missing data value for 1997.

33. *Chocolate Bars.* An experiment is performed to compare the quality of new Swiss chocolate bars being introduced in the United States. People were asked to taste the candies and rate them on a scale of 1 to 10. On the basis of this test, which chocolate bar is better?

Bar A: Swiss Pecan			Bar B: Swiss Hazelnut		
9	10	8	10	6	8
10	9	7	9	10	10
6	9	10	8	7	6
7	8	8	9	10	8

Skill Maintenance

34. Divide and simplify: $\frac{3}{5} \div \frac{12}{125}$.

35. 17 is 25% of what number?

36. On a particular Sunday afternoon, 78% of the television sets that were on were tuned to one of the major networks. Suppose 20,000 TV sets in a town are being watched. How many are tuned to a major network?

37. A baseball player gets 7 hits in the first 20 times at bat. At this rate, how many times at bat will it take to get 119 hits?

Synthesis

38. The ordered set of data 69, 71, 73, a, 78, 98, b has a median of 74 and a mean of 82. Find a and b.

Cumulative Review: Chapters 1–7

1. In 402,513, what does the digit 5 mean?

2. Evaluate: $3 + 5^3$.

3. Find all the factors of 60.

4. Round 52.045 to the nearest tenth.

5. Convert to fractional notation: $3\frac{3}{10}$.

6. Convert from cents to dollars: 210¢.

7. Convert to standard notation: $3.25 billion.

8. Determine whether 11, 30 and 4, 12 are proportional.

Add and simplify.

9. $2\frac{2}{5} + 4\frac{3}{10}$

10. $41.063 + 3.5721$

Subtract and simplify.

11. $\frac{14}{15} - \frac{3}{5}$

12. $350 - 24.57$

Multiply and simplify.

13. $3\frac{3}{7} \cdot 4\frac{3}{8}$

14. $12,456 \times 220$

Divide and simplify.

15. $\frac{13}{15} \div \frac{26}{27}$

16. $104,676 \div 24$

Solve.

17. $\frac{5}{8} = \frac{6}{x}$

18. $\frac{2}{5} \cdot y = \frac{3}{10}$

19. $21.5 \cdot y = 146.2$

20. $x = 398,112 \div 26$

Solve.

21. Tortilla chips cost $2.99 for 14.5 oz. Find the unit price in cents per ounce.

22. A college has a student body of 6000 students. Of these, 55.4% own a car. How many students own a car?

23. *Peanut Products.* In any given year, the average American eats 2.7 lb of peanut butter, 1.5 lb of salted peanuts, 1.2 lb of peanut candy, 0.7 lb of in-shell peanuts, and 0.1 lb of peanuts in other forms. How many pounds of peanuts and products containing peanuts does the average American eat in one year?

24. A piece of fabric $1\frac{3}{4}$ yd long is cut into 7 equal strips. What is the length of each strip?

25. *Energy Consumption.* In a recent year, American utility companies generated 1464 billion kilowatt-hours of electricity using coal, 455 billion using nuclear power, 273 billion using natural gas, 250 billion using hydroelectric plants, 118 billion using petroleum, and 12 billion using geothermal technology and other methods. How many kilowatt-hours of electricity were produced that year?

26. A recipe calls for $\frac{3}{4}$ cup of sugar. How much sugar should be used for an amount that is $\frac{1}{2}$ of the recipe?

FedEx. The following table lists the costs of delivering a package by FedEx Priority Overnight shipping. Use the table for Questions 27–30.

Weight (in pounds)	Cost
1	$18.30
2	19.20
3	21.00
4	22.80
5	24.90
6	27.30
7	29.70
8	31.80
9	34.20
10	36.30
11	37.80
12	?

Source: Federal Express Corporation

27. Find the average and the median of these costs.

28. Make a vertical bar graph of the data.

29. Make a line graph of the data.

30. Use extrapolation to estimate the missing data value for a 12-lb package.

31. A business is owned by four people. One owns $\frac{1}{3}$, the second owns $\frac{1}{4}$, and the third owns $\frac{1}{6}$. How much does the fourth person own?

32. In manufacturing valves for engines, a factory was discovered to have made 4 defective valves in a lot of 18 valves. At this rate, how many defective valves can be expected in a lot of 5049 valves?

33. A landscaper bought 22 evergreen trees for $210. What was the cost of each tree? Round to the nearest cent.

34. A salesperson earns $182 selling $2600 worth of electronic equipment. What is the commission rate?

Voters. The following circle graph shows the percent of 18-year-olds surveyed who planned to vote in an upcoming presidential election. Use the graph for Questions 35 and 36.

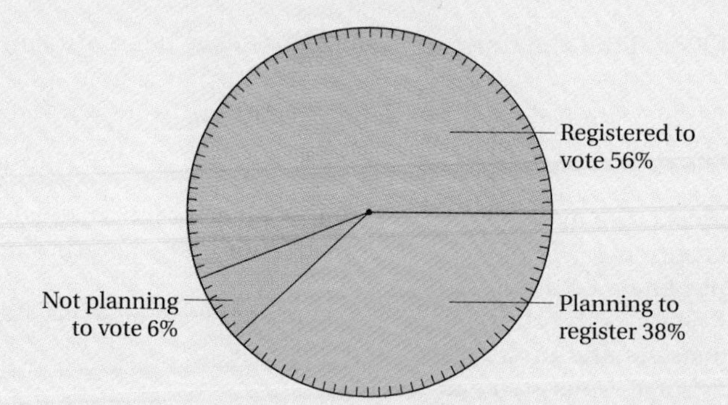

35. What percent of all 18-year-olds were registered to vote?

36. In a class of 250 18-year-old freshmen, how many were not planning to vote?

37. Find the mode of this set of numbers:

3, 5, 2, 5, 1, 3, 5, 2.

Synthesis

38. A photography club meets four times a month. In September, the attendance figures were 28, 23, 26, and 23. In October, the attendance figures were 26, 20, 14, and 28. What was the percent of increase or decrease in average attendance from September to October?

8

Geometry and Measures: Length and Area

Introduction

In this chapter, we introduce American and metric systems used to measure length, and we present conversion from one unit to another within, as well as between, each system. These concepts are then applied to finding areas of squares, rectangles, triangles, parallelograms, trapezoids, and circles. We then study right triangles using square roots and the Pythagorean theorem.

An Application

A standard-sized slow-pitch softball diamond is a square with sides of length 65 ft. What is the perimeter of this softball diamond? (This is the distance you would have to run if you hit a home run.)

This problem appears as Exercise 16 in Exercise Set 8.3.

The Mathematics

We find the perimeter by finding the distance around the square:

$$P = 65 \text{ ft} + 65 \text{ ft} + 65 \text{ ft} + 65 \text{ ft}$$
$$= 4 \cdot (65 \text{ ft}) = \underbrace{260 \text{ ft}}.$$

This is the perimeter.

World Wide Web For more information, visit us at www.mathmax.com

Pretest: Chapter 8

Complete.

1. 8 ft = _______ in.

2. 5 in. = _______ ft

3. 8.46 km = _______ m

4. 9.2 mm = _______ cm

5. Find the perimeter.

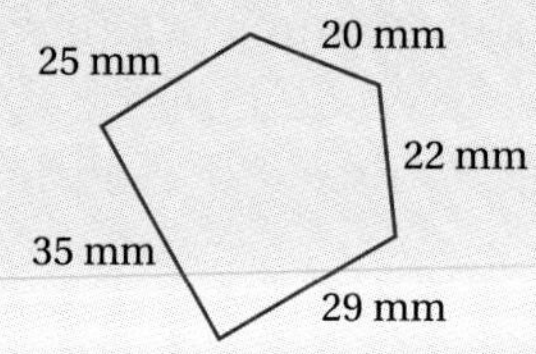

6. Find the area of a square with sides of length 10 ft.

Find the area.

7.

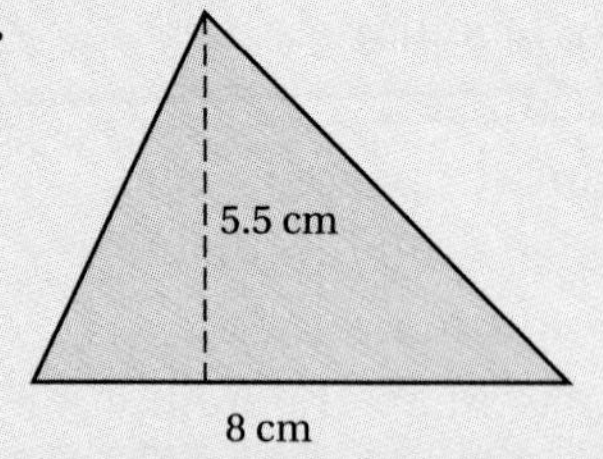

8.

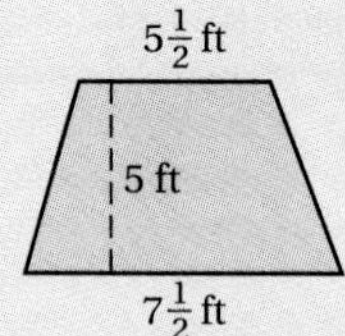

9.

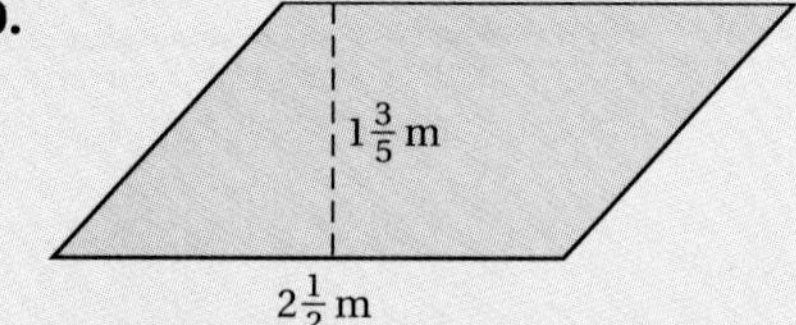

10. Find the length of a diameter of a circle with a radius of 4.8 m.

11. Find the circumference of the circle in Question 10. Use 3.14 for π.

12. Find the area of the circle in Question 10. Use 3.14 for π.

13. Find the area of the shaded region.

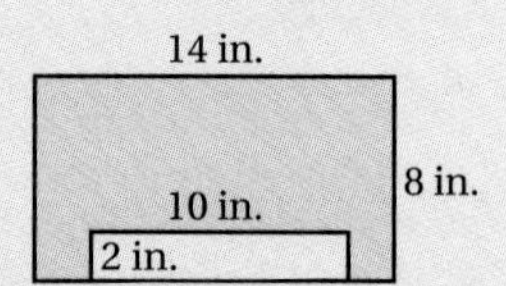

14. Simplify: $\sqrt{81}$.

15. Approximate to three decimal places: $\sqrt{97}$.

In the right triangle, find the length of the side not given. Find an exact answer and an approximation to three decimal places.

16.

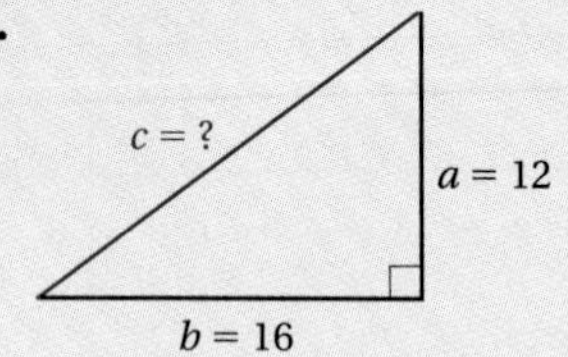

17.

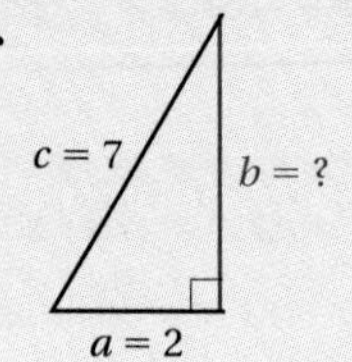

Objectives for Retesting

The objectives to be tested in addition to the material in this chapter are as follows.

[6.1b] Convert from percent notation to decimal notation.
[6.1c] Convert from decimal notation to percent notation.
[6.2a] Convert from fractional notation to percent notation.
[6.2b] Convert from percent notation to fractional notation.

8.1 Linear Measures: American Units

Length, or distance, is one kind of measure. To find lengths, we start with some **unit segment** and assign to it a measure of 1. Suppose $\overline{AB}$ below is a unit segment.

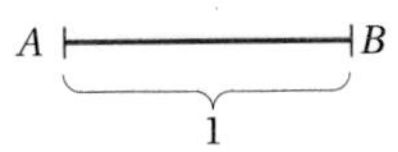

Let's measure segment $\overline{CD}$ below, using $\overline{AB}$ as our unit segment.

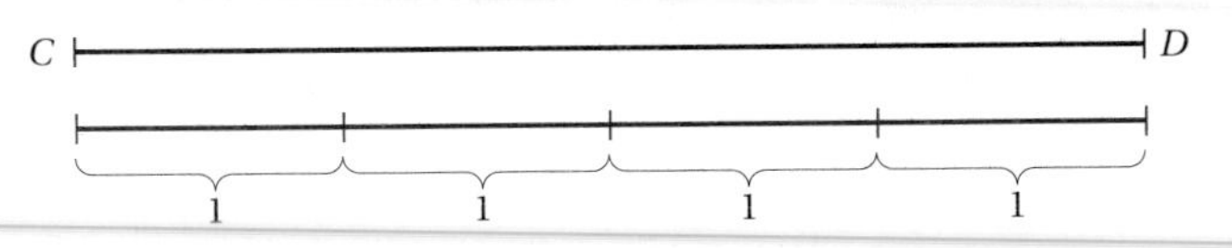

Since we can place 4 unit segments end to end along $\overline{CD}$, the measure of $\overline{CD}$ is 4.

Sometimes we have to use parts of units, called **subunits**. For example, the measure of the segment $\overline{MN}$ below is $1\frac{1}{2}$. We place one unit segment and one half-unit segment end to end.

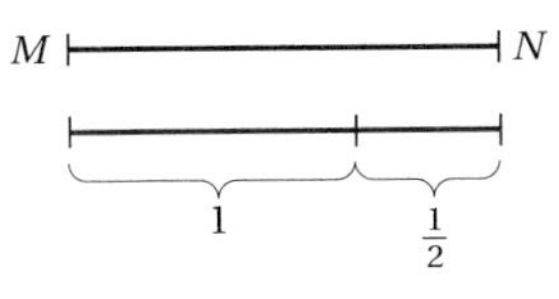

Do Exercises 1–4.

a American Measures

American units of length are related as follows.

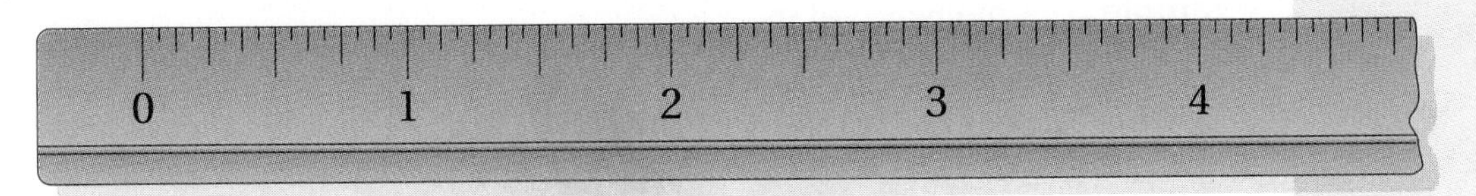

(Actual size, in inches)

> **AMERICAN UNITS OF LENGTH**
>
> 12 inches (in.) = 1 foot (ft) | 3 feet = 1 yard (yd)
>
> 36 inches = 1 yard | 5280 feet = 1 mile (mi)

The symbolism 13 in. = 13″ and 27 ft = 27′ is also used for inches and feet. American units have also been called "English," or "British–American," because at one time they were used by both countries. Today, both Canada and England have officially converted to the metric system. However, if you travel in England, you will still see units such as "miles" on road signs.

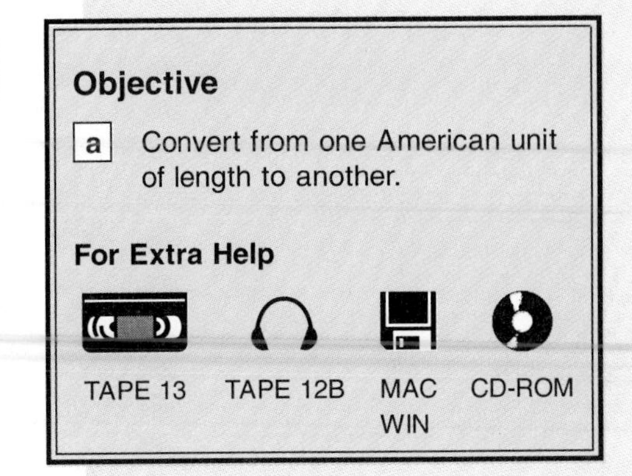

Use the unit below to measure the length of each segment or object.

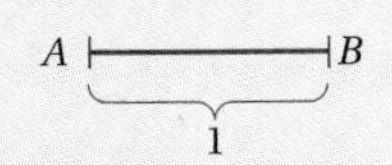

1.

2.

3.

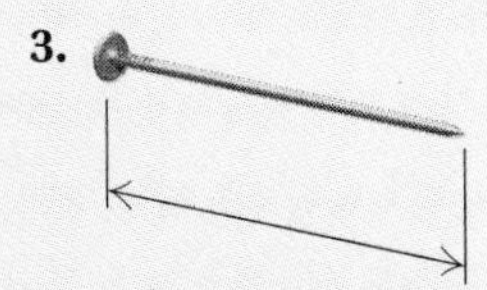

4.

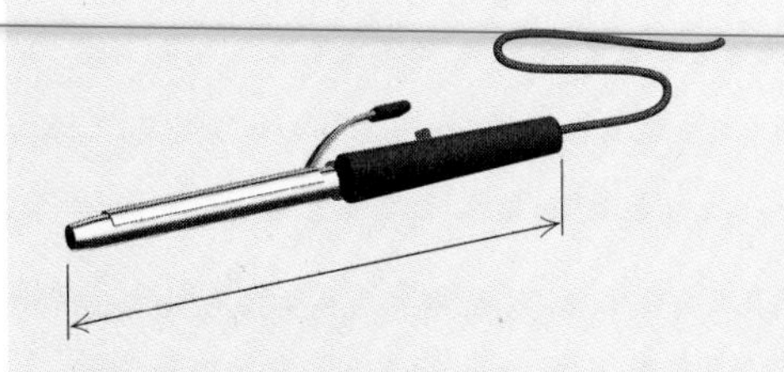

Answers on page A-19

Complete.

5. 8 yd = _______ in.

6. 14.5 yd = _______ ft

7. 3.8 mi = _______ in.

Complete.

8. 72 in. = _______ ft

9. 17 in. = _______ ft

Answers on page A-19

To change from certain American units to others, we make substitutions. Such a substitution is usually helpful when we are converting from a larger unit to a smaller one.

Example 1 Complete: 1 yd = _______ in.

$$\begin{aligned} 1 \text{ yd} &= 3 \text{ ft} \\ &= 3 \times 1 \text{ ft} && \text{We think of 3 ft as } 3 \times \text{ft, or } 3 \times 1 \text{ ft.} \\ &= 3 \times 12 \text{ in.} && \text{Substituting 12 in. for 1 ft} \\ &= 36 \text{ in.} && \text{Multiplying} \end{aligned}$$

Example 2 Complete: 7 yd = _______ in.

$$\begin{aligned} 7 \text{ yd} &= 7 \times 1 \text{ yd} \\ &= 7 \times 3 \text{ ft} && \text{Substituting 3 ft for 1 yd} \\ &= 7 \times 3 \times 1 \text{ ft} \\ &= 7 \times 3 \times 12 \text{ in.} && \text{Substituting 12 in. for 1 ft; } 7 \times 3 = 21;\ 21 \times 12 = 252 \\ &= 252 \text{ in.} \end{aligned}$$

Do Exercises 5–7.

Sometimes it helps to use multiplying by 1 in making conversions. For example, 12 in. = 1 ft, so

$$\frac{12 \text{ in.}}{1 \text{ ft}} = 1 \quad \text{and} \quad \frac{1 \text{ ft}}{12 \text{ in.}} = 1.$$

If we divide 12 in. by 1 ft or 1 ft by 12 in., we get 1 because the lengths are the same. Let's first convert from smaller to larger units.

Example 3 Complete: 48 in. = _______ ft.

We want to convert from "in." to "ft." We multiply by 1 using a symbol for 1 with "in." on the bottom and "ft" on the top to eliminate inches and to convert to feet:

$$\begin{aligned} 48 \text{ in.} &= \frac{48 \text{ in.}}{1} \times \frac{1 \text{ ft}}{12 \text{ in.}} && \text{Multiplying by 1 using } \frac{1 \text{ ft}}{12 \text{ in.}} \text{ to eliminate in.} \\ &= \frac{48 \text{ in.}}{12 \text{ in.}} \times 1 \text{ ft} \\ &= \frac{48}{12} \times \frac{\text{in.}}{\text{in.}} \times 1 \text{ ft} \\ &= 4 \times 1 \text{ ft} && \text{The } \frac{\text{in.}}{\text{in.}} \text{ acts like 1, so we can omit it.} \\ &= 4 \text{ ft.} \end{aligned}$$

We can also look at this conversion as "canceling" units:

$$48 \text{ in.} = \frac{48 \cancel{\text{in.}}}{1} \times \frac{1 \text{ ft}}{12 \cancel{\text{in.}}} = \frac{48}{12} \times 1 \text{ ft} = 4 \text{ ft.}$$

Do Exercises 8 and 9.

Example 4 Complete: 25 ft = _______ yd.

Since we are converting from "ft" to "yd," we choose a symbol for 1 with "yd" on the top and "ft" on the bottom:

$$25\text{ ft} = 25\text{ ft} \times \frac{1\text{ yd}}{3\text{ ft}}$$

3 ft = 1 yd, so $\frac{3\text{ ft}}{1\text{ yd}} = 1$, and $\frac{1\text{ yd}}{3\text{ ft}} = 1$. We use $\frac{1\text{ yd}}{3\text{ ft}}$ to eliminate ft.

$$= \frac{25}{3} \times \frac{\text{ft}}{\text{ft}} \times 1\text{ yd}$$

$$= 8\frac{1}{3} \times 1\text{ yd}$$

The $\frac{\text{ft}}{\text{ft}}$ acts like 1, so we can omit it.

$$= 8\frac{1}{3}\text{ yd, or } 8.\overline{3}\text{ yd.}$$

Again, in this example, we can consider conversion from the point of view of canceling:

$$25\text{ ft} = 25\,\cancel{\text{ft}} \times \frac{1\text{ yd}}{3\,\cancel{\text{ft}}}$$

$$= \frac{25}{3} \times 1\text{ yd} = 8\frac{1}{3}\text{ yd, or } 8.\overline{3}\text{ yd.}$$

Do Exercises 10 and 11.

Complete.

10. 24 ft = _______ yd

11. 35 ft = _______ yd

Example 5 Complete: 23,760 ft = _______ mi.

We choose a symbol for 1 with "mi" on the top and "ft" on the bottom:

$$23{,}760\text{ ft} = 23{,}760\text{ ft} \times \frac{1\text{ mi}}{5280\text{ ft}}$$

5280 ft = 1 mi, so $\frac{1\text{ mi}}{5280\text{ ft}} = 1$.

$$= \frac{23{,}760}{5280} \times \frac{\text{ft}}{\text{ft}} \times 1\text{ mi}$$

$$= 4.5 \times 1\text{ mi}$$

Dividing

$$= 4.5\text{ mi.}$$

Let's also consider this example using canceling:

$$23{,}760\text{ ft} = 23{,}760\,\cancel{\text{ft}} \times \frac{1\text{ mi}}{5280\,\cancel{\text{ft}}}$$

$$= \frac{23{,}760}{5280} \times 1\text{ mi}$$

$$= 4.5 \times 1\text{ mi} = 4.5\text{ mi.}$$

Do Exercises 12 and 13.

Complete.

12. 26,400 ft = _______ mi

13. 2640 ft = _______ mi

We can also use multiplying by 1 to convert from larger to smaller units. Let's redo Example 2.

Example 6 Complete: 7 yd = _______ in.

$$7\text{ yd} = \frac{7\,\cancel{\text{yd}}}{1} \times \frac{3\,\cancel{\text{ft}}}{1\,\cancel{\text{yd}}} \times \frac{12\text{ in.}}{1\,\cancel{\text{ft}}}$$

$$= 7 \times 3 \times 12 \times 1\text{ in.} = 252\text{ in.}$$

Do Exercise 14.

14. Complete. Use multiplying by 1.

8 yd = _______ in.

Answers on page A-19

Improving Your Math Study Skills

Classwork: Before and During Class

Before Class

Textbook

- Check your syllabus (or ask your instructor) to find out which sections will be covered during the next class. Then be sure to read these sections *before* class. Although you may not understand all the concepts, you will at least be familiar with the material, which will help you follow the discussion during class.
- This book makes use of color, shading, and design elements to highlight important concepts, so you do not need to highlight these. Instead, it is more productive for you to note trouble spots with either a highlighter or Post-It notes. Then use these marked points as possible questions for clarification by your instructor at the appropriate time.

Homework

- Review the previous day's homework just before class. This will refresh your memory on the concepts covered in the last class, and again provide you with possible questions to ask your instructor.

During Class

Class Seating

- If possible, choose a seat at the front of the class. In most classes, the more serious students tend to sit there so you will probably be able to concentrate better if you do the same. You should also avoid sitting next to noisy or distracting students.
- If your instructor uses an overhead projector, consider choosing a seat that will give you an unobstructed view of the screen.

Taking Notes

- This textbook has been written and laid out so that it represents a quality set of notes at the same time that it teaches. Thus you might not need to take notes in class. Just watch, listen, and ask yourself questions as the class moves along, rather than racing to keep up your note-taking.

  However, if you still feel more comfortable taking your own notes, consider using the following two-column method. Divide your page in half vertically so that you have two columns side by side. Write down what is on the board in the left column; then, in the right column, write clarifying comments or questions.
- If you have any difficulty keeping up with the instructor, use abbreviations to speed up your note-taking. Consider standard abbreviations like "Ex" for "Example," "$\approx$" for "approximately equal to," or "$\therefore$" for "therefore." Create your own abbreviations as well.
- Another shortcut for note-taking is to write only the beginning of a word, leaving space for the rest. Be sure you write enough of the word to know what it means later on!

Exercise Set 8.1

a Complete.

1. 1 ft = ______ in.
2. 1 yd = ______ ft
3. 1 in. = ______ ft
4. 1 mi = ______ yd
5. 1 mi = ______ ft
6. 1 ft = ______ yd
7. 3 yd = ______ in.
8. 10 yd = ______ ft
9. 84 in. = ______ ft
10. 48 ft = ______ yd
11. 18 in. = ______ ft
12. 29 ft = ______ yd
13. 5 mi = ______ ft
14. 5 mi = ______ yd
15. 36 in. = ______ ft
16. 11,616 ft = ______ mi
17. 10 ft = ______ yd
18. 9.6 yd = ______ ft
19. 10 mi = ______ ft
20. 31,680 ft = ______ mi
21. $4\frac{1}{2}$ ft = ______ yd
22. 48 in. = ______ ft
23. 36 in. = ______ yd
24. 20 yd = ______ in.
25. 330 ft = ______ yd
26. 5280 yd = ______ mi
27. 3520 yd = ______ mi

28. 25 mi = ________ ft

29. 100 yd = ________ ft

30. 480 in. = ________ ft

31. 360 in. = ________ ft

32. 720 in. = ________ yd

33. 1 in. = ________ yd

34. 25 in. = ________ ft

35. 2 mi = ________ in.

36. 63,360 in. = ________ mi

Skill Maintenance

Convert to fractional notation. [6.2b]

37. 9.25%

38. $87\frac{1}{2}\%$

Find fractional notation for the percent notation in the sentence. [6.2b]

39. Of all 18-year-olds, 27.5% are registered to vote.

40. Of all those who buy CDs, 57% are in the 20–39 age group.

Convert to percent notation. [6.2a]

41. $\frac{11}{8}$

42. $\frac{2}{3}$

43. $\frac{1}{4}$

44. $\frac{7}{16}$

Find fractional notation for the ratio. [5.1a]

45. In Washington, D.C., there are 36.1 lawyers for every 1000 people. What is the ratio of lawyers to people? What is the ratio of people to lawyers?

46. In a bread recipe, there are 2 cups of milk to 12 cups of flour. What is the ratio of cups of milk to cups of flour?

Synthesis

47. ◆ A student makes the following error:

23 in. = 23 · (12 ft) = 276 ft.

Explain the error in at least two ways.

48. ◆ Describe two methods of making unit conversions discussed in this section.

49. *National Debt.* Recently the national debt was $5.103 trillion. To get an idea of this amount, picture that if that many $1 bills were stacked on top of each other, they would reach 1.382 times the distance to the moon. The distance to the moon is 238,866 mi. How thick, in inches, is a $1 bill?

Collaborative Learning Manual

Practice conversion with the old British monetary units.

8.2 Linear Measures: The Metric System

Objectives

a Convert from one metric unit of length to another.

b Convert between American and metric units of length.

For Extra Help

TAPE 13 TAPE 13A MAC WIN CD-ROM

The **metric system** is used in most countries of the world, and the United States is now making greater use of it as well. The metric system does not use inches, feet, pounds, and so on, although units for time and electricity are the same as those you use now.

An advantage of the metric system is that it is easier to convert from one unit to another. That is because the metric system is based on the number 10.

The basic unit of length is the **meter**. It is just over a yard. In fact, 1 meter $\approx$ 1.1 yd.

(Comparative sizes are shown.)

1 Meter

1 Yard

The other units of length are multiples of the length of a meter:

10 times a meter, 100 times a meter, 1000 times a meter, and so on,

or fractions of a meter:

$$\frac{1}{10} \text{ of a meter}, \quad \frac{1}{100} \text{ of a meter}, \quad \frac{1}{1000} \text{ of a meter}, \quad \text{and so on.}$$

METRIC UNITS OF LENGTH

1 *kilo*meter (km) = 1000 meters (m)

1 *hecto*meter (hm) = 100 meters (m)

1 *deka*meter (dam) = 10 meters (m)

1 meter (m)

1 *deci*meter (dm) = $\frac{1}{10}$ meter (m)

1 *centi*meter (cm) = $\frac{1}{100}$ meter (m)

1 *milli*meter (mm) = $\frac{1}{1000}$ meter (m)

dam and *dm* are not used often.

You should memorize these names and abbreviations. Think of *kilo*- for 1000, *hecto*- for 100, *deka*- for 10, *deci*- for $\frac{1}{10}$, *centi*- for $\frac{1}{100}$, and *milli*- for $\frac{1}{1000}$. We will also use these prefixes when considering units of area, capacity, and mass.

Thinking Metric

To familiarize yourself with metric units, consider the following.

1 kilometer (1000 meters)	is slightly more than $\frac{1}{2}$ mile (0.6 mi).
1 meter	is just over a yard (1.1 yd).
1 centimeter (0.01 meter)	is a little more than the width of a paperclip (about 0.3937 inch).

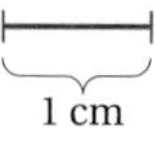

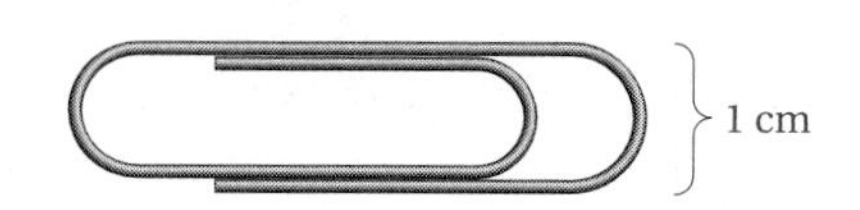

1 inch is about 2.54 centimeters.

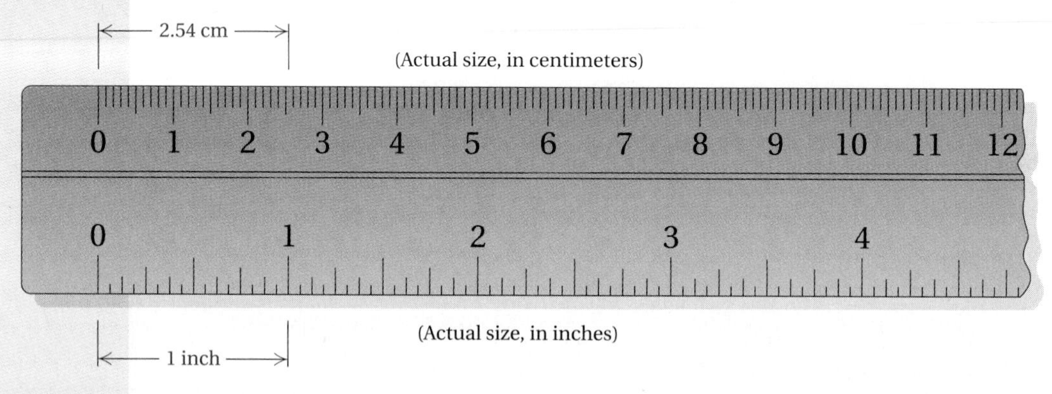

1 millimeter is about the diameter of a paperclip wire.

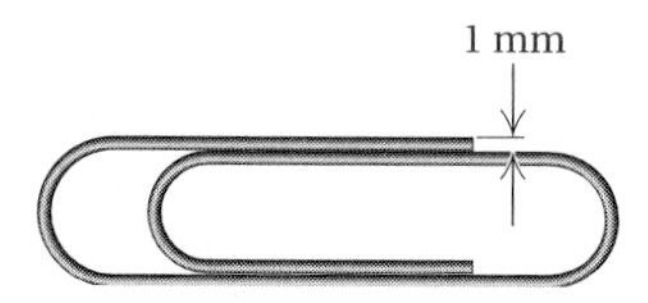

The millimeter (mm) is used to measure small distances, especially in industry.

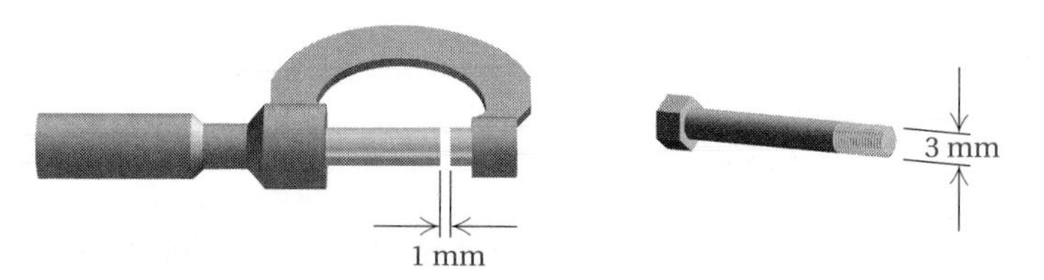

In many countries, the centimeter (cm) is used for body dimensions and clothing sizes.

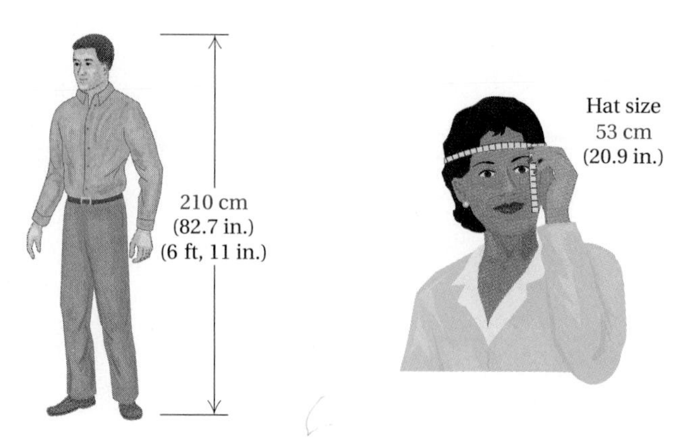

The meter (m) is used for expressing dimensions of larger objects—say, the length of a building—and for shorter distances, such as the length of a rug.

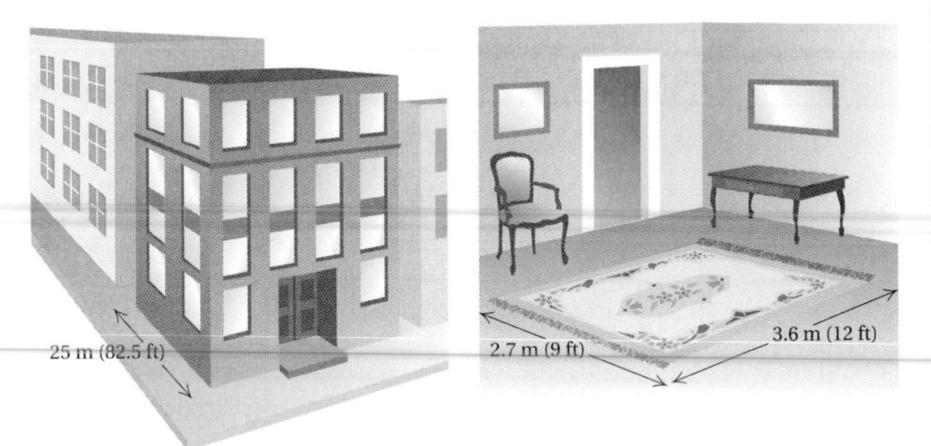

The kilometer (km) is used for longer distances, mostly in cases where miles are now being used.

Do Exercises 1–6.

Complete with mm, cm, m, or km.

1. A stick of gum is 7 ________ long.

2. Minneapolis is 3213 ________ from San Francisco.

3. A penny is 1 ________ thick.

4. The halfback ran 7 ________.

5. The book is 3 ________ thick.

6. The desk is 2 ________ long.

a Changing Metric Units

Example 1 Complete: 4 km = ________ m.

$$\begin{aligned} 4\text{ km} &= 4 \times 1\text{ km} \\ &= 4 \times 1000\text{ m} \quad \textbf{Substituting 1000 m for 1 km} \\ &= 4000\text{ m} \end{aligned}$$

Do Exercises 7 and 8.

Complete.

7. 23 km = ________ m

8. 4 hm = ________ m

Since

$$\frac{1}{10}\text{ m} = 1\text{ dm}, \quad \frac{1}{100}\text{ m} = 1\text{ cm}, \quad \text{and} \quad \frac{1}{1000}\text{ m} = 1\text{ mm},$$

it follows that

> 1 m = 10 dm, 1 m = 100 cm, and 1 m = 1000 mm.

Memorizing these will help you to write forms of 1 when making conversions.

Answers on page A-19

Complete.

9. 1.78 m = ______ cm

10. 9.04 m = ______ mm

Complete.

11. 7814 m = ______ km

12. 7814 m = ______ dam

Answers on page A-19

Example 2 Complete: 93.4 m = ______ cm.

We want to convert from "m" to "cm." We multiply by 1 using a symbol for 1 with "m" on the bottom and "cm" on the top to eliminate meters and convert to centimeters:

$$93.4 \text{ m} = 93.4 \text{ m} \times \frac{100 \text{ cm}}{1 \text{ m}}$$ **Multiplying by 1 using $\frac{100 \text{ cm}}{1 \text{ m}}$**

$$= 93.4 \times 100 \times \frac{\text{m}}{\text{m}} \times 1 \text{ cm}$$ **The $\frac{\text{m}}{\text{m}}$ acts like 1, so we omit it.**

$$= 9340 \text{ cm}.$$ **Multiplying by 100 moves the decimal point two places to the right.**

We can also work this example by canceling:

$$93.4 \text{ m} = 93.4 \cancel{\text{m}} \times \frac{100 \text{ cm}}{1 \cancel{\text{m}}}$$

$$= 93.4 \times 100 \times 1 \text{ cm}$$

$$= 9340 \text{ cm}.$$

Example 3 Complete: 0.248 m = ______ mm.

We are converting from "m" to "mm," so we choose a symbol for 1 with "mm" on the top and "m" on the bottom:

$$0.248 \text{ m} = 0.248 \text{ m} \times \frac{1000 \text{ mm}}{1 \text{ m}}$$ **Multiplying by 1 using $\frac{1000 \text{ mm}}{1 \text{ m}}$**

$$= 0.248 \times 1000 \times \frac{\text{m}}{\text{m}} \times 1 \text{ mm}$$ **The $\frac{\text{m}}{\text{m}}$ acts like 1, so we omit it.**

$$= 248 \text{ mm}.$$ **Multiplying by 1000 moves the decimal point three places to the right.**

Using canceling, we can work this example as follows:

$$0.248 = 0.248 \cancel{\text{m}} \times \frac{1000 \text{ mm}}{1 \cancel{\text{m}}}$$

$$= 0.248 \times 1000 \times 1 \text{ mm} = 248 \text{ mm}.$$

Do Exercises 9 and 10.

Example 4 Complete: 2347 m = ______ km.

$$2347 \text{ m} = 2347 \text{ m} \times \frac{1 \text{ km}}{1000 \text{ m}}$$ **Multiplying by 1 using $\frac{1 \text{ km}}{1000 \text{ m}}$**

$$= \frac{2347}{1000} \times \frac{\text{m}}{\text{m}} \times 1 \text{ km}$$ **The $\frac{\text{m}}{\text{m}}$ acts like 1, so we omit it.**

$$= 2.347 \text{ km}$$ **Dividing by 1000 moves the decimal point three places to the left.**

Using canceling, we can work this example as follows:

$$2347 \text{ m} = 2347 \cancel{\text{m}} \times \frac{1 \text{ km}}{1000 \cancel{\text{m}}} = \frac{2347}{1000} \times 1 \text{ km} = 2.347 \text{ km}.$$

Do Exercises 11 and 12.

Sometimes we multiply by 1 more than once.

Example 5 Complete: 8.42 mm = _______ cm.

$$8.42 \text{ mm} = 8.42 \text{ mm} \times \frac{1 \text{ m}}{1000 \text{ mm}} \times \frac{100 \text{ cm}}{1 \text{ m}}$$

Multiplying by 1 using $\frac{1 \text{ m}}{1000 \text{ mm}}$ and $\frac{100 \text{ cm}}{1 \text{ m}}$

$$= \frac{8.42 \times 100}{1000} \times \frac{\text{mm}}{\text{mm}} \times \frac{\text{m}}{\text{m}} \times 1 \text{ cm}$$

$$= \frac{842}{1000} \text{ cm} = 0.842 \text{ cm}$$

Do Exercises 13 and 14.

Complete.

13. 9.67 mm = _______ cm

14. 89 km = _______ cm

Mental Conversion

Look back over the examples and exercises done so far and you will see that changing from one unit to another in the metric system amounts to only the movement of a decimal point. That is because the metric system is based on 10. Let's find a faster way to convert. Look at the following table.

1000 m	100 m	10 m	1 m	0.1 m	0.01 m	0.001 m
1 km	1 hm	1 dam	1 m	1 dm	1 cm	1 mm

Each place in the table has a value $\frac{1}{10}$ that to the left or 10 times that to the right. Thus moving one place in the table corresponds to one decimal place. Let's convert mentally.

Example 6 Complete: 8.42 mm = _______ cm.

Think: To go from mm to cm in the table is a move of one place to the left. Thus we move the decimal point one place to the left.

1000 m	100 m	10 m	1 m	0.1 m	0.01 m	0.001 m
1 km	1 hm	1 dam	1 m	1 dm	1 cm	1 mm

1 place to the left

8.42 0.8.42 8.42 mm = 0.842 cm

Example 7 Complete: 1.886 km = _______ cm.

Think: To go from km to cm is a move of five places to the right. Thus we move the decimal point five places to the right.

1000 m	100 m	10 m	1 m	0.1 m	0.01 m	0.001 m
1 km	1 hm	1 dam	1 m	1 dm	1 cm	1 mm

5 places to the right

1.886 1.88600. 1.886 km = 188,600 cm

Answers on page A-19

Complete. Try to do this mentally using the table.

15. 6780 m = ________ km

16. 9.74 cm = ________ mm

17. 1 mm = ________ cm

18. 845.1 mm = ________ dm

Complete.

19. 100 yd = ________ m
(The length of a football field)

20. 500 mi = ________ km
(The Indianapolis 500-mile race)

21. 2383 km = ________ mi
(The distance from St. Louis to Phoenix)

Answers on page A-19

Example 8 Complete: 3 m = ________ cm.

Think: To go from m to cm in the table is a move of two places to the right. Thus we move the decimal point two places to the right.

1000 m	100 m	10 m	1 m	0.1 m	0.01 m	0.001 m
1 km	1 hm	1 dam	1 m	1 dm	1 cm	1 mm

2 places to the right

3 3.00. 3 m = 300 cm

You should try to make metric conversions mentally as much as possible.

The fact that conversions can be done so easily is an important advantage of the metric system. The most commonly used metric units of length are km, m, cm, and mm. We have purposely used these more often than the others in the exercises.

Do Exercises 15–18.

b Converting Between American and Metric Units

We can make conversions between American and metric units by using the following table. Again, we either make a substitution or multiply by 1 appropriately.

Metric	American
1 m	39.37 in.
1 m	3.3 ft
0.303 m	1 ft
2.54 cm	1 in.
1 km	0.621 mi
1.609 km	1 mi

Example 9 Complete: 26.2 mi = ________ km. (This is the length of the Olympic marathon.)

$$26.2 \text{ mi} = 26.2 \times 1 \text{ mi}$$
$$\approx 26.2 \times 1.609 \text{ km}$$
$$\approx 42.1558 \text{ km}$$

Example 10 Complete: 100 m = ________ yd. (This is the length of a dash in track.)

$$100 \text{ m} = 100 \times 1 \text{ m} \approx 100 \times 3.3 \text{ ft} \approx 330 \text{ ft}$$
$$\approx 330 \text{ ft} \times \frac{1 \text{ yd}}{3 \text{ ft}} \approx \frac{330}{3} \text{ yd} \approx 110 \text{ yd}$$

Do Exercises 19–21.

Exercise Set 8.2

a Complete. Do as much as possible mentally.

1. **a)** 1 km = 1000 m
b) 1 m = .001 km

2. **a)** 1 hm = 100 m
b) 1 m = .01 hm

3. **a)** 1 dam = 10 m
b) 1 m = .1 dam

4. **a)** 1 dm = .1 m
b) 1 m = 10 dm

5. **a)** 1 cm = .01 m
b) 1 m = 100 cm

6. **a)** 1 mm = .0001 m
b) 1 m = 1000 mm

7. 6.7 km = 6700 m

8. 27 km = 27000 m

9. 98 cm = .98 m

10. 0.789 cm = ______ m

11. 8921 m = ______ km

12. 8664 m = ______ km

13. 56.66 m = ______ km

14. 4.733 m = ______ km

15. 5666 m = ______ cm

16. 869 m = ______ cm

17. 477 cm = ______ m

18. 6.27 mm = ______ m

19. 6.88 m = ______ cm

20. 6.88 m = ______ dm

21. 1 mm = ______ cm

22. 1 cm = ______ km

23. 1 km = ______ cm

24. 2 km = ______ cm

25. 14.2 cm = ______ mm

26. 25.3 cm = ______ mm

27. 8.2 mm = ______ cm

28. 9.7 mm = ______ cm

29. 4500 mm = ______ cm

30. 8,000,000 m = ______ km

31. 0.024 mm = ______ m

32. 60,000 mm = ______ dam

33. 6.88 m = ______ dam

34. 7.44 m = ______ hm

35. 2.3 dam = ______ dm

36. 9 km = ______ hm

37. 392 dam = ______ km

38. 0.056 mm = ______ dm

b Complete.

39. 330 ft = ________ m
(The length of most baseball foul lines)

40. 12 in. = ________ cm

41. 1171.352 km = ________ mi
(The distance from Cleveland to Atlanta)

42. 2 m = ________ ft
(The length of a desk)

43. 65 mph = ________ km/h
(A common speed limit in the United States)

44. 100 km/h = ________ mph
(A common speed limit in Canada)

45. 180 mi = ________ km
(The distance from Indianapolis to Chicago)

46. 141,600,000 mi = ________ km
(The farthest distance of Mars from the sun)

47. 70 mph = ________ km/h
(An interstate speed limit in the United States)

48. 60 km/h = ________ mph
(A city speed limit in Canada)

49. 10 yd = ________ m
(The length needed for a first down in football)

50. 450 ft = ________ m
(The length of a long home run in baseball)

Skill Maintenance

Divide. Find decimal notation for the answer. [4.4a]

51. $21 \div 12$

52. $23.4 \div 10$

53. $23.4 \div 100$

54. $23.4 \div 1000$

55. Multiply 3.14×4.41. Round to the nearest hundredth. [4.3a], [4.1e]

56. Multiply: $4 \times 20\frac{1}{8}$. [3.6a]

57. Multiply: $48 \times \frac{1}{12}$. [2.4a]

Find decimal notation for the percent notation in the sentence.

58. Blood is 90% water. [6.1b]

59. Of those accidents requiring medical attention, 10.8% of them occur on roads. [6.1b]

Synthesis

60. ◆ Explain in your own words why metric units are easier to work with than American units.

61. ◆ Would you expect the world record for the 100-m dash to be longer or shorter than the record for the 100-yd dash? Why?

Complete.

62. 🖩 2 mi = ________ cm

63. 🖩 10 km = ________ in.

64. 🖩 Audio cassettes are generally played at a rate of $1\frac{7}{8}$ in. per second. How many meters of tape are used for a 60-min cassette? (*Note*: A 60-min cassette has 30 min of playing time on each side.)

65. 🖩 In a recent year, the world record for the 100-m dash was 9.86 sec. How fast is this in miles per hour? Round to the nearest tenth of a mile per hour.

8.3 Perimeter

a Finding Perimeters

Objectives

a Find the perimeter of a polygon.

b Solve applied problems involving perimeter.

For Extra Help

TAPE 13 TAPE 13A MAC WIN CD-ROM

A **polygon** is a geometric figure with three or more sides. The **perimeter of a polygon** is the distance around it, or the sum of the lengths of its sides.

Example 1 Find the perimeter of this polygon.

We add the lengths of the sides. Since all the units are the same, we add the numbers, keeping meters (m) as the unit.

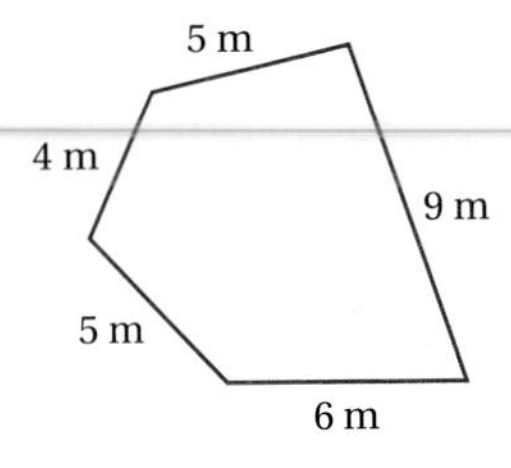

Perimeter $= 6\text{ m} + 5\text{ m} + 4\text{ m} + 5\text{ m} + 9\text{ m}$

$= (6 + 5 + 4 + 5 + 9)\text{ m}$

$= 29\text{ m}$

Do Exercises 1 and 2.

A **rectangle** is a figure with four sides and four 90°-angles, like the one shown in Example 2.

Example 2 Find the perimeter of a rectangle that is 3 cm by 4 cm.

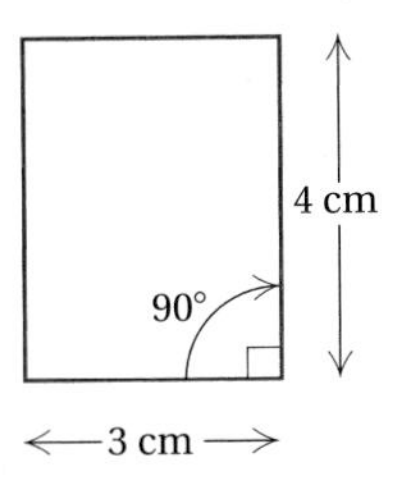

Perimeter $= 3\text{ cm} + 3\text{ cm} + 4\text{ cm} + 4\text{ cm}$

$= (3 + 3 + 4 + 4)\text{ cm}$

$= 14\text{ cm}$

Do Exercise 3.

The **perimeter of a rectangle** is twice the sum of the length and the width, or 2 times the length plus 2 times the width:

$P = 2 \cdot (l + w)$, or $P = 2 \cdot l + 2 \cdot w$.

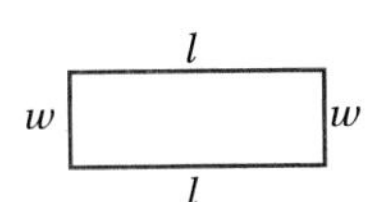

Example 3 Find the perimeter of a rectangle that is 4.3 ft by 7.8 ft.

$P = 2 \cdot (l + w) = 2 \cdot (4.3\text{ ft} + 7.8\text{ ft}) = 2 \cdot (12.1\text{ ft}) = 24.2\text{ ft}$

Do Exercises 4 and 5.

A **square** is a rectangle with all sides the same length.

Example 4 Find the perimeter of a square whose sides are 9 mm long.

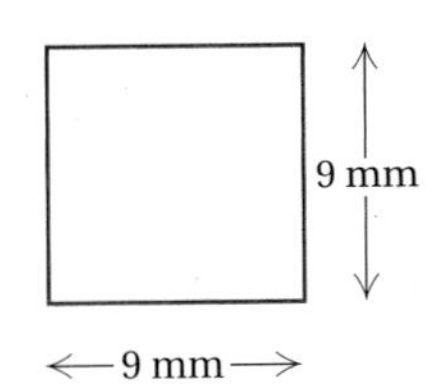

$P = 9\text{ mm} + 9\text{ mm} + 9\text{ mm} + 9\text{ mm}$

$= (9 + 9 + 9 + 9)\text{ mm} = 36\text{ mm}$

Find the perimeter of the polygon.

1.

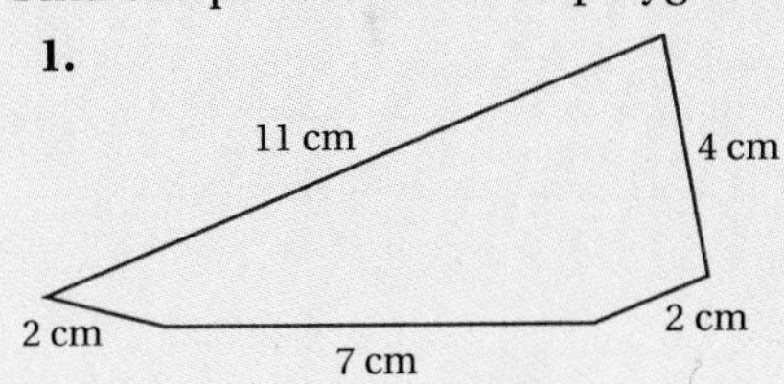

2.

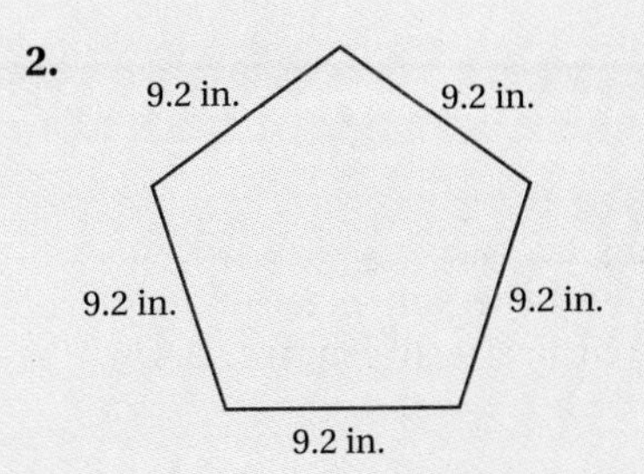

3. Find the perimeter of a rectangle that is 2 cm by 4 cm.

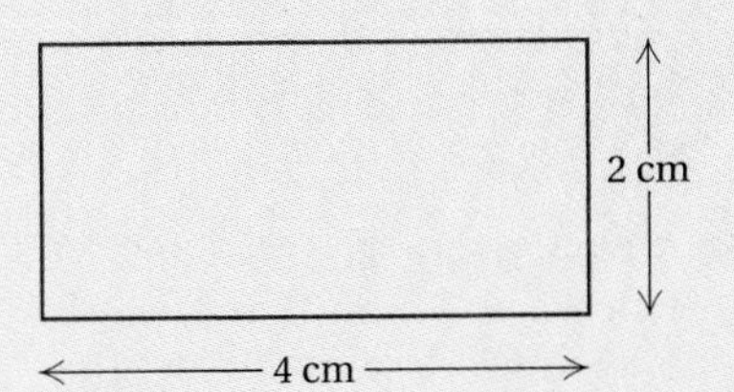

4. Find the perimeter of a rectangle that is 5.25 yd by 3.5 yd.

5. Find the perimeter of a rectangle that is 8 km by 8 km.

Answers on page A-19

6. Find the perimeter of a square with sides of length 10 km.

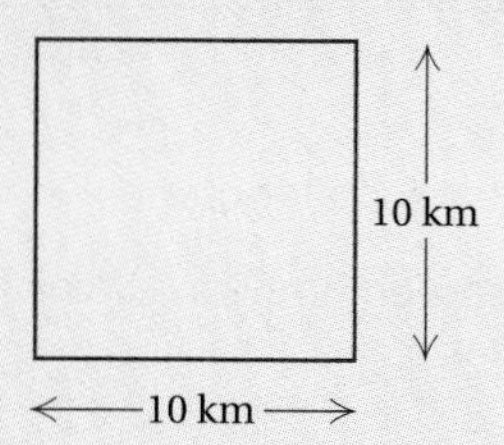

7. Find the perimeter of a square with sides of length $5\frac{1}{4}$ yd.

8. Find the perimeter of a square with sides of length 7.8 km.

9. A play area is 25 ft by 10 ft. A fence is to be built around the play area. How many feet of fencing will be needed? If fencing costs \$4.95 per foot, what will the fencing cost?

Answers on page A-19

Do Exercise 6.

The **perimeter of a square** is four times the length of a side:

$P = 4 \cdot s.$

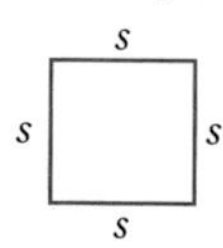

Example 5 Find the perimeter of a square whose sides are $20\frac{1}{8}$ in. long.

$$P = 4 \cdot s = 4 \cdot 20\frac{1}{8} \text{ in.}$$

$$= 4 \cdot \frac{161}{8} \text{ in.} = \frac{4 \cdot 161}{4 \cdot 2} \text{ in.}$$

$$= \frac{161}{2} \cdot \frac{4}{4} \text{ in.} = 80\frac{1}{2} \text{ in.}$$

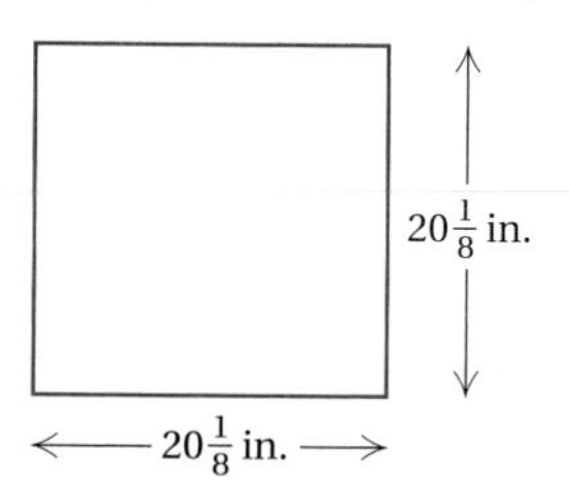

Do Exercises 7 and 8.

b Solving Applied Problems

Example 6 A vegetable garden is 20 ft by 15 ft. A fence is to be built around the garden. How many feet of fence will be needed? If fencing sells for \$2.95 per foot, what will the fencing cost?

1. **Familiarize.** We make a drawing and let $P =$ the perimeter.

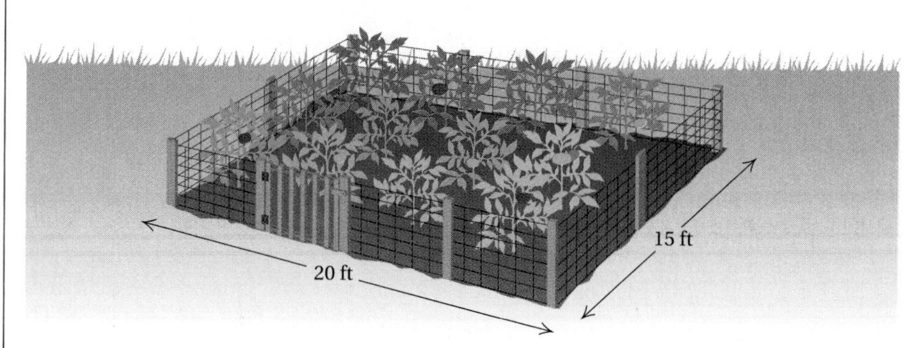

2. **Translate.** The perimeter of the garden is given by

$$P = 2 \cdot (l + w) = 2 \cdot (20 \text{ ft} + 15 \text{ ft}).$$

3. **Solve.** We calculate the perimeter as follows:

$$P = 2 \cdot (20 \text{ ft} + 15 \text{ ft}) = 2 \cdot (35 \text{ ft}) = 70 \text{ ft}.$$

Then we multiply by \$2.95 to find the cost of the fencing:

$$\text{Cost} = \$2.95 \times \text{Perimeter} = \$2.95 \times 70 \text{ ft} = \$206.50.$$

4. **Check.** The check is left to the student.
5. **State.** The 70 ft of fencing that is needed will cost \$206.50.

Do Exercise 9.

Exercise Set 8.3

a Find the perimeter of the polygon.

1. 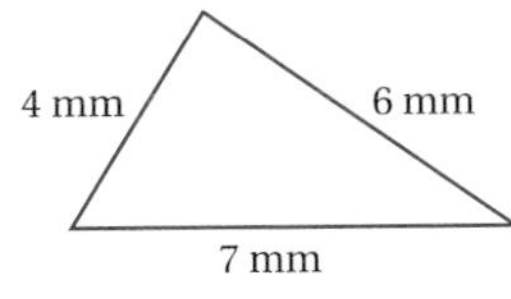

2. 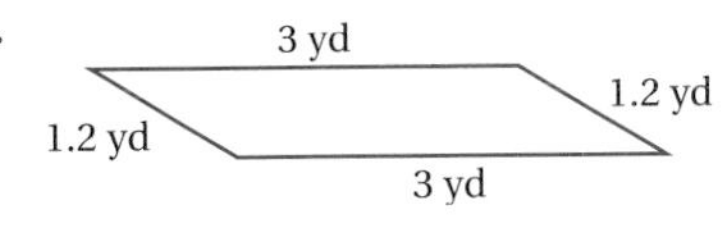

3. 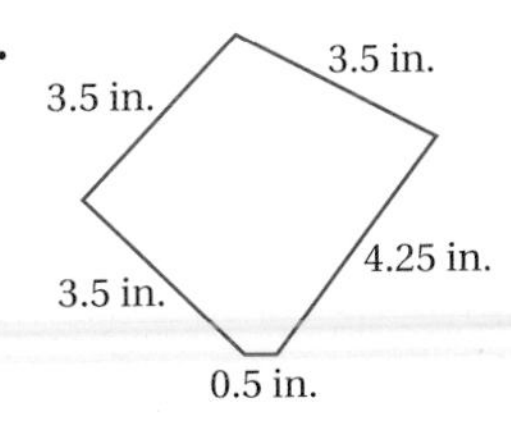

4. 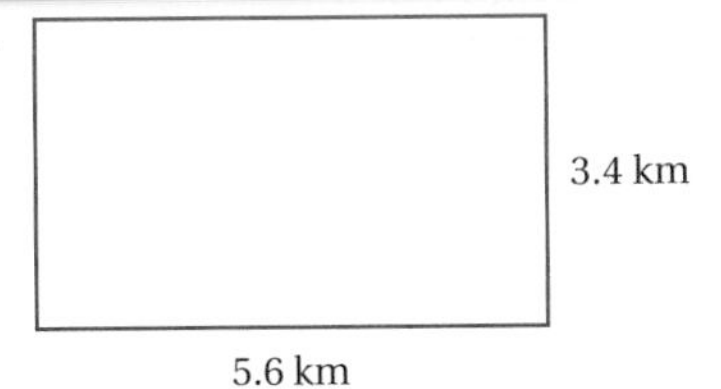

5. 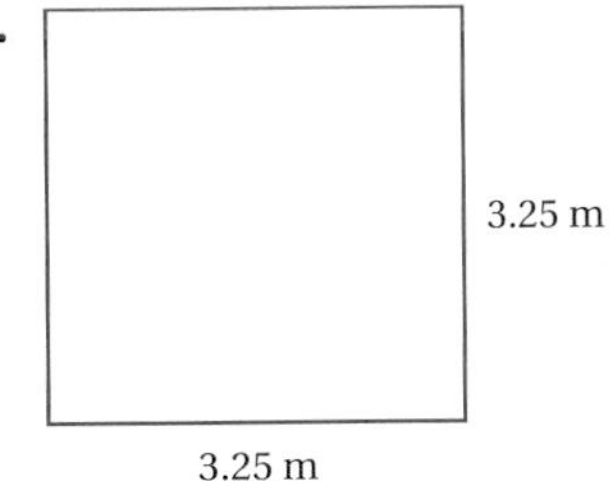

6. 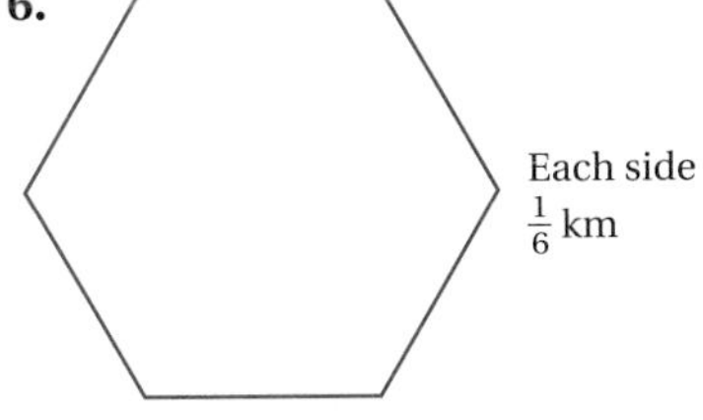

Find the perimeter of the rectangle.

7. 5 ft by 10 ft

8. 2.5 m by 100 m

9. 34.67 cm by 4.9 cm

10. $3\frac{1}{2}$ yd by $4\frac{1}{2}$ yd

Find the perimeter of the square.

11. 22 ft on a side

12. 56.9 km on a side

13. 45.5 mm on a side

14. $3\frac{1}{8}$ yd on a side

b Solve.

15. A security fence is to be built around a 173-m by 240-m field. What is the perimeter of the field? If fence wire costs $1.45 per meter, what will the fencing cost?

16. *Softball Diamond.* A standard-sized slow-pitch softball diamond is a square with sides of length 65 ft. What is the perimeter of this softball diamond? (This is the distance you would have to run if you hit a home run.)

17. A piece of flooring tile is a square with sides of length 30.5 cm. What is the perimeter of a piece of tile?

18. A posterboard is 61.8 cm by 87.9 cm. What is the perimeter of the board?

19. A rain gutter is to be installed around the house shown in the figure.

a) Find the perimeter of the house.
b) If the gutter costs \$4.59 per foot, what is the total cost of the gutter?

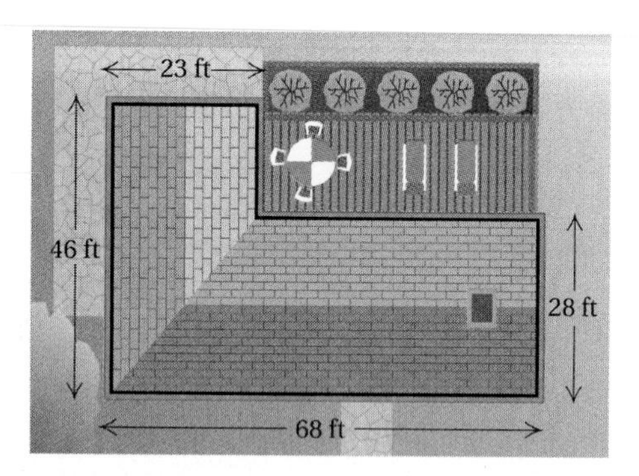

20. A carpenter is to build a fence around a 9-m by 12-m garden.

a) The posts are 3 m apart. How many posts will be needed?
b) The posts cost \$2.40 each. How much will the posts cost?
c) The fence will surround all but 3 m of the garden, which will be a gate. How long will the fence be?
d) The fence costs \$2.85 per meter. What will the cost of the fence be?
e) The gate costs \$9.95. What is the total cost of the materials?

Skill Maintenance

21. Convert to decimal notation: 56.1%. [6.1b]

22. Convert to percent notation: 0.6734. [6.1c]

23. Convert to percent notation: $\frac{9}{8}$. [6.2a]

Evaluate. [1.9b]

24. 5^2

25. 10^2

26. 31^2

Convert the number in the sentence to standard notation. [4.3b]

27. It is estimated that 4.7 million fax machines were sold in a recent year.

28. In a recent year, 4.3 billion CDs were sold.

Synthesis

29. ◆ Create for a fellow student a development of the formula

$$P = 2 \cdot (l + w) = 2 \cdot l + 2 \cdot w$$

for the perimeter of a rectangle.

30. ◆ Create for a fellow student a development of the formula

$$P = 4 \cdot s$$

for the perimeter of a square.

Find the perimeter of the figure in feet.

31.

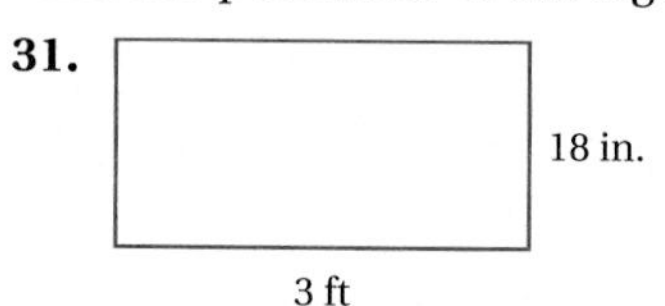

32.

78 in.

5.5 yd

8.4 Area

a Rectangles and Squares

A polygon and its interior form a plane region. We can find the area of a *rectangular region* by filling it with square units. Two such units, a *square inch* and a *square centimeter*, are shown below.

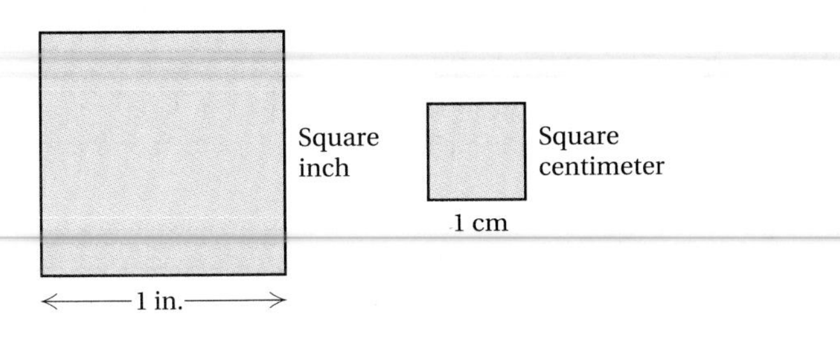

Example 1 What is the area of this region?

We have a rectangular array. Since the region is filled with 12 square centimeters, its area is 12 square centimeters (sq cm), or 12 cm^2. The number of units is 3×4, or 12.

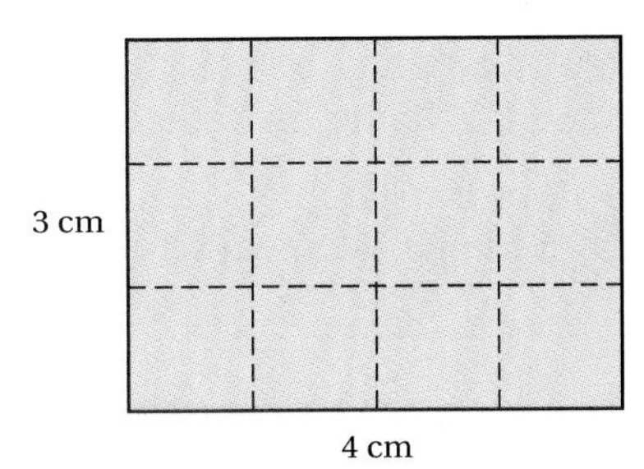

Do Exercise 1.

> The **area of a rectangular region** is the product of the length l and the width w:
>
> $A = l \cdot w.$

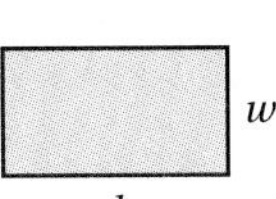

Example 2 Find the area of a rectangle that is 7 yd by 4 yd.

$$A = l \cdot w = 7 \text{ yd} \cdot 4 \text{ yd} = 7 \cdot 4 \cdot \text{yd} \cdot \text{yd} = 28 \text{ yd}^2$$

We think of yd · yd as $(\text{yd})^2$ and denote it yd^2. Thus we read "28 yd^2" as "28 square yards."

Do Exercises 2 and 3.

Example 3 Find the area of a square with sides of length 9 mm.

$$\begin{aligned} A &= (9 \text{ mm}) \cdot (9 \text{ mm}) \\ &= 9 \cdot 9 \cdot \text{mm} \cdot \text{mm} \\ &= 81 \text{ mm}^2 \end{aligned}$$

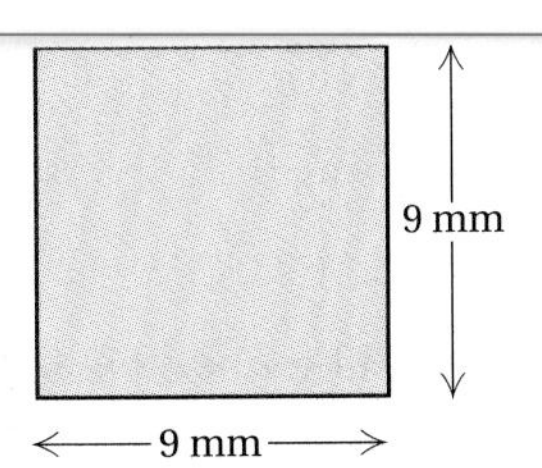

Do Exercise 4.

Objectives

a Find the area of a rectangle or a square.

b Solve applied problems involving areas of rectangles or squares.

For Extra Help

TAPE 14 TAPE 13B MAC WIN CD-ROM

1. What is the area of this region? Count the number of square centimeters.

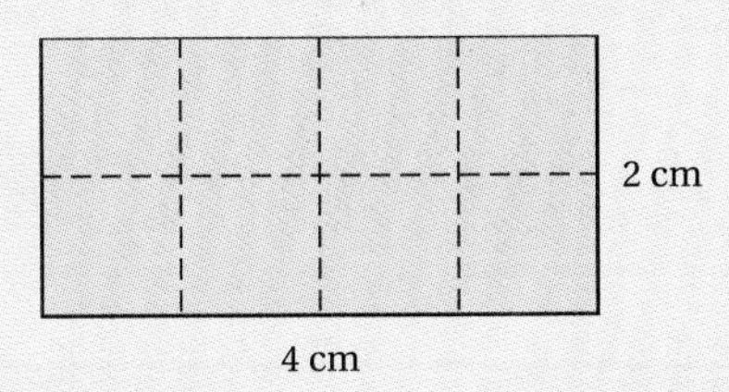

2. Find the area of a rectangle that is 7 km by 8 km.

3. Find the area of a rectangle that is $5\frac{1}{4}$ yd by $3\frac{1}{2}$ yd.

4. Find the area of a square with sides of length 12 km.

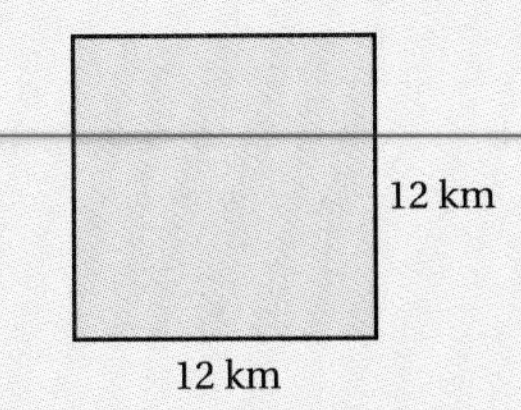

Answers on page A-19

5. Find the area of a square with sides of length 10.9 m.

6. Find the area of a square with sides of length $3\frac{1}{2}$ yd.

7. A square flower bed 3.5 m on a side is dug on a 30-m by 22.4-m lawn. How much area is left over? Draw a picture first.

Answers on page A-19

The **area of a square region** is the square of the length of a side:

$A = s \cdot s,$ or $A = s^2.$

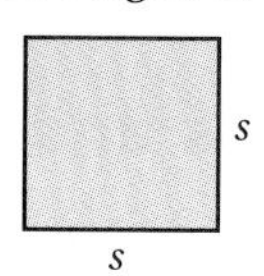

Example 4 Find the area of a square with sides of length 20.3 m.

$$A = s \cdot s = 20.3 \text{ m} \times 20.3 \text{ m}$$
$$= 20.3 \times 20.3 \times \text{m} \times \text{m} = 412.09 \text{ m}^2$$

Do Exercises 5 and 6.

b Solving Applied Problems

Example 5 *Mowing Cost.* A square sandbox 1.5 m on a side is placed on a 20-m by 31.2-m lawn. It costs \$0.04 per square meter to have the lawn mowed. What is the total cost of mowing?

1. **Familiarize.** We make a drawing.

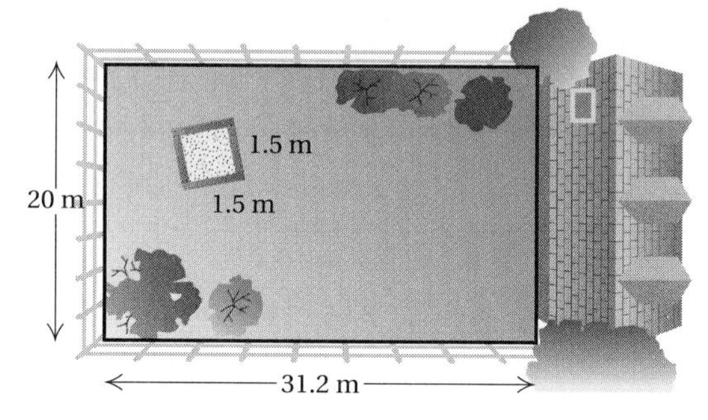

2. **Translate.** This is a two-step problem. We first find the area left over after the area of the sandbox has been subtracted. Then we multiply by the cost per square meter. We let A = the area left over.

Area left over	is	Area of lawn	minus	Area of sandbox
↓	↓	↓	↓	↓
A	$=$	$(20 \text{ m}) \times (31.2 \text{ m})$	$-$	$(1.5 \text{ m}) \times (1.5 \text{ m})$

3. **Solve.** The area of the lawn is

$$(20 \text{ m}) \times (31.2 \text{ m}) = 20 \times 31.2 \times \text{m} \times \text{m} = 624 \text{ m}^2.$$

The area of the sandbox is

$$(1.5 \text{ m}) \times (1.5 \text{ m}) = 1.5 \times 1.5 \times \text{m} \times \text{m} = 2.25 \text{ m}^2.$$

The area left over is

$$A = 624 \text{ m}^2 - 2.25 \text{ m}^2 = 621.75 \text{ m}^2.$$

Then we multiply by \$0.04:

$$\$0.04 \times 621.75 = \$24.87.$$

4. **Check.** The check is left to the student.
5. **State.** The total cost of mowing the lawn is \$24.87.

Do Exercise 7.

Exercise Set 8.4

[a] Find the area.

1. 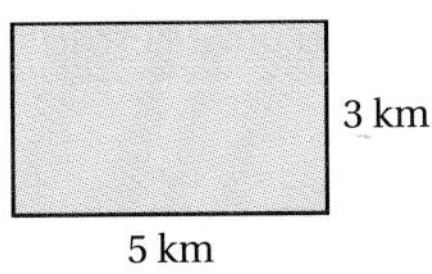

2. 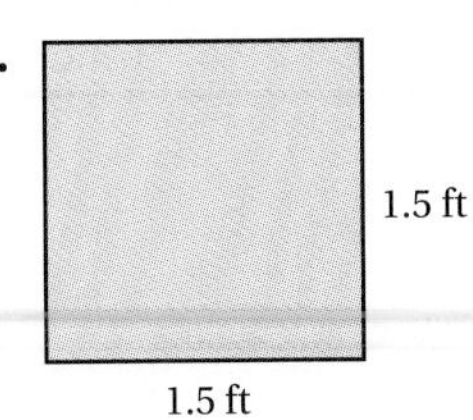

3. 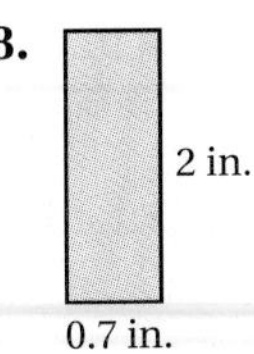

4. 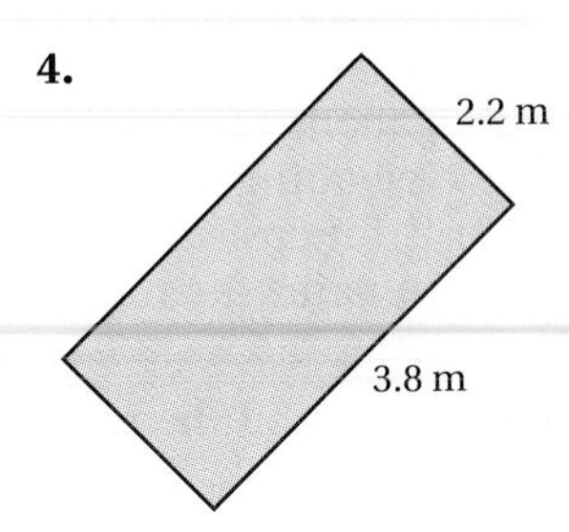

5. 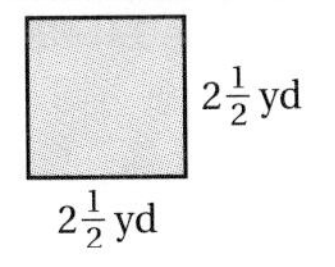

6. 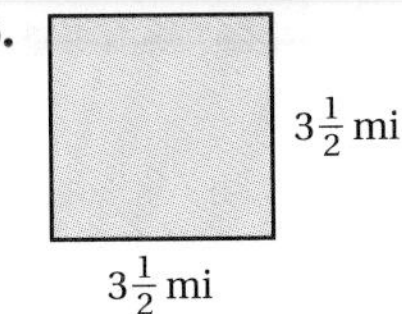

7. 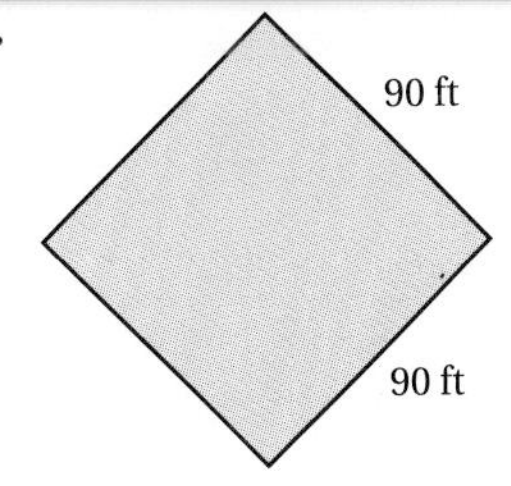

8. 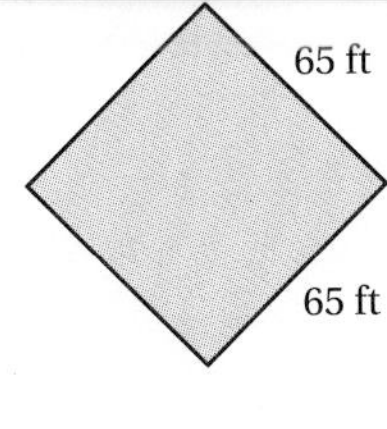

Find the area of the rectangle.

9. 5 ft by 10 ft

10. 14 yd by 8 yd

11. 34.67 cm by 4.9 cm

12. 2.45 km by 100 km

13. $4\frac{2}{3}$ in. by $8\frac{5}{6}$ in.

14. $10\frac{1}{3}$ mi by $20\frac{2}{3}$ mi

Find the area of the square.

15. 22 ft on a side

16. 18 yd on a side

17. 56.9 km on a side

18. 45.5 m on a side

19. $5\frac{3}{8}$ yd on a side

20. $7\frac{2}{3}$ ft on a side

[b] Solve.

21. A lot is 40 m by 36 m. A house 27 m by 9 m is built on the lot. How much area is left over for a lawn?

22. A field is 240.8 m by 450.2 m. Part of the field, 160.4 m by 90.6 m, is paved for a parking lot. How much area is unpaved?

23. Franklin Construction Company builds a sidewalk around two sides of the Municipal Trust Bank building, as shown in the figure. What is the area of the sidewalk?

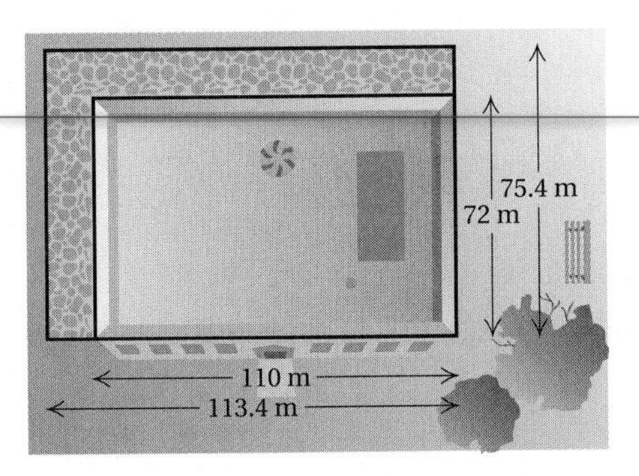

24. A standard sheet of typewriter paper is $8\frac{1}{2}$ in. by 11 in. We generally type on a $7\frac{1}{2}$-in. by 9-in. area of the paper. What is the area of the margin?

25. A room is 15 ft by 20 ft. The ceiling is 8 ft above the floor. There are two windows in the room, each 3 ft by 4 ft. The door is $2\frac{1}{2}$ ft by $6\frac{1}{2}$ ft.

a) What is the total area of the walls and the ceiling?

b) A gallon of paint will cover 86.625 ft^2. How many gallons of paint are needed for the room, including the ceiling?

c) Paint costs $17.95 a gallon. How much will it cost to paint the room?

26. A restaurant owner wants to carpet a 15-yd by 20-yd room.

a) How many square yards of carpeting are needed?

b) The carpeting they want is $18.50 per square yard. How much will it cost to carpet the room?

Find the area of the region.

27.

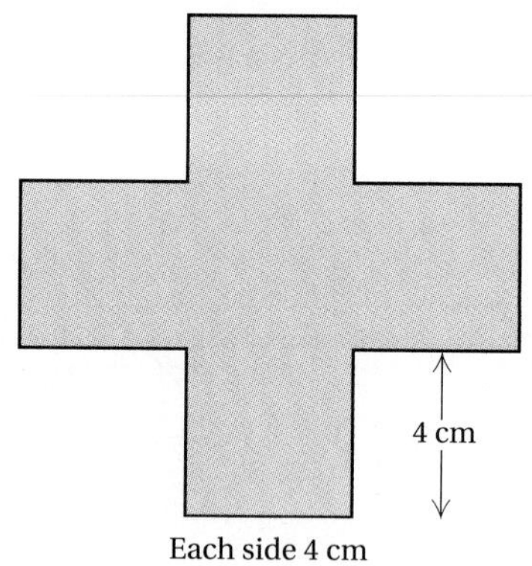

28.

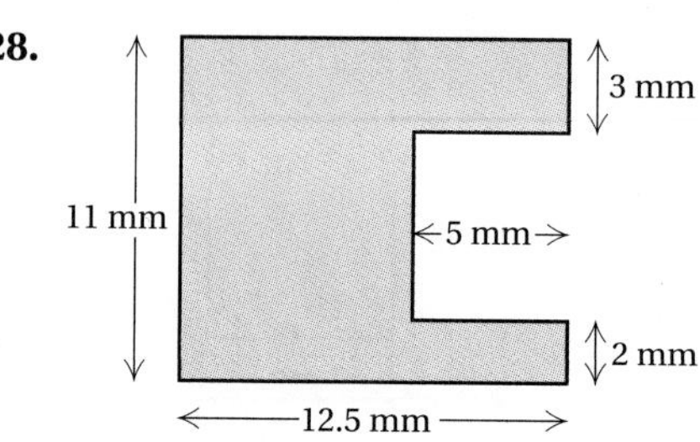

Skill Maintenance

Convert to percent notation. [6.1c], [6.2a]

29. 0.452

30. $\frac{1}{3}$

31. $\frac{11}{20}$

32. $\frac{22}{25}$

33. *Tourist Spending.* Foreign tourists spend $13.1 billion in this country annually. The most money, $2.7 billion, is spent in Florida. What is the ratio of amount spent in Florida to total amount spent? What is the ratio of total amount spent to amount spent in Florida? [5.1a, b]

34. One person in four plays a musical instrument. In a given group of people, what is the ratio of those who play an instrument to total number of people? What is the ratio of those who do not play an instrument to total number of people? [5.1a]

Synthesis

35. ◆ The length and the width of one rectangle are each three times the length and the width of another rectangle. Is the area of the first rectangle three times the area of the other rectangle? Why or why not?

36. ◆ Create for a fellow student a development of the formula

$$A = l \cdot w$$

for the area of a rectangle.

37. Find the area, in square inches, of the shaded region.

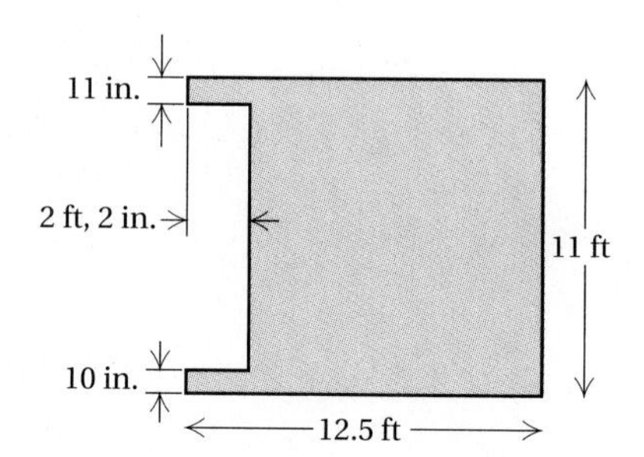

38. Find the area, in square feet, of the shaded region.

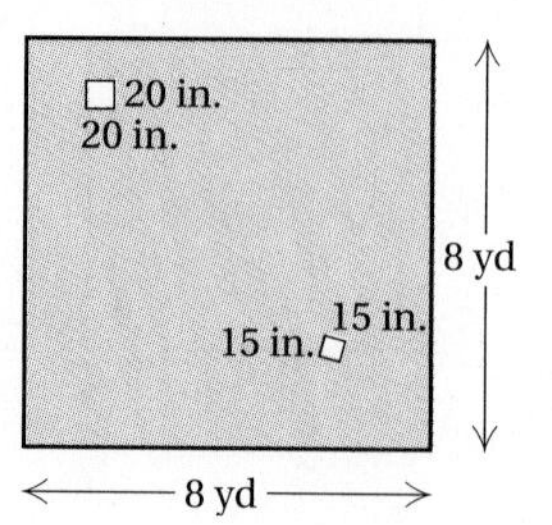

Collaborative Learning Manual

Prepare a budget for redecorating the classroom.

8.5 Areas of Parallelograms, Triangles, and Trapezoids

Objectives

a Find the area of a parallelogram, a triangle, and a trapezoid.

b Solve applied problems involving areas of parallelograms, triangles, and trapezoids.

For Extra Help

TAPE 14 | TAPE 13B | MAC WIN | CD-ROM

a Finding Other Areas

Parallelograms

A **parallelogram** is a four-sided figure with two pairs of parallel sides, as shown below.

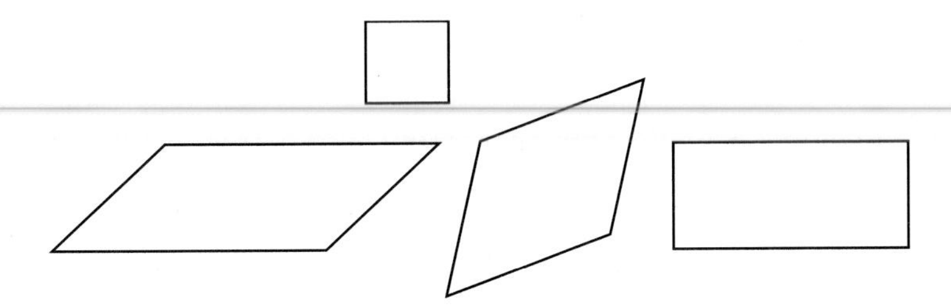

To find the area of a parallelogram, consider the one below.

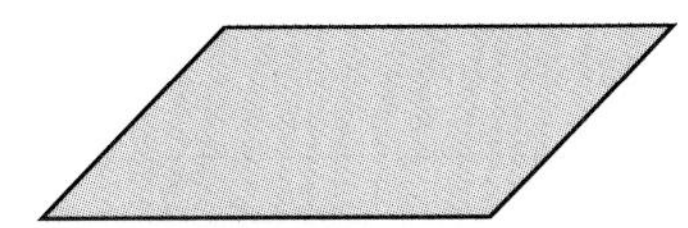

If we cut off a piece and move it to the other end, we get a rectangle.

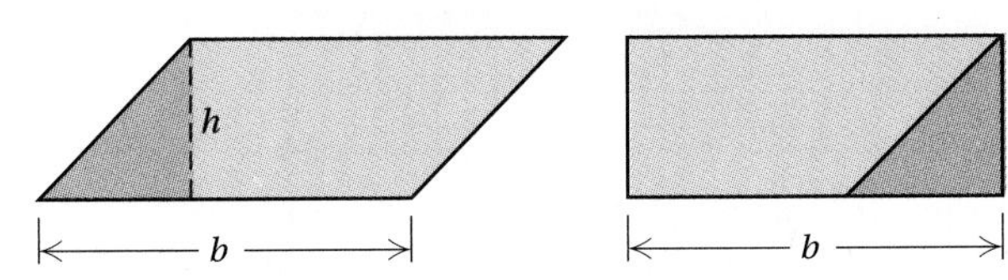

We can find the area by multiplying the length b, called a **base**, by h, called the **height**.

The **area of a parallelogram** is the product of the length of a base b and the height h:

$$A = b \cdot h.$$

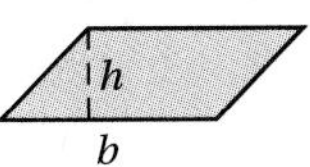

Example 1 Find the area of this parallelogram.

$A = b \cdot h$

$= 7 \text{ km} \cdot 5 \text{ km}$

$= 35 \text{ km}^2$

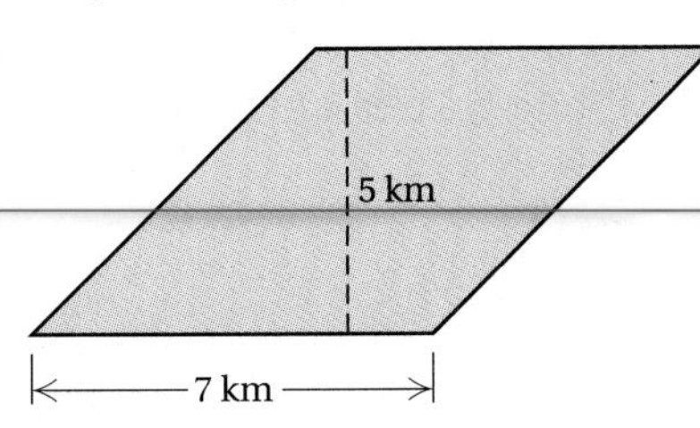

Find the area.

1.

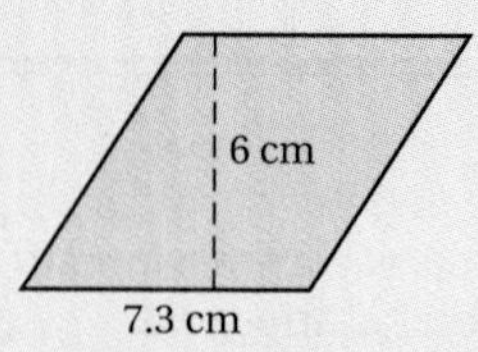

2.

5.5 km

2.25 km

Answers on page A-19

Example 2 Find the area of this parallelogram.

$$A = b \cdot h$$
$$= (1.2 \text{ m}) \times (6 \text{ m})$$
$$= 7.2 \text{ m}^2$$

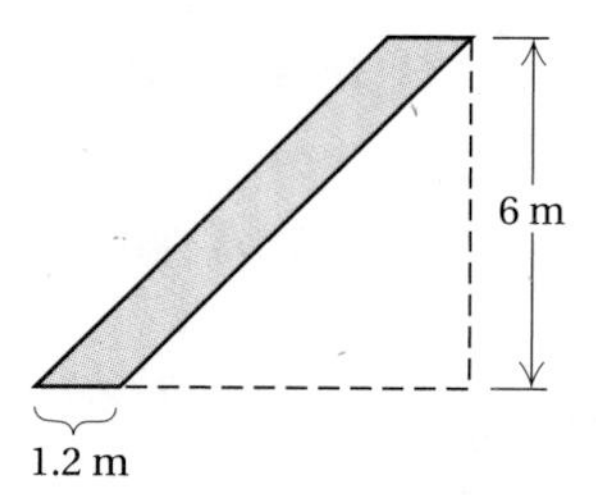

Do Exercises 1 and 2.

Triangles

To find the area of a triangle, think of cutting out another just like it.

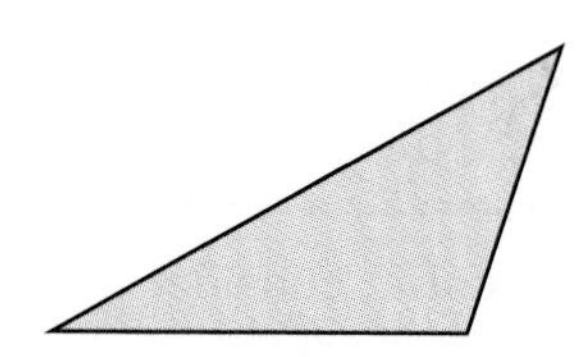

Then place the second one like this.

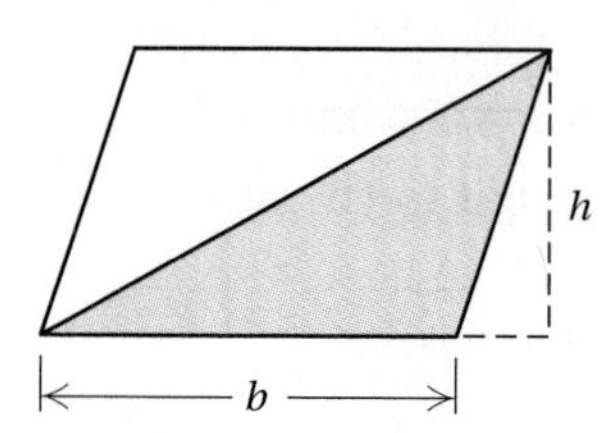

The resulting figure is a parallelogram whose area is

$$b \cdot h.$$

The triangle we started with has half the area of the parallelogram, or

$$\frac{1}{2} \cdot b \cdot h.$$

> The **area of a triangle** is half the length of the base times the height:
>
> $$A = \frac{1}{2} \cdot b \cdot h.$$
>
> 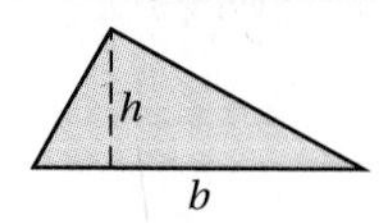
>

Example 3 Find the area of this triangle.

$$A = \frac{1}{2} \cdot b \cdot h$$
$$= \frac{1}{2} \cdot 9 \text{ m} \cdot 6 \text{ m}$$
$$= \frac{9 \cdot 6}{2} \text{ m}^2$$
$$= 27 \text{ m}^2$$

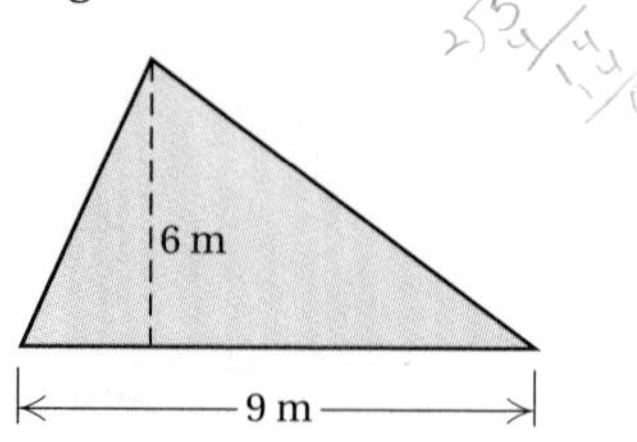

Example 4 Find the area of this triangle.

$$A = \frac{1}{2} \cdot b \cdot h$$
$$= \frac{1}{2} \times 6.25 \text{ cm} \times 5.5 \text{ cm}$$
$$= 0.5 \times 6.25 \times 5.5 \text{ cm}^2$$
$$= 17.1875 \text{ cm}^2$$

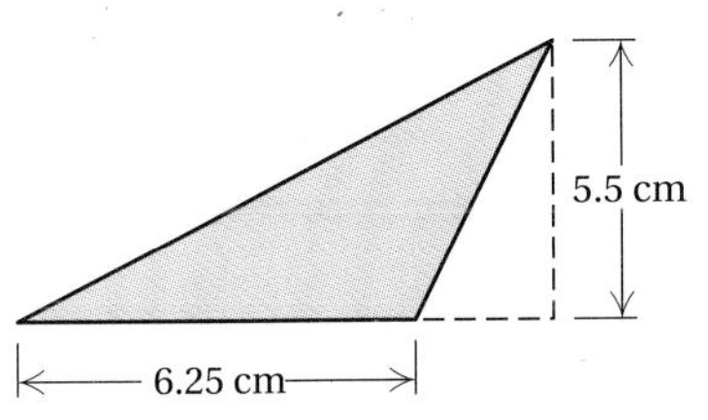

Find the area.

3.

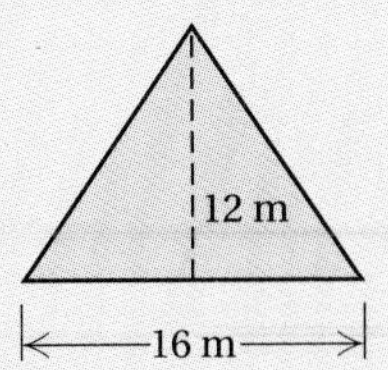

4.

3.4 cm

11 cm

Do Exercises 3 and 4.

Trapezoids

A **trapezoid** is a polygon with four sides, two of which, the **bases**, are parallel to each other.

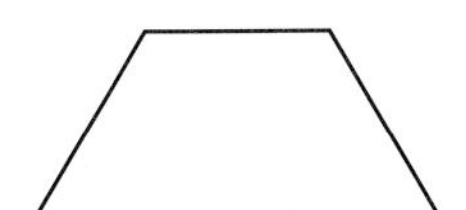
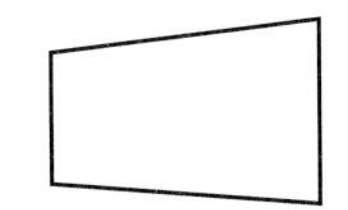
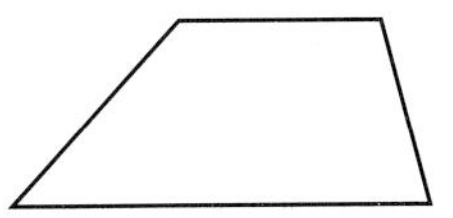

To find the area of a trapezoid, think of cutting out another just like it.

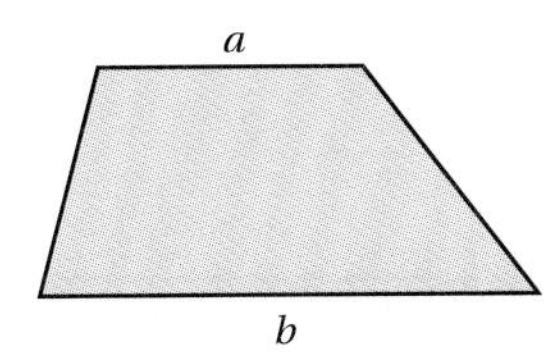

Then place the second one like this.

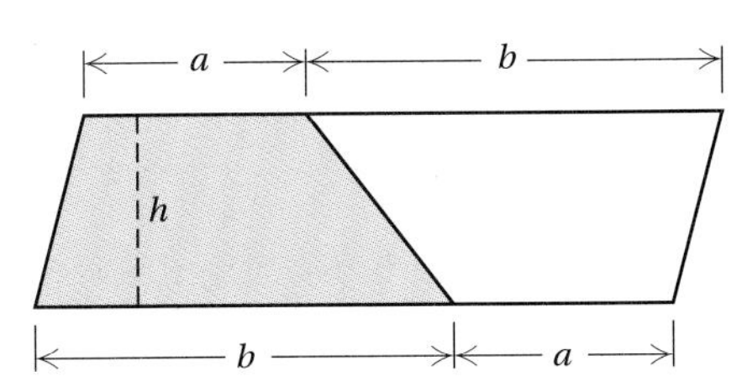

The resulting figure is a parallelogram whose area is

$h \cdot (a + b)$. **The base is $a + b$.**

The trapezoid we started with has half the area of the parallelogram, or

$$\frac{1}{2} \cdot h \cdot (a + b).$$

The **area of a trapezoid** is half the product of the height and the sum of the lengths of the parallel sides, or the product of the height and the average length of the bases:

$$A = \frac{1}{2} \cdot h \cdot (a + b) = h \cdot \frac{a + b}{2}.$$

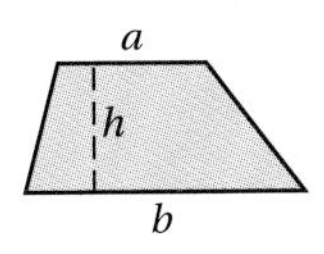

Answers on page A-19

Find the area.

5.

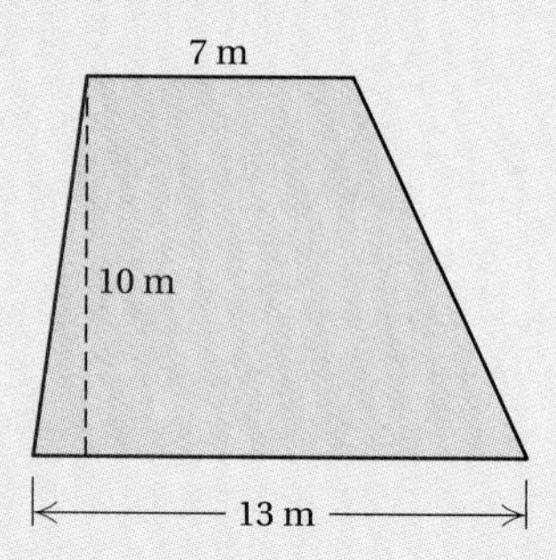

6.

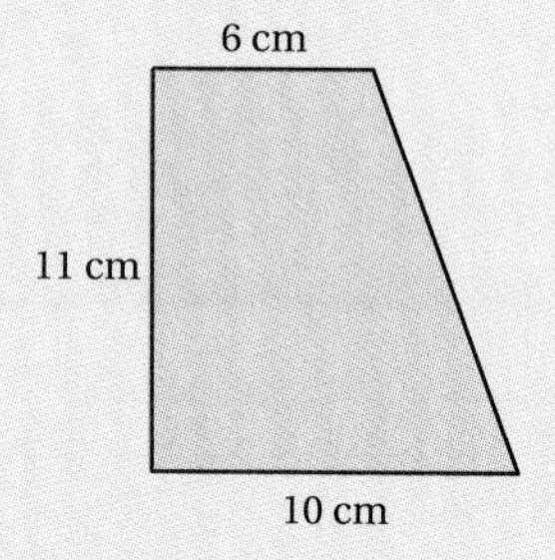

7. Find the area.

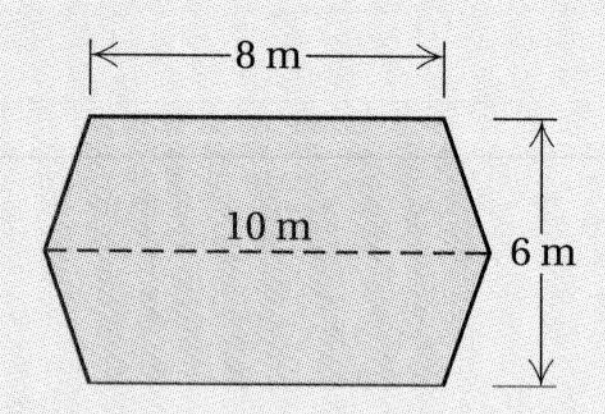

Answers on page A-19

Example 5 Find the area of this trapezoid.

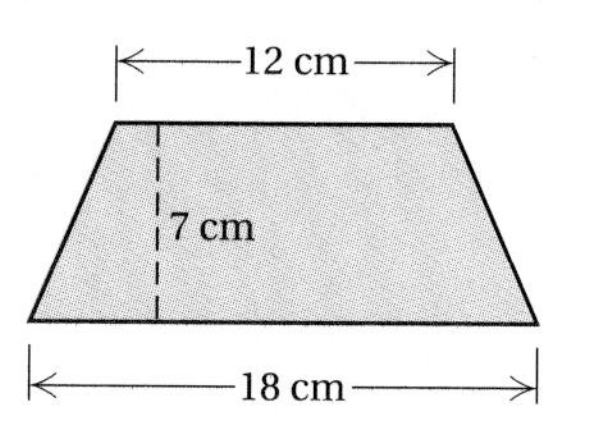

$$A = \frac{1}{2} \cdot h \cdot (a + b)$$
$$= \frac{1}{2} \cdot 7 \text{ cm} \cdot (12 + 18) \text{ cm}$$
$$= \frac{7 \cdot 30}{2} \cdot \text{cm}^2 = \frac{7 \cdot 15 \cdot 2}{1 \cdot 2} \text{cm}^2$$
$$= \frac{7 \cdot 15}{1} \cdot \frac{2}{2} \text{cm}^2$$
$$= 105 \text{ cm}^2$$

Do Exercises 5 and 6.

b Solving Applied Problems

Example 6 Find the area of this kite.

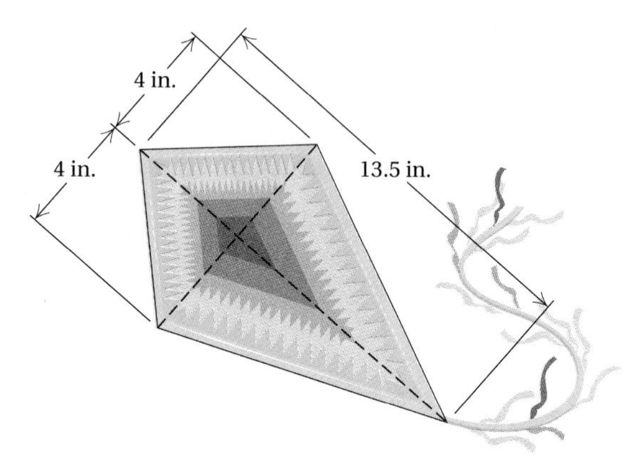

1. **Familiarize.** We look for the kinds of figures whose areas we can calculate using area formulas that we already know.
2. **Translate.** The kite consists of two triangles, each with a base of 13.5 in. and a height of 4 in. We can apply the formula $A = \frac{1}{2} \cdot b \cdot h$ for the area of a triangle and then multiply by 2.
3. **Solve.** We have

   $$A = \frac{1}{2} \cdot (13.5 \text{ in.}) \cdot (4 \text{ in.}) = 27 \text{ in}^2.$$

   Then we multiply by 2:

   $$2 \cdot 27 \text{ in}^2 = 54 \text{ in}^2.$$
4. **Check.** We can check by repeating the calculations.
5. **State.** The area of the kite is 54 in^2.

Do Exercise 7.

Exercise Set 8.5

a Find the area.

1.

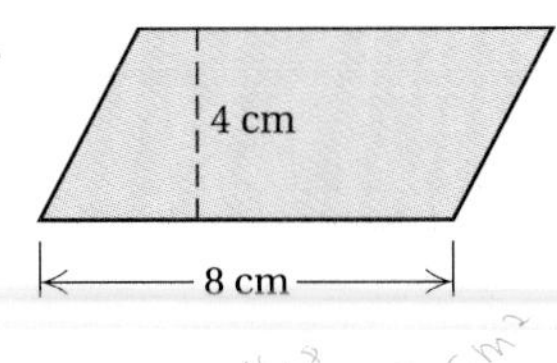

2.

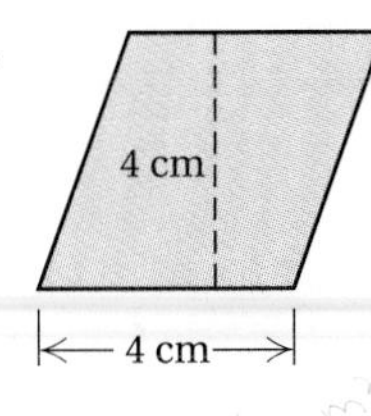

3.

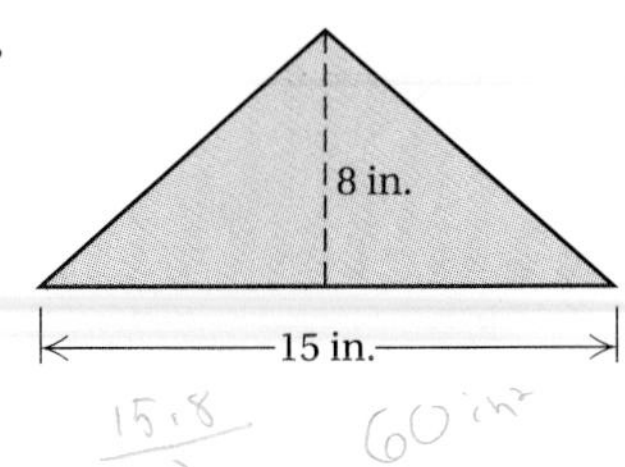

4.

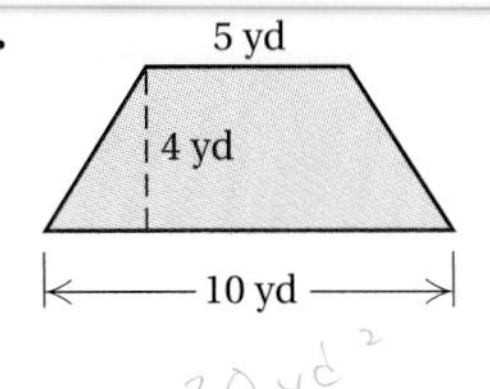

5.

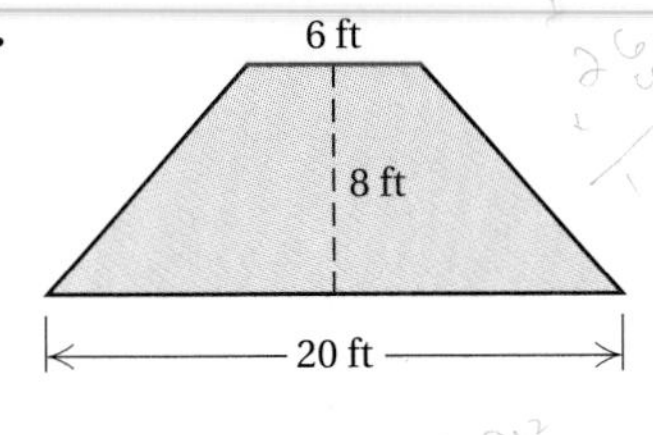

6.

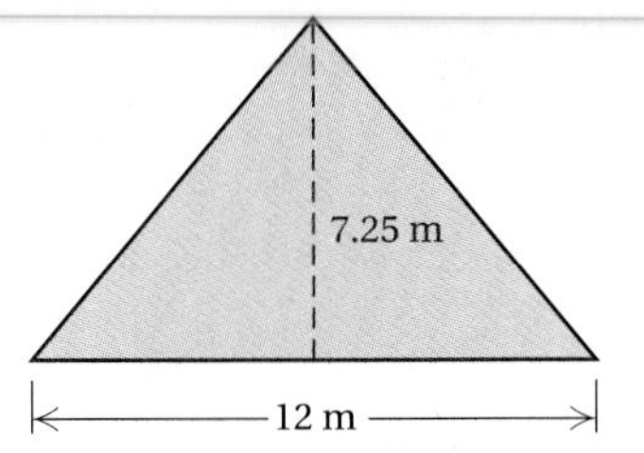

7.

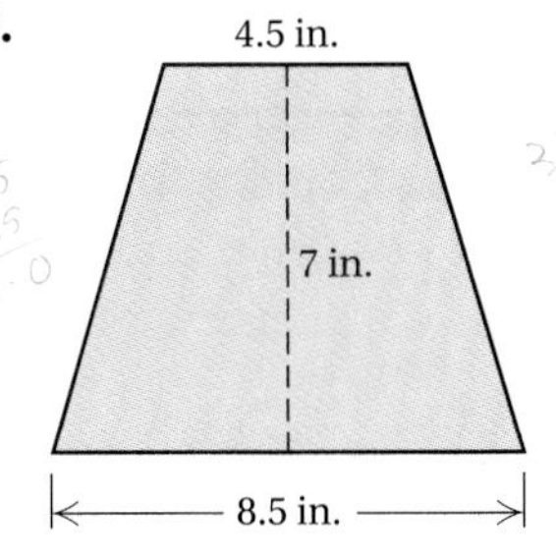

8.

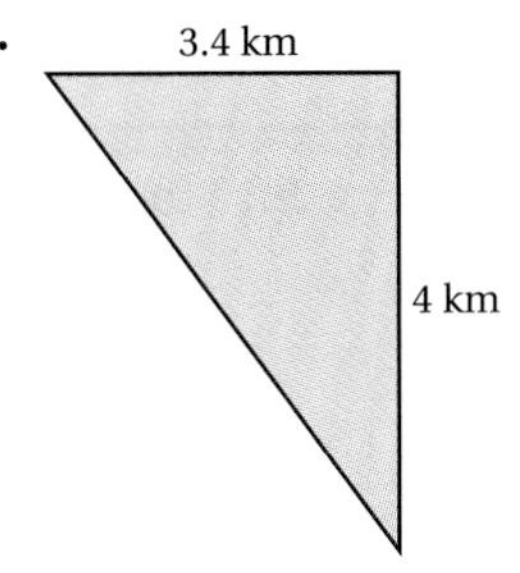

9.

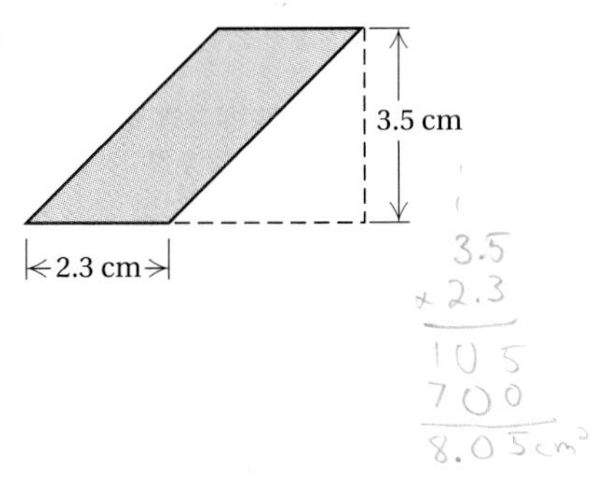

10.

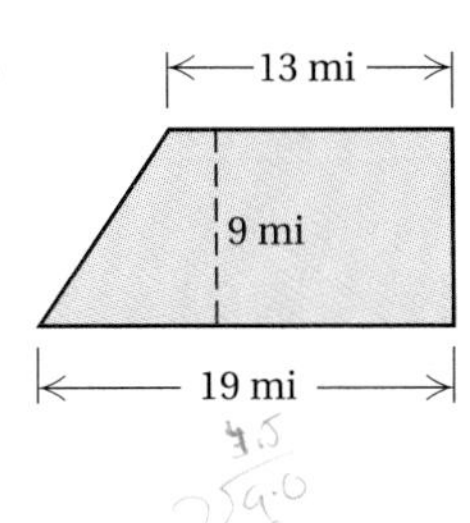

11.

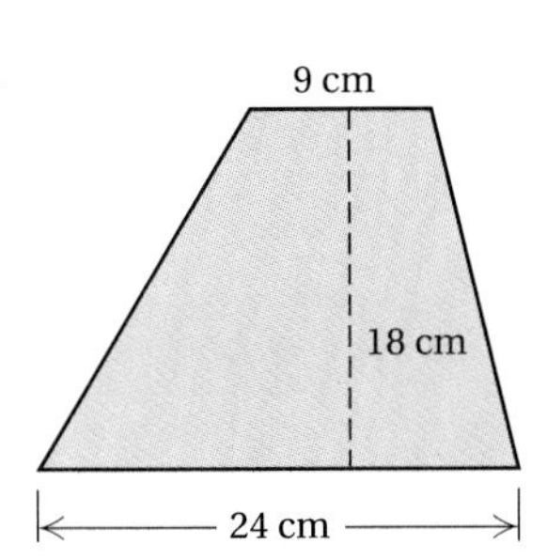

12.

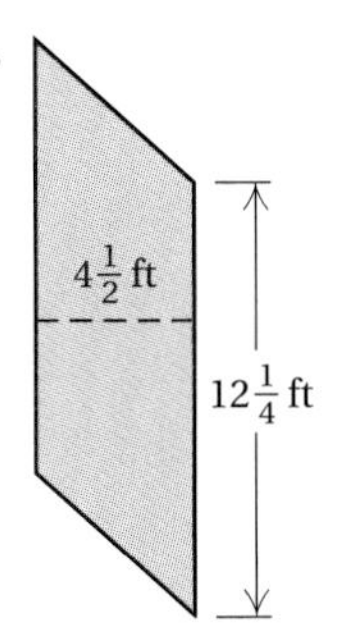

13.

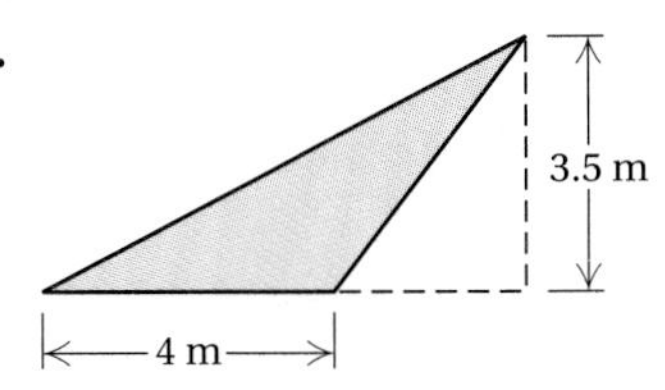

14.

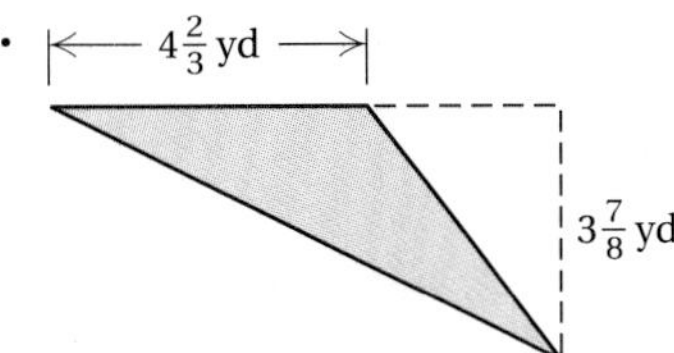

b Find the area of the shaded region.

15.

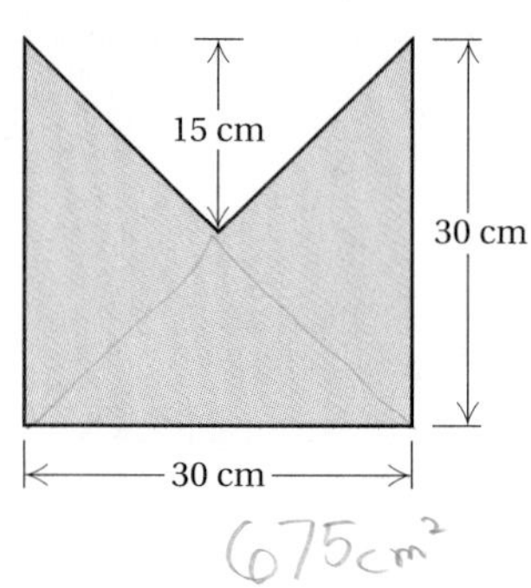

16.

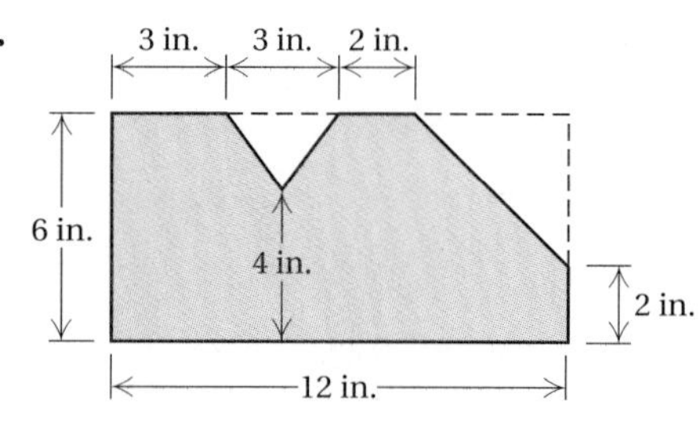

17.

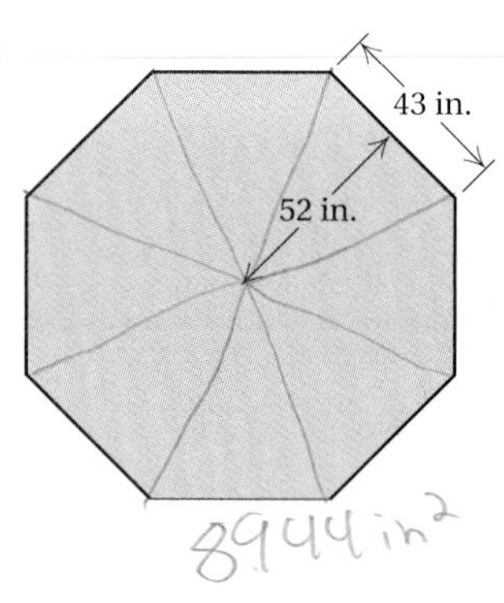

18.

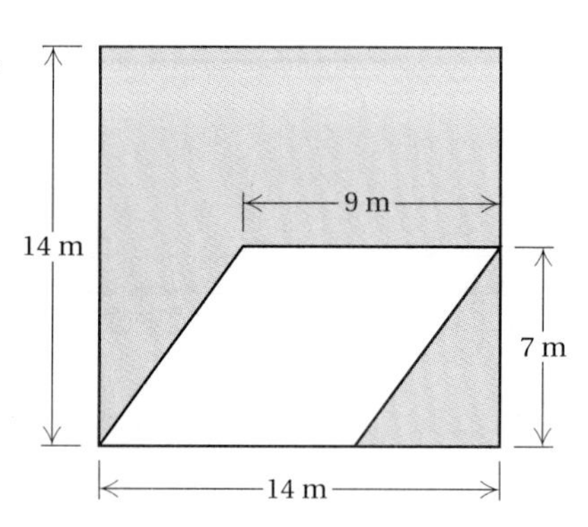

19. A rectangular piece of sailcloth is 36 ft by 24 ft. A triangular area with a height of 4.6 ft and a base of 5.2 ft is cut from the sailcloth. How much area is left over?

20. Find the total area of the sides and ends of the building.

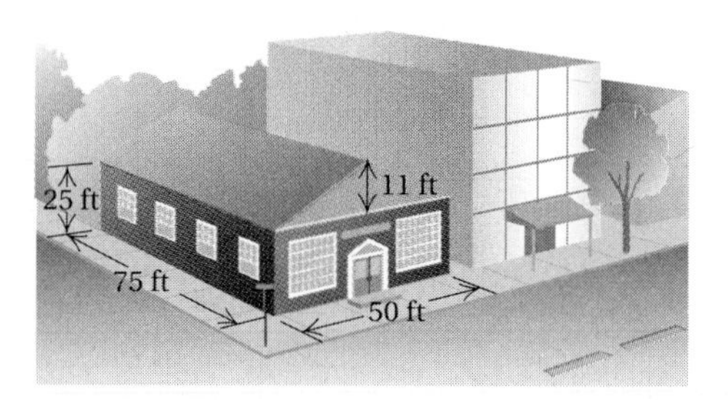

Skill Maintenance

Convert to fractional notation. [6.2b]

21. 35%

22. 85.5%

23. $37\frac{1}{2}\%$

24. $66.\overline{6}\%$

25. $83.\overline{3}\%$

26. $16\frac{2}{3}\%$

Solve. [1.8a]

27. A ream of paper contains 500 sheets. How many sheets are there in 15 reams?

28. A lab technician separates a vial containing 140 cc of blood into test tubes, each of which contains 3 cc of blood. How many test tubes can be filled? How much blood is left over?

Synthesis

29. ◈ Explain how the area of a parallelogram can be found by considering the area of a rectangle.

30. ◈ Explain how the area of a triangle can be found by considering the area of a parallelogram.

Collaborative Learning Manual
Verify the formulas for the area of a parallelogram, a triangle, and a trapezoid.

8.6 Circles

a Radius and Diameter

At the right is a circle with center O. Segment $\overline{AC}$ is a *diameter.* A **diameter** is a segment that passes through the center of the circle and has endpoints on the circle. Segment $\overline{OB}$ is called a *radius.* A **radius** is a segment with one endpoint on the center and the other endpoint on the circle.

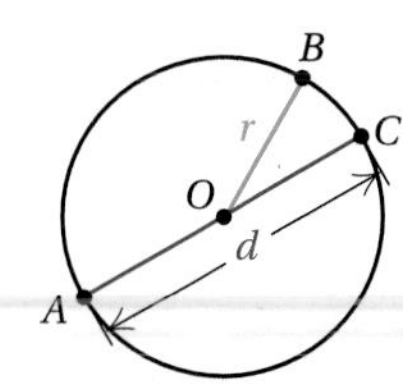

Suppose that d is the diameter of a circle and r is the radius. Then

$$d = 2 \cdot r \quad \text{and} \quad r = \frac{d}{2}.$$

Example 1 Find the length of a radius of this circle.

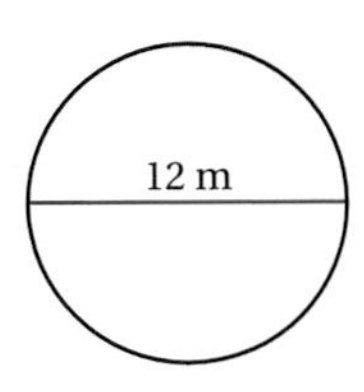

$$r = \frac{d}{2} = \frac{12\text{ m}}{2} = 6\text{ m}$$

The radius is 6 m.

Example 2 Find the length of a diameter of this circle.

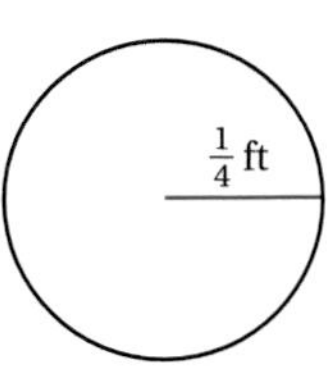

$$d = 2 \cdot r = 2 \cdot \frac{1}{4}\text{ ft} = \frac{1}{2}\text{ ft}$$

The diameter is $\frac{1}{2}$ ft.

Do Exercises 1 and 2.

b Circumference

The **circumference** of a circle is the distance around it. Calculating circumference is similar to finding the perimeter of a polygon.

To find a formula for the circumference of any circle given its diameter, we first need to consider the ratio C/d. Take a 12-oz soda can and measure the circumference C with a tape measure. Also measure the diameter d. The results are shown in the figure. Then

$$\frac{C}{d} = \frac{7.8\text{ in.}}{2.5\text{ in.}} \approx 3.1.$$

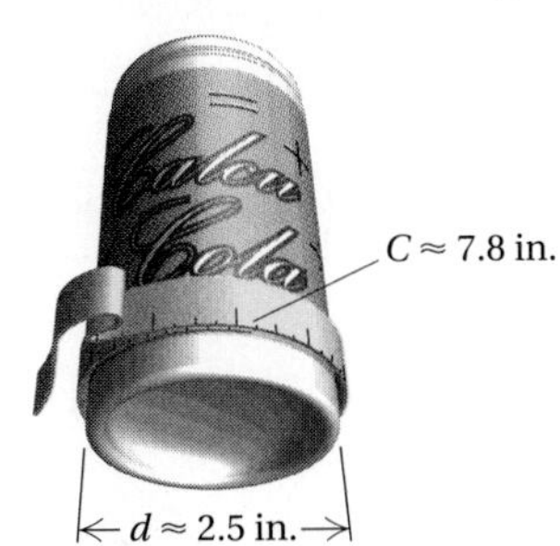

Objectives

a Find the length of a radius of a circle given the length of a diameter, and find the length of a diameter given the length of a radius.

b Find the circumference of a circle given the length of a diameter or a radius.

c Find the area of a circle given the length of a radius.

d Solve applied problems involving circles.

For Extra Help

TAPE 14 | TAPE 14A | MAC WIN | CD-ROM

1. Find the length of a radius.

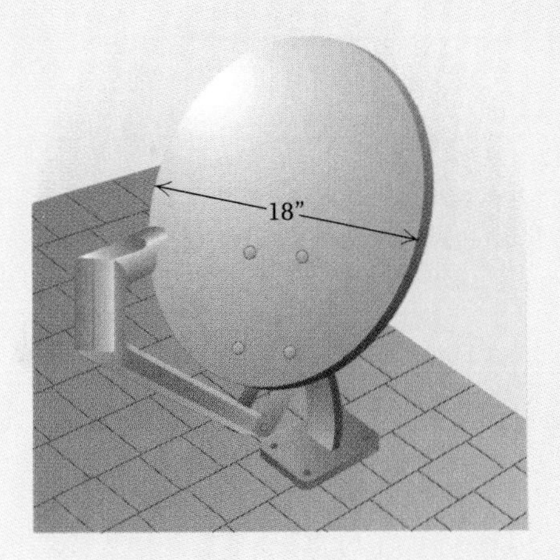

2. Find the length of a diameter.

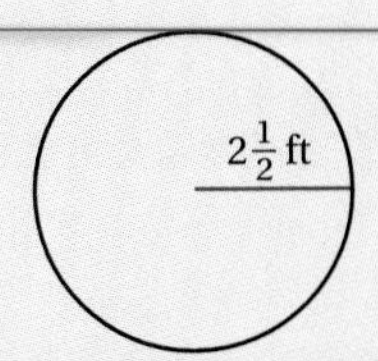

Answers on page A-20

3. Find the circumference of this circle. Use 3.14 for π.

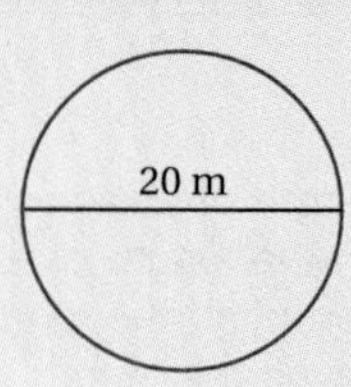

4. Find the circumference of this circle. Use $\frac{22}{7}$ for π.

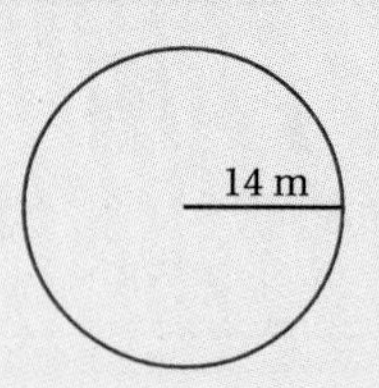

5. Find the perimeter of this figure. Use 3.14 for π.

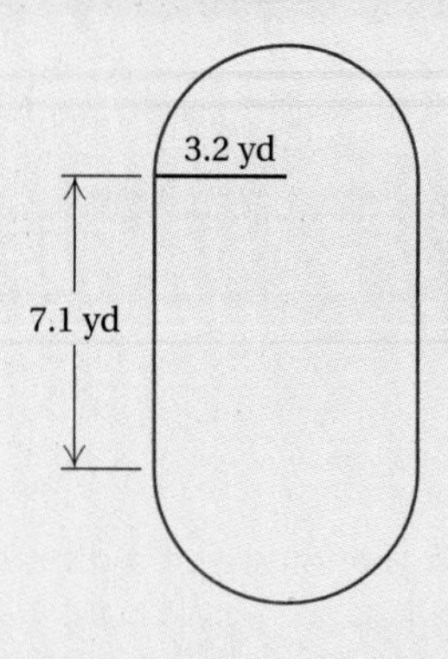

Answers on page A-20

Suppose we did this with cans and circles of several sizes. We would get a number close to 3.1. For any circle, if we divide the circumference C by the diameter d, we get the same number. We call this number π (pi).

> $\frac{C}{d} = \pi$ or $C = \pi \cdot d$. The number π is about 3.14, or about $\frac{22}{7}$.

Example 3 Find the circumference of this circle. Use 3.14 for π.

$$C = \pi \cdot d$$
$$\approx 3.14 \times 6 \text{ cm}$$
$$\approx 18.84 \text{ cm}$$

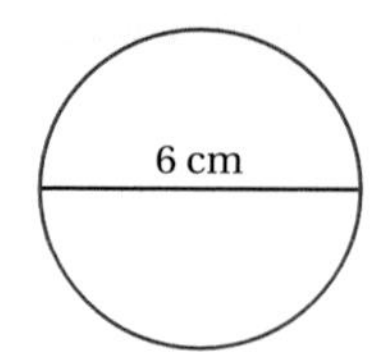

The circumference is about 18.84 cm.

Do Exercise 3.

Since $d = 2 \cdot r$, where r is the length of a radius, it follows that

$$C = \pi \cdot d = \pi \cdot (2 \cdot r).$$

> $C = 2 \cdot \pi \cdot r$

Example 4 Find the circumference of this circle. Use $\frac{22}{7}$ for π.

$$C = 2 \cdot \pi \cdot r$$
$$\approx 2 \cdot \frac{22}{7} \cdot 70 \text{ in.}$$
$$\approx 2 \cdot 22 \cdot \frac{70}{7} \text{ in.}$$
$$\approx 44 \cdot 10 \text{ in.}$$
$$\approx 440 \text{ in.}$$

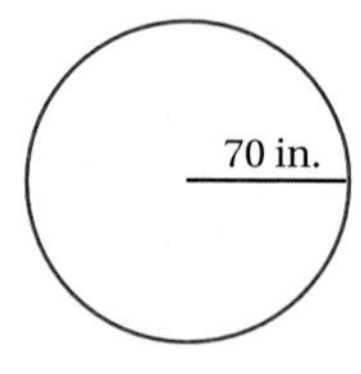

The circumference is about 440 in.

Example 5 Find the perimeter of this figure. Use 3.14 for π.

We let $P =$ the perimeter. We see that we have half a circle attached to a square. Thus we add half the circumference to the lengths of the three line segments.

$$P = 3 \times 9.4 \text{ km} + \frac{1}{2} \times 2 \times \pi \times 4.7 \text{ km}$$
$$\approx 28.2 \text{ km} + 3.14 \times 4.7 \text{ km}$$
$$\approx 28.2 \text{ km} + 14.758 \text{ km}$$
$$\approx 42.958 \text{ km}$$

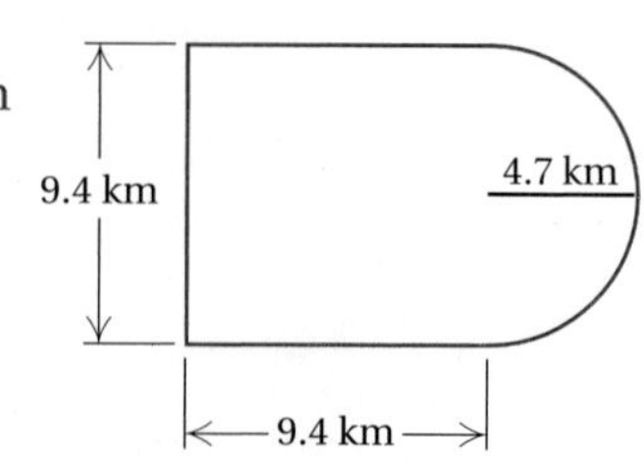

The perimeter is about 42.958 km.

Do Exercises 4 and 5.

C Area

Below is a circle of radius r.

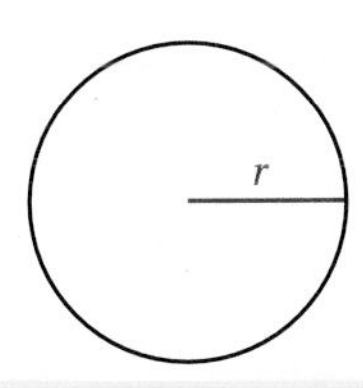

Think of cutting half the circular region into small pieces and arranging them as shown below.

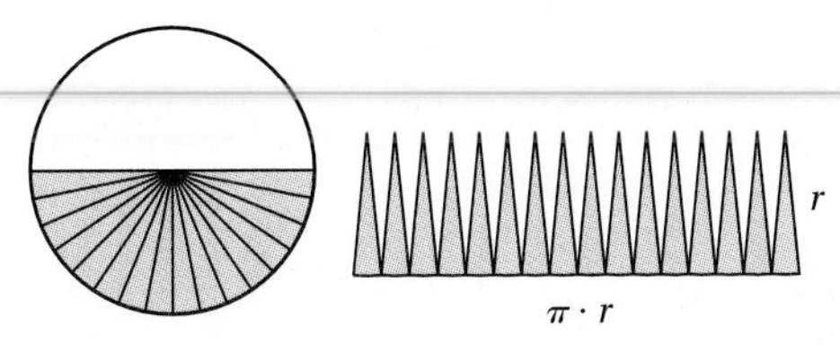

Then imagine cutting the other half of the circular region and arranging the pieces in with the others as shown below.

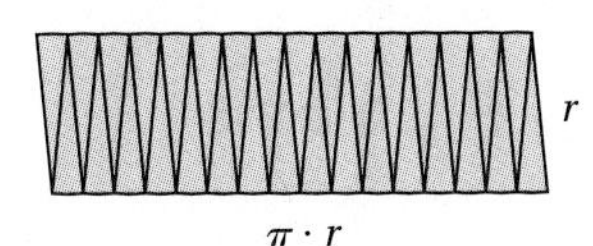

This is almost a parallelogram. The base has length $\frac{1}{2} \cdot 2 \cdot \pi \cdot r$, or $\pi \cdot r$ (half the circumference) and the height is r. Thus the area is

$$(\pi \cdot r) \cdot r.$$

This is the area of a circle.

> The **area of a circle** with radius of length r is given by
>
> $A = \pi \cdot r \cdot r$, or $A = \pi \cdot r^2$.

Example 6 Find the area of this circle. Use $\frac{22}{7}$ for π.

$$A = \pi \cdot r \cdot r$$
$$\approx \frac{22}{7} \cdot 14 \text{ cm} \cdot 14 \text{ cm}$$
$$\approx \frac{22}{7} \cdot 196 \text{ cm}^2$$
$$\approx 616 \text{ cm}^2$$

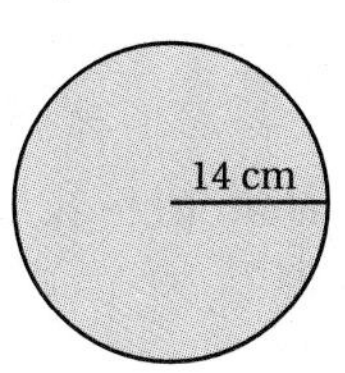

The area is about 616 cm^2.

Do Exercise 6.

6. Find the area of this circle. Use $\frac{22}{7}$ for π.

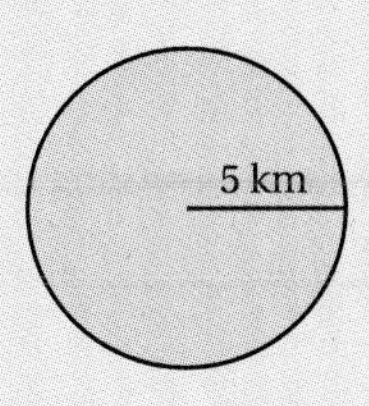

Calculator Spotlight

On certain calculators, there is a pi key, [π]. You can use a [π] key for most computations instead of stopping to round the value of π. Rounding, if necessary, is done at the end.

Exercises

1. If you have a [π] key on your calculator, to how many places does this key give the value of π?
2. Find the circumference and the area of a circle with a radius of 225.68 in.
3. Find the area of a circle with a diameter of $46\frac{12}{13}$ in.
4. Find the area of a large irrigated farming circle with a diameter of 400 ft.

Answer on page A-20

7. Find the area of this circle. Use 3.14 for π.

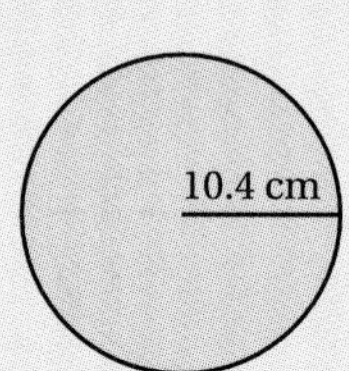

Caution!

Circumference $= \pi \cdot d = \pi \cdot (r + r) = \pi \cdot (2 \cdot r)$,

Area $= \pi \cdot r^2 = \pi \cdot (r \cdot r)$,

and

$r^2 \neq 2 \cdot r$.

Example 7 Find the area of this circle. Use 3.14 for π. Round to the nearest hundredth.

$$\begin{aligned} A &= \pi \cdot r \cdot r \\ &\approx 3.14 \times 2.1 \text{ m} \times 2.1 \text{ m} \\ &\approx 3.14 \times 4.41 \text{ m}^2 \\ &\approx 13.8474 \text{ m}^2 \\ &\approx 13.85 \text{ m}^2 \end{aligned}$$

2.1 m

The area is about 13.85 m^2.

Do Exercise 7.

d Solving Applied Problems

8. Which is larger and by how much: a 10-ft square flower bed or a 12-ft diameter flower bed?

Example 8 *Area of Pizza Pan.* Which makes a larger pizza and by how much: a 16-in. square pizza pan or a 16-in. diameter circular pizza pan?

First, we make a drawing of each.

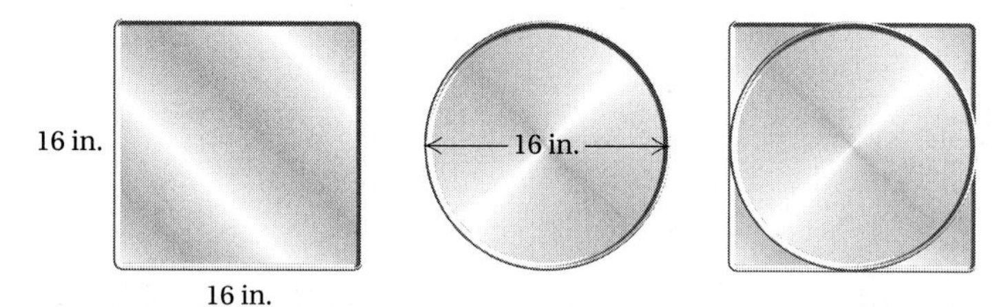

Then we compute areas.

The area of the square is

$$\begin{aligned} A &= s \cdot s \\ &= 16 \text{ in.} \times 16 \text{ in.} = 256 \text{ in}^2. \end{aligned}$$

The diameter of the circle is 16 in., so the radius is 16 in./2, or 8 in. The area of the circle is

$$\begin{aligned} A &= \pi \cdot r \cdot r \\ &\approx 3.14 \times 8 \text{ in.} \times 8 \text{ in.} \approx 200.96 \text{ in}^2. \end{aligned}$$

We see that the square pizza pan is larger by about

$$256 \text{ in}^2 - 200.96 \text{ in}^2, \quad \text{or} \quad 55.04 \text{ in}^2.$$

Thus the square pan makes the larger pizza, by about 55.04 in^2.

Do Exercise 8.

Answers on page A-20

Exercise Set 8.6

a, **b**, **c** For each circle, find the length of a diameter, the circumference, and the area. Use $\frac{22}{7}$ for π.

1.

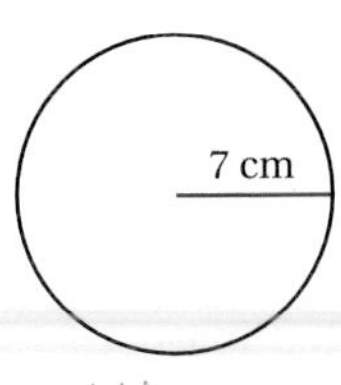

2.

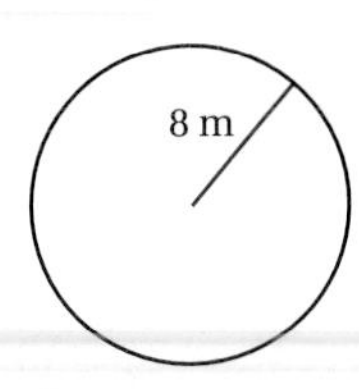

3.

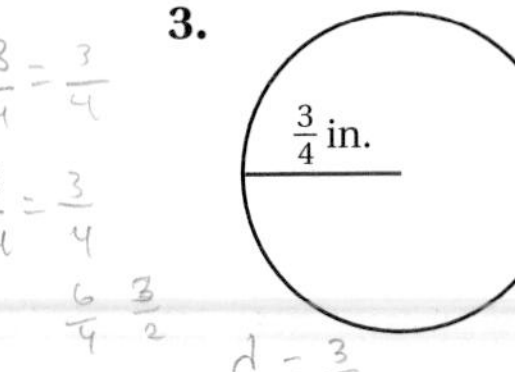

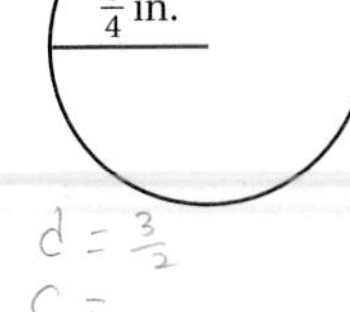

4.

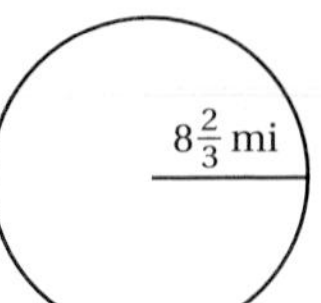

For each circle, find the length of a radius, the circumference, and the area. Use 3.14 for π.

5.

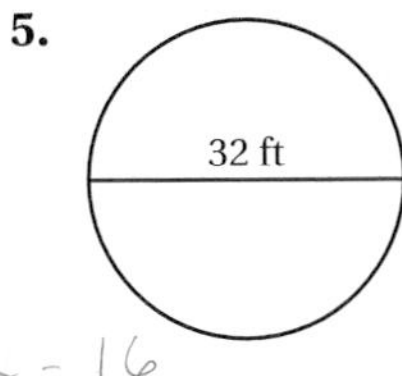

6.

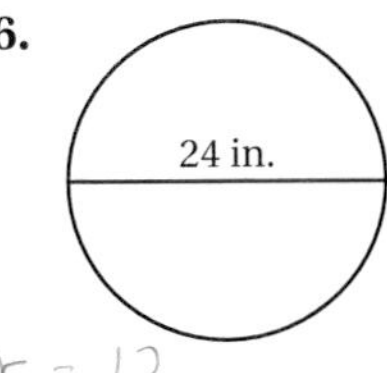

7.

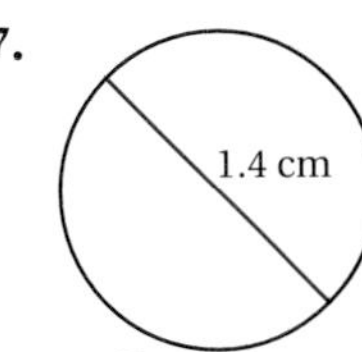

8.

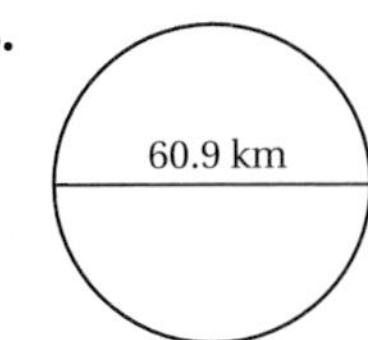

d Solve. Use 3.14 for π.

9. The top of a soda can has a 6-cm diameter. What is its radius? its circumference? its area?

10. A penny has a 1-cm radius. What is its diameter? its circumference? its area?

11. A radio station is allowed by the FCC to broadcast over an area with a radius of 220 mi. How much area is this?

12. *Pizza Areas.* Which is larger and by how much: a 12-in. circular pizza or a 12-in. square pizza?

13. *Dimensions of a Quarter.* The circumference of a quarter is 7.85 cm. What is the diameter? the radius? the area?

14. *Dimensions of a Dime.* The circumference of a dime is 2.23 in. What is the diameter? the radius? the area?

15. *Gypsy-Moth Tape.* To protect an elm tree in your backyard, you need to attach gypsy moth caterpillar tape around the trunk. The tree has a 1.1-ft diameter. What length of tape is needed?

16. *Silo.* A silo has a 10-m diameter. What is its circumference?

17. *Swimming-Pool Walk.* You want to install a 1-yd–wide walk around a circular swimming pool. The diameter of the pool is 20 yd. What is the area of the walk?

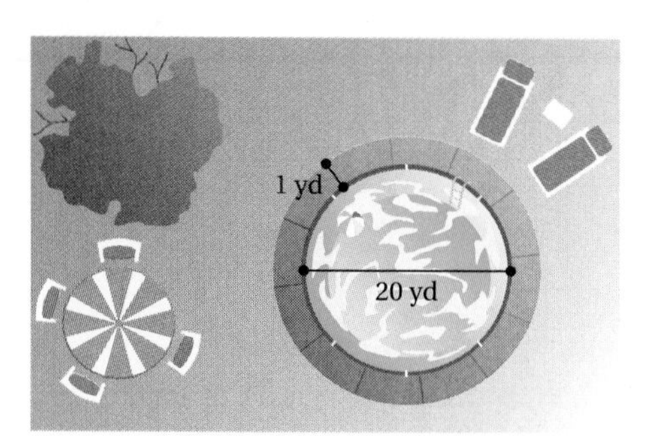

18. *Roller-Rink Floor.* A roller rink floor is shown below. What is its area? If hardwood flooring costs $10.50 per square meter, how much will the flooring cost?

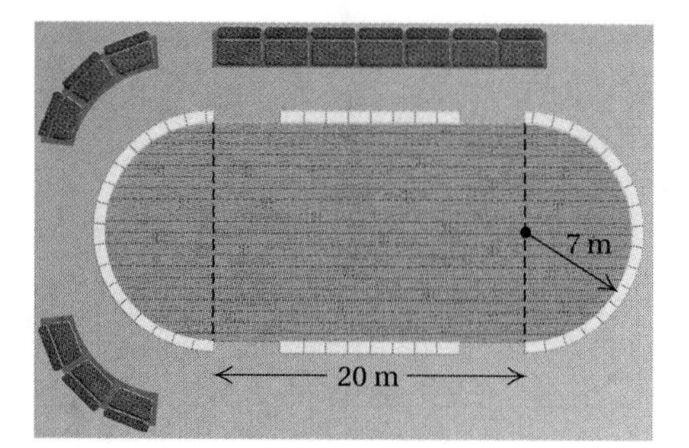

Find the perimeter. Use 3.14 for π.

19.

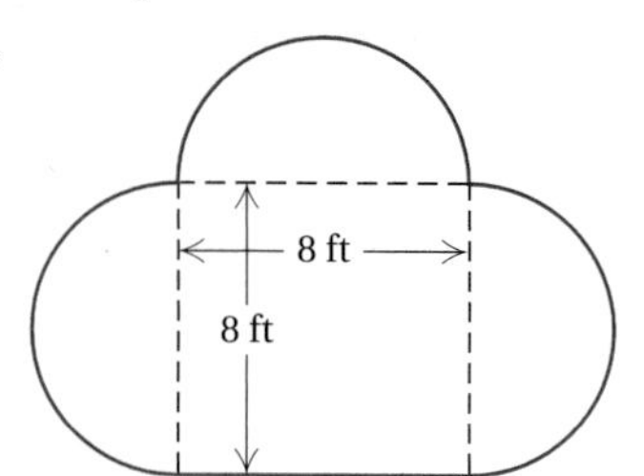

20.

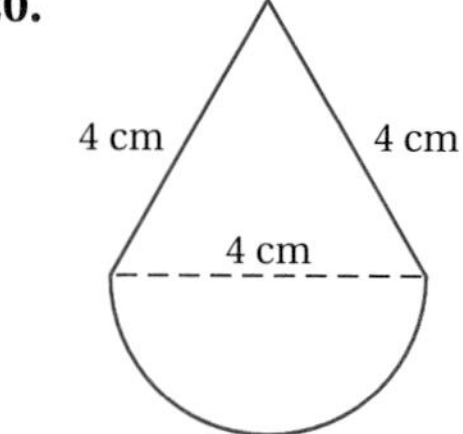

21.

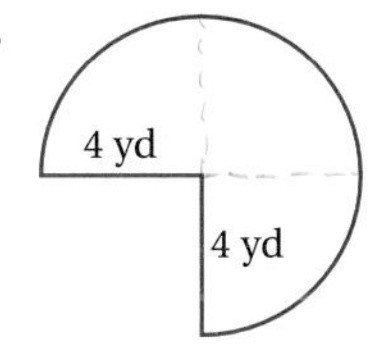

22.

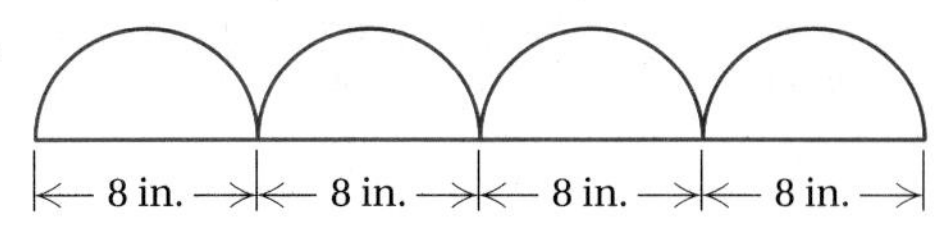

23.

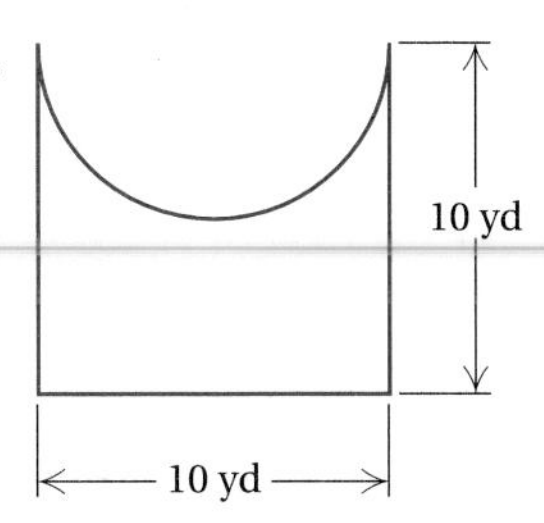

24.

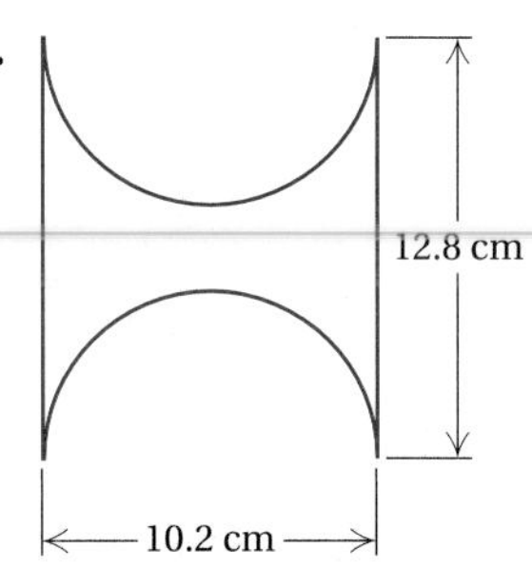

Find the area of the shaded region. Use 3.14 for π.

25.

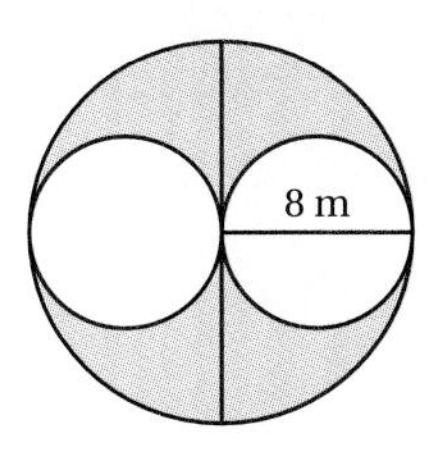

26.

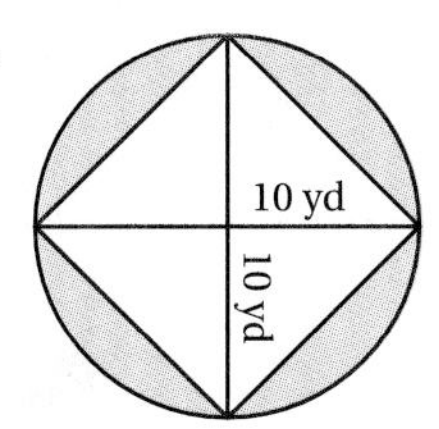

27.

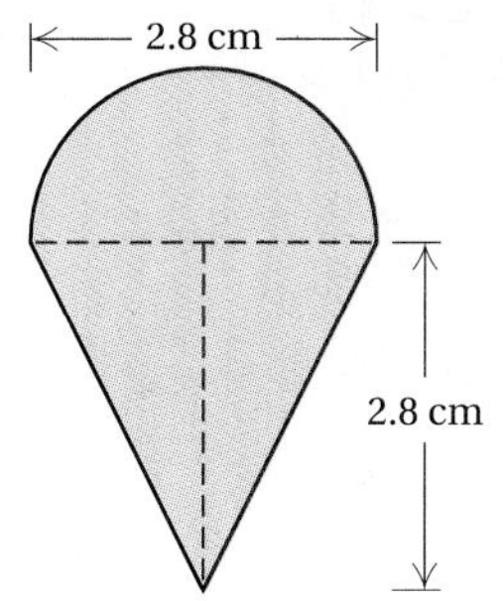

28.

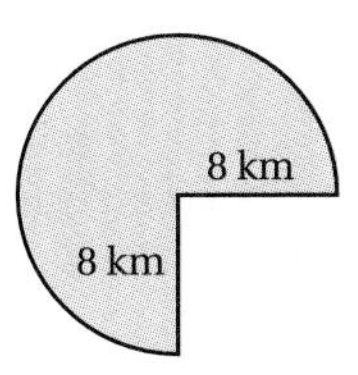

29.

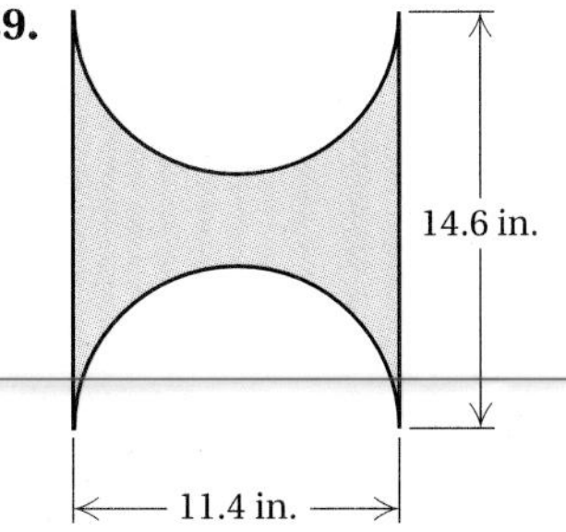

30.

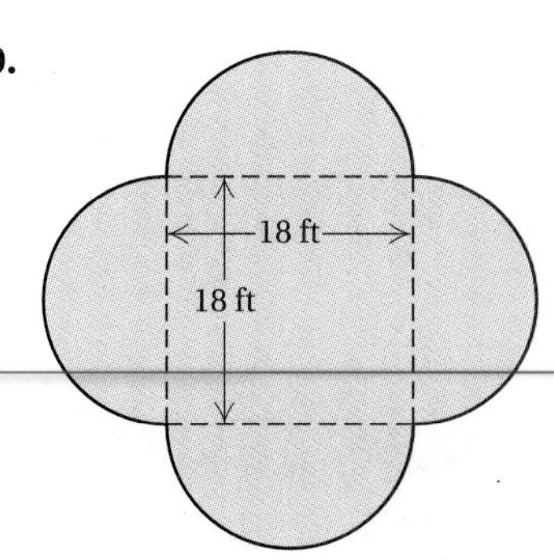

Skill Maintenance

Convert to percent notation. [6.1c]

31. 0.875

32. 0.58

33. $0.\overline{6}$

34. 0.4361

Convert to percent notation. [6.2a]

35. $\frac{3}{8}$

36. $\frac{5}{8}$

37. $\frac{2}{3}$

38. $\frac{1}{5}$

Estimate each of the following as a whole number or as a mixed number where the fractional part is $\frac{1}{2}$. [3.7b]

39. $3\frac{7}{8}$

40. $8\frac{1}{3}$

41. $13\frac{1}{6}$

42. $39\frac{7}{13}$

43. $\frac{4}{5} + 3\frac{7}{8}$

44. $\frac{1}{11} \cdot \frac{7}{15}$

45. $\frac{2}{3} + \frac{7}{15} + \frac{8}{9}$

46. $\frac{8}{9} + \frac{4}{5} + \frac{13}{14}$

47. $\frac{57}{100} - \frac{1}{10} + \frac{9}{1000}$

48. $\frac{23}{24} + \frac{38}{39} + \frac{61}{60}$

49. $11\frac{29}{80} + 10\frac{14}{15} \cdot 24\frac{2}{17}$

50. $\frac{13}{14} + 9\frac{5}{8} - 1\frac{23}{28} \cdot 1\frac{36}{73}$

Synthesis

51. ◆ Explain why a 16-in.–diameter pizza that costs \$16.25 is a better buy than a 10-in.–diameter pizza that costs \$7.85.

52. ◆ The radius of one circle is twice the size of another circle's radius. Is the area of the first circle twice the area of the other circle? Why or why not?

53. 🖩 $\pi \approx \frac{3927}{1250}$ is another approximation for π. Find decimal notation using a calculator. Round to the nearest thousandth.

54. 🖩 The distance from Kansas City to Indianapolis is 500 mi. A car was driven this distance using tires with a radius of 14 in. How many revolutions of each tire occurred on the trip? Use $\frac{22}{7}$ for π.

55. *Tennis Balls.* Tennis balls are usually packed vertically three in a can, one on top of another. Suppose the diameter of a tennis ball is d. Find the height of the stack of balls. Find the circumference of one ball. Which is greater? Explain.

Collaborative Learning Manual

Estimate the value of π.

8.7 Square Roots and the Pythagorean Theorem

Objectives

a. Simplify square roots of squares such as $\sqrt{25}$.

b. Approximate square roots.

c. Given the lengths of any two sides of a right triangle, find the length of the third side.

For Extra Help

TAPE 14

TAPE 14A

MAC WIN

CD-ROM

a Square Roots

If a number is a product of two identical factors, then either factor is called a **square root** of the number. (If $a = c^2$, then c is a square root of a.) The symbol $\sqrt{}$ (called a **radical sign**) is used in naming square roots.

For example, $\sqrt{36}$ is the square root of 36. It follows that

$$\sqrt{36} = \sqrt{6 \cdot 6} = 6 \qquad \text{The square root of 36 is 6.}$$

because $6^2 = 36$.

Example 1 Simplify: $\sqrt{25}$.

$$\sqrt{25} = \sqrt{5 \cdot 5} = 5 \qquad \text{The square root of 25 is 5 because } 5^2 = 25.$$

Example 2 Simplify: $\sqrt{144}$.

$$\sqrt{144} = \sqrt{12 \cdot 12} = 12 \qquad \text{The square root of 144 is 12 because } 12^2 = 144.$$

CAUTION! It is common to confuse squares and square roots. A number squared is that number multiplied by itself. For example, $16^2 = 16 \cdot 16 = 256$. A square root of a number is a number that when multiplied by itself gives the original number. For example, $\sqrt{16} = 4$, because $4 \cdot 4 = 16$.

Examples Simplify.

3. $\sqrt{4} = 2$ **4.** $\sqrt{256} = 16$ **5.** $\sqrt{361} = 19$

Do Exercises 1–24.

b Approximating Square Roots

Square roots of some numbers are not whole numbers or ordinary fractions. For example,

$$\sqrt{2}, \quad \sqrt{3}, \quad \sqrt{39}, \quad \text{and} \quad \sqrt{70}$$

are not whole numbers or ordinary fractions. We can approximate these square roots. For example, consider the following decimal approximations for $\sqrt{2}$. Each gives a closer approximation.

$\sqrt{2} \approx 1.4$ because $(1.4)^2 = 1.96$,

$\sqrt{2} \approx 1.41$ because $(1.41)^2 = 1.9881$,

$\sqrt{2} \approx 1.414$ because $(1.414)^2 = 1.999396$,

$\sqrt{2} \approx 1.4142$ because $(1.4142)^2 = 1.99996164$.

How do we find such approximations? We use a calculator.

Find the square. (See Section 1.9.)

1. 9^2 **2.** 10^2

3. 11^2 **4.** 12^2

It would be helpful to memorize the squares of numbers from 1 to 25.

5. 13^2 **6.** 14^2

7. 15^2 **8.** 16^2

9. 17^2 **10.** 18^2

11. 20^2 **12.** 25^2

Simplify. The results of Exercises 1–12 above may be helpful here.

13. $\sqrt{9}$ **14.** $\sqrt{16}$

15. $\sqrt{121}$ **16.** $\sqrt{100}$

17. $\sqrt{81}$ **18.** $\sqrt{64}$

19. $\sqrt{324}$ **20.** $\sqrt{400}$

21. $\sqrt{225}$ **22.** $\sqrt{169}$

23. $\sqrt{1}$ **24.** $\sqrt{0}$

Answers on page A-20

Approximate to three decimal places.

25. $\sqrt{5}$

26. $\sqrt{78}$

27. $\sqrt{168}$

Example 6 Approximate $\sqrt{3}$, $\sqrt{27}$, and $\sqrt{180}$ to three decimal places. Use a calculator.

We use a calculator to find each square root. Since more than three decimal places are given, we round back to three places.

$$\sqrt{3} \approx 1.732,$$
$$\sqrt{27} \approx 5.196,$$
$$\sqrt{180} \approx 13.416$$

As a check, note that $1 \cdot 1 = 1$ and $2 \cdot 2 = 4$, so we expect $\sqrt{3}$ to be between 1 and 2. Similarly, we expect $\sqrt{27}$ to be between 5 and 6 and $\sqrt{180}$ to be between 13 and 14.

Do Exercises 25–27.

Calculator Spotlight

Most calculators have a square root key, [√‾]. On some calculators, square roots are found by pressing the [x^2] key after first pressing a key labeled [2nd] or [SHIFT]. (On these calculators, finding square roots is a secondary function.)

To find an approximation for $\sqrt{30}$, we simply press

[2nd] [√‾] [3] [0] [ENTER].

The value 5.477225575 appears.

It is always best to wait until calculations are complete before rounding off. For example, to round $9 \cdot \sqrt{5}$ to the nearest tenth, we do *not* first determine that $\sqrt{5} \approx 2.2$. Rather, we press

[9] [×] [2nd] [√‾] [5] [ENTER].

The result, 20.1246118, is then rounded to 20.1.

Exercises

Round to the nearest tenth.

1. $\sqrt{43}$ **2.** $\sqrt{94}$

3. $7 \cdot \sqrt{8}$ **4.** $5 \cdot \sqrt{12}$

5. $\sqrt{35} + 19$ **6.** $17 + \sqrt{57}$

7. $13 \cdot \sqrt{68} + 14$

8. $24 \cdot \sqrt{31} - 18$

9. $5 \cdot \sqrt{30} - 3 \cdot \sqrt{14}$

10. $7 \cdot \sqrt{90} + 3 \cdot \sqrt{40}$

Answers on page A-20

c The Pythagorean Theorem

A **right triangle** is a triangle with a 90° angle, as shown here.

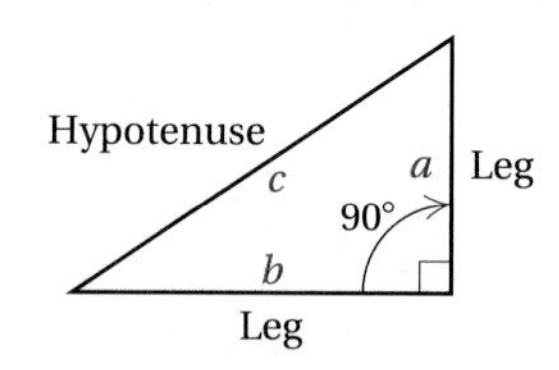

In a right triangle, the longest side is called the **hypotenuse**. It is also the side opposite the right angle. The other two sides are called **legs**. We generally use the letters a and b for the lengths of the legs and c for the length of the hypotenuse. They are related as follows.

THE PYTHAGOREAN THEOREM

In any right triangle, if a and b are the lengths of the legs and c is the length of the hypotenuse, then

$$a^2 + b^2 = c^2, \quad \text{or}$$
$$(\text{Leg})^2 + (\text{Other leg})^2 = (\text{Hypotenuse})^2.$$

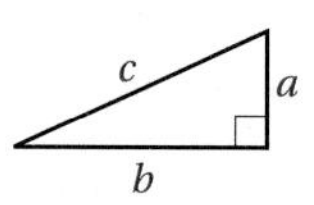

The equation $a^2 + b^2 = c^2$ is called the **Pythagorean equation.**

The Pythagorean theorem is named for the Greek mathematician Pythagoras (569?–500? B.C.). We can think of this relationship as adding areas.

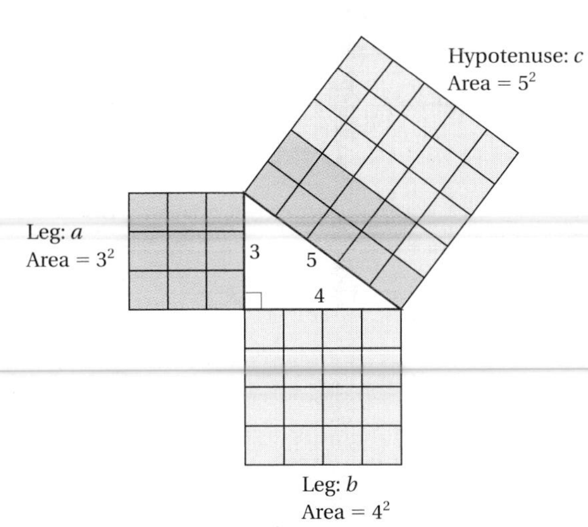

$$a^2 + b^2 = c^2$$
$$3^2 + 4^2 = 5^2$$
$$9 + 16 = 25$$

If we know the lengths of any two sides of a right triangle, we can use the Pythagorean equation to determine the length of the third side.

Example 7 Find the length of the hypotenuse of this right triangle. Give an exact answer and an approximation to three decimal places.

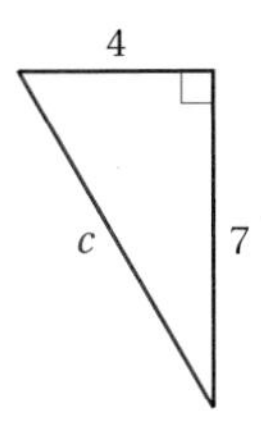

We substitute in the Pythagorean equation:

$$a^2 + b^2 = c^2$$
$$4^2 + 7^2 = c^2 \qquad \textbf{Substituting}$$
$$16 + 49 = c^2$$
$$65 = c^2.$$

The solution of this equation is the square root of 65. We approximate the square root using a calculator.

Exact answer: $c = \sqrt{65}$

Approximate answer: $c \approx 8.062$ **Using a calculator**

Do Exercise 28.

Example 8 Find the length b for the right triangle shown. Give an exact answer and an approximation to three decimal places.

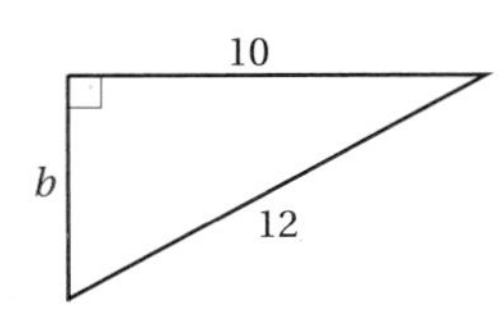

28. Find the length of the hypotenuse of this right triangle. Give an exact answer and an approximation to three decimal places.

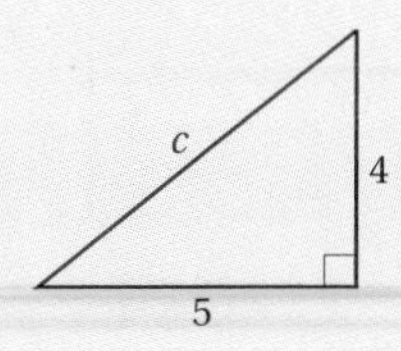

Answer on page A-20

Find the length of the leg of the right triangle. Give an exact answer and an approximation to three decimal places.

29.

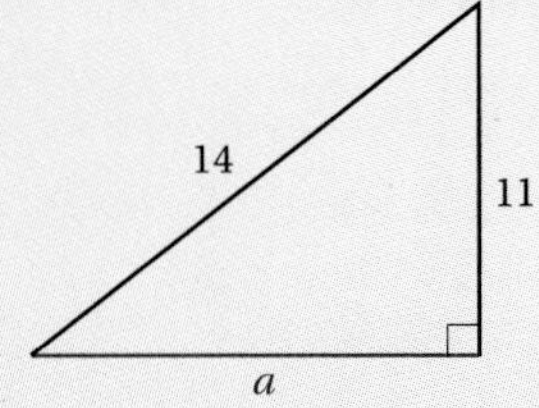

30.

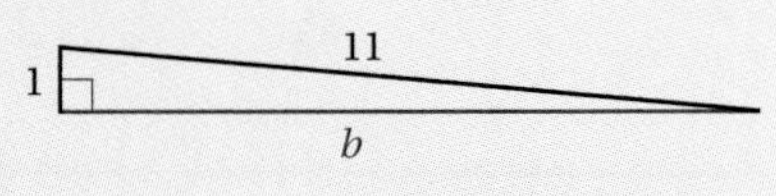

31.

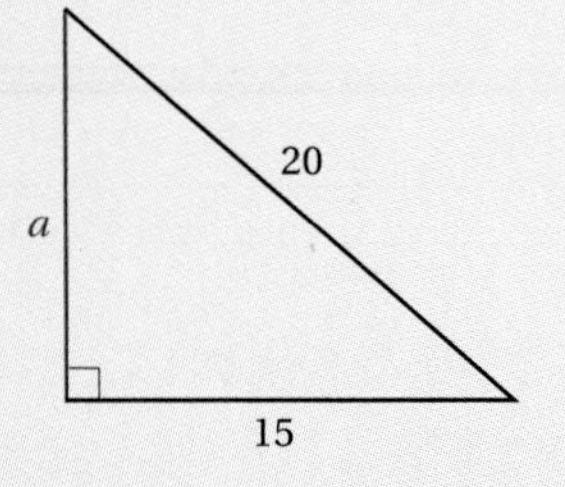

32. How long is a guy wire reaching from the top of an 18-ft pole to a point on the ground 10 ft from the pole? Give an exact answer and an approximation to the nearest tenth of a foot.

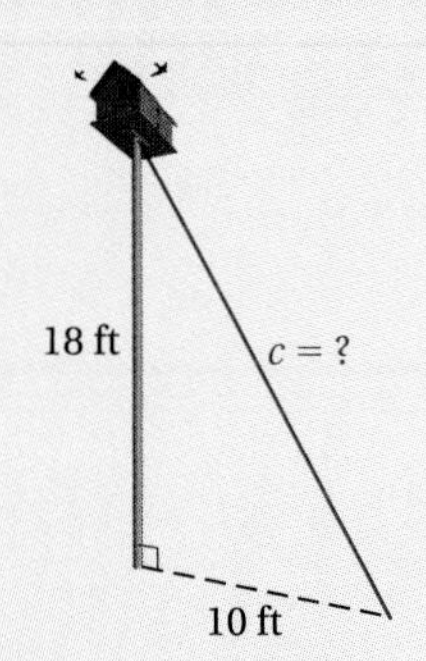

Answers on page A-20

We substitute in the Pythagorean equation. Next, we solve for b^2 and then b, as follows:

$$a^2 + b^2 = c^2$$

$$10^2 + b^2 = 12^2 \quad \text{Substituting}$$

$$100 + b^2 = 144$$

$$100 + b^2 - 100 = 144 - 100 \quad \text{Subtracting 100 on both sides}$$

$$b^2 = 144 - 100$$

$$b^2 = 44$$

Exact answer: $b = \sqrt{44}$

Approximation: $b \approx 6.633$. Using a calculator

Do Exercises 29–31.

Example 9 *Height of Ladder.* A 12-ft ladder leans against a building. The bottom of the ladder is 7 ft from the building. How high is the top of the ladder? Give an exact answer and an approximation to the nearest tenth of a foot.

1. **Familiarize.** We first make a drawing. In it we see a right triangle. We let $h =$ the unknown height.

2. **Translate.** We substitute 7 for a, h for b, and 12 for c in the Pythagorean equation:

$$a^2 + b^2 = c^2 \quad \text{Pythagorean equation}$$

$$7^2 + h^2 = 12^2.$$

3. **Solve.** We solve for h^2 and then h:

$$49 + h^2 = 144$$

$$49 + h^2 - 49 = 144 - 49$$

$$h^2 = 144 - 49$$

$$h^2 = 95$$

Exact answer: $h = \sqrt{95}$

Approximation: $h \approx 9.7$ ft.

4. **Check.** $7^2 + (\sqrt{95})^2 = 49 + 95 = 144 = 12^2$.
5. **State.** The top of the ladder is $\sqrt{95}$, or about 9.7 ft from the ground.

Do Exercise 32.

Exercise Set 8.7

a Simplify.

1. $\sqrt{100}$
10

2. $\sqrt{25}$
5

3. $\sqrt{441}$
21

4. $\sqrt{225}$
15

5. $\sqrt{625}$
25

6. $\sqrt{576}$
24

7. $\sqrt{361}$
19

8. $\sqrt{484}$
22

9. $\sqrt{529}$
23

10. $\sqrt{169}$
13

11. $\sqrt{10{,}000}$
100

12. $\sqrt{4{,}000{,}000}$
2000

b Approximate to three decimal places.

13. $\sqrt{48}$
6.928

14. $\sqrt{17}$
4.123

15. $\sqrt{8}$
2.828

16. $\sqrt{3}$
1.732

17. $\sqrt{18}$
4.24

18. $\sqrt{7}$
2.645

19. $\sqrt{6}$
2.449

20. $\sqrt{61}$
7.810

21. $\sqrt{10}$
3.162

22. $\sqrt{21}$
4.582

23. $\sqrt{75}$
8.660

24. $\sqrt{220}$
14.832

25. $\sqrt{196}$
14

26. $\sqrt{123}$
11.090

27. $\sqrt{183}$
13.528

28. $\sqrt{300}$
17.321

c Find the length of the third side of the right triangle. Give an exact answer and an approximation to three decimal places.

29.

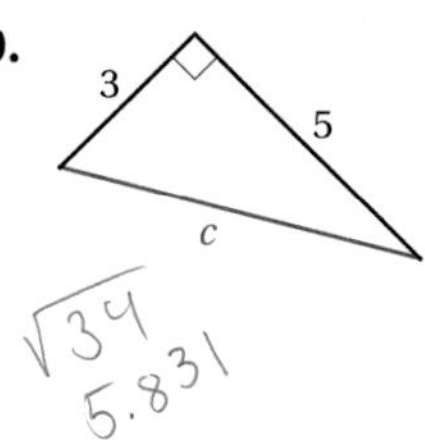

$\sqrt{34}$
5.831

30.

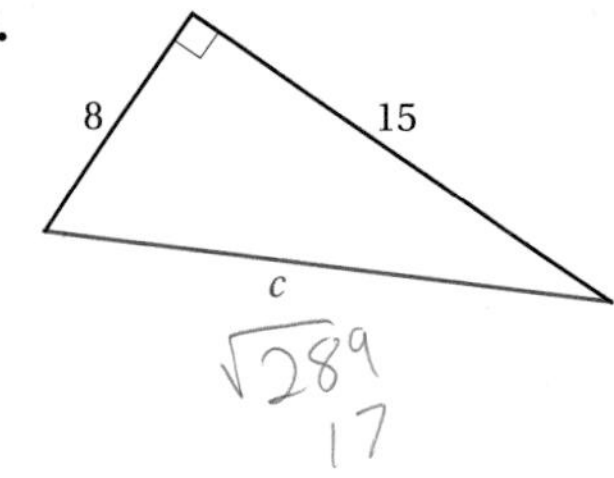

$\sqrt{289}$
17

31.

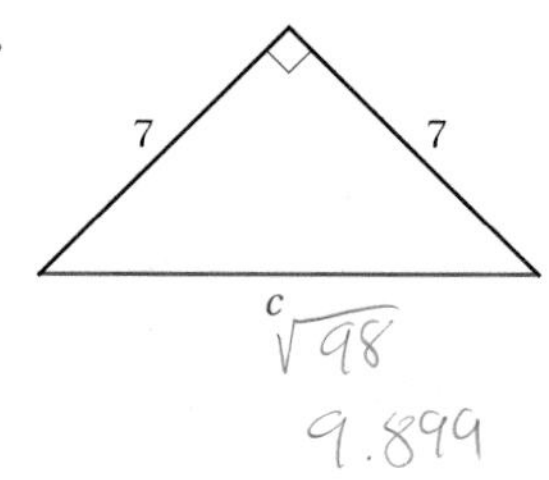

$\sqrt{98}$
9.899

32.

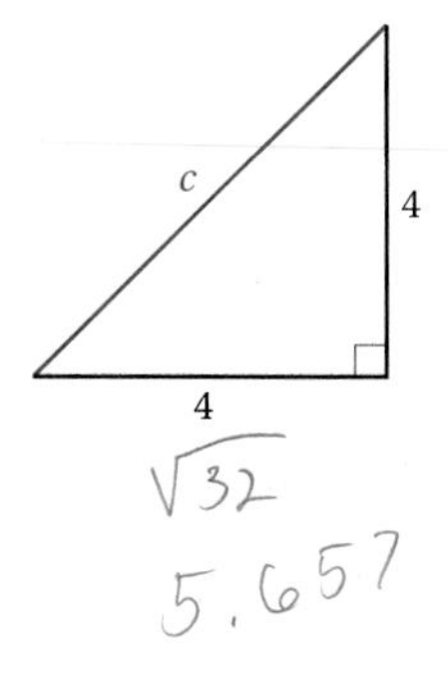

$\sqrt{32}$
5.657

33.

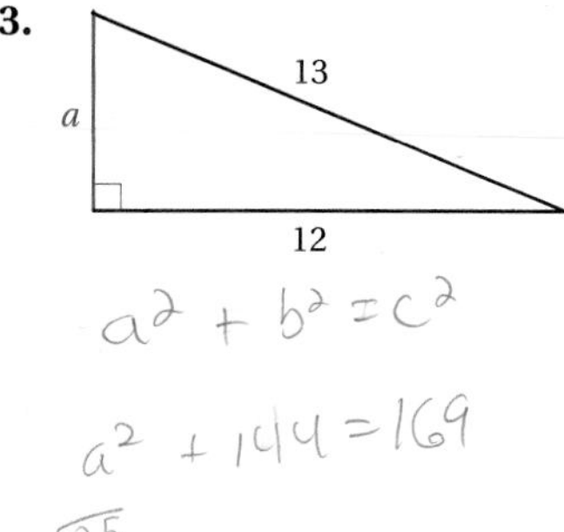

$a^2 + b^2 = c^2$
$a^2 + 144 = 169$
$a = \sqrt{25}$
$a = 5$

34.

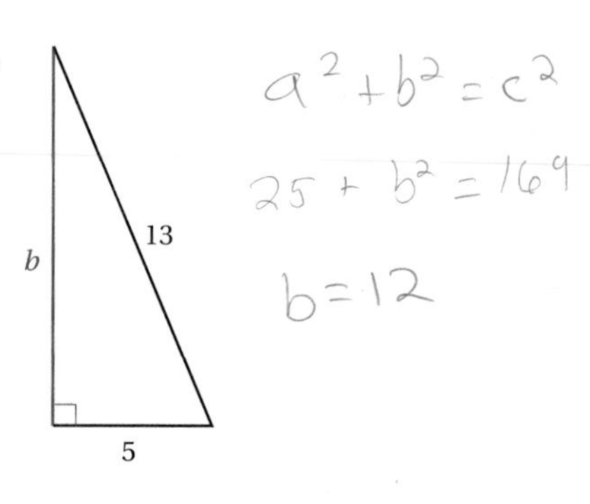

35.

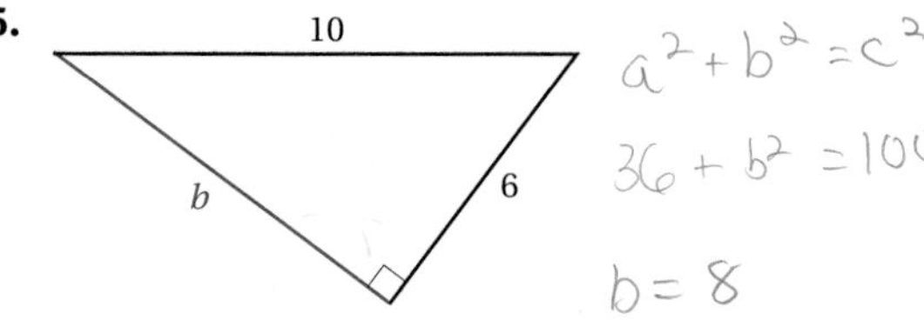

$a^2 + b^2 = c^2$
$36 + b^2 = 100$
$b = 8$

36.

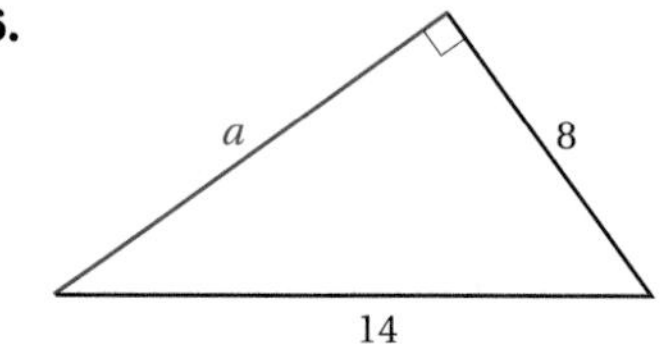

In a right triangle, find the length of the side not given. Give an exact answer and an approximation to three decimal places.

37. $a = 5$, $b = 12$

38. $a = 10$, $b = 24$

39. $a = 18$, $c = 30$

40. $a = 9$, $c = 15$

41. $b = 1$, $c = 20$

42. $a = 1$, $c = 32$

43. $a = 1$, $c = 15$

44. $a = 3$, $b = 4$

In Exercises 45–52, give an exact answer and an approximation to the nearest tenth.

45. How long must a wire be in order to reach from the top of a 13-m telephone pole to a point on the ground 9 m from the base of the pole?

46. How long is a light cord reaching from the top of a 12-ft pole to a point on the ground 8 ft from the base of the pole?

47. *Softball Diamond.* A slow-pitch softball diamond is actually a square 65 ft on a side. How far is it from home plate to second base?

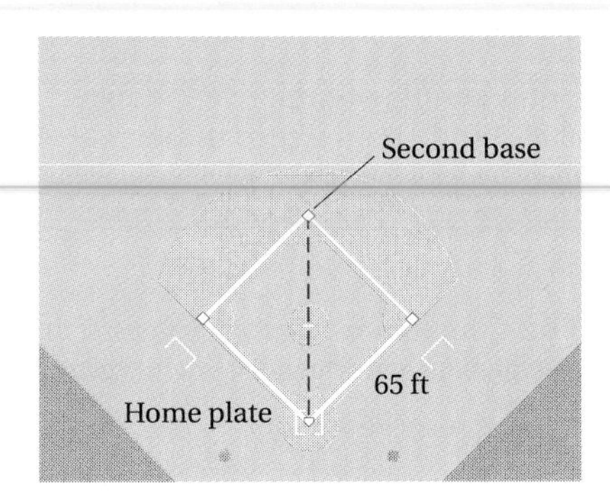

48. *Baseball Diamond.* A baseball diamond is actually a square 90 ft on a side. How far is it from home plate to second base?

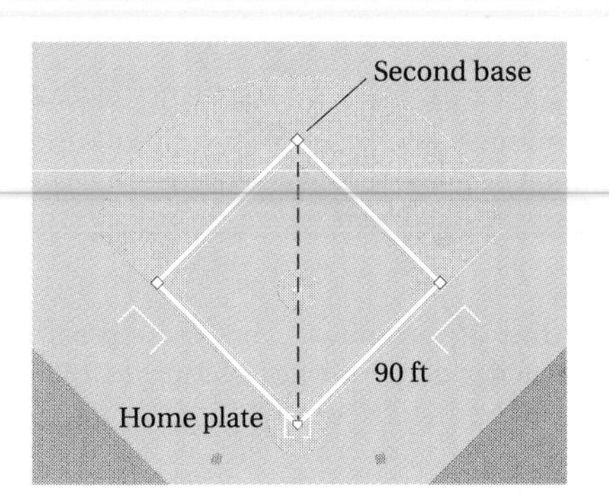

49. How tall is this tree?

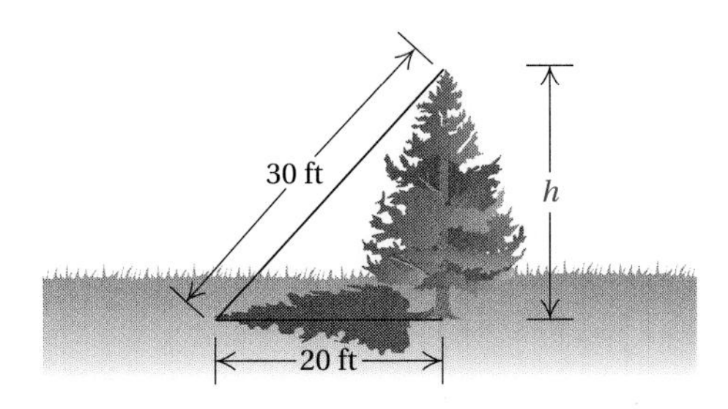

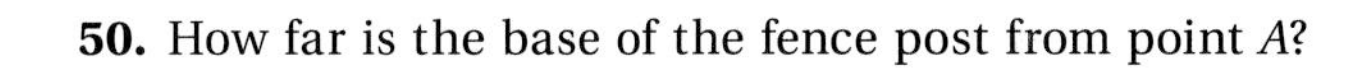

50. How far is the base of the fence post from point A?

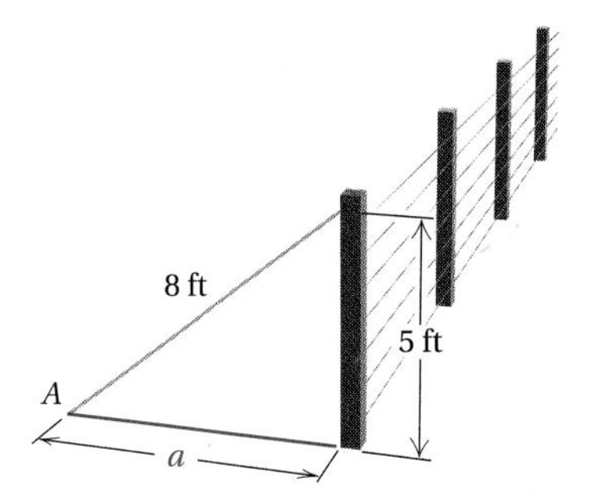

51. An airplane is flying at an altitude of 4100 ft. The slanted distance directly to the airport is 15,100 ft. How far is the airplane horizontally from the airport?

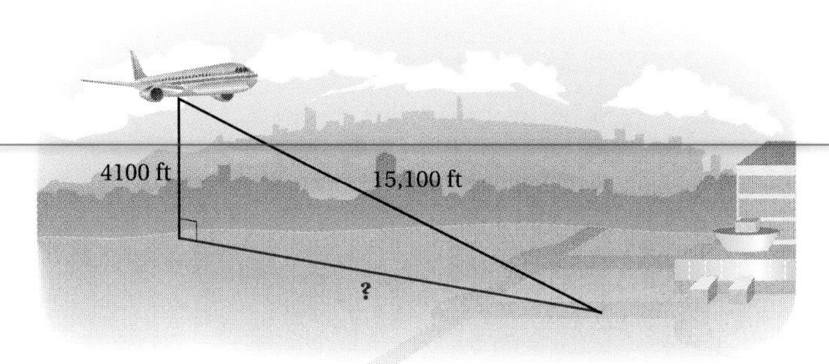

$a^2+b^2=c^2$

$16{,}810{,}000 + b^2 = 228{,}010{,}000$

$b=\sqrt{211200000}$

14532.722

52. A surveyor had poles located at points P, Q, and R around a lake. The distances that the surveyor was able to measure are marked on the drawing. What is the approximate distance from P to R across the lake?

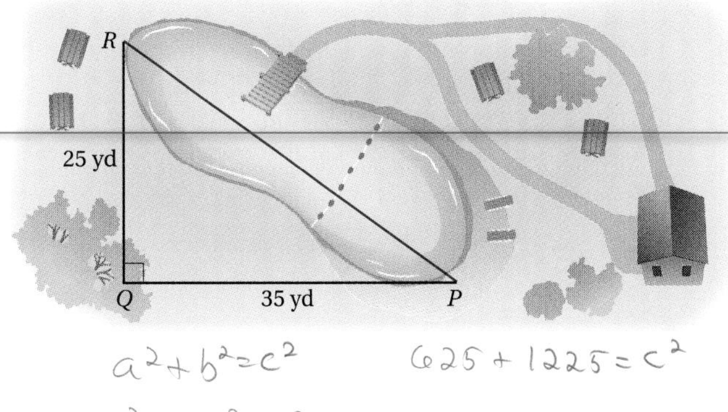

$a^2+b^2=c^2$

$25^2+35^2=c^2$

$625+1225=c^2$

$c=43.012$

Skill Maintenance

Convert to decimal notation. [6.1b]

53. 45.6%

54. 16.34%

55. 123%

56. 99%

57. 0.41%

58. 3%

Solve.

59. Food expenses account for 26% of the average family's budget. A family makes $1800 one month. How much do they spend for food? [6.5a]

60. The price of a cellular phone was reduced from $350 to $308. Find the percent of decrease in price. [6.5b]

61. A county has a population that is increasing by 4% each year. This year the population is 180,000. What will it be next year? [6.5b]

62. The price of a box of cookies increased from $2.85 to $3.99. What was the percent of increase in the price? [6.5b]

63. A college has a student body of 1850 students. Of these, 17.5% are seniors. How many students are seniors? [6.5a]

64. A state charges a meals tax of $4\frac{1}{2}$%. What is the meals tax charged on a dinner party costing $540? [6.6a]

Synthesis

65. ◈ Write a problem similar to Exercises 49–52 for a classmate to solve. Design the problem so that its solution involves the length $\sqrt{58}$ m.

66. ◈ Give an argument that could be used to convince a classmate that $\sqrt{2501}$ is not a whole number. Do not use a calculator.

67. 🖩 Find the area of the trapezoid shown. Round to the nearest hundredth.

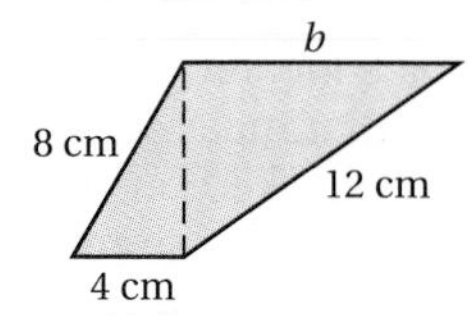

68. Which of the triangles below has the larger area?

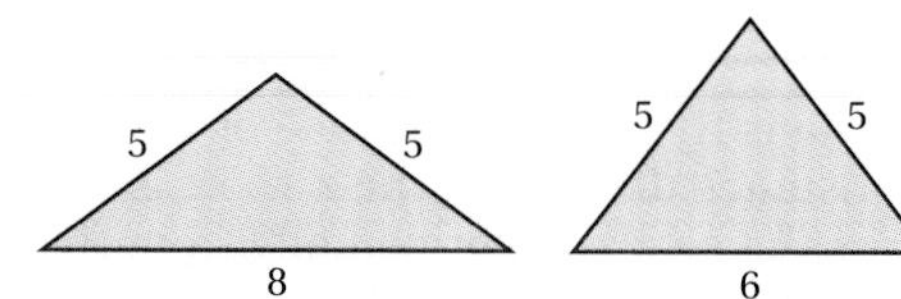

69. A 19-in. television set has a rectangular screen whose diagonal is 19 in. The ratio of length to width in a conventional television set is 4 to 3. Find the length and the width of the screen.

Summary and Review Exercises: Chapter 8

Important Properties and Formulas

American Units of Length:	12 in. = 1 ft; 3 ft = 1 yd; 36 in. = 1 yd; 5280 ft = 1 mi
Metric Units of Length:	1 km = 1000 m; 1 hm = 100 m; 1 dam = 10 m; 1 dm = 0.1 m; 1 cm = 0.01 m; 1 mm = 0.001 m
American–Metric Conversion:	1 m = 39.37 in.; 1 m = 3.3 ft; 0.303 m = 1 ft; 2.54 cm = 1 in.; 1 km = 0.621 mi; 1.609 km = 1 mi
Perimeter of a Rectangle:	$P = 2 \cdot (l + w)$, or $P = 2 \cdot l + 2 \cdot w$
Perimeter of a Square:	$P = 4 \cdot s$
Area of a Rectangle:	$A = l \cdot w$
Area of a Square:	$A = s \cdot s$, or $A = s^2$
Area of a Parallelogram:	$A = b \cdot h$
Area of a Triangle:	$A = \frac{1}{2} \cdot b \cdot h$
Area of a Trapezoid:	$A = \frac{1}{2} \cdot h \cdot (a + b)$
Radius and Diameter of a Circle:	$d = 2 \cdot r$, or $r = \frac{d}{2}$
Circumference of a Circle:	$C = \pi \cdot d$, or $C = 2 \cdot \pi \cdot r$
Area of a Circle:	$A = \pi \cdot r \cdot r$, or $A = \pi \cdot r^2$
Pythagorean Equation:	$a^2 + b^2 = c^2$

The objectives to be tested in addition to the material in this chapter are [6.1b, c] and [6.2a, b].

Complete.

1. 8 ft = ________ yd [8.1a]

2. $\frac{5}{6}$ yd = ________ in. [8.1a]

3. 0.3 mm = ________ cm [8.2a]

4. 4 m = ________ km [8.2a]

5. 2 yd = ________ in. [8.1a]

6. 4 km = ________ cm [8.2a]

7. 14 in. = ________ ft [8.1a]

8. 15 cm = ________ m [8.2a]

9. 200 m = ________ yd [8.2b]

10. 20 mi = ________ km [8.2b]

Find the perimeter. [8.3a]

11.

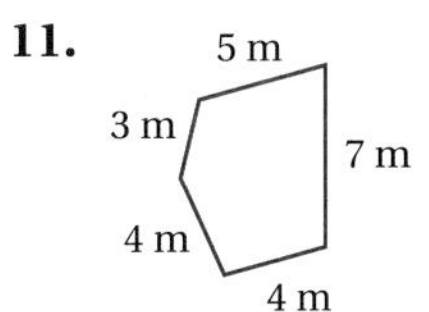

12.

0.5 m
1.9 m
0.8 m
1.2 m

13. The dimensions of a standard-sized tennis court are 78 ft by 36 ft. Find the perimeter and the area of the tennis court. [8.3b], [8.4b]

14. Find the length of a diagonal from one corner to another of the tennis court in Exercise 13. [8.7c]

Find the perimeter and the area. [8.3a], [8.4a]

15.

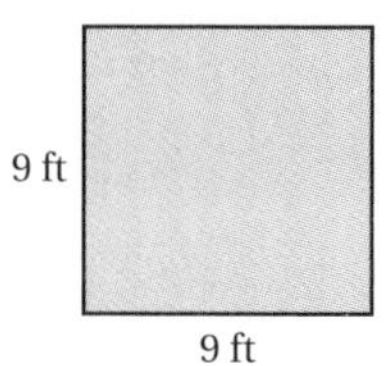

16.

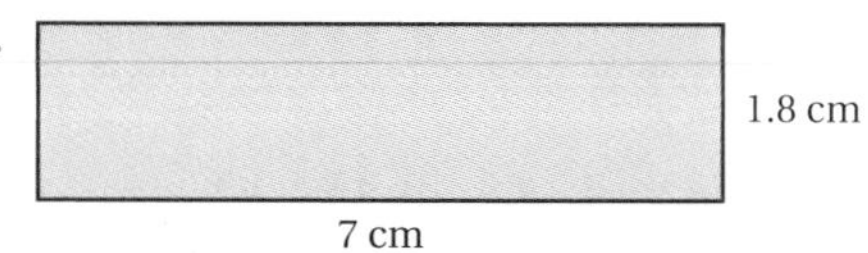

Find the area. [8.5a]

17.

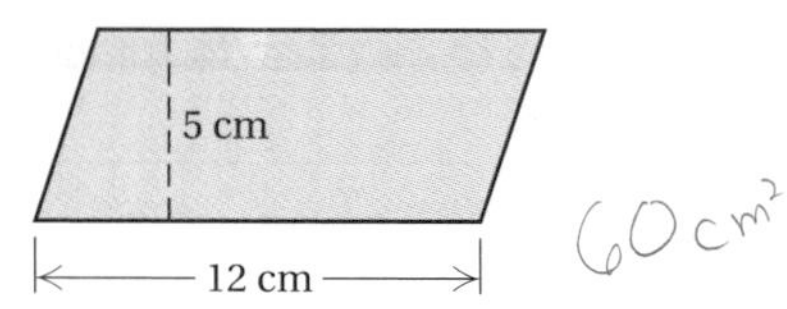

18.

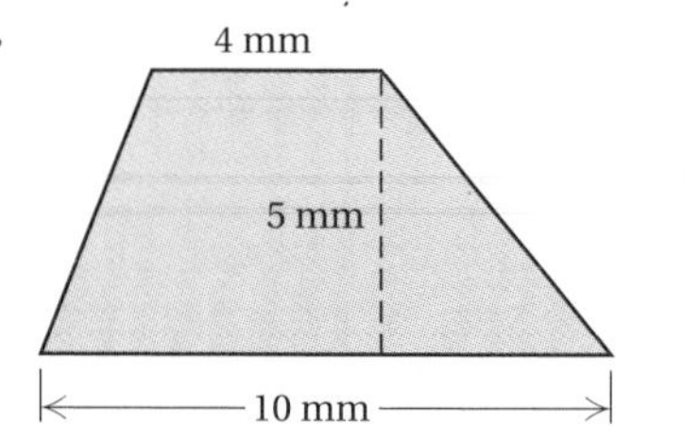

19.

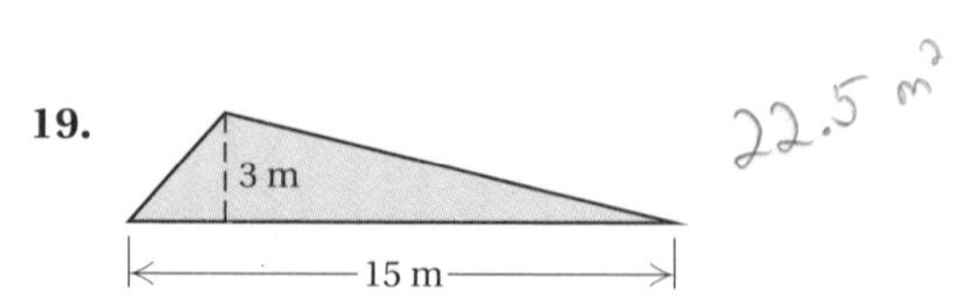

20.

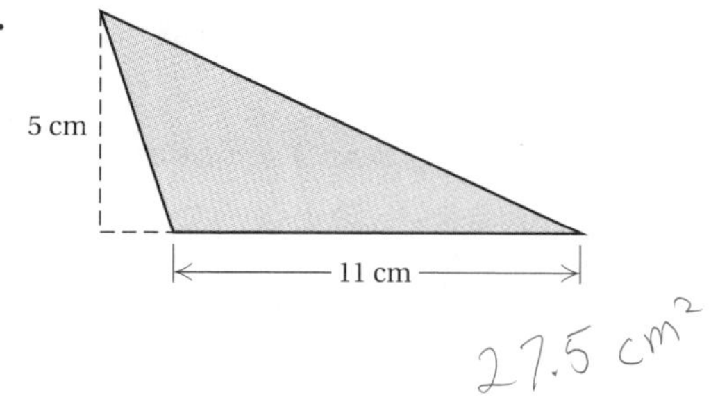

21.

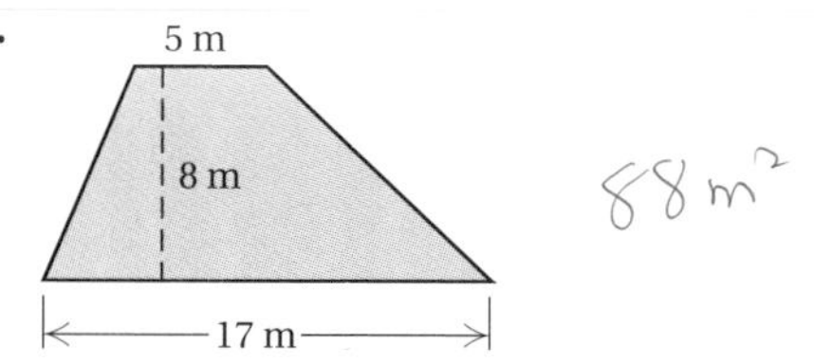

22.

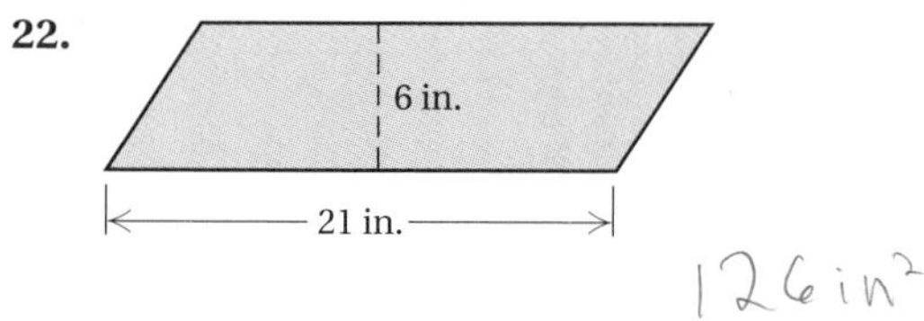

23. A grassy area is to be seeded around three sides of a building and has equal width on the three sides, as shown below. What is the seeded area? [8.4b]

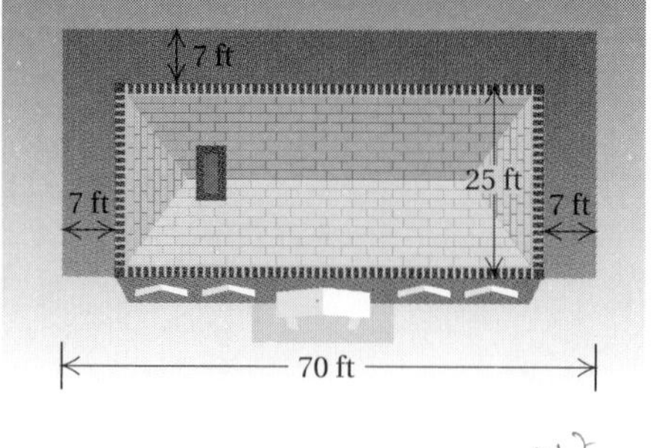

Find the length of a radius of the circle. [8.6a]

24.

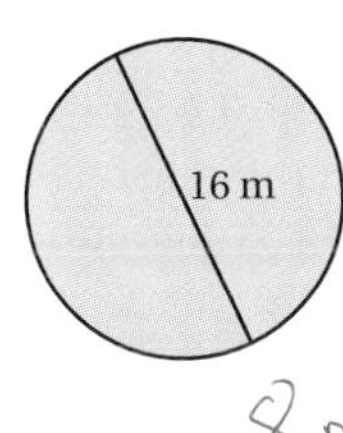

8 m

25.

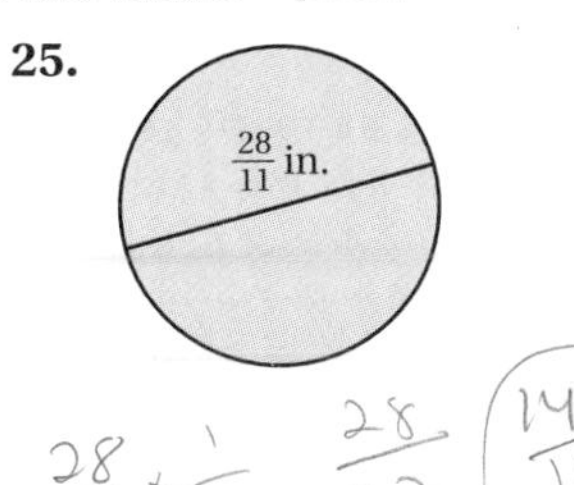

$\frac{28}{11} \times \frac{1}{2}$ $\frac{28}{22}$ $\frac{14}{11}$ in

Find the length of a diameter of the circle. [8.6a]

26.

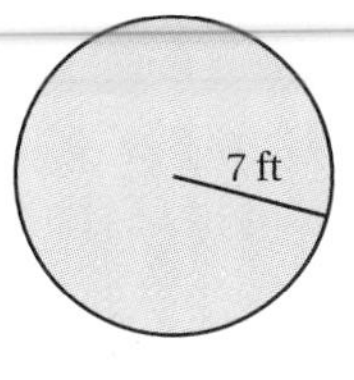

14 ft

27.

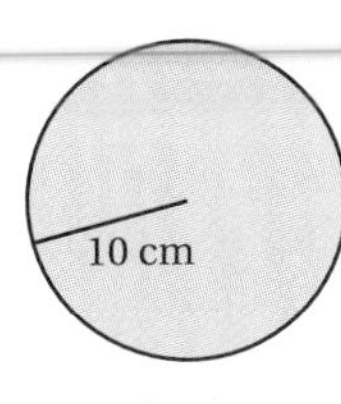

20 cm

28. Find the circumference of the circle in Exercise 24. Use 3.14 for π. [8.6b]

50.24 m

29. Find the circumference of the circle in Exercise 25. Use $\frac{22}{7}$ for π. [8.6b]

30. Find the area of the circle in Exercise 24. Use 3.14 for π. [8.6c]

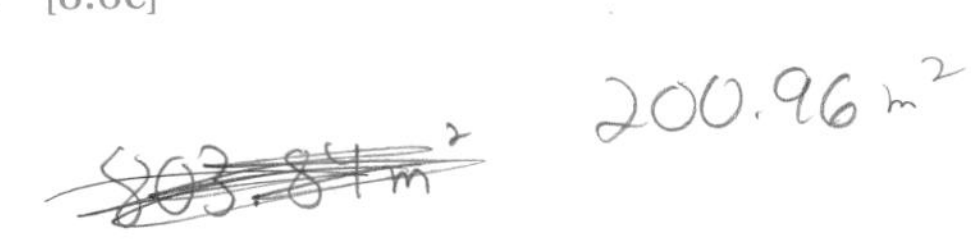

31. Find the area of the circle in Exercise 25. Use $\frac{22}{7}$ for π. [8.6c]

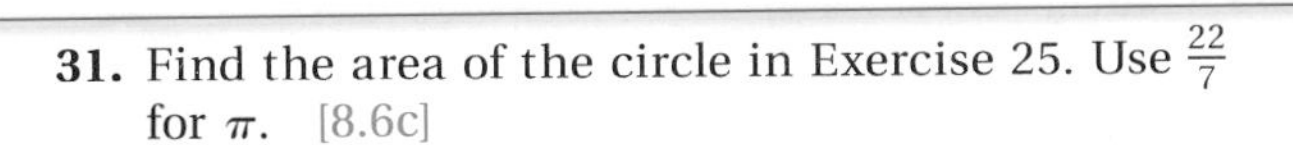

$\frac{14}{11} \times \frac{14}{11}$ $\frac{196}{121} \times \frac{22}{7}$ $\frac{4312}{847}$

32. Find the area of the shaded region. Use 3.14 for π. [8.6d]

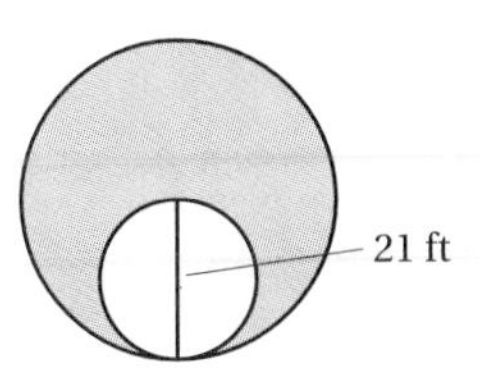

33. Simplify: $\sqrt{64}$. [8.7a]

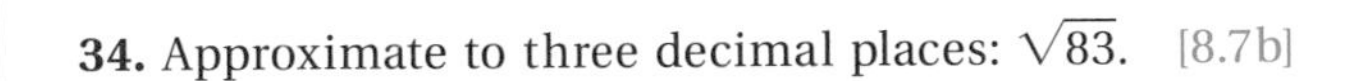

34. Approximate to three decimal places: $\sqrt{83}$. [8.7b]

In a right triangle, find the length of the side not given. Give an exact answer and an approximation to three decimal places. [8.7c]

35. $a = 15$, $b = 25$

36. $a = 7$, $c = 10$

Find the length of the side not given. Give an exact answer and an approximation to three decimal places. [8.7c]

37.

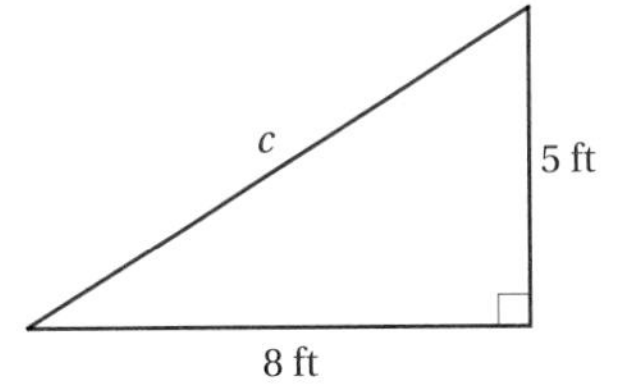

38.

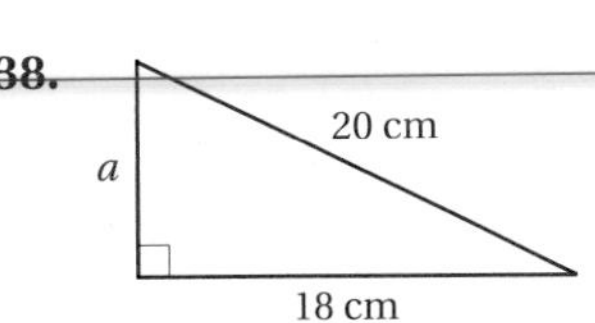

39. How long is a wire reaching from the top of a 24-ft pole to a point on the ground 16 ft from the base of the pole?

40. How tall is this tree?

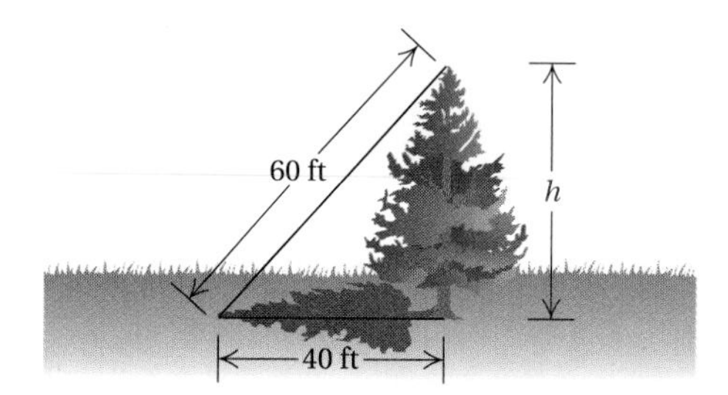

Skill Maintenance

41. Convert to percent notation: 0.47. [6.1c]

42. Convert to percent notation: $\frac{23}{25}$. [6.2a]

43. Convert to decimal notation: 56.7%. [6.1b]

44. Convert to fractional notation: 73%. [6.2b]

Synthesis

45. ◆ List as many reasons as you can for using the metric system exclusively. List as many reasons as you can for continuing our use of the American system. [8.2a, b]

46. ◆ Napoleon is credited with influencing the use of the metric system. Research this possibility and make a report. [8.2a, b]

47. A square is cut in half so that the perimeter of the resulting rectangle is 30 ft. Find the area of the original square. [8.3a], [8.4a]

48. Find the area, in square meters, of the shaded region. [8.2a], [8.4b]

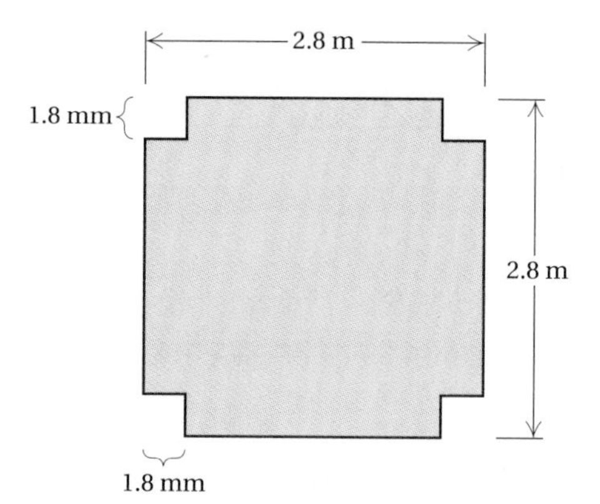

49. Find the area, in square centimeters, of the shaded region. [8.2a], [8.5b]

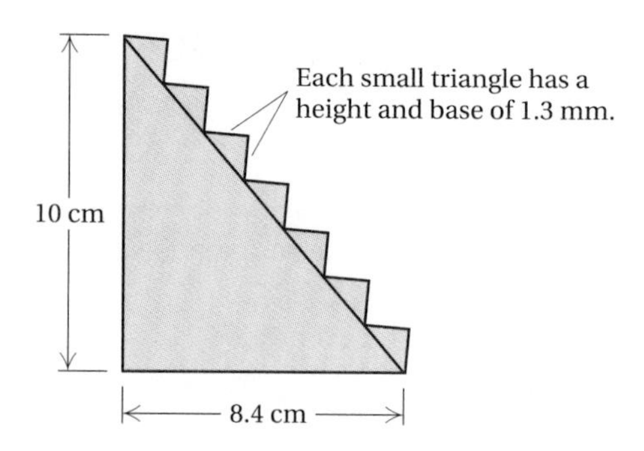

Test: Chapter 8

Complete.

1. 4 ft = _______ in.

2. 4 in. = _______ ft

3. 6 km = _______ m

4. 8.7 mm = _______ cm

5. 200 yd = _______ m

6. 2400 km = _______ mi

Find the perimeter and the area.

7.

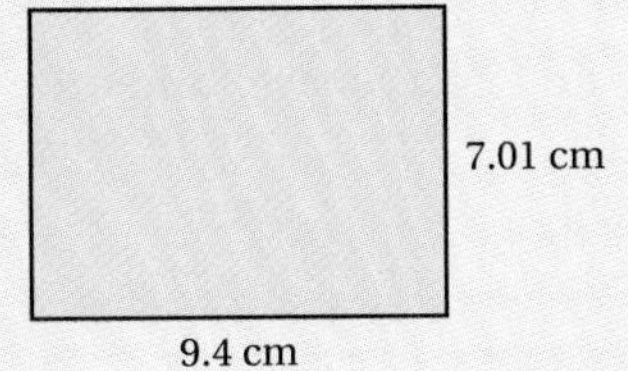

8.

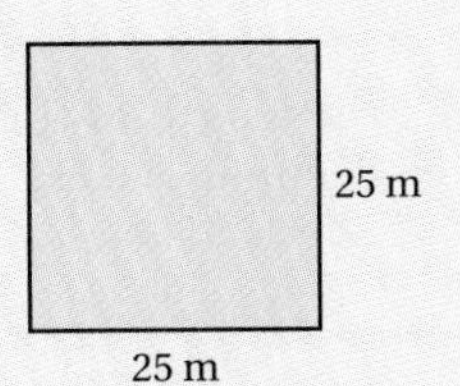

Find the area.

9.

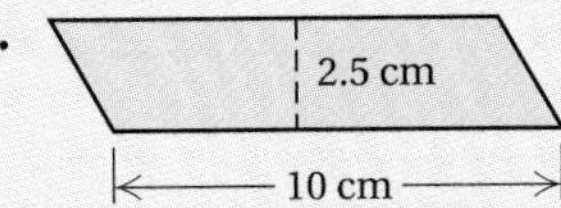

10.

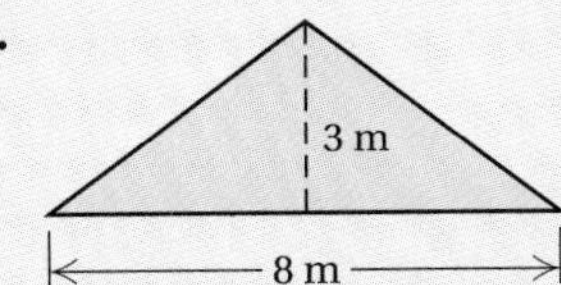

11.

4 ft
3 ft
8 ft

12. Find the length of a diameter of this circle.

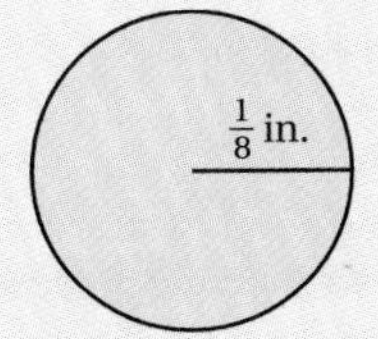

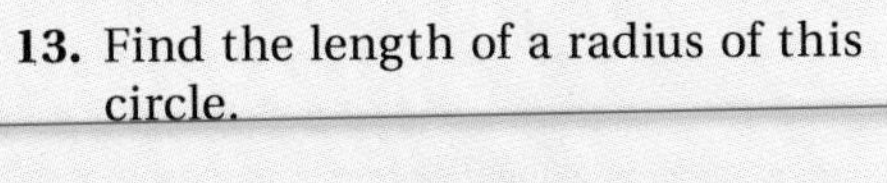

13. Find the length of a radius of this circle.

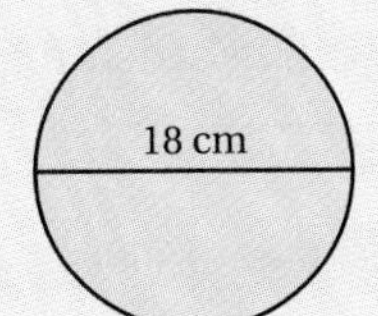

Answers

1. _______
2. _______
3. _______
4. _______
5. _______
6. _______
7. _______
8. _______
9. _______
10. _______
11. _______
12. _______
13. _______

Answers

14. ______

15. ______

16. ______

17. ______

18. ______

19. ______

20. ______

21. ______

22. ______

23. ______

24. ______

25. ______

26. ______

27. ______

28. ______

29. ______

14. Find the circumference of the circle in Question 12. Use $\frac{22}{7}$ for π.

15. Find the area of the circle in Question 13. Use 3.14 for π.

16. Find the area of the shaded region.

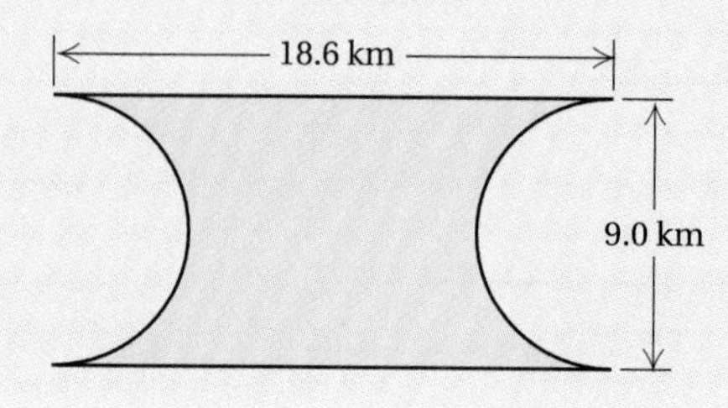

17. Simplify: $\sqrt{225}$.

18. Approximate to three decimal places: $\sqrt{87}$.

In a right triangle, find the length of the side not given. Give an exact answer and an approximation to three decimal places.

19. $a = 24$, $b = 32$

20. $a = 2$, $c = 8$

21.

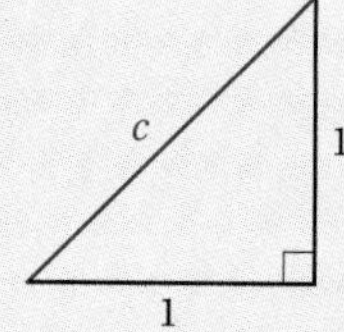

22.

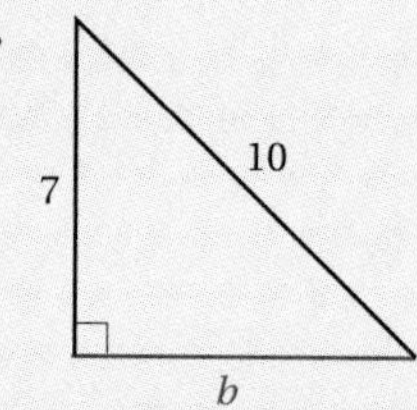

23. How long must a wire be in order to reach from the top of a 13-m antenna to a point on the ground 9 m from the base of the antenna?

Skill Maintenance

24. Convert to percent notation: 0.93.

25. Convert to percent notation: $\frac{13}{16}$.

26. Convert to decimal notation: 93.2%.

27. Convert to fractional notation: $33\frac{1}{3}\%$.

Synthesis

Find the area of the shaded region. Give the answer in square feet.

28.

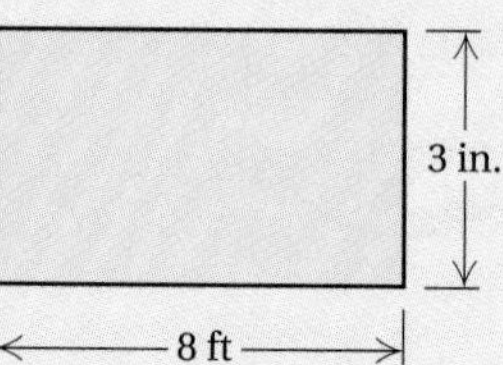

29.

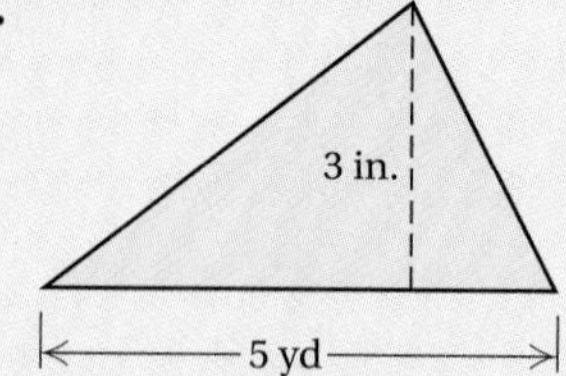

Cumulative Review: Chapters 1–8

Perform the indicated operation and simplify.

1. $46{,}231 \times 1100$

2. $\frac{1}{10} \cdot \frac{5}{6}$

3. $14.5 + \frac{4}{5} - 0.1$

4. $2\frac{3}{5} \div 3\frac{9}{10}$

5. $0.1\overline{)3.56}$

6. $3\frac{1}{2} - 2\frac{2}{3}$

7. Determine whether 1,298,032 is divisible by 8.

8. Determine whether 5,024,120 is divisible by 3.

9. Find the prime factorization of 99.

10. Find the LCM of 35 and 49.

11. Round $35.\overline{7}$ to the nearest tenth.

12. Write a word name for 103.064.

13. Find the average and the median of this set of numbers:

9, 13, 17, 18, 21, 29.

Find percent notation.

14. 0.08

15. $\frac{3}{5}$

16. Simplify: $\sqrt{121}$.

17. Approximate to two decimal places: $\sqrt{29}$.

18. Complete: 2 yd = ______ ft.

19. Find the perimeter.

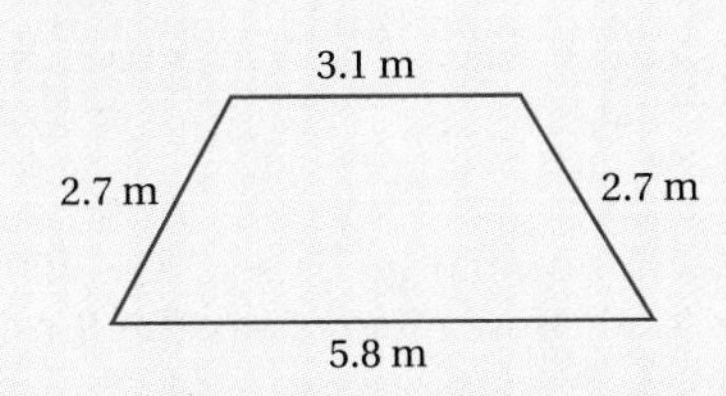

20. Find the area.

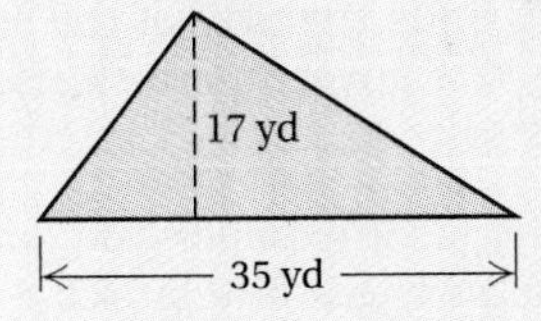

Solve.

21. $0.07 \cdot x = 10.535$

22. $x + 12{,}843 = 32{,}091$

23. $\frac{2}{3} \cdot y = 5$

24. $\frac{4}{5} + y = \frac{6}{7}$

Estimate each of the following as a whole number or as a mixed numeral where the fractional part is $\frac{1}{2}$.

25. $12\frac{3}{8}$

26. $5\frac{1}{4}$

27. $29\frac{15}{16}$

28. $36\frac{2}{13}$

29. $3\frac{4}{5} + 2\frac{7}{8}$

30. $\frac{2}{13} \cdot \frac{7}{15}$

31. $7\frac{2}{3} + \frac{12}{13} + 8\frac{5}{9}$

32. $\frac{19}{20} + \frac{7}{9} + 5\frac{11}{12}$

The following table shows typical sleep requirements in childhood.

Age	Hours of Daytime Sleep	Hours of Nighttime Sleep
1 week	8.0	8.5
1 month	7.0	8.5
3 months	5.5	9.5
6 months	3.3	11.0
9 months	2.5	11.5
12 months	2.3	11.5
18 months	2.0	11.5
2 years	1.5	11.5

33. How many hours of daytime sleep does a 3-month-old child need?

34. How many total hours of sleep does a 1-year-old child need?

35. How many more hours will a 6-month-old child sleep at night than a 1-week-old child?

36. How many hours would you expect a 2-month-old child to sleep at night?

Solve.

37. A microwave oven marked $220 was discounted to a sale price of $194. What was the rate of discount?

38. There are 11 million milk cows in America, each producing, on average, 15,000 lb of milk per year. How many pounds of milk are produced each year in America?

39. The Schwartz family has a rectangular kitchen table measuring 52 in. by 30 in. They replace it with a circular table with a 48-in. diameter. How much bigger is their new table? Use 3.14 for π.

40. A man on a diet loses $3\frac{1}{2}$ lb in 2 weeks. At this rate, how many pounds will he lose in 5 weeks?

41. The U.S. Department of Agriculture requires that 80% of the seeds that a company produces must sprout. To find out about the quality of the seeds it has produced, a company takes 500 seeds and plants them. It finds that 417 of the seeds sprout. Did the seeds pass government standards?

42. A mechanic spent $\frac{1}{3}$ hr changing a car's oil, $\frac{1}{2}$ hr rotating the tires, $\frac{1}{10}$ hr changing the air filter, $\frac{1}{4}$ hr adjusting the idle speed, and $\frac{1}{15}$ hr checking the brake and transmission fluids. How many hours did the mechanic spend working on the car?

43. A driver bought gasoline when the odometer read 86,897.2. At the next gasoline purchase, the odometer read 87,153.0. How many miles had been driven?

44. A family has an annual income of $26,400. Of this, $\frac{1}{4}$ is spent for food. How much does the family spend for food?

Synthesis

45. A homeowner is having a one-story addition built on an existing house. The addition measures 25 ft by 32 ft. The existing house measures 30 ft by 32 ft and has two stories. What is the percent of increase in living area provided by the addition?

46. Mattie is training to run the mile. The first week of her training, she accomplishes the run in 6 min, 58 sec. In the second week, she improves her time by 12 sec. In the third week, she improves the preceding time by 10 sec, and in the fourth week, she improves the preceding time by 8 sec. If this pattern of improvement continues, what will her times be in the fifth, sixth, and seventh weeks? Make a line graph of Mattie's times for the mile. Look for a pattern. Will Mattie eventually "bottom out" to a best time on which she cannot improve? If so, when?

9

More Geometry and Measures

© 1998 AL SATTERWHITE

An Application

How "big" is one million dollars? This photograph shows one million one-dollar bills assembled by the Bureau of Engraving. The width of a dollar is 2.3125 in., the length is 6.0625 in., and the thickness is 0.0041 in. Find the volume occupied by one million one-dollar bills.

This problem appears as Exercise 41 in Exercise Set 9.1.

The Mathematics

To find the volume of a single one-dollar bill, we multiply the length times the width and then times the height, or thickness:

This is a formula for volume.

$$V = l \cdot w \cdot h$$
$$= 6.0625 \times 2.3125 \times 0.0041.$$

Then we multiply the result by one million.

World Wide Web For more information, visit us at www.mathmax.com

Introduction

In this chapter, we consider several other measures: volume, capacity, weight, mass, time, temperature, and units of area. We conclude with angles and triangles and some related concepts.

Any of these topics can be omitted without loss of continuity to the remaining two chapters of the text.

Pretest: Chapter 9

Complete.

1. 2304 mL = _____ L
2. 2.4 L = _____ mL
3. 5 lb = _____ oz
4. 4.4 T = _____ lb
5. 4.8 kg = _____ g
6. 6.2 mg = _____ cg
7. 3400 mg = _____ g
8. 7 hr = _____ min
9. 16 days = _____ hr
10. 128 pt = _____ qt
11. 20 gal = _____ oz
12. 3 cups = _____ oz
13. Convert 77°F to Celsius.
14. Convert 37°C to Fahrenheit.

Complete.

15. 1 ft^2 = _____ in^2
16. 2 km^2 = _____ m^2
17. A bundle of concert programs weighs 1 kg. How many grams does 1 bundle weigh?

Find the volume. Use 3.14 for π.

18.

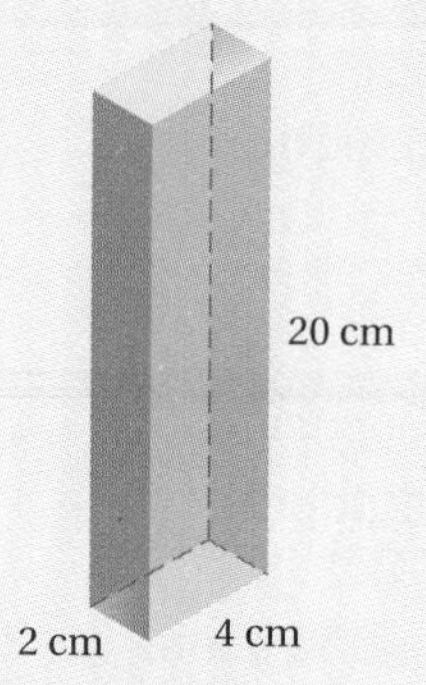

19.

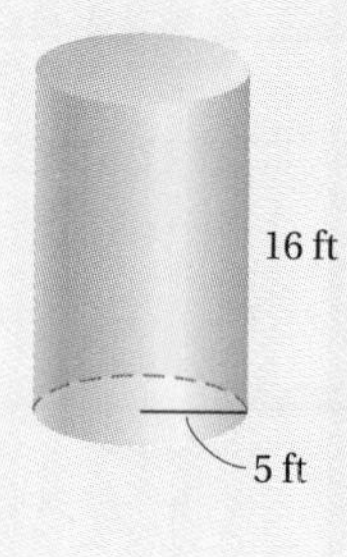

20.

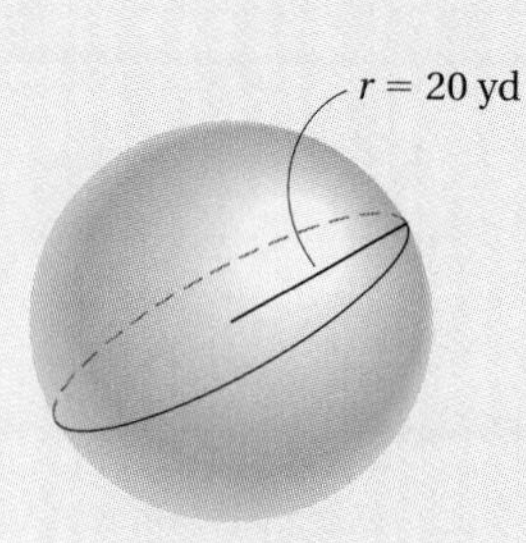

21.

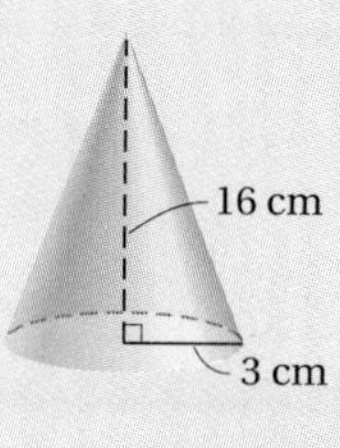

Use a protractor to measure the angle.

22.

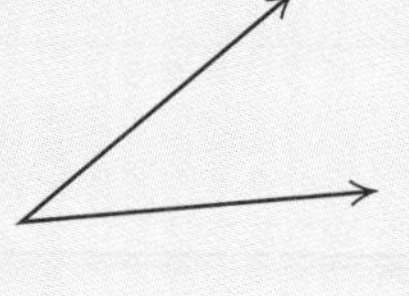

23.

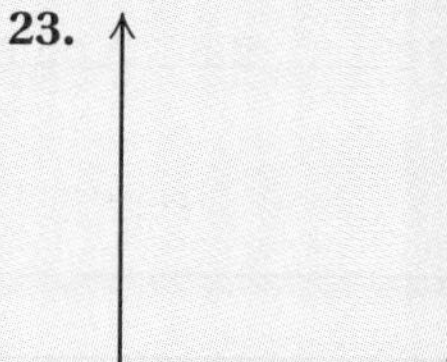

24.

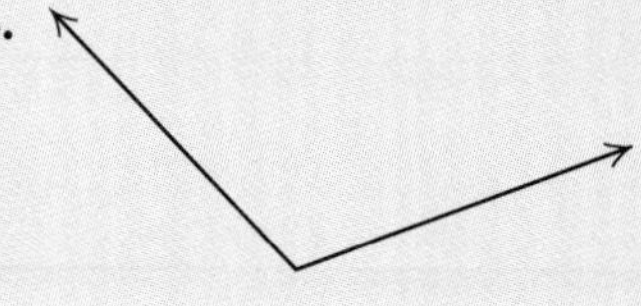

25.

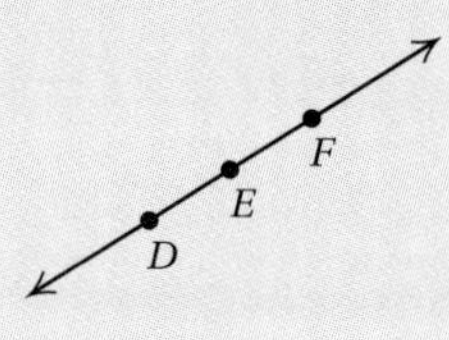

26.–29. Classify each of the angles in Questions 22–25 as right, straight, acute, or obtuse.

Use the triangle shown at right for Questions 30–32.

30. Find the missing angle measure.
31. Classify the triangle as equilateral, isosceles, or scalene.
32. Classify the triangle as right, obtuse, or acute.

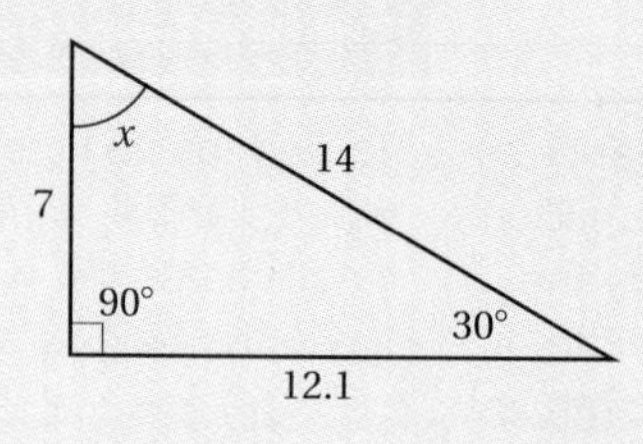

Objectives for Retesting

The objectives to be tested in addition to the material in this chapter are as follows.

[1.9b] Evaluate exponential notation.
[6.7a] Solve applied problems involving simple interest and percent.
[8.1a] Convert from one American unit of length to another.
[8.2a] Convert from one metric unit of length to another.

9.1 Volume and Capacity

a Volume

The **volume** of a **rectangular solid** is the number of unit cubes needed to fill it.

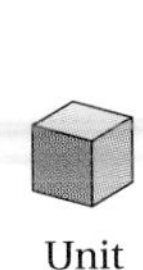

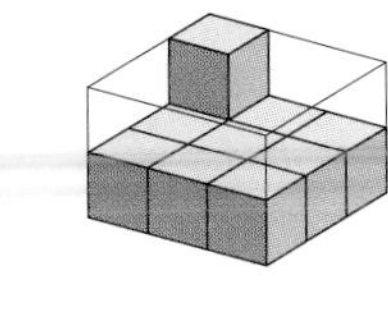

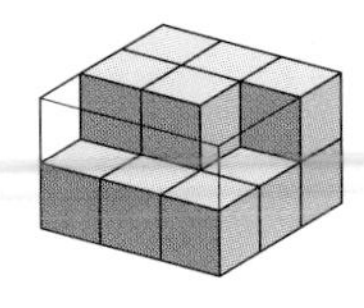

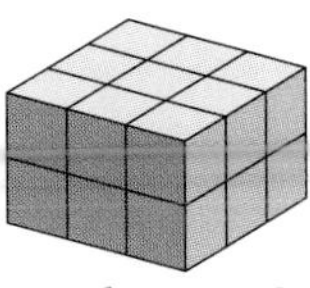

Two other units are shown below.

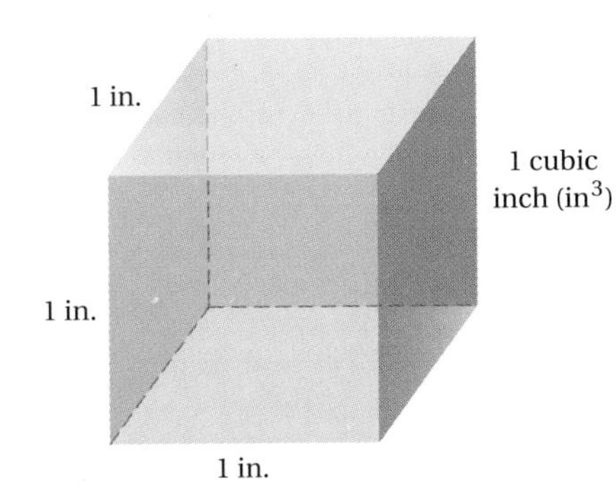

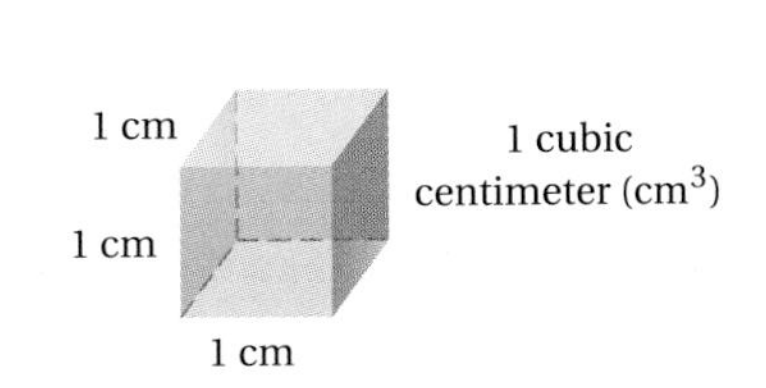

Example 1 Find the volume.

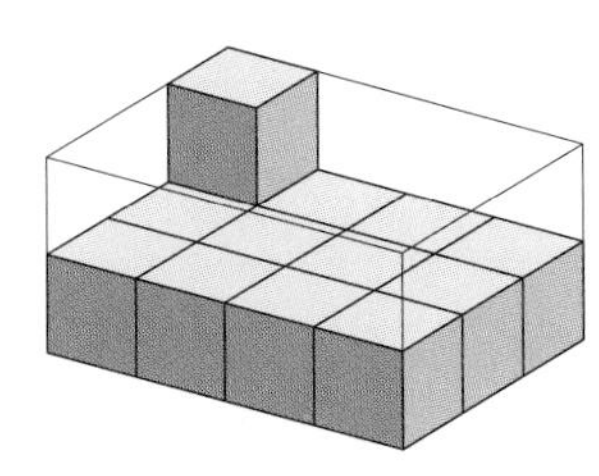

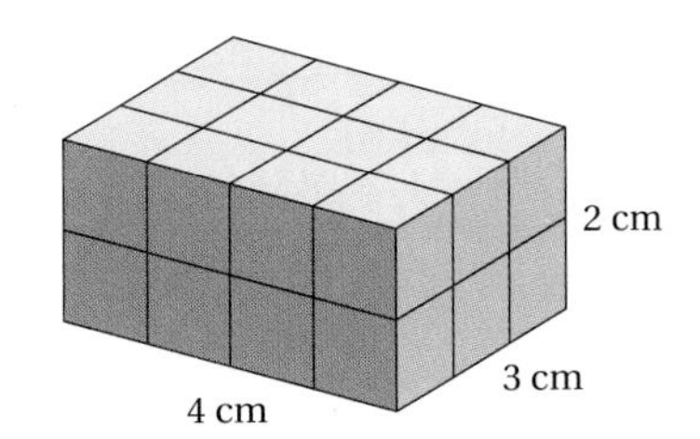

The figure is made up of 2 layers of 12 cubes each, so its volume is 24 cubic centimeters (cm^3).

Do Exercise 1.

The **volume of a rectangular solid** is found by multiplying length by width by height:

$$V = l \cdot w \cdot h.$$

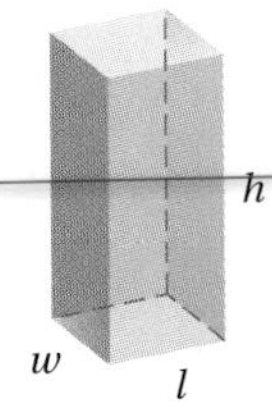

Objectives

a Find the volume of a rectangular solid using the formula $V = l \cdot w \cdot h$.

b Convert from one unit of capacity to another.

c Solve applied problems involving capacity.

For Extra Help

TAPE 15 | TAPE 14B | MAC WIN | CD-ROM

1. Find the volume.

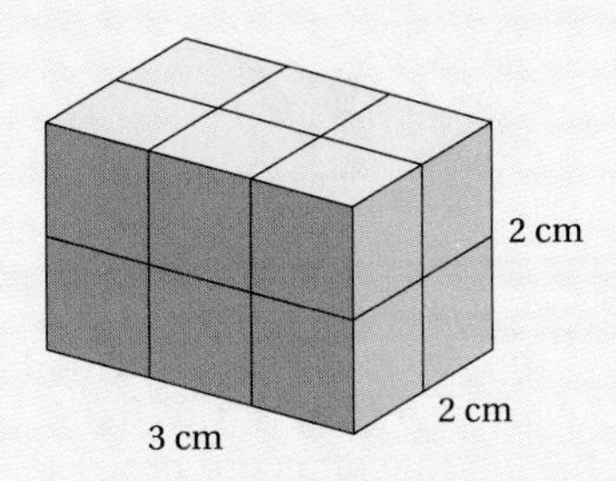

Answer on page A-21

Example 2 The largest piece of luggage that you can carry on an airplane measures 23 in. by 10 in. by 13 in. Find the volume of this solid.

$$V = l \cdot w \cdot h$$
$$= 23 \text{ in.} \cdot 10 \text{ in.} \cdot 13 \text{ in.}$$
$$= 230 \cdot 13 \text{ in}^3$$
$$= 2990 \text{ in}^3$$

Do Exercises 2 and 3.

2. In a recent year, people in the United States bought enough unpopped popcorn to provide every person in the country with a bag of popped corn measuring 2 ft by 2 ft by 5 ft. Find the volume of such a bag.

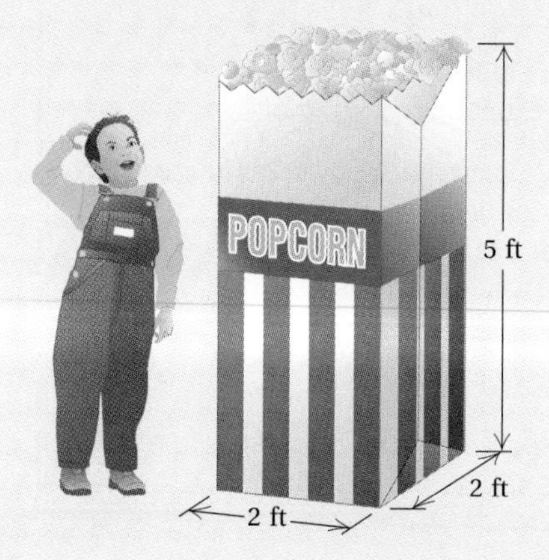

3. *Cord of Wood.* A cord of wood is 4 ft by 4 ft by 8 ft. What is the volume of a cord of wood?

b Capacity

To answer a question like "How much soda is in the can?" we need measures of **capacity**. American units of capacity are ounces, or fluid ounces, cups, pints, quarts, and gallons. These units are related as follows.

AMERICAN UNITS OF CAPACITY

1 gallon (gal) = 4 quarts (qt)
1 qt = 2 pints (pt)
1 pt = 2 cups = 16 ounces (oz)
1 cup = 8 oz

Example 3 Complete: 9 gal = _______ oz.

We convert as follows:

$$9 \text{ gal} = 9 \cdot 1 \text{ gal}$$
$$= 9 \cdot 4 \text{ qt} \qquad \text{Substituting 4 qt for 1 gal}$$
$$= 9 \cdot 4 \cdot 1 \text{ qt}$$
$$= 9 \cdot 4 \cdot 2 \text{ pt} \qquad \text{Substituting 2 pt for 1 qt}$$
$$= 9 \cdot 4 \cdot 2 \cdot 1 \text{ pt}$$
$$= 9 \cdot 4 \cdot 2 \cdot 16 \text{ oz} \qquad \text{Substituting 16 oz for 1 pt}$$
$$= 1152 \text{ oz.}$$

Do Exercise 4.

4. Complete: 5 gal = _______ pt.

Example 4 Complete: 24 qt = _______ gal.

In this case, we multiply by 1 using 1 gal in the numerator, since we are converting to gallons, and 4 qt in the denominator, since we are converting from quarts.

$$24 \text{ qt} = 24 \cancel{\text{qt}} \cdot \frac{1 \text{ gal}}{4 \cancel{\text{qt}}} = \frac{24}{4} \cdot 1 \text{ gal} = 6 \text{ gal}$$

After completing Example 4, we can check whether the answer is reasonable. We are converting from smaller to larger units, so our answer has fewer larger units.

Do Exercise 5.

5. Complete:
80 qt = _______ gal.

Answers on page A-21

Thinking Metric

One unit of capacity in the metric system is a **liter**. A liter is just a bit more than a quart. It is defined as follows.

> **METRIC UNITS OF CAPACITY**
>
> 1 liter (L) = 1000 cubic centimeters (1000 cm^3)
>
> The script letter ℓ is also used for "liter."

The metric prefixes are also used with liters. The most common is **milli-**. The milliliter (mL) is, then, $\frac{1}{1000}$ liter. Thus,

> 1 L = 1000 mL = 1000 cm^3;
>
> 0.001 L = 1 mL = 1 cm^3.

Although the other metric prefixes are rarely used for capacity, we display them in the following table as we did for linear measure.

1000 L	100 L	10 L	1 L	0.1 L	0.01 L	0.001 L
1 kL	1 hL	1 daL	1 L	1 dL	1 cL	1 mL (cc)

A preferred unit for drug dosage is the milliliter (mL) or the cubic centimeter (cm^3). The notation "cc" is also used for cubic centimeter, especially in medicine. The milliliter and the cubic centimeter represent the same measure of capacity. A milliliter is about $\frac{1}{5}$ of a teaspoon.

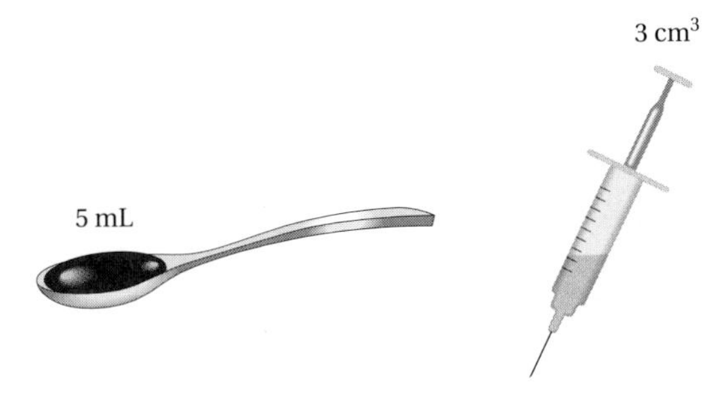

> 1 mL = 1 cm^3 = 1 cc

Volumes for which quarts and gallons are used are expressed in liters. Large volumes in business and industry are expressed using measures of cubic meters (m^3).

Do Exercises 6–9.

Complete with mL or L.

6. The patient received an injection of 2 ________ of penicillin.

7. There are 250________ in a coffee cup.

8. The gas tank holds 80 ________.

9. Bring home 8 ________ of milk.

Example 5 Complete: 4.5 L = ________ mL.

$4.5 \text{ L} = 4.5 \times 1 \text{ L} = 4.5 \times 1000 \text{ mL}$ **Substituting 1000 mL for 1 L**

$= 4500 \text{ mL}$

1000 L	100 L	10 L	1 L	0.1 L	0.01 L	0.001 L
1 kL	1 hL	1 daL	1 L	1 dL	1 cL	1 mL (cc)

3 places to the right

Answers on page A-21

Complete.

10. 0.97 L = ________ mL

11. 8990 mL = ________ L

12. A physician ordered 4800 mL of 0.9% saline solution. How many liters were ordered?

13. A prescription calls for 4 oz of ephedrine.

a) For how many milliliters is the prescription?

b) For how many liters is the prescription?

14. At the same station, the price of premium lead-free gasoline is 43.9 cents a liter. Estimate the price of 1 gal in dollars.

Answers on page A-21

Example 6 Complete: 280 mL = ________ L.

$$280 \text{ mL} = 280 \times 1 \text{ mL}$$
$$= 280 \times 0.001 \text{ L} \quad \text{Substituting 0.001 L for 1 mL}$$
$$= 0.28 \text{ L}$$

1000 L	100 L	10 L	1 L	0.1 L	0.01 L	0.001 L
1 kL	1 hL	1 daL	1 L	1 dL	1 cL	1 mL (cc)

3 places to the left

Do Exercises 10 and 11.

c Solving Applied Problems

The metric system has extensive usage in medicine.

Example 7 *Medical Dosage.* A physician ordered 3.5 L of 5% dextrose in water. How many milliliters were ordered?

We convert 3.5 L to milliliters:

$$3.5 \text{ L} = 3.5 \times 1 \text{ L} = 3.5 \times 1000 \text{ mL} = 3500 \text{ mL}.$$

The physician ordered 3500 mL.

Do Exercise 12.

Example 8 *Medical Dosage.* In pharmaceutical work, liquids at the drugstore are given in liters or milliliters, but a physician's prescription is given in ounces. For conversion, a druggist knows that 1 oz = 29.57 mL. A prescription calls for 3 oz of ephedrine. For how many milliliters is the prescription?

We convert as follows:

$$3 \text{ oz} = 3 \times 1 \text{ oz} = 3 \times 29.57 \text{ mL} = 88.71 \text{ mL}.$$

The prescription calls for 88.71 mL of ephedrine.

Do Exercise 13.

Example 9 *Gasoline Prices.* At a self-service gasoline station, regular lead-free gasoline sells for 29.3¢ a liter. Estimate the cost of 1 gal in dollars.

Since 1 L is about 1 qt and there are 4 qt in a gallon, the price of a gallon is about 4 times the price of a liter:

$$4 \times 29.3¢ = 117.2¢ = \$1.172.$$

Thus regular lead-free gasoline sells for about $1.17 a gallon.

Do Exercise 14.

Exercise Set 9.1

a Find the volume.

1.

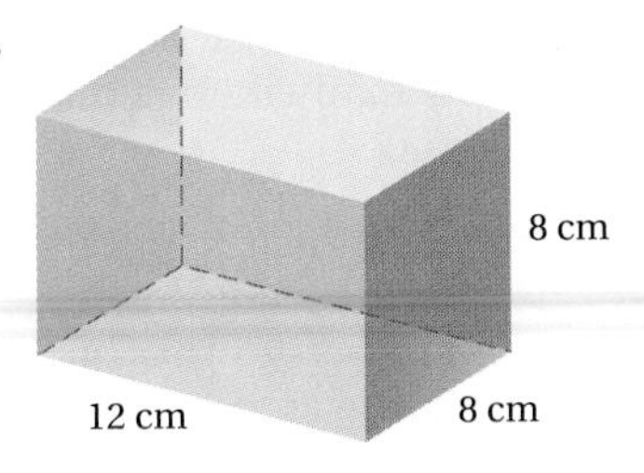

2.

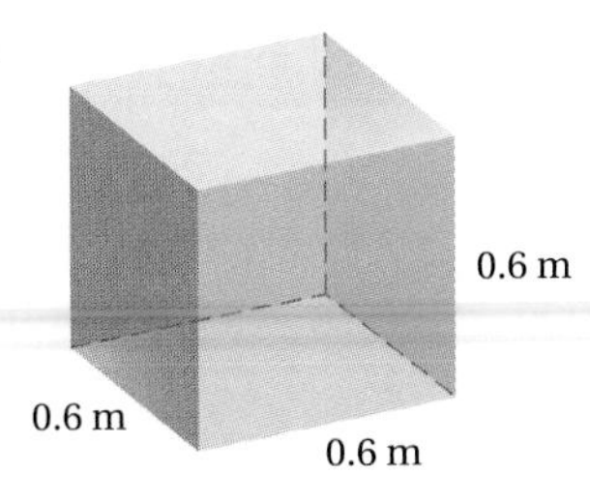

3.

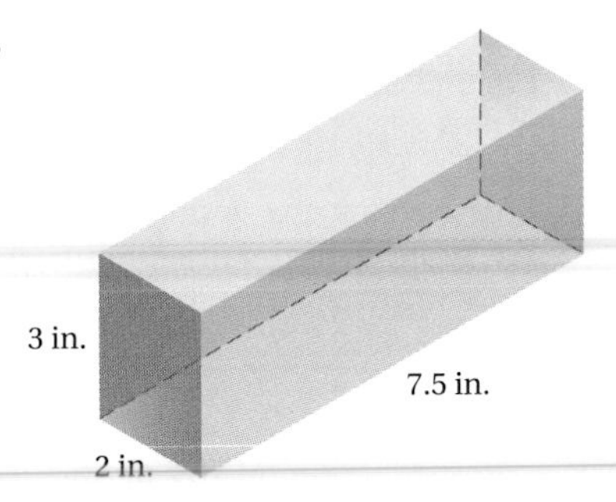

4.

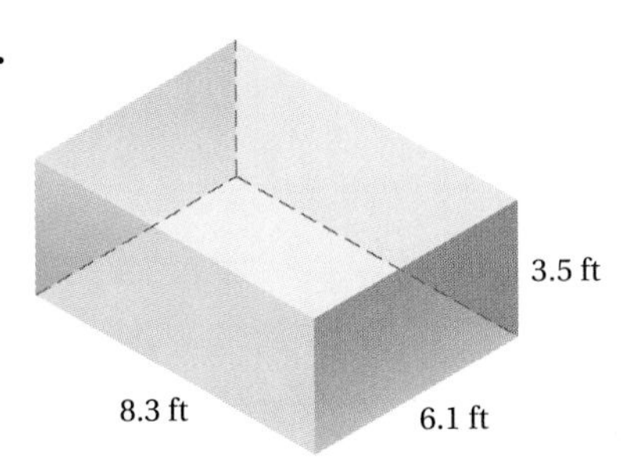

5.

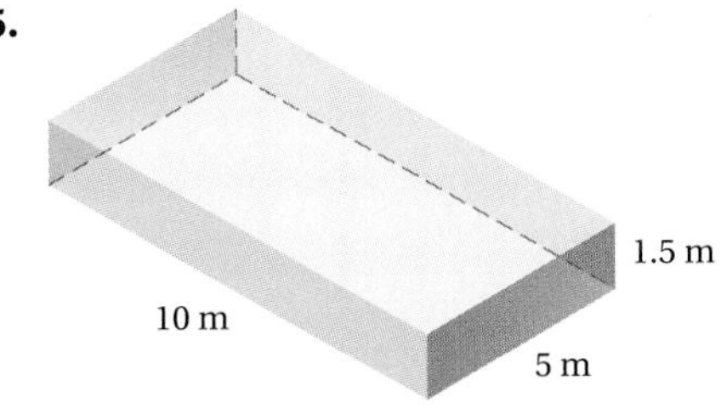

6.

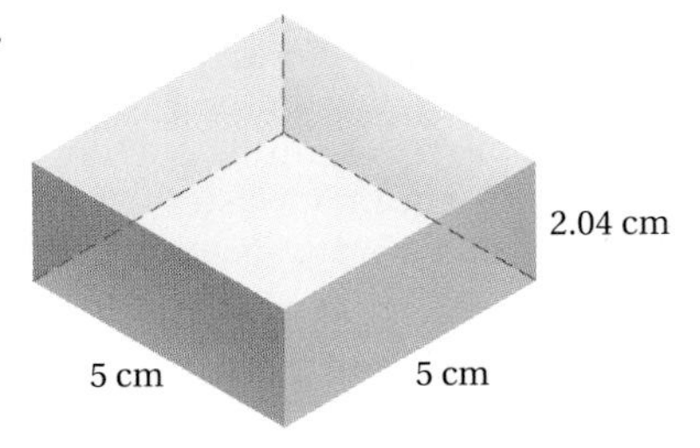

7.

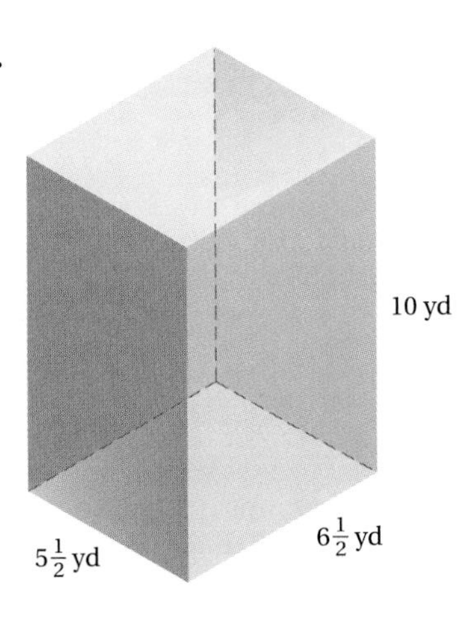

8.

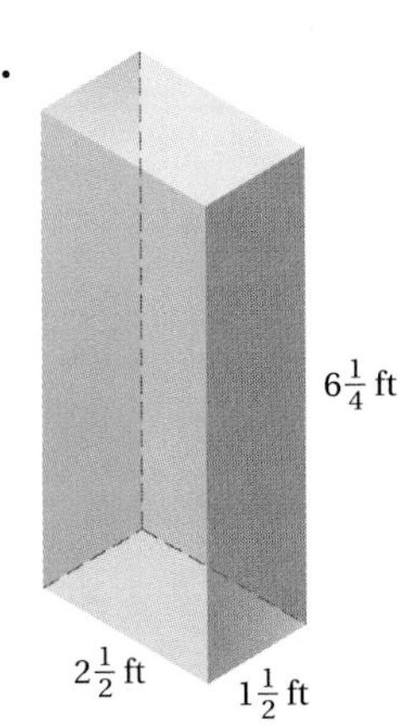

b Complete.

9. 1 L = _______ mL = _______ cm^3

10. _______ L = 1 mL = _______ cm^3

11. 87 L = _______ mL

12. 806 L = _______ mL

13. 49 mL = _______ L

14. 19 mL = _______ L

15. 0.401 mL = _______ L

16. 0.816 mL = _______ L

17. 78.1 L = _______ cm^3

18. 99.6 L = _______ cm^3

19. 10 qt = _______ oz

20. 9.6 oz = _______ pt

21. 20 cups = ________ pt

22. 1 gal = ________ oz

23. 8 gal = ________ qt

24. 1 gal = ________ cups

c Solve.

25. *Medical Dosage.* A physician ordered 0.5 L of normal saline solution. How many milliliters were ordered?

26. *Medical Dosage.* A patient receives 84 mL per hour of normal saline solution. How many liters did the patient receive in a 24-hr period?

27. *Medical Dosage.* A doctor wants a patient to receive 3 L of a normal saline solution in a 24-hr period. How many milliliters per hour must the nurse administer?

28. *Medical Dosage.* A doctor tells a patient to purchase 0.5 L of hydrogen peroxide. Commercially, hydrogen peroxide is found on the shelf in bottles that hold 4 oz, 8 oz, and 16 oz. Which bottle comes closest to filling the prescription? (1 qt = 32 oz)

29. *Wasting Water.* Many people leave the water running while they are brushing their teeth. Suppose that 32 oz of water is wasted in such a way each day by one person. How much water, in gallons, is wasted in a week? in a month (30 days)? in a year? Assuming each of the 261 million people in this country wastes water in this way, estimate how much water is wasted in a day; in a year.

30. *World's Gold.* If all the gold in the world could be gathered together, it would form a cube 18 yd on a side. Find the volume of the world's gold.

Skill Maintenance

31. Find the simple interest on $600 at 6.4% for $\frac{1}{2}$ yr. [6.7a]

32. Find the simple interest on $600 at 8% for 2 yr. [6.7a]

Evaluate. [1.9b]

33. 10^3

34. 15^2

35. 7^2

36. 4^3

Solve.

37. *Sales Tax.* In a certain state, a sales tax of $878 is collected on the purchase of a car for $17,560. What is the sales tax rate? [6.6a]

38. *Commission Rate.* Rich earns $1854.60 selling $16,860 worth of cellular phones. What is the commission rate? [6.6b]

Synthesis

39. ◆ What advantages does the use of metric units of capacity have over that of American units?

40. ◆ How would you go about estimating the volume of an egg? Discuss at least two methods.

41. 🖩 The width of a dollar bill is 2.3125 in., the length is 6.0625 in., and the thickness is 0.0041 in. Find the volume occupied by one million one-dollar bills.

42. 🖩 Audio cassette cases are typically 7 cm by 10.75 cm by 1.5 cm and contain 90 min of music. Compact-disc cases are typically 12.4 cm by 14.1 cm by 1 cm and contain 50 min of music. Which container holds the most music per cubic centimeter?

9.2 Volume of Cylinders, Spheres, and Cones

a Cylinders

A rectangular solid is shown below. Note that we can think of the volume as the product of the area of the base times the height:

$$V = l \cdot w \cdot h$$
$$= (l \cdot w) \cdot h$$
$$= (\text{Area of the base}) \cdot h$$
$$= B \cdot h,$$

Area of base $= B = l \cdot w$

where B represents the area of the base.

Like rectangular solids, **circular cylinders** have bases of equal area that lie in parallel planes. The bases of circular cylinders are circular regions.

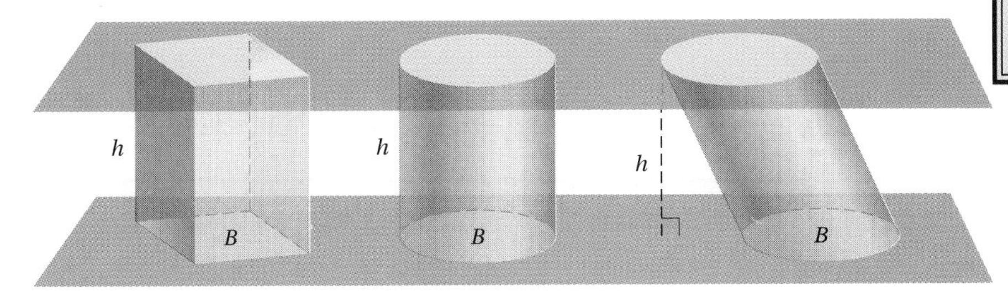

The volume of a circular cylinder is found in a manner similar to finding the volume of a rectangular solid. The volume is the product of the area of the base times the height. The height is always measured perpendicular to the base.

> The **volume of a circular cylinder** is the product of the area of the base B and the height h:
>
> $$V = B \cdot h, \quad \text{or} \quad V = \pi \cdot r^2 \cdot h.$$

Example 1 Find the volume of this circular cylinder. Use 3.14 for π.

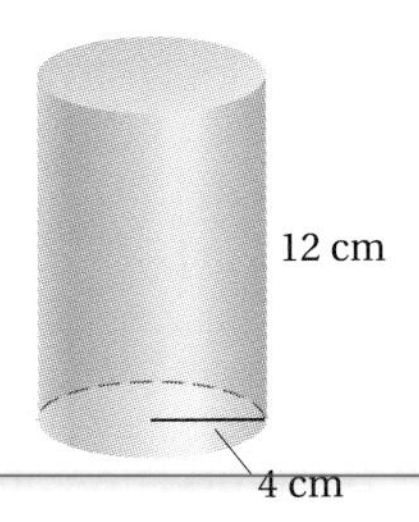

$$V = Bh = \pi \cdot r^2 \cdot h$$
$$\approx 3.14 \times 4 \text{ cm} \times 4 \text{ cm} \times 12 \text{ cm}$$
$$\approx 602.88 \text{ cm}^3$$

Do Exercises 1 and 2.

b Spheres

A **sphere** is the three-dimensional counterpart of a circle. It is the set of all points in space that are a given distance (the radius) from a given point (the center).

Objectives

- a Given the radius and the height, find the volume of a circular cylinder.
- b Given the radius, find the volume of a sphere.
- c Given the radius and the height, find the volume of a circular cone.
- d Solve applied problems involving volume of circular cylinders, spheres, and cones.

For Extra Help

TAPE 15 | TAPE 14B | MAC WIN | CD-ROM

1. Find the volume of the cylinder. Use 3.14 for π.

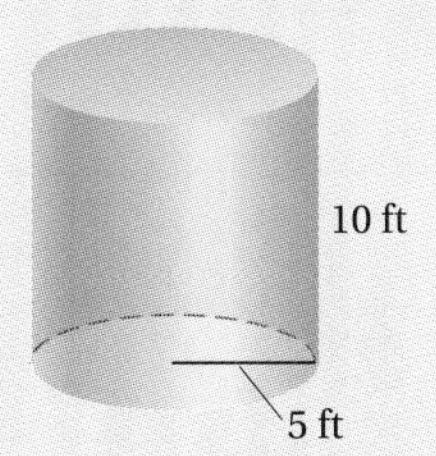

2. Find the volume of the cylinder. Use $\frac{22}{7}$ for π.

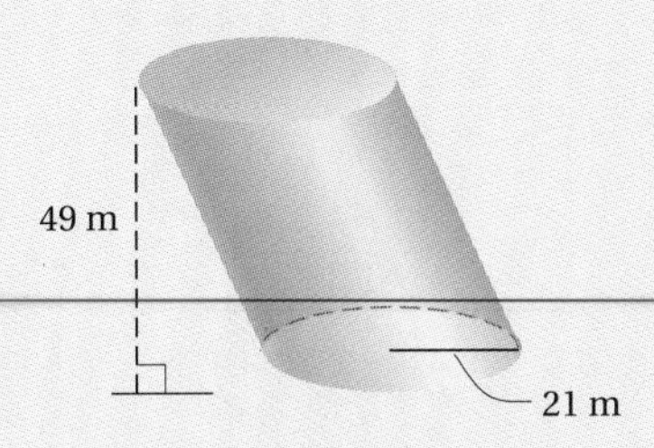

Answers on page A-21

3. Find the volume of the sphere. Use $\frac{22}{7}$ for π.

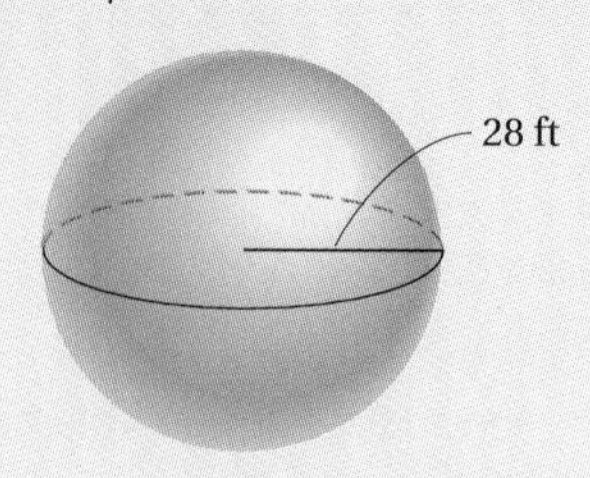

4. The radius of a standard-sized golf ball is 2.1 cm. Find its volume. Use 3.14 for π.

5. Find the volume of this cone. Use 3.14 for π.

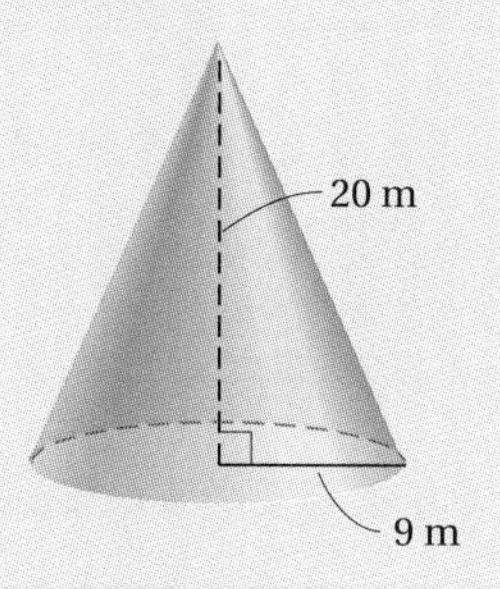

6. Find the volume of this cone. Use $\frac{22}{7}$ for π.

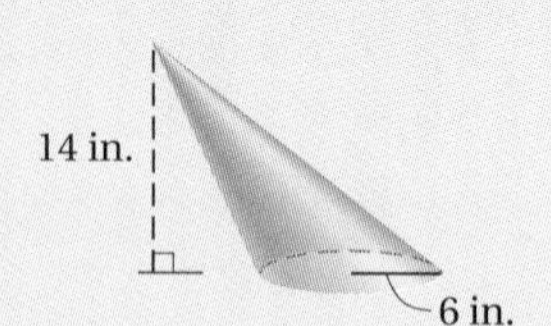

Answers on page A-21

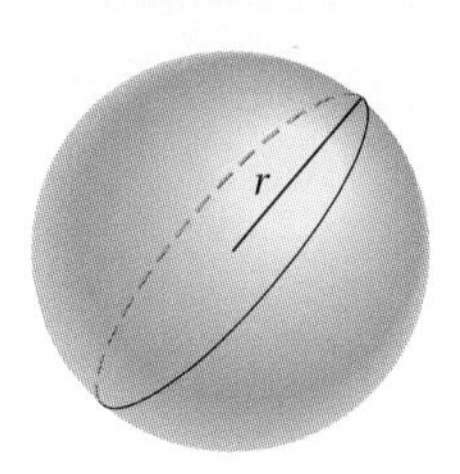

We find the volume of a sphere as follows.

> The **volume of a sphere** of radius r is given by
>
> $$V = \frac{4}{3} \cdot \pi \cdot r^3.$$

Example 2 *Bowling Ball.* The radius of a standard-sized bowling ball is 4.2915 in. Find the volume of a standard-sized bowling ball. Round to the nearest hundredth of a cubic inch. Use 3.14 for π.

$$V = \tfrac{4}{3} \cdot \pi \cdot r^3 \approx \tfrac{4}{3} \times 3.14 \times (4.2915 \text{ in.})^3$$

$$\approx \frac{4 \times 3.14 \times 79.0364 \text{ in}^3}{3} \approx 330.90 \text{ in}^3$$

Do Exercises 3 and 4.

c Cones

Consider a circle in a plane and choose any point P not in the plane. The circular region, together with the set of all segments connecting P to a point on the circle, is called a **circular cone.**

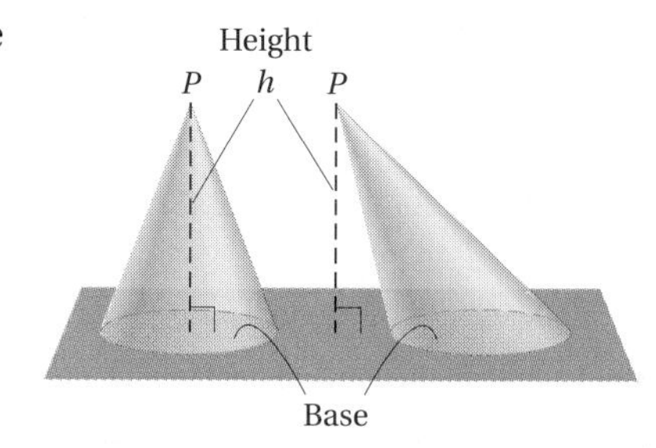

We find the volume of a cone as follows.

> The **volume of a circular cone** with base radius r is one-third the product of the base area and the height:
>
> $$V = \frac{1}{3} \cdot B \cdot h = \frac{1}{3} \pi \cdot r^2 \cdot h.$$

Example 3 Find the volume of this circular cone. Use 3.14 for π.

$$V = \tfrac{1}{3}\pi \cdot r^2 \cdot h$$

$$\approx \tfrac{1}{3} \times 3.14 \times 3 \text{ cm} \times 3 \text{ cm} \times 7 \text{ cm}$$

$$\approx 65.94 \text{ cm}^3$$

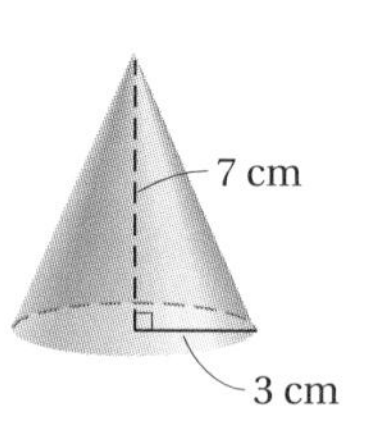

Do Exercises 5 and 6.

d Solving Applied Problems

Example 4 *Propane Gas Tank.* A propane gas tank is shaped like a circular cylinder with half of a sphere at each end. Find the volume of the tank if the cylindrical section is 5 ft long with a 4-ft diameter. Use 3.14 for π.

1. Familiarize. We first make a drawing.

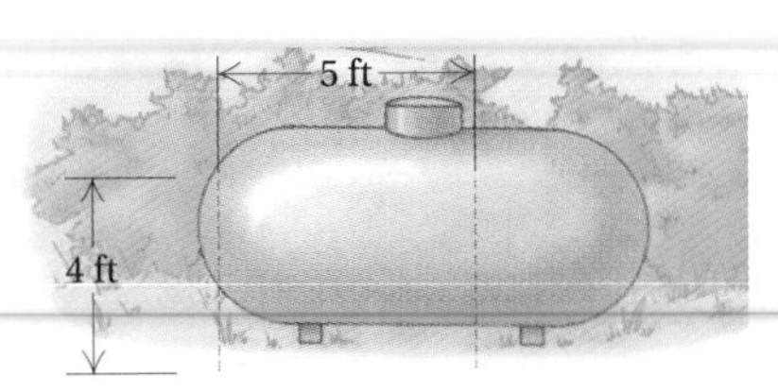

2. Translate. This is a two-step problem. We first find the volume of the cylindrical portion. Then we find the volume of the two ends and add. Note that the radius is 2 ft and that together the two ends make a sphere. We let

$$V = \pi \cdot r^2 \cdot h + \frac{4}{3} \cdot \pi \cdot r^3,$$

where V is the total volume. Then

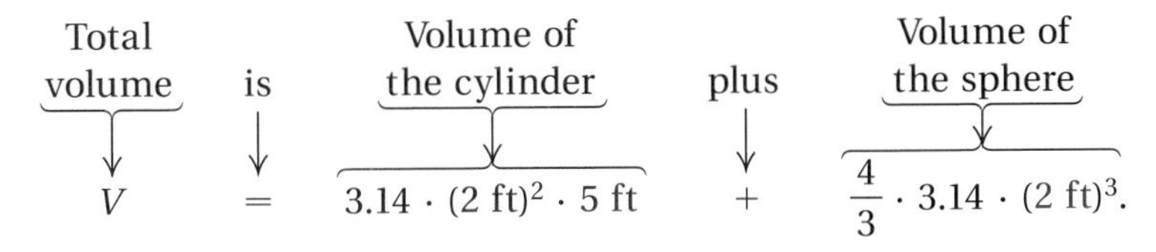

$$V = 3.14 \cdot (2\text{ ft})^2 \cdot 5\text{ ft} + \frac{4}{3} \cdot 3.14 \cdot (2\text{ ft})^3.$$

3. Solve. The volume of the cylinder is approximately

$$\begin{aligned} 3.14 \cdot (2\text{ ft})^2 \cdot 5\text{ ft} &\approx 3.14 \cdot 2\text{ ft} \cdot 2\text{ ft} \cdot 5\text{ ft} \\ &\approx 62.8\text{ ft}^3. \end{aligned}$$

The volume of the two ends is approximately

$$\begin{aligned} \frac{4}{3} \cdot 3.14 \cdot (2\text{ ft})^3 &\approx \frac{4}{3} \cdot 3.14 \cdot 2\text{ ft} \cdot 2\text{ ft} \cdot 2\text{ ft} \\ &\approx 33.5\text{ ft}^3. \end{aligned}$$

The total volume is

$$62.8\text{ ft}^3 + 33.5\text{ ft}^3 = 96.3\text{ ft}^3.$$

4. Check. The check is left to the student.

5. State. The volume of the tank is about 96.3 ft^3.

Do Exercise 7.

7. *Medicine Capsule.* A cold capsule is 8 mm long and 4 mm in diameter. Find the volume of the capsule. Use 3.14 for π. (*Hint*: First find the length of the cylindrical section.)

Calculator Spotlight

Exercises

1. *Measuring the volume of a cloud.* Using a calculator with a $\boxed{\pi}$ key, find the volume of a spherical cloud with a 1000-m diameter.

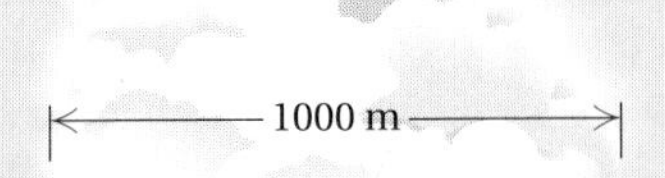

2. The box shown is just big enough to hold 3 golf balls. If the radius of a golf ball is 2.1 cm, how much air surrounds the three balls?

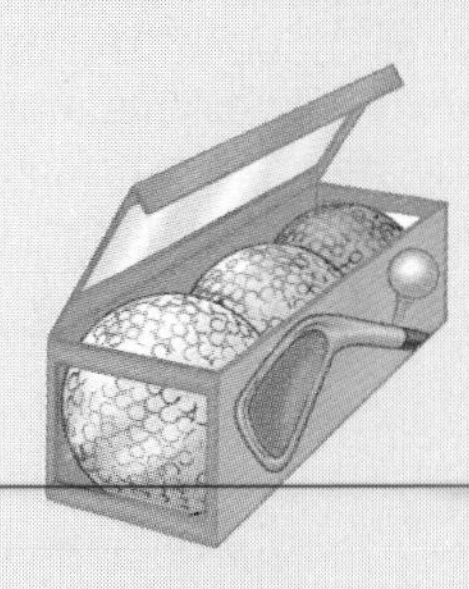

Answer on page A-21

Improving Your Math Study Skills

Classwork: During and After Class

During Class

Asking Questions

Many students are afraid to ask questions in class. You will find that most instructors are not only willing to answer questions during class, but often encourage students to ask questions. In fact, some instructors would like more questions than are offered. Probably your question is one that other students in the class might have been afraid to ask!

Consider waiting for an appropriate time to ask questions. Some instructors will pause to ask the class if they have questions. Use this opportunity to get clarification on any concept you do not understand.

After Class

Restudy Examples and Class Notes

As soon as possible after class, find some time to go over your notes. Read the appropriate sections from the textbook and try to correlate the text with your class notes. You may also want to restudy the examples in the textbook for added comprehension.

Often students make the mistake of doing the homework exercises without reading their notes or textbook. This is not a good idea, since you may lose the opportunity for a complete understanding of the concepts. Simply being able to work the exercises does not ensure that you know the material well enough to work problems on a test.

Videotapes

If you can find the time, visit the library, math lab, or media center to view the videotapes on the textbook. Look on the first page of each section in the textbook for the appropriate tape reference.

The videotapes provide detailed explanations of each objective and they may give you a different presentation than the one offered by your instructor. Being able to pause the tape while you take notes or work the examples or replay the tape as many times as you need are additional advantages to using the videos.

Also, consider studying the special tapes *Math Problem Solving in the Real World* and *Math Study Skills* prepared by the author. If these are not available in the media center, contact your instructor.

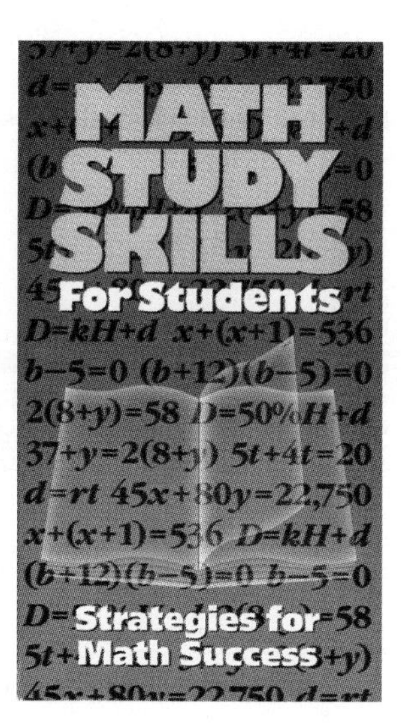

Software

If you would like additional practice on any section of the textbook, you can use the accompanying InterAct Math Tutorial Software. This software can generate many different versions of the basic odd-numbered exercises for added practice. You can also ask the software to work out each problem step by step.

Ask your instructor about the availability of this software.

Exercise Set 9.2

a Find the volume of the circular cylinder. Use 3.14 for π in Exercises 1–4. Use $\frac{22}{7}$ for π in Exercises 5 and 6.

1.

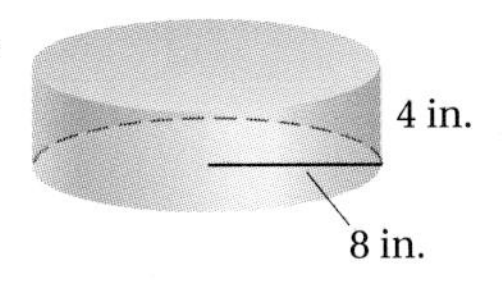

2.

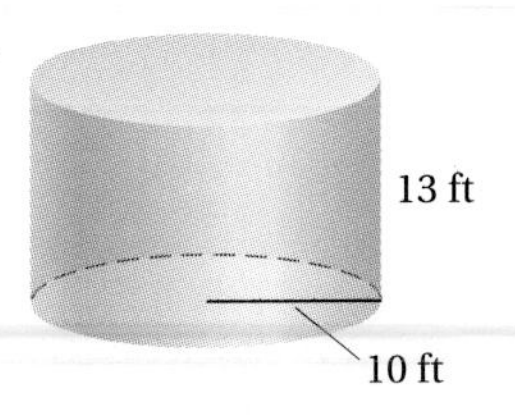

3.

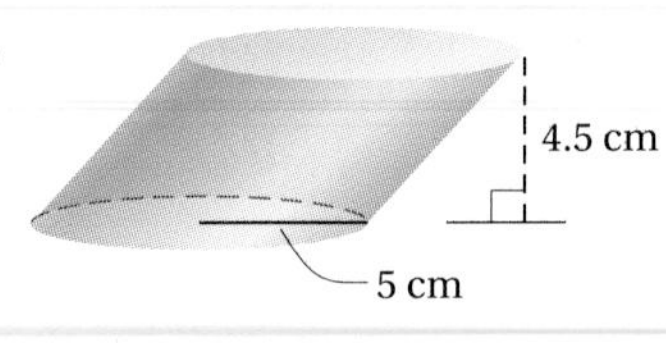

4.

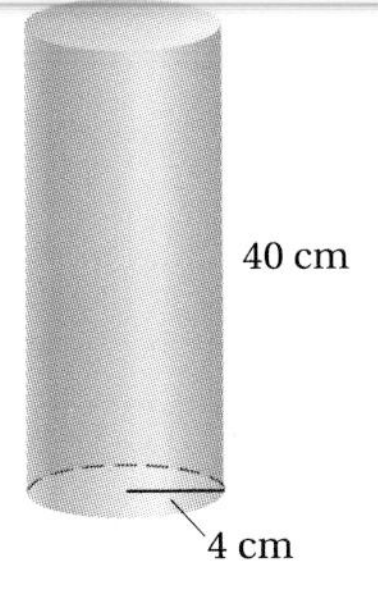

5.

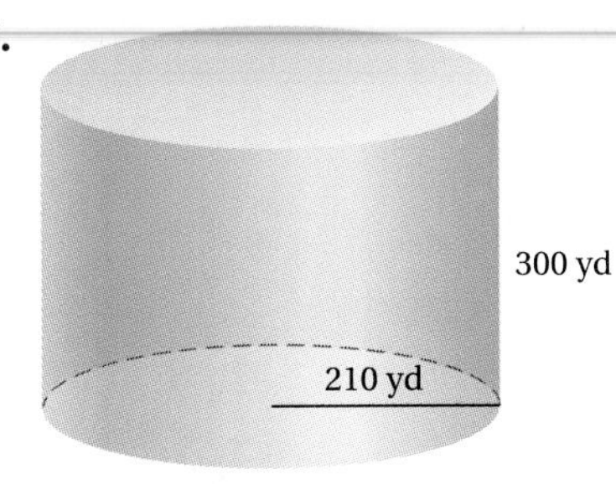

6.

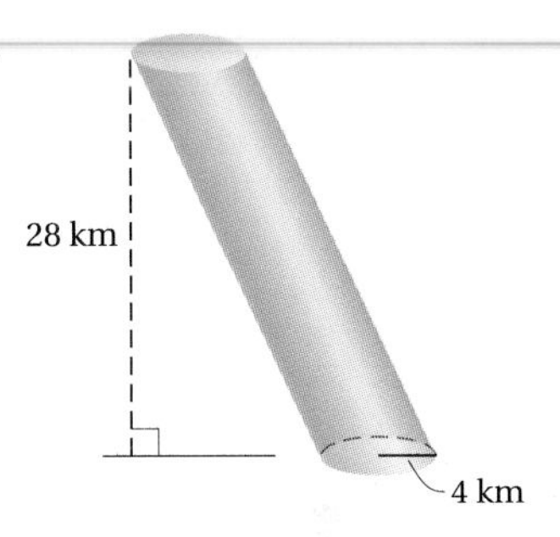

b Find the volume of the sphere. Use 3.14 for π in Exercises 7–10. Use $\frac{22}{7}$ for π in Exercises 11 and 12.

7.

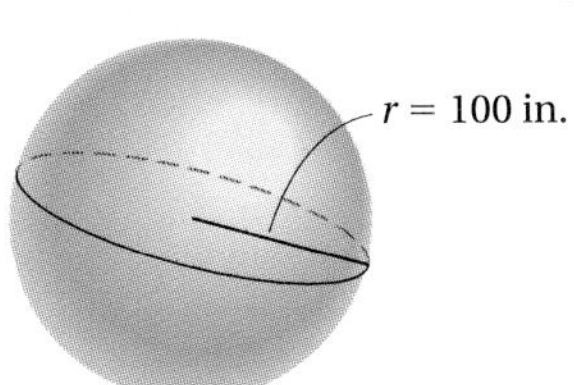

8.

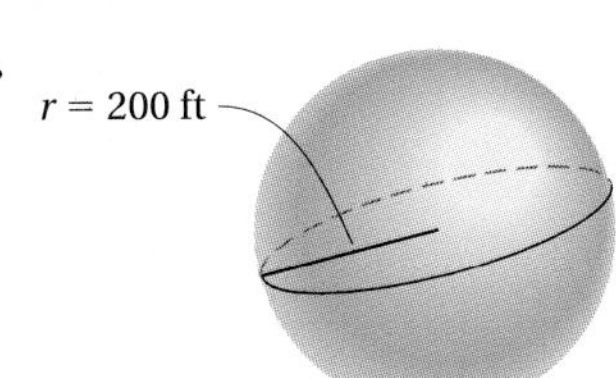

9.

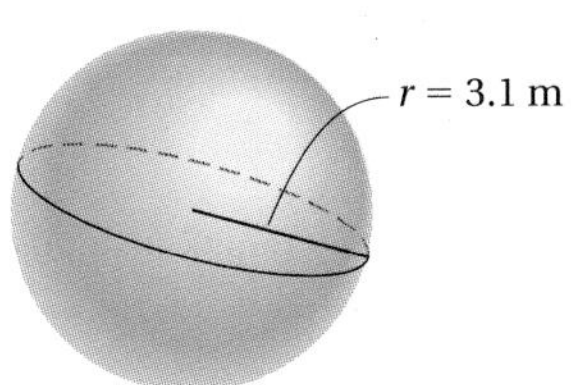

10.

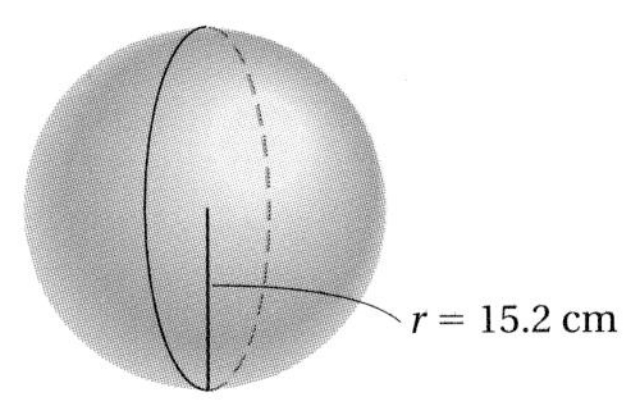

11.

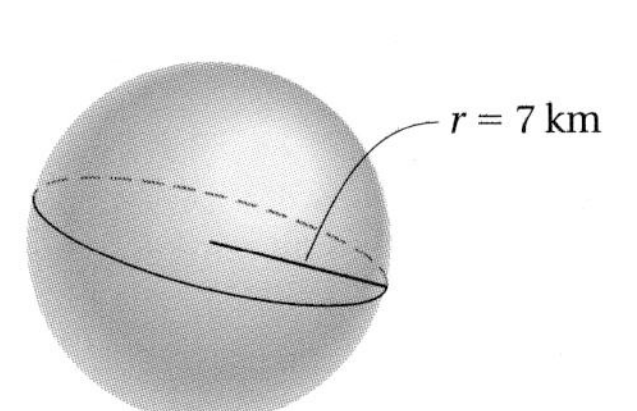

12.

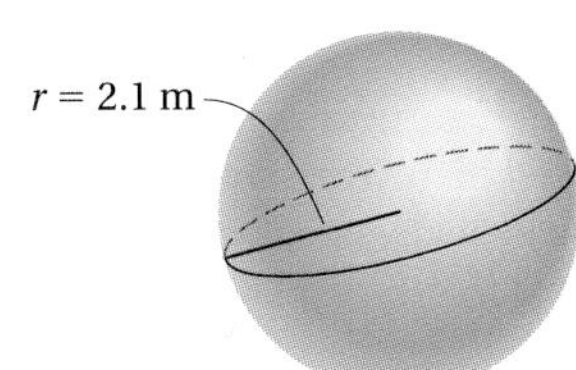

c Find the volume of the circular cone. Use 3.14 for π in Exercises 13 and 14. Use $\frac{22}{7}$ for π in Exercises 15 and 16.

13.

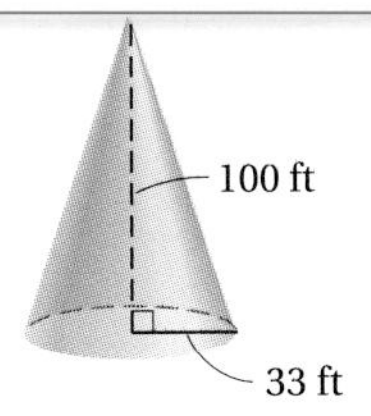

14.

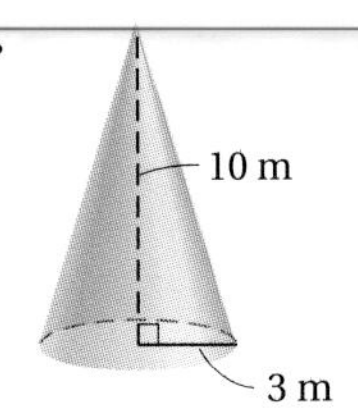

15.

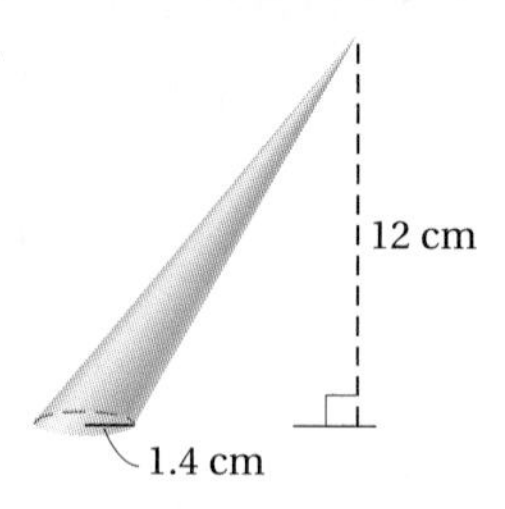

16.

30 mm

35 mm

d Solve.

17. The diameter of the base of a circular cylinder is 14 yd. The height is 220 yd. Find the volume. Use $\frac{22}{7}$ for π.

18. A rung of a ladder is 2 in. in diameter and 16 in. long. Find the volume. Use 3.14 for π.

19. *Barn Silo.* A barn silo, excluding the top, is a circular cylinder. The silo is 6 m in diameter and the height is 13 m. Find the volume. Use 3.14 for π.

20. A log of wood has a diameter of 12 cm and a height of 42 cm. Find the volume. Use 3.14 for π.

21. *Tennis Ball.* The diameter of a tennis ball is 6.5 cm. Find the volume. Use 3.14 for π.

22. *Spherical Gas Tank.* The diameter of a spherical gas tank is 6 m. Find the volume. Use 3.14 for π.

23. *Volume of Earth.* The diameter of the earth is about 3980 mi. Find the volume of the earth. Use 3.14 for π. Round to the nearest ten thousand cubic miles.

24. The volume of a ball is 36π cm^3. Find the dimensions of a rectangular box that is just large enough to hold the ball.

Skill Maintenance

Convert. [8.1a]

25. 11 yd = ________ in.

26. 15,840 ft = ________ mi

27. 42 ft = ________ yd

28. 48 mi = ________ ft

29. 144 in. = ________ ft

30. 5.3 mi = ________ in.

Synthesis

31. ◆ The design of a modern home includes a cylindrical tower that will be capped with either a 10-ft–high dome or a 10-ft–high cone. Which type of cap will be more energy-efficient and why?

32. 🖩 A 2-cm–wide stream of water passes through a 30-m garden hose. At the instant that the water is turned off, how many liters of water are in the hose?

33. 🖩 A hot water tank is a right circular cylinder with a base of diameter 16 in. and height 5 ft. Find the volume of the tank in cubic feet. One cubic foot of water is about 7.5 gal. About how many gallons will the tank hold?

34. 🖩 Find the volume.

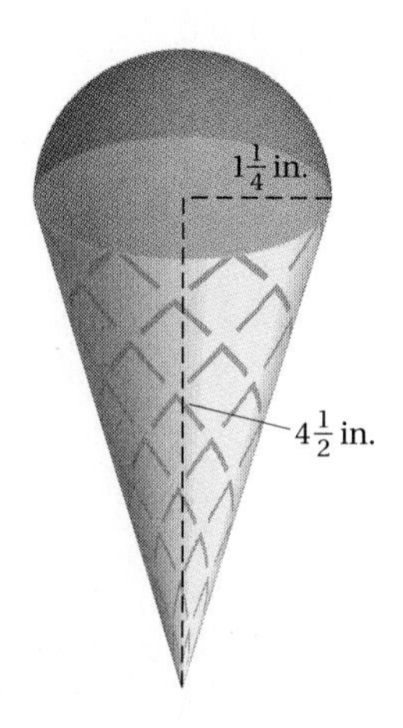

9.3 Weight, Mass, and Time

a Weight: The American System

The American units of weight are as follows.

AMERICAN UNITS OF WEIGHT

1 ton (T) = 2000 pounds (lb)

1 lb = 16 ounces (oz)

The term "ounce" used here for weight is different from the "ounce" we used for capacity in Section 9.1.

Objectives

a Convert from one American unit of weight to another.

b Convert from one metric unit of mass to another.

c Convert from one unit of time to another.

For Extra Help

TAPE 15 TAPE 15A MAC WIN CD-ROM

Example 1 A well-known hamburger is called a "quarter-pounder." Find its name in ounces: a "______ ouncer."

$$\frac{1}{4}\text{ lb} = \frac{1}{4}\cdot 1\text{ lb}$$

$$= \frac{1}{4}\cdot 16\text{ oz} \qquad \textbf{Substituting 16 oz for 1 lb}$$

$$= 4\text{ oz}$$

A "quarter-pounder" can also be called a "four-ouncer."

Example 2 Complete: 15,360 lb = ______ T.

$$15{,}360\text{ lb} = 15{,}360\text{ lb} \times \frac{1\text{ T}}{2000\text{ lb}} \qquad \textbf{Multiplying by 1}$$

$$= \frac{15{,}360}{2000}\text{ T}$$

$$= 7.68\text{ T}$$

Do Exercises 1–3.

Complete.

1. 5 lb = ______ oz

2. 8640 lb = ______ T

3. 1 T = ______ oz

Answers on page A-21

b Mass: The Metric System

There is a difference between **mass** and **weight**, but the terms are often used interchangeably. People sometimes use the word "weight" instead of "mass." Weight is related to the force of gravity. The farther you are from the center of the earth, the less you weigh. Your mass stays the same no matter where you are.

The basic unit of mass is the **gram** (g), which is the mass of 1 cubic centimeter (1 cm^3 or 1 mL) of water. Since a cubic centimeter is small, a gram is a small unit of mass.

1 g = 1 gram = the mass of 1 cm^3 (1 mL) of water

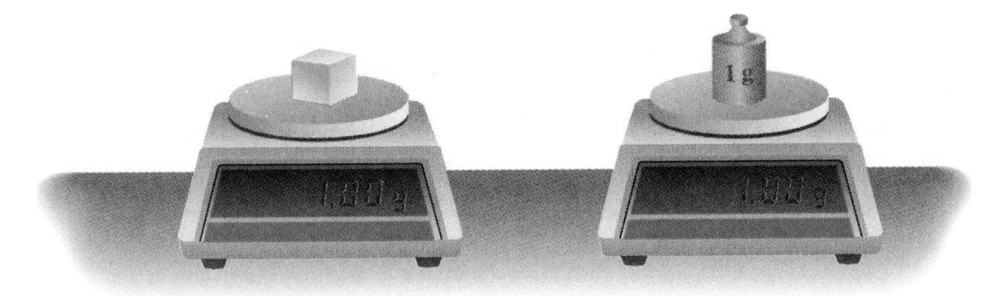

Complete with mg, g, kg, or t.

4. A laptop computer has a mass of 6 ________.

5. That person has a body mass of 85.4 ________.

6. This is a 3-________ vitamin.

7. A pen has a mass of 12 ________.

8. A minivan has a mass of 3 ________.

Answers on page A-21

The following table shows the metric units of mass. The prefixes are the same as those for length.

METRIC UNITS OF MASS

1 metric ton (t) = 1000 kilograms (kg)

1 *kilo*gram (kg) = 1000 grams (g)

1 *hecto*gram (hg) = 100 grams (g)

1 *deka*gram (dag) = 10 grams (g)

1 gram (g)

1 *deci*gram (dg) = $\frac{1}{10}$ gram (g)

1 *centi*gram (cg) = $\frac{1}{100}$ gram (g)

1 *milli*gram (mg) = $\frac{1}{1000}$ gram (g)

Thinking Metric

One gram is about the mass of 1 raisin or 1 paperclip. Since 1 kg is about 2.2 lb, 1000 kg is about 2200 lb, or 1 metric ton (t), which is just a little more than 1 American ton (T).

1 gram

1 kilogram

1 pound

Small masses, such as dosages of medicine and vitamins, may be measured in milligrams (mg). The gram (g) is used for objects ordinarily measured in ounces, such as the mass of a letter, a piece of candy, a coin, or a small package of food.

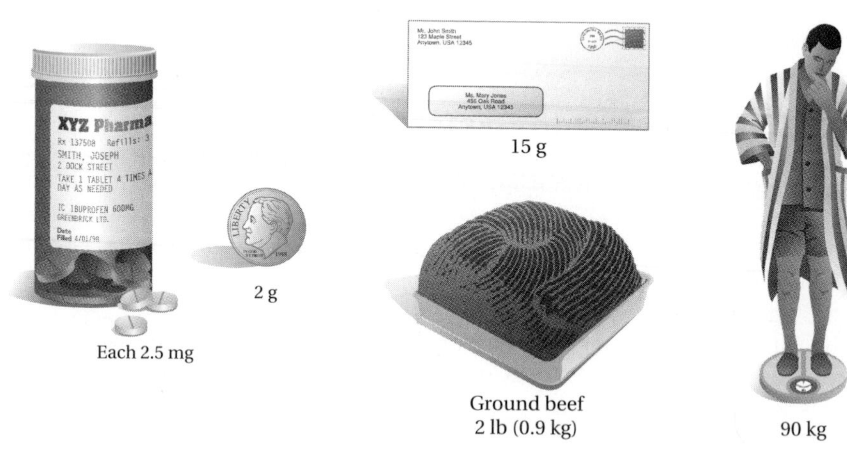

Each 2.5 mg

2 g

15 g

Ground beef
2 lb (0.9 kg)

90 kg

The kilogram (kg) is used for larger food packages, such as meat, or for human body mass. The metric ton (t) is used for very large masses, such as the mass of an automobile, a truckload of gravel, or an airplane.

Do Exercises 4–8.

Changing Units Mentally

As before, changing from one metric unit to another amounts to only the movement of a decimal point. We use this table.

1000 g	100 g	10 g	1 g	0.1 g	0.01 g	0.001 g
1 kg	1 hg	1 dag	1 g	1 dg	1 cg	1 mg

Example 3 Complete: 8 kg = ________ g.

Think: To go from kg to g in the table is a move of three places to the right. Thus we move the decimal point three places to the right.

1000 g	100 g	10 g	1 g	0.1 g	0.01 g	0.001 g
1 kg	1 hg	1 dag	1 g	1 dg	1 cg	1 mg

3 places to the right

8.0   8.000.   8 kg = 8000 g

Example 4 Complete: 4235 g = ________ kg.

Think: To go from g to kg in the table is a move of three places to the left. Thus we move the decimal point three places to the left.

1000 g	100 g	10 g	1 g	0.1 g	0.01 g	0.001 g
1 kg	1 hg	1 dag	1 g	1 dg	1 cg	1 mg

3 places to the left

4235.0   4.235.0   4235 g = 4.235 kg

Do Exercises 9 and 10.

Example 5 Complete: 6.98 cg = ________ mg.

Think: To go from cg to mg is a move of one place to the right. Thus we move the decimal point one place to the right.

1000 g	100 g	10 g	1 g	0.1 g	0.01 g	0.001 g
1 kg	1 hg	1 dag	1 g	1 dg	1 cg	1 mg

1 place to the right

6.98   6.9.8   6.98 cg = 69.8 mg

The most commonly used metric units of mass are kg, g, cg, and mg. We have purposely used those more often than the others in the exercises.

Complete.

9. 6.2 kg = ________ g

10. 304.8 cg = ________ g

Answers on page A-21

Complete.

11. 7.7 cg = _______ mg

12. 2344 mg = _______ cg

13. 67 dg = _______ mg

Complete.

14. 2 hr = _______ min

15. 4 yr = _______ days

16. 1 day = _______ min

17. 168 hr = _______ wk

Answers on page A-21

Example 6 Complete: 89.21 mg = _______ g.

Think: To go from mg to g is a move of three places to the left. Thus we move the decimal point three places to the left.

1000 g	100 g	10 g	1 g	0.1 g	0.01 g	0.001 g
1 kg	1 hg	1 dag	1 g	1 dg	1 cg	1 mg

3 places to the left

89.21   0.089.21   89.21 mg = 0.08921 g

Do Exercises 11–13.

c Time

A table of units of time is shown below. The metric system sometimes uses "h" for hour and "s" for second, but we will use the more familiar "hr" and "sec."

UNITS OF TIME

1 day = 24 hours (hr)
1 hr = 60 minutes (min)
1 min = 60 seconds (sec)

1 year (yr) = $365\frac{1}{4}$ days
1 week (wk) = 7 days

Since we cannot have $\frac{1}{4}$ day on the calendar, we give each year 365 days and every fourth year 366 days (a leap year), unless it is a year at the beginning of a century not divisible by 400.

Example 7 Complete: 1 hr = _______ sec.

$$\begin{aligned} 1 \text{ hr} &= 60 \text{ min} \\ &= 60 \cdot 1 \text{ min} \\ &= 60 \cdot 60 \text{ sec} \quad \text{Substituting 60 sec for 1 min} \\ &= 3600 \text{ sec} \end{aligned}$$

Example 8 Complete: 5 yr = _______ days.

$$\begin{aligned} 5 \text{ yr} &= 5 \cdot 1 \text{ yr} \\ &= 5 \cdot 365\frac{1}{4} \text{ days} \quad \text{Substituting } 365\tfrac{1}{4} \text{ days for 1 yr} \\ &= 1826\frac{1}{4} \text{ days} \end{aligned}$$

Example 9 Complete: 4320 min = _______ days.

$$4320 \text{ min} = 4320 \cancel{\text{min}} \cdot \frac{1 \cancel{\text{hr}}}{60 \cancel{\text{min}}} \cdot \frac{1 \text{ day}}{24 \cancel{\text{hr}}} = \frac{4320}{60 \cdot 24} \text{ days} = 3 \text{ days}$$

Do Exercises 14–17.

Exercise Set 9.3

a Complete.

1. 1 T = ________ lb

2. 1 lb = ________ oz

3. 6000 lb = ________ T

4. 8 T = ________ lb

5. 4 lb = ________ oz

6. 10 lb = ________ oz

7. 6.32 T = ________ lb

8. 8.07 T = ________ lb

9. 3200 oz = ________ T

10. 6400 oz = ________ T

11. 80 oz = ________ lb

12. 960 oz = ________ lb

b Complete.

13. 1 kg = ________ g

14. 1 hg = ________ g

15. 1 dag = ________ g

16. 1 dg = ________ g

17. 1 cg = ________ g

18. 1 mg = ________ g

19. 1 g = ________ mg

20. 1 g = ________ cg

21. 1 g = ________ dg

22. 25 kg = ________ g

23. 234 kg = ________ g

24. 9403 g = ________ kg

25. 5200 g = ________ kg

26. 1.506 kg = ________ g

27. 67 hg = ________ kg

28. 45 cg = ________ g

29. 0.502 dg = ________ g

30. 0.0025 cg = ________ mg

31. 8492 g = ________ kg

32. 9466 g = ________ kg

33. 585 mg = ________ cg

34. 96.1 mg = _______ cg

35. 8 kg = _______ cg

36. 0.06 kg = _______ mg

37. 1 t = _______ kg

38. 2 t = _______ kg

39. 3.4 cg = _______ dag

40. 115 mg = _______ g

c Complete.

41. 1 day = _______ hr

42. 1 hr = _______ min

43. 1 min = _______ sec

44. 1 wk = _______ days

45. 1 yr = _______ days

46. 2 yr = _______ days

47. 180 sec = _______ hr

48. 60 sec = _______ hr

49. 492 sec = _______ min
(the amount of time it takes for the rays of the sun to reach the earth)

50. 18,000 sec = _______ hr

51. 156 hr = _______ days

52. 444 hr = _______ days

53. 645 min = _______ hr

54. 375 min = _______ hr

55. 2 wk = _______ hr

56. 4 hr = _______ sec

57. 756 hr = _______ wk

58. 166,320 min = _______ wk

59. 2922 wk = _______ yr

60. 623 days = _______ wk

Skill Maintenance

Evaluate. [1.9b]

61. 2^4

62. 17^2

63. 5^3

64. 8^2

Complete. [8.2a]

65. 5.43 m = ________ cm

66. 5.43 m = ________ km

Solve. [8.7c]

67. How long is a string of lights reaching from the top of a 24-ft pole to a point on the ground 16 ft from the base of the pole?

68. How long must a wire be in order to reach from the top of a 26-m telephone pole to a point on the ground 18 m from the base of the pole?

69. Find the rate of discount. [6.6c]

70. *Sound Levels.* Make a horizontal bar graph to show the loudness of various sounds, as listed below. A decibel is a measure of the loudness of sounds.

Sound	Loudness (in decibels)
Whisper	15
Tick of watch	30
Speaking aloud	60
Noisy factory	90
Moving car	80
Car horn	98
Subway	104

71. If 2 cans of tomato paste cost $1.49, how many cans of tomato paste can you buy for $7.45? [5.4a]

72. The weight of a human brain is 2.5% of total body weight. A person weighs 200 lb. What does the brain weigh? [6.5a]

Convert to percent notation. [6.1c]

73. 0.0043

74. 2.31

Synthesis

75. ◈ Give at least two reasons why someone might prefer the use of grams to the use of ounces.

76. ◈ Describe a situation in which one object weighs 70 kg, another object weighs 3 g, and a third object weighs 125 mg.

Complete. Use 1 kg = 2.205 lb and 453.5 g = 1 lb. Round to four decimal places.

77. 🖩 1 lb = ________ kg

78. 🖩 1 g = ________ lb

79. At \$0.90 a dozen, the cost of eggs is \$0.60 per pound. How much does an egg weigh?

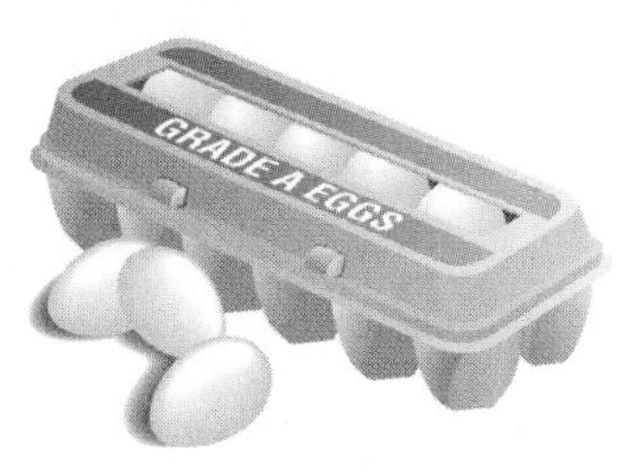

80. Estimate the number of years in one million seconds.

81. Estimate the number of years in one billion seconds.

82. Estimate the number of years in one trillion seconds.

Medical Applications. Another metric unit used in medicine is the microgram (μg). It is defined as follows.

$$1 \text{ microgram} = 1\ \mu\text{g} = \frac{1}{1{,}000{,}000}\ \text{g}; \qquad 1{,}000{,}000\ \mu\text{g} = 1\ \text{g}$$

Thus a microgram is one millionth of a gram, and one million micrograms is one gram.

Complete.

83. 1 mg = ______ μg

84. 1 μg = ______ mg

85. A physician orders 125 μg of digoxin. For how many milligrams is the prescription?

86. A physician orders 0.25 mg of reserpine. For how many micrograms is the prescription?

87. A medicine called sulfisoxazole usually comes in tablets that are 500 mg each. A standard dosage is 2 g. How many tablets would have to be taken in order to achieve this dosage?

88. Quinidine is a liquid mixture, part medicine and part water. There is 80 mg of Quinidine for every milliliter of liquid. A standard dosage is 200 mg. How much of the liquid mixture would be required in order to achieve the dosage?

89. A medicine called cephalexin is obtainable in a liquid mixture, part medicine and part water. There is 250 mg of cephalexin in 5 mL of liquid. A standard dosage is 400 mg. How much of the liquid would be required in order to achieve the dosage?

90. A medicine called Albuterol is used for the treatment of asthma. It typically comes in an inhaler that contains 18 g. One actuation, or inhalation, is 90 mg.

a) How many actuations are in one inhaler?
b) A student is going away for 4 months of college and wants to take enough Albuterol to last for that time. Assuming that she will need 4 actuations per day, estimate about how many inhalers the student will need for the 4-month period.

9.4 Temperature

a Estimated Conversions

Below are two temperature scales: **Fahrenheit** for American measure and **Celsius** for metric measure.

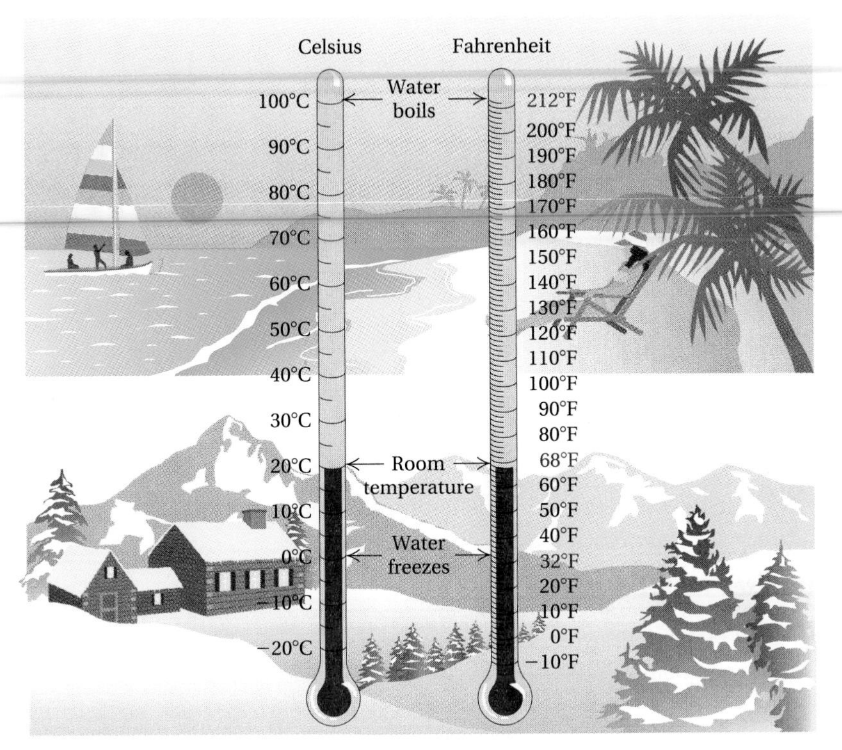

By laying a ruler or a piece of paper horizontally between the scales, we can make an approximate conversion from one measure of temperature to another.

Examples Convert to Celsius (using the scales shown above). Approximate to the nearest ten degrees.

1. 212°F (Boiling point of water)	100°C	This is exact.
2. 32°F (Freezing point of water)	0°C	This is exact.
3. 105°F	40°C	This is approximate.

Do Exercises 1–3.

Examples Make an approximate conversion to Fahrenheit.

4. 44°C (Hot bath)	110°F	This is approximate.
5. 20°C (Room temperature)	68°F	This is exact.
6. 83°C	180°F	This is approximate.

Do Exercises 4–6.

Objectives

a. Make an approximate conversion from Celsius temperature to Fahrenheit, and from Fahrenheit temperature to Celsius.

b. Convert from Celsius temperature to Fahrenheit and from Fahrenheit temperature to Celsius using the formulas $F = \frac{9}{5} \cdot C + 32$ and $C = \frac{5}{9} \cdot (F - 32)$.

For Extra Help

TAPE 15

TAPE 15A

MAC WIN

CD-ROM

Convert to Celsius. Approximate to the nearest ten degrees.

1. 180°F (Brewing coffee)

2. 25°F (Cold day)

3. −10°F (Miserably cold day)

Convert to Fahrenheit. Approximate to the nearest ten degrees.

4. 25°C (Warm day at the beach)

5. 40°C (Temperature of a patient with a high fever)

6. 10°C (A cold bath)

Answers on page A-21

Convert to Fahrenheit.

7. 80°C

8. 35°C

b Exact Conversions

The following formula allows us to make exact conversions from Celsius to Fahrenheit.

$$F = \frac{9}{5} \cdot C + 32, \quad \text{or} \quad F = 1.8 \cdot C + 32$$

(Multiply the Celsius temperature by $\frac{9}{5}$, or 1.8, and add 32.)

Examples Convert to Fahrenheit.

7. 0°C $\quad F = \frac{9}{5} \cdot C + 32$

$$= \frac{9}{5} \cdot 0 + 32 = 0 + 32 = 32°$$

Thus, 0°C = 32°F.

8. 37°C $\quad F = 1.8 \cdot C + 32$

$$= 1.8 \cdot 37 + 32 = 66.6 + 32 = 98.6°$$

Thus, 37°C = 98.6°F. This is normal body temperature.

Do Exercises 7 and 8.

The following formula allows us to make exact conversions from Fahrenheit to Celsius.

$$C = \frac{5}{9} \cdot (F - 32)$$

(Subtract 32 from the Fahrenheit temperature and multiply by $\frac{5}{9}$.)

Convert to Celsius.

9. 95°F

10. 113°F

Examples Convert to Celsius.

9. 212°F $\quad C = \frac{5}{9} \cdot (F - 32)$

$$= \frac{5}{9} \cdot (212 - 32) = \frac{5}{9} \cdot 180 = 100°$$

Thus, 212°F = 100°C.

10. 77°F $\quad C = \frac{5}{9} \cdot (F - 32)$

$$= \frac{5}{9} \cdot (77 - 32) = \frac{5}{9} \cdot 45 = 25°$$

Thus, 77°F = 25°C.

Do Exercises 9 and 10.

Answers on page A-21

Exercise Set 9.4

a Convert to Celsius. Approximate to the nearest ten degrees. Use the scales on p. 493.

1. 178°F

2. 195°F

3. 140°F

4. 107°F

5. 68°F

6. 50°F

7. 10°F

8. 120°F

Convert to Fahrenheit. Approximate to the nearest ten degrees. Use the scales on p. 493.

9. 86°C

10. 93°C

11. 58°C

12. 35°C

13. −10°C

14. −5°C

15. 5°C

16. 15°C

b Convert to Fahrenheit. Use the formula $F = \frac{9}{5} \cdot C + 32$.

17. 25°C

18. 85°C

19. 40°C

20. 90°C

21. 3000°C (melting point of iron)

22. 1000°C (melting point of gold)

Convert to Celsius. Use the formula $C = \frac{5}{9} \cdot (F - 32)$.

23. 86°F

24. 59°F

25. 131°F

26. 140°F

27. 98.6°F (normal body temperature)

28. 104°F (high-fevered body temperature)

Skill Maintenance

Complete. [8.1a], [8.2a]

29. 23.4 cm = ________ mm

30. 0.23 km = ________ m

31. 28 ft = ________ in.

32. 72 ft = ________ yd

33. 72.4 cm = ________ m

34. 72.4 m = ________ km

35. 70 yd = ________ in.

36. 31,680 ft = ________ mi

37. 84 ft = ________ yd

38. $7\frac{1}{2}$ mi = ________ ft

39. 144 in. = ________ ft

40. 0.73 mi = ________ in.

Synthesis

41. ◆ Write a report on the origin of the words *Fahrenheit* and *Celsius*. How is the word *centigrade* related to temperature?

42. Another temperature scale often used is the **Kelvin** scale. Conversions from Celsius to Kelvin can be carried out using the formula

$$K = C + 273.$$

A chemistry textbook describes an experiment in which a reaction takes place at a temperature of 400° Kelvin. A student wishes to perform the experiment, but has only a Fahrenheit thermometer. At what Fahrenheit temperature will the reaction take place?

Convert between American units and metric units of temperature, length, volume, and weight.

9.5 Converting Units of Area

a American Units

Let's do some conversions from one American unit of area to another.

Example 1 Complete: $1\ ft^2 =$ _______ in^2.

$1\ ft^2 = 1 \cdot (12\ in.)^2$ **Substituting 12 in. for 1 ft**

$= 12\ in. \cdot 12\ in. = 144\ in^2$

Example 2 Complete: $8\ yd^2 =$ _______ ft^2.

$8\ yd^2 = 8 \cdot (3\ ft)^2$ **Substituting 3 ft for 1 yd**

$= 8 \cdot 3\ ft \cdot 3\ ft = 8 \cdot 3 \cdot 3 \cdot ft \cdot ft = 72\ ft^2$

Do Exercises 1–3.

American units are related as follows.

1 square yard (yd^2) = 9 square feet (ft^2)
1 square foot (ft^2) = 144 square inches (in^2)
1 square mile (mi^2) = 640 acres
1 acre = 43,560 ft^2

Example 3 Complete: $36\ ft^2 =$ _______ yd^2.

We are converting from "ft^2" to "yd^2". Thus we choose a symbol for 1 with yd^2 on top and ft^2 on the bottom.

$$36\ ft^2 = 36\ ft^2 \times \frac{1\ yd^2}{9\ ft^2} \quad \textbf{Multiplying by 1 using } \frac{1\ yd^2}{9\ ft^2}$$

$$= \frac{36}{9} \times \frac{ft^2}{ft^2} \times 1\ yd^2 = 4\ yd^2$$

Example 4 Complete: $7\ mi^2 =$ _______ acres.

$7\ mi^2 = 7 \cdot 1\ mi^2$

$= 7 \cdot 640$ acres **Substituting 640 acres for 1 mi^2**

$= 4480$ acres

Do Exercises 4 and 5.

b Metric Units

Let's now convert from one metric unit of area to another.

Example 5 Complete: $1\ km^2 =$ _______ m^2.

$1\ km^2 = 1 \cdot (1000\ m)^2$ **Substituting 1000 m for 1 km**

$= 1000\ m \cdot 1000\ m$

$= 1{,}000{,}000\ m^2$

Objectives

a Convert from one American unit of area to another.

b Convert from one metric unit of area to another.

For Extra Help

TAPE 16 TAPE 15B MAC WIN CD-ROM

Complete.

1. $1\ yd^2 =$ _______ ft^2

2. $5\ yd^2 =$ _______ ft^2

3. $20\ ft^2 =$ _______ in^2

Complete.

4. $360\ in^2 =$ _______ ft^2

5. $5\ mi^2 =$ _______ acres

Answers on page A-21

Complete.

6. $1 \text{ m}^2 =$ _______ mm^2

7. $1 \text{ cm}^2 =$ _______ mm^2

Complete.

8. $2.88 \text{ m}^2 =$ _______ cm^2

9. $4.3 \text{ mm}^2 =$ _______ cm^2

10. $678{,}000 \text{ m}^2 =$ _______ km^2

Answers on page A-21

Example 6 Complete: $1 \text{ m}^2 =$ _______ cm^2.

$$1 \text{ m}^2 = 1 \cdot (100 \text{ cm})^2 \quad \text{Substituting 100 cm for 1 m}$$
$$= 100 \text{ cm} \cdot 100 \text{ cm}$$
$$= 10{,}000 \text{ cm}^2$$

Do Exercises 6 and 7.

Mental Conversion

To convert mentally, we first note that $10^2 = 100$, $100^2 = 10{,}000$, and $0.1^2 = 0.01$. We use the diagram as before and multiply the number of moves by 2 to determine the number of moves of the decimal point.

1000 m	100 m	10 m	1 m	0.1 m	0.01 m	0.001 m
1 km	1 hm	1 dam	1 m	1 dm	1 cm	1 mm

Example 7 Complete: $3.48 \text{ km}^2 =$ _______ m^2.

Think: To go from km to m in the table is a move of 3 places to the right.

1000 m	100 m	10 m	1 m	0.1 m	0.01 m	0.001 m
1 km	1 hm	1 dam	1 m	1 dm	1 cm	1 mm

3 moves to the right

So we move the decimal point $2 \cdot 3$, or 6 places to the right.

3.48   3.480000. $3.48 \text{ km}^2 = 3{,}480{,}000 \text{ m}^2$

6 places to the right

Example 8 Complete: $586.78 \text{ cm}^2 =$ _______ m^2.

Think: To go from cm to m in the table is a move of 2 places to the left.

1000 m	100 m	10 m	1 m	0.1 m	0.01 m	0.001 m
1 km	1 hm	1 dam	1 m	1 dm	1 cm	1 mm

2 moves to the left

So we move the decimal point $2 \cdot 2$, or 4 places to the left.

586.78   0.0586.78 $586.78 \text{ cm}^2 = 0.058678 \text{ m}^2$

4 places to the left

Do Exercises 8–10.

Exercise Set 9.5

a Complete.

1. $1\ ft^2 =$ ______ in^2

2. $1\ yd^2 =$ ______ ft^2

3. $1\ mi^2 =$ ______ acres

4. 1 acre = ______ ft^2

5. $1\ in^2 =$ ______ ft^2

6. $1\ ft^2 =$ ______ yd^2

7. $22\ yd^2 =$ ______ ft^2

8. $40\ ft^2 =$ ______ in^2

9. $44\ yd^2 =$ ______ ft^2

10. $144\ ft^2 =$ ______ yd^2

11. $20\ mi^2 =$ ______ acres

12. $576\ in^2 =$ ______ ft^2

13. $1\ mi^2 =$ ______ ft^2

14. $1\ mi^2 =$ ______ yd^2

15. $720\ in^2 =$ ______ ft^2

16. $27\ ft^2 =$ ______ yd^2

17. $144\ in^2 =$ ______ ft^2

18. $72\ in^2 =$ ______ ft^2

19. 1 acre = ______ mi^2

20. 4 acres = ______ ft^2

b Complete.

21. $5.21\ km^2 =$ ______ m^2

22. $65\ km^2 =$ ______ m^2

23. $0.014\ m^2 =$ ______ cm^2

24. $0.028\ m^2 =$ ______ mm^2

25. $2345.6\ mm^2 =$ ______ cm^2

26. $8.38\ cm^2 =$ ______ mm^2

27. $852.14\ cm^2 =$ ______ m^2

28. $125\ mm^2 =$ ______ m^2

29. $250{,}000\ mm^2 =$ ______ cm^2

30. $2400\ mm^2 =$ ______ cm^2

31. $472{,}800\ m^2 =$ ______ km^2

32. $1.37\ cm^2 =$ ______ mm^2

Skill Maintenance

In Exercises 33 and 34, find the simple interest. [6.7a]

33. On $2000 at an interest rate of 8% for 1.5 yr

34. On $2000 at an interest rate of 5.3% for 2 yr

In each of Exercises 35–38, find (a) the amount of simple interest due and (b) the total amount that must be paid back. [6.7a]

35. A firm borrows $15,500 at 9.5% for 120 days.

36. A firm borrows $8500 at 10% for 90 days.

37. A firm borrows $6400 at 8.4% for 150 days.

38. A firm borrows $4200 at 11% for 30 days.

Complete.

39. 22,176 ft = ______ mi [8.1a]

40. 22,176 mm = ______ km [8.2a]

Synthesis

41. ◆ Explain the difference between the way we move the decimal point for area conversion and the way we do for length conversion.

42. *The White House.* The president's family has about 20,175 ft^2 of living area in the White House. Estimate the amount of living area in square meters.

Complete.

43. $1\ m^2 =$ ______ ft^2

44. $1\ in^2 =$ ______ cm^2

45. $2\ yd^2 =$ ______ m^2

46. 1 acre = ______ m^2

Collaborative Learning Manual

Verify the conversions between American units of area.

9.6 Angles and Triangles

a Measuring Angles

An **angle** is a set of points consisting of two **rays**, or half-lines, with a common endpoint. The endpoint is called the **vertex**.

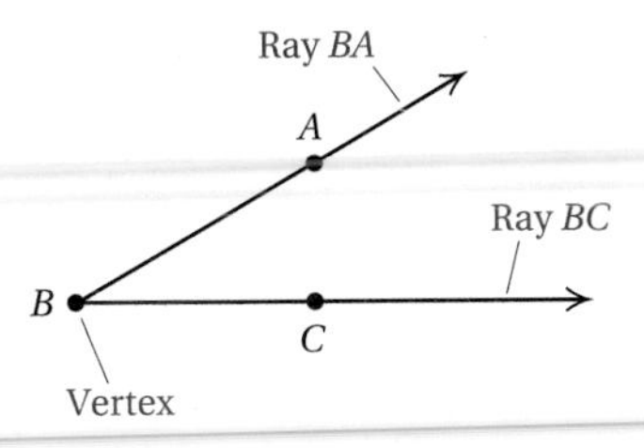

The rays are called the *sides*. The angle above can be named

angle ABC, $\angle ABC$, $\angle CBA$, or $\angle B$.

Note that the name of the vertex is either in the middle or, if no confusion results, listed by itself.

Do Exercises 1 and 2.

Measuring angles is similar to measuring segments. To measure angles, we start with some arbitrary angle and assign to it a measure of 1. We call it a *unit angle*. Suppose that $\angle U$, below, is a unit angle. Let's measure $\angle DEF$. If we made 3 copies of $\angle U$, they would "fill up" $\angle DEF$. Thus the measure of $\angle DEF$ would be 3.

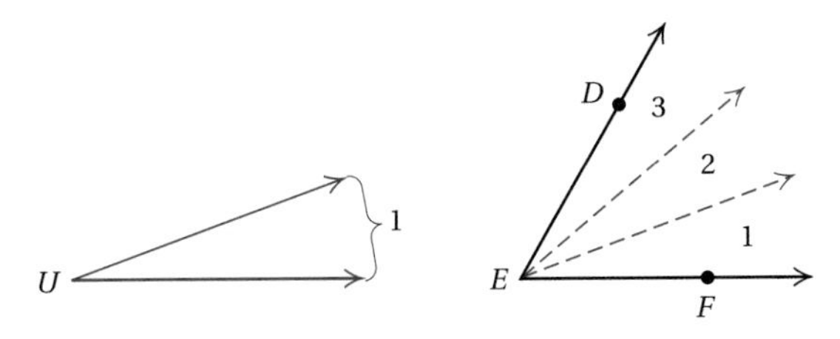

The unit most commonly used for angle measure is the degree. Below is such a unit. Its measure is 1 degree, or 1°.

Here are some other angles with their degree measures.

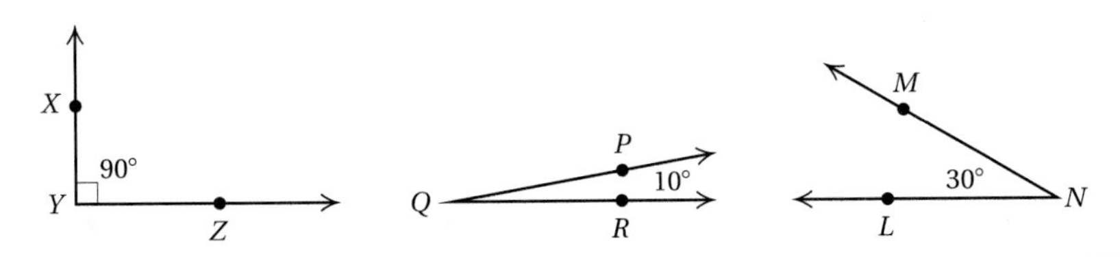

To indicate the *measure* of $\angle XYZ$, we write $m\angle XYZ = 90°$. The symbol ⌉ is sometimes drawn on a figure to indicate a 90° angle.

Objectives

a Name an angle in four different ways and given an angle, measure it with a protractor.

b Classify an angle as right, straight, acute, or obtuse.

c Classify a triangle as equilateral, isosceles, or scalene; and as right, obtuse, or acute.

d Given two of the angle measures of a triangle, find the third.

For Extra Help

TAPE 16 | TAPE 15B | MAC WIN | CD-ROM

Name the angle in four different ways.

1.

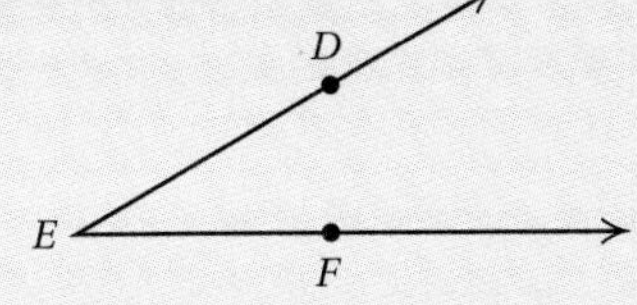

2.

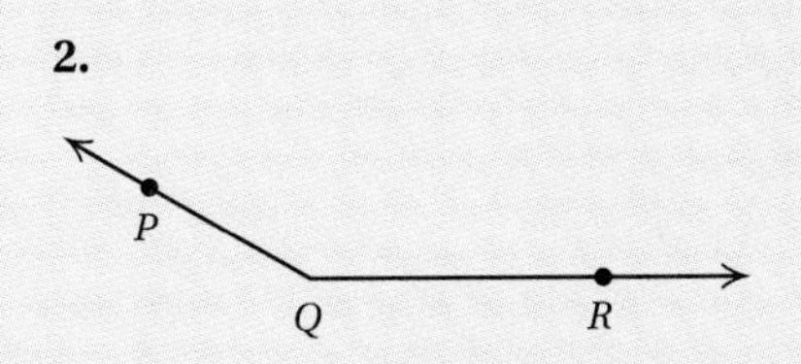

Answers on page A-22

A device called a **protractor** is used to measure angles. Protractors have two scales. To measure an angle like $\angle Q$ below, we place the protractor's ▲ at the vertex and line up one of the angle's sides at 0°. Then we check where the angle's other side crosses the scale. In the figure below, 0° is on the inside scale, so we check where the angle's other side crosses the inside scale. We see that $m\angle Q = 145°$. The notation $m\angle Q$ is read "the measure of angle Q."

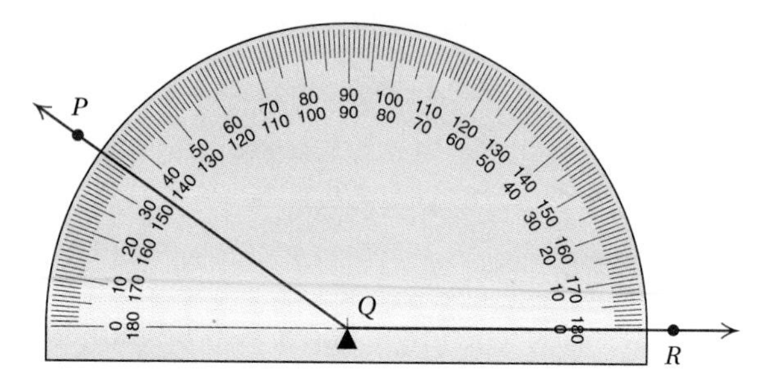

Do Exercise 3.

Let's find the measure of $\angle ABC$. This time we will use the 0° on the outside scale. We see that $m\angle ABC = 42°$.

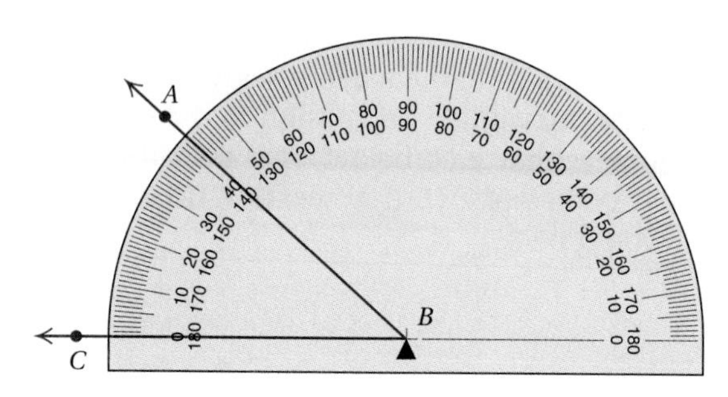

Do Exercise 4.

3. Use a protractor to measure this angle.

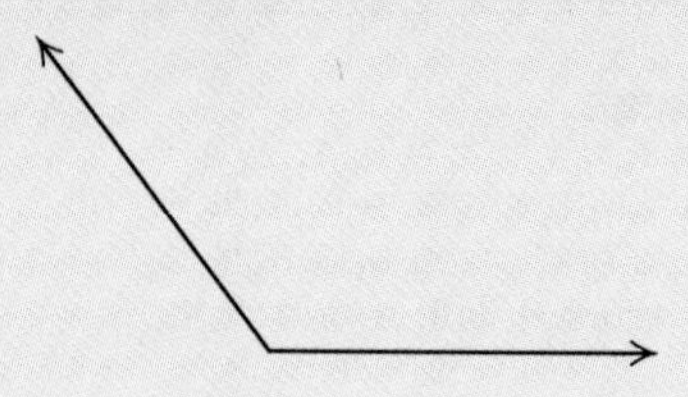

4. Use a protractor to measure this angle.

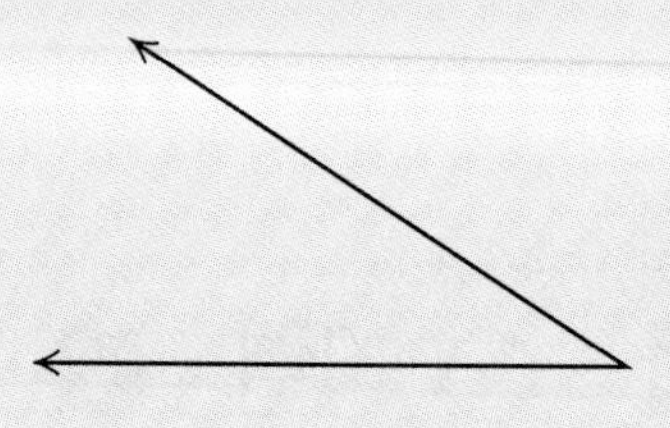

Classify the angle as right, straight, acute, or obtuse. Use a protractor if necessary.

5.

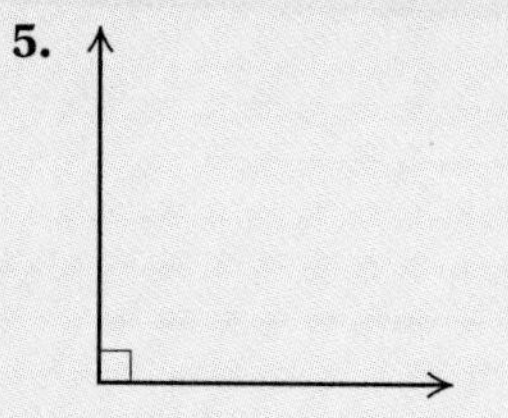

6.

7.

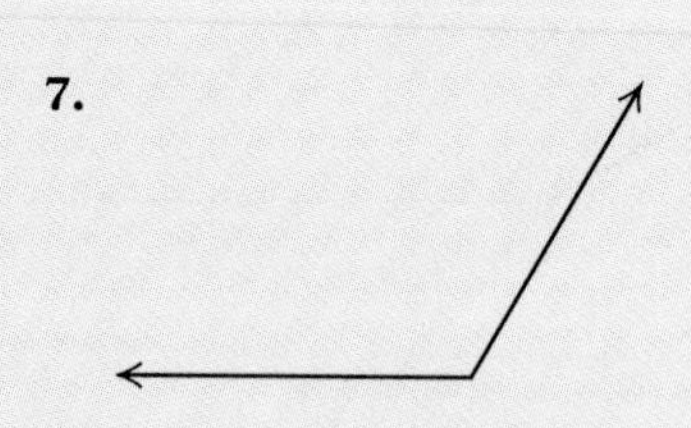

8.

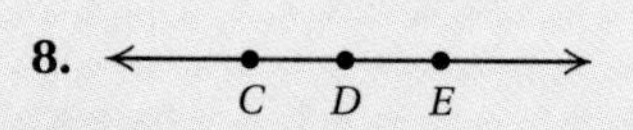

Answers on page A-22

b Classifying Angles

The following are ways in which we classify angles.

Right angle: An angle whose measure is 90°.
Straight angle: An angle whose measure is 180°.
Acute angle: An angle whose measure is greater than 0° and less than 90°.
Obtuse angle: An angle whose measure is greater than 90° and less than 180°.

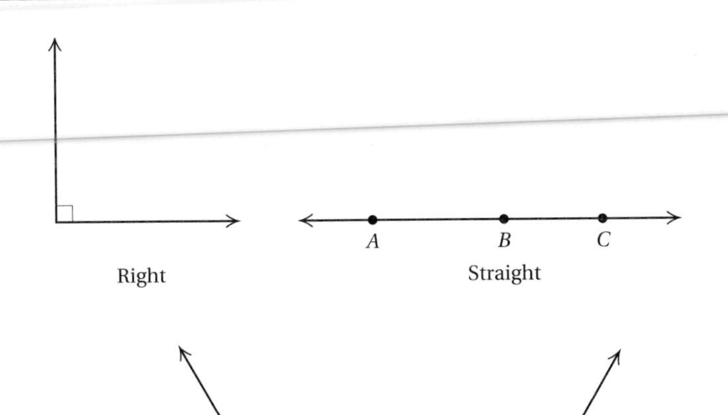

Acute

Obtuse

Do Exercises 5–8 on the preceding page.

c Triangles

A **triangle** is a polygon made up of three segments, or sides. Consider these triangles. The triangle with vertices A, B, and C can be named $\triangle ABC$.

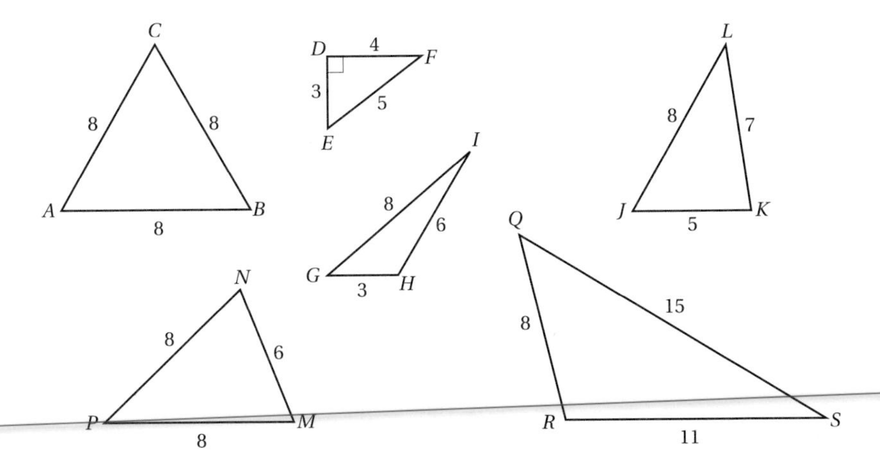

9. Which triangles on this page are:
   a) equilateral?
   b) isosceles?
   c) scalene?

10. Are all equilateral triangles isosceles?

11. Are all isosceles triangles equilateral?

12. Which triangles on this page are:
    a) right triangles?
    b) obtuse triangles?
    c) acute triangles?

Answers on page A-22

13. Find $m(\angle P) + m(\angle Q) + m(\angle R)$.

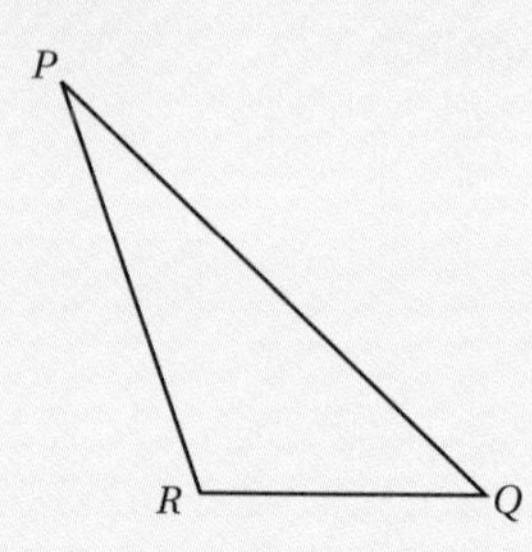

14. Find the missing angle measure.

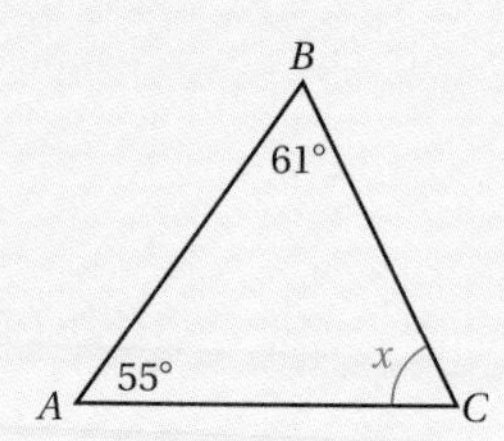

Answers on page A-22

We can classify triangles according to sides and according to angles.

> **Equilateral triangle:** All sides are the same length.
> **Isosceles triangle:** Two or more sides are the same length.
> **Scalene triangle:** All sides are of different lengths.
> **Right triangle:** One angle is a right angle.
> **Obtuse triangle:** One angle is an obtuse angle.
> **Acute triangle:** All three angles are acute.

Do Exercises 9–12 on the preceding page.

d Sum of the Angle Measures of a Triangle

The sum of the angle measures of a triangle is 180°. To see this, note that we can think of cutting apart a triangle as shown on the left below. If we reassemble the pieces, we see that a straight angle is formed.

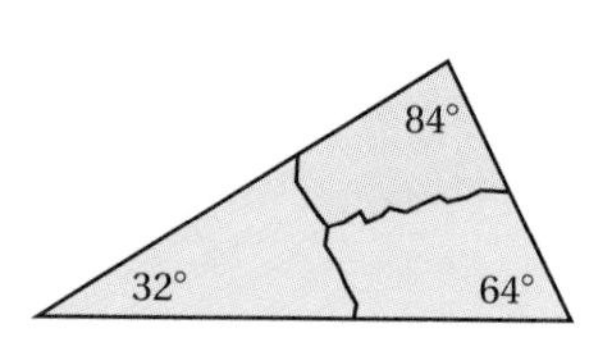

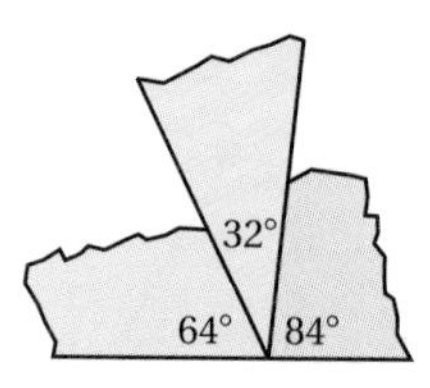

$64° + 32° + 84° = 180°$

> In any triangle ABC, the sum of the measures of the angles is 180°:
> $$m(\angle A) + m(\angle B) + m(\angle C) = 180°.$$

Do Exercise 13.

If we know the measures of two angles of a triangle, we can calculate the third.

Example 1 Find the missing angle measure.

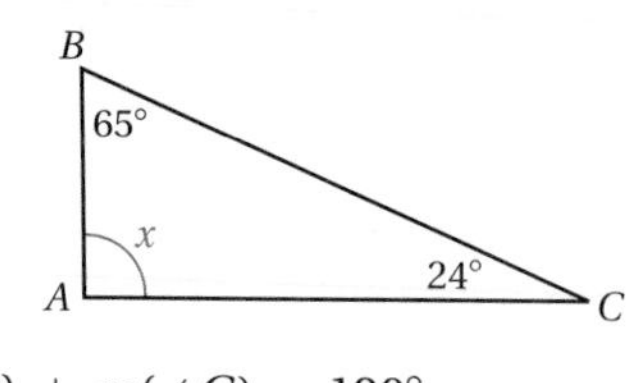

$$m(\angle A) + m(\angle B) + m(\angle C) = 180°$$
$$x + 65° + 24° = 180°$$
$$x + 89° = 180°$$
$$x = 180° - 89°$$
$$x = 91°$$

Do Exercise 14.

Exercise Set 9.6

a Name the angle in four different ways.

1.

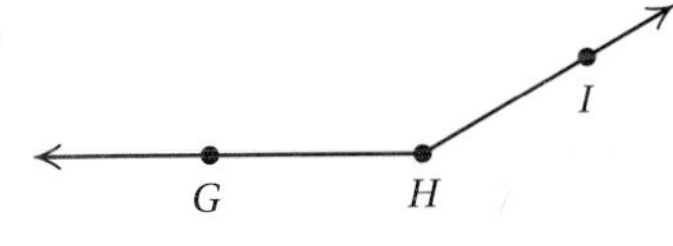

2.

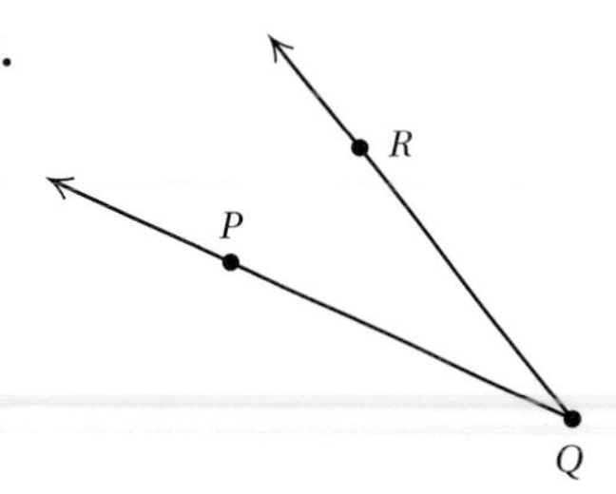

Use a protractor to measure the angle.

3.

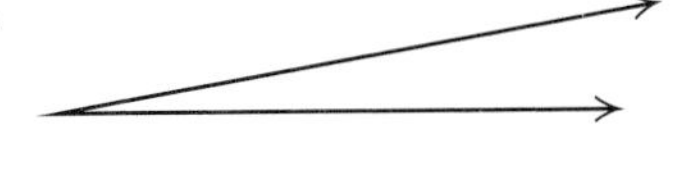

4.

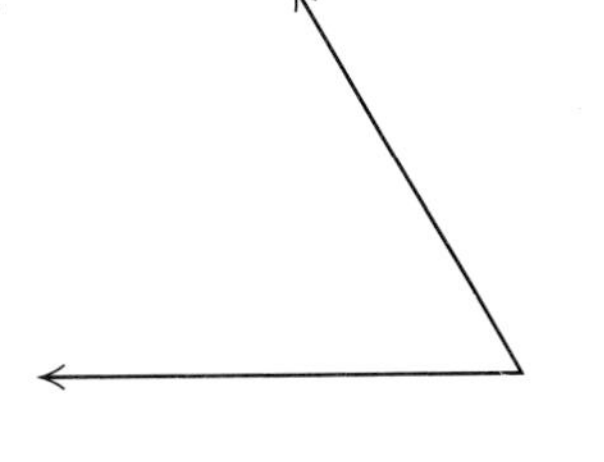

5.

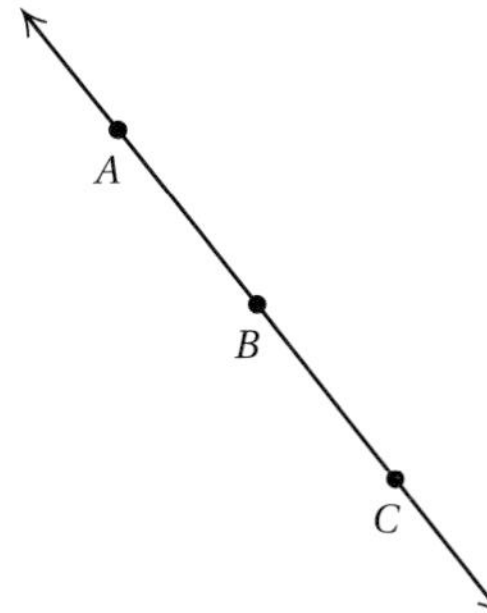

6.

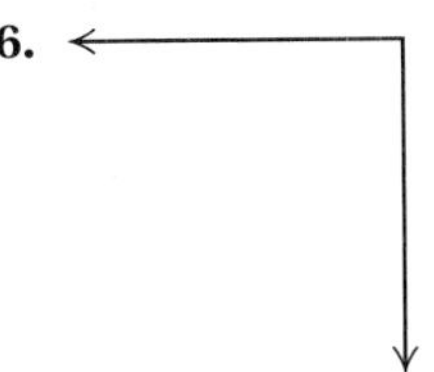

7.

8.

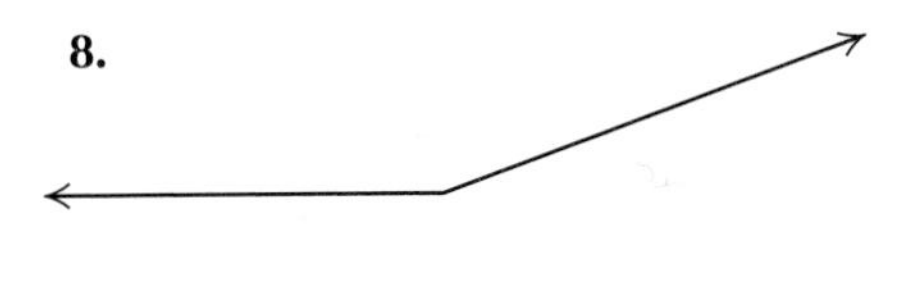

b

9.–16. Classify each of the angles in Exercises 1–8 as right, straight, acute, or obtuse.

17.–20. Classify each of the angles in Margin Exercises 1–4 as right, straight, acute, or obtuse.

c Classify the triangle as equilateral, isosceles, or scalene. Then classify it as right, obtuse, or acute.

21.

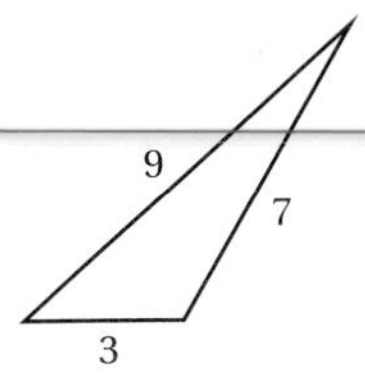

22.

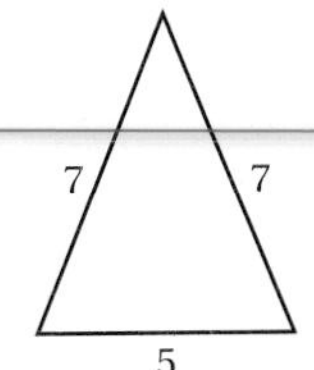

23.

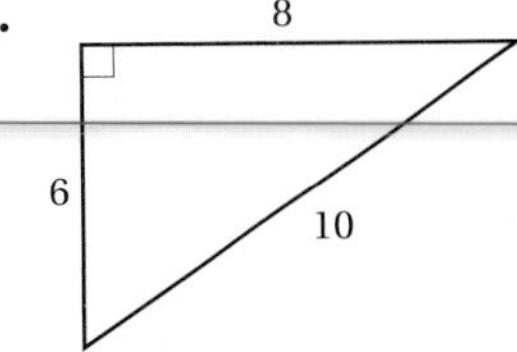

24.

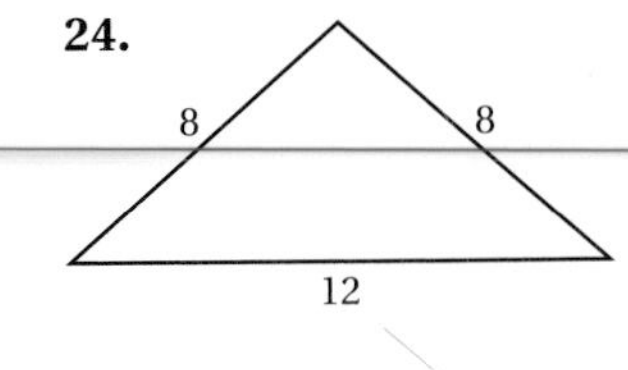

25.

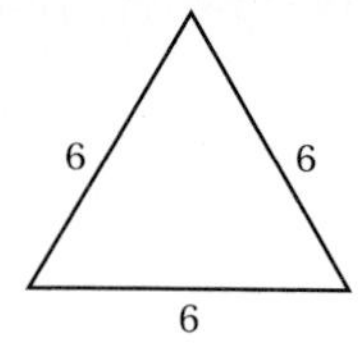

26.

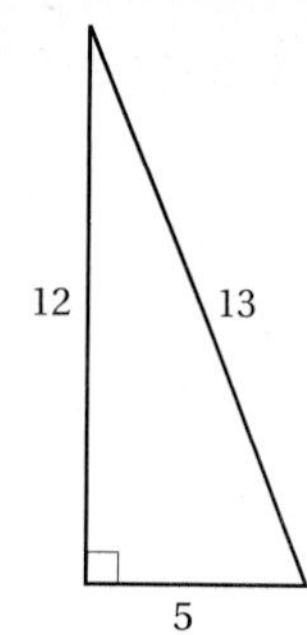

27.

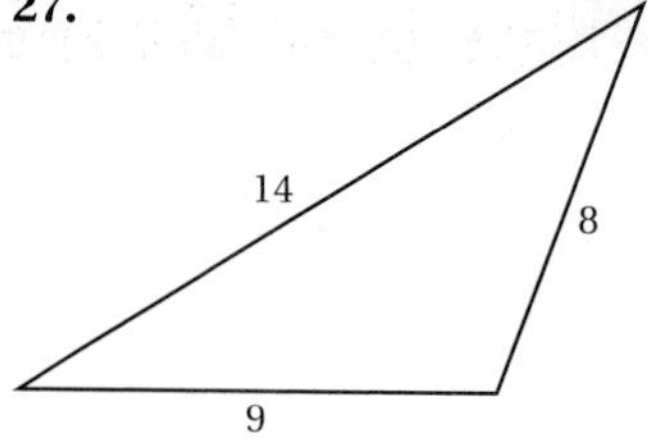

28.

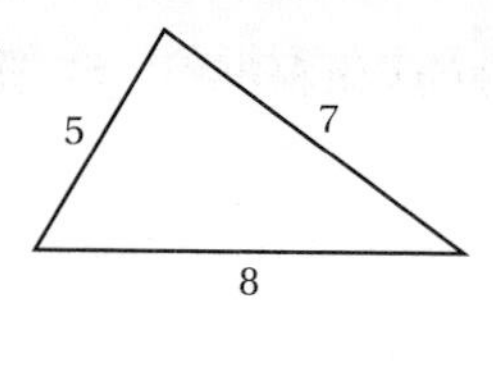

d Find the missing angle measure.

29.

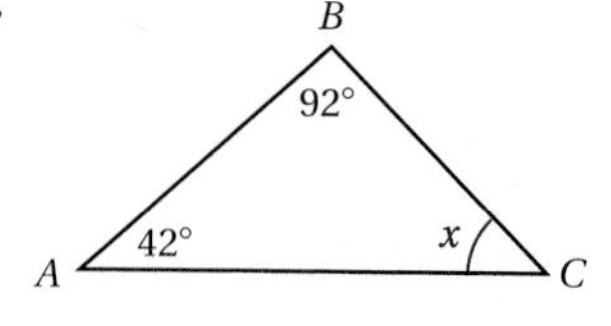

30.

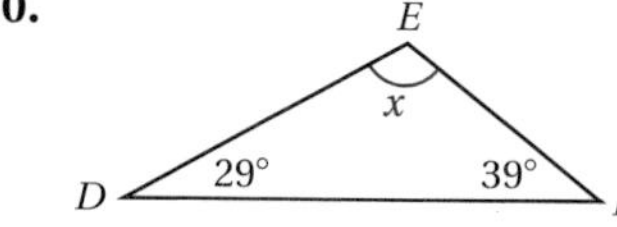

31.

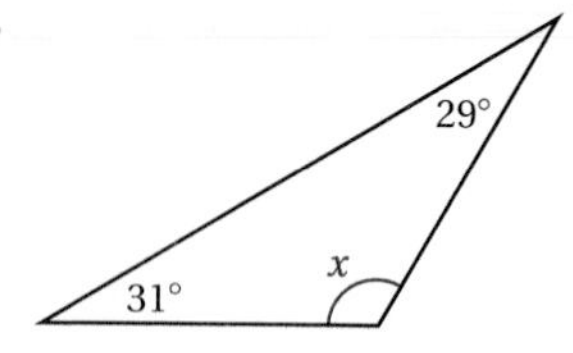

32.

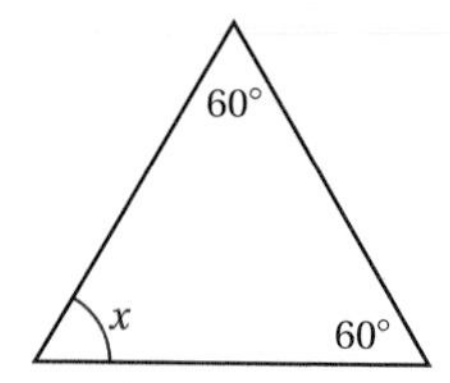

Skill Maintenance

In each of Exercises 33–36, find the simple interest. [6.7a]

33. On $24,000 at an interest rate of 6.4% for 3 yr

34. On $8500 at an interest rate of 10% for 2.5 yr

35. On $6400 at an interest rate of 9.6% for 150 days

36. On $4200 at an interest rate of 10.5% for 60 days

Synthesis

37. ◆ Explain how you might use triangles to find the sum of the angle measures of this figure.

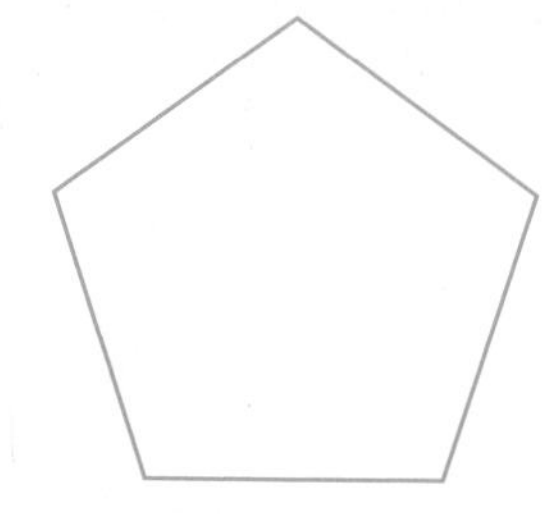

38. Find $m\angle ACB$, $m\angle CAB$, $m\angle EBC$, $m\angle EBA$, $m\angle AEB$, and $m\angle ADB$ in the rectangle shown below.

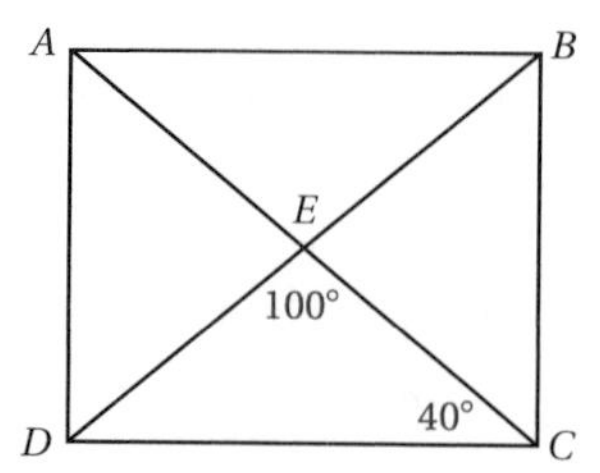

Summary and Review Exercises: Chapter 9

Important Properties and Formulas

Volume of a Rectangular Solid:	$V = l \cdot w \cdot h$
American Units of Capacity:	1 gal = 4 qt; 1 qt = 2 pt; 1 pt = 16 oz; 1 pt = 2 cups; 1 cup = 8 oz
Metric Units of Capacity:	1 L = 1000 mL = 1000 cm^3
Volume of a Circular Cylinder:	$V = \pi \cdot r^2 \cdot h$
Volume of a Sphere:	$V = \frac{4}{3} \cdot \pi \cdot r^3$
Volume of a Cone:	$V = \frac{1}{3} \cdot \pi \cdot r^2 \cdot h$
American System of Weights:	1 T = 2000 lb; 1 lb = 16 oz
Metric System of Mass:	1 t = 1000 kg; 1 kg = 1000 g; 1 hg = 100 g; 1 dag = 10 g; 1 dg = 0.1 g; 1 cg = 0.01 g; 1 mg = 0.001 g
Units of Time:	1 min = 60 sec; 1 hr = 60 min; 1 day = 24 hr; 1 wk = 7 days; 1 yr = 365.25 days
Temperature Conversion:	$F = \frac{9}{5} \cdot C + 32$; $C = \frac{5}{9} \cdot (F - 32)$
Sum of Angle Measures of a Triangle:	$m(\angle A) + m(\angle B) + m(\angle C) = 180°$

The objectives to be tested in addition to the material in this chapter are [1.9b], [6.7a], [8.1a], and [8.2a].

Find the volume. [9.1a]

1.

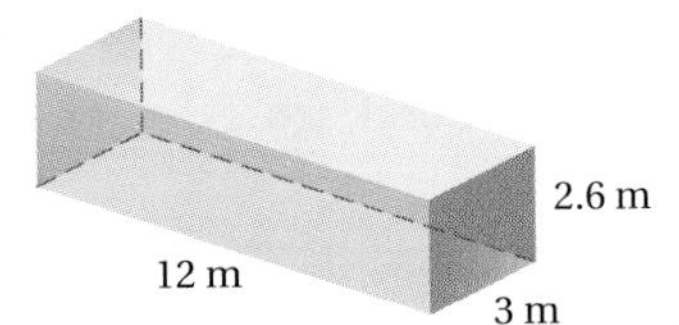

2.

14 cm
3 cm
4.6 cm

Complete. [9.1b], [9.3a, b, c]

3. 7 lb = ______ oz

4. 4 g = ______ kg

5. 16 min = ______ hr

6. 464 mL = ______ L

7. 3 min = ______ sec

8. 4.7 kg = ______ g

9. 8.07 T = ______ lb

10. 0.83 L = ______ mL

11. 6 hr = ______ days

12. 4 cg = ______ g

13. 0.2 g = ______ mg

14. 0.0003 kg = ______ cg

15. 60 mL = ______ L

16. 0.8 T = ______ lb

17. 0.4 L = ______ mL

18. 20 oz = ______ lb

19. $\frac{5}{6}$ min = ______ sec

20. 20 gal = ______ pt

21. 960 oz = ______ gal

22. 54 qt = ______ gal

Solve. [9.1c]

23. A physician prescribed 780 mL per hour of a certain intravenous fluid for a patient. How many liters of fluid did this patient receive in one day?

24. Convert 27°C to Fahrenheit. [9.4b]

25. Convert 68°F to Celsius. [9.4b]

Find the volume. Use 3.14 for π.

26. [9.2a]

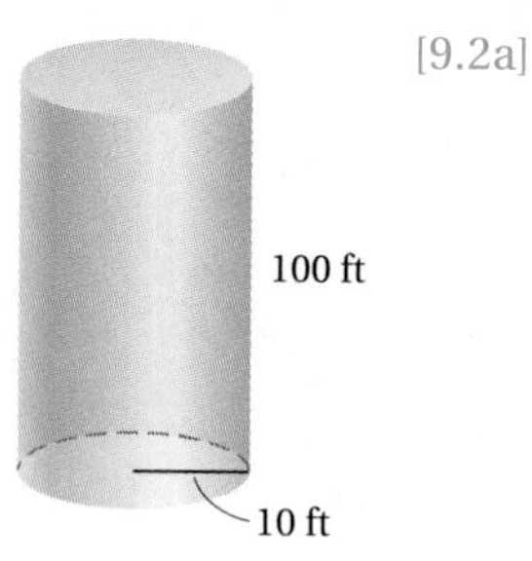

27. [9.2b]

$r = 2$ cm

28. [9.2c]

4.5 in.

1 in.

Complete. [9.5a, b]

29. 4 yd^2 = _______ ft^2

30. 0.3 km^2 = _______ m^2

31. 2070 in^2 = _______ ft^2

32. 600 cm^2 = _______ m^2

Use a protractor to measure each angle. [9.6a]

33.

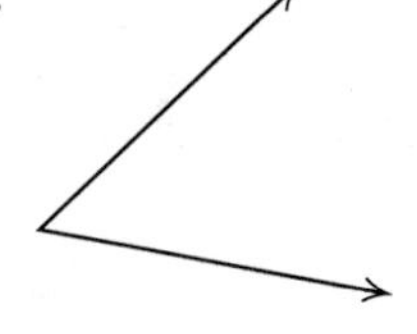

34.

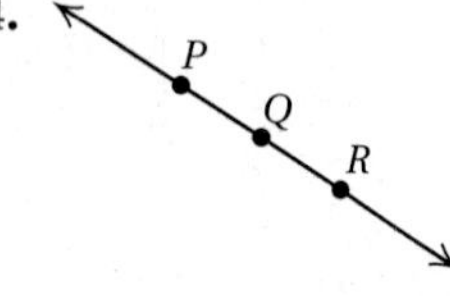

35. **36.**

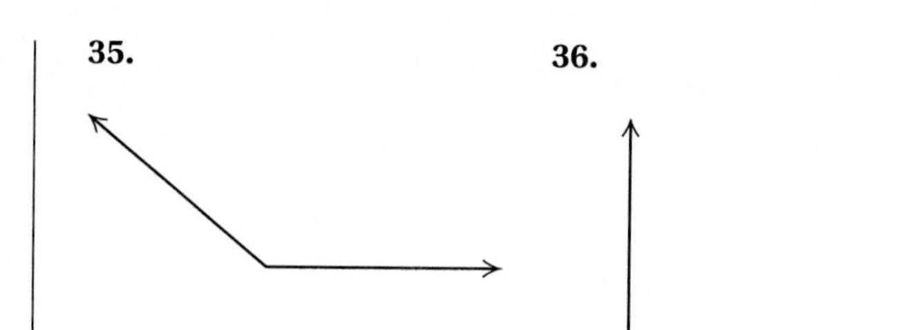

37.–40. Classify each of the angles in Exercises 33–36 as right, straight, acute, or obtuse. [9.6b]

Use the following triangle for Exercises 41–43.

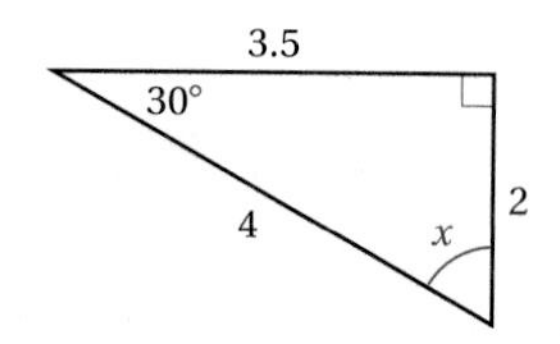

41. Find the missing angle measure. [9.6d]

42. Classify the triangle as equilateral, isosceles, or scalene. [9.6c]

43. Classify the triangle as right, obtuse, or acute. [9.6c]

Skill Maintenance

44. Find the simple interest on $5000 at 9.5% for 30 days. [6.7a]

Evaluate. [1.9b]

45. 3^3

46. $(4.7)^2$

47. 4.7^3

48. $\left(\frac{1}{2}\right)^4$

Complete. [8.1a], [8.2a]

49. 2.5 mi = _______ ft

50. 144 in. = _______ yd

51. 4568 cm = _______ m

52. 4568 cm = _______ mm

Synthesis

53. ◆ It is known that 1 gal of water weighs 8.3453 lb. Which weighs more, an ounce of pennies or an ounce (as capacity) of water? Explain. [9.1b], [9.3b]

54. ◆ List and describe all the volume formulas that you have learned in this chapter. [9.1a], [9.2a, b, c]

Test: Chapter 9

1. Find the volume.

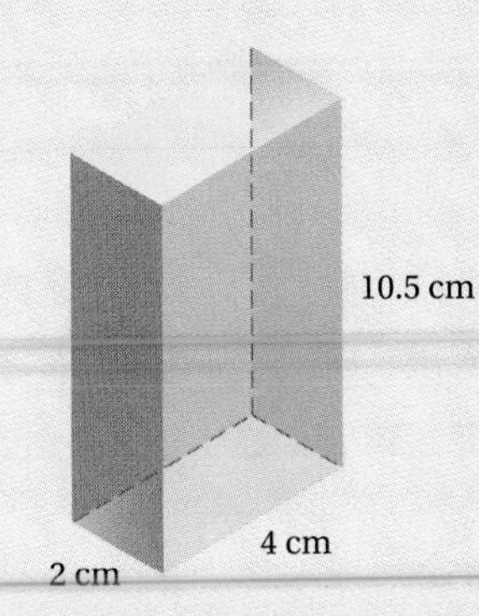

Complete.

2. 3080 mL = _______ L

3. 0.24 L = _______ mL

4. 4 lb = _______ oz

5. 4.11 T = _______ lb

6. 3.8 kg = _______ g

7. 4.325 mg = _______ cg

8. 2200 mg = _______ g

9. 5 hr = _______ min

10. 15 days = _______ hr

11. 64 pt = _______ qt

12. 10 gal = _______ oz

13. 5 cups = _______ oz

14. Convert 95°F to Celsius.

15. Convert 59°C to Fahrenheit.

Complete.

16. 12 ft^2 = _______ in^2

17. 3 cm^2 = _______ m^2

18. A twelve-box carton of 12-oz juice boxes comes in a rectangular box $10\frac{1}{2}$ in. by 8 in. by 5 in. What is the volume of the carton?

Find the volume. Use 3.14 for π.

19.

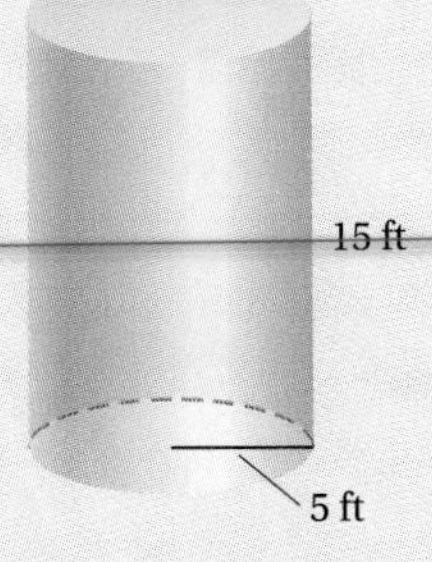

20.

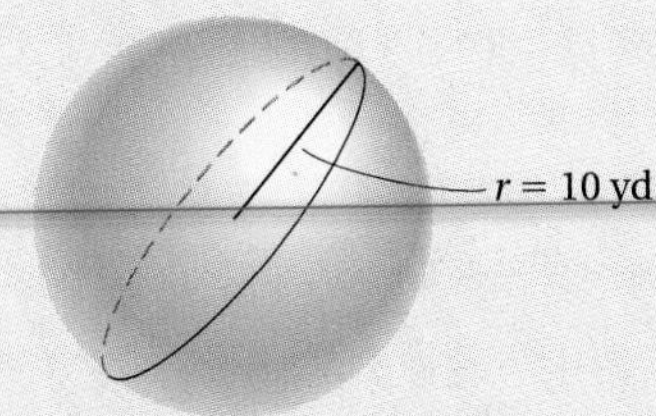

21.

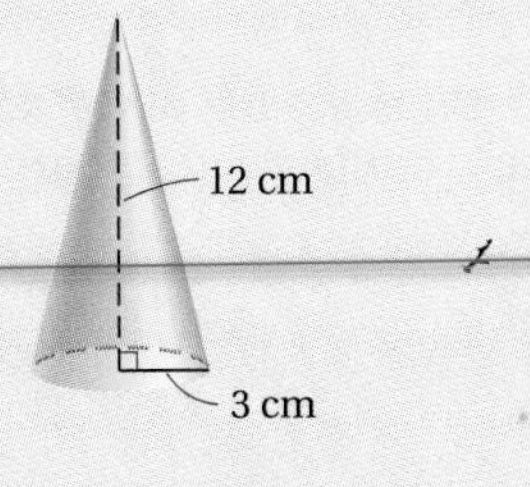

Answers

1. ____________
2. ____________
3. ____________
4. ____________
5. ____________
6. ____________
7. ____________
8. ____________
9. ____________
10. ____________
11. ____________
12. ____________
13. ____________
14. ____________
15. ____________
16. ____________
17. ____________
18. ____________
19. ____________
20. ____________
21. ____________

Answers

22. ______

23. ______

24. ______

25. ______

26. ______

27. ______

28. ______

29. ______

30. ______

31. ______

32. ______

33. ______

34. ______

35. ______

36. ______

37. ______

38. ______

39. ______

40. ______

41. ______

42. ______

43. ______

44. ______

Use a protractor to measure each angle.

22.

23.

24.

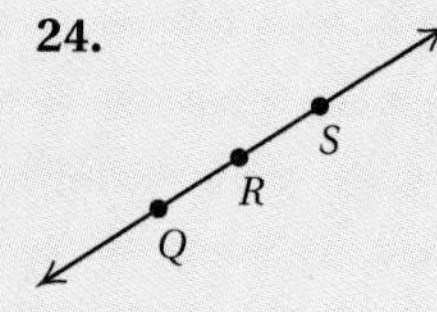

25.

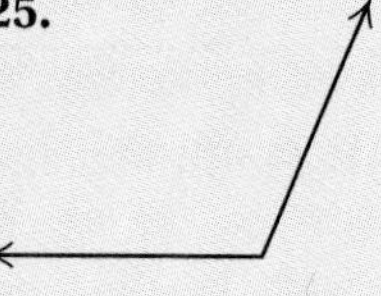

26.–29. Classify each of the angles in Questions 22–25 as right, straight, acute, or obtuse.

Use the following triangle for Questions 30–32.

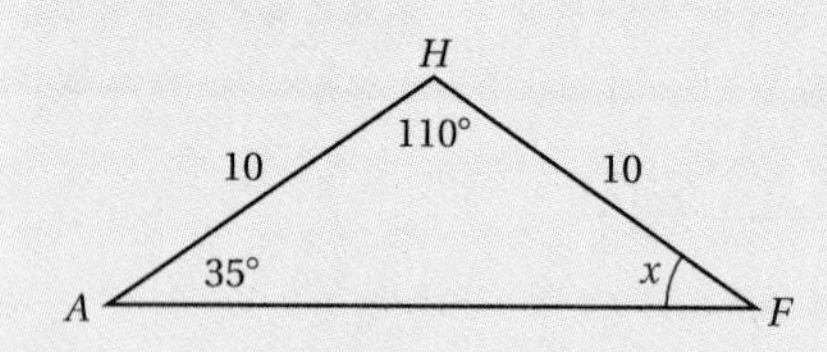

30. Find the missing angle measure.

31. Classify the triangle as equilateral, isosceles, or scalene.

32. Classify the triangle as right, obtuse, or acute.

Skill Maintenance

33. Find the simple interest on \$10,000 at 6.8% for 1 yr.

Evaluate.

34. 10^3

35. $\left(\frac{1}{4}\right)^2$

36. $(3.14)^2$

37. $(0.1)^5$

Complete.

38. 14 yd = ______ ft

39. 3000 in. = ______ ft

40. 2.3 km = ______ m

41. 34,000 mm = ______ cm

Synthesis

Find the volume of the solid. Give the answer in cubic feet. (Note that the solids are not drawn in perfect proportion.

42.

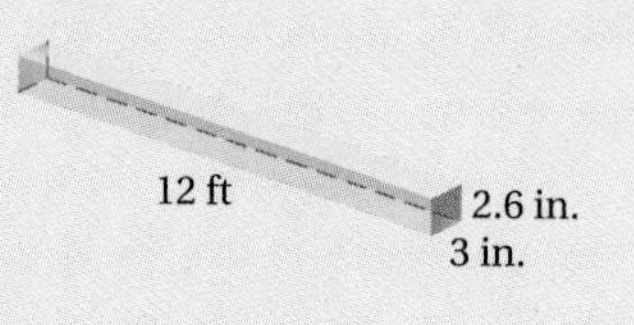

43.

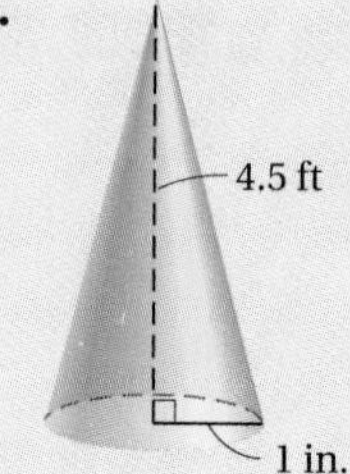

44.

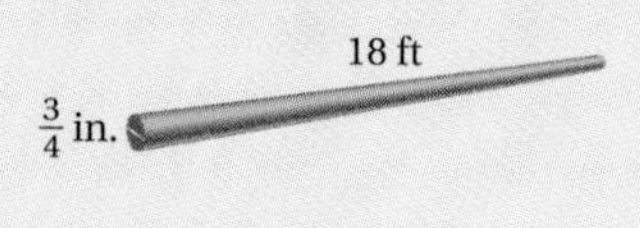

Cumulative Review: Chapters 1–9

Calculate.

1. $1\frac{1}{2} + 2\frac{2}{3}$

2. $\left(\frac{1}{4}\right)^2 \div \left(\frac{1}{2}\right)^3 \times 2^4 + (10.3)(4)$

3. $120.5 - 32.98$

4. $22\overline{)27{,}148}$

5. $14 \div [33 \div 11 + 8 \times 2 - (15 - 3)]$

6. $8^3 + 45 \cdot 24 - 9^2 \div 3$

Find fractional notation.

7. 1.209

8. 17%

Use <, >, or = for ▒ to write a true sentence.

9. $\frac{5}{6}$ ▒ $\frac{7}{8}$

10. $\frac{15}{18}$ ▒ $\frac{10}{12}$

Complete.

11. 6 oz = ______ lb

12. 15°C = ______°F

13. 0.087 L = ______ mL

14. 9 sec = ______ min

15. 3 yd^2 = ______ ft^2

16. 17 cm = ______ m

Find the perimeter and the area.

17.

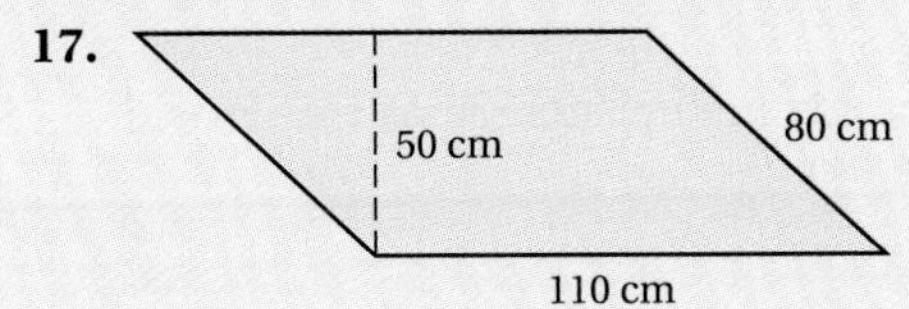

18.

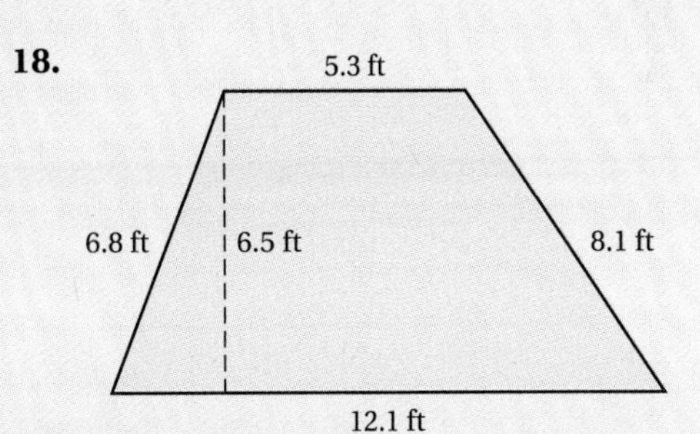

The following table shows the car sales by McQuirk Motors over several months. Use the data in this table for Questions 19 and 20.

Month	Car Sales
1	19
2	17
3	18
4	20
5	22
6	?
7	28

19. Make a line graph of the data.

20. Use interpolation to estimate the number of cars sold in the sixth month.

Estimate each of the following as 0, $\frac{1}{2}$, or 1.

21. $\frac{3}{47}$

22. $\frac{14}{15}$

23. $\frac{2}{13}$

24. $\frac{8}{19}$

Solve.

25. $\frac{12}{15} = \frac{x}{18}$

26. $\frac{3}{x} = \frac{7}{10}$

27. $25 \cdot x = 2835$

28. $x + \frac{3}{4} = \frac{7}{8}$

Solve.

29. To get an A in math, a student must score an average of 90 on five tests. On the first four tests, the scores were 85, 92, 79, and 95. What is the lowest score that the student can get on the last test and still get an A?

30. Americans own 52 million dogs, 56 million cats, 45 million birds, 250 million fish, and 125 million other creatures as house pets. How many pets do Americans own altogether?

31. The diameter of a basketball is 20 cm. What is its volume? Use 3.14 for π.

32. What is the simple interest on \$800 at 12% for $\frac{1}{4}$ year?

33. How long must a rope be in order to reach from the top of an 8-m tree to a point on the ground 15 m from the bottom of the tree?

34. The sales tax on an office supply purchase of \$5.50 is \$0.33. What is the sales tax rate?

35. A bolt of fabric in a fabric store has $10\frac{3}{4}$ yd on it. A customer purchases $8\frac{5}{8}$ yd. How many yards remain on the bolt?

36. What is the cost, in dollars, of 15.6 gal of gasoline at 139.9¢ per gallon? Round to the nearest cent.

37. A box of powdered milk that makes 20 qt costs \$4.99. A box that makes 8 qt costs \$1.99. Which size has the lower unit price?

38. It is $\frac{7}{10}$ km from a student's dormitory to the library. Maria starts to walk there, changes her mind after going $\frac{1}{4}$ of the distance, and returns home. How far did she walk?

39. Find the missing angle measure.

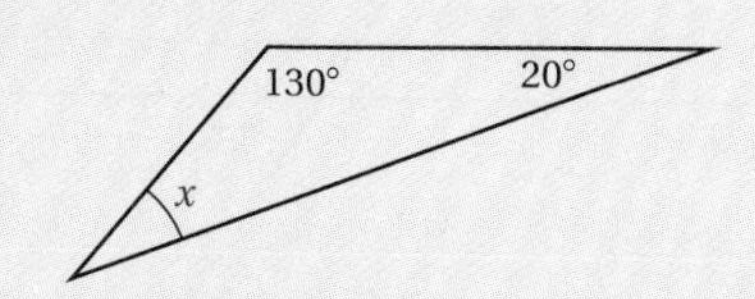

40. Classify the triangle in Question 39 as equilateral, isosceles, or scalene.

41. Classify the triangle in Question 39 as right, obtuse, or acute.

Synthesis

42. Your house sits on a lot measuring 75 ft by 200 ft. The lot is at the intersection of two streets, so there are sidewalks on two sides of the lot. In the winter, you have to shovel the snow off the sidewalks. If the sidewalks are 3 ft wide and the snow is 4 in. deep, what volume of snow must you shovel?

Find the volume in cubic feet. Use 3.14 for π.

43.

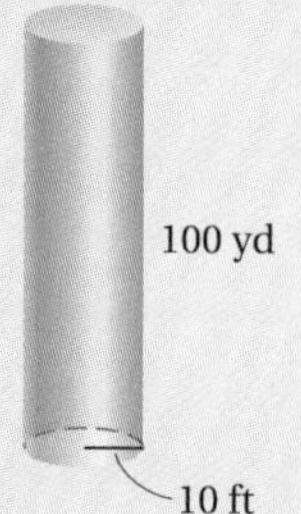

44.

45. J. C. Payne, a 71-year-old rancher at the time, recently asked the *Guinness Book of Records* to accept a world record for constructing a ball of string. The ball was 13.2 ft in diameter. What was the volume of the ball? Assuming the diameter of the string was 0.1 in., how long was the string in feet? in miles? Use the π key on your calculator.

10

The Real-Number System

An Application

A quarterback is sacked for a 5-yd loss. Tell what integer corresponds to this situation.

This problem appears as Margin Exercise 1 in Section 10.1.

The Mathematics

The quarterback being sacked for 5 yd corresponds to

$$\underbrace{-5}.$$

↑ This is a negative number.

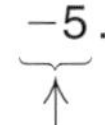

For more information, visit us at www.mathmax.com

Introduction

This chapter is the first of two that form an introduction to algebra. We expand on the arithmetic numbers to include new numbers called *real numbers.* We consider both the rational numbers and the irrational numbers, which make up the real numbers. We will learn to add, subtract, multiply, and divide real numbers.

Pretest: Chapter 10

Use either $<$ or $>$ for ■ to write a true sentence.

1. $0 \ ■ \ -5$

2. $10 \ ■ \ -5$

3. $-35 \ ■ \ -45$

4. $-\frac{2}{3} \ ■ \ \frac{4}{5}$

Find decimal notation.

5. $-\frac{5}{8}$

6. $-\frac{2}{3}$

7. $-\frac{10}{11}$

Find the absolute value.

8. $|-12|$

9. $|2.3|$

10. $|0|$

Find the opposite, or additive inverse.

11. 5.4

12. $-\frac{2}{3}$

Compute and simplify.

13. $-9 + (-8)$

14. $20.2 - (-18.4)$

15. $-\frac{5}{6} - \frac{3}{10}$

16. $-11.5 + 6.5$

17. $-9(-7)$

18. $\frac{5}{8}\left(-\frac{2}{3}\right)$

19. $-19.6 \div 0.2$

20. $-56 \div (-7)$

21. $12 - (-6) + 14 - 8$

22. $20 - 10 \div 5 + 2^3$

The objectives to be tested in addition to the material in this chapter are as follows.

[2.1d] Find the prime factorization of a composite number.
[3.1a] Find the LCM of two or more numbers.
[6.3b] Solve basic percent problems.
[8.4a] Find the area of a rectangle or a square.

10.1 The Real Numbers

In this section, we introduce the *real numbers.* We begin with numbers called *integers* and build up to the real numbers. The integers start with the whole numbers, 0, 1, 2, 3, and so on. For each number 1, 2, 3, and so on, we obtain a new number to the left of zero on the number line:

For the number 1, there will be an *opposite* number -1 (negative 1).

For the number 2, there will be an *opposite* number -2 (negative 2).

For the number 3, there will be an *opposite* number -3 (negative 3), and so on.

The **integers** consist of the whole numbers and these new numbers. We picture them on a number line as follows.

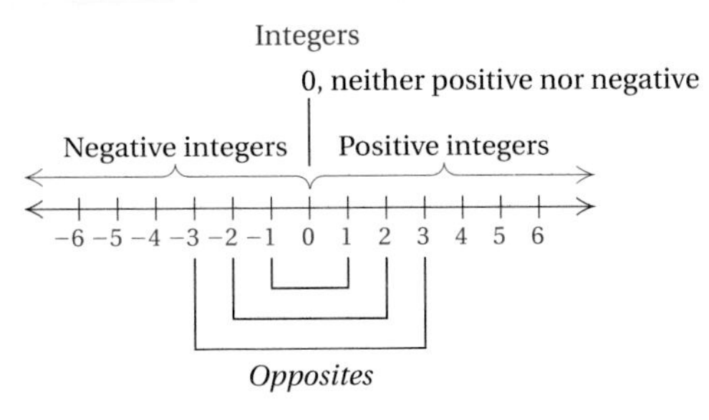

We call the numbers to the left of zero on the number line **negative integers.** The natural numbers are called **positive integers.** Zero is neither positive nor negative. We call -1 and 1 opposites of each other. Similarly, -2 and 2 are opposites, -3 and 3 are opposites, -100 and 100 are opposites, and 0 is its own opposite. Opposite pairs of numbers like -3 and 3 are equidistant from 0. The integers extend infinitely to the left and right of zero.

> The **integers**: $\ldots, -5, -4, -3, -2, -1, 0, 1, 2, 3, 4, 5, \ldots$

Objectives

- **a** Tell which integers correspond to a real-world situation.
- **b** Graph rational numbers on a number line.
- **c** Convert from fractional notation for a rational number to decimal notation.
- **d** Determine which of two real numbers is greater and indicate which, using $<$ or $>$.
- **e** Find the absolute value of a real number.

For Extra Help

TAPE 17 | TAPE 16A | MAC WIN | CD-ROM

a Integers and the Real World

Integers correspond to many real-world problems and situations. The following examples will help you get ready to translate problem situations to mathematical language.

Example 1 Tell which integer corresponds to this situation: The temperature is 3 degrees below zero.

3° below zero is $-3°$

Tell which integers correspond to the situation.

1. The halfback gained 8 yd on first down. The quarterback was sacked for a 5-yd loss on second down.

2. *Highest and Lowest Temperatures.* The highest temperature ever recorded in the United States was 134° in Death Valley on July 10, 1913. The coldest temperature ever recorded in the United States was 80° below zero in Prospect Creek, Alaska, in January 1971.

3. At 10 sec before liftoff, ignition occurs. At 148 sec after liftoff, the first stage is detached from the rocket.

4. A student owes $137 to the bookstore. The student has $289 in a savings account.

Answers on page A-23

Example 2 Tell which integer corresponds to this situation: Death Valley is 280 feet below sea level.

The integer -280 corresponds to the situation. The elevation is -280 ft.

Example 3 Tell which integers correspond to this situation: A business earned $78 on Monday, but lost $57 on Tuesday.

The integers 78 and -57 correspond to the situation. The integer 78 corresponds to the profit on Monday and -57 corresponds to the loss on Tuesday.

Do Exercises 1–4.

b The Rational Numbers

To create a larger number system, called the **rational numbers,** we consider quotients of integers with nonzero divisors. The following are rational numbers:

$$\frac{2}{3},\ -\frac{2}{3},\ \frac{7}{1},\ 4,\ -3,\ 0,\ \frac{23}{-8},\ 2.4,\ -0.17,\ 10\frac{1}{2}.$$

The number $-\frac{2}{3}$ (read "negative two-thirds") can also be named $\frac{2}{-3}$ or $\frac{-2}{3}$. The number 2.4 can be named $\frac{24}{10}$, or $\frac{12}{5}$, and -0.17 can be named $-\frac{17}{100}$.

Note that the rational numbers contain the natural numbers, the whole numbers, the integers, and the arithmetic numbers (also called the nonnegative rational numbers).

> The **rational numbers** consist of all numbers that can be named in the form $\frac{a}{b}$, where a and b are integers and b is not 0.

We picture the rational numbers on a number line, as follows. There is a point on the line for every rational number.

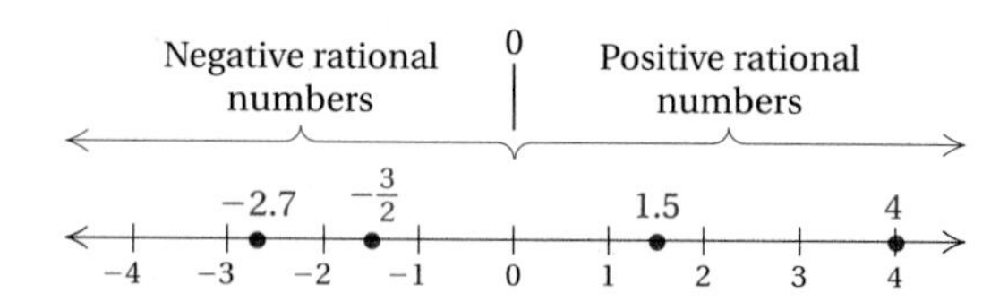

To **graph** a number means to find and mark its point on the number line. Some rational numbers are graphed in the preceding figure.

Example 4 Graph: $\frac{5}{2}$.

The number $\frac{5}{2}$ can be named $2\frac{1}{2}$, or 2.5. Its graph is halfway between 2 and 3.

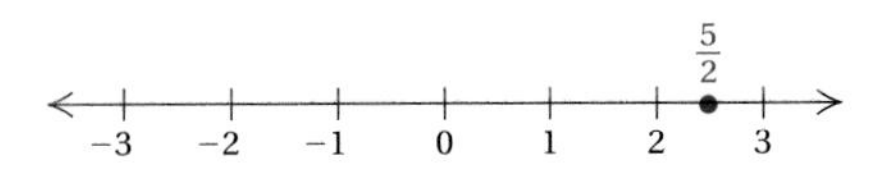

Example 5 Graph: -3.2.

The graph of -3.2 is $\frac{2}{10}$ of the way from -3 to -4.

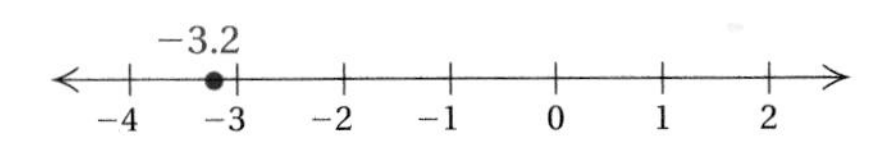

Example 6 Graph: $\frac{13}{8}$.

The number $\frac{13}{8}$ can be named $1\frac{5}{8}$, or 1.625. The graph is about $\frac{6}{10}$ of the way from 1 to 2.

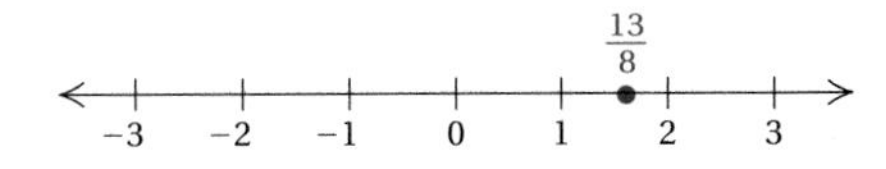

Do Exercises 5–7.

c Notation for Rational Numbers

Each rational number can be named using fractional or decimal notation.

Example 7 Find decimal notation for $-\frac{5}{8}$.

We first find decimal notation for $\frac{5}{8}$. Since $\frac{5}{8}$ means $5 \div 8$, we divide.

$$
\begin{array}{r}
0.625 \\
8\,\overline{)\,5.000} \\
\underline{4\,8} \\
20 \\
\underline{16} \\
40 \\
\underline{40} \\
0
\end{array}
$$

Thus, $\frac{5}{8} = 0.625$, so $-\frac{5}{8} = -0.625$.

Decimal notation for $-\frac{5}{8}$ is -0.625. We consider -0.625 to be a **terminating decimal.** Decimal notation for some numbers repeats.

Graph on a number line.

5. $-\dfrac{7}{2}$

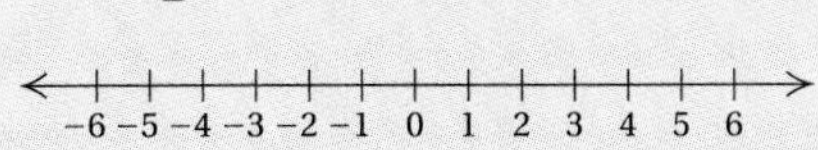

6. 1.4

7. $-\dfrac{11}{4}$

Answers on page A-23

Find decimal notation.

8. $-\frac{3}{8}$

9. $-\frac{6}{11}$

10. $\frac{4}{3}$

Answers on page A-23

Example 8 Find decimal notation for $-\frac{7}{11}$.

We divide to find decimal notation for $\frac{7}{11}$.

$$
\begin{array}{r}
0.6\,3\,6\,3\ldots \\
11\,\overline{)\,7.0\,0\,0\,0} \\
\underline{6\,6} \\
4\,0 \\
\underline{3\,3} \\
7\,0 \\
\underline{6\,6} \\
4\,0 \\
\underline{3\,3} \\
7
\end{array}
$$

Thus, $\frac{7}{11} = 0.6363\ldots$, so $-\frac{7}{11} = -0.6363\ldots$. Repeating decimal notation can be abbreviated by writing a bar over the repeating part; in this case, we write $-0.\overline{63}$.

The following are other examples to show how each rational number can be named using fractional or decimal notation:

$$0 = \frac{0}{6}, \qquad \frac{27}{100} = 0.27, \qquad -8\frac{3}{4} = -8.75, \qquad -\frac{13}{6} = -2.1\overline{6}.$$

Do Exercises 8–10.

d The Real Numbers and Order

The number line has a point for every rational number. However, there are some points on the line for which there are no rational numbers. These points correspond to what are called **irrational numbers.** Some examples of irrational numbers are π and $\sqrt{2}$.

Decimal notation for rational numbers *either* terminates *or* repeats. Decimal notation for irrational numbers *neither* terminates *nor* repeats. Some other examples of irrational numbers are $\sqrt{3}$, $-\sqrt{8}$, $\sqrt{11}$, and 0.121221222122221.... Whenever we take the square root of a number that is not a perfect square, we will get an irrational number.

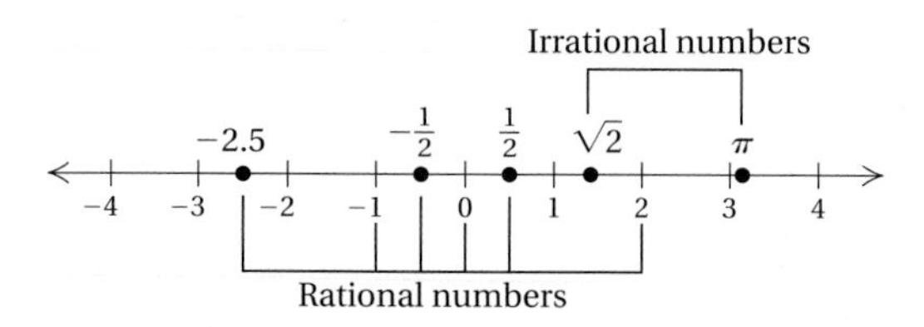

The rational numbers and the irrational numbers together correspond to all the points on a number line and make up what is called the **real-number system**.

The real numbers consist of the rational numbers and the irrational numbers. The following figure shows the relationships among various kinds of numbers.

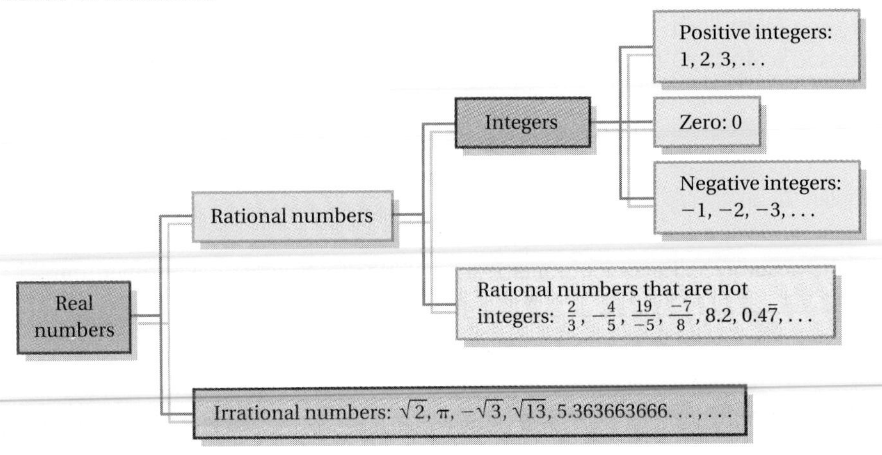

Order

Real numbers are named in order on the number line, with larger numbers named farther to the right. (See Section 1.4.) For any two numbers on the line, the one to the left is less than the one to the right.

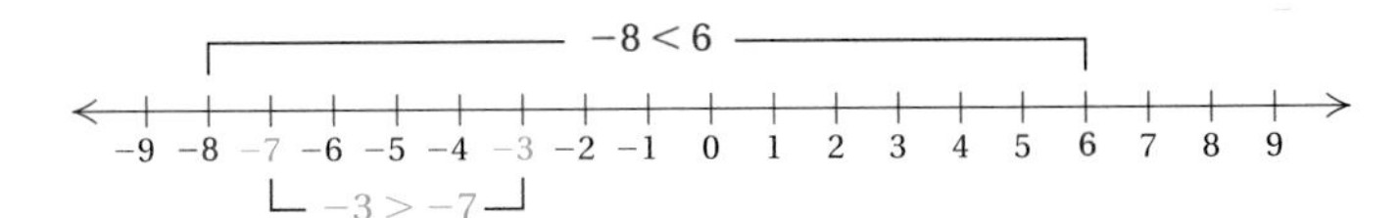

We use the symbol $<$ to mean **"is less than."** The sentence $-8 < 6$ means "-8 is less than 6." The symbol $>$ means **"is greater than."** The sentence $-3 > -7$ means "-3 is greater than -7."

Examples Use either $<$ or $>$ for ▒ to write a true sentence.

9. -7 ▒ 3	Since -7 is to the left of 3, we have $-7 < 3$.
10. 6 ▒ -12	Since 6 is to the right of -12, then $6 > -12$.
11. -18 ▒ -5	Since -18 is to the left of -5, we have $-18 < -5$.
12. -2.7 ▒ $-\frac{3}{2}$	The answer is $-2.7 < -\frac{3}{2}$.
13. 1.5 ▒ -2.7	The answer is $1.5 > -2.7$.
14. -3.45 ▒ 1.32	The answer is $-3.45 < 1.32$.
15. $\frac{5}{8}$ ▒ $\frac{7}{11}$	We convert to decimal notation: $\frac{5}{8} = 0.625$, and $\frac{7}{11} = 0.6363\ldots$. Thus, $\frac{5}{8} < \frac{7}{11}$.
16. -4 ▒ 0	The answer is $-4 < 0$.
17. 5.8 ▒ 0	The answer is $5.8 > 0$.

−2.7 $-\frac{3}{2}$ 1.5
−3 −2 −1 0 1 2 3 4

Do Exercises 11–18.

Use either $<$ or $>$ for ▒ to write a true sentence.

11. -3 ▒ 7

12. -8 ▒ -5

13. 7 ▒ -10

14. 3.1 ▒ -9.5

15. $-\frac{2}{3}$ ▒ -1

16. $-\frac{11}{8}$ ▒ $\frac{23}{15}$

17. $-\frac{2}{3}$ ▒ $-\frac{5}{9}$

18. -4.78 ▒ -5.01

Answers on page A-23

e Absolute Value

From the number line, we see that numbers like 4 and -4 are the same distance from zero. Distance is always a nonnegative number. We call the distance from zero the **absolute value** of the number.

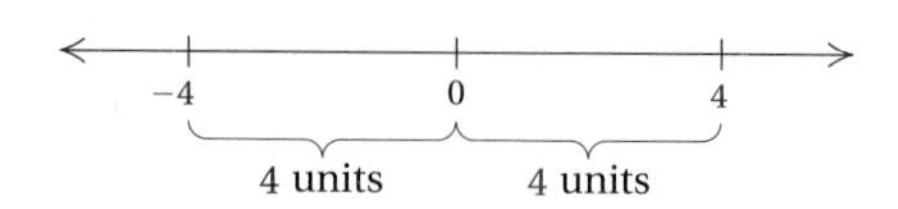

> The **absolute value** of a number is its distance from zero on a number line. We use the symbol $|x|$ to represent the absolute value of a number x.

To find the absolute value of a number:

a) If a number is negative, make it positive.

b) If a number is positive or zero, leave it alone.

Examples Find the absolute value.

18. $|-7|$ The distance of -7 from 0 is 7, so $|-7| = 7$.

19. $|12|$ The distance of 12 from 0 is 12, so $|12| = 12$.

20. $|0|$ The distance of 0 from 0 is 0, so $|0| = 0$.

21. $\left|\frac{3}{2}\right| = \frac{3}{2}$

22. $|-2.73| = 2.73$

Do Exercises 19–23.

Find the absolute value.

19. $|8|$

20. $|0|$

21. $|-9|$

22. $\left|-\frac{2}{3}\right|$

23. $|5.6|$

Answers on page A-23

Exercise Set 10.1

a Tell which real numbers correspond to the situation.

1. In the television game show "Jeopardy," a contestant loses \$200 on the first question. On the second question, she wins \$600.

2. The temperature on Wednesday was 18° above zero. On Thursday, it was 2° below zero.

3. In bowling, after the first game, team A is 34 pins behind team B, and team B is 15 pins ahead of team C.

4. The Dead Sea, between Jordan and Israel, is 1312 ft below sea level, whereas Mt. Whitney in California is 14,494 ft above sea level (***Source***: *World Almanac*).

5. A student deposited \$750 in a savings account. Two weeks later, he withdrew \$125.

6. During a certain time period, the United States had a deficit of \$3 million in foreign trade.

7. During a video game, a player intercepted a missile worth 20 points, lost a starship worth 150 points, and captured a base worth 300 points.

8. It is 3 sec before liftoff of the space shuttle. 128 sec after liftoff, the shuttle loses its solid fuel rockets.

b Graph the number on the number line.

9. $\frac{10}{3}$

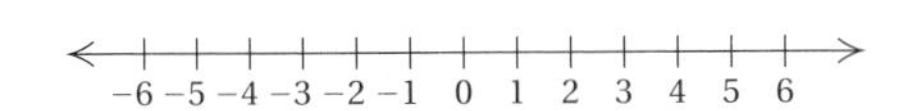

10. $-\frac{12}{5}$

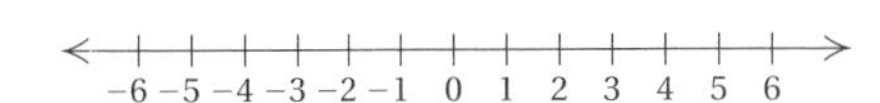

11. -4.3

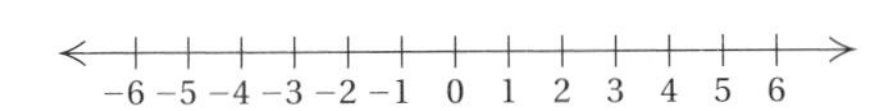

12. 5.78

c Find decimal notation.

13. $-\frac{5}{8}$ **14.** $-\frac{1}{8}$ **15.** $-\frac{5}{3}$ **16.** $-\frac{5}{6}$ **17.** $-\frac{7}{6}$

18. $-\frac{11}{12}$ **19.** $-\frac{7}{8}$ **20.** $-\frac{3}{10}$ **21.** $-\frac{7}{20}$ **22.** $-\frac{7}{3}$

d Use either $<$ or $>$ for $\square$ to write a true sentence.

23. $6 \square 0$ **24.** $9 \square 0$ **25.** $-8 \square 6$ **26.** $8 \square -8$ **27.** $-6 \square 6$

28. $0 \square -7$ **29.** $-6 \square -3$ **30.** $-4 \square -3$ **31.** $-5 \square -11$ **32.** $-3 \square -4$

33. $-6 \square -5$ **34.** $-10 \square -14$ **35.** $2.14 \square 1.24$ **36.** $-3.3 \square -2.2$ **37.** $-14.5 \square 0.011$

38. 17.2 ▭ −1.67

39. −12.88 ▭ −6.45

40. −14.34 ▭ −17.88

41. $\frac{5}{12}$ ▭ $\frac{11}{25}$

42. $-\frac{14}{17}$ ▭ $-\frac{27}{35}$

e Find the absolute value.

43. $|-3|$

44. $|-7|$

45. $|18|$

46. $|0|$

47. $|11|$

48. $|-4|$

49. $|-24|$

50. $|325|$

51. $\left|-\frac{2}{3}\right|$

52. $\left|-\frac{10}{7}\right|$

53. $\left|\frac{0}{4}\right|$

54. $|14.8|$

Skill Maintenance

Find the prime factorization. [2.1d]

55. 54

56. 192

57. 102

58. 260

59. 864

60. 468

Find the LCM. [3.1a]

61. 6, 18

62. 18, 24

63. 6, 24, 32

64. 12, 24, 36

65. 48, 56, 64

66. 12, 36, 84

Synthesis

67. ◆ Give three examples of rational numbers that are not integers. Explain.

68. ◆ Give three examples of irrational numbers. Explain the difference between an irrational number and a rational number.

Use either <, >, or = for ▭ to write a true sentence.

69. $|-5|$ ▭ $|-2|$

70. $|4|$ ▭ $|-7|$

71. $|-8|$ ▭ $|8|$

List in order from the least to the greatest.

72. $-\frac{2}{3}, \frac{1}{2}, -\frac{3}{4}, -\frac{5}{6}, \frac{3}{8}, \frac{1}{6}$

73. $-8\frac{7}{8}, 7, -5, |-6|, 4, |3|, -8\frac{5}{8}, -100, 0, 1^7, \frac{14}{4}, -\frac{67}{8}$

10.2 Addition of Real Numbers

We now consider addition of real numbers. First, to gain an understanding, we add using a number line. Then we consider rules for addition.

Addition on a Number Line

Addition of numbers can be illustrated on a number line. To do the addition $a + b$, we start at a, and then move according to b.

a) If b is positive, we move to the right.

b) If b is negative, we move to the left.

c) If b is 0, we stay at a.

Example 1 Add: $3 + (-5)$.

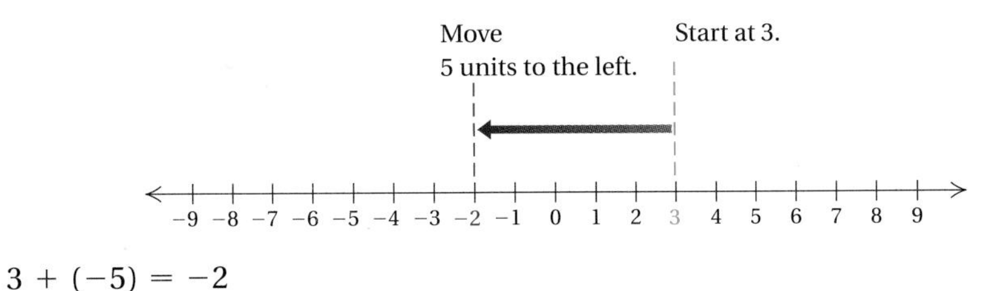

$3 + (-5) = -2$

Example 2 Add: $-4 + (-3)$.

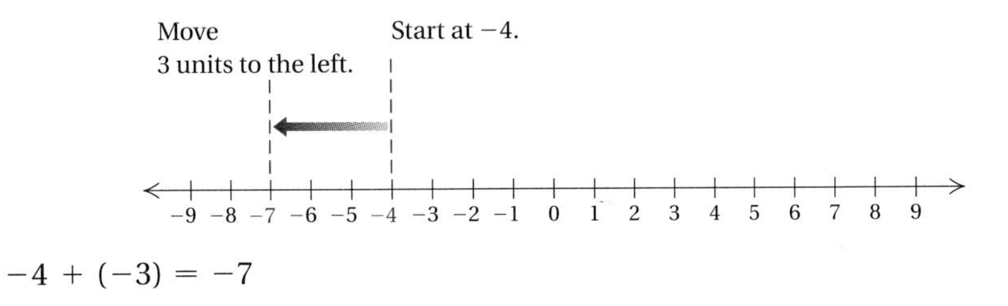

$-4 + (-3) = -7$

Example 3 Add: $-4 + 9$.

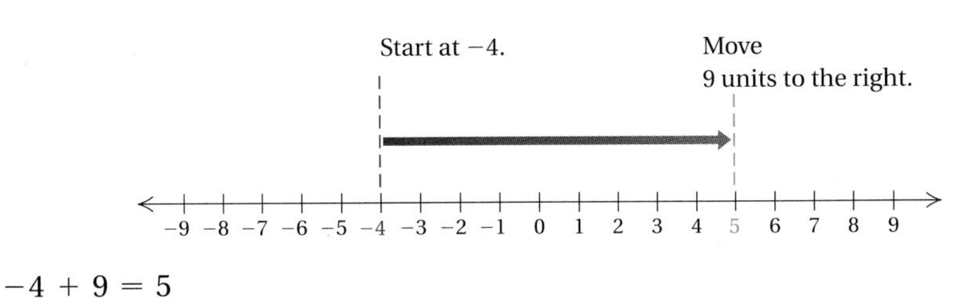

$-4 + 9 = 5$

Example 4 Add: $-5.2 + 0$.

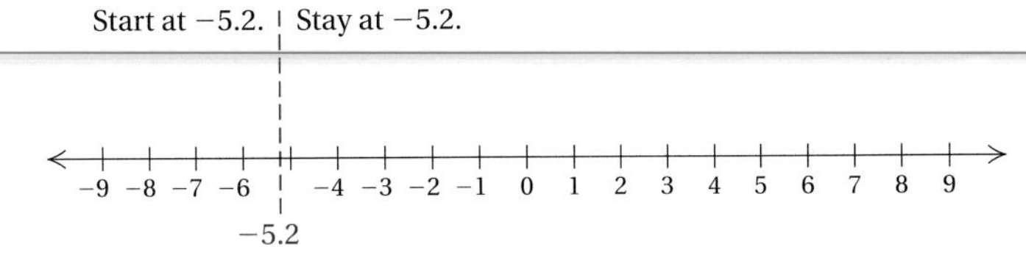

$-5.2 + 0 = -5.2$

Do Exercises 1–6.

Objectives

a — Add real numbers without using a number line.

b — Find the opposite, or additive inverse, of a real number.

For Extra Help

TAPE 17

TAPE 16A

MAC WIN

CD-ROM

Add using a number line.

1. $0 + (-8)$

2. $1 + (-4)$

3. $-3 + (-5)$

4. $-3 + 7$

5. $-5.4 + 5.4$

6. $-\frac{5}{2} + \frac{1}{2}$

Answers on page A-23

Add without using a number line.

7. $-5 + (-6)$

8. $-9 + (-3)$

9. $-4 + 6$

10. $-7 + 3$

11. $5 + (-7)$

12. $-20 + 20$

13. $-11 + (-11)$

14. $10 + (-7)$

15. $-0.17 + 0.7$

16. $-6.4 + 8.7$

17. $-4.5 + (-3.2)$

18. $-8.6 + 2.4$

19. $\frac{5}{9} + \left(-\frac{7}{9}\right)$

20. $-\frac{1}{5} + \left(-\frac{3}{4}\right)$

Answers on page A-23

a Addition Without a Number Line

You may have noticed some patterns in the preceding examples. These lead us to rules for adding without using a number line that are more efficient for adding larger or more complicated numbers.

Rules for Addition of Real Numbers

1. *Positive numbers*: Add the same as arithmetic numbers. The answer is positive.
2. *Negative numbers*: Add absolute values. The answer is negative.
3. *A positive and a negative number*: Subtract the smaller absolute value from the larger. Then:
   a) If the positive number has the greater absolute value, the answer is positive.
   b) If the negative number has the greater absolute value, the answer is negative.
   c) If the numbers have the same absolute value, the answer is 0.
4. *One number is zero*: The sum is the other number.

Rule 4 is known as the **identity property of 0.** It says that for any real number a, $a + 0 = a$.

Examples Add without using a number line.

5. $-12 + (-7) = -19$ — **Two negatives. *Think*: Add the absolute values, 12 and 7, getting 19. Make the answer *negative*, −19.**

6. $-1.4 + 8.5 = 7.1$ — **The absolute values are 1.4 and 8.5. The difference is 7.1. The positive number has the larger absolute value, so the answer is *positive*, 7.1.**

7. $-36 + 21 = -15$ — **The absolute values are 36 and 21. The difference is 15. The negative number has the larger absolute value, so the answer is *negative*, −15.**

8. $1.5 + (-1.5) = 0$ — **The numbers have the same absolute value. The sum is 0.**

9. $-\frac{7}{8} + 0 = -\frac{7}{8}$ — **One number is zero. The sum is $-\frac{7}{8}$.**

10. $-9.2 + 3.1 = -6.1$

11. $-\frac{3}{2} + \frac{9}{2} = \frac{6}{2} = 3$

12. $-\frac{2}{3} + \frac{5}{8} = -\frac{16}{24} + \frac{15}{24} = -\frac{1}{24}$

Do Exercises 7–20.

Suppose we wish to add several numbers, some positive and some negative, as follows. How can we proceed?

$$15 + (-2) + 7 + 14 + (-5) + (-12)$$

We can change grouping and order as we please when adding. For instance, we can group the positive numbers together and the negative numbers together and add them separately. Then we add the two results.

Example 13 Add: $15 + (-2) + 7 + 14 + (-5) + (-12)$.

a) $15 + 7 + 14 = 36$ **Adding the positive numbers**

b) $-2 + (-5) + (-12) = -19$ **Adding the negative numbers**

c) $36 + (-19) = 17$ **Adding the results**

We can also add the numbers in any other order we wish, say, from left to right as follows:

$$\begin{aligned} 15 + (-2) + 7 + 14 + (-5) + (-12) &= 13 + 7 + 14 + (-5) + (-12) \\ &= 20 + 14 + (-5) + (-12) \\ &= 34 + (-5) + (-12) \\ &= 29 + (-12) \\ &= 17 \end{aligned}$$

Do Exercises 21–24.

b Opposites, or Additive Inverses

Suppose we add two numbers that are opposites, such as 6 and -6. The result is 0. When opposites are added, the result is always 0. Such numbers are also called **additive inverses.** Every real number has an opposite, or additive inverse.

> Two numbers whose sum is 0 are called **opposites,** or **additive inverses,** of each other.

Examples Find the opposite.

14. 34 The opposite of 34 is -34 because $34 + (-34) = 0$.

15. -8 The opposite of -8 is 8 because $-8 + 8 = 0$.

16. 0 The opposite of 0 is 0 because $0 + 0 = 0$.

17. $-\frac{7}{8}$ The opposite of $-\frac{7}{8}$ is $\frac{7}{8}$ because $-\frac{7}{8} + \frac{7}{8} = 0$.

Do Exercises 25–30.

To name the opposite, we use the symbol $-$, as follows.

> The opposite, or additive inverse, of a number a can be named $-a$ (read "the opposite of a," or "the additive inverse of a").

Note that if we take a number, say, 8, and find its opposite, -8, and then find the opposite of the result, we will have the original number, 8, again.

> The opposite of the opposite of a number is the number itself. (The additive inverse of the additive inverse of a number is the number itself.) That is, for any number a,
>
> $-(-a) = a$.

Add.

21. $(-15) + (-37) + 25 + 42 + (-59) + (-14)$

22. $42 + (-81) + (-28) + 24 + 18 + (-31)$

23. $-2.5 + (-10) + 6 + (-7.5)$

24.
$$\begin{array}{r} -35 \\ 17 \\ 14 \\ -27 \\ 31 \\ -12 \\ \hline \end{array}$$

Find the opposite.

25. -4

26. 8.7

27. -7.74

28. $-\frac{8}{9}$

29. 0

30. 12

Answers on page A-23

Find $-x$ and $-(-x)$ when x is each of the following.

31. 14

32. 1

33. -19

34. -1.6

35. $\frac{2}{3}$

36. $-\frac{9}{8}$

Change the sign. (Find the opposite.)

37. -4

38. -13.4

39. 0

40. $\frac{1}{4}$

Answers on page A-23

Example 18 Find $-x$ and $-(-x)$ when $x = 16$.

We replace x in each case with 16.

a) If $x = 16$, then $-x = -16 = -16$. **The opposite of 16 is -16.**

b) If $x = 16$, then $-(-x) = -(-16) = 16$. **The opposite of the opposite of 16 is 16.**

Example 19 Find $-x$ and $-(-x)$ when $x = -3$.

We replace x in each case with -3.

a) If $x = -3$, then $-x = -(-3) = 3$.

b) If $x = -3$, then $-(-x) = -(-(-3)) = -3$.

Note that in Example 19 we used an extra set of parentheses to show that we are substituting the negative number -3 for x. Symbolism like $--x$ is not considered meaningful.

Do Exercises 31–36.

A symbol such as -8 is usually read "negative 8." It could be read "the additive inverse of 8," because the additive inverse of 8 is negative 8. It could also be read "the opposite of 8," because the opposite of 8 is -8. Thus a symbol like -8 can be read in more than one way. A symbol like $-x$, which has a variable, should be read "the opposite of x" or "the additive inverse of x" and *not* "negative x," because we do not know whether x represents a positive number, a negative number, or 0. You can verify this by referring to the preceding examples.

We can use the symbolism $-a$ to restate the definition of opposite, or additive inverse.

> For any real number a, the opposite, or additive inverse, of a, $-a$, is such that
>
> $$a + (-a) = (-a) + a = 0.$$

Signs of Numbers

A negative number is sometimes said to have a "negative sign." A positive number is said to have a "positive sign." When we replace a number with its opposite, we can say that we have "changed its sign."

Examples Change the sign. (Find the opposite.)

20. -3 $\quad -(-3) = 3$ **The opposite of -3 is 3.**

21. -10 $\quad -(-10) = 10$

22. 0 $\quad -(0) = 0$

23. 14 $\quad -(14) = -14$

Do Exercises 37–40.

Exercise Set 10.2

a Add. Do not use a number line except as a check.

1. $-9 + 2$

2. $-5 + 2$

3. $-10 + 6$

4. $4 + (-3)$

5. $-8 + 8$

6. $4 + (-4)$

7. $-3 + (-5)$

8. $-6 + (-8)$

9. $-7 + 0$

10. $-10 + 0$

11. $0 + (-27)$

12. $0 + (-36)$

13. $17 + (-17)$

14. $-20 + 20$

15. $-17 + (-25)$

16. $-23 + (-14)$

17. $18 + (-18)$

18. $-13 + 13$

19. $-18 + 18$

20. $11 + (-11)$

21. $8 + (-5)$

22. $-7 + 8$

23. $-4 + (-5)$

24. $10 + (-12)$

25. $13 + (-6)$

26. $-3 + 14$

27. $-25 + 25$

28. $40 + (-40)$

29. $63 + (-18)$

30. $85 + (-65)$

31. $-6.5 + 4.7$

32. $-3.6 + 1.9$

33. $-2.8 + (-5.3)$

34. $-7.9 + (-6.5)$

35. $-\frac{3}{5} + \frac{2}{5}$

36. $-\frac{4}{3} + \frac{2}{3}$

37. $-\frac{3}{7} + \left(-\frac{5}{7}\right)$

38. $-\frac{4}{9} + \left(-\frac{6}{9}\right)$

39. $-\frac{5}{8} + \frac{1}{4}$

40. $-\frac{5}{6} + \frac{2}{3}$

41. $-\frac{3}{7} + \left(-\frac{2}{5}\right)$

42. $-\frac{5}{8} + \left(-\frac{1}{3}\right)$

43. $-\frac{3}{5} + \left(-\frac{2}{15}\right)$

44. $-\frac{5}{9} + \left(-\frac{5}{18}\right)$

45. $-5.7 + (-7.2) + 6.6$

46. $-10.3 + (-7.5) + 3.1$

47. $-\frac{7}{16} + \frac{7}{8}$

48. $-\frac{3}{24} + \frac{7}{36}$

49. $75 + (-14) + (-17) + (-5)$

50. $28 + (-44) + 17 + 31 + (-94)$

51. $-44 + \left(-\frac{3}{8}\right) + 95 + \left(-\frac{5}{8}\right)$

52. $24 + 3.1 + (-44) + (-8.2) + 63$

53. $98 + (-54) + 113 + (-998) + 44 + (-612) + (-18) + 334$

54. $-455 + (-123) + 1026 + (-919) + 213 + 111 + (-874)$

b Find the opposite, or additive inverse.

55. 24

56. -84

57. -26.9

58. 27.4

Find $-x$ when x is each of the following.

59. 9

60. -26

61. $-\frac{14}{3}$

62. $\frac{1}{526}$

Find $-(-x)$ when x is each of the following.

63. -65

64. 31

65. $\frac{5}{3}$

66. -7.8

Change the sign. (Find the opposite.)

67. -14

68. -18.3

69. 10

70. $-\frac{5}{8}$

Skill Maintenance

Find the area.

71. [8.4a]

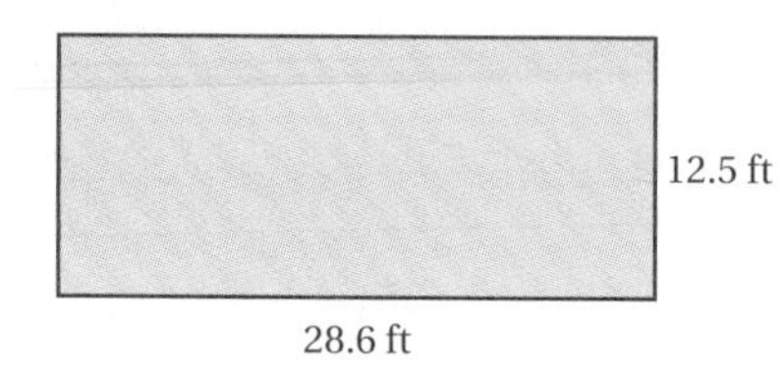

72. [8.5a]

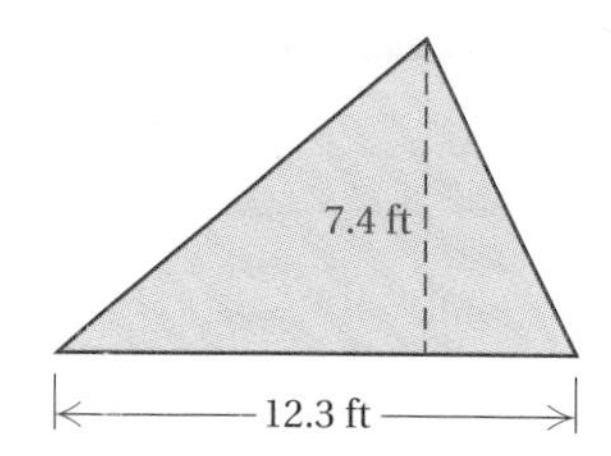

73. [8.4a]

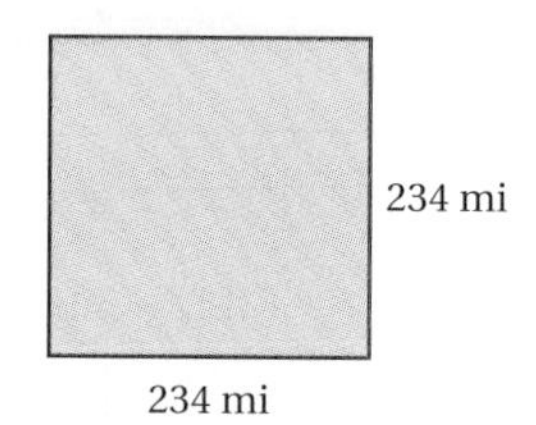

74. [8.5a]

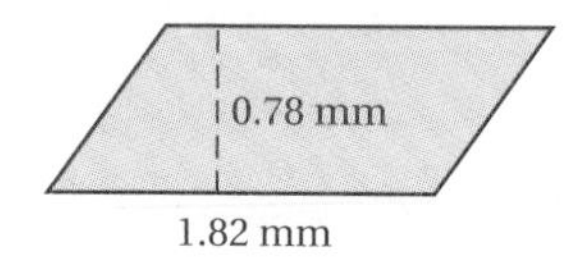

75. [8.5a]

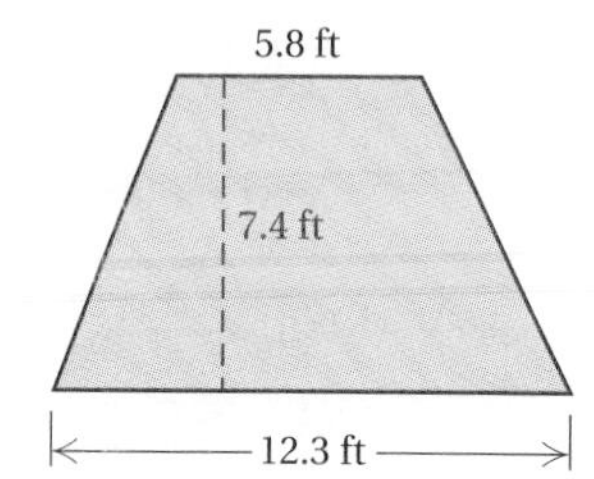

76. [8.6c]

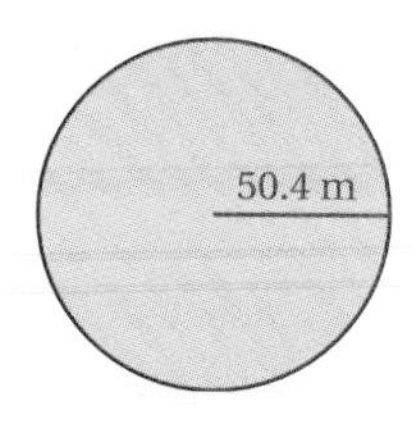

Synthesis

77. ◆ Explain in your own words why the sum of two negative numbers is always negative.

78. ◆ A student states that -93 is "bigger than" -47. What mistake is the student making?

79. For what numbers x is $-x$ negative?

80. For what numbers x is $-x$ positive?

Add.

81. 🖩 $-3496 + (-2987)$

82. 🖩 $497 + (-3028)$

83. For what numbers x is $x + (-7)$ positive?

84. For what numbers x is $-7 + x$ negative?

Tell whether the sum is positive, negative, or zero.

85. If n is positive and m is negative, then $-n + m$ is ________.

86. If $n = m$ and n is negative, then $-n + (-m)$ is ________.

Add integers using a variety of methods.

Collaborative Learning Manual

10.3 Subtraction of Real Numbers

a Subtraction

We now consider subtraction of real numbers. Subtraction is defined as follows.

> The difference $a - b$ is the number that when added to b gives a.

For example, $45 - 17 = 28$ because $28 + 17 = 45$. Let's consider an example whose answer is a negative number.

Example 1 Subtract: $5 - 8$.

Think: $5 - 8$ is the number that when added to 8 gives 5. What number can we add to 8 to get 5? The number must be negative. The number is -3:

$$5 - 8 = -3.$$

That is, $5 - 8 = -3$ because $-3 + 8 = 5$.

Do Exercises 1–3.

The definition above does *not* provide the most efficient way to do subtraction. From that definition, however, a faster way can be developed. Look for a pattern in the following examples.

Subtractions	Adding an Opposite
$5 - 8 = -3$	$5 + (-8) = -3$
$-6 - 4 = -10$	$-6 + (-4) = -10$
$-7 - (-10) = 3$	$-7 + 10 = 3$
$-7 - (-2) = -5$	$-7 + 2 = -5$

Do Exercises 4–7.

Perhaps you have noticed that we can subtract by adding the opposite of the number being subtracted. This can always be done.

> For any real numbers a and b,
>
> $$a - b = a + (-b).$$
>
> (To subtract, add the opposite, or additive inverse, of the number being subtracted.)

This is the method generally used for quick subtraction of real numbers.

Objective

a Subtract real numbers and simplify combinations of additions and subtractions.

For Extra Help

TAPE 17 TAPE 16B MAC WIN CD-ROM

Subtract.

1. $-6 - 4$

Think: What number can be added to 4 to get -6:

$\blacksquare + 4 = -6?$

2. $-7 - (-10)$

Think: What number can be added to -10 to get -7:

$\blacksquare + (-10) = -7?$

3. $-7 - (-2)$

Think: What number can be added to -2 to get -7:

$\blacksquare + (-2) = -7?$

Complete the addition and compare with the subtraction.

4. $4 - 6 = -2$;
$4 + (-6) =$ ______

5. $-3 - 8 = -11$;
$-3 + (-8) =$ ______

6. $-5 - (-9) = 4$;
$-5 + 9 =$ ______

7. $-5 - (-3) = -2$;
$-5 + 3 =$ ______

Answers on page A-23

Subtract.

8. $2 - 8$

9. $-6 - 10$

10. $12.4 - 5.3$

11. $-8 - (-11)$

12. $-8 - (-8)$

13. $\frac{2}{3} - \left(-\frac{5}{6}\right)$

Read each of the following. Then subtract by adding the opposite of the number being subtracted.

14. $3 - 11$

15. $12 - 5$

16. $-12 - (-9)$

17. $-12.4 - 10.9$

18. $-\frac{4}{5} - \left(-\frac{4}{5}\right)$

Simplify.

19. $-6 - (-2) - (-4) - 12 + 3$

20. $9 - (-6) + 7 - 11 - 14 - (-20)$

21. $-9.6 + 7.4 - (-3.9) - (-11)$

Answers on page A-23

Examples Subtract.

2. $2 - 6 = 2 + (-6) = -4$ — **The opposite of 6 is −6. We change the subtraction to addition and add the opposite. *Check:* −4 + 6 = 2.**

3. $4 - (-9) = 4 + 9 = 13$ — **The opposite of −9 is 9. We change the subtraction to addition and add the opposite. *Check:* 13 + (−9) = 4.**

4. $-4.2 - (-3.6) = -4.2 + 3.6 = -0.6$ — **Adding the opposite. *Check:* −0.6 + (−3.6) = −4.2.**

5. $-\frac{1}{2} - \left(-\frac{3}{4}\right) = -\frac{1}{2} + \frac{3}{4} = \frac{1}{4}$ — **Adding the opposite. *Check:* $\frac{1}{4} + \left(-\frac{3}{4}\right) = -\frac{1}{2}$.**

Do Exercises 8–13.

Examples Read each of the following. Then subtract by adding the opposite of the number being subtracted.

6. $3 - 5$;
$3 - 5 = 3 + (-5) = -2$

Read "three minus five is three plus the opposite of five"

7. $\frac{1}{8} - \frac{7}{8}$;
$\frac{1}{8} - \frac{7}{8} = \frac{1}{8} + \left(-\frac{7}{8}\right) = -\frac{6}{8}$, or $-\frac{3}{4}$

Read "one-eighth minus seven-eighths is one-eighth plus the opposite of seven-eighths"

8. $-4.6 - (-9.8)$;
$-4.6 - (-9.8) = -4.6 + 9.8 = 5.2$

Read "negative four point six minus negative nine point eight is negative four point six plus the opposite of negative nine point eight"

9. $-\frac{3}{4} - \frac{7}{5}$;
$-\frac{3}{4} - \frac{7}{5} = -\frac{3}{4} + \left(-\frac{7}{5}\right) = -\frac{15}{20} + \left(-\frac{28}{20}\right) = -\frac{43}{20}$

Read "negative three-fourths minus seven-fifths is negative three-fourths plus the opposite of seven-fifths"

Do Exercises 14–18.

When several additions and subtractions occur together, we can make them all additions.

Examples Simplify.

10. $8 - (-4) - 2 - (-4) + 2 = 8 + 4 + (-2) + 4 + 2 = 16$ — **Adding the opposite where subtractions are indicated**

11. $8.2 - (-6.1) + 2.3 - (-4) = 8.2 + 6.1 + 2.3 + 4 = 20.6$

Do Exercises 19–21.

Exercise Set 10.3

a Subtract.

1. $3 - 7$

2. $5 - 10$

3. $0 - 7$

4. $0 - 8$

5. $-8 - (-2)$

6. $-6 - (-8)$

7. $-10 - (-10)$

8. $-8 - (-8)$

9. $12 - 16$

10. $14 - 19$

11. $20 - 27$

12. $26 - 7$

13. $-9 - (-3)$

14. $-6 - (-9)$

15. $-11 - (-11)$

16. $-14 - (-14)$

17. $8 - (-3)$

18. $-7 - 4$

19. $-6 - 8$

20. $6 - (-10)$

21. $-4 - (-9)$

22. $-14 - 2$

23. $2 - 9$

24. $2 - 8$

25. $0 - 5$

26. $0 - 10$

27. $-5 - (-2)$

28. $-3 - (-1)$

29. $2 - 25$

30. $18 - 63$

31. $-42 - 26$

32. $-18 - 63$

33. $-71 - 2$

34. $-49 - 3$

35. $24 - (-92)$

36. $48 - (-73)$

37. $-2.8 - 0$

38. $6.04 - 1.1$

39. $\frac{3}{8} - \frac{5}{8}$

40. $\frac{3}{9} - \frac{9}{9}$

41. $\frac{3}{4} - \frac{2}{3}$

42. $\frac{5}{8} - \frac{3}{4}$

43. $-\frac{3}{4} - \frac{2}{3}$

44. $-\frac{5}{8} - \frac{3}{4}$

45. $-\frac{5}{8} - \left(-\frac{3}{4}\right)$

46. $-\frac{3}{4} - \left(-\frac{2}{3}\right)$

47. $6.1 - (-13.8)$

48. $1.5 - (-3.5)$

49. $-3.2 - 5.8$

50. $-2.7 - 5.9$

51. $0.99 - 1$

52. $0.87 - 1$

53. $3 - 5.7$

54. $5.1 - 3.02$

55. $7 - 10.53$

56. $8 - (-9.3)$

57. $\frac{1}{6} - \frac{2}{3}$

58. $-\frac{3}{8} - \left(-\frac{1}{2}\right)$

59. $-\frac{4}{7} - \left(-\frac{10}{7}\right)$

60. $\frac{12}{5} - \frac{12}{5}$

61. $-\frac{7}{10} - \frac{10}{15}$

62. $-\frac{4}{18} - \left(-\frac{2}{9}\right)$

63. $\frac{1}{13} - \frac{1}{12}$

64. $-\frac{1}{7} - \left(-\frac{1}{6}\right)$

Simplify.

65. $18 - (-15) - 3 - (-5) + 2$

66. $22 - (-18) + 7 + (-42) - 27$

67. $-31 + (-28) - (-14) - 17$

68. $-43 - (-19) - (-21) + 25$

69. $-93 - (-84) - 41 - (-56)$

70. $84 + (-99) + 44 - (-18) - 43$

71. $-5 - (-30) + 30 + 40 - (-12)$

72. $14 - (-50) + 20 - (-32)$

73. $132 - (-21) + 45 - (-21)$

74. $81 - (-20) - 14 - (-50) + 53$

Skill Maintenance

75. Find the area of a rectangle that is 8.4 cm by 11.5 cm. [8.4a]

76. Find the prime factorization of 750. [2.1d]

77. Find the LCM of 36 and 54. [3.1a]

78. Find the area of a square whose sides are of length 11.2 km. [8.4a]

Evaluate. [1.9b]

79. 4^3

80. 5^3

Solve. [1.8a]

81. How many 12-oz cans of soda can be filled with 96 oz of soda?

82. A case of soda contains 24 bottles. If each bottle contains 12 oz, how many ounces of soda are in the case?

Synthesis

83. ◈ If a negative number is subtracted from a positive number, will the result always be positive? Why or why not?

84. ◈ Write a problem for a classmate to solve. Design the problem so that the solution is "The temperature dropped to $-9°$."

Subtract.

85. ▦ $123{,}907 - 433{,}789$

86. ▦ $23{,}011 - (-60{,}432)$

Tell whether the statement is true or false for all integers m and n. If false, find a number that shows why.

87. $-n = 0 - n$

88. $n - 0 = 0 - n$

89. If $m \neq n$, then $m - n \neq 0$.

90. If $m = -n$, then $m + n = 0$.

91. If $m + n = 0$, then m and n are opposites.

92. If $m - n = 0$, then $m = -n$.

93. $m = -n$ if m and n are opposites.

94. If $m = -m$, then $m = 0$.

95. Velma Quarles is a stockbroker. She kept track of the changes in the stock market over a period of 5 weeks. By how many points had the market risen or fallen over this time?

Week 1	Week 2	Week 3	Week 4	Week 5
Down 13 pts	Down 16 pts	Up 36 pts	Down 11 pts	Up 19 pts

Collaborative Learning Manual

Show how to subtract integers using color tiles.

10.4 Multiplication of Real Numbers

a Multiplication

Multiplication of real numbers is very much like multiplication of arithmetic numbers. The only difference is that we must determine whether the answer is positive or negative.

Multiplication of a Positive Number and a Negative Number

To see how to multiply a positive number and a negative number, consider the pattern of the following.

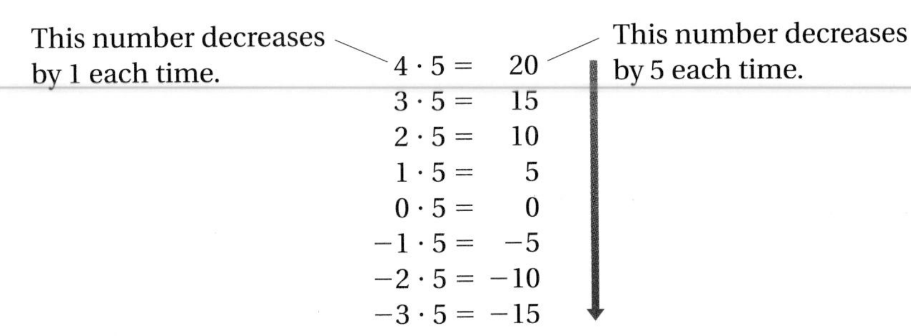

Do Exercise 1.

According to this pattern, it looks as though the product of a negative number and a positive number is negative. That is the case, and we have the first part of the rule for multiplying real numbers.

> To multiply a positive number and a negative number, multiply their absolute values. The answer is negative.

Examples Multiply.

1. $8(-5) = -40$

2. $-\frac{1}{3} \cdot \frac{5}{7} = -\frac{5}{21}$

3. $(-7.2)5 = -36$

Do Exercises 2–7.

Multiplication of Two Negative Numbers

How do we multiply two negative numbers? Again we look for a pattern.

This number decreases by 1 each time.

This number increases by 5 each time.

$$4 \cdot (-5) = -20$$
$$3 \cdot (-5) = -15$$
$$2 \cdot (-5) = -10$$
$$1 \cdot (-5) = -5$$
$$0 \cdot (-5) = 0$$
$$-1 \cdot (-5) = 5$$
$$-2 \cdot (-5) = 10$$
$$-3 \cdot (-5) = 15$$

Do Exercise 8.

Objective

a Multiply real numbers.

For Extra Help

TAPE 18 TAPE 16B MAC WIN CD-ROM

1. Complete, as in the example.

$4 \cdot 10 = 40$
$3 \cdot 10 = 30$
$2 \cdot 10 =$
$1 \cdot 10 =$
$0 \cdot 10 =$
$-1 \cdot 10 =$
$-2 \cdot 10 =$
$-3 \cdot 10 =$

Multiply.

2. $-3 \cdot 6$

3. $20 \cdot (-5)$

4. $4 \cdot (-20)$

5. $-\frac{2}{3} \cdot \frac{5}{6}$

6. $-4.23(7.1)$

7. $\frac{7}{8}\left(-\frac{4}{5}\right)$

8. Complete, as in the example.

$3 \cdot (-10) = -30$
$2 \cdot (-10) = -20$
$1 \cdot (-10) =$
$0 \cdot (-10) =$
$-1 \cdot (-10) =$
$-2 \cdot (-10) =$
$-3 \cdot (-10) =$

Answers on page A-23

Multiply.

9. $-3 \cdot (-4)$

10. $-16 \cdot (-2)$

11. $-7 \cdot (-5)$

12. $-\frac{4}{7}\left(-\frac{5}{9}\right)$

13. $-\frac{3}{2}\left(-\frac{4}{9}\right)$

14. $-3.25(-4.14)$

Multiply.

15. $5(-6)$

16. $(-5)(-6)$

17. $(-3.2) \cdot 10$

18. $\left(-\frac{4}{5}\right)\left(\frac{10}{3}\right)$

Multiply.

19. $5 \cdot (-3) \cdot 2$

20. $-3 \times (-4.1) \times (-2.5)$

21. $-\frac{1}{2} \cdot \left(-\frac{4}{3}\right) \cdot \left(-\frac{5}{2}\right)$

22. $-2 \cdot (-5) \cdot (-4) \cdot (-3)$

23. $(-4)(-5)(-2)(-3)(-1)$

24. $(-1)(-1)(-2)(-3)(-1)(-1)$

Answers on page A-23

According to the pattern, it looks as if the product of two negative numbers is positive. That is actually so, and we have the second part of the rule for multiplying real numbers.

> To multiply two negative numbers, multiply their absolute values. The answer is positive.

Do Exercises 9–14.

The following is another way to consider the rules we have for multiplication.

> To multiply two real numbers:
> a) Multiply the absolute values.
> b) If the signs are the same, the answer is positive.
> c) If the signs are different, the answer is negative.

Examples Multiply.

4. $(-3)(-4) = 12$

5. $-1.6(2) = -3.2$

6. $9 \cdot (-15) = -135$

7. $\left(-\frac{5}{6}\right)\left(-\frac{1}{9}\right) = \frac{5}{54}$

Do Exercises 15–18.

Multiplying More than Two Numbers

When multiplying more than two real numbers, we can choose order and grouping as we please.

Examples Multiply.

8. $-8 \cdot 2(-3) = -16(-3)$ — **Multiplying the first two numbers**
$= 48$ — **Multiplying the results**

9. $-8 \cdot 2(-3) = 24 \cdot 2$ — **Multiplying the negatives. Every pair of negative numbers gives a positive product.**
$= 48$

10. $-3(-2)(-5)(4) = 6(-5)(4)$ — **Multiplying the first two numbers**
$= (-30)4 = -120$

11. $\left(-\frac{1}{2}\right)(8)\left(-\frac{2}{3}\right)(-6) = (-4)4$ — **Multiplying the first two numbers and the last two numbers**
$= -16$

12. $-5 \cdot (-2) \cdot (-3) \cdot (-6) = 10 \cdot 18 = 180$

13. $(-3)(-5)(-2)(-3)(-6) = (-30)(18) = -540$

We can see the following pattern in the results of Examples 12 and 13.

> The product of an even number of negative numbers is positive.
> The product of an odd number of negative numbers is negative.

Do Exercises 19–24.

Exercise Set 10.4

a Multiply.

1. $-8 \cdot 2$

2. $-3 \cdot 5$

3. $8 \cdot (-3)$

4. $-5 \cdot 2$

5. $-9 \cdot 8$

6. $-20 \cdot 3$

7. $-8 \cdot (-2)$

8. $-4 \cdot (-5)$

9. $-7 \cdot (-6)$

10. $-9 \cdot (-2)$

11. $15 \cdot (-8)$

12. $-11 \cdot (-10)$

13. $-14 \cdot 17$

14. $-13 \cdot (-15)$

15. $-25 \cdot (-48)$

16. $39 \cdot (-43)$

17. $-3.5 \cdot (-28)$

18. $97 \cdot (-2.1)$

19. $4 \cdot (-3.1)$

20. $3 \cdot (-2.2)$

21. $-6 \cdot (-4)$

22. $-5 \cdot (-6)$

23. $-7 \cdot (-3.1)$

24. $-4 \cdot (-3.2)$

25. $\frac{2}{3} \cdot \left(-\frac{3}{5}\right)$

26. $\frac{5}{7} \cdot \left(-\frac{2}{3}\right)$

27. $-\frac{3}{8} \cdot \left(-\frac{2}{9}\right)$

28. $-\frac{5}{8} \cdot \left(-\frac{2}{5}\right)$

29. -6.3×2.7

30. -6.2×8.5

31. $-\frac{5}{9} \cdot \frac{3}{4}$

32. $-\frac{8}{3} \cdot \frac{9}{4}$

33. $7 \cdot (-4) \cdot (-3) \cdot 5$

34. $9 \cdot (-2) \cdot (-6) \cdot 7$

35. $-\frac{2}{3} \cdot \frac{1}{2} \cdot \left(-\frac{6}{7}\right)$

36. $-\frac{1}{8} \cdot \left(-\frac{1}{4}\right) \cdot \left(-\frac{3}{5}\right)$

37. $-3 \cdot (-4) \cdot (-5)$

38. $-2 \cdot (-5) \cdot (-7)$

39. $-2 \cdot (-5) \cdot (-3) \cdot (-5)$

40. $-3 \cdot (-5) \cdot (-2) \cdot (-1)$

41. $\frac{1}{5}\left(-\frac{2}{9}\right)$

42. $-\frac{3}{5}\left(-\frac{2}{7}\right)$

43. $-7 \cdot (-21) \cdot 13$

44. $-14 \cdot 34 \cdot 12$

45. $-4 \cdot (-1.8) \cdot 7$

46. $-8 \cdot (-1.3) \cdot (-5)$

47. $-\frac{1}{9}\left(-\frac{2}{3}\right)\left(\frac{5}{7}\right)$

48. $-\frac{7}{2}\left(-\frac{5}{7}\right)\left(-\frac{2}{5}\right)$

49. $4 \cdot (-4) \cdot (-5) \cdot (-12)$

50. $-2 \cdot (-3) \cdot (-4) \cdot (-5)$

51. $0.07 \cdot (-7) \cdot 6 \cdot (-6)$

52. $80 \cdot (-0.8) \cdot (-90) \cdot (-0.09)$

53. $\left(-\frac{5}{6}\right)\left(\frac{1}{8}\right)\left(-\frac{3}{7}\right)\left(-\frac{1}{7}\right)$

54. $\left(\frac{4}{5}\right)\left(-\frac{2}{3}\right)\left(-\frac{15}{7}\right)\left(\frac{1}{2}\right)$

55. $(-14) \cdot (-27) \cdot (-2)$

56. $7 \cdot (-6) \cdot 5 \cdot (-4) \cdot 3 \cdot (-2) \cdot 1 \cdot (-1)$

57. $(-8)(-9)(-10)$

58. $(-7)(-8)(-9)(-10)$

59. $(-6)(-7)(-8)(-9)(-10)$

60. $(-5)(-6)(-7)(-8)(-9)(-10)$

Skill Maintenance

61. Find the prime factorization of 4608. [2.1d]

62. Find the LCM of 36 and 60. [3.1a]

63. 23 is what percent of 69? [6.3b], [6.4b]

64. What is 36% of 729? [6.3b], [6.4b]

65. 40% of what number is 28.8? [6.3b], [6.4b]

66. What percent of 72 is 28.8? [6.3b], [6.4b]

Solve.

67. A rectangular rug measures 5 ft by 8 ft. What is the area of the rug? [8.4b]

68. How many 12-egg cartons can be filled with 2880 eggs? [1.8a]

Synthesis

69. ◆ What rule have we developed that would tell you the sign of $(-7)^8$ and $(-7)^{11}$ without doing the computations? Explain.

70. ◆ Which number is larger, $(-3)^{79}$ or $(-5)^{79}$? Why?

71. Jo wrote seven checks for $13 each. If she had a balance of $68 in her account, what was her balance after writing the checks?

72. After diving 95 m below the surface, a diver rises at a rate of 7 meters per minute for 9 min. What is the diver's new elevation?

73. What must be true of a and b if $-ab$ is to be (a) positive? (b) zero? (c) negative?

10.5 Division and Order of Operations

We now consider division of real numbers. The definition of division results in rules for division that are the same as those for multiplication.

Objectives

a. Divide integers.

b. Find the reciprocal of a real number.

c. Divide real numbers.

d. Simplify expressions using rules for order of operations.

For Extra Help

TAPE 18 TAPE 17A MAC WIN CD-ROM

a Division of Integers

The quotient $\frac{a}{b}$ (or $a \div b$) is the number, if there is one, that when multiplied by b gives a.

Let's use the definition to divide integers.

Examples Divide, if possible. Check your answer.

1. $14 \div (-7) = -2$ — ***Think*: What number multiplied by −7 gives 14? That number is −2. *Check*: $(-2)(-7) = 14$.**

2. $\frac{-32}{-4} = 8$ — ***Think*: What number multiplied by −4 gives −32? That number is 8. *Check*: $8(-4) = -32$.**

3. $\frac{-10}{7} = -\frac{10}{7}$ — ***Think*: What number multiplied by 7 gives −10? That number is $-\frac{10}{7}$. *Check*: $-\frac{10}{7} \cdot 7 = -10$.**

4. $\frac{-17}{0}$ **is undefined.** — ***Think*: What number multiplied by 0 gives −17? There is no such number because the product of 0 and *any* number is 0.**

The rules for division are the same as those for multiplication. We state them together.

To multiply or divide two real numbers:

a) Multiply or divide the absolute values.

b) If the signs are the same, the answer is positive.

c) If the signs are different, the answer is negative.

Do Exercises 1–8.

Divide.

1. $6 \div (-3)$

   Think: What number multiplied by −3 gives 6?

2. $\frac{-15}{-3}$

   Think: What number multiplied by −3 gives −15?

3. $-24 \div 8$

   Think: What number multiplied by 8 gives −24?

4. $\frac{-72}{-8}$

5. $\frac{30}{-5}$

6. $\frac{30}{-7}$

7. $\frac{-5}{0}$

8. $\frac{0}{-3}$

Answers on page A-23

Division by Zero

Example 4 shows why we cannot divide −17 by 0. We can use the same argument to show why we cannot divide any nonzero number b by 0. Consider $b \div 0$. We look for a number that when multiplied by 0 gives b. There is no such number because the product of 0 and any number is 0. Thus we cannot divide a nonzero number b by 0.

On the other hand, if we divide 0 by 0, we look for a number r such that $0 \cdot r = 0$. But $0 \cdot r = 0$ for any number r. Thus it appears that $0 \div 0$ could be any number we choose. Getting any answer we want when we divide 0 by 0 would be very confusing. Thus we agree that division by zero is undefined.

Division by zero is undefined: $a \div 0$, or $\frac{a}{0}$, is undefined for all real numbers a.

Zero divided by a nonzero number a is 0: $0 \div a = 0$, or $\frac{0}{a} = 0$, $a \neq 0$.

b Reciprocals

When two numbers like $\frac{7}{8}$ and $\frac{8}{7}$ are multiplied, the result is 1. Such numbers are called **reciprocals** of each other. Every nonzero real number has a reciprocal, also called a **multiplicative inverse.**

> Two numbers whose product is 1 are called **reciprocals** of each other.

Examples Find the reciprocal.

5. -5 The reciprocal of -5 is $-\frac{1}{5}$ because $-5\left(-\frac{1}{5}\right) = 1$.

6. $-\frac{1}{2}$ The reciprocal of $-\frac{1}{2}$ is -2 because $\left(-\frac{1}{2}\right)(-2) = 1$.

7. $-\frac{2}{3}$ The reciprocal of $-\frac{2}{3}$ is $-\frac{3}{2}$ because $\left(-\frac{2}{3}\right)\left(-\frac{3}{2}\right) = 1$.

> For $a \neq 0$, the reciprocal of a can be named $\frac{1}{a}$ and the reciprocal of $\frac{1}{a}$ is a.
>
> The reciprocal of a nonzero number $\frac{a}{b}$ can be named $\frac{b}{a}$.
>
> The number 0 has no reciprocal.

Do Exercises 9–14.

The reciprocal of a positive number is also a positive number, because their product must be the positive number 1. The reciprocal of a negative number is also a negative number, because their product must be the positive number 1.

> The reciprocal of a number has the same sign as the number itself.

It is important *not* to confuse *opposite* with *reciprocal.* Keep in mind that the opposite, or additive inverse, of a number is what we add to the number to get 0. A reciprocal, or multiplicative inverse, is what we multiply the number by to get 1. Compare the following.

Number	Opposite (Change the Sign.)	Reciprocal (Invert But Do Not Change the Sign.)
$-\frac{3}{8}$	$\frac{3}{8}$	$-\frac{8}{3}$
19	-19	$\frac{1}{19}$
$\frac{18}{7}$	$-\frac{18}{7}$	$\frac{7}{18}$
-7.9	7.9	$-\frac{1}{7.9}$ or $-\frac{10}{79}$
0	0	Undefined

$\left(-\frac{3}{8}\right)\left(-\frac{8}{3}\right) = 1$

$-\frac{3}{8} + \frac{3}{8} = 0$

Do Exercise 15.

Find the reciprocal.

9. $\frac{2}{3}$

10. $-\frac{5}{4}$

11. -3

12. $-\frac{1}{5}$

13. 5.78

14. $-\frac{2}{7}$

15. Complete the following table.

Number	Opposite	Reciprocal
$\frac{2}{3}$		
$-\frac{5}{4}$		
0		
1		
-4.5		

Answers on page A-23

c Division of Real Numbers

We know that we can subtract by adding an opposite. Similarly, we can divide by multiplying by a reciprocal.

> For any real numbers a and b, $b \neq 0$,
>
> $$a \div b = \frac{a}{b} = a \cdot \frac{1}{b}.$$
>
> (To divide, we can multiply by the reciprocal of the divisor.)

Examples Rewrite the division as a multiplication.

8. $-4 \div 3$ $\quad -4 \div 3$ is the same as $-4 \cdot \frac{1}{3}$

9. $\frac{6}{-7}$ $\quad \frac{6}{-7} = 6\left(-\frac{1}{7}\right)$

10. $\frac{3}{5} \div \left(-\frac{9}{7}\right)$ $\quad \frac{3}{5} \div \left(-\frac{9}{7}\right) = \frac{3}{5}\left(-\frac{7}{9}\right)$

Do Exercises 16–20.

When actually doing division calculations, we sometimes multiply by a reciprocal and we sometimes divide directly. With fractional notation, it is generally better to multiply by a reciprocal. With decimal notation, it is generally better to divide directly.

Examples Divide by multiplying by the reciprocal of the divisor.

11. $\frac{2}{3} \div \left(-\frac{5}{4}\right) = \frac{2}{3} \cdot \left(-\frac{4}{5}\right) = -\frac{8}{15}$

12. $-\frac{5}{6} \div \left(-\frac{3}{4}\right) = -\frac{5}{6} \cdot \left(-\frac{4}{3}\right) = \frac{20}{18} = \frac{10 \cdot 2}{9 \cdot 2} = \frac{10}{9} \cdot \frac{2}{2} = \frac{10}{9}$

Caution! Be careful not to change the sign when taking a reciprocal!

13. $-\frac{3}{4} \div \frac{3}{10} = -\frac{3}{4} \cdot \left(\frac{10}{3}\right) = -\frac{30}{12} = -\frac{5}{2} \cdot \frac{6}{6} = -\frac{5}{2}$

With decimal notation, it is easier to carry out long division than to multiply by the reciprocal.

Examples Divide.

14. $-27.9 \div (-3) = \frac{-27.9}{-3} = 9.3$ $\quad$ **Do the long division $3\overline{)27.9}$ = 9.3. The answer is positive.**

15. $-6.3 \div 2.1 = -3$ $\quad$ **Do the long division $2.1\overline{)6.3_{\wedge}0}$ = 3.0. The answer is negative.**

Do Exercises 21–24.

Rewrite the division as a multiplication.

16. $\frac{4}{7} \div \left(-\frac{3}{5}\right)$

17. $\frac{5}{-8}$

18. $\frac{-10}{7}$

19. $-\frac{2}{3} \div \frac{4}{7}$

20. $-5 \div 7$

Divide by multiplying by the reciprocal of the divisor.

21. $\frac{4}{7} \div \left(-\frac{3}{5}\right)$

22. $-\frac{8}{5} \div \frac{2}{3}$

23. $-\frac{12}{7} \div \left(-\frac{3}{4}\right)$

24. Divide: $21.7 \div (-3.1)$.

Answers on page A-24

Simplify.

25. $23 - 42 \cdot 30$

26. $32 \div 8 \cdot 2$

27. $52 \cdot 5 + 5^3 - (4^2 - 48 \div 4)$

28. $\dfrac{5 - 10 - 5 \cdot 23}{2^3 + 3^2 - 7}$

Answers on page A-24

d Order of Operations

When several operations are to be done in a calculation or a problem, we apply the same rules that we did in Sections 1.9, 3.7, and 4.4. We repeat them here for review. If you did not study those sections before, you should do so before continuing.

RULES FOR ORDER OF OPERATIONS

1. Do all calculations within parentheses before operations outside.
2. Evaluate all exponential expressions.
3. Do all multiplications and divisions in order from left to right.
4. Do all additions and subtractions in order from left to right.

These rules are consistent with the way in which most computers do calculations.

Example 16 Simplify: $-34 \cdot 56 - 17$.

There are no parentheses or powers so we start with the third step.

$-34 \cdot 56 - 17 = -1904 - 17$ Carrying out all multiplications and divisions in order from left to right

$= -1921$ Carrying out all additions and subtractions in order from left to right

Example 17 Simplify: $2^4 + 51 \cdot 4 - (37 + 23 \cdot 2)$.

$2^4 + 51 \cdot 4 - (37 + 23 \cdot 2)$

$= 2^4 + 51 \cdot 4 - (37 + 46)$ Carrying out all operations inside parentheses first, multiplying 23 by 2, following the rules for order of operations within the parentheses

$= 2^4 + 51 \cdot 4 - 83$ Completing the addition inside parentheses

$= 16 + 51 \cdot 4 - 83$ Evaluating exponential expressions

$= 16 + 204 - 83$ Doing all multiplications

$= 220 - 83$ Doing all additions and subtractions in order from left to right

$= 137$

A fraction bar can play the role of a grouping symbol, although such a symbol is not as evident as the others.

Example 18 Simplify: $\dfrac{-64 \div (-16) \div (-2)}{2^3 - 3^2}$.

An equivalent expression with brackets as grouping symbols is

$$[-64 \div (-16) \div (-2)] \div [2^3 - 3^2].$$

This shows, in effect, that we can do the calculations in the numerator and then in the denominator, and divide the results:

$$\frac{-64 \div (-16) \div (-2)}{2^3 - 3^2} = \frac{4 \div (-2)}{8 - 9} = \frac{-2}{-1} = 2.$$

Do Exercises 25–28.

Calculator Spotlight

Let's do some calculations with real numbers on a calculator. To enter a negative number on some calculators, we use the [+/−] key. To enter −5, we press [5] and then [+/−]. The display then reads

−5

Some graphing calculators use an opposite key, [(−)]. To enter −5 on such a grapher, we press [(−)] [5].

To do a calculation like −8 − (−2.3), we press the following keys:

[8] [+/−] [−] [2] [.] [3] [+/−] [=]

or

[(−)] [8] [−] [(−)] [2] [.] [3] [ENTER].

The answer is −5.7.

Note that we did not need grouping symbols, or parentheses, in the preceding keystrokes. Many calculators do provide grouping symbols. Such keys may appear as [(] and [)] or [[(. . .] and [. . .)]]. To do a calculation like −7(2 − 9) − 20 on such a calculator, we press the following keys:

[7] [+/−] [×] [(] [2] [−] [9] [)]
[−] [2] [0] [=]

or

[(−)] [7] [(] [2] [−] [9] [)]
[−] [2] [0] [ENTER].

The multiplication key [×] could also be used but often that is not necessary on a graphing calculator. The answer is 29.

If we want to enter a power like $(-39)^4$, the keystrokes are

[3] [9] [+/−] [x^y] [4] [=]

or

[(] [(−)] [3] [9] [)] [^] [4] [ENTER].

The answer is 2,313,441.

To find -39^4, think of the calculation as -1×39^4. The keystrokes are

[1] [+/−] [×] [3] [9] [x^y] [4] [=]

or

[(−)] [3] [9] [^] [4] [ENTER].

The answer is −2,313,441.

To simplify an expression like

$$\frac{38 + 142}{2 - 47},$$

we use grouping symbols to write it as

$(38 + 142) \div (2 - 47)$.

We then press

[(] [3] [8] [+] [1] [4] [2] [)] [÷]
[(] [2] [−] [4] [7] [)] [=].

The answer is −4.

Exercises

Press the appropriate keys so that your calculator displays each of the following numbers.

1. −9

2. −57

3. −1996

4. −24.7

Evaluate.

5. $-8 + 4(7 - 9) + 5$

6. $-3[2 + (-5)]$

7. $7[4 - (-3)] + 5[3^2 - (-4)]$

Evaluate.

8. $(-7)^6$

9. $(-17)^5$

10. $(-104)^3$

11. -7^6

12. -17^5

13. -104^3

Calculate.

14. $\frac{38 - 178}{5 + 30}$

15. $\frac{311 - 17^2}{2 - 13}$

16. $785 - \frac{285 - 5^4}{17 + 3 \cdot 51}$

17. Consider only the numbers 2, 4, 6, and 8. Assume that each can be placed in a blank as follows:

$\square \div \square \cdot \square - \square^2$

What placement of the numbers in the blanks yields the largest number?

18. Consider only the numbers 2, 4, 6, and 8. Assume that each can be placed in a blank as follows:

$\square - \square + \square^2 \div \square$

What placement of the numbers in the blanks yields the largest number?

In Exercises 19 and 20, place one of +, −, ×, and ÷ in each blank to make a true sentence.

19. $-32 \ \square \ (88 \ \square \ 29) = -1888$

20. $3^5 \ \square \ 10^2 \ \square \ 5^2 = -22.57$

Improving Your Math Study Skills

Tips from a Former Student

A former student of Professor Bittinger, Mike Rosenborg earned a master's degree in mathematics and now teaches mathematics. Here are some of his study tips.

- Because working problems is the best way to learn math, instructors generally assign lots of problems. Never let yourself get behind in your math homework.
- If you are struggling with a math concept, do not give up. Ask for help from your friends and your instructor. Since each concept is built on previous concepts, any gaps in your understanding will follow you through the entire course, so make sure you understand each concept as you go along.
- Math contains many rules that cannot be "bent." Don't try inventing your own rules and still expect to get correct answers. Although there is usually more than one way to solve a problem, each method must follow the established rules.
- Read your textbook! It will often contain the tips and help you need to solve any problem with which you're struggling. It may also bring out points that you missed in class or that your instructor may not have covered.
- Learn to use scratch paper to jot down your thoughts and to draw pictures. Don't try to figure everything out "in your head." You will think more clearly and accurately this way.
- When preparing for a test, it is often helpful to work at least two problems per section as practice: one easy and one difficult. Write out all the new rules and procedures your test will cover, and then read through them twice. Doing so will enable you to both learn and retain them better.
- Some people like to work in study groups, while others prefer solitary study. Although it's important to be flexible, it's more important that you be comfortable with your study method, so consider trying both. You may find one or the other or a combination of both effective.
- Most schools have classrooms set up where you can get free help from math tutors. Take advantage of this, but be sure you do the work first. Don't let your tutor do all the work for you—otherwise you'll never learn the material.
- In math, as in many other areas of life, patience and persistence are virtues—cultivate them. "Cramming" for an exam will not help you learn and retain the material.
- Do your work neatly and in pencil. Then if you make a mistake, it will be relatively easy to find and correct. Write out each step in the problem's solution; don't skip steps or take shortcuts. Each step should follow clearly from the preceding step, and the entire solution should be easy to follow. If you understand the concepts and get a wrong answer, the first thing you should look for is a "small" mistake, like writing a "+" instead of a "−."

Exercise Set 10.5

a Divide, if possible. Check each answer.

1. $36 \div (-6)$

2. $\frac{42}{-7}$

3. $\frac{26}{-2}$

4. $24 \div (-12)$

5. $\frac{-16}{8}$

6. $-18 \div (-2)$

7. $\frac{-48}{-12}$

8. $-72 \div (-9)$

9. $\frac{-72}{9}$

10. $\frac{-50}{25}$

11. $-100 \div (-50)$

12. $\frac{-200}{8}$

13. $-108 \div 9$

14. $\frac{-64}{-7}$

15. $\frac{200}{-25}$

16. $-300 \div (-13)$

17. $\frac{75}{0}$

18. $\frac{0}{-5}$

19. $\frac{81}{-9}$

20. $\frac{-145}{-5}$

b Find the reciprocal.

21. $-\frac{15}{7}$

22. $-\frac{5}{8}$

23. 13

24. -8

c Divide.

25. $\frac{3}{4} \div \left(-\frac{2}{3}\right)$

26. $\frac{7}{8} \div \left(-\frac{1}{2}\right)$

27. $-\frac{5}{4} \div \left(-\frac{3}{4}\right)$

28. $-\frac{5}{9} \div \left(-\frac{5}{6}\right)$

29. $-\frac{2}{7} \div \left(-\frac{4}{9}\right)$

30. $-\frac{3}{5} \div \left(-\frac{5}{8}\right)$

31. $-\frac{3}{8} \div \left(-\frac{8}{3}\right)$

32. $-\frac{5}{8} \div \left(-\frac{6}{5}\right)$

33. $-6.6 \div 3.3$

34. $-44.1 \div (-6.3)$

35. $\frac{-11}{-13}$

36. $\frac{-1.7}{20}$

37. $\frac{48.6}{-3}$

38. $\frac{-17.8}{3.2}$

39. $\frac{-9}{17 - 17}$

40. $\frac{-8}{-5 + 5}$

d Simplify.

41. $8 - 2 \cdot 3 - 9$

42. $8 - (2 \cdot 3 - 9)$

43. $(8 - 2 \cdot 3) - 9$

44. $(8 - 2)(3 - 9)$

45. $16 \cdot (-24) + 50$

46. $10 \cdot 20 - 15 \cdot 24$

47. $2^4 + 2^3 - 10$

48. $40 - 3^2 - 2^3$

49. $5^3 + 26 \cdot 71 - (16 + 25 \cdot 3)$

50. $4^3 + 10 \cdot 20 + 8^2 - 23$

51. $4 \cdot 5 - 2 \cdot 6 + 4$

52. $4 \cdot (6 + 8)/(4 + 3)$

53. $4^3/8$

54. $5^3 - 7^2$

55. $8(-7) + 6(-5)$

56. $10(-5) + 1(-1)$

57. $19 - 5(-3) + 3$

58. $14 - 2(-6) + 7$

59. $9 \div (-3) + 16 \div 8$

60. $-32 - 8 \div 4 - (-2)$

61. $6 - 4^2$

62. $(2 - 5)^2$

63. $(3 - 8)^2$

64. $3 - 3^2$

65. $12 - 20^3$

66. $20 + 4^3 \div (-8)$

67. $2 \times 10^3 - 5000$

68. $-7(3^4) + 18$

69. $6[9 - (3 - 4)]$

70. $8[(6 - 13) - 11]$

71. $-1000 \div (-100) \div 10$

72. $256 \div (-32) \div (-4)$

73. $8 - (7 - 9)$

74. $(8 - 7) - 9$

75. $\dfrac{10 - 6^2}{9^2 + 3^2}$

76. $\dfrac{5^2 - 4^3 - 3}{9^2 - 2^2 - 1^5}$

77. $\dfrac{20(8 - 3) - 4(10 - 3)}{10(2 - 6) - 2(5 + 2)}$

78. $\dfrac{(3 - 5)^2 - (7 - 13)}{(12 - 9)^2 + (11 - 14)^2}$

Skill Maintenance

Find the prime factorization. [2.1d]

79. 78

80. 225

81. 960

82. 1025

Find the LCM. [3.1a]

83. 24, 54

84. 56, 63

85. 5, 19, 35

86. 20, 40, 64

Solve. [6.3b], [6.4b]

87. What is 45% of 3800?

88. 344 is what percent of 8600?

Synthesis

89. ◆ Explain how multiplication can be used to justify why the quotient of two negative integers is a positive integer.

90. ◆ Explain how multiplication can be used to justify why a negative integer divided by a positive integer is a negative integer.

Simplify.

91. 🖩 $\dfrac{19 - 17^2}{13^2 - 34}$

92. 🖩 $\dfrac{195 + (-15)^3}{195 - 7 \cdot 5^2}$

Determine the sign of the expression if m is negative and n is positive.

93. $\dfrac{-n}{m}$

94. $\dfrac{-n}{-m}$

95. $-\left(\dfrac{-n}{m}\right)$

96. $-\left(\dfrac{n}{-m}\right)$

97. $-\left(\dfrac{-n}{-m}\right)$

Summary and Review Exercises: Chapter 10

The objectives to be tested in addition to the material in this chapter are [2.1d], [3.1a], [6.3b], and [8.4a].

Find the absolute value. [10.1e]

1. $|-38|$

2. $|7.3|$

3. $\left|\frac{5}{2}\right|$

4. $-|-0.2|$

Find decimal notation. [10.1c]

5. $-\frac{5}{4}$

6. $-\frac{5}{6}$

7. $-\frac{5}{12}$

8. $-\frac{3}{11}$

Graph the number on a number line. [10.1b]

9. -2.5

10. $\frac{8}{9}$

Use either $<$ or $>$ for ▢ to write a true sentence. [10.1d]

11. -3 ▢ 10

12. -1 ▢ -6

13. 0.126 ▢ -12.6

14. $-\frac{2}{3}$ ▢ $-\frac{1}{10}$

Find the opposite, or additive inverse, of the number. [10.2b]

15. 3.8

16. $-\frac{3}{4}$

17. Find $-x$ when x is -34. [10.2b]

18. Find $-(-x)$ when x is 5. [10.2b]

Find the reciprocal. [10.5b]

19. $\frac{3}{8}$

20. -7

21. $-\frac{1}{10}$

Compute and simplify.

22. $4 + (-7)$ [10.2a]

23. $-\frac{2}{3} + \frac{1}{12}$ [10.2a]

24. $6 + (-9) + (-8) + 7$ [10.2a]

25. $-3.8 + 5.1 + (-12) + (-4.3) + 10$ [10.2a]

26. $-3 - (-7)$ [10.3a]

27. $-\frac{9}{10} - \frac{1}{2}$ [10.3a]

28. $-3.8 - 4.1$ [10.3a]

29. $-9 \cdot (-6)$ [10.4a]

30. $-2.7(3.4)$ [10.4a]

31. $\frac{2}{3} \cdot \left(-\frac{3}{7}\right)$ [10.4a]

32. $3 \cdot (-7) \cdot (-2) \cdot (-5)$ [10.4a]

33. $35 \div (-5)$ [10.5a]

34. $-5.1 \div 1.7$ [10.5c]

35. $-\frac{3}{11} \div \left(-\frac{4}{11}\right)$ [10.5c]

36. $(-3.4 - 12.2) - 8(-7)$ [10.5d]

Simplify. [10.5d]

37. $[-12(-3) - 2^3] - (-9)(-10)$

38. $625 \div (-25) \div 5$

Skill Maintenance

39. Find the area of a rectangle of length 10.5 cm and width 20 cm. [8.4a]

40. Find the LCM of 15, 27, and 30. [3.1a]

41. Find the prime factorization of 648. [2.1d]

42. 2016 is what percent of 5600? [6.3b]

Synthesis

43. ◈ Is it possible for a number to be its own reciprocal? Explain. [10.5b]

44. ◈ Write as many arguments as you can to convince a fellow classmate that $-(-a) = a$ for all real numbers a. [10.2b]

45. The sum of two numbers is 800. The difference is 6. Find the numbers. [10.2a], [10.3a]

46. The sum of two numbers is 5. The product is -84. Find the numbers. [10.2a], [10.4a]

47. The following are examples of consecutive integers: 4, 5, 6, 7, 8; and $-13, -12, -11, -10$. Note that consecutive integers can be represented in the form $x, x + 1, x + 2$, and so on. [10.2a], [10.4a]

a) Express the number 8 as the sum of 16 consecutive integers.

b) Find the product of the 16 consecutive integers in part (a).

48. Describe how you might find the following product quickly: [10.4a]

$$\left(-\tfrac{1}{11}\right)\left(-\tfrac{1}{9}\right)\left(-\tfrac{1}{7}\right)\left(-\tfrac{1}{5}\right)\left(-\tfrac{1}{3}\right)(-1)(-3)(-5)(-7)(-9)(-11).$$

49. Simplify: $-\left|\frac{7}{8} - \left(-\frac{1}{2}\right) - \frac{3}{4}\right|$. [10.1e], [10.3a]

50. Simplify: $(|2.7 - 3| + 3^2 - |-3|) \div (-3)$. [10.1e], [10.5d]

Test: Chapter 10

Use either < or > for ▒ to write a true sentence.

1. -4 ▒ 0

2. -3 ▒ -8

3. -0.78 ▒ -0.87

4. $-\frac{1}{8}$ ▒ $\frac{1}{2}$

Find decimal notation.

5. $-\frac{1}{8}$

6. $-\frac{4}{9}$

7. $-\frac{2}{11}$

Find the absolute value.

8. $|-7|$

9. $\left|\frac{9}{4}\right|$

10. $-|-2.7|$

Find the opposite, or additive inverse.

11. $\frac{2}{3}$

12. -1.4

13. Find $-x$ when x is -8.

Find the reciprocal.

14. -2

15. $\frac{4}{7}$

Answers

1. ________

2. ________

3. ________

4. ________

5. ________

6. ________

7. ________

8. ________

9. ________

10. ________

11. ________

12. ________

13. ________

14. ________

15. ________

Answers

16. ______

17. ______

18. ______

19. ______

20. ______

21. ______

22. ______

23. ______

24. ______

25. ______

26. ______

27. ______

28. ______

29. ______

30. ______

31. ______

32. ______

33. ______

34. a) ______

b) ______

c) ______

d) ______

Compute and simplify.

16. $3.1 - (-4.7)$

17. $-8 + 4 + (-7) + 3$

18. $-\frac{1}{5} + \frac{3}{8}$

19. $2 - (-8)$

20. $3.2 - 5.7$

21. $\frac{1}{8} - \left(-\frac{3}{4}\right)$

22. $4 \cdot (-12)$

23. $-\frac{1}{2} \cdot \left(-\frac{3}{8}\right)$

24. $-45 \div 5$

25. $-\frac{3}{5} \div \left(-\frac{4}{5}\right)$

26. $4.864 \div (-0.5)$

27. $-2(16) - [2(-8) - 5^3]$

Skill Maintenance

28. Find the area of a rectangle of length 12.4 ft and width 4.5 ft.

29. 24 is what percent of 50?

30. Find the prime factorization of 280.

31. Find the LCM of 16, 20, and 30.

Synthesis

32. Simplify: $|-27 - 3(4)| - |-36| + |-12|$.

33. The deepest point in the Pacific Ocean is the Marianas Trench with a depth of 11,033 m. The deepest point in the Atlantic Ocean is the Puerto Rico Trench with a depth of 8648 m. (***Source:*** Defense Mapping Agency, Hydrographic/Topographic Center) How much higher is the Puerto Rico Trench than the Marianas Trench?

34. Find the next three numbers in each sequence.

a) 6, 5, 3, 0, __, __, __

b) 14, 10, 6, 2, __, __, __

c) −4, −6, −9, −13, __, __, __

d) 8, −4, 2, −1, 0.5, __, __, __

Cumulative Review: Chapters 1–10

Find decimal notation.

1. 26.3%

2. $-\frac{5}{11}$

Complete.

3. 83.4 cg = _______ mg

4. 2.75 mm^2 = _______ cm^2

5. Find the absolute value: $|-4.5|$.

6. Subtract: $2 - 13$.

7. What is the rate in meters per second?
150 meters, 12 seconds

8. Find the radius, the circumference, and the area of this circle. Use $\frac{22}{7}$ for π.

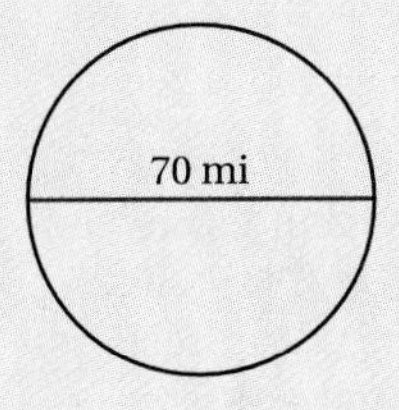

9. Simplify: $\sqrt{225}$.

10. Approximate to two decimal places: $\sqrt{69}$.

11. Multiply: $(-2)(5)$.

12. Divide: $\frac{-48}{-16}$.

13. Add: $-2 + 10$.

14. Draw a pictograph representing the number of hours that each type of farmer works each week using the information given below. Use a clock symbol to represent 10 hr. Be sure to put in all of the appropriate labels.

Dairy	70
Cash grain	40
Tobacco/cotton	35
Beef/hog/sheep	30

Compute and simplify.

15. 14.85×0.001

16. $36 - (-3) + (-42)$

17. $\frac{5}{22} - \frac{4}{11}$

18. $\frac{2}{27} \cdot \left(-\frac{9}{16}\right)$

19. $4\frac{2}{9} - 2\frac{7}{18}$

20. $-\frac{3}{14} \div \frac{6}{7}$

21. $3(-4.5) + (2^2 - 3 \cdot 4^2)$

22. $12{,}854 \cdot 750{,}000$

23. $35.1 + (-2.61)$

24. $32 \div [(-2)(-8) - (15 - (-1))]$

Solve.

25. 7 is what percent of 8?

26. 4 is $12\frac{1}{2}$% of what number?

27. Kerry had $324.98 in a checking account. He wrote a check for $12.76, deposited $35.95, and wrote another check for $213.09. The bank paid $0.97 in interest and deducted a service charge of $3.00. How much is now in his checking account?

28. A can of fruit has a diameter of 7 cm and a height of 8 cm. Find the volume. Use 3.14 for π.

29. The following temperatures were recorded every four hours on a certain day in Seattle: 42°, 40°, 45°, 52°, 50°, 40°. What was the average temperature for the day?

30. Thirteen percent of a student body of 600 received all A's on their grade reports. How many students received all A's?

31. A lot is 125.5 m by 75 m. A house 60 m by 40.5 m and a rectangular swimming pool 10 m by 8 m are built on the lot. How much area is left?

32. A recipe for a pie crust calls for $1\frac{1}{4}$ cups of flour, and a recipe for a cake calls for $1\frac{2}{3}$ cups of flour. How many cups of flour are needed to make both recipes?

33. The four top television game show winners in a recent year won $74,834, $58,253, $57,200, and $49,154. How much did these four win in all? What were the average earnings?

34. A power walker circled a block 6.5 times. If the distance around the block is 0.7 km, how far did the walker go?

The following table shows car sales by McTake Auto Company over several months.

Month	Sales
1	29
2	27
3	28
4	30
5	32
6	34
7	?

35. Make a line graph of the data.

36. Use extrapolation to estimate the number of cars sold in the 7th month.

Estimate each of the following as a whole number or as a mixed numeral where the fractional part is $\frac{1}{2}$.

37. $10\frac{8}{11}$

38. $12\frac{15}{17}$

39. $7\frac{3}{10} + 4\frac{5}{6} - \frac{31}{29}$

40. $33\frac{14}{15} + 27\frac{4}{5} + 8\frac{27}{30} \cdot 8\frac{37}{76}$

41. Find the missing angle measure.

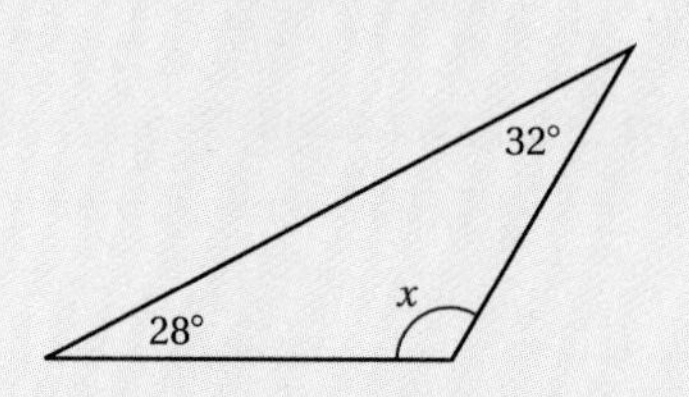

11

Algebra: Solving Equations and Problems

An Application

The Iditarod sled-dog race extends for 1049 mi from Anchorage to Nome. If a musher is twice as far from Anchorage as from Nome, how many miles has the musher traveled?

This problem appears as Exercise 19 in Exercise Set 11.5.

The Mathematics

We let x = the distance that the musher is from Nome. We can then translate the problem to this equation:

$$x + 2x = 1049.$$

World Wide Web For more information, visit us at www.mathmax.com

Introduction

In this chapter, we continue our introduction to algebra. We first consider the manipulation of algebraic expressions and then use the manipulations to solve equations and applied problems.

Pretest: Chapter 11

1. Evaluate $\frac{x}{2y}$ for $x = 5$ and $y = 8$.

2. Write an algebraic expression:

   Seventy-eight percent of some number.

Multiply.

3. $9(z - 2)$

4. $-2(2a + b - 5c)$

Factor.

5. $4x - 12$

6. $6y - 9z - 18$

Collect like terms.

7. $5x - 8x$

8. $6x - 9y - 4x + 11y + 18$

Solve.

9. $-7x = 49$

10. $4y + 9 = 2y + 7$

11. $6a - 2 = 10$

12. $x + (x + 1) + (x + 2) = 12$

Solve.

13. The perimeter of a rectangular peach orchard is 146 m. The width is 5 m less than the length. Find the dimensions of the orchard.

14. Money is invested in a savings account at 6% simple interest. After 1 year, there is $826.80 in the account. How much was originally invested?

The objectives to be tested in addition to the material in this chapter are as follows.

[8.6a, b, c] For circles, find the length of a radius given the length of a diameter, and conversely; find the circumference given the length of a diameter or a radius; and find the area given the length of a radius.

[9.1a] Find the volume of a rectangular solid using the formula $V = l \cdot w \cdot h$.

[10.2a] Add real numbers.

[10.4a] Multiply real numbers.

11.1 Introduction to Algebra

Many types of problems require the use of equations in order to be solved effectively. The study of algebra involves the use of equations to solve problems. Equations are constructed from algebraic expressions.

Objectives

a Evaluate an algebraic expression by substitution.

b Use the distributive laws to multiply expressions like 8 and $x - y$.

c Use the distributive laws to factor expressions like $4x - 12 + 24y$.

d Collect like terms.

For Extra Help

TAPE 19

TAPE 17A

MAC WIN

CD-ROM

a Algebraic Expressions

In arithmetic, you have worked with expressions such as

$$37 + 86, \quad 7 \times 8, \quad 19 - 7, \quad \text{and} \quad \frac{3}{8}.$$

In algebra, we use certain letters for numbers and work with *algebraic expressions* such as

$$x + 86, \quad 7 \times t, \quad 19 - y, \quad \text{and} \quad \frac{a}{b}.$$

Expressions like these should be familiar from the equation and problem solving that we have already done.

Sometimes a letter can stand for various numbers. In that case, we call the letter a **variable**. Let a = your age. Then a is a variable since a changes from year to year. Sometimes a letter can stand for just one number. In that case, we call the letter a **constant**. Let b = your date of birth. Then b is a constant.

An **algebraic expression** consists of variables, constants, numerals, and operation signs. When we replace a variable with a number, we say that we are **substituting** for the variable. This process is called **evaluating the expression.**

Example 1 Evaluate $x + y$ for $x = 37$ and $y = 29$.

We substitute 37 for x and 29 for y and carry out the addition:

$$x + y = 37 + 29 = 66.$$

The number 66 is called the **value** of the expression.

Algebraic expressions involving multiplication can be written in several ways. For example, "8 times a" can be written as $8 \times a$, $8 \cdot a$, $8(a)$, or simply $8a$. Two letters written together without an operation symbol, such as ab, also indicates a multiplication.

Example 2 Evaluate $3y$ for $y = 14$ and for $y = -4.5$.

$$3y = 3(14) = 42;$$
$$3y = 3(-4.5) = -13.5$$

Do Exercises 1–3.

1. Evaluate $a + b$ for $a = 38$ and $b = 26$.

2. Evaluate $x - y$ for $x = 57$ and $y = 29$.

3. Evaluate $4t$ for $t = 15$ and for $t = -6.8$.

Answers on page A-25

4. Evaluate $\frac{a}{b}$ for $a = -200$ and $b = 8$.

5. Evaluate $\frac{10p}{q}$ for $p = 40$ and $q = 25$.

Complete the table by evaluating each expression for the given values.

6.

	$1 \cdot x$	x
$x = 3$		
$x = -6$		
$x = 4.8$		

7.

	$2x$	$5x$
$x = 2$		
$x = -6$		
$x = 4.8$		

Answers on page A-25

Algebraic expressions involving division can also be written in several ways. For example, "8 divided by t" can be written as $8 \div t$, $8/t$, or $\frac{8}{t}$, where the fraction bar is a division symbol.

Example 3 Evaluate $\frac{a}{b}$ for $a = 63$ and $b = -9$.

We substitute 63 for a and -9 for b and carry out the division:

$$\frac{a}{b} = \frac{63}{-9} = -7.$$

Example 4 Evaluate $\frac{12m}{n}$ for $m = 8$ and $n = 16$.

$$\frac{12m}{n} = \frac{12 \cdot 8}{16} = \frac{96}{16} = 6$$

Do Exercises 4 and 5.

b Equivalent Expressions and the Distributive Laws

In solving equations and doing other kinds of work in algebra, we manipulate expressions in various ways. To see how to do this, we consider some examples in which we evaluate expressions.

Example 5 Evaluate $1 \cdot x$ for $x = 5$ and for $x = -8$ and compare the results to x.

We substitute 5 for x:

$$1 \cdot x = 1 \cdot 5 = 5.$$

Then we substitute -8 for x:

$$1 \cdot x = 1 \cdot (-8) = -8.$$

We see that $1 \cdot x$ and x represent the same number.

Do Exercises 6 and 7.

We see in Example 5 and Margin Exercise 6 that the expressions represent the same number for any allowable replacement of x. In that sense, the expressions $1 \cdot x$ and x are **equivalent**.

> Two expressions that have the same value for all allowable replacements are called **equivalent**.

In the expression $3/x$, the number 0 is not allowable because $3/0$ is undefined. Even so, the expressions $3x/x^2$ and $3/x$ are *equivalent* because they represent the same number for any allowable (not 0) replacement of x. We see in Margin Exercise 7 that the expressions $2x$ and $5x$ are *not* equivalent.

The fact that $1 \cdot x$ and x are equivalent is a law of real numbers. It is called the **identity property of 1.** We often refer to the use of the identity property of 1 as "multiplying by 1." We have used multiplying by 1 for understanding many times in this text.

THE IDENTITY PROPERTY OF 1

For any real number a,

$$a \cdot 1 = 1 \cdot a = a.$$

(The number 1 is the *multiplicative identity.*)

We now consider two other laws of real numbers called the **distributive laws.** They are the basis of many procedures in both arithmetic and algebra and are probably the most important laws that we use to manipulate algebraic expressions. The first distributive law involves two operations: addition and multiplication.

Let's begin by considering a multiplication problem from arithmetic:

```
    4 5
  ×   7
  -----
    3 5 ← This is 7 · 5.
  2 8 0 ← This is 7 · 40.
  -----
  3 1 5 ← This is the sum 7 · 40 + 7 · 5.
```

To carry out the multiplication, we actually added two products. That is,

$$7 \cdot 45 = 7(40 + 5) = 7 \cdot 40 + 7 \cdot 5.$$

Let's examine this further. If we wish to multiply a sum of several numbers by a factor, we can either add and then multiply or multiply and then add.

Example 6 Evaluate $5(x + y)$ and $5x + 5y$ for $x = 2$ and $y = 8$ and compare the results.

We substitute 2 for x and 8 for y in each expression. Then we use the rules for order of operations to calculate.

a) $5(x + y) = 5(2 + 8)$

$= 5(10)$ Adding within parentheses first, and then multiplying

$= 50$

b) $5x + 5y = 5 \cdot 2 + 5 \cdot 8$

$= 10 + 40$ Multiplying first and then adding

$= 50$

We see that the expressions $5(x + y)$ and $5x + 5y$ are equivalent.

Do Exercises 8–10.

8. Evaluate $3(x + y)$ and $3x + 3y$ for $x = 5$ and $y = 7$.

9. Evaluate $6x + 6y$ and $6(x + y)$ for $x = 10$ and $y = 5$.

10. Evaluate $4(x + y)$ and $4x + 4y$ for $x = 11$ and $y = 5$.

Answers on page A-25

11. Evaluate $7(x - y)$ and $7x - 7y$ for $x = 9$ and $y = 7$.

12. Evaluate $6x - 6y$ and $6(x - y)$ for $x = 10$ and $y = 5$.

13. Evaluate $2(x - y)$ and $2x - 2y$ for $x = 11$ and $y = 5$.

What are the terms of the expression?

14. $5x - 4y + 3$

15. $-4y - 2x + 3z$

Answers on page A-25

> **THE DISTRIBUTIVE LAW OF MULTIPLICATION OVER ADDITION**
>
> For any numbers a, b, and c,
>
> $a(b + c) = ab + ac.$

In the statement of the distributive law, we know that in an expression such as $ab + ac$, the multiplications are to be done first according to the rules for order of operations. So, instead of writing $(4 \cdot 5) + (4 \cdot 7)$, we can write $4 \cdot 5 + 4 \cdot 7$. However, in $a(b + c)$, we cannot omit the parentheses. If we did we would have $ab + c$, which means $(ab) + c$. For example, $3(4 + 2) = 18$, but $3 \cdot 4 + 2 = 14$.

The second distributive law relates multiplication and subtraction. This law says that to multiply by a difference, we can either subtract and then multiply or multiply and then subtract.

> **THE DISTRIBUTIVE LAW OF MULTIPLICATION OVER SUBTRACTION**
>
> For any numbers a, b, and c,
>
> $a(b - c) = ab - ac.$

We often refer to "*the* distributive law" when we mean *either or both* of these laws.

Do Exercises 11–13.

What do we mean by the *terms* of an expression? **Terms** are separated by addition signs. If there are subtraction signs, we can find an equivalent expression that uses addition signs.

Example 7 What are the terms of $3x - 4y + 2z$?

$$3x - 4y + 2z = 3x + (-4y) + 2z \qquad \text{Separating parts with + signs}$$

The terms are $3x$, $-4y$, and $2z$.

Do Exercises 14 and 15.

The distributive laws are the basis for a procedure in algebra called **multiplying**. In an expression such as $8(a + 2b - 7)$, we multiply each term inside the parentheses by 8:

$$8(a + 2b - 7) = 8 \cdot a + 8 \cdot 2b - 8 \cdot 7 = 8a + 16b - 56.$$

Examples Multiply.

8. $9(x - 5) = 9x - 9(5)$ **Using the distributive law of multiplication over subtraction**

$= 9x - 45$

9. $\frac{2}{3}(w + 1) = \frac{2}{3} \cdot w + \frac{2}{3} \cdot 1$ **Using the distributive law of multiplication over addition**

$= \frac{2}{3}w + \frac{2}{3}$

Example 10 Multiply: $-4(x - 2y + 3z)$.

$$-4(x - 2y + 3z) = -4 \cdot x - (-4)(2y) + (-4)(3z) \quad \text{Using both distributive laws}$$

$$= -4x - (-8y) + (-12z) \quad \text{Multiplying}$$

$$= -4x + 8y - 12z$$

We can also do this problem by first finding an equivalent expression with all plus signs and then multiplying:

$$-4(x - 2y + 3z) = -4[x + (-2y) + 3z]$$

$$= -4 \cdot x + (-4)(-2y) + (-4)(3z) = -4x + 8y - 12z.$$

Do Exercises 16–20.

c Factoring

Factoring is the reverse of multiplying. To factor, we can use the distributive laws in reverse:

$$ab + ac = a(b + c) \quad \text{and} \quad ab - ac = a(b - c).$$

> To **factor** an expression is to find an equivalent expression that is a product.

Look at Example 8. To *factor* $9x - 45$, we find an equivalent expression that is a product, $9(x - 5)$. When all the terms of an expression have a factor in common, we can "factor it out" using the distributive laws. Note the following.

$9x$ has the factors $9, -9, 3, -3, 1, -1, x, -x, 3x, -3x, 9x, -9x$;

-45 has the factors $1, -1, 3, -3, 5, -5, 9, -9, 15, -15, 45, -45$.

We remove the largest common factor. In this case, that factor is 9. Thus,

$$9x - 45 = 9 \cdot x - 9 \cdot 5$$

$$= 9(x - 5).$$

Remember that an expression is factored when we find an equivalent expression that is a product.

Examples Factor.

11. $5x - 10 = 5 \cdot x - 5 \cdot 2$ **Try to do this step mentally.**

$= 5(x - 2)$ ← **You can check by multiplying.**

12. $9x + 27y - 9 = 9 \cdot x + 9 \cdot 3y - 9 \cdot 1$

$= 9(x + 3y - 1)$

CAUTION! Note that although $3(3x + 9y - 3)$ is also equivalent to $9x + 27y - 9$, it is *not* the desired form. However, we can complete the process by factoring out another factor of 3:

$$9x + 27y - 9 = 3(3x + 9y - 3) = 3 \cdot 3(x + 3y - 1) = 9(x + 3y - 1).$$

Remember to factor out the *largest common factor.*

Multiply.

16. $3(x - 5)$

17. $5(x + 1)$

18. $\frac{5}{4}(x - y + 4)$

19. $-2(x - 3)$

20. $-5(x - 2y + 4z)$

Answers on page A-25

Examples Factor. Try to write just the answer, if you can.

13. $5x - 5y = 5(x - y)$

14. $-3x + 6y - 9z = -3 \cdot x - 3(-2y) - 3(3z) = -3(x - 2y + 3z)$

We usually factor out a negative when the first term is negative. The way we factor can depend on the situation in which we are working. We might also factor the expression in Example 14 as follows:

$$-3x + 6y - 9z = 3(-x + 2y - 3z).$$

15. $18z - 12x - 24 = 6(3z - 2x - 4)$

Remember that you can always check such factoring by multiplying. Keep in mind that an expression is factored when it is written as a product.

Do Exercises 21–24.

Factor.

21. $6z - 12$

22. $3x - 6y + 9$

23. $16a - 36b + 42$

24. $-12x + 32y - 16z$

d Collecting Like Terms

Terms such as $5x$ and $-4x$, whose variable factors are exactly the same, are called **like terms.** Similarly, numbers, such as -7 and 13, are like terms. Also, $3y^2$ and $9y^2$ are like terms because the variables are raised to the same power. Terms such as $4y$ and $5y^2$ are not like terms, and $7x$ and $2y$ are not like terms.

The process of **collecting like terms** is based on the distributive laws. We can also apply the distributive law when a factor is on the right.

Examples Collect like terms. Try to write just the answer, if you can.

16. $4x + 2x = (4 + 2)x = 6x$ Factoring out the x using a distributive law

17. $2x + 3y - 5x - 2y = 2x - 5x + 3y - 2y$

$= (2 - 5)x + (3 - 2)y = -3x + y$

18. $3x - x = (3 - 1)x = 2x$

19. $x - 0.24x = 1 \cdot x - 0.24x = (1 - 0.24)x = 0.76x$

20. $x - 6x = 1 \cdot x - 6 \cdot x = (1 - 6)x = -5x$

21. $4x - 7y + 9x - 5 + 3y - 8 = 13x - 4y - 13$

22. $\frac{2}{3}a - b + \frac{4}{5}a + \frac{1}{4}b - 10 = \frac{2}{3}a - 1 \cdot b + \frac{4}{5}a + \frac{1}{4}b - 10$

$$= \left(\frac{2}{3} + \frac{4}{5}\right)a + \left(-1 + \frac{1}{4}\right)b - 10$$

$$= \left(\frac{10}{15} + \frac{12}{15}\right)a + \left(-\frac{4}{4} + \frac{1}{4}\right)b - 10$$

$$= \frac{22}{15}a - \frac{3}{4}b - 10$$

Do Exercises 25–31.

Collect like terms.

25. $6x - 3x$

26. $7x - x$

27. $x - 9x$

28. $x - 0.41x$

29. $5x + 4y - 2x - y$

30. $3x - 7x - 11 + 8y + 4 - 13y$

31. $-\frac{2}{3} - \frac{3}{5}x + y + \frac{7}{10}x - \frac{2}{9}y$

Answers on page A-25

Exercise Set 11.1

a Evaluate.

1. $6x$, for $x = 7$

2. $9t$, for $t = 8$

3. $\frac{x}{y}$, for $x = 9$ and $y = 3$

4. $\frac{m}{n}$, for $m = 18$ and $n = 3$

5. $\frac{3p}{q}$, for $p = -2$ and $q = 6$

6. $\frac{5y}{z}$, for $y = -15$ and $z = -25$

7. $\frac{x + y}{5}$, for $x = 10$ and $y = 20$

8. $\frac{p - q}{2}$, for $p = 17$ and $q = 3$

b Evaluate.

9. $10(x + y)$ and $10x + 10y$, for $x = 20$ and $y = 4$

10. $5(a + b)$ and $5a + 5b$, for $a = 16$ and $b = 6$

11. $10(x - y)$ and $10x - 10y$, for $x = 20$ and $y = 4$

12. $5(a - b)$ and $5a - 5b$, for $a = 16$ and $b = 6$

Multiply.

13. $2(b + 5)$

14. $4(x + 3)$

15. $7(1 - t)$

16. $4(1 - y)$

17. $6(5x + 2)$

18. $9(6m + 7)$

19. $7(x + 4 + 6y)$

20. $4(5x + 8 + 3p)$

21. $-7(y - 2)$

22. $-9(y - 7)$

23. $-9(-5x - 6y + 8)$

24. $-7(-2x - 5y + 9)$

25. $\frac{3}{4}(x - 3y - 2z)$

26. $\frac{2}{5}(2x - 5y - 8z)$

27. $3.1(-1.2x + 3.2y - 1.1)$

28. $-2.1(-4.2x - 4.3y - 2.2)$

c Factor. Check by multiplying.

29. $2x + 4$

30. $5y + 20$

31. $30 + 5y$

32. $7x + 28$

33. $14x + 21y$

34. $18a + 24b$

35. $5x + 10 + 15y$

36. $9a + 27b + 81$

37. $8x - 24$

38. $10x - 50$

39. $32 - 4y$

40. $24 - 6m$

41. $8x + 10y - 22$

42. $9a + 6b - 15$

43. $-18x - 12y + 6$

44. $-14x + 21y + 7$

d Collect like terms.

45. $9a + 10a$

46. $14x + 3x$

47. $10a - a$

48. $-10x + x$

49. $2x + 9z + 6x$

50. $3a - 5b + 4a$

51. $41a + 90 - 60a - 2$

52. $42x - 6 - 4x + 20$

53. $23 + 5t + 7y - t - y - 27$

54. $95 - 90d - 87 - 9d + 3 + 7d$

55. $11x - 3x$

56. $9t - 13t$

57. $6n - n$

58. $10t - t$

59. $y - 17y$

60. $5m - 8m + 4$

61. $-8 + 11a - 5b + 6a - 7b + 7$

62. $8x - 5x + 6 + 3y - 2y - 4$

63. $9x + 2y - 5x$

64. $8y - 3z + 4y$

65. $\frac{11}{4}x + \frac{2}{3}y - \frac{4}{5}x - \frac{1}{6}y + 12$

66. $\frac{13}{2}a + \frac{9}{5}b - \frac{2}{3}a - \frac{3}{10}b - 42$

67. $2.7x + 2.3y - 1.9x - 1.8y$

68. $6.7a + 4.3b - 4.1a - 2.9b$

Skill Maintenance

For a circle with the given radius, find the diameter, the circumference, and the area. Use 3.14 for π. [8.6a, b, c]

69. $r = 15$ yd

70. $r = 8.2$ m

71. $r = 9\frac{1}{2}$ mi

72. $r = 2400$ cm

For a circle with the given diameter, find the radius, the circumference, and the area. Use 3.14 for π. [8.6a, b, c]

73. $d = 20$ mm

74. $d = 264$ km

75. $d = 4.6$ ft

76. $d = 10.3$ m

Synthesis

77. ◆ Determine whether $(a + b)^2$ and $a^2 + b^2$ are equivalent for all real numbers. Explain.

78. ◆ The distributive law is introduced before the material on collecting like terms. Why do you think this is?

11.2 Solving Equations: The Addition Principle

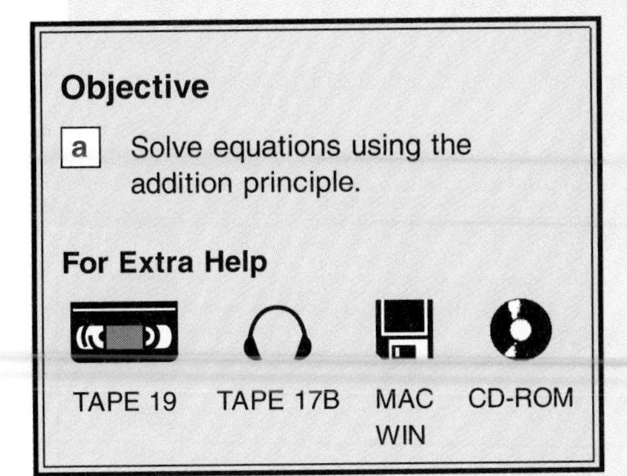

a Using the Addition Principle

Consider the equation

$$x = 7.$$

We can easily "see" that the solution of this equation is 7. If we replace x with 7, we get

$$7 = 7, \quad \text{which is true.}$$

Now consider the equation

$$x + 6 = 13.$$

The solution of this equation is also 7, but the fact that 7 is the solution is not so obvious. We now begin to consider principles that allow us to start with an equation and end up with an equation like $x = 7$, in which the variable is alone on one side and for which the solution is easy to find. The equations $x + 6 = 13$ and $x = 7$ are **equivalent**.

> Equations with the same solutions are called **equivalent equations.**

One principle that we use to solve equations concerns the addition principle, which we have used throughout this text.

> **THE ADDITION PRINCIPLE**
>
> For any real numbers a, b, and c,
>
> $a = b$ is equivalent to $a + c = b + c$.

Let's again solve $x + 6 = 13$ using the addition principle. We want to get x alone on one side. To do so, we use the addition principle, choosing to add -6 on both sides because $6 + (-6) = 0$:

$$x + 6 = 13$$

$$x + 6 + (-6) = 13 + (-6) \quad \text{Using the addition principle; adding } -6 \text{ on both sides}$$

$$x + 0 = 7 \quad \text{Simplifying}$$

$$x = 7. \quad \text{Identity property of 0}$$

Do Exercise 1.

To visualize the addition principle, think of a jeweler's balance. When both sides of the balance hold equal amounts of weight, the balance is level. If weight is added or removed, equally, on both sides, the balance remains level.

1. Solve $x + 2 = 11$ using the addition principle.

   a) First, complete this sentence:

   $2 + \square = 0.$

   b) Then solve the equation.

Answers on page A-25

2. Solve using the addition principle:

$x + 7 = 2$.

Solve.

3. $8.7 = n - 4.5$

4. $y + 17.4 = 10.9$

Answers on page A-25

When we use the addition principle, we sometimes say that we "add the same number on both sides of the equation." This is also true for subtraction, since we can express every subtraction as an addition. That is, since

$$a - c = b - c \quad \text{is equivalent to} \quad a + (-c) = b + (-c),$$

the addition principle tells us that we can "subtract the same number on both sides of an equation."

Example 1 Solve: $x + 5 = -7$.

We have

$$x + 5 = -7$$
$$x + 5 - 5 = -7 - 5 \qquad \text{Using the addition principle: adding } -5 \text{ on both sides or subtracting 5 on both sides}$$
$$x + 0 = -12 \qquad \text{Simplifying}$$
$$x = -12. \qquad \text{Identity property of 0}$$

We can see that the solution of $x = -12$ is the number -12. To check the answer, we substitute -12 in the original equation.

CHECK:

$x + 5 = -7$		
$-12 + 5$	?	-7
-7		TRUE

The solution of the original equation is -12.

In Example 1, to get x alone, we used the addition principle and subtracted 5 on both sides. This eliminated the 5 on the left. We started with $x + 5 = -7$, and using the addition principle we found a simpler equation $x = -12$, for which it was easy to "see" the solution. The equations $x + 5 = -7$ and $x = -12$ are equivalent.

Do Exercise 2.

Now we solve an equation with a subtraction using the addition principle.

Example 2 Solve: $-6.5 = y - 8.4$.

We have

$$-6.5 = y - 8.4$$
$$-6.5 + 8.4 = y - 8.4 + 8.4 \qquad \text{Using the addition principle: adding 8.4 to eliminate } -8.4 \text{ on the right}$$
$$1.9 = y.$$

CHECK:

$-6.5 = y - 8.4$		
-6.5	?	$1.9 - 8.4$
		-6.5 TRUE

The solution is 1.9.

Note that equations are reversible. That is, if $a = b$ is true, then $b = a$ is true. Thus, when we solve $-6.5 = y - 8.4$, we can reverse it and solve $y - 8.4 = -6.5$ if we wish.

Do Exercises 3 and 4.

Exercise Set 11.2

a Solve using the addition principle. Don't forget to check!

1. $x + 5 = 12$

2. $x + 3 = 7$

3. $x + 15 = -5$

4. $y + 8 = 37$

5. $x + 6 = -8$

6. $t + 8 = -14$

7. $x + 16 = -2$

8. $y + 34 = -8$

9. $x - 9 = 6$

10. $x - 9 = 2$

11. $x - 7 = -21$

12. $x - 5 = -16$

13. $5 + t = 7$

14. $6 + y = 22$

15. $-7 + y = 13$

16. $-8 + z = 16$

17. $-3 + t = -9$

18. $-8 + y = -23$

19. $r + \frac{1}{3} = \frac{8}{3}$

20. $t + \frac{3}{8} = \frac{5}{8}$

21. $m + \frac{5}{6} = -\frac{11}{12}$

22. $x + \frac{2}{3} = -\frac{5}{6}$

23. $x - \frac{5}{6} = \frac{7}{8}$

24. $y - \frac{3}{4} = \frac{5}{6}$

25. $-\frac{1}{5} + z = -\frac{1}{4}$

26. $-\frac{1}{8} + y = -\frac{3}{4}$

27. $7.4 = x + 2.3$

28. $9.3 = 4.6 + x$

29. $7.6 = x - 4.8$

30. $9.5 = y - 8.3$

31. $-9.7 = -4.7 + y$

32. $-7.8 = 2.8 + x$

33. $5\frac{1}{6} + x = 7$

34. $5\frac{1}{4} = 4\frac{2}{3} + x$

35. $q + \frac{1}{3} = -\frac{1}{7}$

36. $47\frac{1}{8} = -76 + z$

Skill Maintenance

Add. [10.2a]

37. $-3 + (-8)$

38. $-\frac{2}{3} + \frac{5}{8}$

39. $-14.3 + (-19.8)$

40. $3.2 + (-4.9)$

Subtract. [10.3a]

41. $-3 - (-8)$

42. $-\frac{2}{3} - \frac{5}{8}$

43. $-14.3 - (-19.8)$

44. $3.2 - (-4.9)$

Multiply. [10.4a]

45. $-3(-8)$

46. $-\frac{2}{3} \cdot \frac{5}{8}$

47. $-14.3 \times (-19.8)$

48. $3.2(-4.9)$

Divide. [10.5c]

49. $\frac{-24}{-3}$

50. $-\frac{2}{3} \div \frac{5}{8}$

51. $\frac{283.14}{-19.8}$

52. $\frac{-15.68}{3.2}$

Synthesis

53. ◆ Explain the following mistake made by a fellow student.

$$x + \frac{1}{3} = -\frac{5}{3}$$
$$x = -\frac{4}{3}$$

54. ◆ Explain the role of the opposite of a number when using the addition principle.

Solve.

55. 🖩 $-356.788 = -699.034 + t$

56. $-\frac{4}{5} + \frac{7}{10} = x - \frac{3}{4}$

57. $x + \frac{4}{5} = -\frac{2}{3} - \frac{4}{15}$

58. $8 - 25 = 8 + x - 21$

59. $16 + x - 22 = -16$

60. $x + x = x$

61. $x + 3 = 3 + x$

62. $x + 4 = 5 + x$

63. $-\frac{3}{2} + x = -\frac{5}{17} - \frac{3}{2}$

64. $|x| = 5$

11.3 Solving Equations: The Multiplication Principle

a Using the Multiplication Principle

Suppose that two workers have salaries that are equal. If the salary of each worker were doubled, the salaries would again be equal. Similarly, if each worker's salary were cut by a third, the salaries would remain equal. This brings us to the following.

Suppose that $a = b$ is true and we multiply a by some number c. We get the same answer if we multiply b by c, because a and b are the same number.

THE MULTIPLICATION PRINCIPLE

For any real numbers a, b, and c, $c \neq 0$,

$$a = b \quad \text{is equivalent to} \quad a \cdot c = b \cdot c.$$

When using the multiplication principle, we sometimes say that we "multiply by the same number on both sides."

Example 1 Solve: $\frac{2}{3}x = 18$.

To get x alone, we multiply by the *multiplicative inverse,* or *reciprocal,* of $\frac{2}{3}$. Then we get the *multiplicative identity,* 1, times x, or $1 \cdot x$, which simplifies to x. This allows us to eliminate the $\frac{2}{3}$ on the left.

$$\frac{2}{3}x = 18$$

$$\frac{3}{2} \cdot \frac{2}{3}x = \frac{3}{2} \cdot 18 \qquad \text{Using the multiplication principle: multiplying by } \tfrac{3}{2} \text{ to eliminate } \tfrac{2}{3} \text{ on the left}$$

$$1 \cdot x = 27 \qquad \text{Simplifying}$$

$$x = 27 \qquad \text{Identity property of 1}$$

CHECK:

$$\frac{2}{3}x = 18$$

$$\frac{2}{3} \cdot 27 \;?\; 18$$

$$18 \qquad \text{TRUE}$$

The solution is 27.

Do Exercises 1 and 2.

Objective

a Solve equations using the multiplication principle.

For Extra Help

TAPE 19 TAPE 17B MAC WIN CD-ROM

Solve.

1. $\frac{4}{5}x = 24$

2. $4x = -7$

Answers on page A-25

3. Solve: $5x = 40$.

4. Solve: $108 = -6x$.

Answers on page A-25

The multiplication principle also tells us that we can "divide by a nonzero number on both sides." This is because division is the same as multiplying by a reciprocal. That is,

$$\frac{a}{c} = \frac{b}{c} \quad \text{means} \quad a \cdot \frac{1}{c} = b \cdot \frac{1}{c}, \quad \text{when } c \neq 0.$$

In an expression like $3x$, the number 3 is called the **coefficient**. In practice it is usually more convenient to "divide" on both sides of the equation if the coefficient of the variable is in decimal notation or is an integer. When the coefficient is in fractional notation, it is more convenient to "multiply" by a reciprocal.

Example 2 Solve: $3x = 9$.

We have

$$3x = 9$$

$$\frac{3x}{3} = \frac{9}{3} \qquad \text{Using the multiplication principle: multiplying by } \tfrac{1}{3} \text{ on both sides or dividing by 3 on both sides}$$

$$1 \cdot x = 3 \qquad \text{Simplifying}$$

$$x = 3. \qquad \text{Identity property of 1}$$

It is now easy to see that the solution is 3.

CHECK:

$3x = 9$	
$3 \cdot 3 \;?\; 9$	
9	TRUE

The solution of the original equation is 3.

Do Exercise 3.

Example 3 Solve: $92 = -4x$.

We have

$$92 = -4x$$

$$\frac{92}{-4} = \frac{-4x}{-4} \qquad \text{Using the multiplication principle. Dividing on both sides by } -4 \text{ is the same as multiplying by } -\tfrac{1}{4}.$$

$$-23 = 1 \cdot x \qquad \text{Simplifying. In the next section, we will not include this step.}$$

$$-23 = x. \qquad \text{Identity property of 1}$$

CHECK:

$92 = -4x$	
$92 \;?\; -4(-23)$	
92	TRUE

The solution is -23.

Note that equations are reversible. That is, if $a = b$ is true, then $b = a$ is true. Thus, when we solve $92 = -4x$, we can reverse it and solve $-4x = 92$.

Do Exercise 4.

Exercise Set 11.3

a Solve using the multiplication principle. Don't forget to check!

1. $6x = 36$
2. $4x = 52$
3. $5x = 45$
4. $8x = 56$
5. $84 = 7x$
6. $63 = 7x$
7. $-x = 40$
8. $50 = -x$
9. $-2x = -10$
10. $-78 = -39p$
11. $7x = -49$
12. $9x = -54$
13. $-12x = 72$
14. $-15x = 105$
15. $-21x = -126$
16. $-13x = -104$
17. $\frac{1}{7}t = -9$
18. $-\frac{1}{8}y = 11$
19. $\frac{3}{4}x = 27$
20. $\frac{4}{5}x = 16$
21. $-\frac{1}{3}t = 7$
22. $-\frac{1}{6}x = 9$
23. $-\frac{1}{3}m = \frac{1}{5}$
24. $\frac{1}{5} = -\frac{1}{8}z$
25. $-\frac{3}{5}r = \frac{9}{10}$
26. $\frac{2}{5}y = -\frac{4}{15}$
27. $-\frac{3}{2}r = -\frac{27}{4}$
28. $-\frac{5}{7}x = -\frac{10}{14}$

29. $6.3x = 44.1$

30. $2.7y = 54$

31. $-3.1y = 21.7$

32. $-3.3y = 6.6$

33. $38.7m = 309.6$

34. $29.4m = 235.2$

35. $-\frac{2}{3}y = -10.6$

36. $-\frac{9}{7}y = 12.06$

Skill Maintenance

37. Find the circumference, the diameter, and the area of a circle painted on a driveway. The radius is 10 ft. Use 3.14 for π. [8.6a, b, c]

38. Find the circumference, the radius, and the area of a circle whose diameter is 24 cm. Use 3.14 for π. [8.6a, b, c]

39. Find the volume of a rectangular block of granite of length 25 ft, width 10 ft, and height 32 ft. [9.1a]

40. Find the volume of a rectangular solid of length 1.3 cm, width 10 cm, and height 2.4 cm. [9.1a]

Find the area of the figure. [8.5a]

41.

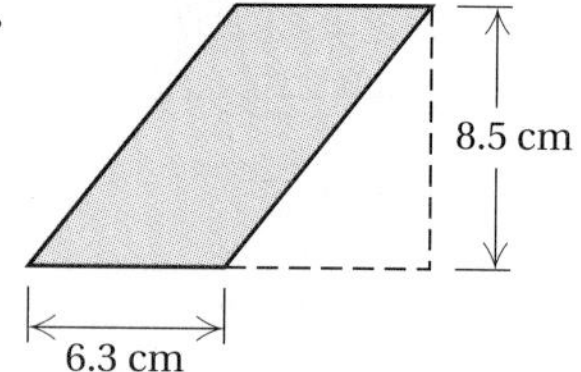

42.

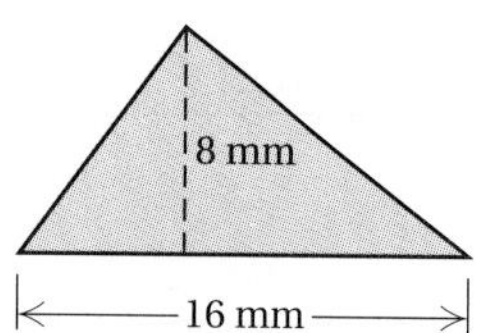

43.

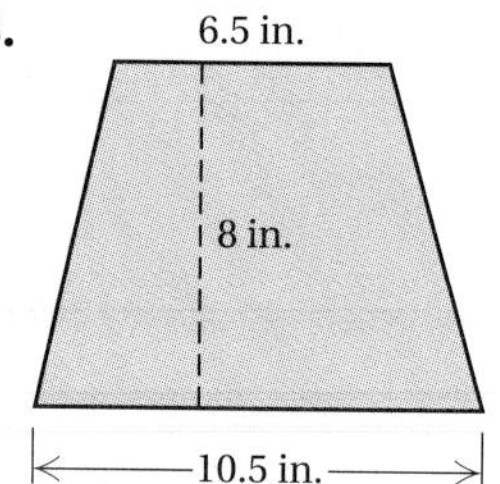

44.

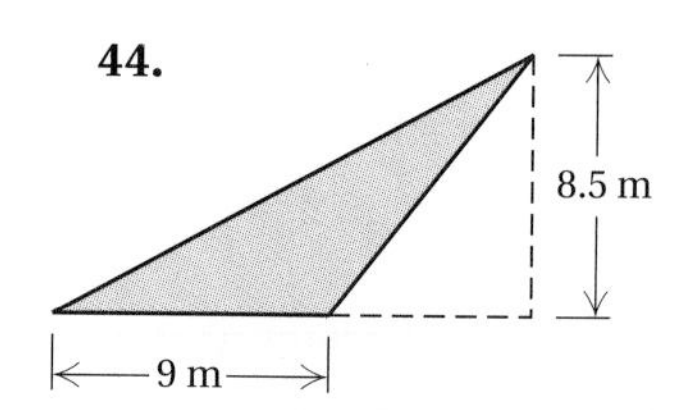

Synthesis

45. ◆ Explain the following mistake made by a fellow student.

$$\frac{2}{3}x = -\frac{5}{3}$$

$$x = \frac{10}{9}$$

46. ◆ Explain the role of the reciprocal of a number when using the multiplication principle.

Solve.

47. 🖩 $-0.2344m = 2028.732$

48. $0 \cdot x = 0$

49. $0 \cdot x = 9$

50. $4|x| = 48$

51. $2|x| = -12$

52. A student makes a calculation and gets an answer of 22.5. On the last step, the student multiplies by 0.3 when a division by 0.3 should have been done. What should the correct answer be?

11.4 Using the Principles Together

a Applying Both Principles

Consider the equation $3x + 4 = 13$. It is more complicated than those in the preceding two sections. In order to solve such an equation, we first isolate the x-term, $3x$, using the addition principle. Then we apply the multiplication principle to get x by itself.

Objectives

a. Solve equations using both the addition and the multiplication principles.

b. Solve equations in which like terms may need to be collected.

c. Solve equations by first removing parentheses and collecting like terms.

For Extra Help

TAPE 20 TAPE 18A MAC WIN CD-ROM

Example 1 Solve: $3x + 4 = 13$.

$$3x + 4 = 13$$

$$3x + 4 - 4 = 13 - 4 \quad \text{Using the addition principle: adding } -4 \text{ or subtracting 4 on both sides}$$

$$3x = 9 \quad \text{Simplifying}$$

$$\frac{3x}{3} = \frac{9}{3} \quad \text{Using the multiplication principle: multiplying by } \tfrac{1}{3} \text{ or dividing by 3 on both sides}$$

$$x = 3 \quad \text{Simplifying}$$

CHECK:

$$\begin{array}{r|l} 3x + 4 & 13 \\ \hline 3 \cdot 3 + 4 & ?\ 13 \\ 9 + 4 & \\ 13 & \quad \text{TRUE} \end{array}$$

The solution is 3.

Do Exercise 1.

1. Solve: $9x + 6 = 51$.

Example 2 Solve: $-5x - 6 = 16$.

$$-5x - 6 = 16$$

$$-5x - 6 + 6 = 16 + 6 \quad \text{Adding 6 on both sides}$$

$$-5x = 22$$

$$\frac{-5x}{-5} = \frac{22}{-5} \quad \text{Dividing by } -5 \text{ on both sides}$$

$$x = -\frac{22}{5}, \text{ or } -4\frac{2}{5} \quad \text{Simplifying}$$

CHECK:

$$\begin{array}{r|l} -5x - 6 & 16 \\ \hline -5\left(-\frac{22}{5}\right) - 6 & ?\ 16 \\ 22 - 6 & \\ 16 & \quad \text{TRUE} \end{array}$$

The solution is $-\frac{22}{5}$.

Do Exercises 2 and 3.

Solve.

2. $8x - 4 = 28$

3. $-\frac{1}{2}x + 3 = 1$

Answers on page A-25

4. Solve: $-18 - x = -57$.

Solve.

5. $-4 - 8x = 8$

6. $41.68 = 4.7 - 8.6y$

Solve.

7. $4x + 3x = -21$

8. $x - 0.09x = 728$

Answers on page A-25

Example 3 Solve: $45 - x = 13$.

$$45 - x = 13$$

$$-45 + 45 - x = -45 + 13 \quad \text{Adding } -45 \text{ on both sides}$$

$$-x = -32$$

$$-1 \cdot x = -32 \quad -x = -1 \cdot x$$

$$\frac{-1 \cdot x}{-1} = \frac{-32}{-1}$$

Dividing by -1 on both sides. (We could have multiplied by -1 on both sides instead. That would also change the sign on both sides.)

$$x = 32$$

CHECK:

$$\begin{array}{c|c} 45 - x & 13 \\ \hline 45 - 32 & ?\ 13 \\ 13 & \end{array} \quad \text{TRUE}$$

The solution is 32.

Do Exercise 4.

Example 4 Solve: $16.3 - 7.2y = -8.18$.

$$16.3 - 7.2y = -8.18$$

$$-16.3 + 16.3 - 7.2y = -16.3 + (-8.18) \quad \text{Adding } -16.3 \text{ on both sides}$$

$$-7.2y = -24.48$$

$$\frac{-7.2y}{-7.2} = \frac{-24.48}{-7.2} \quad \text{Dividing by } -7.2 \text{ on both sides}$$

$$y = 3.4$$

CHECK:

$$\begin{array}{c|c} 16.3 - 7.2y & -8.18 \\ \hline 16.3 - 7.2(3.4) & ?\ -8.18 \\ 16.3 - 24.48 & \\ -8.18 & \end{array} \quad \text{TRUE}$$

The solution is 3.4.

Do Exercises 5 and 6.

b Collecting Like Terms

If there are like terms on one side of the equation, we collect them before using the addition or multiplication principle.

Example 5 Solve: $3x + 4x = -14$.

$$3x + 4x = -14$$

$$7x = -14 \quad \text{Collecting like terms}$$

$$\frac{7x}{7} = \frac{-14}{7} \quad \text{Dividing by 7 on both sides}$$

$$x = -2.$$

The number -2 checks, so the solution is -2.

Do Exercises 7 and 8.

If there are like terms on opposite sides of the equation, we get them on the same side by using the addition principle. Then we collect them. In other words, we get all terms with a variable on one side and all numbers on the other.

Example 6 Solve: $2x - 2 = -3x + 3$.

$2x - 2 = -3x + 3$	
$2x - 2 + 2 = -3x + 3 + 2$	Adding 2
$2x = -3x + 5$	Collecting like terms
$2x + 3x = -3x + 5 + 3x$	Adding $3x$
$5x = 5$	Simplifying
$\frac{5x}{5} = \frac{5}{5}$	Dividing by 5
$x = 1$	Simplifying

CHECK:

$2x - 2$	$= -3x + 3$	
$2 \cdot 1 - 2$	? $-3 \cdot 1 + 3$	
$2 - 2$	$-3 + 3$	
0	0	TRUE

The solution is 1.

Do Exercise 9.

In Example 6, we used the addition principle to get all terms with a variable on one side and all numbers on the other side. Then we collected like terms and proceeded as before. If there are like terms on one side at the outset, they should be collected first.

Example 7 Solve: $6x + 5 - 7x = 10 - 4x + 3$.

$6x + 5 - 7x = 10 - 4x + 3$	
$-x + 5 = 13 - 4x$	Collecting like terms
$4x - x + 5 = 13 - 4x + 4x$	Adding $4x$
$3x + 5 = 13$	Simplifying
$3x + 5 - 5 = 13 - 5$	Subtracting 5
$3x = 8$	Simplifying
$\frac{3x}{3} = \frac{8}{3}$	Dividing by 3
$x = \frac{8}{3}$	Simplifying

The number $\frac{8}{3}$ checks, so $\frac{8}{3}$ is the solution.

Do Exercises 10–12.

9. Solve: $7y + 5 = 2y + 10$.

Solve.

10. $5 - 2y = 3y - 5$

11. $7x - 17 + 2x = 2 - 8x + 15$

12. $3x - 15 = 5x + 2 - 4x$

Answers on page A-25

13. Solve: $\frac{7}{8}x - \frac{1}{4} + \frac{1}{2}x = \frac{3}{4} + x.$

Clearing Fractions and Decimals

For the equations considered thus far, we generally use the addition principle first. There are, however, some situations in which it is to our advantage to use the multiplication principle first. Consider, for example,

$$\frac{1}{2}x = \frac{3}{4}.$$

The LCM of the denominators is 4. If we multiply by 4 on both sides, we get $2x = 3$, which has no fractions. We have "cleared fractions." Now consider

$$2.3x = 4.78.$$

If we multiply by 100 on both sides, we get $230x = 478$, which has no decimal points. We have "cleared decimals." The equations are then easier to solve. It is your choice whether to clear fractions or decimals, but doing so often eases computations.

In what follows, we use the multiplication principle first to "clear," or "eliminate," fractions or decimals. For fractions, the number by which we multiply is the **least common multiple of all the denominators.**

Example 8 Solve:

$$\frac{2}{3}x - \frac{1}{6} + \frac{1}{2}x = \frac{7}{6} + 2x.$$

The number 6 is the least common multiple of all the denominators. We multiply by 6 on both sides:

$$6\left(\frac{2}{3}x - \frac{1}{6} + \frac{1}{2}x\right) = 6\left(\frac{7}{6} + 2x\right)$$ Multiplying by 6 on both sides

$$6 \cdot \frac{2}{3}x - 6 \cdot \frac{1}{6} + 6 \cdot \frac{1}{2}x = 6 \cdot \frac{7}{6} + 6 \cdot 2x$$ Using the distributive laws. (*Caution*! Be sure to multiply all the terms by 6.)

$$4x - 1 + 3x = 7 + 12x$$ Simplifying. Note that the fractions are cleared.

$$7x - 1 = 7 + 12x$$ Collecting like terms

$$7x - 1 - 12x = 7 + 12x - 12x$$ Subtracting $12x$

$$-5x - 1 = 7$$ Simplifying

$$-5x - 1 + 1 = 7 + 1$$ Adding 1

$$-5x = 8$$ Collecting like terms

$$x = -\frac{8}{5}.$$ Multiplying by $-\frac{1}{5}$ or dividing by -5

The number $-\frac{8}{5}$ checks and is the solution.

Do Exercise 13.

Answer on page A-25

To illustrate clearing decimals, we repeat Example 4, but this time we clear the decimals first.

Example 9 Solve: $16.3 - 7.2y = -8.18$.

The greatest number of decimal places in any one number is *two.* Multiplying by 100, which has *two* 0's, will clear the decimals.

$$
\begin{aligned}
100(16.3 - 7.2y) &= 100(-8.18) && \text{Multiplying by 100 on both sides} \\
100(16.3) - 100(7.2y) &= 100(-8.18) && \text{Using a distributive law} \\
1630 - 720y &= -818 && \text{Simplifying} \\
1630 - 720y - 1630 &= -818 - 1630 && \text{Subtracting 1630 on both sides} \\
-720y &= -2448 && \text{Collecting like terms} \\
\frac{-720y}{-720} &= \frac{-2448}{-720} && \text{Dividing by } -720 \text{ on both sides} \\
y &= 3.4
\end{aligned}
$$

The number 3.4 checks and is the solution.

Do Exercise 14.

c Equations Containing Parentheses

To solve certain kinds of equations that contain parentheses, we first use the distributive laws to remove the parentheses. Then we proceed as before.

Example 10 Solve: $4x = 2(12 - 2x)$.

$$
\begin{aligned}
4x &= 2(12 - 2x) \\
4x &= 24 - 4x && \text{Using a distributive law to multiply and remove parentheses} \\
4x + 4x &= 24 - 4x + 4x && \text{Adding } 4x \text{ to get all } x\text{-terms on one side} \\
8x &= 24 && \text{Collecting like terms} \\
\frac{8x}{8} &= \frac{24}{8} && \text{Dividing by 8} \\
x &= 3
\end{aligned}
$$

CHECK:

$4x = 2(12 - 2x)$	
$4 \cdot 3$?	$2(12 - 2 \cdot 3)$
12	$2(12 - 6)$
	$2 \cdot 6$
	12

We use the rules for order of operations to carry out the calculations on each side of the equation.

TRUE

The solution is 3.

Do Exercises 15 and 16.

14. Solve: $41.68 = 4.7 - 8.6y$.

Solve.

15. $2(2y + 3) = 14$

16. $5(3x - 2) = 35$

Answers on page A-25

Solve.

17. $3(7 + 2x) = 30 + 7(x - 1)$

18. $4(3 + 5x) - 4 = 3 + 2(x - 2)$

Answers on page A-25

Here is a procedure for solving the types of equation discussed in this section.

AN EQUATION-SOLVING PROCEDURE

1. Multiply on both sides to clear the equation of fractions or decimals. (This is optional, but it can ease computations.)
2. If parentheses occur, multiply using the *distributive laws* to remove them.
3. Collect like terms on each side, if necessary.
4. Get all terms with variables on one side and all constant terms on the other side, using the *addition principle.*
5. Collect like terms again, if necessary.
6. Multiply or divide to solve for the variable, using the *multiplication principle.*
7. Check all possible solutions in the original equation.

Example 11 Solve: $2 - 5(x + 5) = 3(x - 2) - 1$.

$2 - 5(x + 5) = 3(x - 2) - 1$	
$2 - 5x - 25 = 3x - 6 - 1$	**Using the distributive laws to multiply and remove parentheses**
$-5x - 23 = 3x - 7$	**Collecting like terms**
$-5x - 23 + 5x = 3x - 7 + 5x$	**Adding 5x**
$-23 = 8x - 7$	**Collecting like terms**
$-23 + 7 = 8x - 7 + 7$	**Adding 7**
$-16 = 8x$	**Collecting like terms**
$\frac{-16}{8} = \frac{8x}{8}$	**Dividing by 8**
$-2 = x$	

CHECK:

$2 - 5(x + 5)$	$= 3(x - 2) - 1$	
$2 - 5(-2 + 5)$?	$3(-2 - 2) - 1$	
$2 - 5(3)$	$3(-4) - 1$	
$2 - 15$	$-12 - 1$	
-13	-13	TRUE

The solution is -2.

Note that the solution of $-2 = x$ is -2, which is also the solution of $x = -2$.

Do Exercises 17 and 18.

Exercise Set 11.4

a Solve. Don't forget to check!

1. $5x + 6 = 31$
2. $8x + 6 = 30$
3. $8x + 4 = 68$
4. $8z + 7 = 79$
5. $4x - 6 = 34$
6. $4x - 11 = 21$
7. $3x - 9 = 33$
8. $6x - 9 = 57$
9. $7x + 2 = -54$
10. $5x + 4 = -41$
11. $-45 = 3 + 6y$
12. $-91 = 9t + 8$
13. $-4x + 7 = 35$
14. $-5x - 7 = 108$
15. $-7x - 24 = -129$
16. $-6z - 18 = -132$

b Solve.

17. $5x + 7x = 72$
18. $4x + 5x = 45$
19. $8x + 7x = 60$
20. $3x + 9x = 96$
21. $4x + 3x = 42$
22. $6x + 19x = 100$
23. $-6y - 3y = 27$
24. $-4y - 8y = 48$
25. $-7y - 8y = -15$
26. $-10y - 3y = -39$
27. $10.2y - 7.3y = -58$
28. $6.8y - 2.4y = -88$

29. $x + \frac{1}{3}x = 8$

30. $x + \frac{1}{4}x = 10$

31. $8y - 35 = 3y$

32. $4x - 6 = 6x$

33. $8x - 1 = 23 - 4x$

34. $5y - 2 = 28 - y$

35. $2x - 1 = 4 + x$

36. $5x - 2 = 6 + x$

37. $6x + 3 = 2x + 11$

38. $5y + 3 = 2y + 15$

39. $5 - 2x = 3x - 7x + 25$

40. $10 - 3x = 2x - 8x + 40$

41. $4 + 3x - 6 = 3x + 2 - x$

42. $5 + 4x - 7 = 4x - 2 - x$

43. $4y - 4 + y + 24 = 6y + 20 - 4y$

44. $5y - 7 + y = 7y + 21 - 5y$

Solve. Clear fractions or decimals first.

45. $\frac{7}{2}x + \frac{1}{2}x = 3x + \frac{3}{2} + \frac{5}{2}x$

46. $\frac{7}{8}x - \frac{1}{4} + \frac{3}{4}x = \frac{1}{16} + x$

47. $\frac{2}{3} + \frac{1}{4}t = \frac{1}{3}$

48. $-\frac{3}{2} + x = -\frac{5}{6} - \frac{4}{3}$

49. $\frac{2}{3} + 3y = 5y - \frac{2}{15}$

50. $\frac{1}{2} + 4m = 3m - \frac{5}{2}$

51. $\frac{5}{3} + \frac{2}{3}x = \frac{25}{12} + \frac{5}{4}x + \frac{3}{4}$

52. $1 - \frac{2}{3}y = \frac{9}{5} - \frac{y}{5} + \frac{3}{5}$

53. $2.1x + 45.2 = 3.2 - 8.4x$

54. $0.96y - 0.79 = 0.21y + 0.46$

55. $1.03 - 0.62x = 0.71 - 0.22x$

56. $1.7t + 8 - 1.62t = 0.4t - 0.32 + 8$

57. $\frac{2}{7}x - \frac{1}{2}x = \frac{3}{4}x + 1$

58. $\frac{5}{16}y + \frac{3}{8}y = 2 + \frac{1}{4}y$

c Solve.

59. $3(2y - 3) = 27$

60. $4(2y - 3) = 28$

61. $40 = 5(3x + 2)$

62. $9 = 3(5x - 2)$

63. $2(3 + 4m) - 9 = 45$

64. $3(5 + 3m) - 8 = 88$

65. $5r - (2r + 8) = 16$

66. $6b - (3b + 8) = 16$

67. $6 - 2(3x - 1) = 2$

68. $10 - 3(2x - 1) = 1$

69. $5(d + 4) = 7(d - 2)$

70. $3(t - 2) = 9(t + 2)$

71. $8(2t + 1) = 4(7t + 7)$

72. $7(5x - 2) = 6(6x - 1)$

73. $3(r - 6) + 2 = 4(r + 2) - 21$

74. $5(t + 3) + 9 = 3(t - 2) + 6$

75. $19 - (2x + 3) = 2(x + 3) + x$

76. $13 - (2c + 2) = 2(c + 2) + 3c$

77. $0.7(3x + 6) = 1.1 - (x + 2)$

78. $0.9(2x + 8) = 20 - (x + 5)$

79. $a + (a - 3) = (a + 2) - (a + 1)$

80. $0.8 - 4(b - 1) = 0.2 + 3(4 - b)$

Skill Maintenance

Divide. [10.5c]

81. $22.1 \div 3.4$

82. $-22.1 \div (-3.4)$

83. $22.1 \div (-3.4)$

84. $-22.1 \div 3.4$

Factor. [11.1c]

85. $7x - 21 - 14y$

86. $25a - 625b + 75$

87. $42t + 14m - 56$

88. $16a - 64b + 224 - 32q$

89. Find $-(-x)$ when $x = -8$.

90. Use $<$ or $>$ for ▒ to write a true sentence:

-15 ▒ -13.

Synthesis

91. ◆ A student begins solving the equation $\frac{2}{3}x + 1 = \frac{5}{6}$ by multiplying by 6 on both sides. Is this a wise thing to do? Why or why not?

92. ◆ Describe a procedure that a classmate could use to solve the equation $ax + b = c$ for x.

Solve.

93. $\frac{y - 2}{3} = \frac{2 - y}{5}$

94. $3x = 4x$

95. $\frac{5 + 2y}{3} = \frac{25}{12} + \frac{5y + 3}{4}$

96. ▦ $0.05y - 1.82 = 0.708y - 0.504$

97. $\frac{2}{3}(2x - 1) = 10$

98. $\frac{2}{3}\left(\frac{7}{8} - 4x\right) - \frac{5}{8} = \frac{3}{8}$

99. The perimeter of the figure shown is 15 cm. Solve for x.

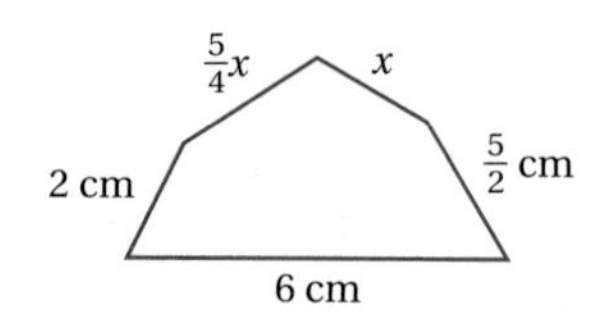

Collaborative Learning Manual

As a group, create and solve equations.

11.5 Solving Applied Problems

a Translating to Algebraic Expressions

In algebra, we translate problems to equations. The different parts of an equation are translations of word phrases to algebraic expressions. To translate, it helps to learn which words translate to certain operation symbols.

KEY WORDS			
Addition (+)	Subtraction (−)	Multiplication (·)	Division (÷)
add sum plus more than increased by	subtract difference minus less than decreased by take from	multiply product times twice of	divide quotient divided by

Objectives

a Translate phrases to algebraic expressions.

b Solve applied problems by translating to equations.

For Extra Help

TAPE 20 TAPE 18A MAC WIN CD-ROM

Example 1 Translate to an algebraic expression:

Twice (or two times) some number.

Think of some number—say, 8. What number is twice 8? It is 16. How did you get 16? You multiplied by 2. Do the same thing using a variable. We can use any variable we wish, such as x, y, m, or n. Let's use y to stand for some number. If we multiply by 2, we get an expression

$$y \times 2, \quad 2 \times y, \quad 2 \cdot y, \quad \text{or} \quad 2y.$$

Example 2 Translate to an algebraic expression:

Seven less than some number.

We let

x = the number.

Now if the number were 23, then the translation would be $23 - 7$. If we knew the number to be 345, then the translation would be $345 - 7$. The translation is found as follows:

Seven less than some number

$$x - 7.$$

Note that $7 - x$ is *not* a correct translation of the expression in Example 2. The expression $7 - x$ is a translation of "seven minus some number" or "some number less than seven."

Translate to an algebraic expression.

1. Twelve less than some number

2. Twelve more than some number

3. Four less than some number

4. Half of some number

5. Six more than eight times some number

6. The difference of two numbers

7. Fifty-nine percent of some number

8. Two hundred less than the product of two numbers

9. The sum of two numbers

Answers on page A-26

Example 3 Translate to an algebraic expression:

Eighteen more than a number.

We let

$t =$ the number.

Now if the number were 26, then the translation would be $18 + 26$. If we knew the number to be 174, then the translation would be $18 + 174$. The translation is

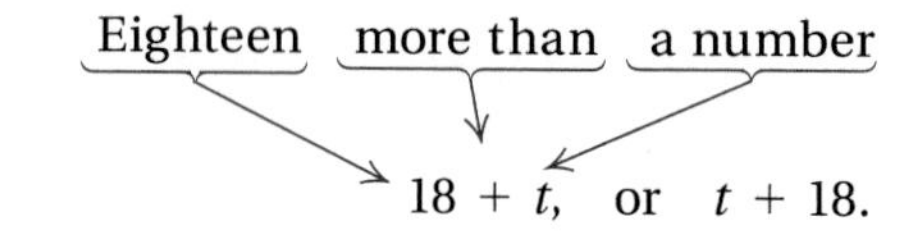

$18 + t$, or $t + 18$.

Example 4 Translate to an algebraic expression:

A number divided by 5.

We let

$m =$ the number.

If the number were 76, then the translation would be $76 \div 5$, or $76/5$, or $\frac{76}{5}$. If the number were 213, then the translation would be $213 \div 5$, or $213/5$, or $\frac{213}{5}$. The translation is found as follows:

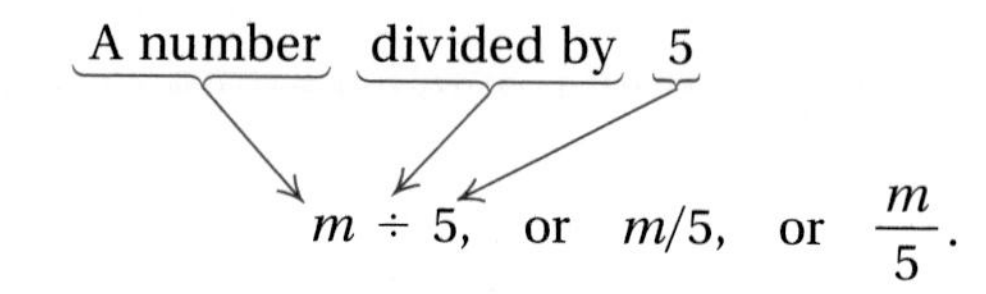

$m \div 5$, or $m/5$, or $\dfrac{m}{5}$.

Example 5 Translate to an algebraic expression.

Phrase	Algebraic Expression
Five more than some number	$5 + n$, or $n + 5$
Half of a number	$\frac{1}{2}t$, or $\frac{t}{2}$
Five more than three times some number	$5 + 3p$, or $3p + 5$
The difference of two numbers	$x - y$
Six less than the product of two numbers	$mn - 6$
Seventy-six percent of some number	$76\%z$, or $0.76z$

Do Exercises 1–9.

b Five Steps for Solving Problems

We have studied many new equation-solving tools in this chapter. We now apply them to problem solving. We have purposely used the following strategy throughout this text in order to introduce you to algebra.

> **FIVE STEPS FOR PROBLEM SOLVING IN ALGEBRA**
>
> 1. *Familiarize* yourself with the problem situation.
> 2. *Translate* to an equation.
> 3. *Solve* the equation.
> 4. *Check* your possible answer in the original problem.
> 5. *State* the answer clearly.

Example 6 Five plus three more than a number is nineteen. What is the number?

1. **Familiarize.** Let $x =$ the number. Then "three more than a number" translates to $x + 3$, and "five plus three more than a number" translates to $5 + (x + 3)$.
2. **Translate.** The familiarization leads us to the following translation:

Five	plus	three more than a number	is	nineteen.
↓	↓	↓	↓	↓
5	+	$(x + 3)$	=	19

3. **Solve.** We solve the equation:

$$5 + (x + 3) = 19$$
$$x + 8 = 19 \quad \text{Collecting like terms}$$
$$x + 8 - 8 = 19 - 8 \quad \text{Subtracting 8}$$
$$x = 11.$$

4. **Check.** Three more than 11 is 14. Adding 5 to 14, we get 19. This checks.
5. **State.** The number is 11.

Do Exercise 10.

10. If 5 is subtracted from three times a certain number, the result is 10. What is the number?

Answer on page A-26

11. *Board Cutting.* A 52-in. board is cut into three pieces. The first piece is one-half the length of the third piece. The second piece is one-eighth the length of the third piece. Find the length of each piece.

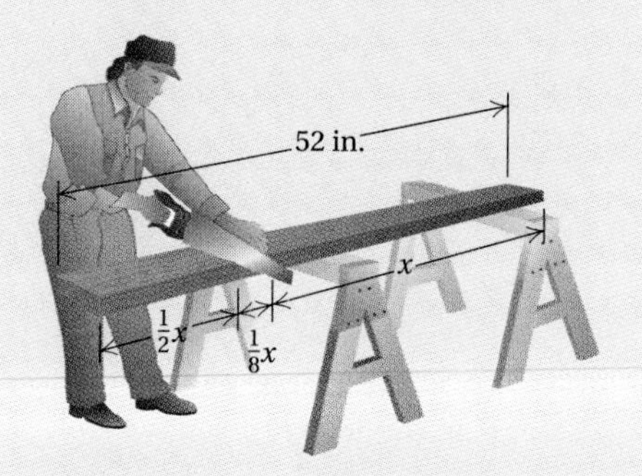

Answer on page A-26

Example 7 *Rocket Sections.* A rocket is divided into three sections: the payload and navigation at the top, the fuel in the middle, and the rocket engine at the bottom. The top section is one-sixth the length of the bottom section. The middle section is one-half the length of the bottom section. The total length of all three is 240 ft. Find the length of each section.

1. **Familiarize.** We first make a drawing. Noting that the lengths of the top and middle sections are expressed in terms of the bottom section, we let

x = the length of the bottom section.

Then $\frac{1}{2}x$ = the length of the middle section

and $\frac{1}{6}x$ = the length of the top section.

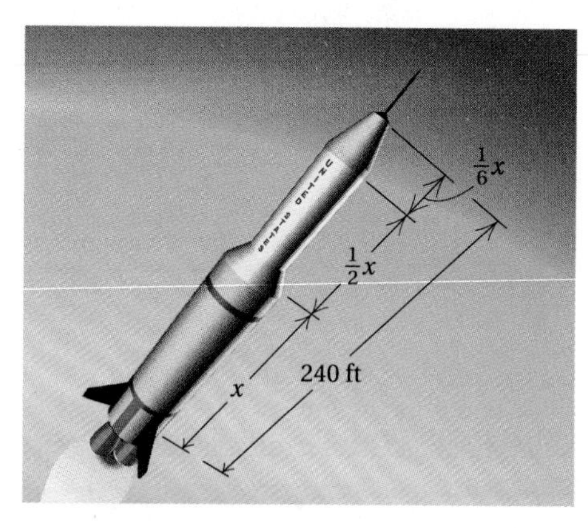

2. **Translate.** From the statement of the problem and the drawing, we see that the lengths add up to 240 ft. This gives us our translation.

Length of top	plus	Length of middle	plus	Length of bottom	is	Total length
↓	↓	↓	↓	↓	↓	↓
$\frac{1}{6}x$	$+$	$\frac{1}{2}x$	$+$	x	$=$	240

3. **Solve.** We solve the equation by clearing fractions, as follows:

$$\frac{1}{6}x + \frac{1}{2}x + x = 240$$ The LCM of all the denominators is 6.

$$6 \cdot \left(\frac{1}{6}x + \frac{1}{2}x + x\right) = 6 \cdot 240$$ Multiplying by the LCM, 6

$$6 \cdot \frac{1}{6}x + 6 \cdot \frac{1}{2}x + 6 \cdot x = 6 \cdot 240$$ Using the distributive law

$$x + 3x + 6x = 1440$$ Simplifying

$$10x = 1440$$ Adding

$$\frac{10x}{10} = \frac{1440}{10}$$ Dividing by 10

$$x = 144.$$

4. **Check.** Do we have an answer *to the problem*? If the rocket engine (bottom section) is 144 ft, then the middle section is $\frac{1}{2} \cdot 144$, or 72 ft, and the top section is $\frac{1}{6} \cdot 144$, or 24 ft. Since these lengths total 240 ft, our answer checks.

5. **State.** The top section is 24 ft, the middle 72 ft, and the bottom 144 ft.

Do Exercise 11.

Example 8 *Truck Rental.* Truck-Rite Rentals rents trucks at a daily rate of \$59.95 plus 49¢ per mile. Concert Productions has budgeted \$400 per day for renting a truck to haul equipment to an upcoming concert. How many miles can a rental truck be driven on a \$400 daily budget?

1. **Familiarize.** Suppose that Concert Productions drives 75 mi. Then the cost is

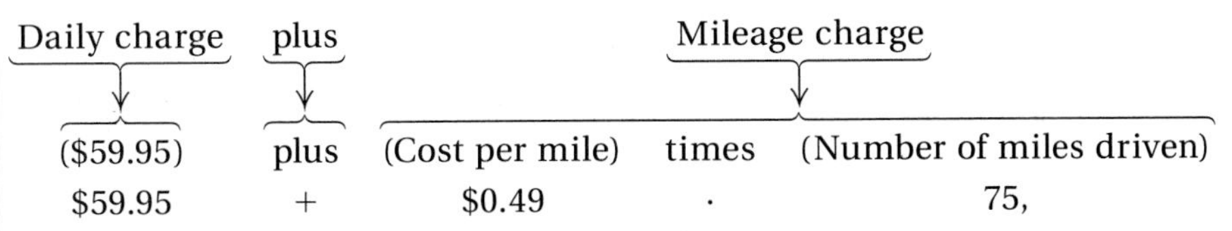

which is \$59.95 + \$36.75, or \$96.70. This familiarizes us with the way in which a calculation is made. Note that we convert 49 cents to \$0.49 so that we are using the same units, dollars, throughout the problem.

Let m = the number of miles that can be driven for \$400.

2. **Translate.** We reword the problem and translate as follows.

Daily rate	plus	Cost per mile	times	Number of miles driven	is	Total cost
↓	↓	↓	↓	↓	↓	↓
\$59.95	+	\$0.49	·	m	=	\$400

3. **Solve.** We solve the equation:

$$59.95 + 0.49m = 400$$

$$100(59.95 + 0.49m) = 100(400) \quad \text{Multiplying by 100 on both sides to clear decimals}$$

$$100(59.95) + 100(0.49m) = 40{,}000 \quad \text{Using the distributive law}$$

$$5995 + 49m = 40{,}000$$

$$49m = 34{,}005 \quad \text{Subtracting 5995}$$

$$\frac{49m}{49} = \frac{34{,}005}{49} \quad \text{Dividing by 49}$$

$$m \approx 694.0. \quad \text{Rounding to the nearest tenth}$$

4. **Check.** We check in the original problem. We multiply 694 by \$0.49, getting \$340.06. Then we add \$340.06 to \$59.95 and get \$400.01, which is just about the \$400 allotted.

5. **State.** The truck can be driven about 694 mi on the rental allotment of \$400.

Do Exercise 12.

12. *Van Rental.* Truck-Rite also rents vans at a daily rate of \$64.95 plus 57 cents per mile. What mileage will allow a salesperson to stay within a daily budget of \$400?

Answer on page A-26

13. *Angles of a Triangle.* The second angle of a triangle is three times as large as the first. The third angle measures 30° more than the first angle. Find the measures of the angles.

Example 9 *Angles of a Triangle.* The second angle of a triangle is twice as large as the first. The measure of the third angle is 20° greater than that of the first angle. How large are the angles?

1. **Familiarize.** We first make a drawing, letting

$$\text{the measure of the first angle} = x.$$

Then $\quad$ the measure of the second angle $= 2x$

and $\quad$ the measure of the third angle $= x + 20$.

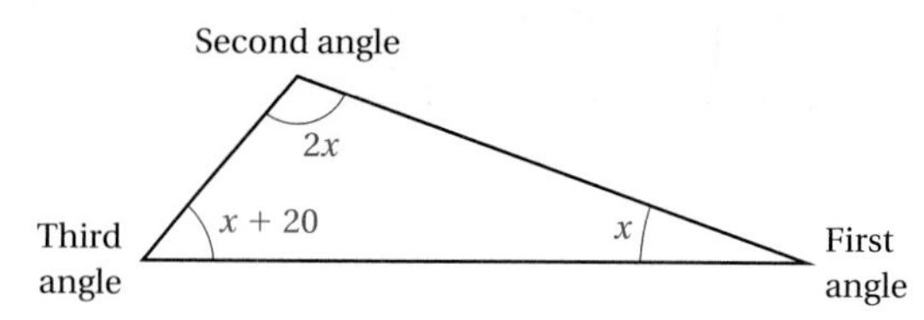

2. **Translate.** To translate, we recall from Section 9.6 that the measures of the angles of any triangle add up to 180°.

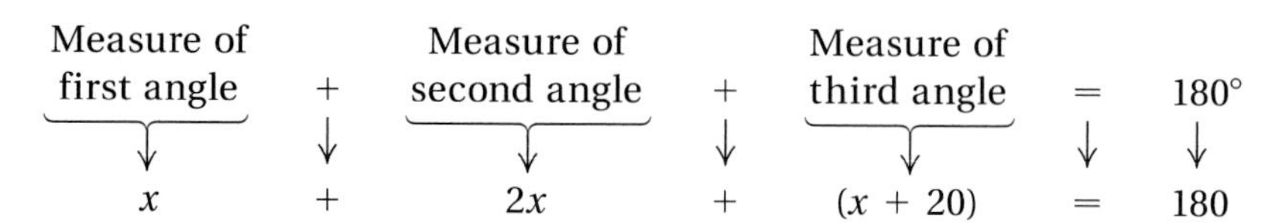

3. **Solve.** We solve the equation:

$$\begin{aligned} x + 2x + (x + 20) &= 180 \\ 4x + 20 &= 180 \\ 4x + 20 - 20 &= 180 - 20 \\ 4x &= 160 \\ \frac{4x}{4} &= \frac{160}{4} \\ x &= 40. \end{aligned}$$

Possible measures for the angles are as follows:

First angle: $x = 40°$;

Second angle: $2x = 2(40) = 80°$;

Third angle: $x + 20 = 40 + 20 = 60°$.

4. **Check.** Consider 40°, 80°, and 60°. The second is twice the first, and the third is 20° greater than the first. The sum is 180°. These numbers check.

5. **State.** The measures of the angles are 40°, 80°, and 60°.

Caution! Units are important in answers. Try to remember to include them, where appropriate.

Do Exercise 13.

Answers on page A-26

Example 10 *Area of Colorado.* The state of Colorado is roughly in the shape of a rectangle whose perimeter is 1300 mi. The length is 110 mi more than the width. Find the dimensions.

1. **Familiarize.** We first make a drawing, letting

$$w = \text{the width of the rectangle.}$$

Then $w + 110 = $ the length.

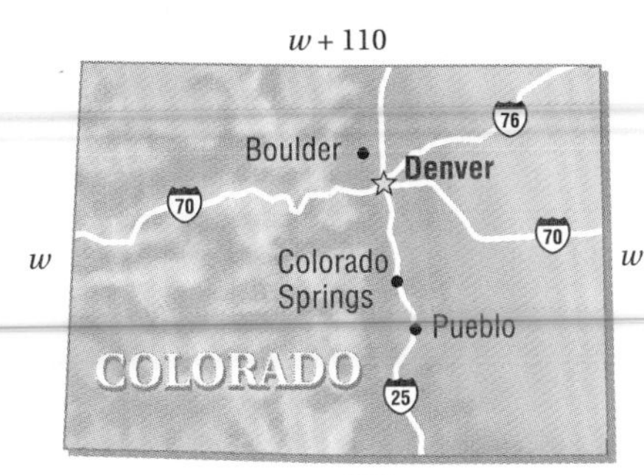

(We can also let $l =$ the length and $l - 110 =$ the width.)

The perimeter P of a rectangle is the distance around it and is given by the formula $2l + 2w = P$, where $l =$ the length and $w =$ the width.

2. **Translate.** To translate the problem, we substitute $w + 110$ for l and 1300 for P, as follows:

$$2l + 2w = P$$
$$2(w + 110) + 2w = 1300.$$

3. **Solve.** We solve the equation:

$$2(w + 110) + 2w = 1300$$
$$2w + 220 + 2w = 1300 \quad \text{Multiplying using a distributive law}$$
$$4w + 220 = 1300$$
$$4w = 1080$$
$$w = 270.$$

Possible dimensions are $w = 270$ mi and $l = w + 110 = 380$ mi.

4. **Check.** If the width is 270 mi and the length is 380 mi, the perimeter is 2(380 mi) + 2(270 mi), or 1300 mi. This checks.

5. **State.** The width is 270 mi, and the length is 380 mi.

Do Exercise 14.

14. *Area of Rug.* A hooked rug has a perimeter of 42 ft. The length is 3 ft more than the width. Find the dimensions of the rug.

Answer on page A-26

We turn now to a different method of solving a problem that we first considered in Sections 7.3 and 7.5.

Example 11 *Movies Released.* The data in the table and the graph show the number of movies released over a period of years.

Years Since 1990	Number of Movies Released
1. 1991	164
2. 1992	150
3. 1993	161
4. 1994	184
5. 1995	234
6. 1996	260
7. 1997	?
10. 2000	?
20. 2010	?

Source: Motion Picture Association of America

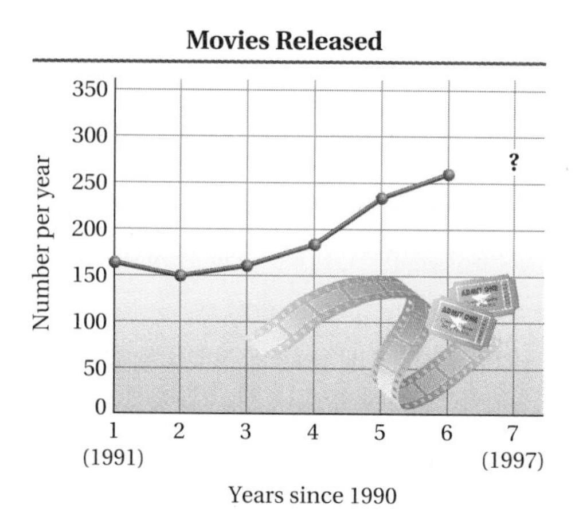

We can use the equation

$$y = 21.571x + 116.667$$

to approximate the number y of movies released in year x, where

$x = 1$ corresponds to 1991,

$x = 2$ corresponds to 1992,

$x = 10$ corresponds to 2000,

and so on.

(This equation was determined from a procedure called *regression.* Its development belongs to a later course.)

a) Find the number of movies to be released in 1997, in 2000, and in 2010.

b) In what year will there be 376 movie releases?

We use the equation to solve the problems.

a) The number of movies to be released in 1997 corresponds to letting $x = 7$. Thus we substitute 7 for x in the equation:

$$y = 21.571x + 116.667 = 21.571(7) + 116.667 \approx 268.$$

We round to a whole number because it would not be meaningful to release a fractional part of a movie. The equation gives us just an estimate.

The number of movies to be released in 2000 corresponds to letting $x = 10$. We substitute 10 for x in the equation:

$$y = 21.571x + 116.667 = 21.571(10) + 116.667 \approx 332.$$

The number of movies to be released in 2010 corresponds to letting $x = 20$. We substitute 20 for x in the equation:

$$y = 21.571x + 116.667 = 21.571(20) + 116.667 \approx 548.$$

Thus the number of movies to be released is 268 in 1997, 332 in 2000, and 548 in 2010.

b) To find the year in which there will be 376 movies released, we substitute 376 for y and solve for x:

$$376 = 21.571x + 116.667$$

$$376 - 116.667 = 21.571x + 116.667 - 116.667 \quad \text{Subtracting 116.667}$$

$$259.333 = 21.571x \quad \text{Collecting like terms}$$

$$\frac{259.333}{21.571} = \frac{21.571x}{21.571} \quad \text{Dividing by 21.571 on both sides}$$

$$12.02 \approx x.$$

The solution to the equation is about 12, which corresponds to the year 1990 + 12, or 2002. Thus in 2002, there will be about 376 movies released.

Do Exercise 15.

Though we are not considering graphs of equations in this text, it turns out that the "graph" of the equation in Example 11 is a straight line, as shown by the dashed line in the following figure. Compare your results to those in Sections 7.3 and 7.5.

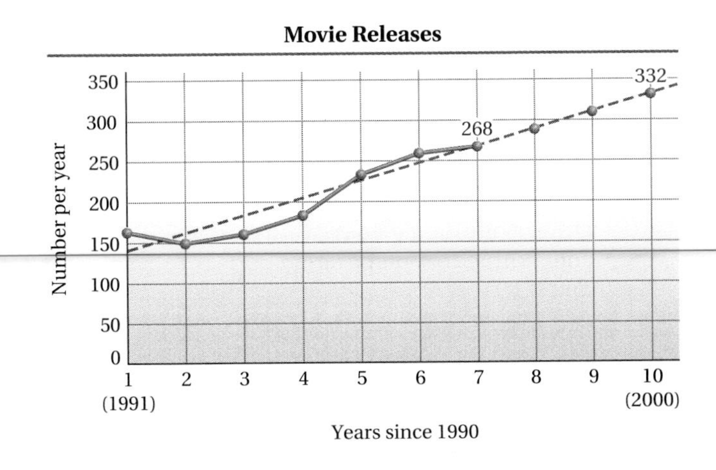

15. *FedEx Mailing Costs.* The cost y, in dollars, of mailing a Priority Overnight package weighing 1 lb or more is given by the equation

$$y = 2.085x + 15.08,$$

where x = the number of pounds (***Source:*** Federal Express Corporation).

a) Find the cost of mailing packages weighing 5 lb, 7 lb, and 20 lb.

b) What is the weight of a package that costs \$177.71 to mail?

(See also Exercises 1–6 in the Chapter 7 Summary and Review.)

Answers on page A-26

Improving Your Math Study Skills

Preparing for a Final Exam

Best Scenario: Two Weeks of Study Time

The best scenario for preparing for a final exam is to do so over a period of at least two weeks. Work in a diligent, disciplined manner, doing some final-exam preparation *each* day. Here is a detailed plan that many find useful.

1. **Begin by browsing through each chapter, reviewing the highlighted or boxed information regarding important formulas in both the text and the Summary and Review.** There may be some formulas that you will need to memorize.
2. **Retake each chapter test that you took in class, assuming your instructor has returned it. Otherwise, use the chapter test in the book.** Restudy the objectives in the text that correspond to each question you missed.
3. **Then work the Cumulative Review that covers all chapters up to that point.** Be careful to avoid any questions corresponding to objectives not covered. Again, restudy the objectives in the text that correspond to each question you missed.
4. **If you are still missing questions, use the supplements for extra review.** For example, you might check out the video- or audiotapes, the *Student's Solutions Manual,* or the InterAct Math Tutorial Software.
5. **For remaining difficulties, see your instructor, go to a tutoring session, or participate in a study group.**
6. **Check for former final exams that may be on file in the math department or a study center, or with students who have already taken the course.** Use them for practice, being alert to trouble spots.
7. **Take the Final Examination in the text during the last couple of days before the final.** Set aside the same amount of time that you will have for the final. See how much of the final exam you can complete under test-like conditions.

Moderate Scenario: Three Days to Two Weeks of Study Time

1. **Begin by browsing through each chapter, reviewing the highlighted or boxed information regarding important formulas in both the text and the Summary and Review.** There may be some formulas that you will need to memorize.
2. **Retake each chapter test that you took in class, assuming your instructor has returned it. Otherwise, use the chapter test in the book.** Restudy the objectives in the text that correspond to each question you missed.
3. **Then work the last Cumulative Review in the text.** Be careful to avoid any questions corresponding to objectives not covered. Again, restudy the objectives in the text that correspond to each question you missed.
4. **For remaining difficulties, see your instructor, go to a tutoring session, or participate in a study group.**
5. **Take the Final Examination in the text during the last couple of days before the final.** Set aside the same amount of time that you will have for the final. See how much of the final exam you can complete under test-like conditions.

Worst Scenario: One or Two Days of Study Time

1. **Begin by browsing through each chapter, reviewing the highlighted or boxed information regarding important formulas in both the text and the Summary and Review.** There may be some formulas that you will need to memorize.
2. **Then work the last Cumulative Review in the text.** Be careful to avoid any questions corresponding to objectives not covered. Restudy the objectives in the text that correspond to each question you missed.
3. **Attend a final-exam review session if one is available.**
4. **Take the Final Examination in the text during the last couple of days before the final.** Set aside the same amount of time that you will have for the final. See how much of the final exam you can complete under test-like conditions.

   Promise yourself that next semester you will allow a more appropriate amount of time for final exam preparation.

Other "Improving Your Math Study Skills" concerning test preparation appear in Sections 2.3 and 3.1.

Exercise Set 11.5

a Translate to an algebraic expression.

1. Three less than twice a number

2. Three times a number divided by a

3. The product of 97% and some number

4. 43% of some number

5. Four more than five times some number

6. Seventy-five less than eight times a number

7. A 240-in. pipe is cut into two pieces. One piece is three times the length of the other. Let x = the length of the longer piece. Write an expression for the length of the shorter piece.

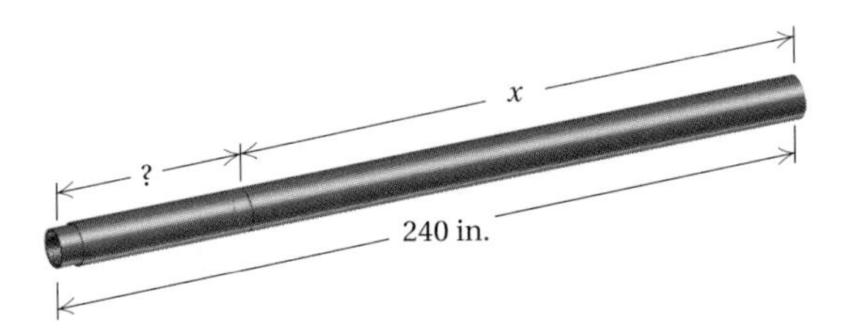

8. The price of a book is decreased by 30% during a sale. Let b = the price of the book before the reduction. Write an expression for the sale price.

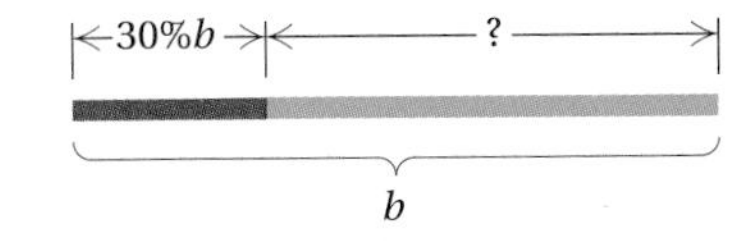

b Solve.

9. What number added to 85 is 117?

10. Eight times what number is 2552?

11. *Statue of Liberty.* The height of the Eiffel Tower is 974 ft, which is about 669 ft higher than the Statue of Liberty. What is the height of the Statue of Liberty?

12. *Area of Lake Ontario.* The area of Lake Superior is about four times the area of Lake Ontario. The area of Lake Superior is 30,172 mi². What is the area of Lake Ontario?

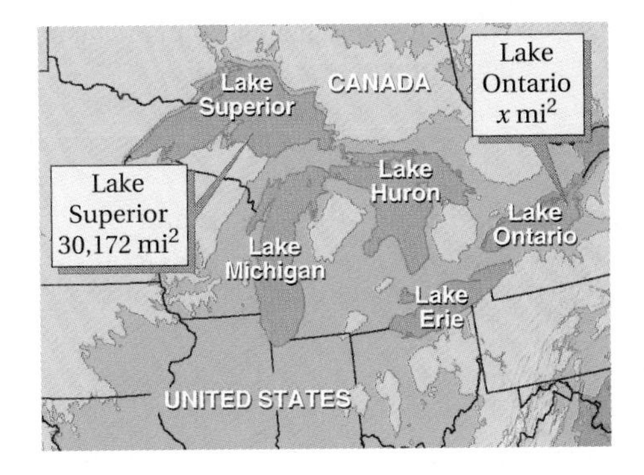

13. *Wheaties.* Recently, the cost of four 18-oz boxes of Wheaties was \$11.56. What was the cost of one box?

14. *Women's Dresses.* The total amount spent on women's blouses in a recent year was \$6.5 billion. This was \$0.2 billion more than was spent on women's dresses. How much was spent on women's dresses?

15. When 17 is subtracted from four times a certain number, the result is 211. What is the number?

16. When 36 is subtracted from five times a certain number, the result is 374. What is the number?

17. If you double a number and then add 16, you get $\frac{2}{3}$ of the original number. What is the original number?

18. If you double a number and then add 85, you get $\frac{3}{4}$ of the original number. What is the original number?

19. *Iditarod Race.* The Iditarod sled-dog race extends for 1049 mi from Anchorage to Nome (***Source:*** Iditarod Trail Commission). If a musher is twice as far from Anchorage as from Nome, how many miles has the musher traveled?

20. *Home Remodeling.* In a recent year, Americans spent a total of approximately $32.8 billion to remodel bathrooms and kitchens. Twice as much was spent on kitchens as on bathrooms. (***Source:*** *Remodeling Magazine*) How much was spent on each?

21. *Pipe Cutting.* A 480-m pipe is cut into three pieces. The second piece is three times as long as the first. The third piece is four times as long as the second. How long is each piece?

22. *Rope Cutting.* A 180-ft rope is cut into three pieces. The second piece is twice as long as the first. The third piece is three times as long as the second. How long is each piece of rope?

23. *Perimeter of NBA Court.* The perimeter of an NBA basketball court is 288 ft. The length is 44 ft longer than the width. Find the dimensions of the court.

24. *Perimeter of High School Court.* The perimeter of a standard high school basketball court is 268 ft. The length is 34 ft longer than the width. Find the dimensions of the court.

25. *Hancock Building Dimensions.* The ground floor of the John Hancock Building in Chicago is in the shape of a rectangle whose length is 100 ft more than the width. The perimeter is 860 ft. Find the length, the width, and the area of the ground floor.

26. *Hancock Building Dimensions.* The top floor of the John Hancock Building in Chicago is in the shape of a rectangle whose length is 60 ft more than the width. The perimeter is 520 ft. Find the length, the width, and the area of the top floor.

27. *Car Rental.* Value Rent-A-Car rents a family-sized car at a daily rate of \$74.95 plus 40 cents per mile. Rick Moneycutt is allotted a daily budget of \$240. How many miles can he drive per day and stay within his budget?

28. *Van Rental.* Value Rent-A-Car rents a van at a daily rate of \$84.95 plus 60 cents per mile. Molly Rickert rents a van to deliver electrical parts to her customers. She is allotted a daily budget of \$320. How many miles can she drive per day and stay within her budget?

29. *Angles of a Triangle.* The second angle of a triangular parking lot is four times as large as the first. The third angle is 45° less than the sum of the other two angles. How large are the angles?

30. *Angles of a Triangle.* The second angle of a triangular field is three times as large as the first. The third angle is 40° greater than the first. How large are the angles?

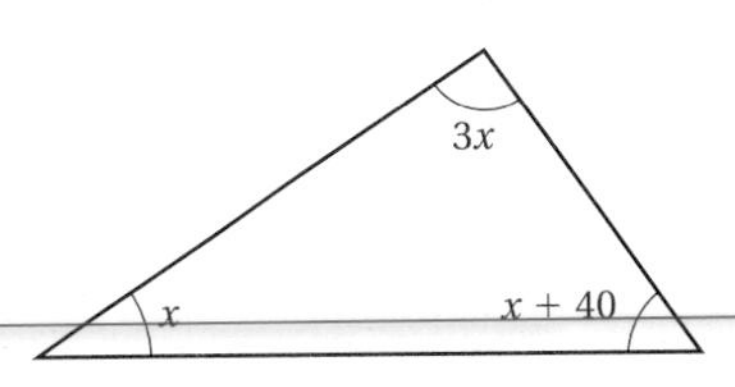

31. *Coca-Cola Co.* The equation

$$y = 66.2x + 460.2$$

can be used to approximate the total revenue y, in millions, of the Coca-Cola Company in year x, where x is the number of years since 1990 and

$x = 1$ corresponds to 1991,
$x = 2$ corresponds to 1992,
$x = 10$ corresponds to 2000,
and so on.

a) Find the total revenue in 1999, in 2000, and in 2010.
b) In what year will the total revenue be about \$1254.6 million?

32. *Nike, Inc.* The equation

$$y = 696.06x + 1687.22$$

can be used to approximate the total revenue y, in millions, of Nike, Inc., in year x, where x is the number of years since 1990 and

$x = 1$ corresponds to 1991,
$x = 2$ corresponds to 1992,
$x = 10$ corresponds to 2000,
and so on.

a) Find the total revenue in 1999, in 2002, and in 2008.
b) In what year will the total revenue be about \$12,824.18 million?

Skill Maintenance

Calculate.

33. $-\frac{4}{5} - \left(\frac{3}{8}\right)$ [10.3a]

34. $-\frac{4}{5} + \frac{3}{8}$ [10.2a]

35. $-\frac{4}{5} \cdot \frac{3}{8}$ [10.4a]

36. $-\frac{4}{5} \div \left(\frac{3}{8}\right)$ [10.5c]

37. $-25.6 \div (-16)$ [10.5c]

38. $-25.6(-16)$ [10.4a]

39. $-25.6 - (-16)$ [10.3a]

40. $-25.6 + (-16)$ [10.2a]

Synthesis

41. ◆ A fellow student claims to be able to solve most of the problems in this section by guessing. Is there anything wrong with this approach? Why or why not?

42. ◆ Write a problem for a classmate to solve so that it can be translated to the equation

$$\tfrac{2}{3}x + (x + 5) + x = 375.$$

43. Abraham Lincoln's 1863 Gettysburg Address refers to the year 1776 as "Four *score* and seven years ago." Write an equation to find out what a score is.

44. A student scored 78 on a test that had 4 seven-point fill-ins and 24 three-point multiple-choice questions. The student had 1 fill-in wrong. How many multiple-choice questions did the student get right?

45. The width of a rectangle is $\frac{3}{4}$ of the length. The perimeter of the rectangle becomes 50 cm when the length and the width are each increased by 2 cm. Find the length and the width.

46. Cookies are set out on a tray for six people to take home. One-third, one-fourth, one-eighth, and one-fifth are given to four people, respectively. The fifth person eats ten cookies on the spot, with one cookie remaining for the sixth person. Find the original number of cookies on the tray.

47. A storekeeper goes to the bank to get \$10 worth of change. She requests twice as many quarters as half dollars, twice as many dimes as quarters, three times as many nickels as dimes, and no pennies or dollars. How many of each coin did the storekeeper receive?

48. A student has an average score of 82 on three tests. His average score on the first two tests is 85. What was the score on the third test?

Summary and Review Exercises: Chapter 11

Important Properties and Formulas

The Addition Principle: If $a = b$ is true, then $a + c = b + c$ is true for any real number c.
The Multiplication Principle: If $a = b$ is true, then $ac = bc$ is true for any real number c.
The Distributive Laws: $a(b + c) = ab + ac$, $a(b - c) = ab - ac$

The objectives to be tested in addition to the material in this chapter are [8.6a, b, c], [9.1a], [10.2a], and [10.4a].

Solve. [11.2a], [11.3a]

1. $x + 5 = -17$

2. $-8x = -56$

3. $-\frac{x}{4} = 48$

4. $n - 7 = -6$

5. $15x = -35$

6. $x - 11 = 14$

7. $-\frac{2}{3} + x = -\frac{1}{6}$

8. $\frac{4}{5}y = -\frac{3}{16}$

9. $y - 0.9 = 9.09$

10. $5 - x = 13$

Solve. [11.4a, b, c]

11. $5t + 9 = 3t - 1$

12. $7x - 6 = 25x$

13. $\frac{1}{4}x - \frac{5}{8} = \frac{3}{8}$

14. $14y = 23y - 17 - 10$

15. $0.22y - 0.6 = 0.12y + 3 - 0.8y$

16. $\frac{1}{4}x - \frac{1}{8}x = 3 - \frac{1}{16}x$

17. $4(x + 3) = 36$

18. $3(5x - 7) = -66$

19. $8(x - 2) = 5(x + 4)$

20. $-5x + 3(x + 8) = 16$

21. Translate to an algebraic expression: [11.5a]
Nineteen percent of some number.

Solve. [11.5b]

22. An HDTV television sold for $829 in May. This was $38 more than the cost in January. Find the cost in January.

23. Selma gets a $4 commission for each appliance that she sells. One week she received $108 in commissions. How many appliances did she sell?

24. An 8-m length of rope is cut into two pieces. One piece is 2 m longer than the other. How long are the pieces?

25. If 14 is added to three times a certain number, the result is 41. Find the number.

26. The perimeter of a rectangular writing pad is 56 cm. The width is 6 cm less than the length. Find the width and the length.

27. *FedEx Mailing Costs.* The cost y, in dollars, of mailing a FedEx 2Day package weighing 1 lb or more is given by the equation

$$y = 1.025x + 8.01,$$

where $x =$ the number of pounds (***Source:*** Federal Express Corporation).

a) Find the cost of mailing packages weighing 5 lb, 7 lb, and 20 lb.

b) What is the weight of a package that costs $59.26 to mail?

Skill Maintenance

28. Find the diameter, the circumference, and the area of a circle when $r = 20$ ft. Use 3.14 for π. [8.6a, b, c]

29. Find the volume of a rectangular solid when the length is 20 cm, the width is 18.5 cm, and the height is 4.6 cm. [9.1a]

30. Add: $-12 + 10 + (-19) + (-24)$. [10.2a]

31. Multiply: $(-2) \cdot (-3) \cdot (-5) \cdot (-2) \cdot (-1)$. [10.4a]

Synthesis

32. ◆ Explain at least three uses of the distributive laws considered in this chapter. [11.1b, c, d]

33. ◆ Explain all the errors in each of the following. [11.4a, c]

a)
$$\begin{aligned} 4 - 3x &= 9 \\ 3x &= 9 \\ x &= 3 \end{aligned}$$

b)
$$\begin{aligned} 2(x - 5) &= 7 \\ 2x - 5 &= 7 \\ 2x &= 12 \\ x &= 6 \end{aligned}$$

34. The total length of the Nile and Amazon Rivers is 13,108 km. If the Amazon were 234 km longer, it would be as long as the Nile. Find the length of each river. [11.5b]

Solve. [10.1e], [11.4a, b]

35. $2|n| + 4 = 50$

36. $|3n| = 60$

Test: Chapter 11

Solve.

1. $x + 7 = 15$

2. $t - 9 = 17$

3. $3x = -18$

4. $-\frac{4}{7}x = -28$

5. $3t + 7 = 2t - 5$

6. $\frac{1}{2}x - \frac{3}{5} = \frac{2}{5}$

7. $8 - y = 16$

8. $-\frac{2}{5} + x = -\frac{3}{4}$

9. $0.4p + 0.2 = 4.2p - 7.8 - 0.6p$

10. $3(x + 2) = 27$

11. $-3x - 6(x - 4) = 9$

Answers

1. ______
2. ______
3. ______
4. ______
5. ______
6. ______
7. ______
8. ______
9. ______
10. ______
11. ______

12. ______

13. ______

14. ______

15. ______

16. a) ______

b) ______

17. ______

18. ______

19. ______

20. ______

21. ______

22. ______

Solve.

12. The perimeter of a rectangular piece of cardboard is 36 cm. The length is 4 cm greater than the width. Find the width and the length.

13. If you triple a number and then subtract 14, you get $\frac{2}{3}$ of the original number. What is the original number?

14. Translate to an algebraic expression:

Nine less than some number.

15. The second angle of a triangle is three times as large as the first. The third angle is 25° less than the sum of the other two angles. Find the measure of the first angle.

16. *FedEx Mailing Costs.* The cost y, in dollars, of mailing a Standard Overnight package weighing 1 lb or more is given by the equation

$$y = 1.489x + 13.52,$$

where $x =$ the number of pounds (***Source:*** Federal Express Corporation).

a) Find the cost of mailing packages weighing 5 lb, 7 lb, and 20 lb.

b) What is the weight of a package that costs $49.26 to mail?

Skill Maintenance

17. Multiply:

$$(-9) \cdot (-2) \cdot (-2) \cdot (-5).$$

18. Add:

$$\frac{2}{3} + \left(-\frac{8}{9}\right).$$

19. Find the diameter, the circumference, and the area of a circle when the radius is 70 yd. Use 3.14 for π.

20. Find the volume of a rectangular solid when the length is 22 ft, the width is 10 ft, and the height is 6 ft.

Synthesis

21. Solve: $3|w| - 8 = 37$.

22. A movie theater had a certain number of tickets to give away. Five people got the tickets. The first got $\frac{1}{3}$ of the tickets, the second got $\frac{1}{4}$ of the tickets, and the third got $\frac{1}{5}$ of the tickets. The fourth person got 8 tickets, and there were 5 tickets left for the fifth person. Find the total number of tickets given away.

Cumulative Review: Chapters 1–11

This is a review of the entire textbook. A question that may occur at this point is what notation to use for a particular problem or exercise. Although there is no hard-and-fast rule, especially as you use mathematics outside the classroom, here is the guideline that we follow: Use the notation given in the problem. That is, if the problem is given using mixed numerals, give the answer in mixed numerals. If the problem is given in decimal notation, give the answer in decimal notation.

1. In 47201, what digit tells the number of thousands?

2. Write expanded notation for 7405.

Add and simplify, if appropriate.

3. $\begin{array}{r} 741 \\ +\ 271 \\ \hline \end{array}$

4. $\begin{array}{r} 4903 \\ 5278 \\ 6391 \\ +\ 4513 \\ \hline \end{array}$

5. $\frac{2}{13} + \frac{1}{26}$

6. $\begin{array}{r} 2\frac{4}{9} \\ +\ 3\frac{1}{3} \\ \hline \end{array}$

7. $\begin{array}{r} 2.048 \\ 63.914 \\ +\ 428.009 \\ \hline \end{array}$

8. $34.56 + 2.783 + 0.433 + 765.1$

Subtract and simplify, if possible.

9. $\begin{array}{r} 674 \\ -\ 522 \\ \hline \end{array}$

10. $\begin{array}{r} 9465 \\ -\ 8791 \\ \hline \end{array}$

11. $\frac{7}{8} - \frac{2}{3}$

12. $\begin{array}{r} 4\frac{1}{3} \\ -\ 1\frac{5}{8} \\ \hline \end{array}$

13. $\begin{array}{r} 20.0 \\ -\ 0.0027 \\ \hline \end{array}$

14. $40.03 - 5.789$

Multiply and simplify, if possible.

15. $\begin{array}{r} 297 \\ \times\ 16 \\ \hline \end{array}$

16. $\begin{array}{r} 349 \\ \times\ 763 \\ \hline \end{array}$

17. $1\frac{3}{4} \cdot 2\frac{1}{3}$

18. $\frac{9}{7} \cdot \frac{14}{15}$

19. $12 \cdot \frac{5}{6}$

20. $\begin{array}{r} 34.09 \\ \times\ 7.6 \\ \hline \end{array}$

Divide and simplify. State the answer using a whole-number quotient and remainder.

21. $6\overline{)3438}$

22. $34\overline{)1914}$

23. Give a mixed numeral for the quotient in Question 22.

24. $\frac{4}{5} \div \frac{8}{15}$

25. $2\frac{1}{3} \div 30$

26. $2.7\overline{)105.3}$

27. Round 68,489 to the nearest thousand.

28. Round 0.4275 to the nearest thousandth.

29. Round $21.\overline{83}$ to the nearest hundredth.

30. Determine whether 1368 is divisible by 8.

31. Find all the factors of 15.

32. Find the LCM of 16, 25, and 32.

Simplify.

33. $\frac{21}{30}$

34. $\frac{275}{5}$

35. Convert to a mixed numeral: $\frac{18}{5}$.

36. Use $=$ or $\neq$ for ▒ to write a true sentence:

$\frac{4}{7}$ ▒ $\frac{3}{5}$.

37. Use $<$ or $>$ for ▒ to write a true sentence:

$\frac{4}{7}$ ▒ $\frac{3}{5}$.

38. Which number is greater, 1.001 or 0.9976?

39. Use $<$ or $>$ for ▒ to write a true sentence:

987 ▒ 879.

40. What part is shaded?

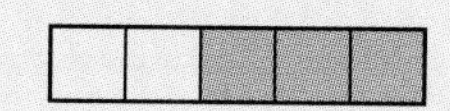

Convert to decimal notation.

41. $\frac{37}{1000}$

42. $\frac{13}{25}$

43. $\frac{8}{9}$

44. 7%

Convert to fractional notation.

45. 4.63

46. $7\frac{1}{4}$

47. 40%

Convert to percent notation.

48. $\frac{17}{20}$

49. 1.5

Solve.

50. $234 + y = 789$

51. $3.9 \times y = 249.6$

52. $\frac{2}{3} \cdot t = \frac{5}{6}$

53. $\frac{8}{17} = \frac{36}{x}$

NBA Three-Point Baskets. The table below lists the average number of three-point baskets made in the NBA in recent years.

Ending Year	Number of Three-Point Baskets per Game
1991	4.6
1992	5.0
1993	6.0
1994	6.6
1995	11.0
1996	11.6
1997	?

Source: National Basketball Association

54. Make a line graph of the data.

55. Use extrapolation to estimate the average number of three-point baskets made in 1997.

Estimate each of the following as a whole number or as a mixed numeral where the fractional part is $\frac{1}{2}$.

56. $24\frac{6}{11}$

57. $28\frac{16}{17}$

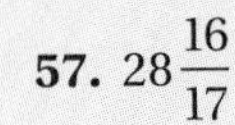

58. $8\frac{3}{11} - 7\frac{5}{6} + \frac{31}{32}$

59. $34\frac{2}{15} - 17\frac{23}{24} + 9\frac{27}{30} \cdot 8\frac{37}{76}$

60. Classify this triangle as right, obtuse, or acute.

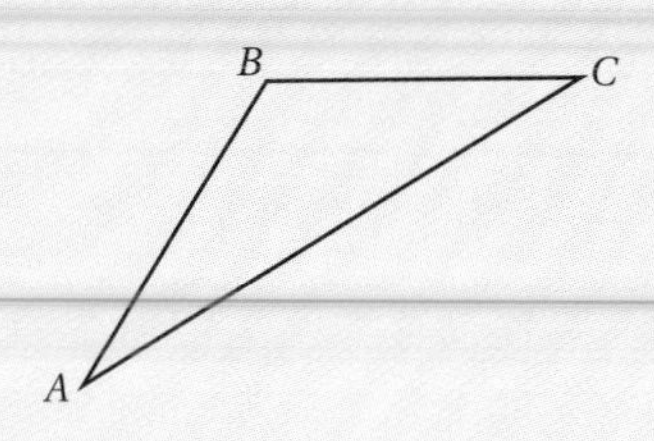

61. Find the missing angle measure.

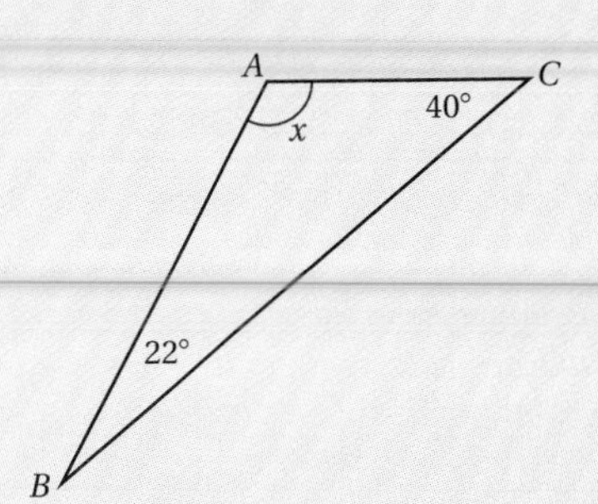

Solve.

62. Lorenzo makes donations of \$627 and \$48. What was the total donation?

63. A machine wraps 134 candy bars per minute. How long does it take this machine to wrap 8710 bars?

64. A share of stock bought for $\$29\frac{5}{8}$ dropped $\$3\frac{7}{8}$ before it was resold. What was the price when it was resold?

65. At the start of a trip, a car's odometer read 27,428.6 mi, and at the end of the trip, the reading was 27,914.5 mi. How long was the trip?

66. From an income of \$12,000, amounts of \$2300 and \$1600 are paid for federal and state taxes. How much remains after these taxes have been paid?

67. A substitute teacher is paid \$47 a day for 9 days. How much was received altogether?

68. A person walks $\frac{3}{5}$ km per hour. At this rate, how far would the person walk in $\frac{1}{2}$ hr?

69. Eight identical sweaters cost a total of \$679.68. What is the cost of each sweater?

70. Eight gallons of exterior paint covers 400 ft^2. How much paint is needed to cover 650 ft^2?

71. Eighteen ounces of a fruit drink costs \$3.06. Find the unit price in cents per ounce.

72. What is the simple interest on \$4000 principal at 5% for $\frac{3}{4}$ year?

73. A real estate agent received \$5880 commission on the sale of an \$84,000 home. What was the rate of commission?

74. The population of a city is 29,000 this year and is increasing at 4% per year. What will the population be next year?

75. The ages of students in a math class at a community college are as follows:

18, 21, 26, 31, 32, 18, 50.

Find the average, the median, and the mode of their ages.

Evaluate.

76. 18^2

77. 20^2

Simplify.

78. $\sqrt{9}$

79. $\sqrt{121}$

80. Approximate to three decimal places: $\sqrt{20}$.

Complete.

81. $\frac{1}{3}$ yd = _____ in. **82.** 4280 mm = _____ cm **83.** 3 days = _____ hr **84.** 20,000 g = _____ kg

85. 5 lb = _____ oz **86.** 0.008 cg = _____ mg **87.** 8190 mL = _____ L **88.** 20 qt = _____ gal

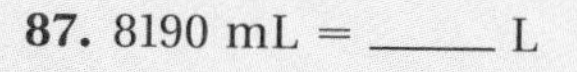

89. Find the length of the third side of this right triangle. Give an exact answer and an approximation to three decimal places.

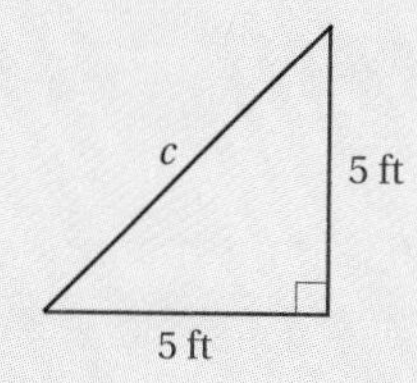

90. Find the perimeter and the area.

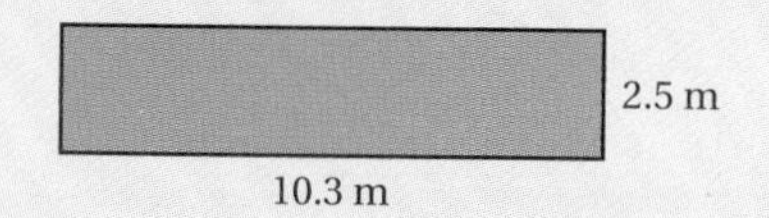

Find the area.

91.

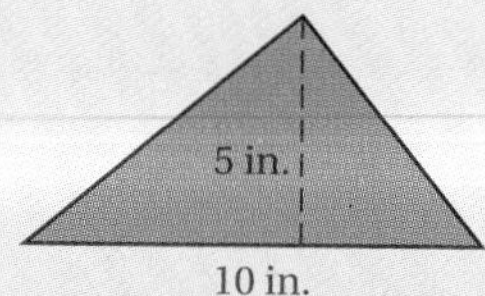

92.

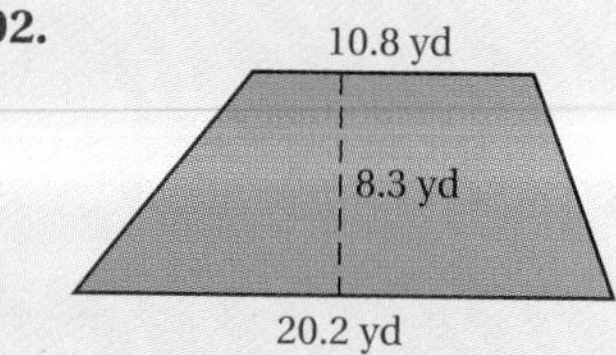

93.

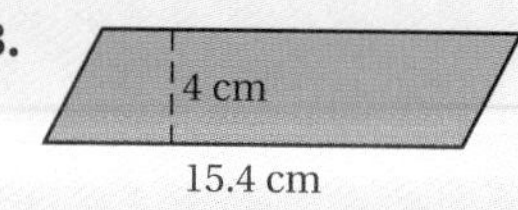

94. Find the diameter, the circumference, and the area of this circle. Use 3.14 for π.

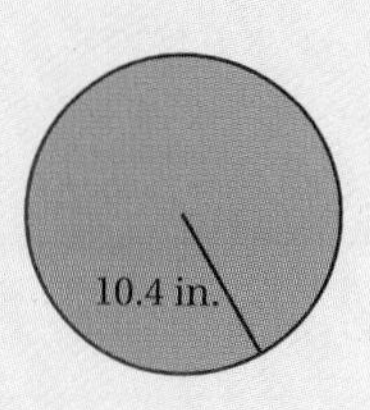

95. Find the volume.

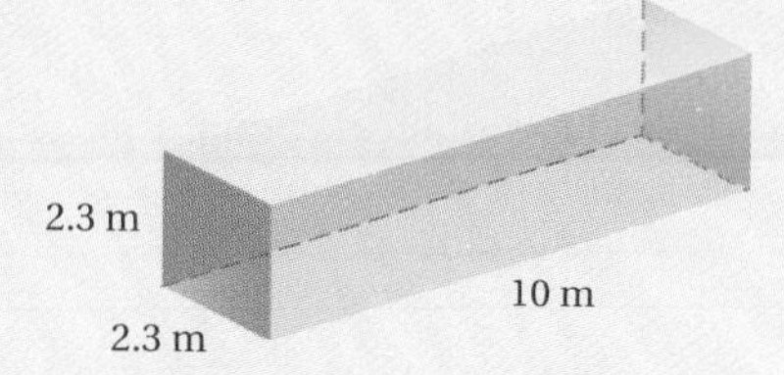

Find the volume. Use 3.14 for π.

96.

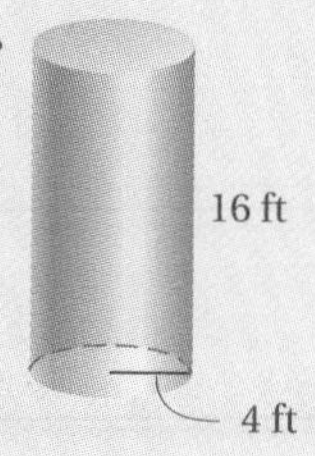

97.

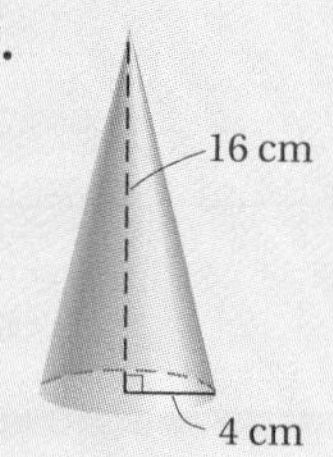

98.

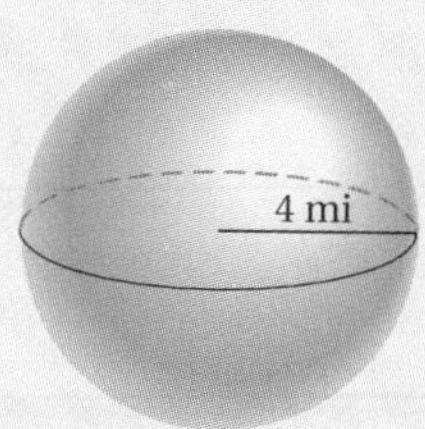

Simplify.

99. $12 \times 20 - 10 \div 5$

100. $4^3 - 5^2 + (16 \cdot 4 + 23 \cdot 3)$

101. $|(-1) \cdot 3|$

102. Add: $17 + (-3)$.

103. Subtract: $\left(-\frac{1}{3}\right) - \left(-\frac{2}{3}\right)$.

104. Multiply: $(-6) \cdot (-5)$.

105. Multiply: $-\frac{5}{7} \cdot \frac{14}{35}$.

106. Divide: $\frac{48}{-6}$.

Solve.

107. $7 - x = 12$

108. $-4.3x = -17.2$

109. $5x + 7 = 3x - 9$

110. $5(x - 2) - 8(x - 4) = 20$

Translate to an algebraic expression.

111. 17 more than some number

112. 38 percent of some number

Solve.

113. A game board has 64 squares. If you win 25 squares and your opponent wins the rest, how many does your opponent get?

114. If you add one-third of a number to the number itself, you get 48. What is the number?

Final Examination

1. Write expanded notation for 8345.

2. In 3784, what digit tells the number of hundreds?

Add and simplify, if possible.

3. $\begin{array}{r} 41.38 \\ 2.013 \\ +172.2247 \\ \hline \end{array}$

4. $\begin{array}{r} 3\frac{1}{4} \\ +5\frac{1}{2} \\ \hline \end{array}$

5. $\frac{7}{5} + \frac{4}{15}$

6. $\begin{array}{r} 432 \\ +327 \\ \hline \end{array}$

7. $\begin{array}{r} 6209 \\ 2134 \\ 9187 \\ +4032 \\ \hline \end{array}$

8. $0.456 + 34.5 + 0.94 + 122.9877$

Subtract and simplify, if possible.

9. $\begin{array}{r} 8987 \\ -3426 \\ \hline \end{array}$

10. $\begin{array}{r} 9006 \\ -3069 \\ \hline \end{array}$

11. $\begin{array}{r} 31.2 \\ -\ 0.808 \\ \hline \end{array}$

12. $\begin{array}{r} 3\frac{1}{2} \\ -2\frac{7}{8} \\ \hline \end{array}$

13. $\frac{3}{4} - \frac{2}{3}$

14. $123.04 - 23.88$

Multiply and simplify, if possible.

15. $3\frac{1}{4} \cdot 7\frac{1}{2}$

16. $\frac{8}{9} \cdot \frac{3}{4}$

17. $\frac{2}{5} \cdot 15$

18. $\begin{array}{r} 342 \\ \times\ 17 \\ \hline \end{array}$

19. $\begin{array}{r} 987 \\ \times 238 \\ \hline \end{array}$

20. $\begin{array}{r} 25.43 \\ \times\ 8.9 \\ \hline \end{array}$

Divide and simplify, if possible.

21. $8\overline{)4137}$
State the answer using a whole-number quotient and remainder.

22. Give a mixed numeral for the quotient in Question 21.

23. $21\overline{)4137}$

24. $\frac{3}{5} \div \frac{9}{10}$

25. $5\frac{2}{3} \div 4\frac{2}{5}$

26. $1.6\overline{)76.8}$

27. Round 42,574 to the nearest thousand.

28. Round 6.7892 to the nearest hundredth.

29. Round $7.\overline{38}$ to the nearest thousandth.

30. Determine whether 3312 is divisible by 9.

Answers

1. ______
2. ______
3. ______
4. ______
5. ______
6. ______
7. ______
8. ______
9. ______
10. ______
11. ______
12. ______
13. ______
14. ______
15. ______
16. ______
17. ______
18. ______
19. ______
20. ______
21. ______
22. ______
23. ______
24. ______
25. ______
26. ______
27. ______
28. ______
29. ______
30. ______

Answers

31. ______

32. ______

33. ______

34. ______

35. ______

36. ______

37. ______

38. ______

39. ______

40. ______

41. ______

42. ______

43. ______

44. ______

45. ______

46. ______

47. ______

48. ______

49. ______

50. ______

51. ______

52. ______

53. ______

54. ______

55. ______

56. ______

57. ______

58. ______

59. ______

31. Find all the factors of 8.

32. Find the LCM of 23, 46, and 10.

Simplify.

33. $\frac{63}{42}$

34. $\frac{100}{10}$

35. Convert to a mixed numeral: $\frac{23}{3}$.

36. Use = or ≠ for ▒ to write a true sentence:

$\frac{3}{5}$ ▒ $\frac{6}{10}$.

37. Use < or > for ▒ to write a true sentence:

$\frac{6}{11}$ ▒ $\frac{5}{9}$.

38. Which is greater, 0.089 or 0.9?

39. Use < or > for ▒ to write a true sentence:

456 ▒ 546.

40. What part is shaded?

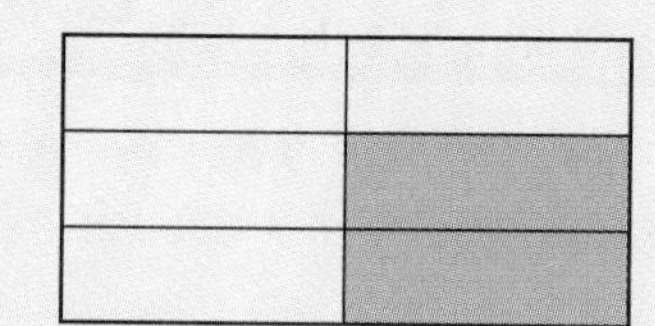

Convert to decimal notation.

41. 49.9%

42. $\frac{6}{25}$

43. $\frac{3}{11}$

44. $\frac{786}{100}$

Convert to fractional notation.

45. $5\frac{3}{4}$

46. 37%

47. 0.897

Convert to percent notation.

48. 0.77

49. $\frac{24}{25}$

Solve.

50. $\frac{25}{12} = \frac{8}{x}$

51. $y + \frac{2}{5} = \frac{11}{25}$

52. $78 \cdot t = 1950$

53. $3.9 + y = 249.6$

Solve.

54. The enrollment in a college increased from 3000 to 3150. Find the percent of increase.

55. A consumer spent $83 for groceries, $204.89 for clothes, and $24.71 for gasoline. How much was spent altogether?

56. How many $\frac{1}{4}$-lb boxes of chocolate can be filled with 20 lb of chocolates?

57. A part-time worker is paid $58 a day for 6 days. How much was received?

58. A $5\frac{1}{2}$-m flagpole was set $1\frac{3}{4}$ m into the ground. How much was above the ground?

59. A person received checks of $324 and $987. What was the total?

60. A student has \$75 in a checking account. Checks of \$17 and \$19 are written. How much is left in the account?

61. A driver travels 325 mi on 25 gal of gasoline. How many miles per gallon did the driver obtain?

62. A lab technician paid \$149.88 for 6 identical lab coats. How much did each lab coat cost?

63. A driver traveled 216 km in 6 hr. At this rate, how far will the driver have traveled in 15 hr?

64. A 3-lb package of meat costs \$11.95. Find the unit cost in dollars per pound.

65. A student got 78% of the questions correct on a test. There were 50 questions. How many of the questions were correct?

66. What is the simple interest on \$2000 principal at 6% for $\frac{1}{2}$ year?

67. Find the average, the median, and the mode of this set of numbers:

\$11, \$12, \$12, \$12, \$19, \$25.

Estimate each of the following as a whole number or as a mixed numeral where the fractional part is $\frac{1}{2}$.

68. $28\frac{16}{17}$

69. $8\frac{3}{11} - 7\frac{5}{6} + \frac{31}{32}$

Evaluate.

70. 25^2

71. 16^2

Simplify.

72. $\sqrt{49}$

73. $\sqrt{625}$

74. Approximate to three decimal places: $\sqrt{24}$.

75. Find the missing angle measure.

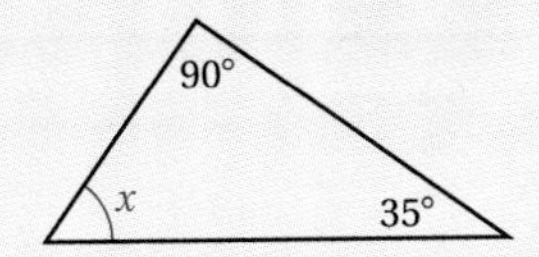

76. Find the length of the third side of this right triangle. Give an exact answer and an approximation to three decimal places.

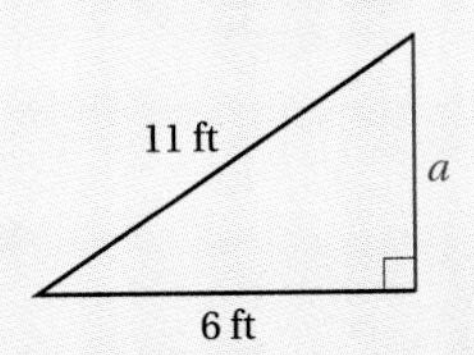

Interactive Technology Spending. The following table lists the projected consumer spending for technology services such as on-line/internet, multimedia software, and video game software for various years.

Year	Spending (in billions)
1994	\$ 5.7
1995	7.2
1996	8.8
1997	10.7
1998	12.6
1999	?

Source: Veronis, Suhler & Associates

77. Make a line graph of the data.

78. Use extrapolation to estimate the projected spending in 1999.

79. The following circle graph shows color preference for a new car.

a) Which is the favorite color?

b) The survey considered 5000 people. How many preferred red?

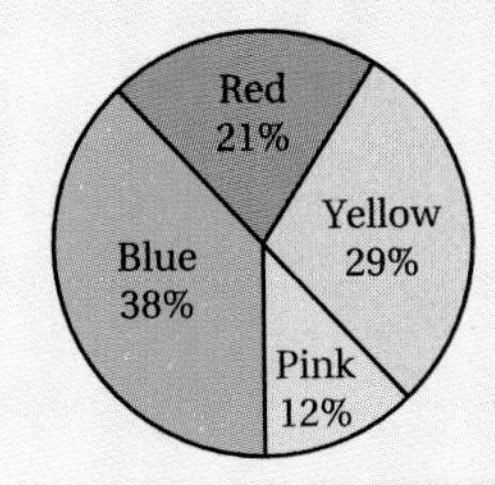

Answers

60. ____

61. ____

62. ____

63. ____

64. ____

65. ____

66. ____

67. ____

68. ____

69. ____

70. ____

71. ____

72. ____

73. ____

74. ____

75. ____

76. ____

77. ____

78. ____

79. a) ____

b) ____

Answers

80. ______
81. ______
82. ______
83. ______
84. ______
85. ______
86. ______
87. ______
88. ______
89. ______
90. ______
91. ______
92. ______
93. ______
94. ______
95. ______
96. ______
97. ______
98. ______
99. ______
100. ______
101. ______
102. ______
103. ______
104. ______
105. ______
106. ______
107. ______
108. ______
109. ______

Complete.

80. 15 ft = ______ yd

81. 2371 m = ______ km

82. 5 L = ______ mL

83. 7 T = ______ lb

84. 24 hr = ______ min

85. 5.34 kg = ______ g

86. 75.4 mg = ______ cg

87. 80 oz = ______ pt

88. Find the perimeter and the area.

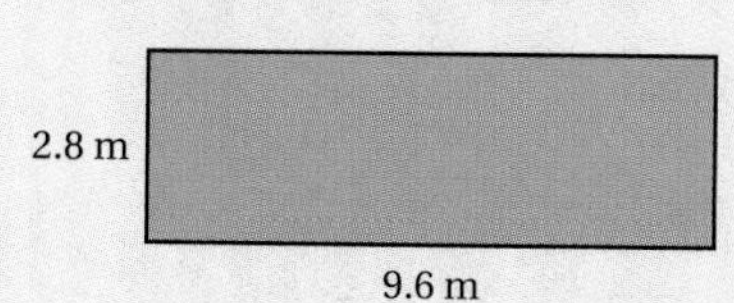

Find the area.

89.

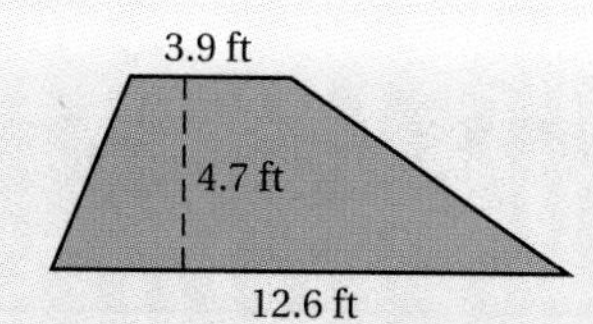

90.

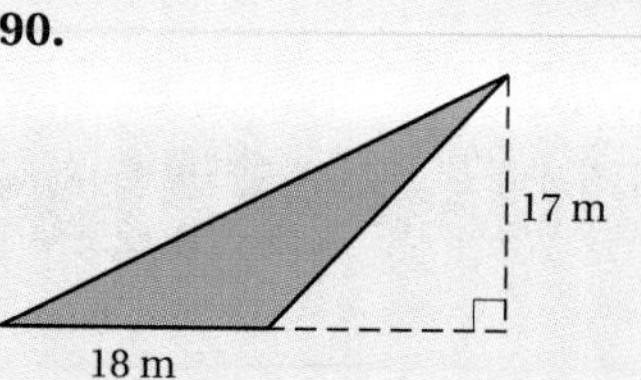

91.

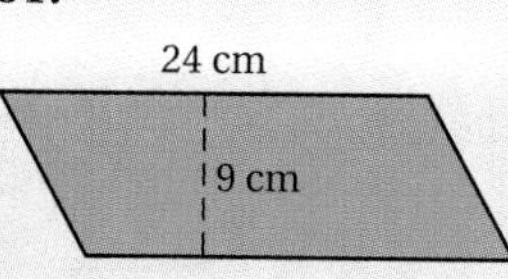

92. Find the radius, the circumference, and the area of this circle. Use 3.14 for π.

8.6 yd

93. Find the volume.

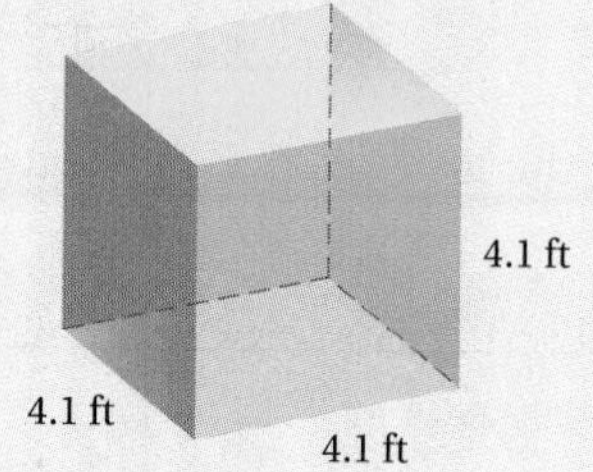

Find the volume. Use 3.14 for π.

94.

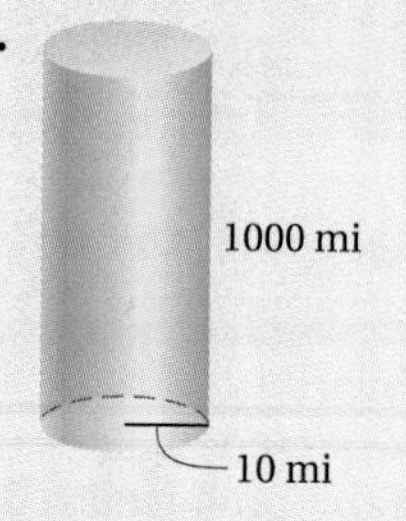

95.

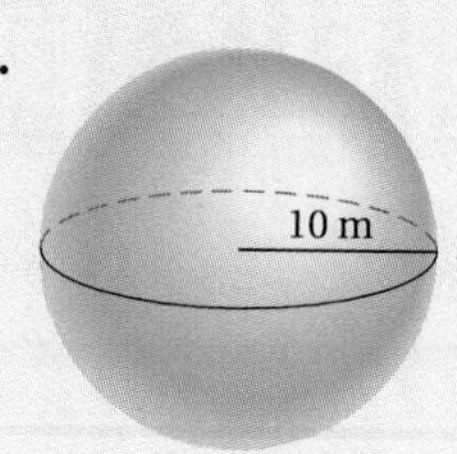

96.

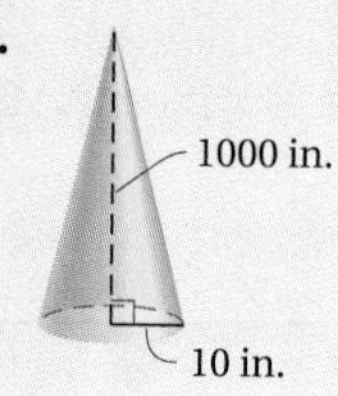

Simplify.

97. $200 \div 25 + 125 \cdot 3$

98. $(2 + 3)^3 - 4^3 + 19 \cdot 2$

99. $|-32|$

100. Subtract: $-7 - 15$.

101. Add: $-7 + (-15)$.

102. Multiply: $-5 \cdot (-6)$.

103. Multiply: $-\frac{2}{3} \cdot \frac{5}{6}$.

104. Divide: $\frac{42}{-7}$.

Solve.

105. $x + 25 = -51.4$

106. $\frac{2}{3}x = 18$

107. $0.5m - 13 = 17 + 2.5m$

Solve.

108. A consultant charges $80 an hour. How many hours did the consultant work in order to earn $42,600?

109. If you add two-fifths of a number to the number itself, you get 56. What is the number?

Developmental Units

Introduction

These developmental units are meant to provide extra instruction for students who have difficulty with any of Sections 1.2, 1.3, 1.5, or 1.6. After reading one of these developmental units and doing the exercises in its exercise set, the student should restudy the appropriate section in Chapter 1.

A Addition
S Subtraction
M Multiplication
D Division

World Wide Web For more information, visit us at www.mathmax.com

A Addition

Objectives

- **a** Add any two of the numbers 0, 1, 2, 3, 4, 5, 6, 7, 8, 9.
- **b** Find certain sums of three numbers such as 1 + 7 + 9.
- **c** Add two whole numbers when carrying is not necessary.
- **d** Add two whole numbers when carrying is necessary.

a Basic Addition

Basic addition can be explained by counting. The sum

$$3 + 4$$

can be found by counting out a set of 3 objects and a separate set of 4 objects, putting them together, and counting all the objects.

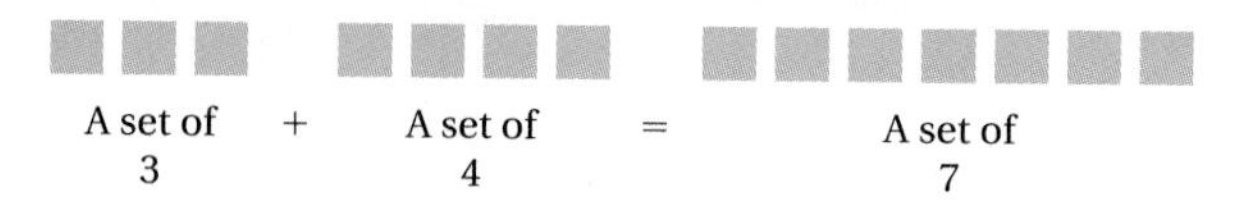

A set of 3 + A set of 4 = A set of 7

The numbers to be added are called **addends**. The result is the **sum**.

$$\underset{\text{Addend}}{3} + \underset{\text{Addend}}{4} = \underset{\text{Sum}}{7}$$

Examples Add. Think of putting sets of objects together.

1. $5 + 6 = 11$

$$\begin{array}{r} 5 \\ +\ 6 \\ \hline 11 \end{array}$$

2. $8 + 5 = 13$

$$\begin{array}{r} 8 \\ +\ 5 \\ \hline 13 \end{array}$$

We can also do these problems by counting up from one of the numbers. For example, in Example 1, we start at 5 and count up 6 times: 6, 7, 8, 9, 10, 11.

Do Exercises 1–6.

What happens when we add 0? Think of a set of 5 objects. If we add 0 objects to it, we still have 5 objects. Similarly, if we have a set with 0 objects in it and add 5 objects to it, we have a set with 5 objects. Thus,

$$5 + 0 = 5 \quad \text{and} \quad 0 + 5 = 5.$$

> Adding 0 to a number does not change the number:
>
> $a + 0 = 0 + a = a.$
>
> We say that 0 is the **additive identity.**

Add; think of joining sets of objects.

1. $4 + 5$ **2.** $3 + 4$

3. $\begin{array}{r} 9 \\ +\ 5 \\ \hline \end{array}$ **4.** $\begin{array}{r} 8 \\ +\ 8 \\ \hline \end{array}$

5. $\begin{array}{r} 9 \\ +\ 7 \\ \hline \end{array}$ **6.** $\begin{array}{r} 7 \\ +\ 9 \\ \hline \end{array}$

The first printed use of the + symbol was in a book by a German, Johann Widmann, in 1498.

Answers on page A-27

Examples Add.

3. $0 + 9 = 9$

$$\begin{array}{r} 0 \\ +\ 9 \\ \hline 9 \end{array}$$

4. $0 + 0 = 0$

$$\begin{array}{r} 0 \\ +\ 0 \\ \hline 0 \end{array}$$

5. $97 + 0 = 97$

$$\begin{array}{r} 97 \\ +\ \ 0 \\ \hline 97 \end{array}$$

Do Exercises 7–12.

Your objective for this part of the section is to be able to add any of the numbers 0, 1, 2, 3, 4, 5, 6, 7, 8, 9. Adding 0 is easy. The rest of the sums are listed in this table. Memorize the table by saying it to yourself over and over or by using flash cards.

+	1	2	3	4	5	6	7	8	9
1	2	3	4	5	6	7	8	9	10
2	3	4	5	6	7	8	9	10	11
3	4	5	6	7	8	9	10	11	12
4	5	6	7	8	9	10	11	12	13
5	6	7	8	9	10	11	12	13	14
6	7	8	9	10	11	12	13	14	15
7	8	9	10	11	12	13	14	15	16
8	9	10	11	12	13	14	15	16	17
9	10	11	12	13	14	15	16	17	18

$6 + 7 = 13$
Find 6 at the left, and 7 at the top.

$7 + 6 = 13$
Find 7 at the left, and 6 at the top.

It is very important that you *memorize* the basic addition facts! If you do not, you will always have trouble with addition.

Note the following.

$3 + 4 = 7$ $\quad 7 + 6 = 13$ $\quad 7 + 2 = 9$

$4 + 3 = 7$ $\quad 6 + 7 = 13$ $\quad 2 + 7 = 9$

We can add whole numbers in any order. This is the *commutative law of addition.* Because of this law, you need to learn only about half the table above, as shown by the shading.

Do Exercises 13 and 14.

b Certain Sums of Three Numbers

To add $3 + 5 + 4$, we can add 3 and 5, then 4:

$3 + 5 + 4$
↘↙
$8 + 4$
↘↙
12.

We can also add 5 and 4, then 3:

$3 + 5 + 4$
↘↙
$3 + 9$
↘↙
12.

Either way we get 12.

Add.

7. $8 + 0$

8. $0 + 8$

9. $\begin{array}{r} 7 \\ +\ 0 \\ \hline \end{array}$

10. $\begin{array}{r} 46 \\ +\ \ 0 \\ \hline \end{array}$

11. $0 + 13$

12. $58 + 0$

Complete the table.

13.

+	1	2	3	4	5
1			4		
2					
3				7	
4					
5					

14.

+	6	5	7	4	9
7			14		
9					
5				9	
8					
4					

Answers on page A-27

Example 6 Add from the top mentally.

$$\begin{array}{r} 1 \\ 7 \\ +\ 9 \\ \hline \end{array}$$

We first add 1 and 7, getting 8. Then we add 8 and 9, getting 17.

$1,\ 7 \rightarrow 8;\quad 8,\ 9 \rightarrow 17$

Example 7 Add from the top mentally.

$$\begin{array}{r} 2 \\ 4 \\ +\ 8 \\ \hline \end{array}$$

$2,\ 4 \rightarrow 6;\quad 6,\ 8 \rightarrow 14$

Do Exercises 15–18.

c Addition (No Carrying)

We now move to a more gradual, conceptual development of the addition procedure you considered in Section 1.2. It is intended to provide you with a greater understanding so that your skill level will increase.

To add larger numbers, we can add the ones first, then the tens, then the hundreds, and so on.

Example 8 Add: 5722 + 3234.

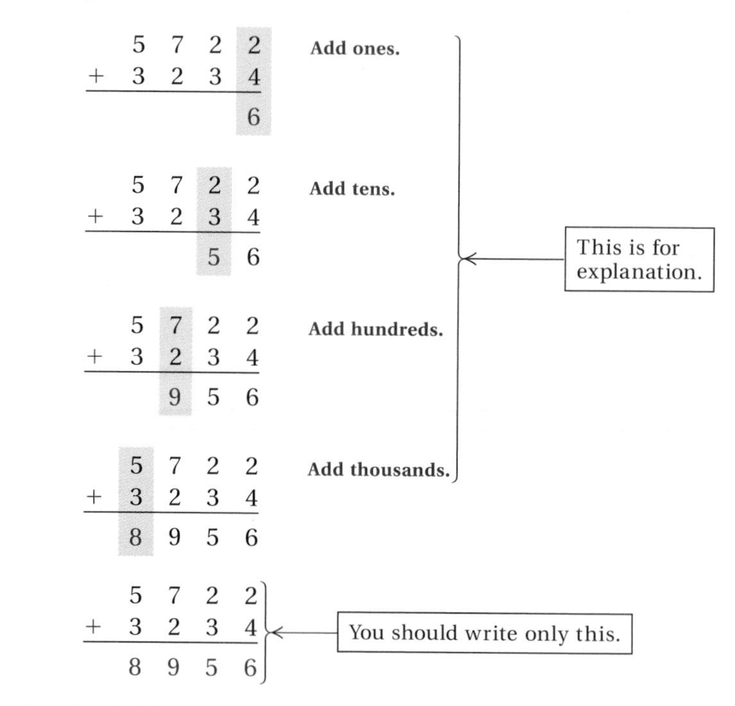

Do Exercises 19–22.

Add from the top mentally.

15. $\begin{array}{r} 1 \\ 6 \\ +\ 9 \\ \hline \end{array}$

16. $\begin{array}{r} 2 \\ 3 \\ +\ 4 \\ \hline \end{array}$

17. $\begin{array}{r} 6 \\ 1 \\ +\ 4 \\ \hline \end{array}$

18. $\begin{array}{r} 5 \\ 2 \\ +\ 8 \\ \hline \end{array}$

Add.

19. $\begin{array}{r} 24 \\ +\ 35 \\ \hline \end{array}$

20. $\begin{array}{r} 346 \\ +\ 203 \\ \hline \end{array}$

21. $\begin{array}{r} 8327 \\ +\ 1652 \\ \hline \end{array}$

22. $\begin{array}{r} 3461 \\ +\ 2035 \\ \hline \end{array}$

Answers on page A-28

d Addition (with Carrying)

Carrying Tens

Example 9 Add: 18 + 27.

$$\begin{array}{r} 1\ 8 \\ +\ 2\ 7 \\ \hline ? \end{array}$$

Add ones.

Think:
$$\begin{array}{r} 8 \\ +\ 7 \\ \hline 1\ 5 \end{array}$$

15 ones = 10 ones + 5 ones
= 1 ten + 5 ones

$$\begin{array}{r} 1 \\ 1\ 8 \\ +\ 2\ 7 \\ \hline 5 \end{array}$$

Write 5 in the ones column.
Write 1 for a reminder above the tens.
This is called *carrying*.

$$\begin{array}{r} 1 \\ 1\ 8 \\ +\ 2\ 7 \\ \hline 4\ 5 \end{array}$$

Add tens.

We can use money to help explain Example 9.

$$\begin{array}{r} 1\ 8¢ \\ +\ 2\ 7¢ \\ \hline 15¢ \end{array}$$

8¢ → 1 dime and 8 pennies
7¢ → 2 dimes and 7 pennies
We first add the pennies.

$$\begin{array}{r} 1 \text{ dime} \\ 1\ 8 \\ +\ 2\ 7 \\ \hline 5 \text{ pennies} \end{array}$$

We exchange ten pennies for a dime.

$$\begin{array}{r} 1 \\ 1\ 8 \\ +\ 2\ 7 \\ \hline 4\ 5 \end{array}$$

We now add the dimes. The result is 4 dimes and 5 pennies.

Do Exercises 23 and 24.

Carrying Hundreds

Example 10 Add: 256 + 391.

$$\begin{array}{r} 2\ 5\ 6 \\ +\ 3\ 9\ 1 \\ \hline 7 \end{array}$$

Add ones.

$$\begin{array}{r} 1 \\ 2\ 5\ 6 \\ +\ 3\ 9\ 1 \\ \hline 4\ 7 \end{array}$$

Add tens. We get 14 tens. Now 14 tens = 10 tens + 4 tens = 1 hundred + 4 tens. Write 4 in the tens column and a 1 above the hundreds.

The carrying here is like exchanging 14 dimes for a 1 dollar bill and 4 dimes.

$$\begin{array}{r} 1 \\ 2\ 5\ 6 \\ +\ 3\ 9\ 1 \\ \hline 6\ 4\ 7 \end{array}$$

Add hundreds.

Add.

23. $\begin{array}{r} 1\ 9 \\ +\ 3\ 7 \\ \hline \end{array}$

24. $\begin{array}{r} 4\ 6 \\ +\ 3\ 9 \\ \hline \end{array}$

Add.

25. $\begin{array}{r} 3\ 4\ 1 \\ +\ 4\ 8\ 8 \\ \hline \end{array}$

26. $\begin{array}{r} 7\ 3\ 0 \\ +\ 2\ 9\ 6 \\ \hline \end{array}$

Answers on page A-28

27. Add.

$$\begin{array}{r} 7850 \\ +\,4848 \\ \hline \end{array}$$

Add.

28.
$$\begin{array}{r} 7989 \\ +\,5672 \\ \hline \end{array}$$

29.
$$\begin{array}{r} 56{,}789 \\ +\,14{,}539 \\ \hline \end{array}$$

To the student: *If you had trouble with Section 1.2 and have studied Developmental Unit A, you should go back and work through Section 1.2 after completing Exercise Set A.*

Answers on page A-28

Do Exercises 25 and 26 on the preceding page.

Carrying Thousands

Example 11 Add: 4803 + 3792.

$$\begin{array}{r} 4803 \\ +\,3792 \\ \hline 5 \end{array}$$

Add ones.

$$\begin{array}{r} 4803 \\ +\,3792 \\ \hline 95 \end{array}$$

Add tens.

$$\begin{array}{r} \scriptstyle 1 \\ 4803 \\ +\,3792 \\ \hline 595 \end{array}$$

Add hundreds. We get 15 hundreds. Now 15 hundreds = 10 hundreds + 5 hundreds = 1 thousand + 5 hundreds. Write 5 in the hundreds column and 1 above the thousands.

$$\begin{array}{r} \scriptstyle 1 \\ 4803 \\ +\,3792 \\ \hline 8595 \end{array}$$

Add thousands.

Do Exercise 27.

Combined Carrying

Example 12 Add: 5767 + 4993.

$$\begin{array}{r} \scriptstyle 1 \\ 5767 \\ +\,4993 \\ \hline 0 \end{array}$$

Add ones. We get 10 ones. Now 10 ones = 1 ten + 0 ones. Write 0 in the ones column and 1 above the tens.

$$\begin{array}{r} \scriptstyle 11 \\ 5767 \\ +\,4993 \\ \hline 60 \end{array}$$

Add tens. We get 16 tens. Now 16 tens = 1 hundred + 6 tens. Write 6 in the tens column and 1 above the hundreds.

$$\begin{array}{r} \scriptstyle 111 \\ 5767 \\ +\,4993 \\ \hline 760 \end{array}$$

Add hundreds. We get 17 hundreds. Now 17 hundreds = 1 thousand + 7 hundreds. Write 7 in the hundreds column and 1 above the thousands.

$$\begin{array}{r} \scriptstyle 111 \\ 5767 \\ +\,4993 \\ \hline 10760 \end{array}$$

Add thousands. We get 10 thousands.

Do Exercises 28 and 29.

Exercise Set A

a Add. Try to do these mentally. If you have trouble, think of putting objects together.

1. $\begin{array}{r} 8 \\ +\ 9 \\ \hline \end{array}$ **2.** $\begin{array}{r} 8 \\ +\ 7 \\ \hline \end{array}$ **3.** $\begin{array}{r} 6 \\ +\ 7 \\ \hline \end{array}$ **4.** $\begin{array}{r} 9 \\ +\ 5 \\ \hline \end{array}$ **5.** $\begin{array}{r} 5 \\ +\ 7 \\ \hline \end{array}$ **6.** $\begin{array}{r} 5 \\ +\ 6 \\ \hline \end{array}$

7. $\begin{array}{r} 9 \\ +\ 8 \\ \hline \end{array}$ **8.** $\begin{array}{r} 9 \\ +\ 7 \\ \hline \end{array}$ **9.** $\begin{array}{r} 8 \\ +\ 4 \\ \hline \end{array}$ **10.** $\begin{array}{r} 9 \\ +\ 1 \\ \hline \end{array}$ **11.** $\begin{array}{r} 8 \\ +\ 2 \\ \hline \end{array}$ **12.** $\begin{array}{r} 3 \\ +\ 8 \\ \hline \end{array}$

13. $\begin{array}{r} 0 \\ +\ 7 \\ \hline \end{array}$ **14.** $\begin{array}{r} 4 \\ +\ 3 \\ \hline \end{array}$ **15.** $\begin{array}{r} 2 \\ +\ 9 \\ \hline \end{array}$ **16.** $\begin{array}{r} 0 \\ +\ 0 \\ \hline \end{array}$ **17.** $\begin{array}{r} 3 \\ +\ 0 \\ \hline \end{array}$ **18.** $\begin{array}{r} 9 \\ +\ 9 \\ \hline \end{array}$

19. $\begin{array}{r} 8 \\ +\ 6 \\ \hline \end{array}$ **20.** $\begin{array}{r} 3 \\ +\ 7 \\ \hline \end{array}$ **21.** $\begin{array}{r} 2 \\ +\ 2 \\ \hline \end{array}$ **22.** $\begin{array}{r} 7 \\ +\ 7 \\ \hline \end{array}$ **23.** $\begin{array}{r} 6 \\ +\ 5 \\ \hline \end{array}$ **24.** $\begin{array}{r} 7 \\ +\ 8 \\ \hline \end{array}$

25. $\begin{array}{r} 8 \\ +\ 8 \\ \hline \end{array}$ **26.** $\begin{array}{r} 8 \\ +\ 1 \\ \hline \end{array}$ **27.** $\begin{array}{r} 5 \\ +\ 8 \\ \hline \end{array}$ **28.** $\begin{array}{r} 5 \\ +\ 9 \\ \hline \end{array}$ **29.** $\begin{array}{r} 4 \\ +\ 7 \\ \hline \end{array}$ **30.** $\begin{array}{r} 6 \\ +\ 1 \\ \hline \end{array}$

31. $6 + 7$ **32.** $7 + 7$ **33.** $3 + 9$ **34.** $6 + 0$ **35.** $6 + 4$

36. $9 + 3$ **37.** $5 + 5$ **38.** $5 + 3$ **39.** $1 + 1$ **40.** $4 + 5$

41. $9 + 4$ **42.** $0 + 8$ **43.** $4 + 6$ **44.** $2 + 7$ **45.** $3 + 7$

46. $3 + 3$ **47.** $5 + 8$ **48.** $3 + 6$ **49.** $4 + 4$ **50.** $4 + 7$

b Add from the top mentally.

51. $\begin{array}{r} 1 \\ 8 \\ +\ 3 \\ \hline \end{array}$ **52.** $\begin{array}{r} 1 \\ 7 \\ +\ 5 \\ \hline \end{array}$ **53.** $\begin{array}{r} 3 \\ 2 \\ +\ 5 \\ \hline \end{array}$ **54.** $\begin{array}{r} 4 \\ 3 \\ +\ 5 \\ \hline \end{array}$ **55.** $\begin{array}{r} 1 \\ 7 \\ +\ 9 \\ \hline \end{array}$

56. $\begin{array}{r} 5 \\ 2 \\ +\ 6 \\ \hline \end{array}$ **57.** $\begin{array}{r} 4 \\ 5 \\ +\ 1 \\ \hline \end{array}$ **58.** $\begin{array}{r} 1 \\ 9 \\ +\ 6 \\ \hline \end{array}$ **59.** $\begin{array}{r} 1 \\ 8 \\ +\ 7 \\ \hline \end{array}$ **60.** $\begin{array}{r} 1 \\ 6 \\ +\ 8 \\ \hline \end{array}$

c Add.

61. 23 + 16
62. 54 + 35
63. 67 + 20
64. 496 + 503
65. 700 + 200

66. 801 + 67
67. 666 + 333
68. 523 + 325
69. 747 + 130
70. 8250 + 9430

71. 6552 + 4321
72. 3406 + 1293
73. 7340 + 3527
74. 4825 + 5070
75. 2073 + 1925

76. 9111 + 9111
77. 7889 + 9000
78. 52,433 + 12,056
79. 43,723 + 56,276
80. 51,670 + 26,107

d Add.

81. 38 + 8
82. 17 + 9
83. 17 + 38
84. 95 + 6
85. 862 + 781

86. 613 + 799
87. 355 + 491
88. 280 + 348
89. 814 + 390
90. 274 + 333

91. 9990 + 10
92. 999 + 11
93. 999 + 111
94. 839 + 388
95. 909 + 202

96. 808 + 909
97. 8718 + 1420
98. 3854 + 2700
99. 4828 + 1283
100. 6995 + 1432

101. 9889 + 1
102. 6889 + 4723
103. 9128 + 1997
104. 8898 + 6645
105. 9989 + 6785

106. 46,889 + 21,786
107. 23,448 + 10,989
108. 67,658 + 98,786
109. 77,548 + 23,767
110. 44,684 + 4,765

S Subtraction

a Basic Subtraction

Subtraction can be explained by taking away part of a set.

Example 1 Subtract: $7 - 3$.

We can do this by counting out 7 objects and then taking away 3 of them. Then we count the number that remain: $7 - 3 = 4$.

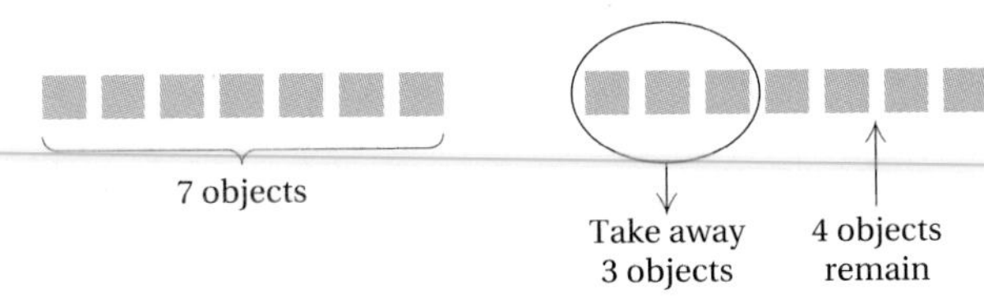

We could also do this mentally by starting at 7 and counting down 3 times: 6, 5, 4.

Examples Subtract. Think of "take away."

2. $11 - 6 = 5$ *Take away*: **"11 take away 6 is 5."**

$$\begin{array}{r} 11 \\ -\ 6 \\ \hline 5 \end{array}$$

3. $17 - 9 = 8$

$$\begin{array}{r} 17 \\ -\ 9 \\ \hline 8 \end{array}$$

Do Exercises 1–4.

In Developmental Unit A, you memorized an addition table. That table will enable you to subtract also. First, let's recall how addition and subtraction are related.

An addition:

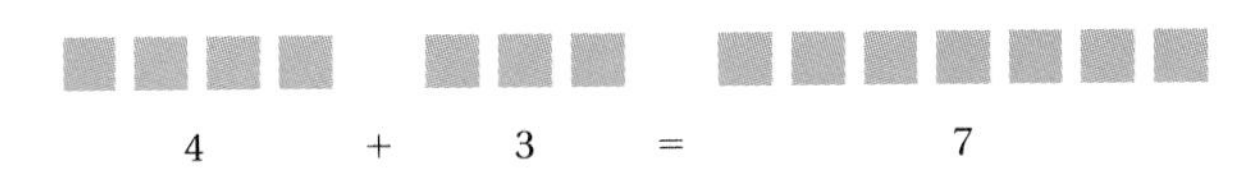

Two related subtractions.

A.

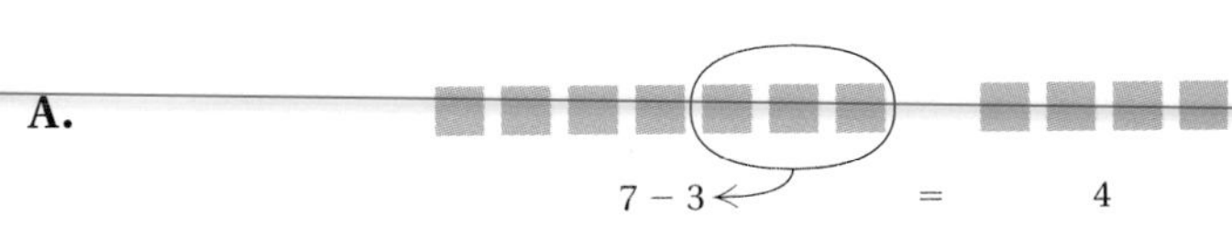

B.

7 − 4 = 3

Objectives

a Find basic differences such as $5 - 3$, $13 - 8$, and so on.

b Subtract one whole number from another when borrowing is not necessary.

c Subtract one whole number from another when borrowing is necessary.

Subtract.

1. $10 - 6$

2. $11 - 4$

3.
$$\begin{array}{r} 16 \\ -\ 8 \\ \hline \end{array}$$

4.
$$\begin{array}{r} 10 \\ -\ 7 \\ \hline \end{array}$$

Answers on page A-28

For each addition fact, write two subtraction facts.

5. $8 + 4 = 12$

6. $6 + 7 = 13$

Subtract. Try to do these mentally.

7. $14 - 6$

8. $12 - 5$

9. $\begin{array}{r} 13 \\ -\ 4 \\ \hline \end{array}$

10. $\begin{array}{r} 11 \\ -\ 7 \\ \hline \end{array}$

Answers on page A-28

Since we know that

$4 + 3 = 7,$ **A basic addition fact**

we also know the two subtraction facts

$7 - 3 = 4$ and $7 - 4 = 3.$

Example 4 From $8 + 9 = 17$, write two subtraction facts.

a) The addend 8 is subtracted from the sum 17.

$8 + 9 = 17$ The related sentence is $17 - 8 = 9$.

b) The addend 9 is subtracted from the sum 17.

$8 + 9 = 17$ The related sentence is $17 - 9 = 8$.

Do Exercises 5 and 6.

We can use the idea that subtraction is defined in terms of addition to think of subtraction as "how much more."

Example 5 Find: $13 - 6$.

To find $13 - 6$, we ask, "6 plus what number is 13?"

$6 + \square = 13$

+	1	2	3	4	5	6	7	8	9
1	2	3	4	5	6	7	8	9	10
2	3	4	5	6	7	8	9	10	11
3	4	5	6	7	8	9	10	11	12
4	5	6	7	8	9	10	11	12	13
5	6	7	8	9	10	11	12	13	14
6	7	8	9	10	11	12	13	14	15
7	8	9	10	11	12	13	14	15	16
8	9	10	11	12	13	14	15	16	17
9	10	11	12	13	14	15	16	17	18

$13 - 6 = 7$

Using the addition table above, we find 13 inside the table and 6 at the left. Then we read the answer 7 from the top. Thus we have $13 - 6 = 7$. Strive to do this kind of thinking mentally as fast as you can, without having to use the table.

Do Exercises 7–10.

b Subtraction (No Borrowing)

We now move to a more gradual, conceptual development of the subtraction procedure you considered in Section 1.3. It is intended to provide you with a greater understanding so that your skill level will increase.

To subtract larger numbers, we can subtract the ones first, then the tens, then the hundreds, and so on.

Example 6 Subtract: 5787 − 3214.

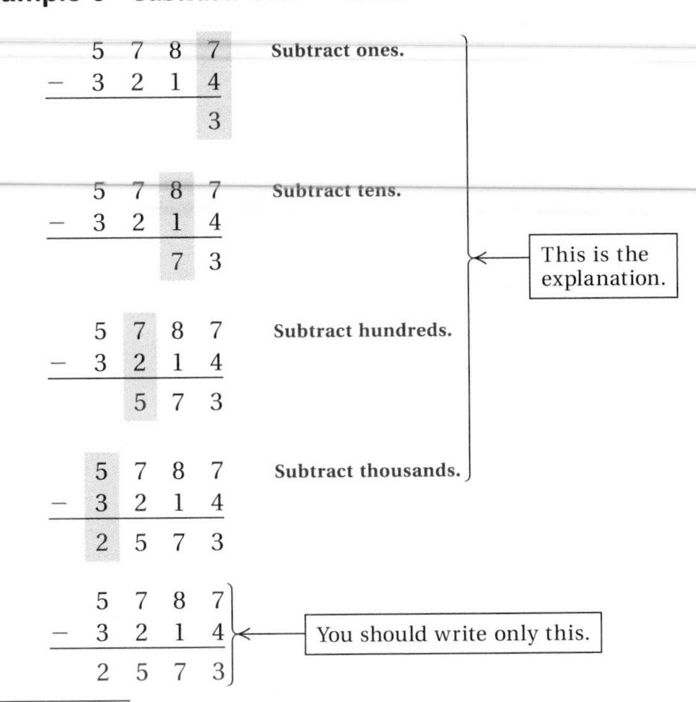

$$\begin{array}{r} 5787 \\ -\ 3214 \\ \hline 2573 \end{array}$$

You should write only this.

Do Exercises 11–14.

c Subtraction (with Borrowing)

We now consider subtraction when borrowing is necessary.

Borrowing from the Tens Place

Example 7 Subtract: 37 − 18.

$$\begin{array}{r} 37 \\ -\ 18 \\ \hline ? \end{array}$$

Try to subtract ones: 7 − 8 is *not* a whole number.

$$\begin{array}{r} \overset{2}{\not{3}}\ \overset{17}{\not{7}} \\ -\ 1\ \ 8 \\ \hline \end{array}$$

Borrow a ten. That is, 1 ten = 10 ones, and 10 ones + 7 ones = 17 ones. Write 2 above the tens column and 17 above the ones.

$$\begin{array}{r} \overset{2}{\not{3}}\ \overset{17}{\not{7}} \\ -\ 1\ \ 8 \\ \hline 9 \end{array}$$

Subtract ones.

The borrowing here is like exchanging 3 dimes and 7 pennies for 2 dimes and 17 pennies.

Subtract.

11. $\begin{array}{r} 78 \\ -\ 64 \\ \hline \end{array}$

12. $\begin{array}{r} 29 \\ -\ 9 \\ \hline \end{array}$

13. $\begin{array}{r} 542 \\ -\ 301 \\ \hline \end{array}$

14. $\begin{array}{r} 6896 \\ -\ 4871 \\ \hline \end{array}$

Answers on page A-28

Subtract.

15. $\begin{array}{r} 46 \\ -29 \\ \hline \end{array}$

16. $\begin{array}{r} 74 \\ -38 \\ \hline \end{array}$

Subtract.

17. $\begin{array}{r} 646 \\ -192 \\ \hline \end{array}$

18. $\begin{array}{r} 733 \\ -483 \\ \hline \end{array}$

Answers on page A-28

$$\begin{array}{rr} \scriptstyle 2 & \scriptstyle 17 \\ \not{3} & \not{7} \\ -\ 1 & 8 \\ \hline 1 & 9 \end{array}$$

Subtract tens.

$$\begin{array}{rr} \scriptstyle 2 & \scriptstyle 17 \\ \not{3} & \not{7} \\ -\ 1 & 8 \\ \hline 1 & 9 \end{array}$$

You should write only this.

Do Exercises 15 and 16.

Borrowing Hundreds

Example 8 Subtract: 538 − 275.

$$\begin{array}{rrr} 5 & 3 & 8 \\ -\ 2 & 7 & 5 \\ \hline & & 3 \end{array}$$

Subtract ones.

$$\begin{array}{rrr} 5 & 3 & 8 \\ -\ 2 & 7 & 5 \\ \hline & ? & 3 \end{array}$$

Try to subtract tens: 30 − 70 is not a whole number.

$$\begin{array}{rrr} \scriptstyle 4 & \scriptstyle 13 & \\ \not{5} & \not{3} & 8 \\ -\ 2 & 7 & 5 \\ \hline & & 3 \end{array}$$

Borrow a hundred. That is, 1 hundred = 10 tens, and 10 tens + 3 tens = 13 tens. Write 4 above the hundreds column and 13 above the tens.

The borrowing is like exchanging 5 dollars, 3 dimes, and 8 pennies for 4 dollars, 13 dimes, and 8 pennies.

$$\begin{array}{rrr} \scriptstyle 4 & \scriptstyle 13 & \\ \not{5} & \not{3} & 8 \\ -\ 2 & 7 & 5 \\ \hline & 6 & 3 \end{array}$$

Subtract tens.

$$\begin{array}{rrr} \scriptstyle 4 & \scriptstyle 13 & \\ \not{5} & \not{3} & 8 \\ -\ 2 & 7 & 5 \\ \hline 2 & 6 & 3 \end{array}$$

Subtract hundreds.

$$\begin{array}{rrr} \scriptstyle 4 & \scriptstyle 13 & \\ \not{5} & \not{3} & 8 \\ -\ 2 & 7 & 5 \\ \hline 2 & 6 & 3 \end{array}$$

You should write only this.

Do Exercises 17 and 18.

Borrowing More Than Once

Sometimes we must borrow more than once.

Example 9 Subtract: 672 − 394.

$$\begin{array}{r} {\scriptstyle 6\ 12} \\ 6\,\not{7}\,\not{2} \\ -\ 3\ 9\ 4 \\ \hline 8 \end{array}$$

Borrowing a ten to subtract ones

$$\begin{array}{r} {\scriptstyle 16} \\ {\scriptstyle 5\ \not{6}\ 12} \\ \not{6}\,\not{7}\,\not{2} \\ -\ 3\ 9\ 4 \\ \hline 2\ 7\ 8 \end{array}$$

Borrowing a hundred to subtract tens

Do Exercises 19 and 20.

Example 10 Subtract: 6357 − 1769.

$$\begin{array}{r} {\scriptstyle 4\ 17} \\ 6\ 3\,\not{5}\,\not{7} \\ -\ 1\ 7\ 6\ 9 \\ \hline 8 \end{array}$$

We cannot subtract 9 from 7. We borrow a ten.

$$\begin{array}{r} {\scriptstyle 14} \\ {\scriptstyle 2\ \not{4}\ 17} \\ 6\,\not{3}\,\not{5}\,\not{7} \\ -\ 1\ 7\ 6\ 9 \\ \hline 8\ 8 \end{array}$$

We cannot subtract 6 tens from 4 tens. We borrow a hundred.

$$\begin{array}{r} {\scriptstyle 12\ 14} \\ {\scriptstyle 5\ \not{2}\ \not{4}\ 17} \\ \not{6}\,\not{3}\,\not{5}\,\not{7} \\ -\ 1\ 7\ 6\ 9 \\ \hline 4\ 5\ 8\ 8 \end{array}$$

We cannot subtract 7 hundreds from 2 hundreds. We borrow a thousand.

We can always check by adding the answer to the number being subtracted.

Example 11 Subtract: 8341 − 2673. Check by adding.

We check by adding 5668 and 2673.

$$\begin{array}{r} {\scriptstyle 12\ 13} \\ {\scriptstyle 7\ \not{2}\ \not{3}\ 11} \\ \not{8}\,\not{3}\,\not{4}\,\not{1} \\ -\ 2\ 6\ 7\ 3 \\ \hline 5\ 6\ 6\ 8 \end{array} \qquad \textit{Check:} \quad \begin{array}{r} {\scriptstyle 1\ 1\ 1} \\ 5\ 6\ 6\ 8 \\ +\ 2\ 6\ 7\ 3 \\ \hline 8\ 3\ 4\ 1 \end{array}$$

Do Exercises 21 and 22.

Zeros in Subtraction

Before subtracting, note the following:

50 is 5 tens;

70 is 7 tens.

Subtract.

19. $\begin{array}{r} 5\ 6\ 3 \\ -\ 1\ 8\ 7 \\ \hline \end{array}$

20. $\begin{array}{r} 7\ 3\ 3 \\ -\ 4\ 8\ 8 \\ \hline \end{array}$

Subtract. Check by adding.

21. $\begin{array}{r} 4\ 2\ 3\ 6 \\ -\ 1\ 6\ 7\ 9 \\ \hline \end{array}$

22. $\begin{array}{r} 7\ 5\ 4\ 1 \\ -\ 3\ 8\ 6\ 7 \\ \hline \end{array}$

Complete.

23. 80 = ______ tens

24. 60 = ______ tens

25. 300 = ______ tens

26. 900 = ______ tens

Answers on page A-28

Complete.

27. 5000 = ________ tens

28. 9000 = ________ tens

29. 5380 = ________ tens

30. 6770 = ________ tens

Subtract.

31. $60 - 18$

32. $480 - 256$

Subtract.

33. $602 - 464$

34. $408 - 364$

Subtract.

35. $4006 - 1238$

36. $9001 - 7804$

Subtract.

37. $3000 - 1754$

38. $8017 - 3289$

To the student: *If you had trouble with Section 1.3 and have studied Developmental Unit S, you should go back and work through Section 1.3 after completing Exercise Set S.*

Answers on page A-28

Then

100 is 10 tens;
200 is 20 tens.

Do Exercises 23–26 on the preceding page.

Also,

230 is 2 hundreds + 3 tens
or 20 tens + 3 tens
or 23 tens.

Similarly,

1000 is 100 tens;
2000 is 200 tens;
4670 is 467 tens.

Do Exercises 27–30.

Example 12 Subtract: 50 − 37.

```
 4 10
 5  0    We have 5 tens.
-3  7    We keep 4 of them in the tens column.
 1  3    We put 1 ten, or 10 ones, with the ones.
```

Do Exercises 31 and 32.

Example 13 Subtract: 803 − 547.

```
 7 9 13
 8 0  3    We have 8 hundreds, or 80 tens.
-5 4  7    We keep 79 tens.
 2 5  6    We put 1 ten, or 10 ones, with the ones.
```

Do Exercises 33 and 34.

Example 14 Subtract: 9003 − 2789.

```
 8 9 9 13
 9 0 0  3    We have 9 thousands, or 900 tens.
-2 7 8  9    We keep 899 tens.
 6 2 1  4    We put 1 ten, or 10 ones, with the ones.
```

Do Exercises 35 and 36.

Examples Subtract.

```
15.  4 9 9 10
     5 0 0  0
    -2 8 6  1
     2 1 3  9
```

```
16.      10
     4 9  0 13
     5 0  1  3
    -1 8  5  7
     3 1  5  6
```

Do Exercises 37 and 38.

Exercise Set S

a Subtract. Try to do these mentally.

1. $\begin{array}{r} 7 \\ -\ 0 \\ \hline \end{array}$ **2.** $\begin{array}{r} 8 \\ -\ 8 \\ \hline \end{array}$ **3.** $\begin{array}{r} 7 \\ -\ 7 \\ \hline \end{array}$ **4.** $\begin{array}{r} 8 \\ -\ 3 \\ \hline \end{array}$ **5.** $\begin{array}{r} 5 \\ -\ 2 \\ \hline \end{array}$

6. $\begin{array}{r} 16 \\ -\ 8 \\ \hline \end{array}$ **7.** $\begin{array}{r} 17 \\ -\ 9 \\ \hline \end{array}$ **8.** $\begin{array}{r} 12 \\ -\ 6 \\ \hline \end{array}$ **9.** $\begin{array}{r} 11 \\ -\ 4 \\ \hline \end{array}$ **10.** $\begin{array}{r} 12 \\ -\ 9 \\ \hline \end{array}$

11. $\begin{array}{r} 14 \\ -\ 7 \\ \hline \end{array}$ **12.** $\begin{array}{r} 18 \\ -\ 9 \\ \hline \end{array}$ **13.** $\begin{array}{r} 13 \\ -\ 7 \\ \hline \end{array}$ **14.** $\begin{array}{r} 15 \\ -\ 9 \\ \hline \end{array}$ **15.** $\begin{array}{r} 9 \\ -\ 7 \\ \hline \end{array}$

16. $7 - 3$ **17.** $4 - 1$ **18.** $2 - 0$ **19.** $3 - 3$ **20.** $6 - 3$

21. $7 - 6$ **22.** $9 - 8$ **23.** $10 - 3$ **24.** $6 - 6$ **25.** $11 - 7$

26. $12 - 8$ **27.** $5 - 0$ **28.** $4 - 0$ **29.** $13 - 9$ **30.** $14 - 9$

31. $11 - 2$ **32.** $12 - 3$ **33.** $16 - 9$ **34.** $18 - 9$ **35.** $11 - 5$

36. $10 - 4$ **37.** $10 - 8$ **38.** $14 - 8$ **39.** $15 - 8$ **40.** $10 - 2$

b Subtract.

41. $\begin{array}{r} 6\ 4 \\ -\ 3\ 1 \\ \hline \end{array}$ **42.** $\begin{array}{r} 5\ 5 \\ -\ 3\ 4 \\ \hline \end{array}$ **43.** $\begin{array}{r} 5\ 4\ 8 \\ -\ 3\ 0\ 1 \\ \hline \end{array}$ **44.** $\begin{array}{r} 5\ 9\ 6 \\ -\ 4\ 0\ 3 \\ \hline \end{array}$ **45.** $\begin{array}{r} 7\ 0\ 0 \\ -\ 2\ 0\ 0 \\ \hline \end{array}$

46. 765 − 111	**47.** 525 − 323	**48.** 747 − 130	**49.** 988 − 700	**50.** 9450 − 8230
51. 6552 − 4321	**52.** 7547 − 3421	**53.** 5875 − 2111	**54.** 38,695 − 37,004	**55.** 67,899 − 66,673
56. 99,999 − 1	**57.** 56,780 − 56,770	**58.** 42,111 − 32,010	**59.** 77,654 − 66,611	**60.** 23,456 − 12,345

c Subtract.

61. 93 − 28	**62.** 42 − 13	**63.** 86 − 78	**64.** 98 − 89	**65.** 625 − 317
66. 735 − 609	**67.** 853 − 236	**68.** 961 − 747	**69.** 787 − 698	**70.** 6769 − 2367
71. 6431 − 2876	**72.** 7654 − 1765	**73.** 5246 − 2859	**74.** 6328 − 2679	**75.** 7641 − 3809
76. 8743 − 599	**77.** 12,647 − 4,897	**78.** 16,222 − 5,777	**79.** 46,781 − 12,988	**80.** 470 − 189
81. 690 − 235	**82.** 703 − 132	**83.** 6406 − 258	**84.** 2309 − 109	**85.** 3406 − 1293
86. 6807 − 3059	**87.** 8000 − 2794	**88.** 8002 − 6543	**89.** 38,000 − 37,695	**90.** 16,043 − 11,588

M Multiplication

a Basic Multiplication

To multiply, we begin with two numbers, called **factors**, and get a third number, called a **product**. Multiplication can be explained by counting. The product 3×5 can be found by counting out 3 sets of 5 objects each, joining them (in a rectangular array if desired), and counting all the objects.

$3 \times 5 = 15$

Factor Factor Product

3

5

We can also think of multiplication as repeated addition.

$3 \times 5 = \underbrace{5 + 5 + 5} = 15$

$\longrightarrow$ 3 addends of 5

Examples Multiply. If you have trouble, think either of putting sets of objects together in a rectangular array or of repeated addition.

1. $5 \times 6 = 30$

$$\begin{array}{r} 6 \\ \times\ 5 \\ \hline 30 \end{array}$$

2. $8 \times 4 = 32$

$$\begin{array}{r} 4 \\ \times\ 8 \\ \hline 32 \end{array}$$

Do Exercises 1–4.

Multiplying by 0

How do we multiply by 0? Consider $4 \cdot 0$. Using repeated addition, we see that

$4 \cdot 0 = \underbrace{0 + 0 + 0 + 0} = 0.$

$\longrightarrow$ 4 addends of 0

We can also think of this using sets. That is, $4 \cdot 0$ is 4 sets with 0 objects in each set, so the total is 0.

Consider $0 \cdot 4$. Using repeated addition, we say that this is 0 addends of 4, which is 0. Using sets, we say that this is 0 sets with 4 objects in each set, which is 0. Thus we have the following.

> Multiplying by 0 gives 0.

Examples Multiply.

3. $13 \times 0 = 0$

$$\begin{array}{r} 0 \\ \times\ 13 \\ \hline 0 \end{array}$$

4. $0 \cdot 11 = 0$

$$\begin{array}{r} 11 \\ \times\ 0 \\ \hline 0 \end{array}$$

5. $0 \cdot 0 = 0$

$$\begin{array}{r} 0 \\ \times\ 0 \\ \hline 0 \end{array}$$

Do Exercises 5 and 6.

Objectives

a Multiply any two of the numbers 0, 1, 2, 3, 4, 5, 6, 7, 8, 9.

b Multiply multiples of 10, 100, and 1000.

c Multiply larger numbers by 0, 1, 2, 3, 4, 5, 6, 7, 8, 9.

d Multiply by multiples of 10, 100, and 1000.

Multiply. Think of joining sets in a rectangular array or of repeated addition.

1. $7 \cdot 8$ (The dot "$\cdot$" means the same as "$\times$".)

2.
$$\begin{array}{r} 9 \\ \times\ 4 \\ \hline \end{array}$$

3. $4 \cdot 7$

4.
$$\begin{array}{r} 7 \\ \times\ 6 \\ \hline \end{array}$$

Multiply.

5. $8 \cdot 0$

6.
$$\begin{array}{r} 17 \\ \times\ 0 \\ \hline \end{array}$$

Answers on page A-28

Multiply.

7. $8 \cdot 1$

8. $\begin{array}{r} 23 \\ \times \ 1 \\ \hline \end{array}$

9. Complete the table.

×	2	3	4	5
2				
3			12	
4				
5		15		
6				

10.

×	6	7	8	9
5				
6			48	
7				
8		56		
9				

Answers on page A-28

Multiplying by 1

How do we multiply by 1? Consider $5 \cdot 1$. Using repeated addition, we see that

$$5 \cdot 1 = \underbrace{1 + 1 + 1 + 1 + 1}_{5 \text{ addends of } 1} = 5.$$

We can also think of this using sets. That is, $5 \cdot 1$ is 5 sets with 1 object in each set, so the total is 5.

Consider $1 \cdot 5$. Using repeated addition, we say that this is 1 addend of 5, which is 5. Using sets, we say that this is 1 set of 5 objects, which is 5. Thus we have the following.

> Multiplying a number by 1 does not change the number:
>
> $a \cdot 1 = 1 \cdot a = a.$
>
> We say that 1 is the **multiplicative identity.**

This is a very important property.

Examples Multiply.

6. $13 \cdot 1 = 13$

$$\begin{array}{r} 1 \\ \times 13 \\ \hline 13 \end{array}$$

7. $1 \cdot 7 = 7$

$$\begin{array}{r} 7 \\ \times 1 \\ \hline 7 \end{array}$$

8. $1 \cdot 1 = 1$

$$\begin{array}{r} 1 \\ \times 1 \\ \hline 1 \end{array}$$

Do Exercises 7 and 8.

You should be able to multiply any of the numbers 0, 1, 2, 3, 4, 5, 6, 7, 8, 9. Multiplying by 0 and 1 is easy. The rest of the products are listed in the following table.

×	2	3	4	5	6	7	8	9
2	4	6	8	10	12	14	16	18
3	6	9	12	15	18	21	24	27
4	8	12	16	20	24	28	32	36
5	10	15	20	25	30	35	40	45
6	12	18	24	30	36	42	48	54
7	14	21	28	35	42	49	56	63
8	16	24	32	40	48	56	64	72
9	18	27	36	45	54	63	72	81

$5 \times 7 = 35$
Find 5 at the left, and 7 at the top.

$8 \cdot 4 = 32$
Find 8 at the left, and 4 at the top.

It is *very* important that you have the basic multiplication facts *memorized.* If you do not, you will always have trouble with multiplication.

The commutative law says that we can multiply numbers in any order. Thus you need to learn only about half the table.

Do Exercises 9 and 10.

b Multiplying Multiples of 10, 100, and 1000

We now move to a more gradual, conceptual development of the multiplication procedure you considered in Section 1.5. It is intended to provide you with a greater understanding so that your skill level will increase.

We begin by considering multiplication by multiples of 10, 100, and 1000. These are numbers such as 10, 20, 30, 100, 400, 1000, and 7000.

Multiplying by a Multiple of 10

We know that

$$50 = 5 \text{ tens} = 5 \cdot 10, \quad 340 = 34 \text{ tens} = 34 \cdot 10, \quad \text{and} \quad 2340 = 234 \text{ tens} = 234 \cdot 10.$$

Turning this around, we see that to multiply any number by 10, all we need do is write a 0 on the end of the number.

> To multiply a number by 10, write 0 on the end of the number.

Examples Multiply.

9. $10 \cdot 6 = 60$

10. $10 \cdot 47 = 470$

11. $10 \cdot 583 = 5830$

Do Exercises 11–15.

Let's find $4 \cdot 90$. This is $4 \cdot (9 \text{ tens})$, or 36 tens. The procedure is the same as multiplying 4 and 9 and writing a 0 on the end. Thus, $4 \cdot 90 = 360$.

Examples Multiply.

12. $5 \cdot 70 = 350$ ← $5 \cdot 7$, then write a 0

13. $8 \cdot 80 = 640$

14. $5 \cdot 60 = 300$

Do Exercises 16 and 17.

Multiplying by a Multiple of 100

Note the following:

$$300 = 3 \text{ hundreds} = 3 \cdot 100, \quad 4700 = 47 \text{ hundreds} = 47 \cdot 100, \quad \text{and} \quad 56{,}800 = 568 \text{ hundreds} = 568 \cdot 100.$$

Turning this around, we see that to multiply any number by 100, all we need do is write two 0's on the end of the number.

> To multiply a number by 100, write two 0's on the end of the number.

Multiply.

11. $10 \cdot 7$

12. $10 \cdot 45$

13. $10 \cdot 273$

14. $10 \cdot 10$

15. $10 \cdot 100$

Multiply.

16. 70×8

17. 60×6

Answers on page A-28

Multiply.

18. 100 · 7

19. 100 · 23

20. 100 · 723

21. 100 · 100

22. 100 · 1000

Multiply.

23. $\begin{array}{r} 700 \\ \times\ \ 8 \\ \hline \end{array}$

24. $\begin{array}{r} 400 \\ \times\ \ 4 \\ \hline \end{array}$

Multiply.

25. 1000 · 9

26. 1000 · 852

27. 1000 · 10

28. 3 · 4000

29. 9 · 8000

Answers on page A-28

Examples Multiply.

15. 100 · 6 = 600

16. 100 · 39 = 3900

17. 100 · 448 = 44,800

Do Exercises 18–22.

Let's find 4 · 900. This is 4 · (9 hundreds). If we use addition, this is

9 hundreds + 9 hundreds + 9 hundreds + 9 hundreds,
or 36 hundreds,

which is the same as multiplying 4 and 9 and writing two 0's on the end. Thus, 4 · 900 = 3600.

Examples Multiply.

18. 6 · 800 = 4800 (6 · 8, then write 00)

19. 9 · 700 = 6300

20. 5 · 500 = 2500

Do Exercises 23 and 24.

Multiplying by a Multiple of 1000

Note the following:

$$6000 = 6 \text{ thousands} = 6 \cdot 1000 \quad \text{and} \quad 19{,}000 = 19 \text{ thousands} = 19 \cdot 1000.$$

Turning this around, we see that to multiply any number by 1000, all we need do is write three 0's on the end of the number.

> To multiply a number by 1000, write three 0's on the end of the number.

Examples Multiply.

21. 1000 · 8 = 8000

22. 2000 · 13 = 26,000

23. 1000 · 567 = 567,000

Do Exercises 25–29.

Multiplying Multiples by Multiples

Let's multiply 50 and 30. This is 50 · (3 tens), or 150 tens, or 1500. The procedure is the same as multiplying 5 and 3 and writing two 0's on the end.

To multiply multiples of tens, hundreds, thousands, and so on:

a) Multiply the one-digit numbers.

b) Count the number of zeros.

c) Write that many 0's on the end.

Examples Multiply.

24.
```
     80      1 zero at end
  ×  60      1 zero at end
  -----
   4800
   ↑———— 6 · 8, then write 00
```

25.
```
     800     2 zeros at end
  ×   60     1 zero at end
  ------
  48,000
  ↑———— 6 · 8, then write 000
```

26.
```
      800    2 zeros at end
  ×   600    2 zeros at end
  -------
  480,000
  ↑———— 6 · 8, then write 0,000
```

27.
```
     800     2 zeros at end
  ×   50     1 zero at end
  ------
  40,000
  ↑———— 5 · 8, then write 000
```

Do Exercises 30–33.

c Multiplying Larger Numbers

The product 3×24 can be represented as

$$
\begin{aligned}
3 \times (2 \text{ tens} + 4) &= (2 \text{ tens} + 4) + (2 \text{ tens} + 4) + (2 \text{ tens} + 4) \\
&= 6 \text{ tens} + 12 \\
&= 6 \text{ tens} + 1 \text{ ten} + 2 \\
&= 7 \text{ tens} + 2 \\
&= 72.
\end{aligned}
$$

We multiply the 4 ones by 3, getting 12

We multiply the 2 tens by 3, getting + 60

Then we add: 72

Example 28 Multiply: 3×24.

```
    2 4
  ×   3
  -----
    1 2 ← Multiply the 4 ones by 3.
    6 0 ← Multiply the 2 tens by 3.
  -----
    7 2 ← Add.
```

Do Exercises 34–36.

Example 29 Multiply: 5×734.

```
      7 3 4
  ×       5
  ---------
        2 0 ← Multiply the 4 ones by 5.
      1 5 0 ← Multiply the 3 tens by 5.
    3 5 0 0 ← Multiply the 7 hundreds by 5.
  ---------
    3 6 7 0 ← Add.
```

Do Exercises 37 and 38.

Multiply.

30. 9000×6

31. 80×70

32. 800×70

33. 600×30

Multiply.

34. 14×2

35. 58×2

36. 37×4

Multiply.

37. 823×6

38. 1348×5

Answers on page A-28

Multiply using the short form.

39. $\begin{array}{r} 58 \\ \times\ 2 \\ \hline \end{array}$

40. $\begin{array}{r} 37 \\ \times\ 4 \\ \hline \end{array}$

41. $\begin{array}{r} 823 \\ \times\ 6 \\ \hline \end{array}$

42. $\begin{array}{r} 1348 \\ \times\ 5 \\ \hline \end{array}$

Multiply.

43. $\begin{array}{r} 746 \\ \times\ 8 \\ \hline \end{array}$

44. $\begin{array}{r} 746 \\ \times\ 80 \\ \hline \end{array}$

45. $\begin{array}{r} 746 \\ \times 800 \\ \hline \end{array}$

To the student: *If you had trouble with Section 1.5 and have studied Developmental Unit M, you should go back and work through Section 1.5 after completing Exercise Set M.*

Answers on page A-28

Let's look at Example 29 again. Instead of writing each product on a separate line, we can use a shorter form.

Example 30 Multiply: 5×734.

$$\begin{array}{r} {}^{2} \\ 734 \\ \times\ \ 5 \\ \hline 0 \end{array}$$

Multiply the ones by 5: 5 · (4 ones) = 20 ones = 2 tens + 0 ones. Write 0 in the ones column and 2 above the tens.

$$\begin{array}{r} {}^{1\,2} \\ 734 \\ \times\ \ 5 \\ \hline 70 \end{array}$$

Multiply the 3 tens by 5 and add 2 tens: 5 · (3 tens) = 15 tens, 15 tens + 2 tens = 17 tens = 1 hundred + 7 tens. Write 7 in the tens column and 1 above the hundreds.

$$\begin{array}{r} {}^{1\,2} \\ 734 \\ \times\ \ 5 \\ \hline 3670 \end{array}$$

Multiply the 7 hundreds by 5 and add 1 hundred: 5 · (7 hundreds) = 35 hundreds, 35 hundreds + 1 hundred = 36 hundreds.

$$\begin{array}{r} {}^{1\,2} \\ 734 \\ \times\ \ 5 \\ \hline 3670 \end{array}$$

You should write only this.

Avoid writing the reminders unless necessary.

Do Exercises 39–42.

d Multiplying by Multiples of 10, 100, and 1000

To multiply 327 by 50, we multiply by 10 (write a 0), and then multiply 327 by 5.

$$\begin{array}{r} 327 \\ \times\ \ 50 \\ \hline 16{,}350 \end{array}$$

Write a 0.

Multiply 5 · 327.

Example 31 Multiply: 400×289.

$$\begin{array}{r} 289 \\ \times 400 \\ \hline 00 \end{array}$$

Write two 0's.

$$\begin{array}{r} 289 \\ \times\ 400 \\ \hline 115{,}600 \end{array}$$

Multiply 4 and 289: $\begin{array}{r} {}^{3\,3} \\ 289 \\ \times\ \ 4 \\ \hline 1156 \end{array}$

$$\begin{array}{r} {}^{3\,3} \\ 289 \\ \times\ 400 \\ \hline 115{,}600 \end{array}$$

You should write only this.

Do Exercises 43–45.

Exercise Set M

a Multiply. Try to do these mentally.

1. $\begin{array}{r} 3 \\ \times 4 \\ \hline \end{array}$ **2.** $\begin{array}{r} 6 \\ \times 0 \\ \hline \end{array}$ **3.** $\begin{array}{r} 7 \\ \times 1 \\ \hline \end{array}$ **4.** $\begin{array}{r} 0 \\ \times 2 \\ \hline \end{array}$ **5.** $\begin{array}{r} 10 \\ \times 1 \\ \hline \end{array}$ **6.** $\begin{array}{r} 6 \\ \times 5 \\ \hline \end{array}$

7. $\begin{array}{r} 5 \\ \times 2 \\ \hline \end{array}$ **8.** $\begin{array}{r} 9 \\ \times 7 \\ \hline \end{array}$ **9.** $\begin{array}{r} 9 \\ \times 6 \\ \hline \end{array}$ **10.** $\begin{array}{r} 2 \\ \times 6 \\ \hline \end{array}$ **11.** $\begin{array}{r} 7 \\ \times 0 \\ \hline \end{array}$ **12.** $\begin{array}{r} 8 \\ \times 9 \\ \hline \end{array}$

13. $\begin{array}{r} 1 \\ \times 8 \\ \hline \end{array}$ **14.** $\begin{array}{r} 8 \\ \times 0 \\ \hline \end{array}$ **15.** $\begin{array}{r} 4 \\ \times 7 \\ \hline \end{array}$ **16.** $\begin{array}{r} 3 \\ \times 8 \\ \hline \end{array}$ **17.** $\begin{array}{r} 5 \\ \times 9 \\ \hline \end{array}$ **18.** $\begin{array}{r} 2 \\ \times 9 \\ \hline \end{array}$

19. $\begin{array}{r} 0 \\ \times 7 \\ \hline \end{array}$ **20.** $\begin{array}{r} 5 \\ \times 7 \\ \hline \end{array}$ **21.** $\begin{array}{r} 9 \\ \times 5 \\ \hline \end{array}$ **22.** $\begin{array}{r} 5 \\ \times 8 \\ \hline \end{array}$ **23.** $\begin{array}{r} 0 \\ \times 0 \\ \hline \end{array}$ **24.** $\begin{array}{r} 2 \\ \times 8 \\ \hline \end{array}$

25. $5 \cdot 5$ **26.** $9 \cdot 9$ **27.** $1 \cdot 1$ **28.** $0 \cdot 0$ **29.** $2 \cdot 2$

30. $6 \cdot 6$ **31.** $1 \cdot 8$ **32.** $0 \cdot 1$ **33.** $3 \cdot 9$ **34.** $2 \cdot 9$

35. $6 \cdot 0$ **36.** $10 \cdot 1$ **37.** $6 \cdot 8$ **38.** $9 \cdot 6$ **39.** $8 \cdot 0$

40. $9 \cdot 8$ **41.** $3 \cdot 5$ **42.** $1 \cdot 8$ **43.** $1 \cdot 9$ **44.** $2 \cdot 1$

45. $8 \cdot 4$ **46.** $3 \cdot 2$ **47.** $5 \cdot 3$ **48.** $1 \cdot 6$ **49.** $4 \cdot 2$

50. $4 \cdot 5$ **51.** $5 \cdot 4$ **52.** $4 \cdot 4$ **53.** $5 \cdot 2$ **54.** $8 \cdot 0$

b Multiply.

55. 10×8
56. 7×10
57. 20×8
58. 30×7
59. 45×10

60. 78×10
61. 80×7
62. 90×4
63. 100×8
64. 100×3

65. 100×9
66. 100×10
67. 3457×100
68. 400×3
69. 700×7

70. 500×8
71. 100×100
72. 1000×7
73. 1000×9
74. 1000×2

75. 457×1000
76. 6769×1000
77. 2000×9
78. 5000×4
79. 6000×8

80. 8000×2
81. 3000×2
82. 1000×1000
83. 40×30
84. 20×10

85. 80×50
86. 50×50
87. 400×30
88. 200×30
89. 700×90

90. 400×300
91. 4000×200
92. 6000×20
93. 4000×4000
94. 8000×10

c Multiply.

95. 49×3
96. 74×6
97. 593×5
98. 609×8
99. 899×7

100. 865×4
101. 8118×2
102. 6754×2
103. $43{,}777 \times 2$
104. $32{,}564 \times 6$

d Multiply.

105. 58×60
106. 93×30
107. 42×80
108. 78×90
109. 346×60

110. 267×40
111. 897×400
112. 366×300
113. 834×700
114. 333×900

115. 5673×2000
116. 4678×5000
117. 6788×9000
118. 9129×8000

D Division

a Basic Division

Division can be explained by arranging a set of objects in a rectangular array. This can be done in two ways.

Example 1 Divide: $18 \div 6$.

METHOD 1 We can do this division by taking 18 objects and determining into how many rows, each with 6 objects, we can arrange the objects.

Since there are 3 rows of 6 objects, we have

$$18 \div 6 = 3.$$

METHOD 2 We can also arrange the objects into 6 rows and determine how many objects are in each row.

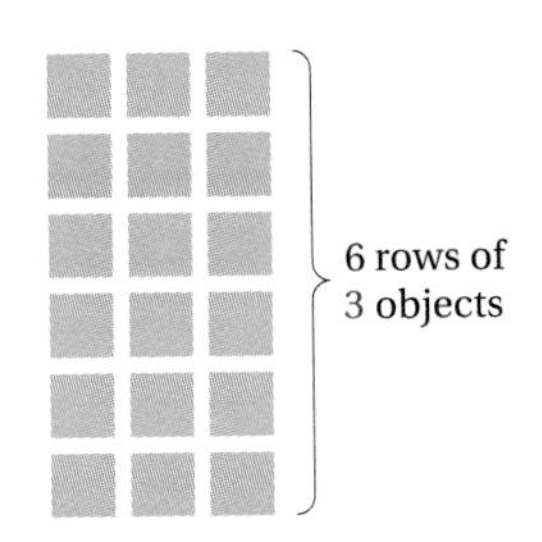

Since there are 3 objects in each of the 6 rows, we have

$$18 \div 6 = 3.$$

We can also use fractional notation for division. That is,

$$18 \div 6 = 18/6 = \frac{18}{6}.$$

Examples Divide.

2. $9\overline{)36}$ with quotient 4   *Think*: 36 objects: How many rows, each with 9 objects? or 36 objects: How many objects in each of 9 rows?

3. $42 \div 7 = 6$

4. $\frac{24}{3} = 8$

Do Exercises 1–4.

Objectives

a. Find basic quotients such as $20 \div 5$, $56 \div 7$, and so on.

b. Divide using the "guess, multiply, and subtract" method.

c. Divide by estimating multiples of thousands, hundreds, tens, and ones.

Divide.

1. $24 \div 6$

2. $64 \div 8$

3. $\frac{63}{7}$

4. $\frac{27}{9}$

Answers on page A-29

For each multiplication fact, write two division facts.

5. $6 \cdot 2 = 12$

6. $7 \times 6 = 42$

Answers on page A-29

In Developmental Unit M, you memorized a multiplication table. That table will enable you to divide as well. First, let's recall how multiplication and division are related.

A multiplication: $5 \cdot 4 = 20$.

Two related divisions:

A.

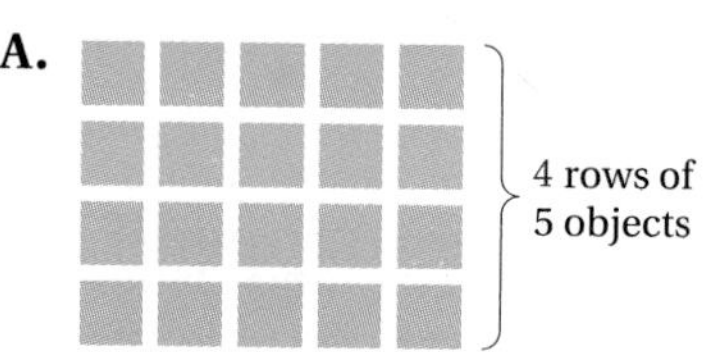

$20 \div 5 = 4$.

B.

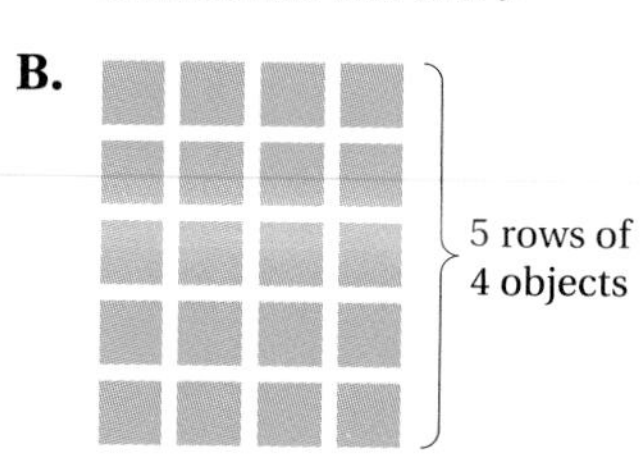

$20 \div 4 = 5$.

Since we know that

$5 \cdot 4 = 20$, **A basic multiplication fact**

we also know the two division facts

$20 \div 5 = 4$ and $20 \div 4 = 5$.

Example 5 From $7 \cdot 8 = 56$, write two division facts.

a) We have

$7 \cdot 8 = 56$ **Division sentence**

$7 = 56 \div 8$. **Related multiplication sentence**

b) We also have

$7 \cdot 8 = 56$ **Division sentence**

$8 = 56 \div 7$. **Related multiplication sentence**

Do Exercises 5 and 6.

We can use the idea that division is defined in terms of multiplication to do basic divisions.

Example 6 Find: $35 \div 5$.

To find $35 \div 5$, we ask, "5 times what number is 35?"

$5 \cdot \square = 35$

×	2	3	4	5	6	7	8	9
2	4	6	8	10	12	14	16	18
3	6	9	12	15	18	21	24	27
4	8	12	16	20	24	28	32	36
5	10	15	20	25	30	35	40	45
6	12	18	24	30	36	42	48	54
7	14	21	28	35	42	49	56	63
8	16	24	32	40	48	56	64	72
9	18	27	36	45	54	63	72	81

$35 \div 5 = 7$

Using the multiplication table above, we find 35 inside the table and 5 at the left. Then we read the answer 7 from the top. Thus we have $35 \div 5 = 7$. Strive to do this kind of thinking mentally as fast as you can, without having to use the table.

Do Exercises 7–10.

Divide.

7. $28 \div 4$

8. $81 \div 9$

9. $\frac{16}{2}$

10. $\frac{54}{6}$

Division by 1

Note that

$$3 \div 1 = 3 \quad \text{because} \quad 3 = 3 \cdot 1; \qquad \frac{14}{1} = 14 \quad \text{because} \quad 14 = 14 \cdot 1.$$

> Any number divided by 1 is that same number:
>
> $$a \div 1 = \frac{a}{1} = a.$$

Examples Divide.

7. $\frac{8}{1} = 8$

8. $6 \div 1 = 6$

9. $34 \div 1 = 34$

Do Exercises 11–13.

Divide.

11. $6 \div 1$

12. $\frac{13}{1}$

13. $1 \div 1$

Answers on page A-29

Division by 0

Why can't we divide by 0? Suppose the number 4 could be divided by 0. Then if $\square$ were the answer,

$$4 \div 0 = \square$$

and since 0 times any number is 0, we would have

$$4 = \square \cdot 0 = 0. \quad \text{False!}$$

Divide, if possible. If not possible, write "not defined."

14. $\frac{8}{4}$

15. $\frac{5}{0}$

16. $\frac{0}{5}$

17. $\frac{0}{0}$

18. $12 \div 0$

19. $100 \div 10$

20. $\frac{5}{3-3}$

21. $\frac{8-8}{4}$

Answers on page A-29

Suppose 12 could be divided by 0. If $\square$ were the answer,

$$12 \div 0 = \square$$

and since 0 times any number is 0, we would have

$$12 = \square \cdot 0 = 0. \qquad \text{False!}$$

Thus, $a \div 0$ would be some number $\square$ such that $a = \square \cdot 0 = 0$. So the only possible number that could be divided by 0 would be 0 itself.

But such a division would give us any number we wish, for

$$\left.\begin{array}{lll} 0 \div 0 = 8 & \text{because} & 0 = 8 \cdot 0; \\ 0 \div 0 = 3 & \text{because} & 0 = 3 \cdot 0; \\ 0 \div 0 = 7 & \text{because} & 0 = 7 \cdot 0. \end{array}\right\} \quad \text{All true!}$$

We avoid the preceding difficulties by agreeing to exclude division by 0.

> Division by 0 is not defined. (We agree not to divide by 0.)

Dividing 0 by Other Numbers

Note that

$$0 \div 3 = 0 \quad \text{because} \quad 0 = 0 \cdot 3; \qquad \frac{0}{12} = 0 \quad \text{because } 0 = 0 \cdot 12.$$

> Zero divided by any number greater than 0 is 0:
>
> $$\frac{0}{a} = 0, \quad a > 0.$$

Examples Divide.

10. $0 \div 8 = 0$

11. $0 \div 22 = 0$

12. $\frac{0}{9} = 0$

Do Exercises 14–21.

Division of a Number by Itself

Note that

$$3 \div 3 = 1 \quad \text{because} \quad 3 = 1 \cdot 3; \qquad \frac{34}{34} = 1 \quad \text{because} \quad 34 = 1 \cdot 34.$$

> Any number greater than 0 divided by itself is 1:
>
> $$\frac{a}{a} = 1, \quad a > 0.$$

Examples Divide.

13. $8 \div 8 = 1$

14. $27 \div 27 = 1$

15. $\frac{32}{32} = 1$

Do Exercises 22–27.

b Dividing by "Guess, Multiply, and Subtract"

To understand the process of division, we use a method known as "guess, multiply, and subtract."

Example 16 Divide $275 \div 4$. Use "guess, multiply, and subtract."

We *guess* a partial quotient of 35. We could guess *any* number—say, 4, 16, or 30. We *multiply* and *subtract* as follows:

```
      3 5 <— Partial quotient
4 ) 2 7 5
    1 4 0 <— 35 · 4
    1 3 5 <— Remainder
```

Next, we look at 135 and *guess* another partial quotient—say, 20. Then we *multiply* and *subtract*:

```
      2 0 <— Second partial quotient
      3 5
4 ) 2 7 5
    1 4 0
    1 3 5
      8 0 <— 20 · 4
      5 5 <— Remainder
```

Next, we look at 55 and *guess* another partial quotient—say, 13. Then we *multiply* and *subtract*:

```
      1 3 <— Third partial quotient
      2 0
      3 5
4 ) 2 7 5
    1 4 0
    1 3 5
      8 0
      5 5
      5 2 <— 13 · 4
        3 <— Remainder is less than 4
```

Divide.

22. $23 \div 23$

23. $\frac{67}{67}$

24. $\frac{41}{41}$

25. $17 \div 17$

26. $17 \div 1$

27. $\frac{54}{54}$

Divide using the "guess, multiply, and subtract" method.

28. $6 \overline{)\,4\,5\,4}$

29. $3\,2 \overline{)\,7\,4\,7}$

Answers on page A-29

Divide using the "guess, multiply, and subtract" method.

30. $7\overline{)6789}$

31. $64\overline{)3012}$

Answers on page A-29

Since we cannot subtract any more 4's, the division is finished. We add our partial quotients.

```
      6 8 <—— Quotient (sum of guesses)
     ____
      1 3
      2 0
      3 5
    _____
 4 ) 2 7 5        CHECK:   275 = (4 × 68) + 3
     1 4 0                 275 ? 272 + 3
     _____                     | 275
     1 3 5
       8 0
     _____
       5 5
       5 2
     _____
         3
```

The answer is 68 R 3. This tells us that with 275 objects, we could make 68 rows of 4 and have 3 left over.

The partial quotients (guesses) can be made in any manner so long as subtraction is possible.

Do Exercises 28 and 29 on the preceding page.

Example 17 Divide: 1506 ÷ 32.

```
          4 7 <—— Quotient (sum of guesses)
        _____
          2 0 ]
            2 ]
          2 0 ] <—— Guesses
            5 ]
        _______
 3 2 ) 1 5 0 6
         1 6 0 <—— 5 · 32
       _______
       1 3 4 6
         6 4 0 <—— 20 · 32
       _______
         7 0 6
           6 4 <—— 2 · 32
       _______
         6 4 2
         6 4 0 <—— 20 · 32
       _______
             2 <—— Remainder: smaller than the divisor
```

The answer is 47 R 2.

Remember, you can *guess any partial quotient* so long as subtraction is possible.

Do Exercises 30 and 31.

c Dividing by Estimating Multiples

Let's refine the guessing process. We guess multiples of 10, 100, and 1000, and so on.

Example 18 Divide: $7643 \div 3$.

a) Are there any thousands in the quotient? Yes, $3 \cdot 1000 = 3000$, which is less than 7643. To find how many thousands, we find products of 3 and multiples of 1000.

$3 \cdot 1000 = 3000$
$3 \cdot 2000 = 6000$ ← 7643 is here, so there are 2000 threes in the quotient.
$3 \cdot 3000 = 9000$

```
     2 0 0 0
3 ) 7 6 4 3
    6 0 0 0
    1 6 4 3
```

b) Now go to the hundreds place. Are there any hundreds in the quotient?

$3 \cdot 100 = 300$
$3 \cdot 200 = 600$
$3 \cdot 300 = 900$
$3 \cdot 400 = 1200$
$3 \cdot 500 = 1500$ ← 1643
$3 \cdot 600 = 1800$

```
       5 0 0
     2 0 0 0
3 ) 7 6 4 3
    6 0 0 0
    1 6 4 3
    1 5 0 0
      1 4 3
```

c) Now go to the tens place. Are there any tens in the quotient?

$3 \cdot 10 = 30$
$3 \cdot 20 = 60$
$3 \cdot 30 = 90$
$3 \cdot 40 = 120$ ← 143
$3 \cdot 50 = 150$

```
         4 0
       5 0 0
     2 0 0 0
3 ) 7 6 4 3
    6 0 0 0
    1 6 4 3
    1 5 0 0
      1 4 3
      1 2 0
        2 3
```

d) Now go to the ones place. Are there any ones in the quotient?

$3 \cdot 1 = 3$
$3 \cdot 2 = 6$
$3 \cdot 3 = 9$
$3 \cdot 4 = 12$
$3 \cdot 5 = 15$
$3 \cdot 6 = 18$
$3 \cdot 7 = 21$ ← 23
$3 \cdot 8 = 24$

```
    2 5 4 7
          7
        4 0
      5 0 0
    2 0 0 0
3 ) 7 6 4 3
    6 0 0 0
    1 6 4 3
    1 5 0 0
      1 4 3
      1 2 0
        2 3
        2 1
          2
```

The answer is 2547 R 2.

Do Exercises 32 and 33.

Divide.

32. $4 \overline{)\,3\,8\,5}$

33. $7 \overline{)\,8\,8\,4\,6}$

Answers on page A-29

Divide using the short form.

34. $2\,\overline{)\,648}$

35. $9\,\overline{)\,3758}$

Divide.

36. $11\,\overline{)\,415}$

37. $46\,\overline{)\,1075}$

To the student: *If you had trouble with Section 1.6 and have studied Developmental Unit D, you should go back and work through Section 1.6 after completing Exercise Set D.*

Answers on page A-29

A Short Form

Here is a shorter way to write Example 18.

Instead of this,

```
       2 5 4 7
       -------
             7
           4 0
         5 0 0
       2 0 0 0
     ---------
 3 ) 7 6 4 3
     6 0 0 0
     -------
     1 6 4 3
     1 5 0 0
     -------
       1 4 3
       1 2 0
     -------
         2 3
         2 1
     -------
           2
```

Short form

we write this.

```
     2 5 4 7
   ---------
3 ) 7 6 4 3
    6 0 0 0
    -------
    1 6 4 3
    1 5 0 0
    -------
      1 4 3
      1 2 0
    -------
        2 3
        2 1
    -------
          2
```

We write a 2 above the thousands digit in the dividend to record 2000.
We write a 5 to record 500.
We write a 4 to record 40.
We write a 7 to record 7.

Do Exercises 34 and 35.

Example 19 Divide 2637 ÷ 41. Use the short form.

```
          6
     -------
4 1 ) 2 6 3 7
      2 4 6 0
      -------
        1 7 7
```

```
        6 4
     -------
4 1 ) 2 6 3 7
      2 4 6 0
      -------
        1 7 7
        1 6 4
      -------
          1 3
```

The answer is 64 R 13.

Do Exercises 36 and 37.

In Section 1.6, the process of long division was refined with an estimation method. After doing Exercise Set D, you should restudy that procedure.

Exercise Set D

a Divide, if possible.

1. $24 \div 8$ **2.** $72 \div 9$ **3.** $28 \div 7$ **4.** $22 \div 22$ **5.** $32 \div 1$

6. $45 \div 5$ **7.** $14 \div 2$ **8.** $40 \div 8$ **9.** $37 \div 1$ **10.** $10 \div 2$

11. $36 \div 4$ **12.** $12 \div 3$ **13.** $54 \div 9$ **14.** $18 \div 2$ **15.** $20 \div 4$

16. $16 \div 2$ **17.** $72 \div 8$ **18.** $42 \div 7$ **19.** $12 \div 4$ **20.** $8 \div 4$

21. $54 \div 6$ **22.** $18 \div 9$ **23.** $9 \div 3$ **24.** $28 \div 4$ **25.** $56 \div 7$

26. $24 \div 6$ **27.** $14 \div 2$ **28.** $14 \div 7$ **29.** $21 \div 7$ **30.** $36 \div 6$

31. $8 \div 8$ **32.** $32 \div 8$ **33.** $30 \div 5$ **34.** $18 \div 6$ **35.** $49 \div 7$

36. $81 \div 9$ **37.** $0 \div 7$ **38.** $9 \div 0$ **39.** $16 \div 0$ **40.** $42 \div 6$

41. $\frac{48}{6}$ **42.** $\frac{35}{5}$ **43.** $\frac{9}{9}$ **44.** $\frac{45}{9}$ **45.** $\frac{0}{5}$ **46.** $\frac{0}{8}$

47. $\frac{6}{2}$ **48.** $\frac{3}{3}$ **49.** $\frac{8}{2}$ **50.** $\frac{7}{1}$ **51.** $\frac{5}{5}$ **52.** $\frac{6}{1}$

53. $\frac{2}{2}$ **54.** $\frac{25}{5}$ **55.** $\frac{4}{2}$ **56.** $\frac{24}{3}$ **57.** $\frac{0}{9}$ **58.** $\frac{0}{4}$

59. $\frac{40}{5}$ **60.** $\frac{3}{1}$ **61.** $\frac{16}{4}$ **62.** $\frac{9}{0}$ **63.** $\frac{32}{8}$ **64.** $\frac{9}{9}$

b Divide using the "guess, multiply, and subtract" method.

65. $4\overline{)277}$

66. $2\overline{)399}$

67. $8\overline{)737}$

68. $6\overline{)831}$

69. $5\overline{)8619}$

70. $3\overline{)8775}$

71. $9\overline{)7777}$

72. $8\overline{)4179}$

73. $7\overline{)3691}$

74. $2\overline{)5794}$

75. $20\overline{)875}$

76. $30\overline{)987}$

77. $21\overline{)999}$

78. $23\overline{)975}$

79. $85\overline{)7757}$

80. $54\overline{)2821}$

81. $111\overline{)3219}$

82. $102\overline{)5612}$

83. $346\overline{)78{,}910}$

84. $781\overline{)15{,}999}$

c Divide.

85. $5\overline{)105}$

86. $6\overline{)708}$

87. $9\overline{)820}$

88. $3\overline{)965}$

89. $5\overline{)4823}$

90. $8\overline{)5437}$

91. $7\overline{)9298}$

92. $41\overline{)1115}$

93. $46\overline{)1058}$

94. $24\overline{)7722}$

95. $38\overline{)8522}$

96. $81\overline{)2247}$

97. $94\overline{)2153}$

98. $82\overline{)4064}$

99. $117\overline{)44{,}902}$

100. $740\overline{)55{,}200}$

Answers

Diagnostic Pretest, p. xxi

1. [1.2b] 1807 **2.** [1.7b] 29 **3.** [1.5b] 15,087 **4.** [1.8a] 21 cups, 2 oz left over **5.** [2.7c] $\frac{3}{10}$ **6.** [2.4a] $\frac{3}{2}$ **7.** [2.1d] $2 \cdot 2 \cdot 2 \cdot 2 \cdot 3 \cdot 3$ **8.** [2.6b] $\frac{1}{2}$ cup **9.** [3.1a] 48 **10.** [3.2b] $\frac{13}{24}$ **11.** [3.6a] $15\frac{2}{5}$ **12.** [3.5c] $1\frac{3}{5}$ m **13.** [3.6c] 30 **14.** [4.1d] 0.0001 **15.** [4.1e] 25.6 **16.** [4.2a] 39.815 **17.** [4.7a] 186.9 mi **18.** [4.3a] 0.03 **19.** [4.5a] $2.\overline{3}$ **20.** [4.4b] 2.4 **21.** [4.7a] $119.95 **22.** [4.6a] 160 **23.** [5.3b] 4.5 **24.** [5.4a] 12 **25.** [5.2b] 17.2¢/oz **26.** [6.2a] 12.5% **27.** [6.1b] 0.0135 **28.** [6.5b] 40% **29.** [6.7a] $19.55 **30.** [7.1a, b, c] Average: $24.\overline{3}$; median: 25; mode: 25 **31.** [7.1a] 18.5 **32.** [7.1a] 85 **33.** [8.1a] 72 **34.** [8.2a] 0.00004 **35.** [8.6a, b, c] 113.04 cm^2; 37.68 cm **36.** [8.3a], [8.4a] 7.5 ft^2; 11 ft **37.** [8.7a] 11 **38.** [9.3c] 120 **39.** [9.3b] 0.005 **40.** [9.5a] 18 **41.** [9.1b] 2.5 **42.** [9.2b] 3052.08 cm^3 **43.** [10.1e] 4.2 **44.** [10.1c] $-0.\overline{4}$ **45.** [10.3a] -0.1 **46.** [10.4a] $-\frac{1}{12}$ **47.** [11.4b] -3 **48.** [11.4b] $\frac{4}{5}$ **49.** [11.5b] $51.56 **50.** [11.5b] 8

Chapter 1

Pretest: Chapter 1, p. 2

1. [1.1c] Three million, seventy-eight thousand, fifty-nine **2.** [1.1a] 6 thousands + 9 hundreds + 8 tens + 7 ones **3.** [1.1d] 2,047,398,589 **4.** [1.1e] 6 ten thousands **5.** [1.4a] 956,000 **6.** [1.5c] 60,000 **7.** [1.2b] 10,216 **8.** [1.3d] 4108 **9.** [1.5b] 22,976 **10.** [1.6c] 503 R 11 **11.** [1.4c] $<$ **12.** [1.4c] $>$ **13.** [1.7b] 5542 **14.** [1.7b] 22 **15.** [1.7b] 34 **16.** [1.7b] 25 **17.** [1.8a] 12 lb **18.** [1.8a] 126 **19.** [1.8a] 22,981,000 **20.** [1.8a] 2292 sq ft **21.** [1.9b] 25 **22.** [1.9b] 64 **23.** [1.9c] 0 **24.** [1.9d] 0

Margin Exercises, Section 1.1, pp. 3–6

1. 1 thousand + 8 hundreds + 0 tens + 5 ones, or 1 thousand + 8 hundreds + 5 ones **2.** 3 ten thousands + 6 thousands + 2 hundreds + 2 tens + 3 ones **3.** 3 thousands + 2 hundreds + 1 ten **4.** 2 thousands + 9 ones **5.** 5 thousands + 7 hundreds **6.** 5689 **7.** 87,128 **8.** 9003 **9.** Fifty-seven **10.** Twenty-nine **11.** Eighty-eight **12.** Two hundred four **13.** Seventy-nine thousand, two hundred four **14.** One million, eight hundred seventy-nine thousand, two hundred four **15.** Twenty-two billion, three hundred one million, eight hundred seventy-nine thousand, two hundred four **16.** 213,105,329 **17.** 2 ten thousands **18.** 2 hundred thousands **19.** 2 millions **20.** 2 ten millions **21.** 6 **22.** 8 **23.** 5 **24.** 5

Exercise Set 1.1, p. 7

1. 5 thousands + 7 hundreds + 4 tens + 2 ones **3.** 2 ten thousands + 7 thousands + 3 hundreds + 4 tens + 2 ones **5.** 5 thousands + 6 hundreds + 9 ones **7.** 2 thousands + 3 hundreds **9.** 2475 **11.** 68,939 **13.** 7304 **15.** 1009 **17.** Eighty-five **19.** Eighty-eight thousand **21.** One hundred twenty-three thousand, seven hundred sixty-five **23.** Seven billion, seven hundred fifty-four million, two hundred eleven thousand, five hundred seventy-seven **25.** One million, eight hundred sixty-seven thousand **27.** One billion, five hundred eighty-three million, one hundred forty-one thousand **29.** 2,233,812 **31.** 8,000,000,000 **33.** 9,460,000,000,000 **35.** 2,974,600 **37.** 5 thousands **39.** 5 hundreds **41.** 3 **43.** 0 **45.** ◈ **47.** All 9's as digits. Answers may vary. For an 8-digit readout, it would be 99,999,999. This number has three periods.

Margin Exercises, Section 1.2, pp. 9–13

1. 8 + 2 = 10 **2.** $45 + $33 = $78 **3.** 100 mi + 93 mi = 193 mi **4.** 5 ft + 7 ft = 12 ft **5.** 4 in. + 5 in. + 9 in. + 6 in. + 5 in. = 29 in. **6.** 5 ft + 6 ft + 5 ft + 6 ft = 22 ft **7.** 30,000 sq ft + 40,000 sq ft = 70,000 sq ft **8.** 8 sq yd + 9 sq yd = 17 sq yd **9.** 6 cu yd + 8 cu yd = 14 cu yd **10.** 80 gal + 56 gal = 136 gal **11.** 9745 **12.** 13,465 **13.** 16,182 **14.** 27 **15.** 34 **16.** 27 **17.** 38 **18.** 47 **19.** 61 **20.** 27,474

Exercise Set 1.2, p. 15

1. 7 + 8 = 15 **3.** 500 acres + 300 acres = 800 acres **5.** 114 mi **7.** 52 in. **9.** 1300 ft **11.** 387 **13.** 5198 **15.** 164 **17.** 100 **19.** 900 **21.** 1010 **23.** 8503 **25.** 5266 **27.** 4466 **29.** 8310 **31.** 6608 **33.** 16,784 **35.** 34,432 **37.** 101,310 **39.** 100,111 **41.** 28 **43.** 26 **45.** 67 **47.** 230 **49.** 130 **51.** 1349 **53.** 36,926 **55.** 18,424 **57.** 2320 **59.** 31,685 **61.** 11,679 **63.** 22,654 **65.** 12,765,097 **67.** 7992 **68.** Nine hundred twenty-four million, six hundred thousand **69.** 8 ten thousands **70.** 23,000,000 **71.** ◈ **73.** 56,055,667 **75.** 1 + 99 = 100, 2 + 98 = 100, . . . , 49 + 51 = 100. Then 49 · 100 = 4900 and 4900 + 50 + 100 = 5050.

Margin Exercises, Section 1.3, pp. 19–22

1. 67 cu yd − 5 cu yd = 62 cu yd **2.** 20,000 sq ft − 12,000 sq ft = 8,000 sq ft **3.** 7 = 2 + 5, or 7 = 5 + 2 **4.** 17 = 9 + 8, or 17 = 8 + 9 **5.** 5 = 13 − 8; 8 = 13 − 5 **6.** 11 = 14 − 3; 3 = 14 − 11 **7.** 67 + $\square$ = 348; $\square$ = 348 − 67 **8.** 800 + $\square$ = 1200; $\square$ = 1200 − 800 **9.** 3801 **10.** 6328 **11.** 4747 **12.** 56 **13.** 205 **14.** 658 **15.** 2851 **16.** 1546

Exercise Set 1.3, p. 23

1. \$1260 − \$450 = $\square$ **3.** 16 oz − 5 oz = $\square$ **5.** 7 = 3 + 4, or 7 = 4 + 3 **7.** 13 = 5 + 8, or 13 = 8 + 5 **9.** 23 = 14 + 9, or 23 = 9 + 14 **11.** 43 = 27 + 16, or 43 = 16 + 27 **13.** 6 = 15 − 9; 9 = 15 − 6 **15.** 8 = 15 − 7; 7 = 15 − 8 **17.** 17 = 23 − 6; 6 = 23 − 17 **19.** 23 = 32 − 9; 9 = 32 − 23 **21.** 17 + $\square$ = 32; $\square$ = 32 − 17 **23.** 10 + $\square$ = 23; $\square$ = 23 − 10 **25.** 12 **27.** 44 **29.** 533 **31.** 1126 **33.** 39 **35.** 298 **37.** 226 **39.** 234 **41.** 5382 **43.** 1493 **45.** 2187 **47.** 3831 **49.** 7748 **51.** 33,794 **53.** 2168 **55.** 43,028 **57.** 56 **59.** 36 **61.** 84 **63.** 454 **65.** 771 **67.** 2191 **69.** 3749 **71.** 7019 **73.** 5745 **75.** 95,974 **77.** 9989 **79.** 83,818 **81.** 4206 **83.** 10,305 **85.** 7 ten thousands **86.** Six million, three hundred seventy-five thousand, six hundred two **87.** 29,708 **88.** 22,692 **89.** ◈ **91.** 2,829,177 **93.** 3; 4

Margin Exercises, Section 1.4, pp. 27–30

1. 40 **2.** 50 **3.** 70 **4.** 100 **5.** 40 **6.** 80 **7.** 90 **8.** 140 **9.** 470 **10.** 240 **11.** 290 **12.** 600 **13.** 800 **14.** 800 **15.** 9300 **16.** 8000 **17.** 8000 **18.** 19,000 **19.** 69,000 **20.** 200 **21.** 1800 **22.** 2600 **23.** 11,000 **24.** $<$ **25.** $>$ **26.** $>$ **27.** $<$ **28.** $<$ **29.** $>$

Exercise Set 1.4, p. 31

1. 50 **3.** 70 **5.** 730 **7.** 900 **9.** 100 **11.** 1000 **13.** 9100 **15.** 32,900 **17.** 6000 **19.** 8000 **21.** 45,000 **23.** 373,000 **25.** 180 **27.** 5720 **29.** 220; incorrect **31.** 890; incorrect **33.** 16,500 **35.** 5200 **37.** 1600 **39.** 1500 **41.** 31,000 **43.** 69,000 **45.** $<$ **47.** $>$ **49.** $<$ **51.** $>$ **53.** $>$ **55.** $>$ **57.** 86,754 **58.** 13,589 **59.** 48,824 **60.** 4415 **61.** ◈ **63.** 30,411 **65.** 69,594

Margin Exercises, Section 1.5, pp. 34–38

1. 8 · 7 = 56 **2.** 10 · 75 = 750 mL **3.** 8 · 8 = 64 **4.** 4 · 6 = 24 sq ft **5.** 1035 **6.** 3024 **7.** 46,252 **8.** 205,065 **9.** 144,432 **10.** 287,232 **11.** 14,075,720 **12.** 391,760 **13.** 17,345,600 **14.** 56,200 **15.** 562,000 **16.** **(a)** 1081; **(b)** 1081; **(c)** same **17.** 40 **18.** 15 **19.** 210,000; 160,000

Exercise Set 1.5, p. 39

1. 21 · 21 = 441 **3.** 8 · 12 oz = 96 oz **5.** 18 sq ft **7.** 121 sq yd **9.** 144 sq mm **11.** 870 **13.** 2,340,000 **15.** 520 **17.** 564 **19.** 65,200 **21.** 4,371,000 **23.** 1527 **25.** 64,603 **27.** 4770 **29.** 3995 **31.** 46,080 **33.** 14,652 **35.** 207,672 **37.** 798,408 **39.** 166,260 **41.** 11,794,332 **43.** 20,723,872 **45.** 362,128 **47.** 20,064,048 **49.** 25,236,000 **51.** 302,220 **53.** 49,101,136 **55.** 30,525 **57.** 298,738 **59.** 50 · 70 = 3500 **61.** 30 · 30 = 900 **63.** 900 · 300 = 270,000 **65.** 400 · 200 = 80,000 **67.** 6000 · 5000 = 30,000,000 **69.** 8000 · 6000 = 48,000,000 **71.** 4370 **72.** 3109 **73.** 2350; 2300; 2000 **75.** ◈

Margin Exercises, Section 1.6, pp. 44–49

1. 112 ÷ 14 = $\square$ **2.** 112 ÷ 8 = $\square$ **3.** 15 = 5 · 3, or 15 = 3 · 5 **4.** 72 = 9 · 8, or 72 = 8 · 9 **5.** 6 = 12 ÷ 2; 2 = 12 ÷ 6 **6.** 6 = 42 ÷ 7; 7 = 42 ÷ 6 **7.** 6; 6 · 9 = 54 **8.** 6 R 7; 6 · 9 = 54, 54 + 7 = 61 **9.** 4 R 5; 4 · 12 = 48, 48 + 5 = 53 **10.** 6 R 13; 6 · 24 = 144, 144 + 13 = 157 **11.** 59 R 3 **12.** 1475 R 5 **13.** 1015 **14.** 134 **15.** 63 R 12 **16.** 807 R 4 **17.** 1088 **18.** 360 R 4 **19.** 800 R 47

Calculator Spotlight, p. 48

1. 1475 R 5 **2.** 360 R 4 **3.** 800 R 47 **4.** 134

Exercise Set 1.6, p. 51

1. 760 ÷ 4 = $\square$ **3.** 455 ÷ 5 = $\square$ **5.** 18 = 3 · 6, or 18 = 6 · 3 **7.** 22 = 22 · 1, or 22 = 1 · 22 **9.** 54 = 6 · 9, or 54 = 9 · 6 **11.** 37 = 1 · 37, or 37 = 37 · 1 **13.** 9 = 45 ÷ 5; 5 = 45 ÷ 9 **15.** 37 = 37 ÷ 1; 1 = 37 ÷ 37 **17.** 8 = 64 ÷ 8 **19.** 11 = 66 ÷ 6; 6 = 66 ÷ 11 **21.** 55 R 2 **23.** 108 **25.** 307 **27.** 753 R 3 **29.** 74 R 1 **31.** 92 R 2 **33.** 1703 **35.** 987 R 5 **37.** 12,700 **39.** 127 **41.** 52 R 52 **43.** 29 R 5 **45.** 40 R 12 **47.** 90 R 22 **49.** 29 **51.** 105 R 3 **53.** 1609 R 2

55. 1007 R 1 **57.** 23 **59.** 107 R 1 **61.** 370 **63.** 609 R 15 **65.** 304 **67.** 3508 R 219 **69.** 8070 **71.** 7 thousands + 8 hundreds + 8 tens + 2 ones **72.** > **73.** 21 = 16 + 5, or 21 = 5 + 16 **74.** 56 = 14 + 42, or 56 = 42 + 14 **75.** 47 = 56 − 9; 9 = 56 − 47 **76.** 350 = 414 − 64; 64 = 414 − 350 **77.** ◈ **79.** 30

Margin Exercises, Section 1.7, pp. 55–58

1. 7 **2.** 5 **3.** No **4.** Yes **5.** 5 **6.** 10 **7.** 5 **8.** 22 **9.** 22,490 **10.** 9022 **11.** 570 **12.** 3661 **13.** 8 **14.** 45 **15.** 77 **16.** 3311 **17.** 6114 **18.** 8 **19.** 16 **20.** 644 **21.** 96 **22.** 94

Exercise Set 1.7, p. 59

1. 14 **3.** 0 **5.** 29 **7.** 0 **9.** 8 **11.** 14 **13.** 1035 **15.** 25 **17.** 450 **19.** 90,900 **21.** 32 **23.** 143 **25.** 79 **27.** 45 **29.** 324 **31.** 743 **33.** 37 **35.** 66 **37.** 15 **39.** 48 **41.** 175 **43.** 335 **45.** 104 **47.** 45 **49.** 4056 **51.** 17,603 **53.** 18,252 **55.** 205 **57.** 7 = 15 − 8; 8 = 15 − 7 **58.** 6 = 48 ÷ 8; 8 = 48 ÷ 6 **59.** < **60.** > **61.** 142 R 5 **62.** 334 R 11 **63.** ◈ **65.** 347

Margin Exercises, Section 1.8, pp. 62–68

1. 1,424,000 **2.** $369 **3.** $18 **4.** $38,988 **5.** 9180 sq in. **6.** 378 packages; 1 can left over **7.** 37 gal **8.** 70 min, or 1 hr, 10 min **9.** 106

Exercise Set 1.8, p. 69

1. $12,276 **3.** $7,004,000,000 **5.** $64,000,000 **7.** 4007 mi **9.** 384 in. **11.** 4500 **13.** 7280 **15.** $247 **17.** 54 weeks; 1 episode left over **19.** 168 hr **21.** $400 **23.** **(a)** 4700 ft^2; **(b)** 288 ft **25.** 35 **27.** 56 cartons; 11 books left over **29.** 1600 mi; 27 in. **31.** 18 **33.** 22 **35.** $704 **37.** 525 min, or 8 hr, 45 min **39.** 3000 in^2 **41.** 234,600 **42.** 235,000 **43.** 22,000 **44.** 16,000 **45.** 320,000 **46.** 720,000 **47.** ◈ **49.** 792,000 mi; 1,386,000 mi

Margin Exercises, Section 1.9, pp. 73–76

1. 5^4 **2.** 5^5 **3.** 10^2 **4.** 10^4 **5.** 10,000 **6.** 100 **7.** 512 **8.** 32 **9.** 51 **10.** 30 **11.** 584 **12.** 84 **13.** 4; 1 **14.** 52; 52 **15.** 29 **16.** 1880 **17.** 253 **18.** 93 **19.** 1880 **20.** 305 **21.** 93 **22.** 87 in. **23.** 46 **24.** 4

Calculator Spotlight, p. 75

1. 1024 **2.** 40,353,607 **3.** 1,048,576 **4.** 49 **5.** 85 **6.** 135 **7.** 176

Exercise Set 1.9, p. 77

1. 3^4 **3.** 5^2 **5.** 7^5 **7.** 10^3 **9.** 49 **11.** 729 **13.** 20,736 **15.** 121 **17.** 22 **19.** 20 **21.** 100 **23.** 1 **25.** 49 **27.** 5 **29.** 434 **31.** 41 **33.** 88 **35.** 4 **37.** 303 **39.** 20 **41.** 70 **43.** 295 **45.** 32 **47.** 906 **49.** 62 **51.** 102 **53.** $94 **55.** 110 **57.** 7 **59.** 544 **61.** 708 **63.** 452 **64.** 13 **65.** 102,600 mi^2 **66.** 98 gal **67.** ◈ **69.** 675 **71.** 24; 1 + 5 · (4 + 3) = 36 **73.** 7; 12 ÷ (4 + 2) · 3 − 2 = 4

Summary and Review: Chapter 1, p. 79

1. 2 thousands + 7 hundreds + 9 tens + 3 ones **2.** 5 ten thousands + 6 thousands + 7 tens + 8 ones **3.** 8669 **4.** 90,844 **5.** Sixty-seven thousand, eight hundred nineteen **6.** Two million, seven hundred eighty-one thousand, four hundred twenty-seven **7.** 476,588 **8.** 36,260,064 **9.** 8 thousands **10.** 3 **11.** $406 + $78 = $484, or $78 + $406 = $484 **12.** 986 yd **13.** 14,272 **14.** 66,024 **15.** 22,098 **16.** 98,921 **17.** 151 − 12 = 139 **18.** $196 + □ = $340, or □ = $340 − $196 **19.** 10 = 6 + 4, or 10 = 4 + 6 **20.** 8 = 11 − 3; 3 = 11 − 8 **21.** 5148 **22.** 1153 **23.** 2274 **24.** 17,757 **25.** 345,800 **26.** 345,760 **27.** 346,000 **28.** 41,300 + 19,700 = 61,000 **29.** 38,700 − 24,500 = 14,200 **30.** 400 · 700 = 280,000 **31.** > **32.** < **33.** 32 · 15 = 480 **34.** 125 · 368 = 46,000 yd^2 **35.** 420,000 **36.** 6,276,800 **37.** 506,748 **38.** 27,589 **39.** 5,331,810 **40.** 176 ÷ 4 = □ **41.** 222 ÷ 6 = □ **42.** 56 = 8 · 7, or 56 = 7 · 8 **43.** 4 = 52 ÷ 13; 13 = 52 ÷ 4 **44.** 12 R 3 **45.** 5 **46.** 913 R 3 **47.** 384 R 1 **48.** 4 R 46 **49.** 54 **50.** 452 **51.** 5008 **52.** 4389 **53.** 8 **54.** 45 **55.** 546 **56.** $2413 **57.** 1982 **58.** $19,748 **59.** 137 beakers filled; 13 mL of alcohol left over **60.** 4^3 **61.** 10,000 **62.** 36 **63.** 65 **64.** 233 **65.** 56 **66.** 32 **67.** 260 **68.** 165 **69.** ◈ A vat contains 1152 oz of hot sauce. If 144 bottles are to be filled equally, how much will each bottle contain? Answers may vary.
70. ◈ No; if subtraction were associative, then $a - (b - c) = (a - b) - c$ for any a, b, and c. But, for example,

$$12 - (8 - 4) = 12 - 4 = 8,$$

whereas

$$(12 - 8) - 4 = 4 - 4 = 0.$$

Since $8 \neq 0$, this example shows that subtraction is not associative.
71. $d = 8$ **72.** $a = 8, b = 4$ **73.** 7 days

Test: Chapter 1, p. 81

1. [1.1a] 8 thousands + 8 hundreds + 4 tens + 3 ones **2.** [1.1c] Thirty-eight million, four hundred three thousand, two hundred seventy-seven **3.** [1.1e] 5 **4.** [1.2b] 9989 **5.** [1.2b] 63,791 **6.** [1.2b] 34 **7.** [1.2b] 10,515 **8.** [1.3d] 3630 **9.** [1.3d] 1039 **10.** [1.3d] 6848 **11.** [1.3d] 5175 **12.** [1.5b] 41,112

13. [1.5b] 5,325,600 **14.** [1.5b] 2405 **15.** [1.5b] 534,264 **16.** [1.6c] 3 R 3 **17.** [1.6c] 70 **18.** [1.6c] 97 **19.** [1.6c] 805 R 8 **20.** [1.8a] 1955 **21.** [1.8a] 92 packages, 3 cans left over **22.** [1.8a] 62,811 mi^2 **23.** [1.8a] 120,000 m^2; 1600 m **24.** [1.8a] 1808 lb **25.** [1.8a] 20 **26.** [1.7b] 46 **27.** [1.7b] 13 **28.** [1.7b] 14 **29.** [1.4a] 35,000 **30.** [1.4a] 34,580 **31.** [1.4a] 34,600 **32.** [1.4b] $23{,}600 + 54{,}700 = 78{,}300$ **33.** [1.4b] $54{,}800 - 23{,}600 = 31{,}200$ **34.** [1.5c] $800 \cdot 500 = 400{,}000$ **35.** [1.4c] $>$ **36.** [1.4c] $<$ **37.** [1.9a] 12^4 **38.** [1.9b] 343 **39.** [1.9b] 8 **40.** [1.9c] 64 **41.** [1.9c] 96 **42.** [1.9c] 2 **43.** [1.9d] 216 **44.** [1.9c] 18 **45.** [1.9c] 92 **46.** [1.5a], [1.8a] 336 in^2 **47.** [1.8a] 80 **48.** [1.9c] 83 **49.** [1.9c] 9

Chapter 2

Pretest: Chapter 2, p. 84

1. [2.1c] Prime **2.** [2.1d] $2 \cdot 2 \cdot 5 \cdot 7$ **3.** [2.2a] Yes **4.** [2.2a] No **5.** [2.3c] 1 **6.** [2.3c] 68 **7.** [2.3c] 0 **8.** [2.5b] $\frac{1}{4}$ **9.** [2.6a] $\frac{6}{5}$ **10.** [2.6a] 20 **11.** [2.6a] $\frac{5}{4}$ **12.** [2.7a] $\frac{8}{7}$ **13.** [2.7a] $\frac{1}{11}$ **14.** [2.7b] 24 **15.** [2.7b] $\frac{3}{4}$ **16.** [2.7c] 30 **17.** [2.5c] $\neq$ **18.** [2.6b] \$36 **19.** [2.7d] $\frac{1}{24}$ m

Margin Exercises, Section 2.1, pp. 85–90

1. 1, 2, 3, 6 **2.** 1, 2, 4, 8 **3.** 1, 2, 5, 10 **4.** 1, 2, 4, 8, 16, 32 **5.** $5 = 1 \cdot 5$, $45 = 9 \cdot 5$, $100 = 20 \cdot 5$ **6.** $10 = 1 \cdot 10$, $60 = 6 \cdot 10$, $110 = 11 \cdot 10$ **7.** 5, 10, 15, 20, 25, 30, 35, 40, 45, 50 **8.** Yes **9.** Yes **10.** No **11.** 13, 19, 41 are prime; 4, 6, 8 are composite; 1 is neither **12.** $2 \cdot 3$ **13.** $2 \cdot 2 \cdot 3$ **14.** $3 \cdot 3 \cdot 5$ **15.** $2 \cdot 7 \cdot 7$ **16.** $2 \cdot 3 \cdot 3 \cdot 7$ **17.** $2 \cdot 2 \cdot 2 \cdot 2 \cdot 3 \cdot 3$

Calculator Spotlight, p. 87

1. Yes **2.** No **3.** No **4.** Yes

Calculator Spotlight, p. 90

1. No **2.** Yes **3.** Yes **4.** No **5.** Yes **6.** No

Exercise Set 2.1, p. 91

1. 1, 2, 3, 6, 9, 18 **3.** 1, 2, 3, 6, 9, 18, 27, 54 **5.** 1, 2, 4 **7.** 1, 7 **9.** 1 **11.** 1, 2, 7, 14, 49, 98 **13.** 4, 8, 12, 16, 20, 24, 28, 32, 36, 40 **15.** 20, 40, 60, 80, 100, 120, 140, 160, 180, 200 **17.** 3, 6, 9, 12, 15, 18, 21, 24, 27, 30 **19.** 12, 24, 36, 48, 60, 72, 84, 96, 108, 120 **21.** 10, 20, 30, 40, 50, 60, 70, 80, 90, 100 **23.** 9, 18, 27, 36, 45, 54, 63, 72, 81, 90 **25.** No **27.** Yes **29.** Yes **31.** No **33.** No **35.** Neither **37.** Composite **39.** Prime **41.** Prime **43.** $2 \cdot 2 \cdot 2$ **45.** $2 \cdot 7$ **47.** $2 \cdot 3 \cdot 7$ **49.** $5 \cdot 5$ **51.** $2 \cdot 5 \cdot 5$ **53.** $13 \cdot 13$ **55.** $2 \cdot 2 \cdot 5 \cdot 5$ **57.** $5 \cdot 7$ **59.** $2 \cdot 2 \cdot 2 \cdot 3 \cdot 3$ **61.** $7 \cdot 11$ **63.** $2 \cdot 2 \cdot 7 \cdot 103$ **65.** $3 \cdot 17$ **67.** 26 **68.** 256 **69.** 425 **70.** 4200 **71.** 0 **72.** 22 **73.** 1 **74.** 3 **75.** \$612 **76.** 201 min, or 3 hr, 21 min **77.** ◆ **79.** Row 1: 48, 90, 432, 63; row 2: 7, 2, 2, 10, 8, 6, 21, 10; row 3: 9, 18, 36, 14, 12, 11, 21; row 4: 29, 19, 42

Margin Exercises, Section 2.2, pp. 93–96

1. Yes **2.** No **3.** Yes **4.** No **5.** Yes **6.** No **7.** Yes **8.** No **9.** Yes **10.** No **11.** No **12.** Yes **13.** No **14.** Yes **15.** No **16.** Yes **17.** No **18.** Yes **19.** No **20.** Yes **21.** Yes **22.** No **23.** No **24.** Yes **25.** Yes **26.** No **27.** No **28.** Yes **29.** No **30.** Yes **31.** Yes **32.** No

Exercise Set 2.2, p. 97

1. 46, 224, 300, 36, 45,270, 4444, 256, 8064, 21,568 **3.** 224, 300, 36, 4444, 256, 8064, 21,568 **5.** 300, 36, 45,270, 8064 **7.** 36, 45,270, 711, 8064 **9.** 324, 42, 501, 3009, 75, 2001, 402, 111,111, 1005 **11.** 55,555, 200, 75, 2345, 35, 1005 **13.** 324 **15.** 200 **17.** 138 **18.** 139 **19.** 56 **20.** 26 **21.** 234 **22.** 4003 **23.** 45 gal **24.** 4320 min **25.** ◆ **27.** $2 \cdot 2 \cdot 2 \cdot 3 \cdot 5 \cdot 5 \cdot 13$ **29.** $2 \cdot 2 \cdot 3 \cdot 3 \cdot 7 \cdot 11$ **31.** 7,652,341

Margin Exercises, Section 2.3, pp. 99–102

1. 1, numerator; 6, denominator **2.** 5, numerator; 7, denominator **3.** 22, numerator; 3, denominator **4.** $\frac{1}{2}$ **5.** $\frac{1}{3}$ **6.** $\frac{1}{3}$ **7.** $\frac{1}{6}$ **8.** $\frac{5}{8}$ **9.** $\frac{2}{3}$ **10.** $\frac{3}{4}$ **11.** $\frac{4}{6}$ **12.** $\frac{5}{4}$ **13.** $\frac{7}{4}$ **14.** $\frac{2}{5}$ **15.** $\frac{2}{3}$ **16.** $\frac{2}{6}, \frac{4}{6}$ **17.** 1 **18.** 1 **19.** 1 **20.** 1 **21.** 1 **22.** 1 **23.** 0 **24.** 0 **25.** 0 **26.** 0 **27.** Not defined **28.** Not defined **29.** 8 **30.** 10 **31.** 346 **32.** 23

Exercise Set 2.3, p. 103

1. 3, numerator; 4, denominator **3.** 11, numerator; 20, denominator **5.** $\frac{2}{4}$ **7.** $\frac{1}{8}$ **9.** $\frac{4}{3}$ **11.** $\frac{3}{4}$ **13.** $\frac{4}{8}$ **15.** $\frac{6}{12}$ **17.** $\frac{5}{8}$ **19.** $\frac{3}{5}$ **21.** 0 **23.** 7 **25.** 1 **27.** 1 **29.** 0 **31.** 1 **33.** 1 **35.** 1 **37.** 1 **39.** 18 **41.** Not defined **43.** Not defined **45.** 34,560 **46.** 34,600 **47.** 35,000 **48.** \$733 **49.** 3728 lb **50.** 848 **51.** 2203 **52.** 37,239 **53.** ◆ **55.** $\frac{3}{4}; \frac{1}{4}$

Margin Exercises, Section 2.4, pp. 105–108

1. $\frac{2}{3}$ **2.** $\frac{5}{8}$ **3.** $\frac{10}{3}$ **4.** $\frac{33}{8}$ **5.** $\frac{46}{5}$ **6.** $\frac{15}{56}$ **7.** $\frac{32}{15}$ **8.** $\frac{3}{100}$ **9.** $\frac{14}{3}$

10.

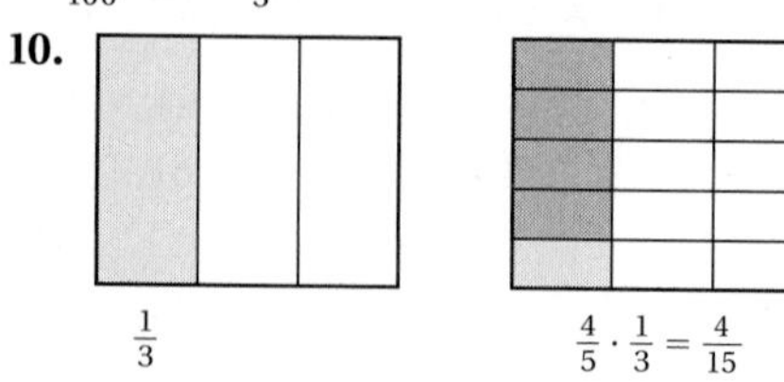

$\frac{1}{3}$ $\qquad$ $\frac{4}{5} \cdot \frac{1}{3} = \frac{4}{15}$

11. $\frac{3}{8}$ **12.** $\frac{63}{100}$ cm^2 **13.** $\frac{3}{40}$

Exercise Set 2.4, p. 109

1. $\frac{3}{5}$ **3.** $\frac{5}{8}$ **5.** $\frac{8}{11}$ **7.** $\frac{70}{9}$ **9.** $\frac{2}{5}$ **11.** $\frac{6}{5}$ **13.** $\frac{21}{4}$ **15.** $\frac{85}{6}$ **17.** $\frac{1}{6}$ **19.** $\frac{1}{40}$ **21.** $\frac{2}{15}$ **23.** $\frac{4}{15}$ **25.** $\frac{9}{16}$ **27.** $\frac{14}{39}$ **29.** $\frac{7}{100}$ **31.** $\frac{49}{64}$ **33.** $\frac{1}{1000}$ **35.** $\frac{182}{285}$ **37.** $\frac{12}{25}\ \text{m}^2$ **39.** $\frac{1}{1521}$ **41.** $\frac{3}{8}$ cup **43.** $\frac{56}{100}$ **45.** 204 **46.** 700 **47.** 3001 **48.** 204 R 8 **49.** 8 thousands **50.** 8 millions **51.** 8 ones **52.** 8 hundreds **53.** 3 **54.** 81 **55.** 50 **56.** 6399 **57.** ◆ **59.** $\frac{71{,}269}{180{,}433}$ **61.** $\frac{56}{1125}$

Margin Exercises, Section 2.5, pp. 111–114

1. $\frac{8}{16}$ **2.** $\frac{30}{50}$ **3.** $\frac{52}{100}$ **4.** $\frac{200}{75}$ **5.** $\frac{12}{9}$ **6.** $\frac{18}{24}$ **7.** $\frac{90}{100}$ **8.** $\frac{9}{45}$ **9.** $\frac{56}{49}$ **10.** $\frac{1}{4}$ **11.** $\frac{5}{6}$ **12.** 5 **13.** $\frac{4}{3}$ **14.** $\frac{7}{8}$ **15.** $\frac{89}{78}$ **16.** $\frac{8}{7}$ **17.** $\frac{1}{4}$ **18.** $\frac{2}{100} = \frac{1}{50}$; $\frac{4}{100} = \frac{1}{25}$; $\frac{32}{100} = \frac{8}{25}$; $\frac{44}{100} = \frac{11}{25}$; $\frac{18}{100} = \frac{9}{50}$ **19.** $=$ **20.** $\neq$

Calculator Spotlight, p. 113

1. $\frac{14}{15}$ **2.** $\frac{7}{8}$ **3.** $\frac{138}{167}$ **4.** $\frac{7}{25}$

Exercise Set 2.5, p. 115

1. $\frac{5}{10}$ **3.** $\frac{20}{32}$ **5.** $\frac{27}{30}$ **7.** $\frac{28}{32}$ **9.** $\frac{20}{48}$ **11.** $\frac{51}{54}$ **13.** $\frac{75}{45}$ **15.** $\frac{42}{132}$ **17.** $\frac{1}{2}$ **19.** $\frac{3}{4}$ **21.** $\frac{1}{5}$ **23.** 3 **25.** $\frac{3}{4}$ **27.** $\frac{7}{8}$ **29.** $\frac{6}{5}$ **31.** $\frac{1}{3}$ **33.** 6 **35.** $\frac{1}{3}$ **37.** $=$ **39.** $\neq$ **41.** $=$ **43.** $\neq$ **45.** $=$ **47.** $\neq$ **49.** $=$ **51.** $\neq$ **53.** $4992\ \text{ft}^2$ **54.** \$928 **55.** 11 **56.** 32 **57.** 186 **58.** 2737 **59.** 5 **60.** 3520 **61.** ◆ **63.** $\frac{137}{149}$ **65.** $\frac{2}{5}$; $\frac{3}{5}$ **67.** No. $\frac{168}{551} \neq \frac{165}{545}$ because $168 \cdot 545 \neq 551 \cdot 165$.

Margin Exercises, Section 2.6, p. 118

1. $\frac{7}{12}$ **2.** $\frac{1}{3}$ **3.** 6 **4.** $\frac{5}{2}$ **5.** 14 lb

Exercise Set 2.6, p. 119

1. $\frac{1}{3}$ **3.** $\frac{1}{8}$ **5.** $\frac{1}{10}$ **7.** $\frac{1}{6}$ **9.** $\frac{27}{10}$ **11.** $\frac{14}{9}$ **13.** 1 **15.** 1 **17.** 1 **19.** 1 **21.** 2 **23.** 4 **25.** 9 **27.** 9 **29.** $\frac{26}{5}$ **31.** $\frac{98}{5}$ **33.** 60 **35.** 30 **37.** $\frac{1}{5}$ **39.** $\frac{9}{25}$ **41.** $\frac{11}{40}$ **43.** $\frac{5}{14}$ **45.** \$27 **47.** 625 **49.** $\frac{1}{3}$ cup **51.** \$93,000 **53.** 160 mi **55.** Food: \$6750; housing: \$5400; clothing: \$2700; savings: \$3000; taxes: \$6750; other: \$2400 **57.** 35 **58.** 85 **59.** 4989 **60.** 8546 **61.** 4673 **62.** 5338 **63.** ◆ **65.** $\frac{129}{485}$ **67.** $\frac{1}{12}$ **69.** $\frac{1}{168}$

Margin Exercises, Section 2.7, pp. 123–126

1. $\frac{5}{2}$ **2.** $\frac{7}{10}$ **3.** $\frac{1}{9}$ **4.** 5 **5.** $\frac{8}{7}$ **6.** $\frac{8}{3}$ **7.** $\frac{1}{10}$ **8.** 100 **9.** 1 **10.** $\frac{14}{15}$ **11.** $\frac{4}{5}$ **12.** 32 **13.** 320 **14.** 200 gal

Exercise Set 2.7, p. 127

1. $\frac{6}{5}$ **3.** $\frac{1}{6}$ **5.** 6 **7.** $\frac{3}{10}$ **9.** $\frac{4}{5}$ **11.** $\frac{4}{15}$ **13.** 4 **15.** 2 **17.** $\frac{1}{8}$ **19.** $\frac{3}{7}$ **21.** 8 **23.** 35 **25.** 1 **27.** $\frac{2}{3}$ **29.** $\frac{9}{4}$ **31.** 144 **33.** 75 **35.** 2 **37.** $\frac{3}{5}$ **39.** 315 **41.** 75 **43.** 32 **45.** 24 **47.** 16 L **49.** 288 km; 108 km **51.** 67 **52.** 33 R 4 **53.** 285 R 2 **54.** 103 R 10 **55.** 67 **56.** 264 **57.** 8499 **58.** 4368 **59.** ◆ **61.** $\frac{9}{19}$ **63.** 36 **65.** $\frac{3}{8}$

Summary and Review: Chapter 2, p. 129

1. $2 \cdot 5 \cdot 7$ **2.** $2 \cdot 3 \cdot 5$ **3.** $3 \cdot 3 \cdot 5$ **4.** $2 \cdot 3 \cdot 5 \cdot 5$ **5.** No **6.** No **7.** No **8.** Yes **9.** Prime **10.** 2, numerator; 7, denominator **11.** $\frac{3}{5}$ **12.** $\frac{3}{100}$; $\frac{8}{100} = \frac{2}{25}$; $\frac{10}{100} = \frac{1}{10}$; $\frac{15}{100} = \frac{3}{20}$; $\frac{21}{100}$; $\frac{43}{100}$ **13.** 0 **14.** 1 **15.** 48 **16.** 6 **17.** $\frac{2}{3}$ **18.** $\frac{1}{4}$ **19.** 1 **20.** 0 **21.** $\frac{2}{5}$ **22.** 18 **23.** 4 **24.** $\frac{1}{3}$ **25.** Not defined **26.** Not defined **27.** $\neq$ **28.** $=$ **29.** $\neq$ **30.** $=$ **31.** $\frac{3}{2}$ **32.** 56 **33.** $\frac{5}{2}$ **34.** 24 **35.** $\frac{2}{3}$ **36.** $\frac{1}{14}$ **37.** $\frac{2}{3}$ **38.** $\frac{1}{22}$ **39.** $\frac{5}{4}$ **40.** $\frac{1}{3}$ **41.** 9 **42.** $\frac{36}{47}$ **43.** $\frac{9}{2}$ **44.** 2 **45.** $\frac{11}{6}$ **46.** $\frac{1}{4}$ **47.** $\frac{9}{4}$ **48.** 300 **49.** 1 **50.** $\frac{4}{9}$ **51.** $\frac{3}{10}$ **52.** 240 **53.** 224 days **54.** 160 km **55.** $\frac{1}{3}$ cup **56.** \$6 **57.** 24 **58.** 469 **59.** \$912 **60.** 774 mi **61.** 408 R 9 **62.** 3607

63. ◆ To simplify fractional notation, first factor the numerator and the denominator into prime numbers. Examine the factorizations for factors common to both the numerator and the denominator. Change the order of the factorizations, if necessary, so that pairs of like factors are above and below each other. Factor the fraction, with each pair of like factors forming a factor of 1. Remove the factors of 1, and multiply the remaining factors in the numerator and in the denominator, if necessary.

64. ◆ Taking $\frac{1}{2}$ of a number is equivalent to multiplying the number by $\frac{1}{2}$. Dividing by $\frac{1}{2}$ is equivalent to multiplying by the reciprocal of $\frac{1}{2}$, or 2. Thus taking $\frac{1}{2}$ of a number is not the same as dividing by $\frac{1}{2}$.

65. $a = 11{,}176$; $b = 9887$ **66.** 13, 11, 101

Test: Chapter 2, p. 131

1. [2.1d] $2 \cdot 3 \cdot 3$ **2.** [2.1d] $2 \cdot 2 \cdot 3 \cdot 5$ **3.** [2.2a] Yes **4.** [2.2a] No **5.** [2.3a] 4, numerator; 9, denominator **6.** [2.3b] $\frac{3}{4}$ **7.** [2.3c] 26 **8.** [2.3c] 1 **9.** [2.3c] 0 **10.** [2.5b] $\frac{1}{2}$ **11.** [2.5b] 6 **12.** [2.5b] $\frac{1}{14}$ **13.** [2.3c] Not defined **14.** [2.3c] Not defined **15.** [2.5c] $=$ **16.** [2.5c] $\neq$ **17.** [2.6a] 32 **18.** [2.6a] $\frac{3}{2}$ **19.** [2.6a] $\frac{5}{2}$ **20.** [2.6a] $\frac{1}{10}$ **21.** [2.7a] $\frac{8}{5}$ **22.** [2.7a] 4 **23.** [2.7a] $\frac{1}{18}$ **24.** [2.7b] $\frac{3}{10}$ **25.** [2.7b] $\frac{8}{5}$ **26.** [2.7b] 18 **27.** [2.7c] 64 **28.** [2.7c] $\frac{7}{4}$ **29.** [2.6b] 28 lb **30.** [2.7d] $\frac{3}{40}$ m **31.** [1.7b] 1805 **32.** [1.7b] 101 **33.** [1.8a] 3635 mi **34.** [1.6c] 380 R 7 **35.** [1.3d] 4434 **36.** [2.6b] $\frac{15}{8}$ tsp **37.** [2.6b] $\frac{7}{48}$ acre **38.** [2.6a], [2.7b] $\frac{7}{960}$ **39.** [2.7c] $\frac{7}{5}$

Cumulative Review: Chapters 1–2, p. 133

1. [1.1d] 584,017,800 **2.** [1.1c] Five million, three hundred eighty thousand, six hundred twenty-one **3.** [1.1e] 0 **4.** [1.2b] 17,797 **5.** [1.2b] 8866 **6.** [1.3d] 4946 **7.** [1.3d] 1425 **8.** [1.5b] 16,767 **9.** [1.5b] 8,266,500 **10.** [2.4a] $\frac{3}{20}$ **11.** [2.6a] $\frac{1}{6}$ **12.** [1.6c] 241 R 1 **13.** [1.6c] 62 **14.** [2.7b] $\frac{3}{50}$ **15.** [2.7b] $\frac{16}{45}$ **16.** [1.4a] 428,000 **17.** [1.4a] 5300 **18.** [1.4b] 749,600 + 301,400 = 1,051,000 **19.** [1.5c] 700 · 500 = 350,000 **20.** [1.4c] $>$ **21.** [2.5c] $\neq$ **22.** [1.9b] 81 **23.** [1.9c] 36 **24.** [1.9d] 2 **25.** [2.1a] 1, 2, 4, 7, 14, 28 **26.** [2.1d] 2 · 2 · 7 **27.** [2.1c] Composite **28.** [2.2a] Yes **29.** [2.2a] No **30.** [2.3c] 35 **31.** [2.5b] 7 **32.** [2.5b] $\frac{2}{7}$ **33.** [2.3c] 0 **34.** [1.7b] 37 **35.** [2.7c] $\frac{3}{2}$ **36.** [1.7b] 3 **37.** [1.7b] 24 **38.** [1.8a] $664,163 **39.** [1.8a] 11,719 **40.** [1.8a] $75 **41.** [2.4c] 3 cups **42.** [2.7d] 8 days **43.** [1.8a] Westside **44.** [1.8a], [2.6b] Yes; $950 **45.** [2.3b] $\frac{3}{6}$, or $\frac{1}{2}$

Chapter 3

Pretest: Chapter 3, p. 136

1. [3.1a] 120 **2.** [3.3b] $<$ **3.** [3.4a] $\frac{61}{8}$ **4.** [3.4b] $5\frac{1}{2}$ **5.** [3.4c] $399\frac{1}{12}$ **6.** [3.5a] $11\frac{31}{60}$ **7.** [3.5b] $6\frac{1}{6}$ **8.** [3.6a] $13\frac{3}{5}$ **9.** [3.6a] $21\frac{2}{3}$ **10.** [3.6b] 6 **11.** [3.6b] $1\frac{2}{3}$ **12.** [3.3c] $\frac{2}{9}$ **13.** [3.5c] $21\frac{1}{4}$ lb **14.** [3.6c] $4\frac{1}{4}$ cu ft **15.** [3.5c] $351\frac{1}{5}$ mi **16.** [3.6c] $22\frac{1}{2}$ cups **17.** [3.7b] 1 **18.** [3.7b] 0 **19.** [3.7b] $\frac{1}{2}$ **20.** [3.7b] 10 **21.** [3.7b] 2 **22.** [3.7b] 58

Margin Exercises, Section 3.1, pp. 137–142

1. 45 **2.** 40 **3.** 30 **4.** 24 **5.** 10 **6.** 80 **7.** 40 **8.** 360 **9.** 864 **10.** 2520 **11.** 18 **12.** 24 **13.** 36 **14.** 210 **15.** 2520 **16.** 3780

Exercise Set 3.1, p. 143

1. 4 **3.** 50 **5.** 40 **7.** 54 **9.** 150 **11.** 120 **13.** 72 **15.** 420 **17.** 144 **19.** 288 **21.** 30 **23.** 105 **25.** 72 **27.** 60 **29.** 36 **31.** 900 **33.** 48 **35.** 50 **37.** 143 **39.** 420 **41.** 378 **43.** 810 **45.** 60 yr **47.** 90 days **48.** 250 **49.** 7935 **50.** $\frac{2}{3}$ **51.** $\frac{8}{7}$ **52.** 6939 **53.** ◆ **55.** 2592 **57.** 18,900 **59.** 5 in. by 24 in.

Margin Exercises, Section 3.2, pp. 145–148

1. $\frac{4}{5}$ **2.** 1 **3.** $\frac{1}{2}$ **4.** $\frac{3}{4}$ **5.** $\frac{5}{6}$ **6.** $\frac{29}{24}$ **7.** $\frac{5}{9}$ **8.** $\frac{413}{1000}$ **9.** $\frac{759}{1000}$ **10.** $\frac{197}{210}$ **11.** $\frac{11}{10}$ lb

Exercise Set 3.2, p. 149

1. 1 **3.** $\frac{3}{4}$ **5.** $\frac{3}{2}$ **7.** $\frac{7}{24}$ **9.** $\frac{3}{2}$ **11.** $\frac{19}{24}$ **13.** $\frac{9}{10}$ **15.** $\frac{29}{18}$ **17.** $\frac{31}{100}$ **19.** $\frac{41}{60}$ **21.** $\frac{189}{100}$ **23.** $\frac{7}{8}$ **25.** $\frac{13}{24}$ **27.** $\frac{17}{24}$ **29.** $\frac{3}{4}$ **31.** $\frac{437}{500}$ **33.** $\frac{53}{40}$ **35.** $\frac{391}{144}$ **37.** $\frac{5}{6}$ lb **39.** $\frac{23}{12}$ mi **41.** 690 kg; $\frac{14}{23}$ cement; $\frac{5}{23}$ stone; $\frac{4}{23}$ sand; 1 **43.** $\frac{51}{32}$ in. **45.** 210,528 **46.** 4,194,000 **47.** 3,387,807 **48.** 352,350 **49.** $3077 **50.** $739 **51.** 12 **52.** $\frac{3}{64}$ acre **53.** ◆ **55.** $\frac{4}{15}$; $320

Margin Exercises, Section 3.3, pp. 151–154

1. $\frac{1}{2}$ **2.** $\frac{3}{8}$ **3.** $\frac{1}{2}$ **4.** $\frac{1}{12}$ **5.** $\frac{13}{18}$ **6.** $\frac{1}{2}$ **7.** $\frac{9}{112}$ **8.** $<$ **9.** $>$ **10.** $>$ **11.** $>$ **12.** $<$ **13.** $\frac{1}{6}$ **14.** $\frac{11}{40}$ **15.** $\frac{11}{20}$ cup

Exercise Set 3.3, p. 155

1. $\frac{2}{3}$ **3.** $\frac{3}{4}$ **5.** $\frac{5}{8}$ **7.** $\frac{1}{24}$ **9.** $\frac{1}{2}$ **11.** $\frac{9}{14}$ **13.** $\frac{3}{5}$ **15.** $\frac{7}{10}$ **17.** $\frac{17}{60}$ **19.** $\frac{53}{100}$ **21.** $\frac{26}{75}$ **23.** $\frac{9}{100}$ **25.** $\frac{13}{24}$ **27.** $\frac{1}{10}$ **29.** $\frac{1}{24}$ **31.** $\frac{13}{16}$ **33.** $\frac{31}{75}$ **35.** $\frac{13}{75}$ **37.** $<$ **39.** $>$ **41.** $<$ **43.** $<$ **45.** $>$ **47.** $>$ **49.** $<$ **51.** $\frac{1}{15}$ **53.** $\frac{2}{15}$ **55.** $\frac{1}{15}$ **57.** $\frac{5}{12}$ hr **59.** $\frac{1}{32}$ in. **61.** $\frac{1}{4}$ **63.** $\frac{4}{21}$ **64.** $\frac{3}{2}$ **65.** 21 **66.** $\frac{1}{32}$ **67.** 6 lb **68.** 9 cups **69.** ◆ **71.** $\frac{14}{3553}$ **73.** $\frac{227}{420}$ km **75.** $\frac{19}{24}$ **77.** $\frac{145}{144}$ **79.** $\frac{21}{40}$ km **81.** $>$

Margin Exercises, Section 3.4, pp. 157–160

1. $1\frac{2}{3}$ **2.** $8\frac{3}{4}$ **3.** $12\frac{2}{3}$ **4.** $\frac{22}{5}$ **5.** $\frac{61}{10}$ **6.** $\frac{29}{6}$ **7.** $\frac{37}{4}$ **8.** $\frac{62}{3}$ **9.** $2\frac{1}{3}$ **10.** $1\frac{1}{10}$ **11.** $18\frac{1}{3}$ **12.** $807\frac{2}{3}$ **13.** $134\frac{23}{45}$

Calculator Spotlight, p. 160

1. $1476\frac{1}{6}$ **2.** $676\frac{4}{9}$ **3.** $800\frac{51}{56}$ **4.** $13{,}031\frac{1}{2}$ **5.** $51{,}626\frac{9}{11}$ **6.** $7330\frac{7}{32}$ **7.** $134\frac{1}{15}$ **8.** $2666\frac{130}{213}$ **9.** $3571\frac{51}{112}$ **10.** $12\frac{169}{454}$

Exercise Set 3.4, p. 161

1. $\frac{17}{3}$ **3.** $\frac{13}{4}$ **5.** $\frac{81}{8}$ **7.** $\frac{51}{10}$ **9.** $\frac{103}{5}$ **11.** $\frac{59}{6}$ **13.** $\frac{73}{10}$ **15.** $\frac{13}{8}$ **17.** $\frac{51}{4}$ **19.** $\frac{43}{10}$ **21.** $\frac{203}{100}$ **23.** $\frac{200}{3}$ **25.** $\frac{279}{50}$ **27.** $3\frac{3}{5}$ **29.** $4\frac{2}{3}$ **31.** $4\frac{1}{2}$ **33.** $5\frac{7}{10}$ **35.** $7\frac{4}{7}$ **37.** $7\frac{1}{2}$ **39.** $11\frac{1}{2}$ **41.** $1\frac{1}{2}$ **43.** $7\frac{57}{100}$ **45.** $43\frac{1}{8}$ **47.** $108\frac{5}{8}$ **49.** $618\frac{1}{5}$ **51.** $40\frac{4}{7}$ **53.** $55\frac{1}{51}$ **55.** 18 **56.** $\frac{5}{2}$ **57.** $\frac{1}{4}$ **58.** $\frac{2}{5}$ **59.** 24 **60.** 49 **61.** $\frac{2560}{3}$ **62.** $\frac{4}{3}$ **63.** ◆ **65.** $237\frac{19}{541}$ **67.** $8\frac{2}{3}$ **69.** $52\frac{2}{7}$

Margin Exercises, Section 3.5, pp. 163–166

1. $7\frac{2}{5}$ **2.** $12\frac{1}{10}$ **3.** $13\frac{7}{12}$ **4.** $1\frac{1}{2}$ **5.** $3\frac{1}{6}$ **6.** $3\frac{1}{3}$ **7.** $3\frac{2}{3}$ **8.** $17\frac{1}{12}$ yd **9.** $\frac{3}{8}$ in. **10.** $23\frac{1}{4}$ gal

Exercise Set 3.5, p. 167

1. $6\frac{1}{2}$ **3.** $2\frac{11}{12}$ **5.** $14\frac{7}{12}$ **7.** $12\frac{1}{10}$ **9.** $16\frac{5}{24}$ **11.** $21\frac{1}{2}$

13. $27\frac{7}{8}$ **15.** $27\frac{13}{24}$ **17.** $1\frac{3}{5}$ **19.** $4\frac{1}{10}$ **21.** $21\frac{17}{24}$ **23.** $12\frac{1}{4}$ **25.** $15\frac{3}{8}$ **27.** $7\frac{5}{12}$ **29.** $13\frac{3}{8}$ **31.** $11\frac{5}{18}$ **33.** $7\frac{5}{12}$ lb **35.** $6\frac{5}{12}$ in. **37.** $19\frac{1}{16}$ ft **39.** $95\frac{1}{5}$ mi **41.** $36\frac{1}{2}$ in. **43.** $\$17\frac{1}{8}$ **45.** $20\frac{1}{8}$ in. **47.** $78\frac{1}{12}$ in. **49.** $3\frac{4}{5}$ hr **51.** $28\frac{3}{4}$ yd **53.** $7\frac{3}{8}$ ft **55.** $1\frac{9}{16}$ in. **56.** 286 cartons; 2 oz left over **57.** 16 **58.** $\frac{1}{10}$ **59.** $\frac{10}{13}$ **61.** ◆ **63.** $8568\frac{786}{1189}$ **65.** $5\frac{3}{4}$ ft

Margin Exercises, Section 3.6, pp. 171–174

1. 20 **2.** $1\frac{7}{8}$ **3.** $12\frac{4}{5}$ **4.** $8\frac{1}{3}$ **5.** 16 **6.** $7\frac{3}{7}$ **7.** $1\frac{7}{8}$ **8.** $\frac{7}{10}$ **9.** $227\frac{1}{2}$ mi **10.** 20 mpg **11.** $240\frac{3}{4}$ ft^2

Exercise Set 3.6, p. 175

1. $22\frac{2}{3}$ **3.** $2\frac{5}{12}$ **5.** $8\frac{1}{6}$ **7.** $9\frac{31}{40}$ **9.** $24\frac{91}{100}$ **11.** $975\frac{4}{5}$ **13.** $6\frac{1}{4}$ **15.** $1\frac{1}{5}$ **17.** $3\frac{9}{16}$ **19.** $1\frac{1}{8}$ **21.** $1\frac{8}{43}$ **23.** $\frac{9}{40}$ **25.** 24 **27.** 7 oz **29.** $343\frac{3}{4}$ lb **31.** 1 chicken bouillon cube, $\frac{3}{4}$ cup hot water, $1\frac{1}{2}$ tbsp margarine, $1\frac{1}{2}$ tbsp flour, $1\frac{1}{4}$ cups diced cooked chicken, $\frac{1}{2}$ cup cooked peas, 2 oz sliced mushrooms (drained), $\frac{1}{6}$ cup sliced cooked carrots, $\frac{1}{8}$ cup chopped onion, 1 tbsp chopped pimiento, $\frac{1}{2}$ tsp salt; 6 chicken bouillon cubes, $4\frac{1}{2}$ cups hot water, 9 tbsp margarine, 9 tbsp flour, $7\frac{1}{2}$ cups diced cooked chicken, 3 cups cooked peas, 3 4-oz cans (12 oz) sliced mushrooms (drained), 1 cup sliced cooked carrots, $\frac{3}{4}$ cup chopped onion, 6 tbsp chopped pimiento, 3 tsp salt **33.** 68°F **35.** $82\frac{1}{2}$ in. **37.** 15 mpg **39.** 4 cu ft **41.** 16 **43.** $35\frac{115}{256}$ sq in. **45.** $59{,}538\frac{1}{8}$ sq ft **47.** 1,429,017 **48.** 45,800 **49.** 588 **50.** $\frac{2}{3}$ **51.** $\frac{1}{6}$ **52.** 45,770 **53.** ◆ **55.** $360\frac{60}{473}$ **57.** $35\frac{57}{64}$ **59.** $\frac{4}{9}$ **61.** $\frac{9}{5}$, or $1\frac{4}{5}$

Margin Exercises, Section 3.7, pp. 179–181

1. $\frac{1}{2}$ **2.** $\frac{3}{10}$ **3.** $20\frac{2}{3}$ **4.** $\frac{5}{9}$ **5.** $\frac{31}{40}$ **6.** $\frac{27}{56}$ **7.** 0 **8.** 1 **9.** $\frac{1}{2}$ **10.** 1 **11.** 12; answers may vary **12.** 32; answers may vary **13.** $22\frac{1}{2}$ **14.** 132 **15.** 37

Calculator Spotlight, p. 182

1. $\frac{5}{8}$ **2.** $\frac{7}{10}$ **3.** $\frac{5}{7}$ **4.** $\frac{133}{68}$ **5.** $\frac{73}{150}$ **6.** $\frac{97}{116}$ **7.** $24\frac{3}{7}$ **8.** $\frac{115}{147}$ **9.** $13\frac{41}{63}$ **10.** $5\frac{59}{72}$ **11.** $3\frac{5}{8}$ **12.** $19\frac{5}{8}$

Exercise Set 3.7, p. 183

1. $\frac{1}{24}$ **3.** $\frac{2}{5}$ **5.** $\frac{4}{7}$ **7.** $\frac{59}{30}$, or $1\frac{29}{30}$ **9.** $\frac{3}{20}$ **11.** $\frac{211}{8}$, or $26\frac{3}{8}$ **13.** $\frac{7}{16}$ **15.** $\frac{1}{36}$ **17.** $\frac{3}{8}$ **19.** $\frac{37}{48}$ **21.** $\frac{25}{72}$ **23.** $\frac{103}{16}$, or $6\frac{7}{16}$ **25.** $\frac{17}{6}$, or $2\frac{5}{6}$ **27.** $\frac{8395}{84}$, or $99\frac{79}{84}$ **29.** 0 **31.** 0 **33.** $\frac{1}{2}$ **35.** $\frac{1}{2}$ **37.** 0 **39.** 1 **41.** 6 **43.** 12 **45.** 19 **47.** 15 **49.** 6 **51.** 12 **53.** 16 **55.** 19 **57.** 3 **59.** 13 **61.** 2 **63.** 2 **65.** $\frac{1}{2}$ **67.** $271\frac{1}{2}$ **69.** 3 **71.** 100 **73.** $29\frac{1}{2}$ **75.** 3402 **76.** 1,038,180 **77.** 59 R 77 **78.** 42 **79.** $\frac{8}{3}$, or $2\frac{2}{3}$ **80.** Prime: 5, 7, 23, 43; composite: 9, 14; neither: 1 **81.** 16 **82.** 43 mg **83.** ◆ **85.** **(a)** $13 \cdot 9\frac{1}{4} + 8\frac{1}{4} \cdot 7\frac{1}{4}$; **(b)** $\frac{2881}{16}$, or $180\frac{1}{16}$ in^2; **(c)** Multiply before adding. **87.** $a = 2$; $b = 8$ **89.** The largest is $\frac{4}{3} + \frac{5}{2} = \frac{23}{6}$.

Summary and Review: Chapter 3, p. 187

1. 36 **2.** 90 **3.** 30 **4.** 1404 **5.** $\frac{63}{40}$ **6.** $\frac{19}{48}$ **7.** $\frac{29}{15}$ **8.** $\frac{7}{16}$ **9.** $\frac{1}{3}$ **10.** $\frac{1}{8}$ **11.** $\frac{5}{27}$ **12.** $\frac{11}{18}$ **13.** $>$ **14.** $>$ **15.** $\frac{19}{40}$ **16.** $\frac{2}{5}$ **17.** $\frac{15}{2}$ **18.** $\frac{67}{8}$ **19.** $\frac{13}{3}$ **20.** $\frac{75}{7}$ **21.** $2\frac{1}{3}$ **22.** $6\frac{3}{4}$ **23.** $12\frac{3}{5}$ **24.** $3\frac{1}{2}$ **25.** $877\frac{1}{3}$ **26.** $456\frac{5}{23}$ **27.** $10\frac{2}{5}$ **28.** $11\frac{11}{15}$ **29.** $10\frac{2}{3}$ **30.** $8\frac{1}{4}$ **31.** $7\frac{7}{9}$ **32.** $4\frac{11}{15}$ **33.** $4\frac{3}{20}$ **34.** $13\frac{3}{8}$ **35.** 16 **36.** $3\frac{1}{2}$ **37.** $2\frac{21}{50}$ **38.** 6 **39.** 12 **40.** $1\frac{7}{17}$ **41.** $\frac{1}{8}$ **42.** $\frac{9}{10}$ **43.** 15 **44.** $\$70\frac{3}{8}$ **45.** $1\frac{73}{100}$ in. **46.** $177\frac{3}{4}$ in^2 **47.** $50\frac{1}{4}$ in^2 **48.** $8\frac{3}{8}$ cups **49.** 1 **50.** $\frac{77}{240}$ **51.** $\frac{1}{2}$ **52.** 0 **53.** 1 **54.** 7 **55.** 10 **56.** 2 **57.** $28\frac{1}{2}$ **58.** $\frac{6}{5}$ **59.** $\frac{3}{2}$ **60.** 708,048 **61.** 17 days
62. ◆ It might be necessary to find the least common denominator before adding or subtracting. The least common denominator is the least common multiple of the denominators.
63. ◆ Suppose that a room has dimensions $15\frac{3}{4}$ ft by $28\frac{5}{8}$ ft. The equation $2 \cdot 15\frac{3}{4} + 2 \cdot 28\frac{5}{8} = 88\frac{3}{4}$ gives the perimeter of the room, in feet. Answers may vary.
64. 12 min

Test: Chapter 3, p. 189

1. [3.1a] 48 **2.** [3.2a] 3 **3.** [3.2b] $\frac{37}{24}$ **4.** [3.2b] $\frac{79}{100}$ **5.** [3.3a] $\frac{1}{3}$ **6.** [3.3a] $\frac{1}{12}$ **7.** [3.3a] $\frac{1}{12}$ **8.** [3.3b] $>$ **9.** [3.3c] $\frac{1}{4}$ **10.** [3.4a] $\frac{7}{2}$ **11.** [3.4a] $\frac{79}{8}$ **12.** [3.4b] $4\frac{1}{2}$ **13.** [3.4b] $8\frac{2}{9}$ **14.** [3.4c] $162\frac{7}{11}$ **15.** [3.5a] $14\frac{1}{5}$ **16.** [3.5a] $14\frac{5}{12}$ **17.** [3.5b] $4\frac{7}{24}$ **18.** [3.5b] $6\frac{1}{6}$ **19.** [3.6a] 39 **20.** [3.6a] $4\frac{1}{2}$ **21.** [3.6a] $5\frac{5}{6}$ **22.** [3.6b] 6 **23.** [3.6b] 2 **24.** [3.6b] $\frac{1}{36}$ **25.** [3.6c] $17\frac{1}{2}$ cups **26.** [3.6c] 80 **27.** [3.5c] $360\frac{5}{12}$ lb **28.** [3.5c] $2\frac{1}{2}$ in. **29.** [3.7a] $3\frac{1}{2}$ **30.** [3.7a] $\frac{11}{20}$ **31.** [3.7b] $\frac{1}{2}$ **32.** [3.7b] 0 **33.** [3.7b] 1 **34.** [3.7b] 4 **35.** [3.7b] $18\frac{1}{2}$ **36.** [3.7b] 16 **37.** [3.7b] $1214\frac{1}{2}$ **38.** [1.5b] 346,636 **39.** [2.7b] $\frac{8}{5}$ **40.** [2.6a] $\frac{10}{9}$ **41.** [1.8a] 535 bottles; 10 oz **42.** [3.1a] **(a)** 24, 48, 72; **(b)** 24 **43.** [3.3b], [3.5c] Dolores runs $\frac{17}{56}$ mi farther.

Cumulative Review: Chapters 1–3, p. 191

1. [1.1e] 5 **2.** [1.1a] 6 thousands + 7 tens + 5 ones **3.** [1.1c] Twenty-nine thousand, five hundred **4.** [1.2b] 899 **5.** [1.2b] 8982 **6.** [3.2b] $\frac{5}{12}$ **7.** [3.5a] $8\frac{1}{4}$ **8.** [1.3d] 5124 **9.** [1.3d] 4518

10. [3.3a] $\frac{5}{12}$ **11.** [3.5b] $1\frac{1}{6}$ **12.** [1.5b] 5004
13. [1.5b] 293,232 **14.** [2.6a] $\frac{3}{2}$ **15.** [2.6a] 15
16. [3.6a] $7\frac{1}{3}$ **17.** [1.6c] 715 **18.** [1.6c] 56 R 11
19. [3.4c] $56\frac{11}{45}$ **20.** [2.7b] $\frac{4}{7}$ **21.** [3.6b] $7\frac{1}{3}$
22. [1.4a] 38,500 **23.** [3.1a] 72 **24.** [2.2a] No
25. [2.1a] 1, 2, 4, 8, 16 **26.** [2.3b] $\frac{1}{4}$ **27.** [3.3b] $>$
28. [3.3b] $<$ **29.** [2.5b] $\frac{4}{5}$ **30.** [2.5b] 32
31. [3.4a] $\frac{37}{8}$ **32.** [3.4b] $5\frac{2}{3}$ **33.** [1.7b] 93
34. [3.3c] $\frac{5}{9}$ **35.** [2.7c] $\frac{12}{7}$ **36.** [1.7b] 905
37. [1.8a] $235 **38.** [1.8a] $108 **39.** [1.8a] 297 sq ft
40. [1.8a] 31 **41.** [2.6b] $\frac{2}{5}$ tsp **42.** [3.6c] 39 lb
43. [3.6c] 16 **44.** [3.2c] $\frac{33}{20}$ mi **45.** [3.7b] 1
46. [3.7b] $\frac{1}{2}$ **47.** [3.7b] 0 **48.** [3.7b] 30 **49.** [3.7b] 1
50. [3.7b] 42 **51.** [3.2b] **(a)** $\frac{1}{2}, \frac{2}{3}, \frac{3}{4}, \frac{4}{5}$; **(b)** $\frac{9}{10}$

Chapter 4

Pretest: Chapter 4, p. 194

1. [4.1a] Two and three hundred forty-seven thousandths **2.** [4.1a] Three thousand, two hundred sixty-four and $\frac{78}{100}$ dollars **3.** [4.1b] $\frac{21}{100}$
4. [4.1b] $\frac{5408}{1000}$ **5.** [4.1c] 0.379 **6.** [4.1c] 28.439
7. [4.1d] 3.2 **8.** [4.1d] 0.099 **9.** [4.1e] 21.0
10. [4.1e] 21.045 **11.** [4.2a] 607.219
12. [4.2b] 39.0901 **13.** [4.3a] 0.6179
14. [4.3a] 0.32456 **15.** [4.4a] 30.4
16. [4.4a] 0.57698 **17.** [4.4b] 84.26
18. [4.2c] 6345.157 **19.** [4.7a] 1081.6 mi
20. [4.7a] $285.95 **21.** [4.7a] $89.70
22. [4.7a] $3397.71 **23.** [4.6a] 224 **24.** [4.5a] $1.\overline{4}$
25. [4.5a] 0.925 **26.** [4.5a] 2.75 **27.** [4.5a] $4.\overline{142857}$
28. [4.5b] 4.1 **29.** [4.5b] 4.14 **30.** [4.5b] 4.143
31. [4.3b] $9.49 **32.** [4.3b] 490,000,000,000,000
33. [4.4c] 1548.8836 **34.** [4.5c] 58.17

Margin Exercises, Section 4.1, pp. 196–200

1. Twenty-one and one tenth **2.** Two and four thousand five hundred thirty-three ten-thousandths **3.** Two hundred forty-five and eighty-nine hundredths **4.** Thirty-one thousand, seventy-nine and seven hundred sixty-four thousandths **5.** Four thousand, two hundred seventeen and $\frac{56}{100}$ dollars **6.** Thirteen and $\frac{98}{100}$ dollars **7.** $\frac{896}{1000}$ **8.** $\frac{2378}{100}$ **9.** $\frac{56,789}{10,000}$ **10.** $\frac{19}{10}$
11. 7.43 **12.** 0.406 **13.** 6.7089 **14.** 0.9 **15.** 0.057
16. 0.083 **17.** 4.3 **18.** 283.71 **19.** 456.013
20. 2.04 **21.** 0.06 **22.** 0.58 **23.** 1 **24.** 0.8989
25. 21.05 **26.** 2.8 **27.** 13.9 **28.** 234.4 **29.** 7.0
30. 0.64 **31.** 7.83 **32.** 34.68 **33.** 0.03 **34.** 0.943
35. 8.004 **36.** 43.112 **37.** 37.401 **38.** 7459.355
39. 7459.35 **40.** 7459.4 **41.** 7459 **42.** 7460
43. 7500 **44.** 7000

Exercise Set 4.1, p. 201

1. Four hundred forty-nine and six hundredths **3.** One and five thousand five hundred ninety-nine ten-thousandths **5.** Thirty-four and eight hundred ninety-one thousandths **7.** Three hundred twenty-six and $\frac{48}{100}$ dollars **9.** Thirty-six and $\frac{72}{100}$ dollars
11. $\frac{83}{10}$ **13.** $\frac{356}{100}$ **15.** $\frac{4603}{100}$ **17.** $\frac{13}{100,000}$ **19.** $\frac{10,008}{10,000}$
21. $\frac{20,003}{1000}$ **23.** 0.8 **25.** 8.89 **27.** 3.798 **29.** 0.0078
31. 0.00019 **33.** 0.376193 **35.** 99.44 **37.** 3.798
39. 2.1739 **41.** 8.953073 **43.** 0.58 **45.** 0.91
47. 0.001 **49.** 235.07 **51.** $\frac{4}{100}$ **53.** 0.4325 **55.** 0.1
57. 0.5 **59.** 2.7 **61.** 123.7 **63.** 0.89 **65.** 0.67
67. 1.00 **69.** 0.09 **71.** 0.325 **73.** 17.002
75. 10.101 **77.** 9.999 **79.** 800 **81.** 809.473
83. 809 **85.** 34.5439 **87.** 34.54 **89.** 35 **91.** 6170
92. 6200 **93.** 6000 **94.** 830 **95.** $\frac{830}{1000}$, or $\frac{83}{100}$
96. 182 **97.** $\frac{182}{100}$, or $\frac{91}{50}$ **99.** ◆ **101.** 6.78346
103. 0.03030

Margin Exercises, Section 4.2, pp. 203–206

1. 10.917 **2.** 34.2079 **3.** 4.969 **4.** 3.5617
5. 9.40544 **6.** 912.67 **7.** 2514.773 **8.** 10.754
9. 0.339 **10.** 0.5345 **11.** 0.5172 **12.** 7.36992
13. 1194.22 **14.** 4.9911 **15.** 38.534 **16.** 14.164
17. 2133.5

Calculator Spotlight, p. 206

$8744.16 should be $8744.17; $8764.65 should be $8723.68; $8848.65 should be $8808.68; $8801.05 should be $8760.08; $8533.09 should be $8492.13

Exercise Set 4.2, p. 207

1. 334.37 **3.** 1576.215 **5.** 132.560 **7.** 84.417
9. 50.0248 **11.** 40.007 **13.** 771.967 **15.** 20.8649
17. 227.4680 **19.** 8754.8221 **21.** 1.3 **23.** 49.02
25. 45.61 **27.** 85.921 **29.** 2.4975 **31.** 3.397
33. 8.85 **35.** 3.37 **37.** 1.045 **39.** 3.703
41. 0.9902 **43.** 99.66 **45.** 4.88 **47.** 0.994
49. 17.802 **51.** 51.13 **53.** 2.491 **55.** 32.7386
57. 1.6666 **59.** 2344.90886 **61.** 11.65 **63.** 19.251
65. 384.68 **67.** 582.97 **69.** 15,335.3 **71.** 35,000
72. 34,000 **73.** $\frac{1}{6}$ **74.** $\frac{34}{45}$ **75.** 6166 **76.** 5366
77. $16\frac{1}{2}$ **78.** $60\frac{1}{5}$ mi **79.** ◆ **81.** 345.8

Margin Exercises, Section 4.3, pp. 212–215

1. 529.48 **2.** 5.0594 **3.** 34.2906 **4.** 0.348
5. 0.0348 **6.** 0.00348 **7.** 0.000348 **8.** 34.8
9. 348 **10.** 3480 **11.** 34,800 **12.** $938,000,000
13. $44,100,000,000 **14.** 1569¢ **15.** 17¢ **16.** $0.35
17. $5.77

Exercise Set 4.3, p. 217

1. 60.2 **3.** 6.72 **5.** 0.252 **7.** 0.522 **9.** 237.6 **11.** 583,686.852 **13.** 780 **15.** 8.923 **17.** 0.09768 **19.** 0.782 **21.** 521.6 **23.** 3.2472 **25.** 897.6 **27.** 322.07 **29.** 55.68 **31.** 3487.5 **33.** 50.0004 **35.** 114.42902 **37.** 13.284 **39.** 90.72 **41.** 0.0028728 **43.** 0.72523 **45.** 1.872115 **47.** 45,678 **49.** 2888¢ **51.** 66¢ **53.** $0.34 **55.** $34.45 **57.** $3,600,000,000 **59.** 196,800,000 **61.** $11\frac{1}{5}$ **62.** $\frac{35}{72}$ **63.** 342 **64.** 87 **65.** 4566 **66.** 1257 **67.** ◆ **69.** 10^{21}

Margin Exercises, Section 4.4, pp. 219–224

1. 0.6 **2.** 1.5 **3.** 0.47 **4.** 0.32 **5.** 3.75 **6.** 0.25 **7.** **(a)** 375; **(b)** 15 **8.** 4.9 **9.** 12.8 **10.** 15.625 **11.** 12.78 **12.** 0.001278 **13.** 0.09847 **14.** 67.832 **15.** 0.78314 **16.** 1105.6 **17.** 0.2426 **18.** 593.44 **19.** 1.2825 billion

Exercise Set 4.4, p. 225

1. 2.99 **3.** 23.78 **5.** 7.48 **7.** 7.2 **9.** 1.143 **11.** 4.041 **13.** 0.07 **15.** 70 **17.** 20 **19.** 0.4 **21.** 0.41 **23.** 8.5 **25.** 9.3 **27.** 0.625 **29.** 0.26 **31.** 15.625 **33.** 2.34 **35.** 0.47 **37.** 0.2134567 **39.** 21.34567 **41.** 1023.7 **43.** 9.3 **45.** 0.0090678 **47.** 45.6 **49.** 2107 **51.** 303.003 **53.** 446.208 **55.** 24.14 **57.** 13.0072 **59.** 19.3204 **61.** 473.188278 **63.** 10.49 **65.** 911.13 **67.** 205 **69.** $1288.36 **71.** 59.49° **73.** $15\frac{1}{8}$ **74.** $5\frac{7}{8}$ **75.** $\frac{6}{7}$ **76.** $2 \cdot 3 \cdot 3 \cdot 3 \cdot 3$ **77.** $2 \cdot 2 \cdot 3 \cdot 3 \cdot 19$ **78.** $\frac{7}{8}$ **79.** ◆ **81.** 6.254194585 **83.** 1000 **85.** 100

Margin Exercises, Section 4.5, pp. 229–232

1. 0.8 **2.** 0.45 **3.** 0.275 **4.** 1.32 **5.** 0.4 **6.** 0.375 **7.** $0.1\overline{6}$ **8.** $0.\overline{6}$ **9.** $0.\overline{45}$ **10.** $1.\overline{09}$ **11.** $0.\overline{428571}$ **12.** 0.7; 0.67; 0.667 **13.** 0.8; 0.81; 0.808 **14.** 6.2; 6.25; 6.245 **15.** 0.72 **16.** 0.552 **17.** 9.6575

Exercise Set 4.5, p. 233

1. 0.6 **3.** 0.325 **5.** 0.2 **7.** 0.85 **9.** 0.475 **11.** 0.975 **13.** 0.52 **15.** 20.016 **17.** 0.25 **19.** 0.575 **21.** 0.72 **23.** 1.1875 **25.** $0.2\overline{6}$ **27.** $0.\overline{3}$ **29.** $1.\overline{3}$ **31.** $1.1\overline{6}$ **33.** $0.\overline{571428}$ **35.** $0.91\overline{6}$ **37.** 0.3; 0.27; 0.267 **39.** 0.3; 0.33; 0.333 **41.** 1.3; 1.33; 1.333 **43.** 1.2; 1.17; 1.167 **45.** 0.6; 0.57; 0.571 **47.** 0.9; 0.92; 0.917 **49.** 0.2; 0.18; 0.182 **51.** 0.3; 0.28; 0.278 **53.** 11.06 **55.** 8.4 **57.** $417.51\overline{6}$ **59.** 0 **61.** 2.8125 **63.** 0.20425 **65.** 317.14 **67.** 0.1825 **69.** 18 **71.** 2.736 **73.** 21 **74.** 10 **75.** $3\frac{2}{5}$ **76.** $30\frac{7}{10}$ **77.** $1\frac{1}{24}$ cups **78.** $1\frac{73}{100}$ in. **79.** ◆ **81.** $0.\overline{142857}$ **83.** $0.\overline{428571}$ **85.** $0.\overline{714285}$ **87.** $0.\overline{1}$ **89.** $0.\overline{001}$

Margin Exercises, Section 4.6, pp. 235–237

1. (b) **2.** (a) **3.** (d) **4.** (b) **5.** (a) **6.** (d) **7.** (b) **8.** (c) **9.** (b) **10.** (b) **11.** (c) **12.** (a) **13.** (c) **14.** (c)

Calculator Spotlight, p. 238

1. **(a)** [+] [×]; **(b)** [+] [×] [−] **2.** $a = 5$, $b = 9$ **3.** $66.70, $77.82, $88.94, $100.06, $111.18, $122.30, $133.42 **4.** $2029.66, $1950.88, $1872.10, $1793.32

5.

4.55	1.3	1.95
0	2.6	5.2
3.25	3.9	0.65

Magic sum = 7.8

6.

2.16	0.81	1.08
0.27	1.35	2.43
1.62	1.89	0.54

Magic sum = 4.05

7.

6.16	43.12	34.72	16.24
38.64	12.32	9.52	39.76
15.12	34.16	44.24	6.72
40.32	10.64	11.76	37.52

Magic sum = 100.24

Exercise Set 4.6, p. 239

1. (d) **3.** (c) **5.** (a) **7.** (c) **9.** 1.6 **11.** 6 **13.** 60 **15.** 2.3 **17.** 180 **19.** (a) **21.** (c) **23.** (b) **25.** (b) **27.** 7700 **29.** $2 \cdot 2 \cdot 3 \cdot 3 \cdot 3$ **30.** $2 \cdot 2 \cdot 2 \cdot 2 \cdot 5 \cdot 5$ **31.** $5 \cdot 5 \cdot 13$ **32.** $2 \cdot 3 \cdot 3 \cdot 37$ **33.** $\frac{5}{16}$ **34.** $\frac{129}{251}$ **35.** $\frac{8}{9}$ **36.** $\frac{13}{25}$ **37.** ◆ **39.** Yes **41.** No

Margin Exercises, Section 4.7, pp. 241–248

1. 8.4° **2.** 148.1 gal **3.** $55.92 **4.** $368.75 **5.** 96.52 cm^2 **6.** $0.89 **7.** 28.6 miles per gallon **8.** $221,519 **9.** $594,444

Exercise Set 4.7, p. 249

1. $39.60 **3.** $21.22 **5.** $3.01 **7.** 102.8° **9.** $21,219.17 **11.** 250,205.04 ft^2 **13.** 22,691.5 mi **15.** 8.9 billion **17.** mach 0.3 **19.** 20.2 mpg **21.** 11.9752 cu ft **23.** $10 **25.** 78.1 cm **27.** 2.31 cm **29.** 876 calories **31.** $1171.74 **33.** 227.75 ft^2 **35.** 0.305 **37.** $57.35 **39.** $349.44 **41.** 5.8¢, or $0.058 **43.** 2152.56 yd^2 **45.** $316,987.20; $196,987.20 **47.** 31 million **49.** 90.6 million **51.** 1.4°F **53.** No **55.** $435,976 **57.** $87,494 **59.** $67,972 **61.** 6335 **62.** $\frac{31}{24}$ **63.** $6\frac{5}{6}$ **64.** $\frac{23}{15}$ **65.** 13,766 **66.** 2803 **67.** $\frac{1}{24}$ **68.** $1\frac{5}{6}$ **69.** $\frac{2}{15}$ **70.** 2432 **71.** 28 min **72.** $7\frac{1}{5}$ min **73.** ◆ **75.** $1.44

Summary and Review: Chapter 4, p. 255

1. Three and forty-seven hundredths **2.** Thirty-one thousandths **3.** Five hundred ninety-seven and $\frac{25}{100}$ dollars **4.** Zero and $\frac{96}{100}$ dollars **5.** $\frac{9}{100}$
6. $\frac{4561}{1000}$ **7.** $\frac{89}{1000}$ **8.** $\frac{30,227}{10,000}$ **9.** 0.034 **10.** 4.2603
11. 27.91 **12.** 867.006 **13.** 0.034 **14.** 0.91
15. 0.741 **16.** 1.041 **17.** 17.4 **18.** 17.43 **19.** 17.429
20. 17 **21.** 574.519 **22.** 0.6838 **23.** 229.1
24. 45.551 **25.** 29.2092 **26.** 790.29 **27.** 29.148
28. 70.7891 **29.** 12.96 **30.** 0.14442 **31.** 4.3
32. 0.02468 **33.** 7.5 **34.** 0.45 **35.** 45.2
36. 1.022 **37.** 0.2763 **38.** 1389.2 **39.** 496.2795
40. 6.95 **41.** 42.54 **42.** 4.9911 **43.** $7.76
44. $5888.74 **45.** 24.36; 104.4 **46.** $239.80
47. 11.16 **48.** 6365.1 bu **49.** 272 **50.** 216 **51.** 4
52. $125 **53.** 2.6 **54.** 1.28 **55.** 2.75 **56.** 3.25
57. $1.1\overline{6}$ **58.** $1.\overline{54}$ **59.** 1.5 **60.** 1.55 **61.** 1.545
62. $82.73 **63.** $4.87 **64.** 2493¢ **65.** 986¢
66. 5,500,000,000,000 **67.** 1,200,000 **68.** 1.8045
69. 57.1449 **70.** 15.6375 **71.** $41.537\overline{3}$ **72.** $43\frac{3}{4}$
73. $3\frac{3}{4}$ **74.** $19\frac{4}{5}$ **75.** $6\frac{3}{5}$ **76.** $\frac{1}{2}$
77. $2 \cdot 2 \cdot 2 \cdot 2 \cdot 2 \cdot 2 \cdot 3$
78. ◆ Multiply by 1 to get a denominator that is a power of 10:

$$\frac{44}{125} = \frac{44}{125} \cdot \frac{8}{8} = \frac{352}{1000} = 0.352.$$

We can also divide to find that $\frac{44}{125} = 0.352$.
79. ◆ Each decimal place in the decimal notation corresponds to one zero in the power of ten in the fractional notation. When the fractions are multiplied, the number of zeros in the denominator of the product is the sum of the number of zeros in the denominators of the factors. So the number of decimal places in the product is the sum of the number of decimal places in the factors.
80. **(a)** $2.56 \cdot 6.4 \div 51.2 - 17.4 + 89.7 = 72.62$;
(b) $(0.37 + 18.78) \cdot 2^{13} = 156,876.8$
81. $\frac{1}{3} + \frac{2}{3} = 0.33333333\ldots + 0.66666666 = 0.99999999\ldots$. Therefore, $1 = 0.99999999\ldots$ because $\frac{1}{3} + \frac{2}{3} = 1$. **82.** $2 = 1.\overline{9}$

Test: Chapter 4, p. 257

1. [4.1a] Two and thirty-four hundredths **2.** [4.1a] One thousand, two hundred thirty-four and $\frac{78}{100}$ dollars
3. [4.1b] $\frac{91}{100}$ **4.** [4.1b] $\frac{2769}{1000}$ **5.** [4.1c] 0.074
6. [4.1c] 3.7047 **7.** [4.1c] 756.09 **8.** [4.1c] 91.703
9. [4.1d] 0.162 **10.** [4.1d] 0.078 **11.** [4.1d] 0.9
12. [4.1e] 6 **13.** [4.1e] 5.68 **14.** [4.1e] 5.678
15. [4.1e] 5.7 **16.** [4.2a] 405.219 **17.** [4.2a] 0.7902
18. [4.2a] 186.5 **19.** [4.2a] 1033.23 **20.** [4.2b] 48.357
21. [4.2b] 19.0901 **22.** [4.2b] 1.9946
23. [4.2b] 152.8934 **24.** [4.3a] 0.03 **25.** [4.3a] 8
26. [4.3a] 0.21345 **27.** [4.3a] 73,962 **28.** [4.4a] 4.75
29. [4.4a] 0.24 **30.** [4.4a] 30.4 **31.** [4.4a] 0.19
32. [4.4a] 0.34689 **33.** [4.4a] 34,689 **34.** [4.4b] 84.26
35. [4.2c] 8.982 **36.** [4.7a] 4.97 km
37. [4.7a] $6572.45 **38.** [4.7a] $1675.50
39. [4.7a] $479.70 **40.** [4.6a] 198 **41.** [4.6a] 4
42. [4.5a] 1.6 **43.** [4.5a] 0.88 **44.** [4.5a] 5.25
45. [4.5a] 0.75 **46.** [4.5a] $1.\overline{2}$ **47.** [4.5a] $2.\overline{142857}$
48. [4.5b] 2.1 **49.** [4.5b] 2.14 **50.** [4.5b] 2.143
51. [4.3b] $9.49 **52.** [4.3b] $2,800,000,000
53. [4.4c] 40.0065 **54.** [4.4c] 384.8464
55. [4.5c] 302.4 **56.** [4.5c] $52.339\overline{4}$ **57.** [3.6a] 14
58. [3.5a] $2\frac{11}{16}$ **59.** [3.5b] $26\frac{1}{2}$ **60.** [3.6b] $1\frac{1}{8}$
61. [2.5b] $\frac{11}{18}$ **62.** [2.1d] $2 \cdot 2 \cdot 2 \cdot 3 \cdot 3 \cdot 5$
63. [4.7a] $35 **64.** [4.1c, d] $\frac{2}{3}, \frac{5}{7}, \frac{15}{19}, \frac{11}{13}, \frac{17}{20}, \frac{13}{15}$

Cumulative Review: Chapters 1–4, p. 259

1. [3.4a] $\frac{20}{9}$ **2.** [4.1b] $\frac{3052}{1000}$ **3.** [4.5a] 1.4
4. [4.5a] $0.5\overline{4}$ **5.** [2.1c] Prime **6.** [2.2a] Yes
7. [1.9c] 1754 **8.** [4.4c] 4.364 **9.** [4.1e] 584.90
10. [4.5b] 218.56 **11.** [4.6a] 160 **12.** [4.6a] 4
13. [1.5c] 12,800,000 **14.** [4.6a] 6 **15.** [3.5a] $6\frac{1}{20}$
16. [1.2b] 139,116 **17.** [3.2b] $\frac{31}{18}$ **18.** [4.2a] 145.953
19. [1.3d] 710,137 **20.** [4.2b] 13.097 **21.** [3.5b] $\frac{5}{7}$
22. [3.3a] $\frac{1}{110}$ **23.** [2.6a] $\frac{1}{6}$ **24.** [1.5b] 5,317,200
25. [4.3a] 4.78 **26.** [4.3a] 0.0279431 **27.** [4.4a] 2.122
28. [1.6c] 1843 **29.** [4.4a] 13,862.1 **30.** [2.7b] $\frac{5}{6}$
31. [4.2c] 0.78 **32.** [1.7b] 28 **33.** [4.4b] 8.62
34. [1.7b] 367,251 **35.** [3.3c] $\frac{1}{18}$ **36.** [2.7c] $\frac{1}{2}$
37. [1.8a] 14,274 **38.** [2.7d] $500 **39.** [1.8a] 86,400
40. [2.6b] $2400 **41.** [4.7a] $258.77 **42.** [3.5c] $6\frac{1}{2}$ lb
43. [3.2c] 2 lb **44.** [4.7a] 467.28 ft^2 **45.** [3.7a] $\frac{9}{32}$
46. [4.7a] $0.91 **47.** [3.6c] 144

Chapter 5

Pretest: Chapter 5, p. 262

1. [5.1a] $\frac{35}{43}$ **2.** [5.1a] $\frac{0.079}{1.043}$ **3.** [5.3b] 22.5
4. [5.3b] 0.75 **5.** [5.3b] $\frac{117}{14}$, or $8\frac{5}{14}$
6. [5.2a] 25.5 mi/gal **7.** [5.2a] $\frac{2}{9}$ qt/min
8. [5.2b] 5.79 ¢/oz **9.** [5.2b] Brand B
10. [5.4a] 1944 km **11.** [5.4a] 22
12. [5.4a] 12 min **13.** [5.4a] 393.75 mi
14. [5.5a] $x = 15$, $y = 9$

Margin Exercises, Section 5.1, pp. 263–264

1. $\frac{5}{11}$, or 5:11 **2.** $\frac{57.3}{86.1}$, or 57.3:86.1 **3.** $\frac{6\frac{3}{4}}{7\frac{2}{5}}$, or $6\frac{3}{4}:7\frac{2}{5}$
4. $\frac{21.1}{182.5}$ **5.** $\frac{2.5}{0.8}$ **6.** $\frac{4}{7\frac{2}{3}}$ **7.** $\frac{38.2}{56.1}$ **8.** 18 is to 27 as 2 is to 3 **9.** 3.6 is to 12 as 3 is to 10 **10.** 1.2 is to 1.5 as 4 is to 5 **11.** $\frac{3}{4}$

Exercise Set 5.1, p. 265

1. $\frac{4}{5}$ **3.** $\frac{178}{572}$ **5.** $\frac{0.4}{12}$ **7.** $\frac{3.8}{7.4}$ **9.** $\frac{56.78}{98.35}$ **11.** $\frac{8\frac{3}{4}}{9\frac{5}{6}}$

13. $\frac{1}{4}; \frac{3}{1}$ **15.** $\frac{4}{1}$ **17.** $\frac{478}{213}; \frac{213}{478}$ **19.** $\frac{2}{3}$ **21.** $\frac{3}{4}$
23. $\frac{12}{25}$ **25.** $\frac{7}{9}$ **27.** $\frac{2}{3}$ **29.** $\frac{14}{25}$ **31.** $\frac{1}{2}$ **33.** $\frac{3}{4}$ **35.** $\frac{27}{50}$
37. $\frac{32}{101}$ **39.** = **40.** ≠ **41.** 50 **42.** 9.5 **43.** 14.5
44. 152 **45.** $6\frac{7}{20}$ cm **46.** $17\frac{11}{20}$ cm **47.** ◆
49. 0.065584309 **51.** $\frac{30}{47}$ **53.** 1:2:3

Margin Exercises, Section 5.2, pp. 267–268

1. 5 mi/hr **2.** 12 mi/hr **3.** 0.3 mi/hr
4. 1100 ft/sec **5.** 4 ft/sec **6.** 14.5 ft/sec
7. 250 ft/sec **8.** 2 gal/day **9.** 20.64 ¢/oz **10.** Jar A

Exercise Set 5.2, p. 269

1. 40 km/h **3.** 11 m/sec **5.** 152 yd/day
7. 25 mi/hr; 0.04 hr/mi **9.** 57.5 ¢/min
11. 0.623 gal/ft^2 **13.** 186,000 mi/sec **15.** 124 km/h
17. 560 mi/hr **19.** $9.50/yd **21.** 20.59 ¢/oz
23. $4.34/lb **25.** A **27.** A **29.** A **31.** B
33. Four 12-oz bottles **35.** B **37.** Bert's
39. 1.7 million **40.** $37\frac{1}{2}$ **41.** 109.608 **42.** 67,819
43. 5833.56 **44.** 466,190.4 **45.** ◆ **47.** **(a)** 10.83¢, 10.91¢; **(b)** 1.14¢, 1.22¢ **49.** The unit price increases by 0.76 $\frac{¢}{oz}$.

Margin Exercises, Section 5.3, pp. 273–276

1. Yes **2.** No **3.** No **4.** Yes **5.** No **6.** Yes
7. 14 **8.** $11\frac{1}{4}$ **9.** 10.5 **10.** 2.64 **11.** 10.8
12. $\frac{125}{42}$, or $2\frac{41}{42}$

Calculator Spotlight, p. 276

1. $\frac{35}{3}$, 1.1, 4.8, 14.5 **2.** 14, 11.25, 10.5, 2.64, 10.8

Exercise Set 5.3, p. 277

1. No **3.** Yes **5.** Yes **7.** No **9.** 45 **11.** 12
13. 10 **15.** 20 **17.** 5 **19.** 18 **21.** 22 **23.** 28
25. $9\frac{1}{3}$ **27.** $2\frac{8}{9}$ **29.** 0.06 **31.** 5 **33.** 1 **35.** 1
37. 14 **39.** $2\frac{3}{16}$ **41.** $\frac{51}{16}$, or $3\frac{3}{16}$ **43.** 12.5725
45. $\frac{1748}{249}$, or $7\frac{5}{249}$ **47.** ≠ **48.** = **49.** ≠ **50.** ≠
51. 65 **52.** 39.5 **53.** 290.5 **54.** $1523.\overline{1}$ **55.** ◆
57. Approximately 2731.4

Margin Exercises, Section 5.4, pp. 279–283

1. 445 **2.** Approximately 14.1 gal **3.** 8
4. 38 in. or less **5.** 9.5 in. **6.** 2074

Exercise Set 5.4, p. 285

1. 702 mi **3.** $84.60 **5.** 8 ft **7.** $61\frac{2}{3}$ lb **9.** 171 gal
11. 309 **13.** 4 **15.** 6 gal **17.** 954 **19.** 100 oz
21. 1980 **23.** 58.1 km **25.** $7\frac{1}{3}$ in. **27.** No. The money will be gone in 30 weeks; $53.33 more.
29. 2.72 **31.** 5.28 **33.** **(a)** 33.7 in.; **(b)** 85.598 cm
34. **(a)** 2438 mi; **(b)** 3900.8 km **35.** ◆ **37.** 17
39. 7 gal

Margin Exercises, Section 5.5, pp. 289–290

1. 15 **2.** 24.75 ft **3.** 34.9 ft

Exercise Set 5.5, p. 291

1. 25 **3.** $\frac{4}{3}$, or $1\frac{1}{3}$ **5.** $x = \frac{27}{4}$, or $6\frac{3}{4}$; $y = 9$
7. $x = 7.5$; $y = 7.2$ **9.** 1.25 m **11.** 36 ft **13.** 7 ft
15. 100 ft **17.** 4 **19.** $10\frac{1}{2}$ **21.** $x = 6$, $y = 5.25$, $z = 3$ **23.** $x = 5\frac{1}{3}$, or $5.\overline{3}$; $y = 4\frac{2}{3}$ or $4.\overline{6}$; $z = 5\frac{1}{3}$, or $5.\overline{3}$ **25.** 20 ft **27.** 145 ft **29.** $59.81 **30.** 9.63
31. 679.4928 **32.** 2.74568 **33.** 27,456.8
34. 0.549136 **35.** ◆ **37.** 13.75 ft **39.** 1.034 cm
41. 3681.437 **43.** $x \approx 0.35$, $y = 0.4$

Summary and Review: Chapter 5, p. 295

1. $\frac{47}{84}$ **2.** $\frac{46}{1.27}$ **3.** $\frac{83}{100}$ **4.** $\frac{0.72}{197}$ **5.** $\frac{537{,}969}{2{,}312{,}203}$ **6.** $\frac{3}{4}$
7. $\frac{9}{16}$ **8.** 23.54 mi/hr **9.** 0.638 gal/ft^2
10. $25.36/kg **11.** 0.72 serving/lb **12.** 5.7 ¢/oz
13. 5.45 ¢/oz **14.** B **15.** B **16.** No **17.** No
18. 32 **19.** 7 **20.** $\frac{1}{40}$ **21.** 24 **22.** $4.45 **23.** 351
24. 832 mi **25.** 27 acres **26.** 2,990,000 kg
27. 6 in. **28.** Approximately 9906 **29.** $x = 20$, $y = 16$ **30.** $1630.40 **31.** $2799.60 **32.** = **33.** ≠
34. 10,672.74 **35.** 45.5 **36.** ◆ In terms of cost, a low faculty-to-student ratio is less expensive than a high faculty-to-student ratio. In terms of quality of education and student satisfaction, a high faculty-to-student ratio is more desirable. A college president must balance the cost and quality issues. **37.** ◆ Leslie used 4 gal of gasoline to drive 92 mi. At the same rate, how many gallons would be needed to travel 368 mi?
38. 105 min, or 1 hr, 45 min **39.** $x = 4258.5$, $z \approx 10{,}094.3$

Test: Chapter 5, p. 297

1. [5.1a] $\frac{85}{97}$ **2.** [5.1a] $\frac{0.34}{124}$ **3.** [5.1b] $\frac{9}{10}$ **4.** [5.1b] $\frac{25}{32}$
5. [5.2a] 0.625 ft/sec **6.** [5.2a] $1\frac{1}{3}$ servings/lb
7. [5.2b] 19.39 ¢/oz **8.** [5.2b] B **9.** [5.3a] Yes
10. [5.3a] No **11.** [5.3b] 12 **12.** [5.3b] 360
13. [5.4a] 1512 km **14.** [5.4a] 44 **15.** [5.4a] 4.8 min
16. [5.4a] 525 mi **17.** [5.5a] 66 m
18. [4.7a] 25.8 million lb **19.** [2.5c] ≠
20. [4.3a] 17,324.14 **21.** [4.4a] 0.9944
22. [5.4a] New York: 112; Portland: 92 **23.** [5.4a] 5888

Cumulative Review: Chapters 1–5, p. 299

1. [4.2a] 513.996 **2.** [3.5a] $6\frac{3}{4}$ **3.** [3.2b] $\frac{7}{20}$
4. [4.2b] 30.491 **5.** [4.2b] 72.912 **6.** [3.3a] $\frac{7}{60}$
7. [4.3a] 222.076 **8.** [4.3a] 567.8 **9.** [3.6a] 3
10. [4.4a] 43 **11.** [1.6c] 899 **12.** [2.7b] $\frac{3}{2}$
13. [1.1a] 3 ten thousands + 7 tens + 4 ones
14. [4.1a] One hundred twenty and seven hundredths
15. [4.1d] 0.7 **16.** [4.1d] 0.8
17. [2.1d] $2 \cdot 2 \cdot 2 \cdot 2 \cdot 3 \cdot 3$ **18.** [3.1a] 140

19. [2.3b] $\frac{5}{8}$ **20.** [2.5b] $\frac{5}{8}$ **21.** [4.5c] 5.718
22. [4.5c] 0.179 **23.** [5.1a] $\frac{0.3}{15}$ **24.** [5.3a] Yes
25. [5.2a] 55 m/sec **26.** [5.2b] The 14-oz jar
27. [5.3b] 30.24 **28.** [1.7b] 26.4375 **29.** [2.7c] $\frac{8}{9}$
30. [5.3b] 128 **31.** [4.2c] 33.34 **32.** [3.3c] $\frac{76}{175}$
33. [2.6b] 390 cal **34.** [1.8a] 178,666
35. [4.7a] 976.9 mi **36.** [5.4a] 7 min
37. [3.5c] $2\frac{1}{4}$ cups **38.** [2.7d] 12
39. [4.7a], [5.2a] 42.2025 mi **40.** [4.7a] 132
41. [5.1b] $\frac{10}{11}$ **42.** [5.2b] The 12-oz bag

Chapter 6

Pretest: Chapter 6, p. 302

1. [6.1b] 0.87 **2.** [6.1c] 53.7% **3.** [6.2a] 75%
4. [6.2b] $\frac{37}{100}$ **5.** [6.3a, b] $x = 60\% \times 75$; 45
6. [6.4a, b] $\frac{n}{100} = \frac{35}{50}$; 70% **7.** [6.5a] 90 lb
8. [6.6b] 20% **9.** [6.6a] \$14.30; \$300.30
10. [6.6b] \$5152 **11.** [6.6c] \$112.50 discount; \$337.50 sale price **12.** [6.7a] \$99.60 **13.** [6.7a] \$20
14. [6.7b] \$7128.60 **15.** [6.7b] \$7689.38

Margin Exercises, Section 6.1, pp. 304–306

1. $\frac{70}{100}$; $70 \times \frac{1}{100}$; 70×0.01 **2.** $\frac{23.4}{100}$; $23.4 \times \frac{1}{100}$; 23.4×0.01 **3.** $\frac{100}{100}$; $100 \times \frac{1}{100}$; 100×0.01 **4.** 0.34
5. 0.789 **6.** 0.83 **7.** 0.042 **8.** 24% **9.** 347%
10. 100% **11.** 90% **12.** 10.8%

Exercise Set 6.1, p. 307

1. $\frac{90}{100}$; $90 \times \frac{1}{100}$; 90×0.01 **3.** $\frac{12.5}{100}$; $12.5 \times \frac{1}{100}$; 12.5×0.01 **5.** 0.67 **7.** 0.456 **9.** 0.5901 **11.** 0.1
13. 0.01 **15.** 2 **17.** 0.001 **19.** 0.0009 **21.** 0.0018
23. 0.2319 **25.** 0.4 **27.** 0.025 **29.** 0.622 **31.** 47%
33. 3% **35.** 870% **37.** 33.4% **39.** 75% **41.** 40%
43. 0.6% **45.** 1.7% **47.** 27.18% **49.** 2.39%
51. 0.0104% **53.** 24% **55.** 58.1% **57.** $33\frac{1}{3}$
58. $37\frac{1}{2}$ **59.** $9\frac{3}{8}$ **60.** $18\frac{9}{16}$ **61.** $0.\overline{6}$ **62.** $0.\overline{3}$
63. $0.8\overline{3}$ **64.** $1.41\overline{6}$ **65.** ◆

Margin Exercises, Section 6.2, pp. 310–311

1. 25% **2.** 62.5%, or $62\frac{1}{2}\%$ **3.** $66.\overline{6}\%$, or $66\frac{2}{3}\%$
4. $83.\overline{3}\%$, or $83\frac{1}{3}\%$ **5.** 57% **6.** 76% **7.** $\frac{3}{5}$ **8.** $\frac{13}{400}$
9. $\frac{2}{3}$
10.

$\frac{1}{5}$	$\frac{5}{6}$	$\frac{3}{8}$
0.2	$0.8\overline{33}$	0.375
20%	$83.\overline{3}\%$, or $83\frac{1}{3}\%$	$37\frac{1}{2}\%$

Calculator Spotlight, p. 309

1. 52% **2.** 38.46% **3.** 107.69% **4.** 171.43%
5. 59.62% **6.** 28.31%

Calculator Spotlight, p. 312

1. 25.35; 1.68% **2.** 11.95; 0% **3.** 57.14; 2.5%
4. 39.41; 3.42%

Exercise Set 6.2, p. 313

1. 41% **3.** 5% **5.** 20% **7.** 30% **9.** 50%
11. 62.5%, or $62\frac{1}{2}\%$ **13.** 80% **15.** $66.\overline{6}\%$, or $66\frac{2}{3}\%$
17. $16.\overline{6}\%$, or $16\frac{2}{3}\%$ **19.** 16% **21.** 5% **23.** 34%
25. 36% **27.** 21% **29.** 24% **31.** $\frac{17}{20}$ **33.** $\frac{5}{8}$ **35.** $\frac{1}{3}$
37. $\frac{1}{6}$ **39.** $\frac{29}{400}$ **41.** $\frac{1}{125}$ **43.** $\frac{203}{800}$ **45.** $\frac{176}{225}$ **47.** $\frac{711}{1100}$
49. $\frac{3}{2}$ **51.** $\frac{13}{40{,}000}$ **53.** $\frac{1}{3}$ **55.** $\frac{11}{20}$ **57.** $\frac{19}{50}$ **59.** $\frac{11}{100}$
61. $\frac{1}{4}$ **63.** $\frac{569}{10{,}000}$
65.

Fractional Notation	Decimal Notation	Percent Notation
$\frac{1}{8}$	0.125	$12\frac{1}{2}\%$, or 12.5%
$\frac{1}{6}$	$0.1\overline{6}$	$16\frac{2}{3}\%$, or $16.\overline{6}\%$
$\frac{1}{5}$	0.2	20%
$\frac{1}{4}$	0.25	25%
$\frac{1}{3}$	$0.\overline{3}$	$33\frac{1}{3}\%$, or $33.\overline{3}\%$
$\frac{3}{8}$	0.375	$37\frac{1}{2}\%$, or 37.5%
$\frac{2}{5}$	0.4	40%
$\frac{1}{2}$	0.5	50%

67.

Fractional Notation	Decimal Notation	Percent Notation
$\frac{1}{2}$	0.5	50%
$\frac{1}{3}$	$0.\overline{3}$	$33\frac{1}{3}\%$, or $33.\overline{3}\%$
$\frac{1}{4}$	0.25	25%
$\frac{1}{6}$	$0.1\overline{6}$	$16\frac{2}{3}\%$, or $16.\overline{6}\%$
$\frac{1}{8}$	0.125	$12\frac{1}{2}\%$, or 12.5%
$\frac{3}{4}$	0.75	75%
$\frac{5}{6}$	$0.8\overline{3}$	$83\frac{1}{3}\%$, or $83.\overline{3}\%$
$\frac{3}{8}$	0.375	$37\frac{1}{2}\%$, or 37.5%

69. 70 **70.** 5 **71.** 400 **72.** 18.75 **73.** 23.125 **74.** 25.5 **75.** $33\frac{1}{3}$ **76.** $37\frac{1}{2}$ **77.** $83\frac{1}{3}$ **78.** $20\frac{1}{2}$ **79.** $43\frac{1}{8}$ **80.** $62\frac{1}{6}$ **81.** $18\frac{3}{4}$ **82.** $7\frac{4}{9}$ **83.** ◆ **85.** $11.\overline{1}\%$ **87.** $0.01\overline{5}$

Margin Exercises, Section 6.3, pp. 317–320

1. $12\% \times 50 = a$ **2.** $a = 40\% \times 60$ **3.** $45 = 20\% \times t$ **4.** $120\% \times y = 60$ **5.** $16 = n \times 40$ **6.** $b \times 84 = 10.5$ **7.** 6 **8.** $35.20 **9.** 225 **10.** $50 **11.** 40% **12.** 12.5%

Exercise Set 6.3, p. 321

1. $y = 32\% \times 78$ **3.** $89 = a \times 99$ **5.** $13 = 25\% \times y$ **7.** 234.6 **9.** 45 **11.** $18 **13.** 1.9 **15.** 78% **17.** 200% **19.** 50% **21.** 125% **23.** 40 **25.** $40 **27.** 88 **29.** 20 **31.** 6.25 **33.** $846.60 **35.** $\frac{9}{100}$ **36.** $\frac{179}{100}$ **37.** $\frac{875}{1000}$, or $\frac{7}{8}$ **38.** $\frac{9375}{10,000}$, or $\frac{15}{16}$ **39.** 0.89 **40.** 0.07 **41.** 0.3 **42.** 0.017 **43.** ◆ **45.** $880 (can vary); $843.20 **47.** 108 to 135 tons

Margin Exercises, Section 6.4, pp. 324–326

1. $\frac{12}{100} = \frac{a}{50}$ **2.** $\frac{40}{100} = \frac{a}{60}$ **3.** $\frac{130}{100} = \frac{a}{72}$ **4.** $\frac{20}{100} = \frac{45}{b}$ **5.** $\frac{120}{100} = \frac{60}{b}$ **6.** $\frac{N}{100} = \frac{16}{40}$ **7.** $\frac{N}{100} = \frac{10.5}{84}$ **8.** $225 **9.** 35.2 **10.** 6 **11.** 50 **12.** 30% **13.** 12.5%

Exercise Set 6.4, p. 327

1. $\frac{37}{100} = \frac{a}{74}$ **3.** $\frac{N}{100} = \frac{4.3}{5.9}$ **5.** $\frac{25}{100} = \frac{14}{b}$ **7.** 68.4 **9.** 462 **11.** 40 **13.** 2.88 **15.** 25% **17.** 102% **19.** 25% **21.** 93.75% **23.** $72 **25.** 90 **27.** 88 **29.** 20 **31.** 25 **33.** $780.20 **35.** 8 **36.** 4000 **37.** 8 **38.** 2074 **39.** 100 **40.** 15 **41.** $8.0\overline{4}$ **42.** $\frac{3}{16}$, or 0.1875 **43.** $\frac{43}{48}$ qt **44.** $\frac{1}{8}$ T **45.** ◆ **47.** $1134 (can vary); $1118.64

Margin Exercises, Section 6.5, pp. 329–334

1. 16% **2.** 50 **3.** 9% **4.** 4% **5.** $10,682

Calculator Spotlight, p. 330

1. 50 **2.** 7001.88 **3.** 83,931.456 **4.** 36,458.03724

Calculator Spotlight, p. 334

1. $10,682 **2.** $8918

Exercise Set 6.5, p. 335

1. 20; 100 **3.** 536; 264 **5.** 32.5%; 67.5% **7.** 20.4 mL; 659.6 mL **9.** 25% **11.** 166; 156; 146; 140; 122 **13.** 8% **15.** 20% **17.** $30,030 **19.** $12,600 **21.** 6.096 billion; 6.194 billion; 6.293 billion **23.** $36,400 **25.** $17.25; $39.10; $56.35 **27.** **(a)** 4.1%; **(b)** 31.0% **29.** 35.9% **31.** $2.\overline{27}$ **32.** $0.\overline{44}$ **33.** 3.375 **34.** $4.\overline{7}$ **35.** 0.92 **36.** $0.8\overline{3}$ **37.** 0.4375 **38.** 2.317 **39.** 3.4809 **40.** 0.675 **41.** ◆ **43.** 19% **45.** About 5 ft, 6 in. **47.** $83\frac{1}{3}\%$

Margin Exercises, Section 6.6, pp. 339–343

1. $53.52; $722.47 **2.** $1.03; $15.78 **3.** 6% **4.** $420 **5.** $5628 **6.** 25% **7.** $1675 **8.** $33.60; $106.40 **9.** 20%

Exercise Set 6.6, p. 345

1. $29.30; $615.30 **3.** $16.56; $281.56 **5.** 5% **7.** 4% **9.** $2000 **11.** $800 **13.** $711.55 **15.** 5.6% **17.** $2700 **19.** 5% **21.** $980 **23.** $5880 **25.** 12% **27.** $420 **29.** $30; $270 **31.** $2.55; $14.45 **33.** $125; $112.50 **35.** 40%; $360 **37.** $387; 30.4% **39.** $460; 18.043% **41.** 18 **42.** $\frac{22}{7}$ **43.** 265.625 **44.** 1.113 **45.** $0.\overline{5}$ **46.** $2.\overline{09}$ **47.** $0.91\overline{6}$ **48.** $1.\overline{857142}$ **49.** $2.\overline{142857}$ **50.** $1.58\overline{3}$ **51.** 4,030,000,000,000 **52.** 5,800,000 **53.** 42,700,000 **54.** 6,090,000,000,000 **55.** ◆ **57.** ◆ **59.** $2.69 **61.** He bought the plaques for $\$166\frac{2}{3} + \$250$, or $\$416\frac{2}{3}$, and sold them for $400, so he lost money.

Margin Exercises, Section 6.7, pp. 349–352

1. $602 **2.** $451.50 **3.** $27.62; $4827.62 **4.** $2464.20 **5.** $8103.38

Calculator Spotlight, p. 352

1. \$16,357.18 **2.** \$12,764.72

Exercise Set 6.7, p. 353

1. \$26 **3.** \$124 **5.** \$150.50 **7. (a)** \$147.95; **(b)** \$10,147.95 **9. (a)** \$128.22; **(b)** \$6628.22 **11. (a)** \$46.03; **(b)** \$5646.03 **13.** \$484 **15.** \$236.75 **17.** \$4284.90 **19.** \$2604.52 **21.** \$4101.01 **23.** \$1324.58 **25.** 4.5 **26.** $8\frac{3}{4}$ **27.** 8 **28.** 87.5 **29.** $33\frac{1}{3}$ **30.** $3\frac{13}{17}$ **31.** $12\frac{2}{3}$ **32.** $3\frac{5}{11}$ **33.** $\frac{18}{17}$ **34.** $\frac{209}{10}$ **35.** $\frac{203}{2}$ **36.** $\frac{259}{8}$ **37.** ◈ **39.** \$1434.53 **41.** \$33,981.98 **43.** 9.38% **45.** \$7883.24

Summary and Review: Chapter 6, p. 357

1. 48.3% **2.** 36% **3.** 37.5%, or $37\frac{1}{2}\%$ **4.** $33.\overline{3}\%$, or $33\frac{1}{3}\%$ **5.** 0.735 **6.** 0.065 **7.** $\frac{6}{25}$ **8.** $\frac{63}{1000}$ **9.** $30.6 = x\% \times 90$; 34% **10.** $63 = 84\% \times n$; 75 **11.** $y = 38\frac{1}{2}\% \times 168$; 64.68 **12.** $\frac{24}{100} = \frac{16.8}{b}$; 70 **13.** $\frac{42}{30} = \frac{N}{100}$; 140% **14.** $\frac{10.5}{100} = \frac{a}{84}$; 8.82 **15.** \$598 **16.** 12% **17.** 82,400 **18.** 20% **19.** 168 **20.** \$14.40 **21.** 5% **22.** 11% **23.** \$42; \$308 **24.** \$42.70; \$262.30 **25.** \$2940 **26.** \$36 **27. (a)** \$394.52; **(b)** \$24,394.52 **28.** \$121 **29.** \$7727.26 **30.** \$9504.80 **31.** Approximately 25% **32.** $\frac{56}{3}$ **33.** 42 **34.** 64 **35.** 7.6123 **36.** $3.\overline{6}$ **37.** $1.\overline{571428}$ **38.** $3\frac{2}{3}$ **39.** $17\frac{2}{7}$ **40.** ◈ No; the 10% discount was based on the original price rather than on the sale price. **41.** ◈ A 40% discount is better. When successive discounts are taken, each is based on the previous discounted price rather than on the original price. A 20% discount followed by a 22% discount is the same as a 37.6% discount off the original price. **42.** 19.5% increase **43.** $66\frac{2}{3}\%$ **44.** \$168

Test: Chapter 6, p. 359

1. [6.1b] 0.89 **2.** [6.1c] 67.4% **3.** [6.2a] 137.5% **4.** [6.2b] $\frac{13}{20}$ **5.** [6.3a, b] $a = 40\% \cdot 55$; 22 **6.** [6.4a, b] $\frac{N}{100} = \frac{65}{80}$; 81.25% **7.** [6.5a] 50 lb **8.** [6.5b] 140% **9.** [6.6a] \$16.20; \$340.20 **10.** [6.6b] \$630 **11.** [6.6c] \$40; \$160 **12.** [6.7a] \$8.52 **13.** [6.7a] \$5356 **14.** [6.7b] \$1102.50 **15.** [6.7b] \$10,226.69 **16.** [6.6c] \$131.95; 52.8% **17.** [4.4b] 222 **18.** [5.3b] 16 **19.** [4.5a] $1.41\overline{6}$ **20.** [3.4b] $3\frac{21}{44}$ **21.** [6.6b] \$117,800 **22.** [6.6b], [6.7b] \$2546.16 **23.** [6.2a], [6.5a] **(a)** #1 pays $\frac{1}{25}$, #2 pays $\frac{9}{100}$, #3 pays $\frac{47}{300}$, #4 pays $\frac{77}{300}$, #5 pays $\frac{137}{300}$; **(b)** 4%, 9%, $15\frac{2}{3}\%$, $25\frac{2}{3}\%$, $45\frac{2}{3}\%$; **(c)** 87%

Cumulative Review: Chapters 1–6, p. 361

1. [4.1b] $\frac{91}{1000}$ **2.** [4.5a] $2.1\overline{6}$ **3.** [6.1b] 0.03 **4.** [6.2a] 112.5% **5.** [5.1a] $\frac{10}{1}$ **6.** [5.2a] $23\frac{1}{3}$ km/h **7.** [3.3b] $<$ **8.** [3.3b] $<$ **9.** [4.6a] 296,200 **10.** [1.4b] 50,000 **11.** [1.9c, d] 13 **12.** [4.4c] 1.5 **13.** [3.5a] $3\frac{1}{30}$ **14.** [4.2a] 49.74 **15.** [1.2b] 515,150 **16.** [4.2b] 0.02 **17.** [3.5b] $\frac{2}{3}$ **18.** [3.3a] $\frac{2}{63}$ **19.** [2.4b] $\frac{1}{6}$ **20.** [1.5b] 853,142,400 **21.** [4.3a] 1.38036 **22.** [3.6b] $1\frac{1}{2}$ **23.** [4.4a] 12.25 **24.** [3.4c] $123\frac{1}{3}$ **25.** [1.7b] 95 **26.** [4.2c] 8.13 **27.** [2.7c] 9 **28.** [3.3c] $\frac{1}{12}$ **29.** [5.3b] 40 **30.** [5.3b] $8\frac{8}{21}$ **31.** [4.7a] \$6878.84 **32.** [4.7a] \$11.50 **33.** [5.2b] 31 ¢/oz **34.** [5.4a] 608 km **35.** [6.5b] 30% **36.** [6.5a] 1,033,710 mi^2 **37.** [1.8a] 1887 **38.** [3.6c] 5 **39.** [3.2c] $1\frac{1}{2}$ mi **40.** [2.6b], [5.4a] 60 mi **41.** [6.7b] \$8787.14 **42.** [6.5b], [6.7a] \$10,560; 5.6% **43.** [6.7a, b] First National Bank **44.** [3.6b], [6.5b] 12.5% increase

Chapter 7

Pretest: Chapter 7, p. 364

1. [7.1a, b, c] **(a)** 51; **(b)** 51.5; **(c)** no mode exists
2. [7.1a, b, c] **(a)** 3; **(b)** 3; **(c)** no mode exists
3. [7.1a, b, c] **(a)** 12.75; **(b)** 17; **(c)** 4
4. [7.1a] 55 mi/hr **5.** [7.1a] 76
6. [7.4b]

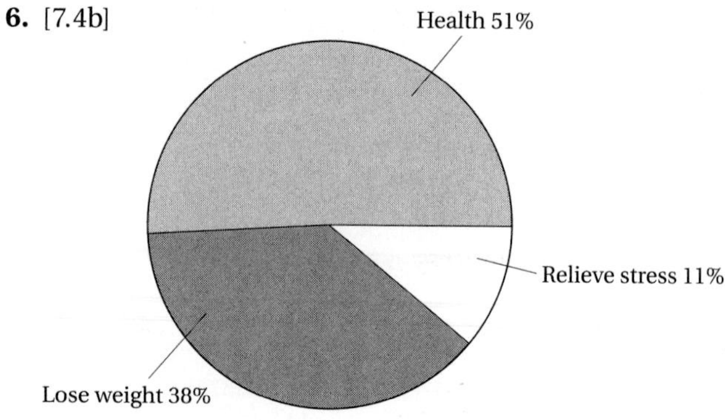

7. [7.2a] **(a)** \$208; **(b)** \$92
8. [7.3b]

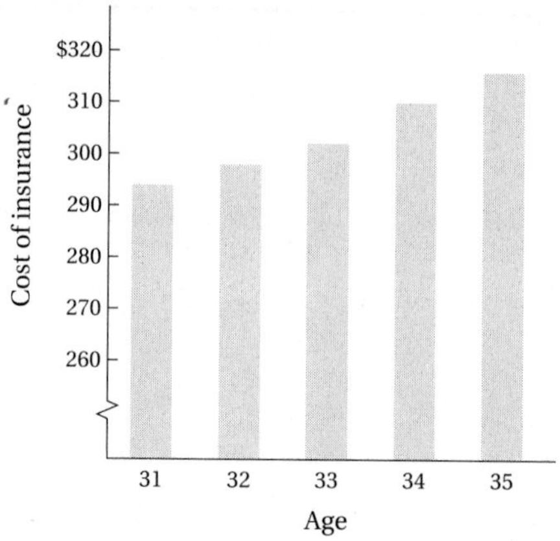

\$320
310
300
290
280
270
260
Cost of insurance
31 32 33 34 35
Age

9. [7.3d]

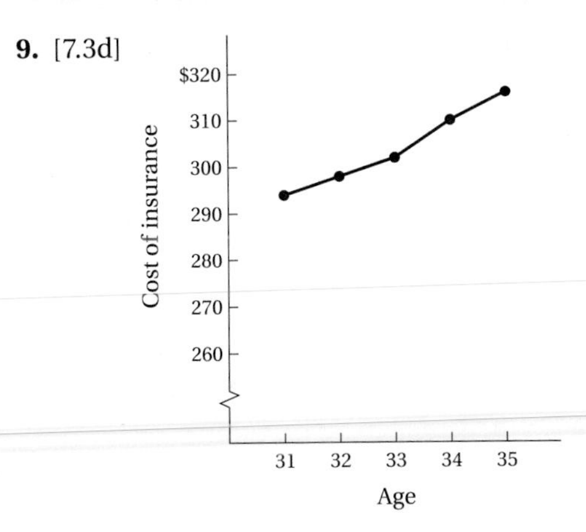

10. [7.3c] 260 **11.** [7.3c] 160
12. [7.3d], [7.5b]
(a)

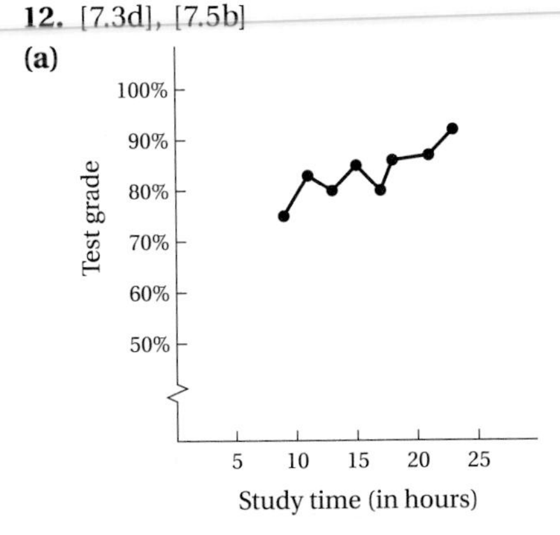

(b) Answers may vary; 94%

Margin Exercises, Section 7.1, pp. 367–370

1. 75 **2.** 54.9 **3.** 81 **4.** 19.4 **5.** 0.4 lb **6.** 19 mpg **7.** 2.5 **8.** 94 **9.** 17 **10.** 17 **11.** 91 **12.** $1700 **13.** 67.5 **14.** 45 **15.** 34, 67 **16.** No mode exists. **17.** **(a)** 17 g; **(b)** 18 g; **(c)** 19 g

Calculator Spotlight, p. 369

1. Only 79 would be divided by 3. The result would be $203\frac{1}{3}$. **2.** The answers are the same.

Exercise Set 7.1, p. 371

1. Average: 20; median: 18; mode: 29 **3.** Average: 20; median: 20; mode: 5, 20 **5.** Average: 5.2; median: 5.7; mode: 7.4 **7.** Average: 239.5; median: 234; mode: 234 **9.** Average: 40°; median: 40°; no mode exists **11.** 33 mpg **13.** 2.8 **15.** Average: $9.19; median: $9.49; mode: $7.99 **17.** 90 **19.** 263 days **21.** 196 **22.** $\frac{4}{9}$ **23.** 1.96 **24.** 1.999396 **25.** $1139.05 **26.** 3360 mi **27.** ◆ **29.** 182 **31.** 10

Margin Exercises, Section 7.2, pp. 373–377

1. Kellogg's Complete Bran Flakes **2.** Kellogg's Special K **3.** Kellogg's Special K, Wheaties **4.** Ralston Rice Chex, Kellogg's Complete Bran Flakes, Honey Nut Cheerios **5.** 510.66 mg **6.** 500 mg **7.** Mean: 232; median: 240; mode: 240 **8.** 20 **9.** Yes **10.** 440 mg **11.** 24% **12.** 6 g **13.** 60,000 **14.** Two and one half times as many in Zimbabwe as in Cameroon **15.** 55,000 **16.** 795; answers may vary **17.** 750; answers may vary **18.** 1830; answers may vary

19. Total Gross Revenue

Jimmy Buffett
Reba McEntire
Alanis Morissette
Hootie & the Blowfish
Ozzy Osbourne
= $10,000,000

Exercise Set 7.2, p. 379

1. 483,612,200 mi **3.** Neptune **5.** All **7.** 11 **9.** Average: $27{,}884.\overline{1}$ mi; median: 7926 mi; no mode exists **11.** 92° **13.** 108° **15.** 3 **17.** 90° and higher **19.** 30% and higher **21.** 50% **23.** 59.29°, 59.58°; 0.5% **25.** 59.50°; 59.62°; 0.12° **27.** 1.0 billion **29.** 1999 **31.** 1650 and 1850 **33.** 2.0 billion; 50% **35.** 1998 **37.** 1994 and 1995 **39.** 7000 **41.** 1997

43.

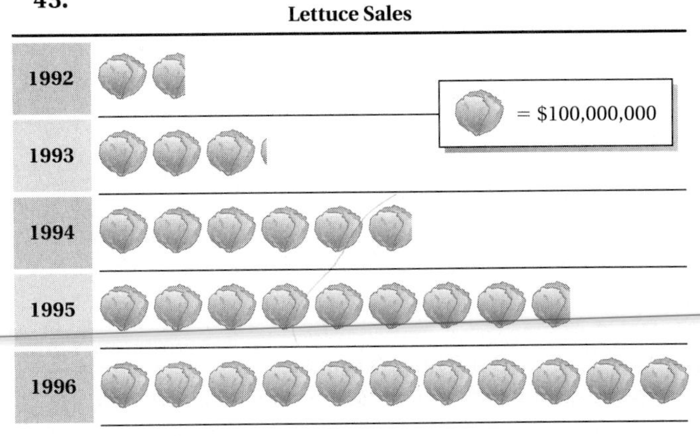

44. 12 **45.** 27,859.5 mi^2 **46.** $\frac{24}{100}$, or $\frac{6}{25}$ **47.** $\frac{4.8}{100}$, or $\frac{48}{1000}$, or $\frac{6}{125}$ **49.** ◈
50.

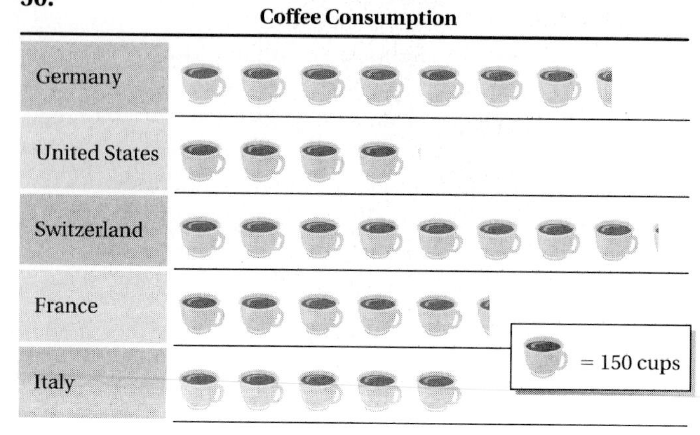

Margin Exercises, Section 7.3, pp. 385–390

1. 16 g **2.** Big Bacon Classic **3.** Single with Everything, chicken club, Big Bacon Classic **4.** 60 **5.** 85+ **6.** 60–64 **7.** Yes
8.

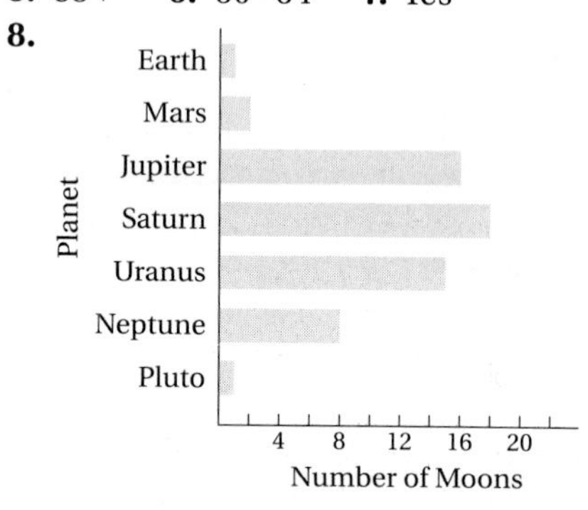

9. Month 7 **10.** Months 1 and 2, 4 and 5, 6 and 7, 11 and 12 **11.** Months 2, 5, 6, 7, 8, 9, 12 **12.** \$920 **13.** 40 yr **14.** \$1300
15.

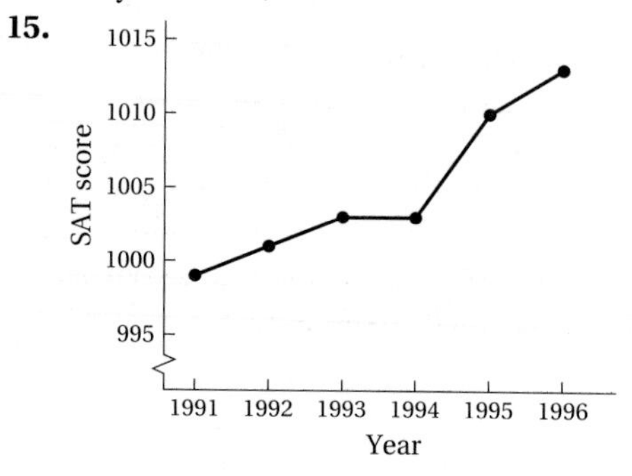

Exercise Set 7.3, p. 391

1. 190 **3.** 1 slice of chocolate cake with fudge frosting **5.** 1 cup premium chocolate ice cream **7.** 120 calories **9.** 920 calories **11.** 28 lb **13.** 920,000 hectares **15.** Latin America **17.** Africa **19.** 880,000 hectares

21.

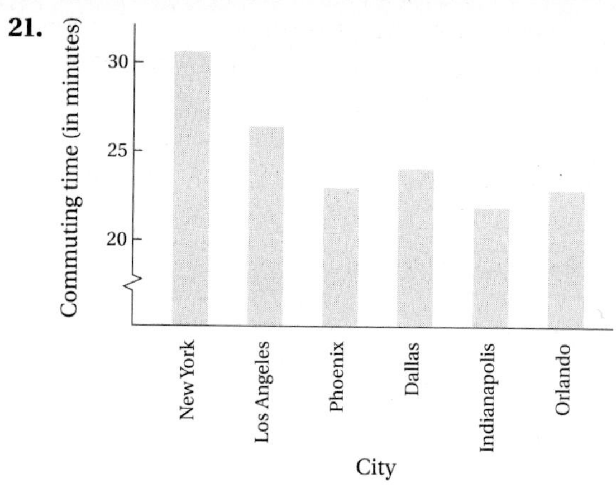

23. Indianapolis **25.** 23.55 min
26.

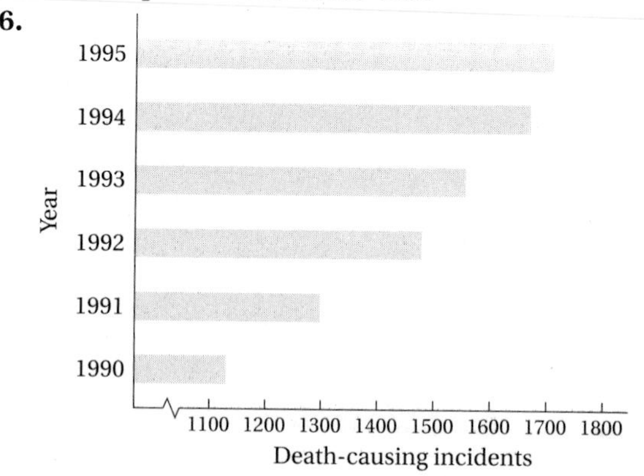

27. 1991 and 1992 **29.** $1472.\overline{6}$ **31.** 1997 **33.** About \$0.5 million **35.** 1994 and 1995
37.

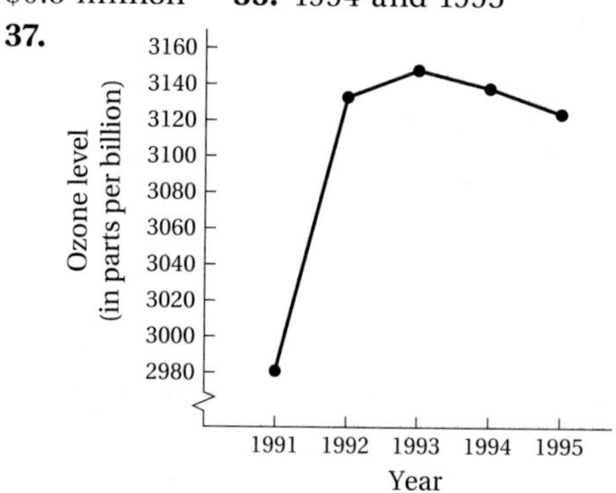

39. 1994 and 1995
41. 3133 parts per billion

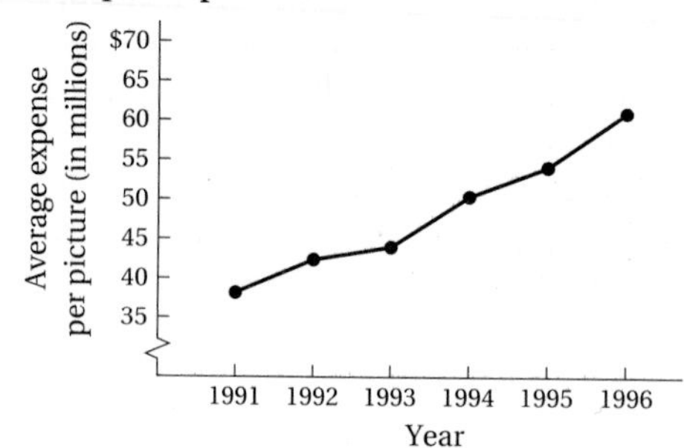

43. 1993 and 1994 **45.** $48.35 million **47.** $41.5 million **49.** 18 min **50.** 18% **51.** 82.5 **52.** $66\frac{2}{3}\%$ **53.** ◈ **55.** $65.6

Margin Exercises, Section 7.4, pp. 397–398

1. Spaying **2.** 5% **3.** $1122 **4.** 8% + 3%, or 11%
5.

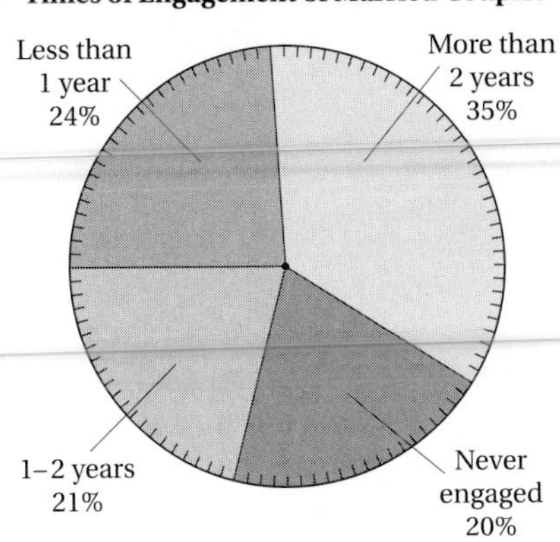

Exercise Set 7.4, p. 399

1. 3.7% **3.** 270 **5.** 6.8% **7.** Food **9.** 14%
11.

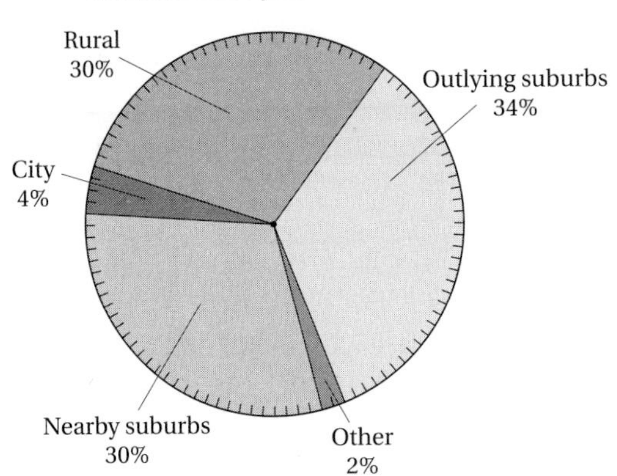

13.

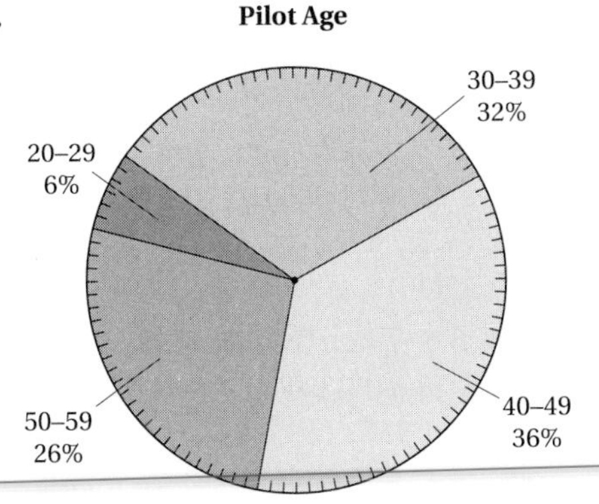

15.

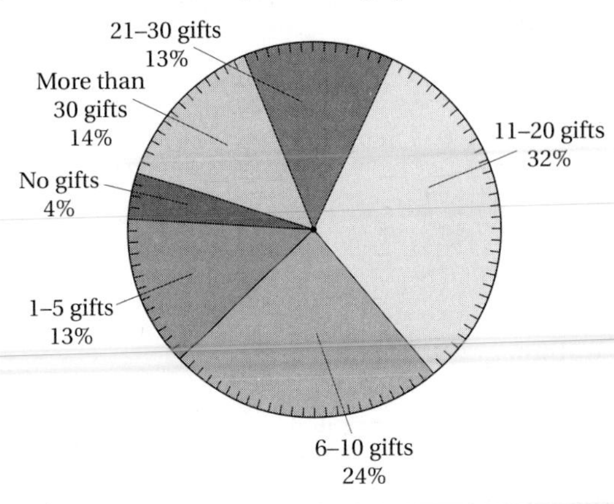

17. 300.6 **18.** 25% **19.** 115

Margin Exercises, Section 7.5, pp. 401–403

1. Wheat A: average stalk height ≈ 25.21 in.; wheat B: average stalk height ≈ 22.54 in.; wheat B is better. **2.** 111 million **3.** 94%

Exercise Set 7.5, p. 405

1. Bulb A: average time = 1171.25 hr; bulb B: average time ≈ 1251.58 hr; bulb B is better **3.** 83 **5.** 3112 parts per billion **7.** $148.8 billion **9.** $128,727 **10.** 0.625 cubic centimeter **11.** $1378.85 **12.** 5040 mi **13.** $\frac{15}{7}$, or $2\frac{1}{7}$ **14.** $\frac{1408}{3}$, or $469\frac{1}{3}$ **15.** $\frac{17}{25{,}000}$ **16.** $\frac{11}{12}$ **17.** ◈

Summary and Review: Chapter 7, p. 407

1. $21.00 **2.** $28.75 **3.** $10.00 **4.** $10.50 **5.** No **6.** $12.00 **7.** 5500 **8.** 1997 **9.** 1999 **10.** 4075 **11.** 38.5 **12.** 13.4 **13.** 1.55 **14.** 1840 **15.** $\$16.\overline{6}$ **16.** $321.\overline{6}$ **17.** 38.5 **18.** 14 **19.** 1.8 **20.** 1900 **21.** $17 **22.** 375 **23.** 26 **24.** 11; 17 **25.** 0.2 **26.** 700; 800 **27.** $17 **28.** 20 **29.** $110.50; $107 **30.** $66.1\overline{6}°$ **31.** 96 **32.** 420 **33.** 440 **34.** Big Bacon Classic **35.** Grilled chicken **36.** Plain single **37.** Chicken club **38.** 80 **39.** 250 **40.** Under 20 **41.** 12 **42.** 13 **43.** 45–74 **44.** 11 **45.** Under 20 **46.** 22% **47.** 11% **48.** 1600 **49.** 25%
50.

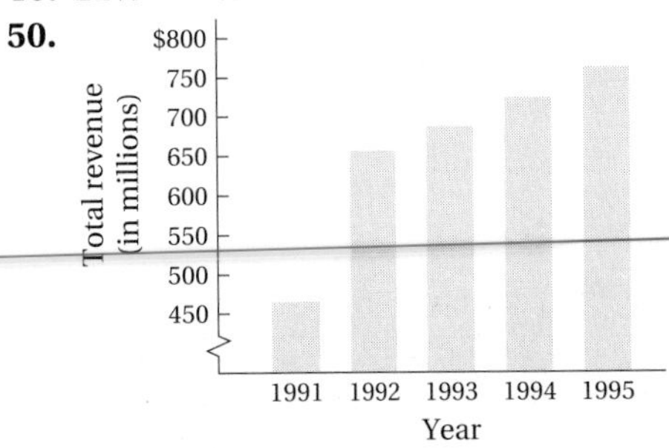

51.

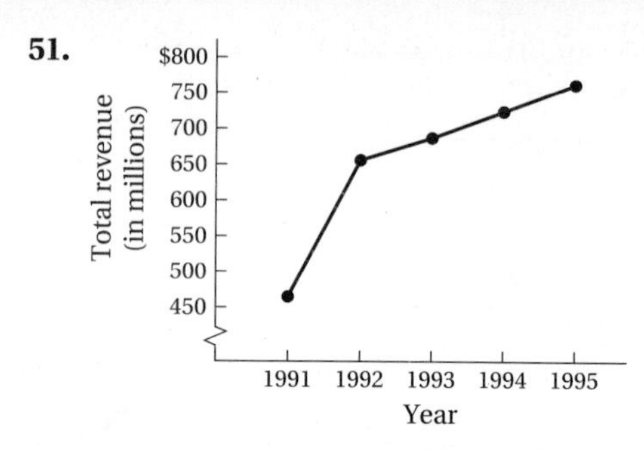

52. $800 million **53.** $806 million **54.** Battery A: average ≈ 43.04 hr; battery B: average = 41.55 hr; battery A is better. **55.** 12,600 mi **56.** 5.428 billion **57.** $222.\overline{2}\%$ **58.** 50% **59.** $\frac{9}{10}$ **60.** $\frac{5}{12}$ **61.** ◆ The average, the median, and the mode are "center points" that characterize a set of data. You might use the average to find a center point that is midway between the extreme values of the data. The median is a center point that is in the middle of all the data. That is, there are as many values less than the median than there are values greater than the median. The mode is a center point that represents the value or values that occur most frequently. **62.** ◆ The equation could represent a person's average income during a 4-yr period. Answers may vary. **63.** $a = 316$, $b = 349$

Test: Chapter 7, p. 411

1. [7.2a] $341,413 **2.** [7.2a] $2558 **3.** [7.2a] Couple, age 65; yearly income $100,000 **4.** [7.2a] Single female, age 45; yearly income $150,000
5. [7.2a] $78,169 **6.** [7.2a] $2264 **7.** [7.2b] 2003
8. [7.2b] 2002 and 2003 **9.** [7.2b] 7000
10. [7.2b] 2002 **11.** [7.1a] 50 **12.** [7.1a] 3
13. [7.1a] 15.5 **14.** [7.1b, c] Median: 50.5; no mode exists **15.** [7.1b, c] Median: 3; no mode exists
16. [7.1b, c] Median: 17.5; mode: 17, 18
17. [7.1a] 58 km/h **18.** [7.1a] 76
19. [7.3c] $6.5 billion **20.** [7.3c] $4.4 billion
21. [7.3c] $3.1 billion **22.** [7.3c] 1996
23. [7.3c] $3.9 billion **24.** [7.5b] $7.9 billion
25. [7.3b]

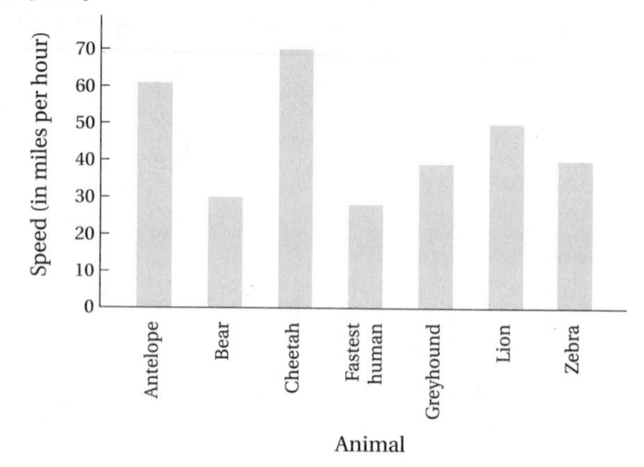

26. [7.3a] 42 mph **27.** [7.3a] No. The maximum speed of the fastest human is 22 mph slower than that of the lion. **28.** [7.3a] About 45.4 mph **29.** [7.3a] 40 mph
30. [7.4b]

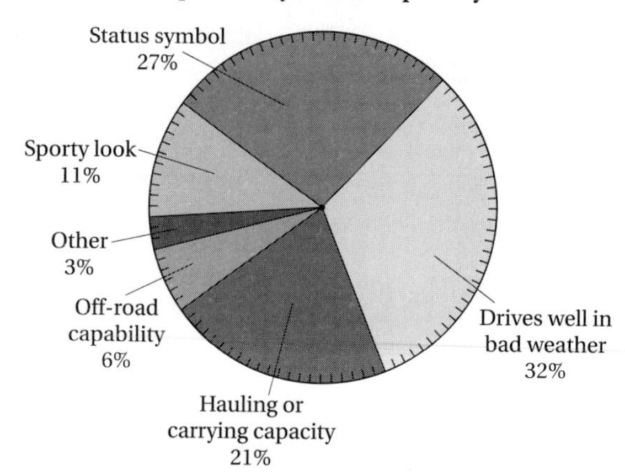

31. [7.3d]

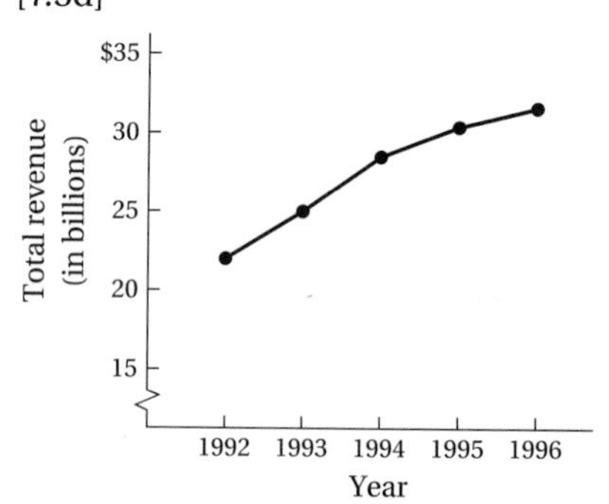

32. [7.5b] $34.4 billion **33.** [7.5a] Bar A: average ≈ 8.417; bar B: average ≈ 8.417; equal quality.
34. [2.7b] $\frac{25}{4}$, or $6\frac{1}{4}$ **35.** [6.3b], [6.4b] 68
36. [6.5a] 15,600 **37.** [5.4a] 340 **38.** [7.1a, b] $a = 74$, $b = 111$

Cumulative Review: Chapters 1–7, p. 415

1. [1.1e] 5 hundreds **2.** [1.9c] 128 **3.** [2.1a] 1, 2, 3, 4, 5, 6, 10, 12, 15, 20, 30, 60 **4.** [4.1e] 52.0 **5.** [3.4a] $\frac{33}{10}$
6. [4.3b] $2.10 **7.** [4.3b] $3,250,000,000 **8.** [5.3a] No
9. [3.5a] $6\frac{7}{10}$ **10.** [4.2a] 44.6351 **11.** [3.3a] $\frac{1}{3}$
12. [4.2b] 325.43 **13.** [3.6a] 15 **14.** [1.5b] 2,740,320
15. [2.7b] $\frac{9}{10}$ **16.** [3.4c] $4361\frac{1}{2}$ **17.** [5.3b] $9\frac{3}{5}$
18. [2.7c] $\frac{3}{4}$ **19.** [4.4b] 6.8 **20.** [1.7b] 15,312
21. [5.2b] 20.6¢/oz **22.** [6.5a] 3324 **23.** [4.7a] 6.2 lb
24. [3.6c] $\frac{1}{4}$ yd **25.** [1.8a] 2572 billion
26. [2.6b] $\frac{3}{8}$ cup **27.** [7.1a, b] Average: $27.57; median: $27.30

28. [7.3b]

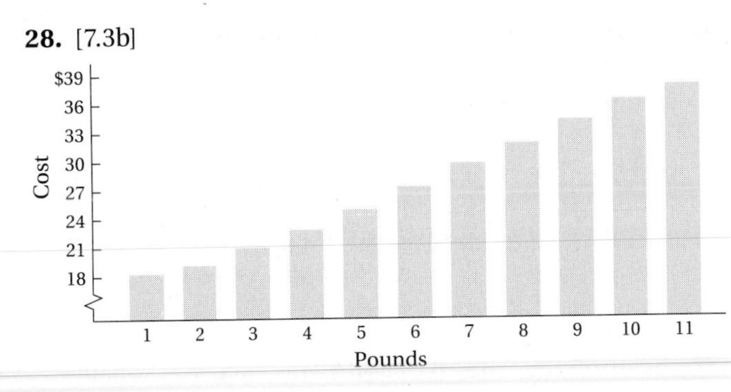

29. [7.3d]

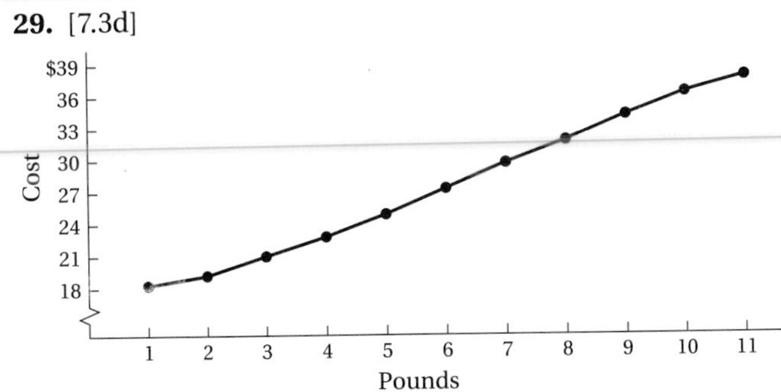

30. [7.5b] $39.60 **31.** [3.2c], [3.3d] $\frac{1}{4}$ **32.** [5.4a] 1122 **33.** [4.7a] $9.55 **34.** [6.6b] 7% **35.** [7.4a] 56% **36.** [7.4a] 15 **37.** [7.1c] 5 **38.** [6.5b], [7.1a] 12% decrease

Chapter 8

Pretest: Chapter 8, p. 418

1. [8.1a] 96 **2.** [8.1a] $\frac{5}{12}$ **3.** [8.2a] 8460 **4.** [8.2a] 0.92 **5.** [8.3a] 131 mm **6.** [8.4a] 100 ft^2 **7.** [8.5a] 22 cm^2 **8.** [8.5a] $32\frac{1}{2}$ ft^2 **9.** [8.5a] 4 m^2 **10.** [8.6a] 9.6 m **11.** [8.6b] 30.144 m **12.** [8.6c] 72.3456 m^2 **13.** [8.4b] 92 in^2 **14.** [8.7a] 9 **15.** [8.7b] 9.849 **16.** [8.7c] $c = 20$ **17.** [8.7c] $b = \sqrt{45}$; $b \approx 6.708$

Margin Exercises, Section 8.1, pp. 419–421

1. 2 **2.** 3 **3.** $1\frac{1}{2}$ **4.** $2\frac{1}{2}$ **5.** 288 **6.** 43.5 **7.** 240,768 **8.** 6 **9.** $1\frac{5}{12}$ **10.** 8 **11.** $11\frac{2}{3}$, or $11.\overline{6}$ **12.** 5 **13.** 0.5 **14.** 288

Exercise Set 8.1, p. 423

1. 12 **3.** $\frac{1}{12}$ **5.** 5280 **7.** 108 **9.** 7 **11.** $1\frac{1}{2}$ **13.** 26,400 **15.** 3 **17.** $3\frac{1}{3}$ **19.** 52,800 **21.** $1\frac{1}{2}$ **23.** 1 **25.** 110 **27.** 2 **29.** 300 **31.** 30 **33.** $\frac{1}{36}$ **35.** 126,720 **37.** $\frac{37}{400}$ **38.** $\frac{7}{8}$ **39.** $\frac{275}{1000}$, or $\frac{11}{40}$ **40.** $\frac{57}{100}$ **41.** 137.5% **42.** $66.\overline{6}$% **43.** 25% **44.** 43.75% **45.** $\frac{361}{10,000}$; $\frac{10,000}{361}$ **46.** $\frac{1}{6}$ **47.** ◆ **49.** 0.0041 in.

Margin Exercises, Section 8.2, pp. 427–430

1. cm **2.** km **3.** mm **4.** m **5.** cm **6.** m **7.** 23,000 **8.** 400 **9.** 178 **10.** 9040 **11.** 7.814 **12.** 781.4 **13.** 0.967 **14.** 8,900,000 **15.** 6.78 **16.** 97.4 **17.** 0.1 **18.** 8.451 **19.** 90.909 **20.** 804.5 **21.** 1479.843

Exercise Set 8.2, p. 431

1. **(a)** 1000; **(b)** 0.001 **3.** **(a)** 10; **(b)** 0.1 **5.** **(a)** 0.01; **(b)** 100 **7.** 6700 **9.** 0.98 **11.** 8.921 **13.** 0.05666 **15.** 566,600 **17.** 4.77 **19.** 688 **21.** 0.1 **23.** 100,000 **25.** 142 **27.** 0.82 **29.** 450 **31.** 0.000024 **33.** 0.688 **35.** 230 **37.** 3.92 **39.** 100 **41.** 727.409592 **43.** 104.585 **45.** 289.62 **47.** 112.63 **49.** 9.09 **51.** 1.75 **52.** 2.34 **53.** 0.234 **54.** 0.0234 **55.** 13.85 **56.** $80\frac{1}{2}$ **57.** 4 **58.** 0.9 **59.** 0.108 **61.** ◆ **63.** 393,700 **65.** 22.7 mph

Margin Exercises, Section 8.3, pp. 433–434

1. 26 cm **2.** 46 in. **3.** 12 cm **4.** 17.5 yd **5.** 32 km **6.** 40 km **7.** 21 yd **8.** 31.2 km **9.** 70 ft; $346.50

Exercise Set 8.3, p. 435

1. 17 mm **3.** 15.25 in. **5.** 13 m **7.** 30 ft **9.** 79.14 cm **11.** 88 ft **13.** 182 mm **15.** 826 m; $1197.70 **17.** 122 cm **19.** **(a)** 228 ft; **(b)** $1046.52 **21.** 0.561 **22.** 67.34% **23.** 112.5%, or $112\frac{1}{2}$% **24.** 25 **25.** 100 **26.** 961 **27.** 4,700,000 **28.** 4,300,000,000 **29.** ◆ **31.** 9 ft

Margin Exercises, Section 8.4, pp. 437–438

1. 8 cm^2 **2.** 56 km^2 **3.** $18\frac{3}{8}$ yd^2 **4.** 144 km^2 **5.** 118.81 m^2 **6.** $12\frac{1}{4}$ yd^2 **7.** 659.75 m^2

Exercise Set 8.4, p. 439

1. 15 km^2 **3.** 1.4 in^2 **5.** $6\frac{1}{4}$ yd^2 **7.** 8100 ft^2 **9.** 50 ft^2 **11.** 169.883 cm^2 **13.** $41\frac{2}{9}$ in^2 **15.** 484 ft^2 **17.** 3237.61 km^2 **19.** $28\frac{57}{64}$ yd^2 **21.** 1197 m^2 **23.** 630.36 m^2 **25.** **(a)** 819.75 ft^2; **(b)** 10 gal; **(c)** $179.50 **27.** 80 cm^2 **29.** 45.2% **30.** $33\frac{1}{3}$%, or $33.\overline{3}$% **31.** 55% **32.** 88% **33.** $\frac{27}{131}$; $\frac{131}{27}$ **34.** $\frac{1}{4}$; $\frac{3}{4}$ **35.** ◆ **37.** 16,914 in^2

Margin Exercises, Section 8.5, pp. 442–444

1. 43.8 cm^2 **2.** 12.375 km^2 **3.** 96 m^2 **4.** 18.7 cm^2 **5.** 100 m^2 **6.** 88 cm^2 **7.** 54 m^2

Exercise Set 8.5, p. 445

1. 32 cm^2 **3.** 60 in^2 **5.** 104 ft^2 **7.** 45.5 in^2 **9.** 8.05 cm^2 **11.** 297 cm^2 **13.** 7 m^2 **15.** 675 cm^2 **17.** 8944 in^2 **19.** 852.04 ft^2 **21.** $\frac{7}{20}$ **22.** $\frac{171}{200}$ **23.** $\frac{3}{8}$ **24.** $\frac{2}{3}$ **25.** $\frac{5}{6}$ **26.** $\frac{1}{6}$ **27.** 7500 **28.** 46; 2 cc left over **29.** ◆

Margin Exercises, Section 8.6, pp. 447–450

1. 9 in. **2.** 5 ft **3.** 62.8 m **4.** 88 m **5.** 34.296 yd **6.** $78\frac{4}{7}$ km^2 **7.** 339.62 cm^2 **8.** 12-ft diameter flower bed, by about 13.04 ft^2

Calculator Spotlight, p. 449

1. Answers may vary; 9 **2.** 1417.99 in.; 160,005.91 in^2 **3.** 1729.27 in^2 **4.** 125,663.71 ft^2

Exercise Set 8.6, p. 451

1. 14 cm; 44 cm; 154 cm^2 **3.** $1\frac{1}{2}$ in.; $4\frac{5}{7}$ in.; $1\frac{43}{56}$ in^2 **5.** 16 ft; 100.48 ft; 803.84 ft^2 **7.** 0.7 cm; 4.396 cm; 1.5386 cm^2 **9.** 3 cm; 18.84 cm; 28.26 cm^2 **11.** 151,976 mi^2 **13.** 2.5 cm; 1.25 cm; 4.90625 cm^2 **15.** 3.454 ft **17.** 65.94 yd^2 **19.** 45.68 ft **21.** 26.84 yd **23.** 45.7 yd **25.** 100.48 m^2 **27.** 6.9972 cm^2 **29.** 64.4214 in^2 **31.** 87.5% **32.** 58% **33.** $66.\overline{6}$% **34.** 43.61% **35.** 37.5% **36.** 62.5% **37.** $66.\overline{6}$% **38.** 20% **39.** 4 **40.** $8\frac{1}{2}$ **41.** 13 **42.** $39\frac{1}{2}$ **43.** 5 **44.** 0 **45.** 2 **46.** 3 **47.** $\frac{1}{2}$ **48.** 3 **49.** $275\frac{1}{2}$ **50.** $7\frac{1}{2}$ **51.** ◆ **53.** 3.142 **55.** $3d$; πd; circumference of one ball, since $\pi > 3$

Margin Exercises, Section 8.7, pp. 455–458

1. 81 **2.** 100 **3.** 121 **4.** 144 **5.** 169 **6.** 196 **7.** 225 **8.** 256 **9.** 289 **10.** 324 **11.** 400 **12.** 625 **13.** 3 **14.** 4 **15.** 11 **16.** 10 **17.** 9 **18.** 8 **19.** 18 **20.** 20 **21.** 15 **22.** 13 **23.** 1 **24.** 0 **25.** 2.236 **26.** 8.832 **27.** 12.961 **28.** $c = \sqrt{41}$; $c \approx 6.403$ **29.** $a = \sqrt{75}$; $a \approx 8.660$ **30.** $b = \sqrt{120}$; $b \approx 10.954$ **31.** $a = \sqrt{175}$; $a \approx 13.229$ **32.** $\sqrt{424}$ ft $\approx$ 20.6 ft

Calculator Spotlight, p. 456

1. 6.6 **2.** 9.7 **3.** 19.8 **4.** 17.3 **5.** 24.9 **6.** 24.5 **7.** 121.2 **8.** 115.6 **9.** 16.2 **10.** 85.4

Exercise Set 8.7, p. 459

1. 10 **3.** 21 **5.** 25 **7.** 19 **9.** 23 **11.** 100 **13.** 6.928 **15.** 2.828 **17.** 4.243 **19.** 2.449 **21.** 3.162 **23.** 8.660 **25.** 14 **27.** 13.528 **29.** $c = \sqrt{34}$; $c \approx 5.831$ **31.** $c = \sqrt{98}$; $c \approx 9.899$ **33.** $a = 5$ **35.** $b = 8$ **37.** $c = 13$ **39.** $b = 24$ **41.** $a = \sqrt{399}$; $a \approx 19.975$ **43.** $b = \sqrt{224}$; $b \approx 14.967$ **45.** $\sqrt{250}$ m $\approx$ 15.8 m **47.** $\sqrt{8450}$ ft $\approx$ 91.9 ft **49.** $h = \sqrt{500}$ ft $\approx$ 22.4 ft **51.** $\sqrt{211{,}200{,}000}$ ft $\approx$ 14,532.7 ft **53.** 0.456 **54.** 0.1634 **55.** 1.23 **56.** 0.99 **57.** 0.0041 **58.** 0.03 **59.** $468 **60.** 12% **61.** 187,200 **62.** 40% **63.** About 324 **64.** $24.30 **65.** ◆ **67.** 47.80 cm^2 **69.** Length: 15.2 in.; width: 11.4 in.

Summary and Review: Chapter 8, p. 463

1. $2\frac{2}{3}$ **2.** 30 **3.** 0.03 **4.** 0.004 **5.** 72 **6.** 400,000 **7.** $1\frac{1}{6}$ **8.** 0.15 **9.** 220 **10.** 32.18 **11.** 23 m **12.** 4.4 m **13.** 228 ft; 2808 ft^2 **14.** 85.9 ft **15.** 36 ft; 81 ft^2 **16.** 17.6 cm; 12.6 cm^2 **17.** 60 cm^2 **18.** 35 mm^2 **19.** 22.5 m^2 **20.** 27.5 cm^2 **21.** 88 m^2 **22.** 126 in^2 **23.** 840 ft^2 **24.** 8 m **25.** $\frac{14}{11}$ in., or $1\frac{3}{11}$ in. **26.** 14 ft **27.** 20 cm **28.** 50.24 m **29.** 8 in. **30.** 200.96 m^2 **31.** $5\frac{1}{11}$ in^2 **32.** 1038.555 ft^2 **33.** 8 **34.** 9.110 **35.** $c = \sqrt{850}$; $c \approx 29.155$ **36.** $b = \sqrt{51}$; $b \approx 7.141$ **37.** $c = \sqrt{89}$ ft; $c \approx 9.434$ ft **38.** $a = \sqrt{76}$ cm; $a \approx 8.718$ cm **39.** 28.8 ft **40.** 44.7 ft **41.** 47% **42.** 92% **43.** 0.567 **44.** $\frac{73}{100}$ **45.** ◆ Answers might include the following: The metric system is used in most countries. It is easier to convert from one metric unit to another than from one American unit to another, because the metric system is based on the number 10. **46.** ◆ The metric system was adopted by law in France in about 1790, during the rule of Napoleon I. **47.** 100 ft^2 **48.** 7.83998704 m^2 **49.** 42.05915 cm^2

Test: Chapter 8, p. 467

1. [8.1a] 48 **2.** [8.1a] $\frac{1}{3}$ **3.** [8.2a] 6000 **4.** [8.2a] 0.87 **5.** [8.2b] 181.8 **6.** [8.2b] 1490.4 **7.** [8.3a], [8.4a] 32.82 cm; 65.894 cm^2 **8.** [8.3a], [8.4a] 100 m; 625 m^2 **9.** [8.5a] 25 cm^2 **10.** [8.5a] 12 m^2 **11.** [8.5a] 18 ft^2 **12.** [8.6a] $\frac{1}{4}$ in. **13.** [8.6a] 9 cm **14.** [8.6b] $\frac{11}{14}$ in. **15.** [8.6c] 254.34 cm^2 **16.** [8.6d] 103.815 km^2 **17.** [8.7a] 15 **18.** [8.7b] 9.327 **19.** [8.7c] $c = 40$ **20.** [8.7c] $b = \sqrt{60}$; $b \approx 7.746$ **21.** [8.7c] $c = \sqrt{2}$; $c \approx 1.414$ **22.** [8.7c] $b = \sqrt{51}$; $b \approx 7.141$ **23.** [8.7c] 15.8 m **24.** [6.1c] 93% **25.** [6.2a] 81.25% **26.** [6.1b] 0.932 **27.** [6.2b] $\frac{1}{3}$ **28.** [8.1a], [8.4a] 2 ft^2 **29.** [8.1a], [8.5a] 1.875 ft^2

Cumulative Review: Chapters 1–8, p. 469

1. [1.5b] 50,854,100 **2.** [2.6a] $\frac{1}{12}$ **3.** [4.5c] 15.2 **4.** [3.6b] $\frac{2}{3}$ **5.** [4.4a] 35.6 **6.** [3.5b] $\frac{5}{6}$ **7.** [2.2a] Yes **8.** [2.2a] No **9.** [2.1d] $3 \cdot 3 \cdot 11$ **10.** [3.1a] 245 **11.** [4.5b] 35.8 **12.** [4.1a] One hundred three and sixty-four thousandths **13.** [7.1a, b] $17.8\overline{3}$, 17.5 **14.** [6.1c] 8% **15.** [6.2a] 60% **16.** [8.7a] 11 **17.** [8.7b] 5.39 **18.** [8.1a] 6 **19.** [8.3a] 14.3 m **20.** [8.5a] 297.5 yd^2 **21.** [4.4b] 150.5 **22.** [1.7b] 19,248 **23.** [2.7c] $\frac{15}{2}$ **24.** [3.3c] $\frac{2}{35}$ **25.** [3.7b] $12\frac{1}{2}$ **26.** [3.7b] 5 **27.** [3.7b] 30 **28.** [3.7b] 36 **29.** [3.7b] 7 **30.** [3.7b] 0 **31.** [3.7b] $17\frac{1}{2}$ **32.** [3.7b] 8 **33.** [7.2a] 5.5 hr **34.** [7.2a] 13.8 hr **35.** [7.2a] 2.5 hr **36.** [7.5b] 9 hr **37.** [6.6c] $11\frac{9}{11}$% **38.** [1.8a] 165,000,000,000 lb

39. [8.4a], [8.6c] 248.64 in^2 **40.** [5.4a] $8\frac{3}{4}$ lb **41.** [6.5a] Yes **42.** [3.2c] $1\frac{1}{4}$ hr **43.** [4.7a] 255.8 mi **44.** [2.6b] $6600 **45.** [8.4a], [6.5b] $41\frac{2}{3}$% **46.** [7.3c, d] The times, in seconds, are 418, 406, 396, and 388. If the pattern continues, the times, in seconds, are 382, 378, 376, 376, and so on. The times eventually bottom out to 376 seconds by the seventh week, a time on which she cannot improve.

Chapter 9

Pretest: Chapter 9, p. 472

1. [9.1b] 2.304 **2.** [9.1b] 2400 **3.** [9.3a] 80 **4.** [9.3a] 8800 **5.** [9.3b] 4800 **6.** [9.3b] 0.62 **7.** [9.3b] 3.4 **8.** [9.3c] 420 **9.** [9.3c] 384 **10.** [9.1b] 64 **11.** [9.1b] 2560 **12.** [9.1b] 24 **13.** [9.4b] 25°C **14.** [9.4b] 98.6°F **15.** [9.5a] 144 **16.** [9.5b] 2,000,000 **17.** [9.3b] 1000 g **18.** [9.1a] 160 cm^3 **19.** [9.2a] 1256 ft^3 **20.** [9.2b] 33,493.$\overline{3}$ yd^3 **21.** [9.2c] 150.72 cm^3 **22.** [9.6a] 35° **23.** [9.6a] 90° **24.** [9.6a] 113° **25.** [9.6a] 180° **26.** [9.6b] Acute **27.** [9.6b] Right **28.** [9.6b] Obtuse **29.** [9.6b] Straight **30.** [9.6d] 60° **31.** [9.6c] Scalene **32.** [9.6c] Right

Margin Exercises, Section 9.1, pp. 473–476

1. 12 cm^3 **2.** 20 ft^3 **3.** 128 ft^3 **4.** 40 **5.** 20 **6.** mL **7.** mL **8.** L **9.** L **10.** 970 **11.** 8.99 **12.** 4.8 L **13.** **(a)** 118.28 mL; **(b)** 0.11828 L **14.** $1.76

Exercise Set 9.1, p. 477

1. 768 cm^3 **3.** 45 in^3 **5.** 75 m^3 **7.** $357\frac{1}{2}$ yd^3 **9.** 1000; 1000 **11.** 87,000 **13.** 0.049 **15.** 0.000401 **17.** 78,100 **19.** 320 **21.** 10 **23.** 32 **25.** 500 mL **27.** 125 mL **29.** 1.75 gal/week; 7.5 gal/month; 91.25 gal/year; 65,250,000 gal/day; 23,816,250,000 gal/year **31.** $19.20 **32.** $96 **33.** 1000 **34.** 225 **35.** 49 **36.** 64 **37.** 5% **38.** 11% **39.** ◆ **41.** About 57,480 in^3

Margin Exercises, Section 9.2, pp. 479–481

1. 785 ft^3 **2.** 67,914 m^3 **3.** $91,989\frac{1}{3}$ ft^3 **4.** 38.77272 cm^3 **5.** 1695.6 m^3 **6.** 528 in^3 **7.** 83.7$\overline{3}$ mm^3

Calculator Spotlight, p. 481

1. 523,598,776 m^3 **2.** 105.87 cm^3

Exercise Set 9.2, p. 483

1. 803.84 in^3 **3.** 353.25 cm^3 **5.** 41,580,000 yd^3 **7.** $4,186,666\frac{2}{3}$ in^3 **9.** 124.72 m^3 **11.** $1437\frac{1}{3}$ km^3 **13.** 113,982 ft^3 **15.** 24.64 cm^3 **17.** 33,880 yd^3 **19.** 367.38 m^3 **21.** 143.72 cm^3 **23.** 32,993,440,000 mi^3 **25.** 396 **26.** 3 **27.** 14 **28.** 253,440 **29.** 12 **30.** 335,808 **31.** ◆ **33.** 6.9$\overline{7}$ ft^3; 52.$\overline{3}$ gal

Margin Exercises, Section 9.3, pp. 485–488

1. 80 **2.** 4.32 **3.** 32,000 **4.** kg **5.** kg **6.** mg **7.** g **8.** t **9.** 6200 **10.** 3.048 **11.** 77 **12.** 234.4 **13.** 6700 **14.** 120 **15.** 1461 **16.** 1440 **17.** 1

Exercise Set 9.3, p. 489

1. 2000 **3.** 3 **5.** 64 **7.** 12,640 **9.** 0.1 **11.** 5 **13.** 1000 **15.** 10 **17.** $\frac{1}{100}$, or 0.01 **19.** 1000 **21.** 10 **23.** 234,000 **25.** 5.2 **27.** 6.7 **29.** 0.0502 **31.** 8.492 **33.** 58.5 **35.** 800,000 **37.** 1000 **39.** 0.0034 **41.** 24 **43.** 60 **45.** $365\frac{1}{4}$ **47.** 0.05 **49.** 8.2 **51.** 6.5 **53.** 10.75 **55.** 336 **57.** 4.5 **59.** 56 **61.** 16 **62.** 289 **63.** 125 **64.** 64 **65.** 543 **66.** 0.00543 **67.** About 28.8 ft **68.** About 31.6 m **69.** 50.1%

70.

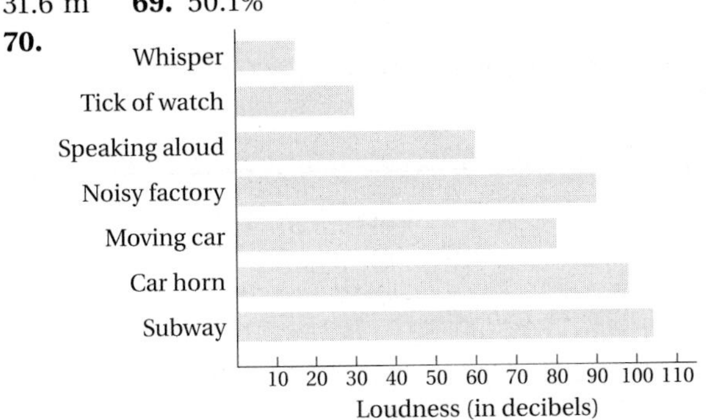

71. 10 **72.** 5 lb **73.** 0.43% **74.** 231% **75.** ◆ **77.** 0.4535 **79.** 2 oz **81.** About 31.7 yr **83.** 1000 **85.** 0.125 mg **87.** 4 **89.** 8 mL

Margin Exercises, Section 9.4, pp. 493–494

1. 80°C **2.** 0°C **3.** −20°C **4.** 80°F **5.** 100°F **6.** 50°F **7.** 176°F **8.** 95°F **9.** 35°C **10.** 45°C

Exercise Set 9.4, p. 495

1. 80°C **3.** 60°C **5.** 20°C **7.** −10°C **9.** 190°F **11.** 140°F **13.** 10°F **15.** 40°F **17.** 77°F **19.** 104°F **21.** 5432°F **23.** 30°C **25.** 55°C **27.** 37°C **29.** 234 **30.** 230 **31.** 336 **32.** 24 **33.** 0.724 **34.** 0.0724 **35.** 2520 **36.** 6 **37.** 28 **38.** 39,600 **39.** 12 **40.** 46,252.8 **41.** ◆

Margin Exercises, Section 9.5, pp. 497–498

1. 9 **2.** 45 **3.** 2880 **4.** 2.5 **5.** 3200 **6.** 1,000,000 **7.** 100 **8.** 28,800 **9.** 0.043 **10.** 0.678

Exercise Set 9.5, p. 499

1. 144 **3.** 640 **5.** $\frac{1}{144}$ **7.** 198 **9.** 396 **11.** 12,800 **13.** 27,878,400 **15.** 5 **17.** 1 **19.** $\frac{1}{640}$, or 0.0015625 **21.** 5,210,000 **23.** 140 **25.** 23.456 **27.** 0.085214 **29.** 2500 **31.** 0.4728 **33.** $240 **34.** $212 **35. (a)** $484.11; **(b)** $15,984.11 **36. (a)** $209.59; **(b)** $8709.59 **37. (a)** $220.93; **(b)** $6620.93 **38. (a)** $37.97; **(b)** $4237.97 **39.** 4.2 **40.** 0.022176 **41.** ◈ **43.** 10.89 **45.** 1.65

Margin Exercises, Section 9.6, pp. 501–504

1. Angle *DEF*, angle *FED*, ∠*DEF*, ∠*FED*, or ∠*E* **2.** Angle *PQR*, angle *RQP*, ∠*PQR*, ∠*RQP*, or ∠*Q* **3.** 127° **4.** 33° **5.** Right **6.** Acute **7.** Obtuse **8.** Straight **9. (a)** △*ABC*; **(b)** △*ABC*, △*MPN*; **(c)** △*DEF*, △*GHI*, △*JKL*, △*QRS* **10.** Yes **11.** No **12. (a)** △*EDF*; **(b)** △*GHI*, △*QRS*; **(c)** △*ABC*, △*PMN*, △*JKL* **13.** 180° **14.** 64°

Exercise Set 9.6, p. 505

1. Angle *GHI*, angle *IHG*, ∠*GHI*, ∠*IHG*, or ∠*H* **3.** 10° **5.** 180° **7.** 130° **9.** Obtuse **11.** Acute **13.** Straight **15.** Obtuse **17.** Acute **19.** Obtuse **21.** Scalene; obtuse **23.** Scalene; right **25.** Equilateral; acute **27.** Scalene; obtuse **29.** 46° **31.** 120° **33.** $4608 **34.** $2125 **35.** $252.49 **36.** $72.49 **37.** ◈

Summary and Review: Chapter 9, p. 507

1. 93.6 m^3 **2.** 193.2 cm^3 **3.** 112 **4.** 0.004 **5.** $\frac{4}{15}$ **6.** 0.464 **7.** 180 **8.** 4700 **9.** 16,140 **10.** 830 **11.** $\frac{1}{4}$ **12.** 0.04 **13.** 200 **14.** 30 **15.** 0.06 **16.** 1600 **17.** 400 **18.** $1\frac{1}{4}$ **19.** 50 **20.** 160 **21.** 7.5 **22.** 13.5 **23.** 18.72 L **24.** 80.6°F **25.** 20°C **26.** 31,400 ft^3 **27.** $33.49\overline{3}$ cm^3 **28.** 4.71 in^3 **29.** 36 **30.** 300,000 **31.** 14.375 **32.** 0.06 **33.** 54° **34.** 180° **35.** 140° **36.** 90° **37.** Acute **38.** Straight **39.** Obtuse **40.** Right **41.** 60° **42.** Scalene **43.** Right **44.** $39.04 **45.** 27 **46.** 22.09 **47.** 103.823 **48.** $\frac{1}{16}$ **49.** 13,200 **50.** 4 **51.** 45.68 **52.** 45,680 **53.** ◈ 1 gal = 128 oz, so 1 oz of water (as capacity) weighs $\frac{8.3453}{128}$ lb, or about 0.0652 lb. An ounce of pennies weighs $\frac{1}{16}$ lb, or 0.0625 lb. Thus an ounce of water (as capacity) weighs more than an ounce of pennies. **54.** ◈ See the volume formulas listed at the beginning of the Summary and Review Exercises for Chapter 9.

Test: Chapter 9, p. 509

1. [9.1a] 84 cm^3 **2.** [9.1b] 3.08 **3.** [9.1b] 240 **4.** [9.3a] 64 **5.** [9.3a] 8220 **6.** [9.3b] 3800 **7.** [9.3b] 0.4325 **8.** [9.3b] 2.2 **9.** [9.3c] 300 **10.** [9.3c] 360 **11.** [9.1b] 32 **12.** [9.1b] 1280 **13.** [9.1b] 40 **14.** [9.4b] 35°C **15.** [9.4b] 138.2°F **16.** [9.5a] 1728 **17.** [9.5b] 0.0003 **18.** [9.1a] 420 in^3 **19.** [9.2a] 1177.5 ft^3 **20.** [9.2b] $4186.\overline{6}$ yd^3 **21.** [9.2c] 113.04 cm^3 **22.** [9.6a] 90° **23.** [9.6a] 35° **24.** [9.6a] 180° **25.** [9.6a] 113° **26.** [9.6b] Right **27.** [9.6b] Acute **28.** [9.6b] Straight **29.** [9.6b] Obtuse **30.** [9.6d] 35° **31.** [9.6c] Isosceles **32.** [9.6c] Obtuse **33.** [6.7a] $680 **34.** [1.9b] 1000 **35.** [1.9b] $\frac{1}{16}$ **36.** [1.9b] 9.8596 **37.** [1.9b] 0.00001 **38.** [8.1a] 42 **39.** [8.1a] 250 **40.** [8.2a] 2300 **41.** [8.2a] 3400 **42.** [9.1a] 0.65 ft^3 **43.** [9.2c] 0.033 ft^3 **44.** [9.2a] 0.055 ft^3

Cumulative Review: Chapters 1–9, p. 511

1. [3.5a] $4\frac{1}{6}$ **2.** [4.5c] 49.2 **3.** [4.2b] 87.52 **4.** [1.6c] 1234 **5.** [1.9d] 2 **6.** [1.9c] 1565 **7.** [4.1b] $\frac{1209}{1000}$ **8.** [6.2b] $\frac{17}{100}$ **9.** [3.3b] < **10.** [3.3b] = **11.** [9.3a] $\frac{3}{8}$ **12.** [9.4b] 59 **13.** [9.1b] 87 **14.** [9.3c] $\frac{3}{20}$ **15.** [9.5a] 27 **16.** [8.2a] 0.17 **17.** [8.3a], [8.5a] 380 cm; 5500 cm^2 **18.** [8.3a], [8.5a] 32.3 ft; 56.55 ft^2 **19.** [7.3d]

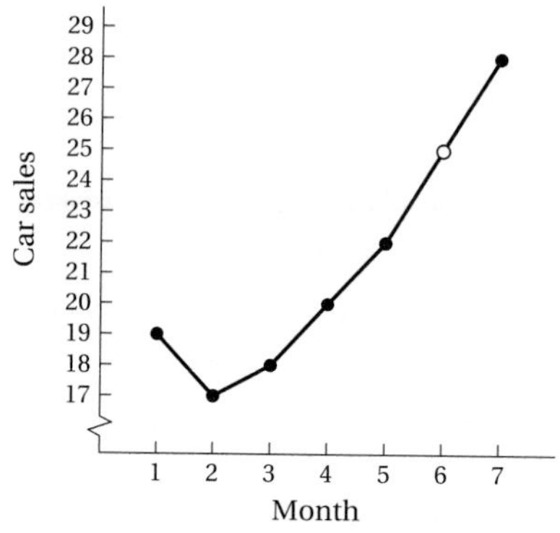

Car sales
29 28 27 26 25 24 23 22 21 20 19 18 17
1 2 3 4 5 6 7
Month

20. [7.5b] 25 **21.** [3.7b] 0 **22.** [3.7b] 1 **23.** [3.7b] 0 **24.** [3.7b] $\frac{1}{2}$ **25.** [5.3b] $14\frac{2}{5}$ **26.** [5.3b] $4\frac{2}{7}$ **27.** [4.4b] 113.4 **28.** [3.3c] $\frac{1}{8}$ **29.** [7.1a] 99 **30.** [1.8a] 528 million **31.** [9.2b] 4187 cm^3 **32.** [6.7a] $24 **33.** [8.7a] 17 m **34.** [6.6a] 6% **35.** [3.5c] $2\frac{1}{8}$ yd **36.** [4.7a] $21.82 **37.** [5.2b] The 8-qt box **38.** [2.6b] $\frac{7}{20}$ km **39.** [9.6d] 30° **40.** [9.6c] Scalene **41.** [9.6c] Obtuse **42.** [8.1a], [9.1a] 272 ft^3 **43.** [9.2a] 94,200 ft^3 **44.** [9.1a] 1.342 ft^3 **45.** [9.2a, b] 1204.260429 ft^3; about 22,079,692.8 ft, or 4181.76 mi

Chapter 10

Pretest: Chapter 10, p. 514

1. [10.1d] > **2.** [10.1d] > **3.** [10.1d] > **4.** [10.1d] < **5.** [10.1c] −0.625 **6.** [10.1c] $-0.\overline{6}$ **7.** [10.1c] $-0.\overline{90}$ **8.** [10.1e] 12 **9.** [10.1e] 2.3 **10.** [10.1e] 0 **11.** [10.2b] −5.4 **12.** [10.2b] $\frac{2}{3}$ **13.** [10.2a] −17 **14.** [10.3a] 38.6 **15.** [10.3a] $-\frac{17}{15}$ **16.** [10.2a] −5 **17.** [10.4a] 63 **18.** [10.4a] $-\frac{5}{12}$ **19.** [10.5c] −98 **20.** [10.5a] 8 **21.** [10.3a] 24 **22.** [10.5d] 26

Margin Exercises, Section 10.1, pp. 516–520

1. 8; −5 **2.** 134; −80 **3.** −10; 148 **4.** −137; 289

5. $-\frac{7}{2}$

−6 −5 −4 −3 −2 −1 0 1 2 3 4 5 6

6. 1.4

−6 −5 −4 −3 −2 −1 0 1 2 3 4 5 6

7. $-\frac{11}{4}$

−6 −5 −4 −3 −2 −1 0 1 2 3 4 5 6

8. −0.375 **9.** $-0.\overline{54}$ **10.** $1.\overline{3}$ **11.** $<$ **12.** $<$ **13.** $>$ **14.** $>$ **15.** $>$ **16.** $<$ **17.** $<$ **18.** $>$ **19.** 8 **20.** 0 **21.** 9 **22.** $\frac{2}{3}$ **23.** 5.6

Exercise Set 10.1, p. 521

1. −200; 600 **3.** −34; 15 **5.** 750; −125 **7.** 20; −150; 300

9. $\frac{10}{3}$

−6 −5 −4 −3 −2 −1 0 1 2 3 4 5 6

11. −4.3

−6 −5 −4 −3 −2 −1 0 1 2 3 4 5 6

13. −0.625 **15.** $-1.\overline{6}$ **17.** $-1.1\overline{6}$ **19.** −0.875 **21.** −0.35 **23.** $>$ **25.** $<$ **27.** $<$ **29.** $<$ **31.** $>$ **33.** $<$ **35.** $>$ **37.** $<$ **39.** $<$ **41.** $<$ **43.** 3 **45.** 18 **47.** 11 **49.** 24 **51.** $\frac{2}{3}$ **53.** 0 **55.** $2 \cdot 3 \cdot 3 \cdot 3$ **56.** $2 \cdot 2 \cdot 2 \cdot 2 \cdot 2 \cdot 2 \cdot 3$ **57.** $2 \cdot 3 \cdot 17$ **58.** $2 \cdot 2 \cdot 5 \cdot 13$ **59.** $2 \cdot 2 \cdot 2 \cdot 2 \cdot 2 \cdot 3 \cdot 3 \cdot 3$ **60.** $2 \cdot 2 \cdot 3 \cdot 3 \cdot 13$ **61.** 18 **62.** 72 **63.** 96 **64.** 72 **65.** 1344 **66.** 252 **67.** ◈ **69.** $>$ **71.** $=$ **73.** $-100, -8\frac{7}{8}, -8\frac{5}{8}, -\frac{67}{8}, -5, 0, 1^7, |3|, \frac{14}{4}, 4, |-6|, 7$

Margin Exercises, Section 10.2, pp. 523–526

1. −8 **2.** −3 **3.** −8 **4.** 4 **5.** 0 **6.** −2 **7.** −11 **8.** −12 **9.** 2 **10.** −4 **11.** −2 **12.** 0 **13.** −22 **14.** 3 **15.** 0.53 **16.** 2.3 **17.** −7.7 **18.** −6.2 **19.** $-\frac{2}{9}$ **20.** $-\frac{19}{20}$ **21.** −58 **22.** −56 **23.** −14 **24.** −12 **25.** 4 **26.** −8.7 **27.** 7.74 **28.** $\frac{8}{9}$ **29.** 0 **30.** −12 **31.** −14; 14 **32.** −1; 1 **33.** 19; −19 **34.** 1.6; −1.6 **35.** $-\frac{2}{3}; \frac{2}{3}$ **36.** $\frac{9}{8}; -\frac{9}{8}$ **37.** 4 **38.** 13.4 **39.** 0 **40.** $-\frac{1}{4}$

Exercise Set 10.2, p. 527

1. −7 **3.** −4 **5.** 0 **7.** −8 **9.** −7 **11.** −27 **13.** 0 **15.** −42 **17.** 0 **19.** 0 **21.** 3 **23.** −9 **25.** 7 **27.** 0 **29.** 45 **31.** −1.8 **33.** −8.1 **35.** $-\frac{1}{5}$ **37.** $-\frac{8}{7}$ **39.** $-\frac{3}{8}$ **41.** $-\frac{29}{35}$ **43.** $-\frac{11}{15}$ **45.** −6.3 **47.** $\frac{7}{16}$ **49.** 39 **51.** 50 **53.** −1093 **55.** −24 **57.** 26.9 **59.** −9 **61.** $\frac{14}{3}$ **63.** −65 **65.** $\frac{5}{3}$ **67.** 14 **69.** −10 **71.** 357.5 ft^2 **72.** 45.51 ft^2 **73.** 54,756 mi^2 **74.** 1.4196 mm^2 **75.** 66.97 ft^2 **76.** 7976.1 m^2 **77.** ◈ **79.** All positive **81.** −6483 **83.** All numbers greater than 7 **85.** Negative

Margin Exercises, Section 10.3, pp. 529–530

1. −10 **2.** 3 **3.** −5 **4.** −2 **5.** −11 **6.** 4 **7.** −2 **8.** −6 **9.** −16 **10.** 7.1 **11.** 3 **12.** 0 **13.** $\frac{3}{2}$ **14.** −8 **15.** 7 **16.** −3 **17.** −23.3 **18.** 0 **19.** −9 **20.** 17 **21.** 12.7

Exercise Set 10.3, p. 531

1. −4 **3.** −7 **5.** −6 **7.** 0 **9.** −4 **11.** −7 **13.** −6 **15.** 0 **17.** 11 **19.** −14 **21.** 5 **23.** −7 **25.** −5 **27.** −3 **29.** −23 **31.** −68 **33.** −73 **35.** 116 **37.** −2.8 **39.** $-\frac{1}{4}$ **41.** $\frac{1}{12}$ **43.** $-\frac{17}{12}$ **45.** $\frac{1}{8}$ **47.** 19.9 **49.** −9 **51.** −0.01 **53.** −2.7 **55.** −3.53 **57.** $-\frac{1}{2}$ **59.** $\frac{6}{7}$ **61.** $-\frac{41}{30}$ **63.** $-\frac{1}{156}$ **65.** 37 **67.** −62 **69.** 6 **71.** 107 **73.** 219 **75.** 96.6 cm^2 **76.** $2 \cdot 3 \cdot 5 \cdot 5 \cdot 5$ **77.** 108 **78.** 125.44 km^2 **79.** 64 **80.** 125 **81.** 8 **82.** 288 oz **83.** ◈ **85.** −309,882 **87.** True **89.** True **91.** True **93.** True **95.** Rose by 15 points

Margin Exercises, Section 10.4, pp. 533–534

1. 20; 10; 0; −10; −20; −30 **2.** −18 **3.** −100 **4.** −80 **5.** $-\frac{5}{9}$ **6.** −30.033 **7.** $-\frac{7}{10}$ **8.** −10; 0; 10; 20; 30 **9.** 12 **10.** 32 **11.** 35 **12.** $\frac{20}{63}$ **13.** $\frac{2}{3}$ **14.** 13.455 **15.** −30 **16.** 30 **17.** −32 **18.** $-\frac{8}{3}$ **19.** −30 **20.** −30.75 **21.** $-\frac{5}{3}$ **22.** 120 **23.** −120 **24.** 6

Exercise Set 10.4, p. 535

1. −16 **3.** −24 **5.** −72 **7.** 16 **9.** 42 **11.** −120 **13.** −238 **15.** 1200 **17.** 98 **19.** −12.4 **21.** 24 **23.** 21.7 **25.** $-\frac{2}{5}$ **27.** $\frac{1}{12}$ **29.** −17.01 **31.** $-\frac{5}{12}$ **33.** 420 **35.** $\frac{2}{7}$ **37.** −60 **39.** 150 **41.** $-\frac{2}{45}$ **43.** 1911 **45.** 50.4 **47.** $\frac{10}{189}$ **49.** −960 **51.** 17.64 **53.** $-\frac{5}{784}$ **55.** −756 **57.** −720 **59.** −30,240 **61.** $2 \cdot 2 \cdot 2 \cdot 2 \cdot 2 \cdot 2 \cdot 2 \cdot 2 \cdot 2 \cdot 3 \cdot 3$ **62.** 180 **63.** $33\frac{1}{3}\%$, or $33.\overline{3}\%$ **64.** 262.44 **65.** 72 **66.** 40% **67.** 40 ft^3 **68.** 240 **69.** ◈ **71.** −$23 **73. (a)** One must be negative and one must be positive. **(b)** Either or both must be zero. **(c)** Both must be negative or both must be positive.

Margin Exercises, Section 10.5, pp. 537–540

1. −2 **2.** 5 **3.** −3 **4.** 9 **5.** −6 **6.** $-\frac{30}{7}$ **7.** Undefined **8.** 0 **9.** $\frac{3}{2}$ **10.** $-\frac{4}{5}$ **11.** $-\frac{1}{3}$ **12.** −5 **13.** $\frac{1}{5.78}$ **14.** $-\frac{7}{2}$

15.

Number	Opposite	Reciprocal
$\frac{2}{3}$	$-\frac{2}{3}$	$\frac{3}{2}$
$-\frac{5}{4}$	$\frac{5}{4}$	$-\frac{4}{5}$
0	0	Undefined
1	-1	1
-4.5	4.5	$-\frac{1}{4.5}$

16. $\frac{4}{7}\cdot\left(-\frac{5}{3}\right)$ **17.** $5\cdot\left(-\frac{1}{8}\right)$ **18.** $-10\cdot\left(\frac{1}{7}\right)$ **19.** $-\frac{2}{3}\cdot\frac{7}{4}$ **20.** $-5\cdot\left(\frac{1}{7}\right)$ **21.** $-\frac{20}{21}$ **22.** $-\frac{12}{5}$ **23.** $\frac{16}{7}$ **24.** -7 **25.** -1237 **26.** 8 **27.** 381 **28.** -12

Calculator Spotlight, p. 541

1. -9 **2.** -57 **3.** -1996 **4.** -24.7 **5.** -11 **6.** 9 **7.** 114 **8.** 117,649 **9.** $-1{,}419{,}857$ **10.** $-1{,}124{,}864$ **11.** $-117{,}649$ **12.** $-1{,}419{,}857$ **13.** $-1{,}124{,}864$ **14.** -4 **15.** -2 **16.** 787 **17.** The largest value is 8 for either $6 \div 4 \cdot 8 - 2^2$, or $8 \div 4 \cdot 6 - 2^2$. **18.** The largest value is 34 for $6 - 4 + 8^2 \div 2$. **19.** $-32 \cdot (88 - 29) = -1888$ **20.** $3^5 \div 10^2 - 5^2 = -22.57$

Exercise Set 10.5, p. 543

1. -6 **3.** -13 **5.** -2 **7.** 4 **9.** -8 **11.** 2 **13.** -12 **15.** -8 **17.** Undefined **19.** -9 **21.** $-\frac{7}{15}$ **23.** $\frac{1}{13}$ **25.** $-\frac{9}{8}$ **27.** $\frac{5}{3}$ **29.** $\frac{9}{14}$ **31.** $\frac{9}{64}$ **33.** -2 **35.** $\frac{11}{13}$ **37.** -16.2 **39.** Undefined **41.** -7 **43.** -7 **45.** -334 **47.** 14 **49.** 1880 **51.** 12 **53.** 8 **55.** -86 **57.** 37 **59.** -1 **61.** -10 **63.** 25 **65.** -7988 **67.** -3000 **69.** 60 **71.** 1 **73.** 10 **75.** $-\frac{13}{45}$ **77.** $-\frac{4}{3}$ **79.** $2\cdot3\cdot13$ **80.** $3\cdot3\cdot5\cdot5$ **81.** $2\cdot2\cdot2\cdot2\cdot2\cdot2\cdot3\cdot5$ **82.** $5\cdot5\cdot41$ **83.** 216 **84.** 504 **85.** 665 **86.** 320 **87.** 1710 **88.** 4% **89.** **91.** -2 **93.** Positive **95.** Negative **97.** Positive

Summary and Review: Chapter 10, p. 545

1. 38 **2.** 7.3 **3.** $\frac{5}{2}$ **4.** -0.2 **5.** -1.25 **6.** $-0.8\overline{3}$ **7.** $-0.41\overline{6}$ **8.** $-0.\overline{27}$

9.

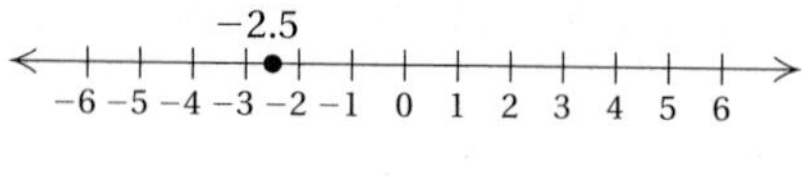

10.

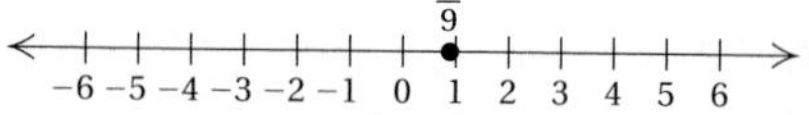

11. < **12.** > **13.** > **14.** < **15.** -3.8 **16.** $\frac{3}{4}$ **17.** 34 **18.** 5 **19.** $\frac{8}{3}$ **20.** $-\frac{1}{7}$ **21.** -10 **22.** -3 **23.** $-\frac{7}{12}$ **24.** -4 **25.** -5 **26.** 4 **27.** $-\frac{7}{5}$ **28.** -7.9 **29.** 54 **30.** -9.18 **31.** $-\frac{2}{7}$ **32.** -210 **33.** -7 **34.** -3 **35.** $\frac{3}{4}$ **36.** 40.4 **37.** -62 **38.** -5 **39.** 210 cm^2 **40.** 270 **41.** $2\cdot2\cdot2\cdot3\cdot3\cdot3\cdot3$ **42.** 36% **43.** Yes; the numbers 1 and -1 are their own reciprocals: $1\cdot1=1$ and $-1(-1)=1$. **44.** We know that $a + (-a) = 0$, so the opposite of $-a$ is a. That is, $-(-a) = a$. **45.** 403 and 397 **46.** 12 and -7 **47.** **(a)** $-7 + (-6) + (-5) + (-4) + (-3) + (-2) + (-1) + 0 + 1 + 2 + 3 + 4 + 5 + 6 + 7 + 8 = 8$; **(b)** 0 **48.** -1; Consider reciprocals and pairs of products of negative numbers. **49.** $-\frac{5}{8}$ **50.** -2.1

Test: Chapter 10, p. 547

1. [10.1d] < **2.** [10.1d] > **3.** [10.1d] > **4.** [10.1d] < **5.** [10.1c] -0.125 **6.** [10.1c] $-0.\overline{4}$ **7.** [10.1c] $-0.\overline{18}$ **8.** [10.1e] 7 **9.** [10.1e] $\frac{9}{4}$ **10.** [10.1e] -2.7 **11.** [10.2b] $-\frac{2}{3}$ **12.** [10.2b] 1.4 **13.** [10.2b] 8 **14.** [10.5b] $-\frac{1}{2}$ **15.** [10.5b] $\frac{7}{4}$ **16.** [10.3a] 7.8 **17.** [10.2a] -8 **18.** [10.2a] $\frac{7}{40}$ **19.** [10.3a] 10 **20.** [10.3a] -2.5 **21.** [10.3a] $\frac{7}{8}$ **22.** [10.4a] -48 **23.** [10.4a] $\frac{3}{16}$ **24.** [10.5a] -9 **25.** [10.5c] $\frac{3}{4}$ **26.** [10.5c] -9.728 **27.** [10.5d] 109 **28.** [8.4a] 55.8 ft^2 **29.** [6.3b] 48% **30.** [2.1d] $2\cdot2\cdot2\cdot5\cdot7$ **31.** [3.1a] 240 **32.** [10.1e], [10.3a] 15 **33.** [10.3a] 2385 m **34.** [10.3a], [10.5c] **(a)** -4, -9, -15; **(b)** -2, -6, -10; **(c)** -18, -24, -31; **(d)** -0.25, 0.125, -0.0625

Cumulative Review: Chapters 1–10, p. 549

1. [6.1b] 0.263 **2.** [10.1c] $-0.\overline{45}$ **3.** [9.3b] 834 **4.** [9.5b] 0.0275 **5.** [10.1e] 4.5 **6.** [10.3a] -11 **7.** [5.2a] 12.5 m/sec **8.** [8.6a, b, c] 35 mi; 220 mi; 3850 mi^2 **9.** [8.7a] 15 **10.** [8.7b] 8.31 **11.** [10.4a] -10 **12.** [10.5a] 3 **13.** [10.2a] 8 **14.** [7.2c]

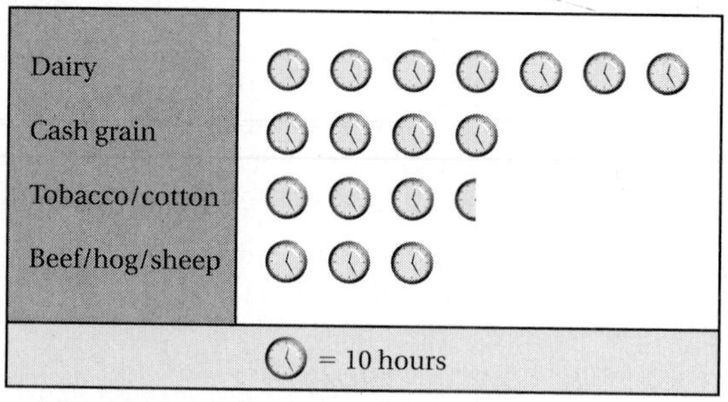

15. [4.3a] 0.01485 **16.** [10.3a] -3 **17.** [10.3a] $-\frac{3}{22}$ **18.** [10.4a] $-\frac{1}{24}$ **19.** [3.5b] $1\frac{5}{6}$ **20.** [10.5c] $-\frac{1}{4}$ **21.** [10.5d] -57.5 **22.** [1.5b] 9,640,500,000 **23.** [10.2a] 32.49 **24.** [10.5d] Undefined **25.** [6.3b] 87.5% **26.** [6.3b] 32 **27.** [4.7a] \$133.05 **28.** [9.2a] 307.72 cm^3 **29.** [7.1a] $44.8\overline{3}^\circ$ **30.** [6.5a] 78 **31.** [8.4b] 6902.5 m^2 **32.** [3.5c] $2\frac{11}{12}$ cups **33.** [1.8a], [7.1a] \$239,441; \$59,860.25

34. [4.7a] 4.55 km

35. [7.3d]

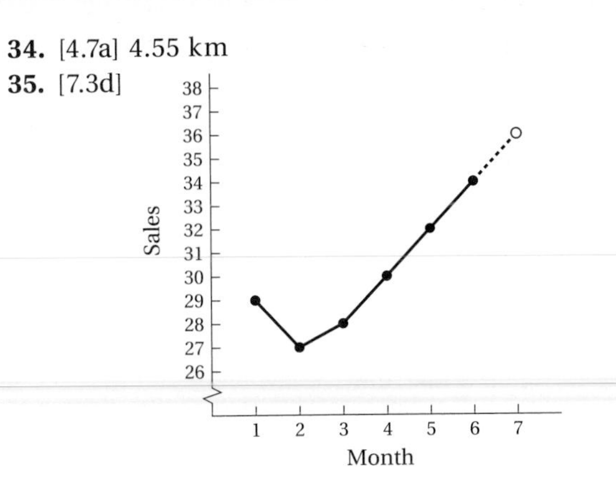

36. [7.5b] 36 **37.** [3.7b] 11 **38.** [3.7b] 13 **39.** [3.7b] 11 **40.** [3.7b] $138\frac{1}{2}$ **41.** [9.6d] 120°

Chapter 11

Pretest: Chapter 11, p. 552

1. [11.1a] $\frac{5}{16}$ **2.** [11.5a] $78\%x$, or $0.78x$ **3.** [11.1b] $9z - 18$ **4.** [11.1b] $-4a - 2b + 10c$ **5.** [11.1c] $4(x - 3)$ **6.** [11.1c] $3(2y - 3z - 6)$ **7.** [11.1d] $-3x$ **8.** [11.1d] $2x + 2y + 18$ **9.** [11.3a] -7 **10.** [11.4b] -1 **11.** [11.4a] 2 **12.** [11.4c] 3 **13.** [11.5b] Width: 34 m; length: 39 m **14.** [11.5b] $780

Margin Exercises, Section 11.1, pp. 553–558

1. 64 **2.** 28 **3.** 60; -27.2 **4.** -25 **5.** 16

6.

	$1 \cdot x$	x
$x = 3$	3	3
$x = -6$	-6	-6
$x = 4.8$	4.8	4.8

7.

	$2x$	$5x$
$x = 2$	4	10
$x = -6$	-12	-30
$x = 4.8$	9.6	24

8. 36; 36 **9.** 90; 90 **10.** 64; 64 **11.** 14; 14 **12.** 30; 30 **13.** 12; 12 **14.** $5x$; $-4y$; 3 **15.** $-4y$; $-2x$; $3z$ **16.** $3x - 15$ **17.** $5x + 5$ **18.** $\frac{5}{4}x - \frac{5}{4}y + 5$ **19.** $-2x + 6$ **20.** $-5x + 10y - 20z$ **21.** $6(z - 2)$ **22.** $3(x - 2y + 3)$ **23.** $2(8a - 18b + 21)$ **24.** $-4(3x - 8y + 4z)$ **25.** $3x$ **26.** $6x$ **27.** $-8x$ **28.** $0.59x$ **29.** $3x + 3y$ **30.** $-4x - 5y - 7$ **31.** $\frac{1}{10}x + \frac{7}{9}y - \frac{2}{3}$

Exercise Set 11.1, p. 559

1. 42 **3.** 3 **5.** -1 **7.** 6 **9.** 240; 240 **11.** 160; 160 **13.** $2b + 10$ **15.** $7 - 7t$ **17.** $30x + 12$ **19.** $7x + 28 + 42y$ **21.** $-7y + 14$ **23.** $45x + 54y - 72$ **25.** $\frac{3}{4}x - \frac{9}{4}y - \frac{3}{2}z$ **27.** $-3.72x + 9.92y - 3.41$ **29.** $2(x + 2)$ **31.** $5(6 + y)$ **33.** $7(2x + 3y)$ **35.** $5(x + 2 + 3y)$ **37.** $8(x - 3)$ **39.** $4(8 - y)$ **41.** $2(4x + 5y - 11)$ **43.** $6(-3x - 2y + 1)$, or $-6(3x + 2y - 1)$ **45.** $19a$ **47.** $9a$ **49.** $8x + 9z$ **51.** $-19a + 88$ **53.** $4t + 6y - 4$ **55.** $8x$ **57.** $5n$ **59.** $-16y$ **61.** $17a - 12b - 1$ **63.** $4x + 2y$ **65.** $\frac{39}{20}x + \frac{1}{2}y + 12$ **67.** $0.8x + 0.5y$ **69.** 30 yd; 94.2 yd; 706.5 yd^2 **70.** 16.4 m; 51.496 m; 211.1336 m^2 **71.** 19 mi; 59.66 mi; 283.385 mi^2 **72.** 4800 cm; 15,072 cm; 18,086,400 cm^2 **73.** 10 mm; 62.8 mm; 314 mm^2 **74.** 132 km; 828.96 km; 54,711.36 km^2 **75.** 2.3 ft; 14.444 ft; 16.6106 ft^2 **76.** 5.15 m; 32.342 m; 83.28065 m^2 **77.** ◆

Margin Exercises, Section 11.2, pp. 561–562

1. (a) $2 + -2 = 0$; (b) 9 **2.** -5 **3.** 13.2 **4.** -6.5

Exercise Set 11.2, p. 563

1. 7 **3.** -20 **5.** -14 **7.** -18 **9.** 15 **11.** -14 **13.** 2 **15.** 20 **17.** -6 **19.** $\frac{7}{3}$ **21.** $-\frac{7}{4}$ **23.** $\frac{41}{24}$ **25.** $-\frac{1}{20}$ **27.** 5.1 **29.** 12.4 **31.** -5 **33.** $1\frac{5}{6}$ **35.** $-\frac{10}{21}$ **37.** -11 **38.** $-\frac{1}{24}$ **39.** -34.1 **40.** -1.7 **41.** 5 **42.** $-\frac{31}{24}$ **43.** 5.5 **44.** 8.1 **45.** 24 **46.** $-\frac{5}{12}$ **47.** 283.14 **48.** -15.68 **49.** 8 **50.** $-\frac{16}{15}$ **51.** -14.3 **52.** -4.9 **53.** ◆ **55.** 342.246 **57.** $-\frac{26}{15}$ **59.** -10 **61.** All real numbers **63.** $-\frac{5}{17}$

Margin Exercises, Section 11.3, pp. 565–566

1. 30 **2.** $-\frac{7}{4}$ **3.** 8 **4.** -18

Exercise Set 11.3, p. 567

1. 6 **3.** 9 **5.** 12 **7.** -40 **9.** 5 **11.** -7 **13.** -6 **15.** 6 **17.** -63 **19.** 36 **21.** -21 **23.** $-\frac{3}{5}$ **25.** $-\frac{3}{2}$ **27.** $\frac{9}{2}$ **29.** 7 **31.** -7 **33.** 8 **35.** 15.9 **37.** 62.8 ft; 20 ft; 314 ft^2 **38.** 75.36 cm; 12 cm; 452.16 cm^2 **39.** 8000 ft^3 **40.** 31.2 cm^3 **41.** 53.55 cm^2 **42.** 64 mm^2 **43.** 68 in^2 **44.** 38.25 m^2 **45.** ◆ **47.** -8655 **49.** No solution **51.** No solution

Margin Exercises, Section 11.4, pp. 569–574

1. 5 **2.** 4 **3.** 4 **4.** 39 **5.** $-\frac{3}{2}$ **6.** -4.3 **7.** -3 **8.** 800 **9.** 1 **10.** 2 **11.** 2 **12.** $\frac{17}{2}$ **13.** $\frac{8}{3}$ **14.** -4.3 **15.** 2 **16.** 3 **17.** -2 **18.** $-\frac{1}{2}$

Exercise Set 11.4, p. 575

1. 5 **3.** 8 **5.** 10 **7.** 14 **9.** −8 **11.** −8 **13.** −7 **15.** 15 **17.** 6 **19.** 4 **21.** 6 **23.** −3 **25.** 1 **27.** −20 **29.** 6 **31.** 7 **33.** 2 **35.** 5 **37.** 2 **39.** 10 **41.** 4 **43.** 0 **45.** −1 **47.** $-\frac{4}{3}$ **49.** $\frac{2}{5}$ **51.** −2 **53.** −4 **55.** $\frac{4}{5}$ **57.** $-\frac{28}{27}$ **59.** 6 **61.** 2 **63.** 6 **65.** 8 **67.** 1 **69.** 17 **71.** $-\frac{5}{3}$ **73.** −3 **75.** 2 **77.** $-\frac{51}{31}$ **79.** 2 **81.** 6.5 **82.** 6.5 **83.** −6.5 **84.** −6.5 **85.** $7(x - 3 - 2y)$ **86.** $25(a - 25b + 3)$ **87.** $14(3t + m - 4)$ **88.** $16(a - 4b + 14 - 2q)$ **89.** −8 **90.** < **91.** ◈ **93.** 2 **95.** −2 **97.** 8 **99.** 2 cm

Margin Exercises, Section 11.5, pp. 580–587

1. $x - 12$ **2.** $y + 12$, or $12 + y$ **3.** $m - 4$ **4.** $\frac{1}{2} \cdot p$ **5.** $6 + 8x$, or $8x + 6$ **6.** $a - b$ **7.** 59%x, or $0.59x$ **8.** $xy - 200$ **9.** $p + q$ **10.** 5 **11.** 4 in., 16 in., 32 in. **12.** 587.8 mi **13.** 30°, 90°, 60° **14.** Width: 9 ft; length: 12 ft **15.** **(a)** $25.51, $29.68, $56.78; **(b)** 78 lb

Exercise Set 11.5, p. 589

1. $2x - 3$ **3.** 97%y, or $0.97y$ **5.** $5x + 4$, or $4 + 5x$ **7.** $\frac{1}{3}x$, or $240 - x$ **9.** 32 **11.** 305 ft **13.** $2.89 **15.** 57 **17.** −12 **19.** $699\frac{1}{3}$ mi **21.** First: 30 m; second: 90 m; third: 360 m **23.** Length: 94 ft; width: 50 ft **25.** Length: 265 ft; width: 165 ft; area: 43,725 ft^2 **27.** About 412.6 mi **29.** First: 22.5°; second: 90°; third: 67.5° **31.** **(a)** $1056 million, $1122.2 million, $1784.2 million; **(b)** 2002 **33.** $-\frac{47}{40}$ **34.** $-\frac{17}{40}$ **35.** $-\frac{3}{10}$ **36.** $-\frac{32}{15}$ **37.** 1.6 **38.** 409.6 **39.** −9.6 **40.** −41.6 **41.** ◈ **43.** $1863 = 1776 + (4s + 7)$ **45.** Length: 12 cm; width: 9 cm **47.** 5 half dollars, 10 quarters, 20 dimes, 60 nickels

Summary and Review: Chapter 11, p. 593

1. −22 **2.** 7 **3.** −192 **4.** 1 **5.** $-\frac{7}{3}$ **6.** 25 **7.** $\frac{1}{2}$ **8.** $-\frac{15}{64}$ **9.** 9.99 **10.** −8 **11.** −5 **12.** $-\frac{1}{3}$ **13.** 4 **14.** 3 **15.** 4 **16.** 16 **17.** 6 **18.** −3 **19.** 12 **20.** 4 **21.** 19%x, or $0.19x$ **22.** $791 **23.** 27 **24.** 3 m, 5 m **25.** 9 **26.** Width: 11 cm; length: 17 cm **27.** **(a)** $13.14, $15.19, $28.51; **(b)** 50 lb **28.** 40 ft; 125.6 ft; 1256 ft^2 **29.** 1702 cm^3 **30.** −45 **31.** −60 **32.** ◈ The distributive laws are used to multiply, factor, and collect like terms in this chapter. **33.** ◈ **(a)** $4 - 3x = 9$

$$3x = 9 \quad (1)$$
$$x = 3 \quad (2)$$

1. 4 was subtracted on the left side but not on the right side. Also, the minus sign preceding $3x$ has been dropped.
2. This step would give the correct result if the preceding step were correct.

The correct steps are

$$4 - 3x = 9$$
$$-3x = 5 \quad (1)$$
$$x = -\tfrac{5}{3}. \quad (2)$$

(b) $2(x - 5) = 7$

$$2x - 5 = 7 \quad (1)$$
$$2x = 12 \quad (2)$$
$$x = 6 \quad (3)$$

1. When a distributive law was used to remove parentheses, x was multiplied by 2 but 5 was not.
2. and 3. These steps would give the correct result if the preceding step were correct.

The correct steps are

$$2(x - 5) = 7$$
$$2x - 10 = 7 \quad (1)$$
$$2x = 17 \quad (2)$$
$$x = \tfrac{17}{2}. \quad (3)$$

34. Amazon: 6437 km; Nile: 6671 km **35.** 23, −23 **36.** 20, −20

Test: Chapter 11, p. 595

1. [11.2a] 8 **2.** [11.2a] 26 **3.** [11.3a] −6 **4.** [11.3a] 49 **5.** [11.4b] −12 **6.** [11.4a] 2 **7.** [11.4a] −8 **8.** [11.2a] $-\frac{7}{20}$ **9.** [11.4b] 2.5 **10.** [11.4c] 7 **11.** [11.4c] $\frac{5}{3}$ **12.** [11.5b] Width: 7 cm; length: 11 cm **13.** [11.5b] 6 **14.** [11.5a] $x - 9$ **15.** [11.5b] 25.625° **16.** [11.5b] **(a)** $20.97, $23.94, $43.30; **(b)** about 24 lb **17.** [10.4a] 180 **18.** [10.2a] $-\frac{2}{9}$ **19.** [8.6a, b, c] 140 yd; 439.6 yd; 15,386 yd^2 **20.** [9.1a] 1320 ft^3 **21.** [10.1e], [11.4b] 15, −15 **22.** [11.5b] 60

Cumulative Review: Chapters 1–11, p. 597

1. [1.1e] 7 **2.** [1.1a] 7 thousands + 4 hundreds + 5 ones **3.** [1.2b] 1012 **4.** [1.2b] 21,085 **5.** [3.2b] $\frac{5}{26}$ **6.** [3.5a] $5\frac{7}{9}$ **7.** [4.2a] 493.971 **8.** [4.2a] 802.876 **9.** [1.3d] 152 **10.** [1.3d] 674 **11.** [3.3a] $\frac{5}{24}$ **12.** [3.5b] $2\frac{17}{24}$ **13.** [4.2b] 19.9973 **14.** [4.2b] 34.241 **15.** [1.5b] 4752 **16.** [1.5b] 266,287 **17.** [3.6a] $4\frac{1}{12}$ **18.** [2.4b] $\frac{6}{5}$ **19.** [2.4a] 10 **20.** [4.3a] 259.084 **21.** [1.6c] 573 **22.** [1.6c] 56 R 10 **23.** [3.4c] $56\frac{5}{17}$ **24.** [2.7b] $\frac{3}{2}$ **25.** [3.6b] $\frac{7}{90}$ **26.** [4.4a] 39 **27.** [1.4a] 68,000 **28.** [4.1e] 0.428 **29.** [4.5b] 21.84 **30.** [2.2a] Yes **31.** [2.1a] 1, 3, 5, 15 **32.** [3.1a] 800 **33.** [2.5b] $\frac{7}{10}$ **34.** [2.5b] 55 **35.** [3.4b] $3\frac{3}{5}$ **36.** [2.5c] ≠ **37.** [3.3b] < **38.** [4.1d] 1.001 **39.** [1.4c] > **40.** [2.3b] $\frac{3}{5}$ **41.** [4.1c] 0.037 **42.** [4.5a] 0.52 **43.** [4.5a] $0.\overline{8}$ **44.** [6.1b] 0.07 **45.** [4.1b] $\frac{463}{100}$ **46.** [3.4a] $\frac{29}{4}$ **47.** [6.2b] $\frac{2}{5}$ **48.** [6.2a] 85% **49.** [6.1c] 150% **50.** [1.7b] 555 **51.** [4.4b] 64 **52.** [2.7c] $\frac{5}{4}$ **53.** [5.3b] $76\frac{1}{2}$, or 76.5

54. [7.3d]

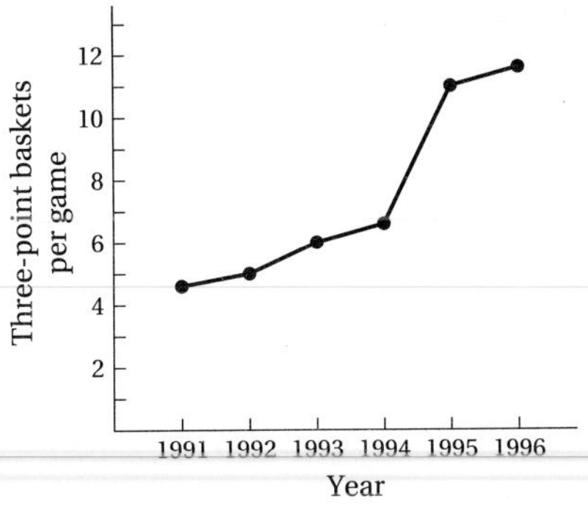

55. [7.5b] 14.5 **56.** [3.7b] $24\frac{1}{2}$ **57.** [3.7b] 29
58. [3.7b] 1 **59.** [3.7b] 101 **60.** [9.6c] Obtuse
61. [9.6d] 118° **62.** [1.8a] $675 **63.** [1.8a] 65 min
64. [3.5c] $\$25\frac{3}{4}$ **65.** [4.7a] 485.9 mi **66.** [1.8a] $8100
67. [1.8a] $423 **68.** [2.4c] $\frac{3}{10}$ km **69.** [4.7a] $84.96
70. [5.4a] 13 gal **71.** [5.2b] 17 cents/oz
72. [6.7a] $150 **73.** [6.6b] 7% **74.** [6.5b] 30,160
75. [7.1a, b, c] 28; 26; 18 **76.** [1.9b] 324
77. [1.9b] 400 **78.** [8.7a] 3 **79.** [8.7a] 11
80. [8.7b] 4.472 **81.** [8.1a] 12 **82.** [8.2a] 428
83. [9.3c] 72 **84.** [9.3b] 20 **85.** [9.3a] 80
86. [9.3b] 0.08 **87.** [9.1b] 8.19 **88.** [9.1b] 5
89. [8.7c] $c = \sqrt{50}$ ft; $c \approx 7.071$ ft
90. [8.3a], [8.4a] 25.6 m; 25.75 m^2 **91.** [8.5a] 25 in^2
92. [8.5a] 128.65 yd^2 **93.** [8.5a] 61.6 cm^2
94. [8.6a, b, c] 20.8 in.; 65.312 in.; 339.6224 in^2
95. [9.1a] 52.9 m^3 **96.** [9.2a] 803.84 ft^3
97. [9.2c] $267.94\overline{6}$ cm^3 **98.** [9.2b] $267.94\overline{6}$ mi^3
99. [1.9c] 238 **100.** [1.9c] 172 **101.** [10.1e], [10.4a] 3
102. [10.2a] 14 **103.** [10.3a] $\frac{1}{3}$ **104.** [10.4a] 30
105. [10.4a] $-\frac{2}{7}$ **106.** [10.5a] −8 **107.** [11.4a] −5
108. [11.3a] 4 **109.** [11.4b] −8 **110.** [11.4c] $\frac{2}{3}$
111. [11.5a] $y + 17$ **112.** [11.5a] 38%x, or $0.38x$
113. [11.5b] 39 **114.** [11.5b] 36

Final Examination, p. 601

1. [1.1a] 8 thousands + 3 hundreds + 4 tens + 5 ones **2.** [1.1e] 7 **3.** [4.2a] 215.6177 **4.** [3.5a] $8\frac{3}{4}$
5. [3.2b] $\frac{5}{3}$ **6.** [1.2b] 759 **7.** [1.2b] 21,562
8. [4.2a] 158.8837 **9.** [1.3d] 5561 **10.** [1.3d] 5937
11. [4.2b] 30.392 **12.** [3.5b] $\frac{5}{8}$ **13.** [3.3a] $\frac{1}{12}$
14. [4.2b] 99.16 **15.** [3.6a] $24\frac{3}{8}$ **16.** [2.4b] $\frac{2}{3}$
17. [2.4a] 6 **18.** [1.5b] 5814 **19.** [1.5b] 234,906
20. [4.3a] 226.327 **21.** [1.6c] 517 R 1 **22.** [3.4c] $517\frac{1}{8}$
23. [1.6c] 197 **24.** [2.7b] $\frac{2}{3}$ **25.** [3.6b] $1\frac{19}{66}$
26. [4.4a] 48 **27.** [1.4a] 43,000 **28.** [4.1e] 6.79
29. [4.5b] 7.384 **30.** [2.2a] Yes **31.** [2.1a] 1, 2, 4, 8
32. [3.1a] 230 **33.** [2.5b] $\frac{3}{2}$ **34.** [2.5b] 10
35. [3.4b] $7\frac{2}{3}$ **36.** [2.5c] = **37.** [3.3b] <
38. [4.1d] 0.9 **39.** [1.4c] < **40.** [2.3b] $\frac{2}{6}$, or $\frac{1}{3}$
41. [6.1b] 0.499 **42.** [4.5a] 0.24 **43.** [4.5a] $0.\overline{27}$
44. [4.1c] 7.86 **45.** [3.4a] $\frac{23}{4}$ **46.** [6.2b] $\frac{37}{100}$
47. [4.1b] $\frac{897}{1000}$ **48.** [6.1c] 77% **49.** [6.2a] 96%
50. [5.3b] 3.84, or $3\frac{21}{25}$ **51.** [3.3c] $\frac{1}{25}$ **52.** [1.7b] 25
53. [4.2c] 245.7 **54.** [6.5b] 5% **55.** [4.7a] $312.60
56. [2.7d] 80 **57.** [1.8a] $348 **58.** [3.5c] $3\frac{3}{4}$ m
59. [1.8a] $1311 **60.** [1.8a] $39 **61.** [5.2a] 13 mpg
62. [4.7a] $24.98 **63.** [5.4a] 540 km **64.** [5.2b] $3.98
65. [6.5a] 39 **66.** [6.7a] $60 **67.** [7.1a, b, c] $15.17; $12; $12 **68.** [3.7b] 29 **69.** [3.7b] $1\frac{1}{2}$ **70.** [1.9b] 625
71. [1.9b] 256 **72.** [8.7a] 7 **73.** [8.7a] 25
74. [8.7b] 4.899 **75.** [9.6d] 55°
76. [8.7c] $a = \sqrt{85}$ ft; $a \approx 9.220$ ft
77. [7.3d]

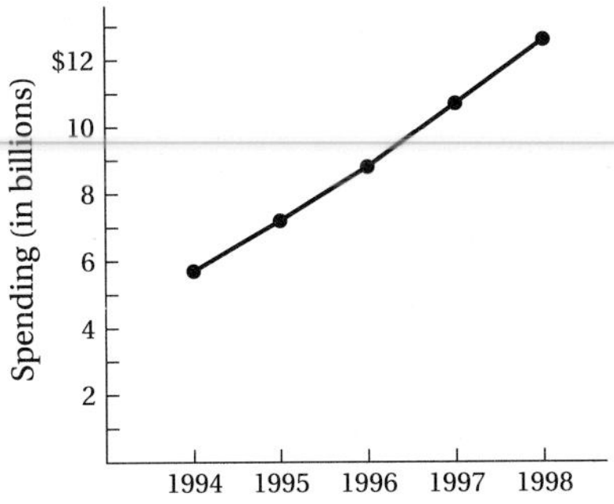

78. [7.5b] $14.5 billion **79.** [7.4a] **(a)** Blue; **(b)** 1050
80. [8.1a] 5 **81.** [8.2a] 2.371 **82.** [9.1b] 5000
83. [9.3a] 14,000 **84.** [9.3c] 1440 **85.** [9.3b] 5340
86. [9.3b] 7.54 **87.** [9.1b] 5 **88.** [8.3a], [8.4a] 24.8 m; 26.88 m^2 **89.** [8.5a] 38.775 ft^2 **90.** [8.5a] 153 m^2
91. [8.5a] 216 cm^2 **92.** [8.6a, b, c] 4.3 yd; 27.004 yd; 58.0586 yd^2 **93.** [9.1a] 68.921 ft^3
94. [9.2a] 314,000 mi^3 **95.** [9.2b] $4186.\overline{6}$ m^3
96. [9.2c] $104{,}666.\overline{6}$ in^3 **97.** [1.9c] 383 **98.** [1.9c] 99
99. [10.1e] 32 **100.** [10.3a] −22 **101.** [10.2a] −22
102. [10.4a] 30 **103.** [10.4a] $-\frac{5}{9}$ **104.** [10.5a] −6
105. [11.2a] −76.4 **106.** [11.3a] 27 **107.** [11.4b] −15
108. [11.5b] 532.5 **109.** [11.5b] 40

Developmental Units

Margin Exercises, Section A, pp. 606–610

1. 9 **2.** 7 **3.** 14 **4.** 16 **5.** 16 **6.** 16 **7.** 8
8. 8 **9.** 7 **10.** 46 **11.** 13 **12.** 58
13.

+	1	2	3	4	5
1	2	3	4	5	6
2	3	4	5	6	7
3	4	5	6	7	8
4	5	6	7	8	9
5	6	7	8	9	10

14.

+	6	5	7	4	9
7	13	12	14	11	16
9	15	14	16	13	18
5	11	10	12	9	14
8	14	13	15	12	17
4	10	9	11	8	13

15. 16 **16.** 9 **17.** 11 **18.** 15 **19.** 59 **20.** 549 **21.** 9979 **22.** 5496 **23.** 56 **24.** 85 **25.** 829 **26.** 1026 **27.** 12,698 **28.** 13,661 **29.** 71,328

Exercise Set A, p. 611

1. 17 **2.** 15 **3.** 13 **4.** 14 **5.** 12 **6.** 11 **7.** 17 **8.** 16 **9.** 12 **10.** 10 **11.** 10 **12.** 11 **13.** 7 **14.** 7 **15.** 11 **16.** 0 **17.** 3 **18.** 18 **19.** 14 **20.** 10 **21.** 4 **22.** 14 **23.** 11 **24.** 15 **25.** 16 **26.** 9 **27.** 13 **28.** 14 **29.** 11 **30.** 7 **31.** 13 **32.** 14 **33.** 12 **34.** 6 **35.** 10 **36.** 12 **37.** 10 **38.** 8 **39.** 2 **40.** 9 **41.** 13 **42.** 8 **43.** 10 **44.** 9 **45.** 10 **46.** 6 **47.** 13 **48.** 9 **49.** 8 **50.** 11 **51.** 12 **52.** 13 **53.** 10 **54.** 12 **55.** 17 **56.** 13 **57.** 10 **58.** 16 **59.** 16 **60.** 15 **61.** 39 **62.** 89 **63.** 87 **64.** 999 **65.** 900 **66.** 868 **67.** 999 **68.** 848 **69.** 877 **70.** 17,680 **71.** 10,873 **72.** 4699 **73.** 10,867 **74.** 9895 **75.** 3998 **76.** 18,222 **77.** 16,889 **78.** 64,489 **79.** 99,999 **80.** 77,777 **81.** 46 **82.** 26 **83.** 55 **84.** 101 **85.** 1643 **86.** 1412 **87.** 846 **88.** 628 **89.** 1204 **90.** 607 **91.** 10,000 **92.** 1010 **93.** 1110 **94.** 1227 **95.** 1111 **96.** 1717 **97.** 10,138 **98.** 6554 **99.** 6111 **100.** 8427 **101.** 9890 **102.** 11,612 **103.** 11,125 **104.** 15,543 **105.** 16,774 **106.** 68,675 **107.** 34,437 **108.** 166,444 **109.** 101,315 **110.** 49,449

Margin Exercises, Section S, pp. 613–618

1. 4 **2.** 7 **3.** 8 **4.** 3 **5.** 12 − 8 = 4; 12 − 4 = 8 **6.** 13 − 6 = 7; 13 − 7 = 6 **7.** 8 **8.** 7 **9.** 9 **10.** 4 **11.** 14 **12.** 20 **13.** 241 **14.** 2025 **15.** 17 **16.** 36 **17.** 454 **18.** 250 **19.** 376 **20.** 245 **21.** 2557 **22.** 3674 **23.** 8 **24.** 6 **25.** 30 **26.** 90 **27.** 500 **28.** 900 **29.** 538 **30.** 677 **31.** 42 **32.** 224 **33.** 138 **34.** 44 **35.** 2768 **36.** 1197 **37.** 1246 **38.** 4728

Exercise Set S, p. 619

1. 7 **2.** 0 **3.** 0 **4.** 5 **5.** 3 **6.** 8 **7.** 8 **8.** 6 **9.** 7 **10.** 3 **11.** 7 **12.** 9 **13.** 6 **14.** 6 **15.** 2 **16.** 4 **17.** 3 **18.** 2 **19.** 0 **20.** 3 **21.** 1 **22.** 1 **23.** 7 **24.** 0 **25.** 4 **26.** 4 **27.** 5 **28.** 4 **29.** 4 **30.** 5 **31.** 9 **32.** 9 **33.** 7 **34.** 9 **35.** 6 **36.** 6 **37.** 2 **38.** 6 **39.** 7 **40.** 8 **41.** 33 **42.** 21 **43.** 247 **44.** 193 **45.** 500 **46.** 654 **47.** 202 **48.** 617 **49.** 288 **50.** 1220 **51.** 2231 **52.** 4126 **53.** 3764 **54.** 1691 **55.** 1226 **56.** 99,998 **57.** 10 **58.** 10,101 **59.** 11,043 **60.** 11,111 **61.** 65 **62.** 29 **63.** 8 **64.** 9 **65.** 308 **66.** 126 **67.** 617 **68.** 214 **69.** 89 **70.** 4402 **71.** 3555 **72.** 5889 **73.** 2387 **74.** 3649 **75.** 3832 **76.** 8144 **77.** 7750 **78.** 10,445 **79.** 33,793 **80.** 281 **81.** 455 **82.** 571 **83.** 6148 **84.** 2200 **85.** 2113 **86.** 3748 **87.** 5206 **88.** 1459 **89.** 305 **90.** 4455

Margin Exercises, Section M, pp. 621–626

1. 56 **2.** 36 **3.** 28 **4.** 42 **5.** 0 **6.** 0 **7.** 8 **8.** 23

9.

×	2	3	4	5
2	4	6	8	10
3	6	9	12	15
4	8	12	16	20
5	10	15	20	25
6	12	18	24	30

10.

×	6	7	8	9
5	30	35	40	45
6	36	42	48	54
7	42	49	56	63
8	48	56	64	72
9	54	63	72	81

11. 70 **12.** 450 **13.** 2730 **14.** 100 **15.** 1000 **16.** 560 **17.** 360 **18.** 700 **19.** 2300 **20.** 72,300 **21.** 10,000 **22.** 100,000 **23.** 5600 **24.** 1600 **25.** 9000 **26.** 852,000 **27.** 10,000 **28.** 12,000 **29.** 72,000 **30.** 54,000 **31.** 5600 **32.** 56,000 **33.** 18,000 **34.** 28 **35.** 116 **36.** 148 **37.** 4938 **38.** 6740 **39.** 116 **40.** 148 **41.** 4938 **42.** 6740 **43.** 5968 **44.** 59,680 **45.** 596,800

Exercise Set M, p. 627

1. 12 **2.** 0 **3.** 7 **4.** 0 **5.** 10 **6.** 30 **7.** 10 **8.** 63 **9.** 54 **10.** 12 **11.** 0 **12.** 72 **13.** 8 **14.** 0 **15.** 28 **16.** 24 **17.** 45 **18.** 18 **19.** 0 **20.** 35 **21.** 45 **22.** 40 **23.** 0 **24.** 16 **25.** 25 **26.** 81 **27.** 1 **28.** 0 **29.** 4 **30.** 36 **31.** 8 **32.** 0 **33.** 27 **34.** 18 **35.** 0 **36.** 10 **37.** 48 **38.** 54 **39.** 0 **40.** 72 **41.** 15 **42.** 8 **43.** 9 **44.** 2 **45.** 32 **46.** 6 **47.** 15 **48.** 6 **49.** 8 **50.** 20 **51.** 20 **52.** 16 **53.** 10 **54.** 0 **55.** 80 **56.** 70 **57.** 160 **58.** 210 **59.** 450 **60.** 780 **61.** 560 **62.** 360 **63.** 800 **64.** 300 **65.** 900 **66.** 1000 **67.** 345,700 **68.** 1200 **69.** 4900 **70.** 4000 **71.** 10,000 **72.** 7000 **73.** 9000 **74.** 2000 **75.** 457,000 **76.** 6,769,000 **77.** 18,000 **78.** 20,000

79. 48,000 **80.** 16,000 **81.** 6000 **82.** 1,000,000 **83.** 1200 **84.** 200 **85.** 4000 **86.** 2500 **87.** 12,000 **88.** 6000 **89.** 63,000 **90.** 120,000 **91.** 800,000 **92.** 120,000 **93.** 16,000,000 **94.** 80,000 **95.** 147 **96.** 444 **97.** 2965 **98.** 4872 **99.** 6293 **100.** 3460 **101.** 16,236 **102.** 13,508 **103.** 87,554 **104.** 195,384 **105.** 3480 **106.** 2790 **107.** 3360 **108.** 7020 **109.** 20,760 **110.** 10,680 **111.** 358,800 **112.** 109,800 **113.** 583,800 **114.** 299,700 **115.** 11,346,000 **116.** 23,390,000 **117.** 61,092,000 **118.** 73,032,000

Margin Exercises, Section D, pp. 629–636

1. 4 **2.** 8 **3.** 9 **4.** 3 **5.** $12 \div 2 = 6$; $12 \div 6 = 2$ **6.** $42 \div 6 = 7$; $42 \div 7 = 6$ **7.** 7 **8.** 9 **9.** 8 **10.** 9 **11.** 6 **12.** 13 **13.** 1 **14.** 2 **15.** Not defined **16.** 0 **17.** Not defined **18.** Not defined **19.** 10 **20.** Not defined **21.** 0 **22.** 1 **23.** 1 **24.** 1 **25.** 1 **26.** 17 **27.** 1 **28.** 75 R 4 **29.** 23 R 11 **30.** 969 R 6 **31.** 47 R 4 **32.** 96 R 1 **33.** 1263 R 5 **34.** 324 **35.** 417 R 5 **36.** 37 R 8 **37.** 23 R 17

Exercise Set D, p. 637

1. 3 **2.** 8 **3.** 4 **4.** 1 **5.** 32 **6.** 9 **7.** 7 **8.** 5 **9.** 37 **10.** 5 **11.** 9 **12.** 4 **13.** 6 **14.** 9 **15.** 5 **16.** 8 **17.** 9 **18.** 6 **19.** 3 **20.** 2 **21.** 9 **22.** 2 **23.** 3 **24.** 7 **25.** 8 **26.** 4 **27.** 7 **28.** 2 **29.** 3 **30.** 6 **31.** 1 **32.** 4 **33.** 6 **34.** 3 **35.** 7 **36.** 9 **37.** 0 **38.** Not defined **39.** Not defined **40.** 7 **41.** 8 **42.** 7 **43.** 1 **44.** 5 **45.** 0 **46.** 0 **47.** 3 **48.** 1 **49.** 4 **50.** 7 **51.** 1 **52.** 6 **53.** 1 **54.** 5 **55.** 2 **56.** 8 **57.** 0 **58.** 0 **59.** 8 **60.** 3 **61.** 4 **62.** Not defined **63.** 4 **64.** 1 **65.** 69 R 1 **66.** 199 R 1 **67.** 92 R 1 **68.** 138 R 3 **69.** 1723 R 4 **70.** 2925 **71.** 864 R 1 **72.** 522 R 3 **73.** 527 R 2 **74.** 2897 **75.** 43 R 15 **76.** 32 R 27 **77.** 47 R 12 **78.** 42 R 9 **79.** 91 R 22 **80.** 52 R 13 **81.** 29 **82.** 55 R 2 **83.** 228 R 22 **84.** 20 R 379 **85.** 21 **86.** 118 **87.** 91 R 1 **88.** 321 R 2 **89.** 964 R 3 **90.** 679 R 5 **91.** 1328 R 2 **92.** 27 R 8 **93.** 23 **94.** 321 R 18 **95.** 224 R 10 **96.** 27 R 60 **97.** 22 R 85 **98.** 49 R 46 **99.** 383 R 91 **100.** 74 R 440

Index

A

B

C

D

M

N

O